www.wadsworth.com

POPULATION

An Introduction to Concepts and Issues
Ninth Edition

John R. Weeks
San Diego State University

WADSWORTH

THOMSON LEARNING

Australia • Canada • Mexico • Singapore • Spain • United Kingdom • United States

WADSWORTH

THOMSON LEARNING™

Acquisitions Editor: Jay Whitney
Assistant Editor: Stephanie Monzon
Editorial Assistant: Jennifer Walsh
Technology Project Manager: Dee Dee Zobian
Marketing Manager: Matthew Wright
Marketing Assistant: Tara Pierson
Advertising Project Manager: Linda Yip
Project Manager, Editorial Production: Matt Ballantyne
Art Director: Rob Hugel

Print/Media Buyer: Doreen Suruki
Permissions Editor: Kiely Sexton
Production Service: Stratford Publishing Services
Copy Editor: Mary Jean Frank
Cover Designer: Margarite Reynolds
Cover Printer: Phoenix Color Corp
Compositor: Stratford Publishing Services
Text Printer: Courier – Westford

For more information about our products,
contact us at:
Thomson Learning Academic Resource Center
1-800-423-0563

For permission to use material from this text
or product, submit a request online at
http://www.thomsonrights.com.

Any additional questions about permissions
can be submitted by email to
thomsonrights@thomson.com.

Library of Congress Control Number: 2004106583
Student Edition: ISBN 0-534-62769-2
Loose Leaf Edition: ISBN 0-534-62779-X

Wadsworth/Thomson Learning
10 Davis Drive
Belmont, CA 94002-3098
USA

Asia
Thomson Learning
5 Shenton Way #01-01
UIC Building
Singapore 068808

Australia/New Zealand
Thomson Learning
102 Dodds Street
Southbank, Victoria 3006
Australia

Canada
Nelson
1120 Birchmount Road
Toronto, Ontario M1K 5G4
Canada

Europe/Middle East/Africa
Thomson Learning
High Holborn House
50/51 Bedford Row
London WC1R 4LR
United Kingdom

Latin America
Thomson Learning
Seneca, 53
Colonia Polanco
11560 Mexico D.F.
Mexico

Spain/Portugal
Paraninfo
Calle Magallanes, 25
28015 Madrid, Spain

To Deanna

BRIEF TABLE OF CONTENTS

DETAILED TABLE OF CONTENTS

PREFACE

Growth and transition. These are the key demographic trends as we move through the twenty-first century, and they both will have a huge impact on your life. When I think about population growth in the world, I conjure up an image of a bus hurtling down the highway toward what appears to be a cliff. The bus is semiautomatic and has no driver in charge of its progress. Some of the passengers on the bus are ignorant of what seems to lie ahead and are more worried about whether or not the air-conditioning is turned up high enough or wondering how many snacks they have left for the journey. Other more alert passengers are looking down the road, but some of them think that what seems like a cliff is really just an optical illusion and is nothing to worry about; some think it may just be a dip, not really a cliff. Those who think it is a cliff are trying to figure out how to apply the brakes, knowing that a big bus takes a long time to slow down even after the brakes are put on. Are we headed toward a disastrous scenario? We don't really know for sure, but we simply can't afford the luxury of hoping for the best. The population bus is causing damage and creating vortexes of change as it charges down the highway, whether or not we are on the cliff route; and the better we understand its speed and direction, the better we will be at steering it and managing it successfully. No matter how many stories you have heard about the rate of population growth coming down or about the end of the population explosion, the world *will* continue to add billions more to the current six billion before it stops growing. Huge implications for the future lie in that growth in numbers.

The transition I mentioned is closely related to population growth, but also leads to its own set of complications. The world's population is growing because death rates have declined over the past several decades at a much faster pace than have birth rates, and as we go from the historical pattern of high birth and death rates to the increasingly common pattern of low birth and death rates, we pass through the demographic transition. This is actually a whole set of transitions relating to changes in mortality, fertility, migration, age structure, urbanization, and family and household structure. Each of these separate, but interrelated, changes has serious consequences for the way societies and economies work, and for that reason they have big implications for your personally.

The growth in numbers (the bus hurtling toward what we hope is *not* a cliff) and the transitions created in the process (the vortex created by the passing bus) have to be dealt with simultaneously and our success as a human civilization depends on how well we do in this project. A lot is at stake here and my goal in this book is to provide you with as much insight as possible into the ways in which these demographic trends of growth and transition affect your life in small and large ways.

Over the years, I have found that most people are either blissfully unaware of the enormous impact of population growth and change on their lives, or they are nearly overwhelmed whenever they think of population growth because they have heard so many horror stories about impending doom, or, increasingly, they have heard that population growth is ending and thus assume that the story has a happy ending. This latter belief is in many ways the scariest, because the lethargy that develops from thinking that the impact of population growth is a thing of the past is exactly what will lead us to doom. My purpose in this book is to shake you out of your lethargy (if you are one of those types), without necessarily scaring you in the process. I will introduce you to the basic concepts of population studies and help you develop your own demographic perspective, enabling you to understand some of the most important issues confronting the world. My intention is to sharpen your perception of population growth and change, to increase your awareness of what is happening and why, and to help prepare you to cope with (and help shape) a future that will be shared with billions more people than there are today.

I wrote this book with a wide audience in mind because I find that students in my classes come from a wide range of academic disciplines and bring with them an incredible variety of viewpoints and backgrounds. No matter who you are, demographic events are influencing your life, and the more you know about them, the better off you will be.

What Is New in this Ninth Edition

This edition represents a significant revision from the eighth edition. Users of previous editions will immediately notice that the book has been slimmed down to 13 chapters to make it easier to get through the material in a single semester. Furthermore, the life table and TFR calculations that used to be in the Appendix have now been incorporated into the appropriate chapters, so that teaching these techniques will flow more easily than in the past. In the process of revising the book, I rewrote major sections of every chapter to more clearly tell the story of population growth and change. And, of course, I have scoured the literature and data to incorporate the latest trends and thinking in the field of population studies.

Organization of the Book

The book is organized into four parts, each building on the previous one.

Part One "A Demographic Perspective" (Chapters 1–4) begins with an introduction to the field of population studies and illustrates why this is such an important topic for students. The second chapter reviews world population trends (past, present, and future projections), so that you have a good idea of what is happening in the world demographically, how we got to this point, and where we seem to be heading. The third chapter introduces you to the major perspectives or ways of thinking about population growth and change, and the fourth chapter reviews the sources of data that form the basis of our understanding of demographic trends.

Part Two "Population Processes" (Chapters 5–7) discusses the three basic demographic processes whose transitions are transforming the world. Chapter 5 covers the mortality transition, Chapter 6 deals with the fertility transition, and Chapter 7 reviews the migration transition. Knowledge of these three population processes and transitions provides you with the foundation you need to understand why changes occur and what might be done about them.

Part Three "Population Structure and Characteristics" (Chapters 8-11) is devoted to studying the interaction of the population processes and societal change according to demographic characteristics such as the age transition (Chapter 8), population aging (Chapter 9), the family and household transition (Chapter 10), and the urban transition (Chapter 11).

Part Four "Using the Demographic Perspective" (Chapters 12-13) is the final section. Chapter 12 explores the relationship between population and the environment: Can economic growth and development be sustained in the face of continued population growth? There are no simple answers, but we are faced with a future in which we will have to deal with the global and local consequences of a larger and constantly changing population. Chapter 13 reviews the ways in which we can cope with these changes as they alter the fabric of human society all across the globe.

Special Features of the Book

To help increase your understanding of the basic concepts and issues of population studies, the book contains the following special features.

Short Essays Each chapter contains a short essay on a particular population concept, designed to help you better understand current demographic issues, such as the one in Chapter 8 on "Oil Boom, Baby Boom, and Social Change in Saudi Arabia," or the one in Chapter 11 on "NIMBY and BNANA—The Politics of Urban Sprawl in America."

Main Points A list of main points appears at the end of each chapter, following the summary, to help you review chapter highlights.

Suggested Readings At the end of each chapter, I have listed five of the most important and readable references for additional review of the topics covered in that chapter.

Web Sites of Interest The Internet has become a useful supplement to books and journals, and there is a great deal of relevant information on population studies available on the World Wide Web. At the end of each chapter, I list and annotate five Web sites that I have found to be particularly interesting and helpful to students. These suggested Internet resources are tied to classroom and homework exercises I have prepared that are available through the Wadsworth home page: *http://www.wadsworth.com.*

Glossary A Glossary in the back of the book defines key population terms. These terms are in boldface type when introduced in the text to signal that they also appear in the Glossary.

Complete Bibliography This is a fully referenced book and all of the publications and data sources I have used are included in the Bibliography at the end of the book.

A Thorough Index To help you find what you need in the book, we have built as complete an index as possible.

Personal Acknowledgments

Like most authors, I have an intellectual lineage that I feel is worth tracing. In particular, I would like to acknowledge the late Kingsley Davis, whose standards as a teacher and scholar will always keep me reaching; Eduardo Arriaga; the late Judith Blake; Thomas Burch; Carlo Cipolla; Murray Gendell; Nathan Keyfitz; and Samuel Preston. Individually and collectively, they have helped me unravel the mysteries of how the world operates demographically. Thanks are due also to Steve Rutter, formerly of Wadsworth Publishing Company, who first suggested that I write this book. Jay Whitney at Wadsworth provided key insights for updating this ninth edition, and I am grateful for the excellent production work of Linda DeMasi at Stratford Publishing Services.

Special thanks go to John, Gregory, Jennifer, Suzanne, and Amy for teaching me the costs and benefits of children. They have instructed me, respectively, in the advantages of being first-born, in the coziness of the middle child, in the joys that immigration can bring to a family, and in the wonderful gifts (including Andrew, Sophie, and Benjamin) that daughters-in-law can bring to a family.

However, the one person who is directly responsible for the fact that the first, second, third, fourth, fifth, updated fifth, sixth, seventh, eighth, and now ninth editions were written, and who deserves credit for the book's strengths, is my wife Deanna. Her creativity, good judgment, and hard work in reviewing and editing the manuscript benefited virtually every page, and I have dedicated the book to her.

Other Acknowledgments

I would also like to thank the users of the earlier editions, including professors, their students, and my own students, for their comments and suggestions. In particular, for the ninth edition, I appreciate the important input and advice provided by Tim Liao, University of Illinois–Urbana; Jack Carter, University of New Orleans; Jianjun Ji, University of Wisconsin–Eau Claire; Scott Meyers, Washington State; Ted Fuller, Virginia Tech; Dudley Poston, Texas A&M University; and Gillian Stevens, University of Illinois–Urbana. Many, many other people have helped since the first edition went into production almost 30 years ago, and I am naturally very grateful for all of their assistance. The essay in Chapter 13 was co-authored by John R. Weeks, Jr., INSEAD, and Gregory B. Weeks, University of North Carolina at Charlotte.

PART ONE
A Demographic Perspective

CHAPTER 1
Introduction to Demography

Figure 1.1 The World's Population Distribution Defined by Nighttime Lights

Source: Marc Imhoff (NASA/GSFC) and Christopher Elvidge (NOAA/NGDC): NASA/ Goddard Space Flight Center Scientific Visualization Studio, Images collected in 1994–95.

WHAT IS DEMOGRAPHY?

WHY IS DEMOGRAPHY IMPORTANT?
  Terrorism and Regional Conflict
  Armies of Children in Sub-Saharan Africa
  Embracing and Hating Immigration
  Globalization
  Degradation of the Environment
  The Danger of Demographic Fatigue

HOW WILL YOU USE THIS INFORMATION?
  Demographics of Politics
  Demographics of Social Planning
  Demographics of Business

ESSAY: The Demography of Conflict in the
  Middle East

Population growth is the single most important set of events ever to occur in human history. It has changed—and continues to alter—the way of life in even the most remote corners of the earth. No matter how much you may have heard about declining birth rates, it is still true that the number of people added to the world each day is unprecedented in history and unparalleled in its consequences. We now live in a world crowded not only with people but also with contradiction. There are more highly educated people than ever before, yet also more illiterates; more rich people, but also more poor; more well-fed children, but also more hunger-ravaged babies whose images haunt us. We have better control over the environment than ever before, but we are damaging our living space in ways we are afraid to imagine.

Our partial mastery of the environment is, indeed, key to understanding why the population is growing, because we have learned how to conquer more and more of the diseases that once routinely killed us. Although the rapid, dramatic drop in mortality all over the world is certainly one of humanity greatest triumphs, it seems that no good deed goes unpunished, even an altruistic one like conquering (or at least delaying) death. Because birth rates almost never go down in tandem with the decline in the death rate, the result is rapid population growth. This relentless increase in the population has fueled both environmental damage and social upheaval.

Not all demographic change is bad, of course, but population growth does make implacable demands on natural and societal resources that will certainly be detrimental to the long-term health of the planet and its inhabitants, human and otherwise. A baby born this year may not create much of an immediate stir, but in a few years that child will be eating more and needing clothes, an education, and then a job and place of her own to live.

Population growth is an irresistible force. Virtually every social, political, and economic problem facing the world has demographic change as one of its root causes. Furthermore, I can guarantee you that it is a force that will increasingly affect you personally, in ways both large and small—population change is not just something that happens to other people. It is taking place all around us, and we each make our own contribution to it. Whether your concern with demography is personal or global, unraveling the "whys" of population growth and change will provide you with a better perspective on how the world works. Understanding these and a wide range of related issues is the business of demography.

What is Demography?

Demography is the scientific study of human populations. The term itself was coined in 1855 by Achille Guillard, who used it in the title of his book *Éléments de Statistique Humaine ou Démographie Comparée*. The word he invented is a combination of two Greek roots: *demos,* which means people, and *graphy,* which means the branch of knowledge relating to a particular science (in this instance, population).

Guillard defined demography as "the mathematical knowledge of populations, their general movements, and their physical, civil, intellectual and moral state" (Guillard 1855:xxvi). This is generally in tune with how we use the term today in that modern demography is the study of the determinants and consequences of population change and is concerned with virtually everything that influences or can be influenced by:

- **population size** (how many people there are in a given place)
- **population growth or decline** (how the number of people in that place is changing over time)
- **population processes** (the levels and trends in mortality, fertility, and migration that are determining population size and change)
- **population distribution** (where people are located and why)
- **population structure** (how many males and females there are of each age)
- **population characteristics** (what people are like in a given place, in terms of variables such as education, income, occupation, family and household relationships, immigrant and refugee status, and the many other characteristics that add up to who we are as individuals or groups of people)

It has been said that "the past is a foreign country; they do things differently there" (Hartley 1967:3). Population change and all that goes with it is an integral part of creating a present that seems foreign by comparison to the past, and it will create a future that will make today seem strange to those who look back on it several decades from now. Table 1.1 illustrates this idea, comparing the United States in the year 1900 with the year 2000. Although the population of the United States grew considerably during that century, from 76 million to 281 million, it did not keep pace with overall world population growth and so accounted for a slightly smaller fraction of the world's population in 2000 than it had in 1900 (more on this in the next chapter). Mortality levels dropped substantially over the century, leading to a truly amazing 30-year rise in life expectancy, from 47 in 1900 to 77 in 2000 (the reasons for this are laid out in Chapter 5).

Fertility also declined, although by world standards fertility was already fairly low (3.5 children per woman) in 1900. Still, the drop from 3.5 to 2.1 clearly makes a huge difference in the composition of families, and I discuss this more in Chapter 6. Americans rearranged themselves considerably within the country over that century, and the westward movement is exemplified by the change in the fraction of the population living in California. It went from only 2 percent in 1900 to 12 percent in 2000. Consider that in 1900 Los Angeles had about the same population as Buffalo, New York, but in 2000, Los Angeles was 35 times more populous than Buffalo. In the latter part of the twentieth century, much of that growth in Los Angeles was fueled by immigrants from Mexico and Central America, but over the course of the century the composition of international immigrants had shifted substantially. In the decade following the 1900 census, there were about 50,000 Mexican immigrants to the U.S., compared to two million Italian immigrants in the same time period. By contrast, in the decade prior to Census 2000, the numbers were essentially reversed, with 63,000 Italian immigrants, and at least 2.2 million Mexican immigrants. Yet, as strange as it might seem in an era when there is so much talk about immigrants, the data in Table 1.1 show that the foreign-born population actually represented a

Table 1.1   The Past Is a Foreign Country

	1900	2000
World population (billions)	1.6	6.1
U.S. population (millions)	76	281
U.S. percent of world total	4.8%	4.6%
Life expectancy	47	77
Children per woman	3.5	2.1
% of U.S. population in California	2%	12%
Population of Buffalo, NY, compared to Los Angeles	About the same	LA is 35 times more populous
Immigrants from Italy (1900–1910); (1990–2000)	2 million	63,000
Immigrants from Mexico (1900–1910); (1990–2000)	50,000	2.2 million
% foreign-born	13.6%	11.1%
% urban	40%	80%
Number of passenger cars	8,000	130 million
% of population under 15	34.4%	21.4%
% of population 65+	4.1%	12.4%
Average persons per household	4.76	2.59
% high school graduates	10%	80%

Source: Data for 1900 from http://www.census.gov/prod/99pubs/99statab/sec31.pdf, accessed 2003; data for 2000 from http://www.census.gov, accessed 2003.

greater fraction of the nation in 1900 than it did a century later. We'll explore the reasons for that in Chapter 7.

The past was predominantly rural and the present is predominantly urban (as I discuss in Chapter 11); the past was predominantly pedestrian (there were only 8,000 passenger cars in 1900) and the present is heavily dependent on the automobile (with more than 130 million passenger cars being driven around the country). The past was young, with 34 percent of the population under the age of 15 and only 4 percent aged 65 and older; whereas the present is older, with only 21 percent under 15 and 12 percent aged 65 and older (see Chapter 8 for more on this). In the past, people lived in households with more people than today (4.76 compared to 2.59), and the average person in each household was considerably less well educated than today, with only 10 percent of those in 1900 achieving a high school education, compared to 80 percent now. These trends are discussed more in Chapter 11.

The world of 1900 was very different from the world of 2000, and the demographics represent an important part of that difference. The future will be different, in its turn, partly because of demographic changes taking place even as you read this page. The study of demography is thus an integral part of understanding human society.

Why is Demography Important?

The demographic foundation of our lives is deep and broad. As you will see in this book, demography affects nearly every facet of your life in some way or another. Population change is one of the prime forces behind social and technological change all over the world. As population size and composition changes in an area—whether it be growth or decline—people have to adjust, and from those adjustments radiate innumerable alterations to the way society operates. Very few acres of this earth await human discovery, as more and more people with more and better machines traverse the planet, coming into contact (and often conflict) with one another. A century and a half ago the American philosopher and naturalist Henry David Thoreau wrote, "I would rather sit on a pumpkin and have it all to myself than be crowded on a velvet cushion." By now, population growth has nearly swallowed up his famous Walden Pond in the Boston metropolitan area, and his ghost shares even his pumpkin, never mind a velvet cushion, with thousands of visitors each year.

The population of the world is increasing by more than 200,000 people per day, as I will discuss in more detail in the next chapter, and that is reflected in the political instability of several regions of the world. Population growth is, of course, not the only source of trouble in the world, but its impact is often incendiary, igniting other dilemmas that face human society (Davis 1984a). Without knowledge of population dynamics, we cannot fully understand the roots of terrorism and conflict from the Middle East to Southeast Asia; nor can we understand why there is a simultaneous acceptance of and a backlash against immigrants in the United States and Europe; nor can we understand why the world is globalizing at such a rapid pace. And, we cannot begin to imagine our future without taking into account the fact that the population of the world at the middle of this century is expected to be half again larger than it is now, since the health of the planet depends upon being able to sustain a much larger number of people than are currently alive. Because so much that happens in your life will be influenced by the consequences of population change, it behooves you to understand the causes and mechanisms of those changes. Let's look at some examples.

Terrorism and Regional Conflict

One reaction to population growth is to embrace, or at least accept, the change and seek positive solutions to the dilemmas presented by a rapidly growing younger population. Another reaction, of course, is to reject change and to try to stuff the genie back into the bottle. This is what the Taliban tried to do in Afghanistan—to forcibly prevent a society from modernizing and, in the process, put a damper on population growth by keeping death rates higher than they might otherwise be (you will learn in Chapter 5 that Afghanistan has the highest rate of maternal mortality in the world). Less extreme, but still very dangerous, forms of this sort of religious fundamentalism have arisen in the Indian subcontinent, where rapidly growing Muslim populations confront rapidly growing Hindu populations.

Throughout the Middle East we can see with special clarity the crucial role that demography has played. The migration of poor rural peasants to the cities of Iran,

THE DEMOGRAPHY OF CONFLICT IN THE MIDDLE EAST

At the end of World War I, the British took control of Palestine from the remnants of the Ottoman Empire, which had shrunk to its base in Turkey. At that time, Palestine under the British mandate included the territory of modern Israel and Jordan. In 1922, it was divided into the mandates of Transjordan, east of the Jordan River, and Palestine, west of the river. Under the Balfour Declaration of 1917, the British had already agreed to help establish a Jewish national home in Palestine, but later were sorry they had done so (Oren 2002). In the 1930s and 1940s, European anti-Semitism encouraged the migration of Jews to Palestine and despite the drop in support from the British the changing demographics were leading inexorably in the direction of a Jewish state. Not unexpectedly, this trend was resisted first by Palestinian Arabs, and subsequently by virtually all Arab states. In 1946, at the end of World War II, Transjordan was granted full independence and became the modern state of Jordan, which had in fact been ruled by the great-grandfather of the current King Abdullah II since the end of World War I.

Britain handed the decision about Palestine to the United Nations, and in 1947 the UN passed General Assembly Resolution 181. "This provided for the creation of two states, one Arab and the

other Jewish, in Palestine, and an international regime for Jerusalem. The Zionists approved of the plan but the Arabs, having already rejected an earlier, more favorable (for them) partition offer from Britain, stood firm in their demand for sovereignty over Palestine in full" (Oren 2002:4). The stage was thus set for the continuing struggle for control of the region. The nascent state of Israel was immediately attacked by armies from all surrounding Arab nations, but managed to prevail, and when hostilities ended in 1949, Israel had claimed more territory than originally allotted to it by the UN. Because as many as 750,000 of Palestine's Arabs (who came to be known simply as Palestinians) had fled the area when fighting broke out, the Jewish population emerged as the demographic majority. The Palestinian population was effectively cordoned into the Gaza Strip (controlled by Egypt) and the West Bank (controlled by Jordan), as shown on the accompanying map, with the remainder of what had been Palestine being the state of Israel.

As of 1950, then, the population of Israel accounted for 56 percent of the population of the former Palestine, and 45 percent of the original British mandate that had included Jordan, as you can see in the accompanying table. Israel accepted large numbers of immigrants early in its existence and by 1970, shortly after the 1967 war (the "Six Day War" to Israelis and the "June 1967 War" to Arabs), you can see in the table that Israel's population accounted for nearly three-fourths of the population of the former Palestine, and more than half of the area that included both Israel and Jordan. Israel had once again defeated the invading Arab armies in 1967, and as a consequence took control of both Gaza and the West Bank.

After each round of hostilities, Jordan received a new flood of Palestinian refugees, and in 1970 a group of Palestinians attempted to overthrow the government of Jordan in order to establish it as the Palestinian state. Jordan is thus a very important part of the picture and, in fact, in 1983, Benjamin Netanyahu, then a member of the Israeli mission to the U.S., but later Prime Minister of Israel, wrote that "the Arabs of Palestine already have a state, called Jordan, in eastern Palestine" (Netanyahu 1983:30). Jordanians do not share that view, of course. Nonetheless, as of the year 2000, the

United Nations Relief and Works Agency for Palestine Refugees estimated that there were 3.7 million Palestinian refugees in the Middle East (this does not include those who had gone to countries outside the region, such as the United States or even Denmark, which has several thousand), of whom 1.6 million (43 percent) are in Jordan.

In the meantime, Yasser Arafat, who emerged as the primary leader and spokesperson of the Palestinian population through his role as Chairman of the Palestinian Liberation Organization, was applauding the high birth rate among Palestinians, calling it their "secret weapon" to beat back the Israeli challenge (Waldman 1991), and he urged women to have babies for the cause (the *Intifada* or uprising against Israel) to replace the young male fighters who were being killed fighting the Israelis (Sontag 2000).

To be sure, Palestinians in the region have consistently had fertility rates that are the highest, or close to the highest, of any group in the world. As you can see in the accompanying table, in the Gaza Strip women were averaging more than eight children each as recently as 1990. Because of low infant mortality, most of these children survive to adulthood and so the Palestinian population is very young. In the year 2000, it was estimated that 47 percent of the population of Gaza and the West Bank combined was under the age of 15, compared to 29 percent of the Israeli population.

The fact that there are more young Palestinians each year than there are new jobs being created means that there is a pool of discontent waiting to be tapped by any group interested in overturning the present social order (Fargues 1995). Although the fertility levels in Israel are generally lower than among Palestinians, fertility rates among the ultra-orthodox Jewish population are nearly as high as for Palestinians (Courbage 2000). The high fertility and steadily declining mortality will produce a projected population for the region (including Jordan) in 2030 of 24 million—nearly nine times what it was in the middle of the twentieth century. The demographics of the region are thus not very encouraging, and keep in mind that we haven't even talked about the water scarcity already existing in that part of the world . . .

				Year			
Total Population:	1950	1970	1990	2000	2010	2020	2030
Israel	1,286	2,903	4,512	5,842	6,645	7,315	7,873
Gaza	245	343	643	1,132	1,651	2,260	2,921
West Bank	771	690	1,255	2,020	2,765	3,501	4,258
Total	2,302	3,936	6,410	8,994	11,061	13,076	15,052
Jordan	561	1,503	3,262	4,999	6,486	7,920	9,373
Total including Jordan	2,863	5,439	9,672	13,993	17,547	20,996	24,425
Israel as % of Israel, Gaza, and West Bank	56%	74%	70%	65%	60%	56%	52%
Israel as % of Gaza, West Bank, and Jordan	25%	31%	28%	25%	23%	21%	20%
Total Fertility Rate:							
Israel			3.0	2.6	2.3	2.1	1.9
Gaza			8.1	6.6	5.2	4.0	3.1
West Bank			5.9	5.0	3.8	3.0	2.5
Jordan			5.3	3.6	2.3	2.1	2.0

Source: U.S. Census Bureau, International Programs Center, International Database, http://www.census.gov/ipc/www/idbnew.html, accessed 2003.

especially Tehran, contributed to the political revolution in that country in 1978 by creating a pool of young, unemployed men who were ready recruits to the cause of overthrowing the existing government (Kazemi 1980). It has been said that the "dogs of war" are young and male (Mesquida and Wiener 1999), and this description applies especially to the Middle East, where large fractions of the population are young, and where males are routinely accorded higher status than females. This does not mean that having a large population of young men automatically leads to conflict. Rather, it suggests that such a demographic situation is incendiary (Choucri 1984). And, of course, the very existence of Israel is openly threatened by the more rapid growth rate of the Arab Palestinian population, as I discuss in the accompanying essay.

Armies of Children in Sub-Saharan Africa

Sub-Saharan Africa is another part of the world where population growth has been increasing faster than resources can be generated to support it—despite the devastation caused by HIV/AIDS—increasing the level of poverty and disease, and encouraging child labor, slavery, and despair. Rwanda and Burundi, two land-locked countries in central sub-Saharan Africa, both reached the exploding point in the mid-1990s. These two countries, which in decades past were likened to the countryside of Switzerland, have now been almost completely deforested, as a population that doubles every 30 years harvests firewood and seeks land on which to grow food. The vast majority of people are in grinding poverty with little hope of improvement and it is a recipe for disaster that has produced interethnic violence and mass murders. In 1993, a genocidal civil war erupted between the rival Tutsi and Hutu ethnic groups, leaving at least 100,000 Burundis dead. The following year a similar eruption in Rwanda killed an estimated 1 million Rwandans. Throughout sub-Saharan Africa, the large number of children, enmeshed in poverty and often orphaned because their parents have died of AIDS, provides recruits for rebel armies waging warfare against one government or another. Those children who resist the army recruiters may find themselves being sold into slavery (*The Economist* 2001), which is part of a larger global problem of child trafficking (International Labour Office 2003). This kind of abuse of children is not caused by demographic trends, but the demographic structure of society contributes to the problem.

Embracing and Hating Immigration

Low fertility is not without its problems. The transition from higher to lower fertility in North America, Europe, and East Asia, as well as Australia and New Zealand, has created a situation in all of these parts of the world in which the younger population is declining as a fraction of the total population, creating holes in the labor force and concerns about who will pay the taxes needed to fund the pensions of the older population. For a variety of reasons that I will discuss in Chapter 6, women in the richer countries are choosing to have fewer children than required to replace the population. But, since each of the rich countries has jobs available and needs people to pay taxes, the only answer in the short term is immigration. Canada has been the country most welcoming of immigrants, at least on the basis of immigrants per resident population, with Asians being the largest group entering Canada (a pattern followed also in Australia). The United States has been the most accepting of all countries in the world in terms of absolute numbers of immigrants, including both legal and undocumented ones, with Mexico leading the list of countries from which immigrants to the U.S. come.

Like it or not, the countries sending migrants have their own demographics issues that complement those of the richer countries. For example, in Mexico, fertility has not declined as quickly as mortality, and the resulting high rates of population growth have made it impossible for the economy to generate enough jobs for each year's crop of new workers. Underemployment in Mexico naturally increases the attractiveness of migrating to where better jobs are. This happens especially to be the United States, not just because the United States is next door, but because low rates of population growth have left many jobs open, particularly at the lower end of the economic ladder. These positions provide foreign laborers with a higher standard of living than they could have had in Mexico. The North American Free Trade Agreement, passed in 1994, promised new and higher-wage jobs in Mexico that would eventually reduce the need for Mexicans to leave their country to find work (Cornelius 2002). That promise has been sideswiped, however, by the fact that rising wages in Mexico have led many companies to take their manufacturing businesses to other countries where wages are still low, particularly China, whose population outnumbers Mexico's by a ratio of 10 to 1 (*The Economist* 2003).

Since the terrorist attacks of September 11, 2001, it has been increasingly difficult for undocumented immigrants to enter the United States. As a consequence, many Latin American migrants have been going to Europe instead, both legally and illegally (Millman and Vitzthum 2003). The open-border policy within the European Union (EU) means that once a person enters Europe they are free to travel to any of the other EU countries in search of a job, even if they arrived without a work permit. Not surprisingly, Spain is the largest recipient of these predominantly Spanish-speaking immigrants, but there are growing communities of Latin Americans in Switzerland and Italy, as well. There is a certain amount of symmetry, one might say, in the fact that the migration of Spaniards to the New World created "Latin America" from the mixing of Europeans with the indigenous population. Now, five centuries later, the tide is turning.

There is, in fact, a bigger vacuum of laborers in Europe than in North America, because birth rates there have declined dramatically in the past two decades, and are now considerably lower than in the United States. There is thus the "sucking sound" of people from developing nations, notably former European colonies, filling the jobs in Europe that would otherwise go begging. These immigrants are most likely to come from former colonies, and so the United Kingdom has large immigrant populations from India, Pakistan, and the Caribbean, while France has immigrants from Algeria and Senegal, Germany has immigrants from Turkey (not a former colony, but a sympathizer in both World Wars), the Netherlands has immigrants from Indonesia, and Spain has immigrants from Morocco (along with those from Latin America). Europeans, however, are not necessarily in favor of this trend. Several elections in Europe in 2002 focused on immigrant issues and a reporter for the *The Wall Street Journal* suggested that "across Europe, unassimilated immigrant populations from former colonies have become political headaches for governments that never thought their 'guest workers' would stay" (Champion 2002:A20).

Given the complementary needs of European countries for laborers and younger people in developing countries for jobs, we can expect that immigration will quite literally change the face of Europe in your lifetime. Each of these countries needs immigrants in lieu of a replacement birth rate, but the problem is always that immigrants tend to be different. They may look different, have a different language, a different religion, and differ in their expectations about how society operates. Furthermore, since the immigrants tend to be young adults, they will be contributing disproportionately to the birth rate in their new countries, leading to a rapid and profound shift in the ethnic composition of the younger population. These differences create problems for all societies, and create situations of backlash against immigrants.

American history is replete with stories of discrimination against immigrant groups for one or more generations until the children and grandchildren of immigrants finally are accepted as part of mainstream society. This process produces children who would not be recognizable to their ancestors and a society that is in certain ways a foreign country relative to the past. Just as in the United States, European nations have highly visible anti-immigrant groups, but the immigrants keep coming anyway because jobs are available and they need the jobs. This is less true in Japan because the level of antiforeign sentiment is so high. The Japanese simply take it for granted that people from other countries will not be able to become permanent members of Japanese society. This means that Japan has had fewer immigrant workers per person than in North America or Europe, and it is not unreasonable to think that the Japanese economy has been moribund for several years now because it has not been invigorated by immigration.

Globalization

Most broadly, globalization can be thought of as an increasing level of connectedness among and between people and places all over the world. The term has taken on a more politically charged meaning, however, of the penetration of less developed nations by multinational companies from the more developed nations. This trend is promoted by the removal of trade barriers that protect local industries and by the integration of local and regional economies into a larger world arena. The pros and cons of this process invite heated debate, but an important, yet generally ignored, element of globalization is that it is closely related to the enormous increase in worldwide population growth that took place after the end of World War II. Control over mortality, which has permitted the growth of population, occurred first in the countries of Europe and North America and it was there that population first began to grow rapidly in the modern world. However, after World War II, death control technology was spread globally, especially through the work of various United Nations agencies, funded by the governments of the richer countries. Since declines in mortality initially affect infants more than any other age group, there tends to be a somewhat delayed reaction in the realization of the effects of a mortality decline until those people who would otherwise have died reach an age where they must be educated, clothed, fed, and jobs and homes must be created for them on a scale never before imagined. As huge new cohorts of young people have come of age and needed jobs in developing countries, their willingness to work for relatively low wages has not gone unnoticed by manufacturers in North America, Europe, and Japan. Nor have big companies failed to notice the growing number of potential consumers for products, especially those aimed at younger people, who represent the bulk of the population in developing countries. Given the demographics, it should not be surprising to us that jobs have moved to the developing countries and that younger consumers in those countries have been encouraged to spend their new wages on products that are popular with younger people in the richer countries, including music, fast food, cars, and mobile phones. Globalization exists, in essence, because of the nature of world demographic trends.

Degradation of the Environment

As the human population has increased, its potential for disrupting the earth's biosphere has grown. We are polluting the **atmosphere** (producing problems such as acid rain, global warming, and holes in the ozone layer); the **hydrosphere** (contaminating the fresh water supply, destroying coral reefs, and fishing out the ocean); and the **lithosphere** (degrading the land with toxic waste and permitting topsoil loss, desertification, and deforestation). This degradation is caused by our intensive use of resources, which has dramatically increased our standard of living over the past 200 years. That very same use of resources has permitted us to bring mortality under control, launching the world on its course of a huge increase in the number of people alive on the planet. The task that we confront in the future is to maintain our standard of living while using many fewer resources per person. Keep in mind that international agencies such as the United Nations and the World Bank have suggested that long-term sustainability of the planet requires that we lift all people out of poverty so that everyone can be a better steward of the planet. So, the task is really to maintain our standard of living, while increasing that of people everywhere else in the world, and doing so while not using resources in an unsustainable fashion. This is not going to be a simple project, based on data from the United Nations Human Development Report released in 2003 indicating that the richer countries became even richer in the 1990s, but 54 less developed countries actually saw their living standards decline in that same decade (United Nations Development Program 2003).

The number of people living in extreme poverty has increased in tandem with population growth, and poverty is one of the major reasons why millions of children in less developed nations are forced to work, jeopardizing their health, their education, and their future. Worldwide, the gap has been increasing between rich and poor countries, especially between the very richest and the very poorest. Reducing poverty is a worthwhile goal in all events, but these plans rarely reckon with how this can be accomplished given the huge increase in numbers that the United Nations itself (through its Population Division) projects will be sharing the planet by the middle of this century. Can we reduce worldwide poverty and not further degrade our environment in the face of continued population pressure? No one really knows the answer to that question.

None of the basic resources required to expand food output—land, water, energy—can be considered abundant today. This especially affects the food security of people in less developed countries with rapidly rising food demands and small energy reserves. Even now in sub-Saharan Africa, food production is not keeping pace with population growth and this raises the fear that the world may have surpassed its ability to sustain current levels of food production. Genetically modified foods (GMF) may not be the answer, but some other solutions that we currently don't know about will have to come forth if we are not only to feed nine billion people, but to meet the United Nations Food and Agricultural Organization's goal of reducing malnutrition throughout the world. In other words, we want to feed more people, and feed them better than we are currently doing. It is often argued that there is, in fact, plenty of food in the world, but it is just not properly distributed to the places that need it. However, Jenkins and Scanlan (2001) have argued convincingly that even when these other factors are taken into account, population pressure has undermined the supply of food and the access to food in less developed nations, thus aggravating problems of food security in much of the world.

Keep in mind that every person added to the world's population requires energy to prepare food, to provide clothing and shelter, and to fuel economic life. Each increment in demand is another claim on energy resources (and creates problems of what to do with the by-products of energy use), forcing further global adjustments in the use of energy. Water is also an issue. More than half a billion people already face water scarcity, and an additional two to three billion live in areas that are water-stressed.

Paradoxically, it turns out that in order to grow enough food for the expected party of nine billion, we need fewer people in agricultural areas, not more. Humans are poor converters of energy, and machines are required to increase production. This means that cities will be absorbing an increasing fraction of the world's population. A safe and healthy urban environment requires the infrastructure necessary to provide clean water and to treat sewage and waste, and a growing economy requires transportation and communications infrastructure. Rapidly growing countries routinely find that their "demographic overhead" is too great to keep up with the demand for not only new infrastructure but also just to maintain what they already have.

The Danger of Demographic Fatigue

In the face of nearly relentless population pressure, as we continue to add tens of millions of people per year to the world's total, there is the danger that those of us who live in more developed countries will tire of and turn our backs on the less developed countries, where population growth continues to create one problem after another. After all, much of the population news that we hear about concerns declining birth rates and laments how low fertility in rich countries creates problems such as the need for immigrants. Our rates of growth may be low, but the "silent explosion" (Appleman 1965) continues in the rest of the world. Although we live in a demographically divided world, we in the rich countries were the ones who lit the fuse on the population bomb, and increasing globalization demands that we remember that there is no such thing as "their" problem—all problems are ours (and theirs) to deal with together.

How Will You Use This Information?

It is my hope that you will use the demographic insights in this book to improve your understanding of how the world works. In that general way, a demographic perspective can be personally empowering for you. But it is also no exaggeration to say that much of what you will glean from the book is potentially very useful in more mundane, but not necessarily less important, ways. **Demographics** is the term that has become associated with practical applications of population information and this can be a veritable treasure trove for you to mine.

Because you live in a social world, many if not most of the decisions that you have to make about life involve people, and when the issues relate to how many people there are, and where they live and work, and what they are like, demographics become part of the decision-making process, allowing you to lay out a systematic strategy for achieving your goal whether your problems relate to business, social, or political planning.

Demographics of Politics

Demographics are central to the political process in the United States. The constitutional basis of the Census of Population is to provide data for the **apportionment** of seats in the House of Representatives, and this process reaches down to the local level. Legislators also ask questions about how population growth and distribution influence the allocation of tax dollars. Will the increase in the older population bankrupt the Social Security system? Would federal subsidies to inner-city areas help lower the unemployment rate? Are illegal aliens creating an undue burden on the criminal justice systems in border counties? Politicians are not completely altruistic, of course, and have been known to apply demographics for their own personal use when campaign time rolls around. What are the characteristics of a candidate's supporters and where are these voters located?

The original purpose of the Census of Population in the United States was to determine how membership in the House of Representatives should be distributed. For example, the U.S. Bureau of Census was required by law to deliver total population counts for all 50 states from Census 2000 to the President on or before December 31, 2000. These data were then sent to the House of Representatives, where they were used to determine the number of representatives to which each state is entitled (Poston 1990; U.S. Census Bureau 2000a). Since the 1910 census, the total number of House seats has been fixed at 435, and the Constitution requires that every state get at least one seat. The first 50 House seats are thus used up. The question remains of how to allocate or apportion the remaining 385 seats, remembering that Congressional districts cannot cross state boundaries and there can be neither partial districts nor sharing of seats. Since 1940, this number has been based on a formula called the method of equal proportions, which rank-orders a state's priority for each of those 385 seats based on the total population of a state compared with all other states. The calculations themselves are cumbersome (Poston 1990; U.S. Census Bureau 2000b), but they produce an allocation of House seats that is now accepted without much criticism or controversy.

The reapportionment based on Census 2000 caused little stir for the 32 states that neither lost nor gained seats. However, there were changes in 18 states and, in general, the Sun Belt gained at the expense of the Rust Belt, repeating the pattern from the 1990 census. No state gained more than two seats (they included Arizona, Georgia, Florida, and Texas), and no state lost more than two (New York and Pennsylvania), as you can see in Figure 1.2. Once the number of seats per state has been reapportioned, the real fight begins. This involves **redistricting**, the reconfiguration of Congressional districts (geographic areas) that each seat will represent. In the 1960s, a series of Supreme Court decisions extended to the state and local levels the requirement that legislative districts be drawn in such a way as to ensure demographically equal representation ("one person, one vote"). Public Law 94-171 mandates that the Census Bureau provide population counts by age (under 18 and 18 and over) and by race and ethnicity down to the block level for local communities. These data are then used to redefine legislative district boundaries (Williams 1994). State legislatures are charged with redistricting Congressional seats, and local governments may also use the census data to redistrict city council or other boundaries.

The redistricting process is rooted in the U.S. Constitution, which requires a census to be taken every 10 years. After the census is complete, each state must change its political boundaries to make sure each district is equally populated. There are very few rules that govern the formation of a district. Newly drawn districts do not have to take into account existing political boundaries, such as cities and counties, nor do they have to take into account natural geographic boundaries, such as mountains and rivers. They need only be equally populated. When political boundaries are drawn solely for partisan gain, the process is called "gerrymandering," after early-nineteenth-century Massachusetts Governor (and later vice president of the United States) Elbridge Gerry. Gerry attempted to draw political districts to favor his own Federalist Party over the opposing Democrat-Republicans. Since that time, gerrymandering has become a common weapon that a majority party uses to increase its majority without increasing its share of the vote (Leonard 2000). This mix of demography, geography, and politics does not end with redistricting. Indeed, that is only part of the interplay. The candidates running for elected office in these and other places increasingly depend on demographic intelligence.

Demographics are used extensively by politicians seeking office. After researching the sociodemographic characteristics of voters in their constituency, they must either match their election platform to the political attitudes of such voters or realize that they will face an uphill battle to convince people to vote for them. "The demography of the voting-age population within a political district is a crucial ingredient in winning an election because demographic characteristics such as age, sex, race, and education are strongly related to the likelihood of voting and, to some extent, to political preferences" (Hill and Kent 1988:2). In 1969, the political analyst Kevin Phillips used the demographic characteristics of voters in the United States to

Figure 1.2 Apportionment of the U.S. House of Representatives Based on Census 2000 Results

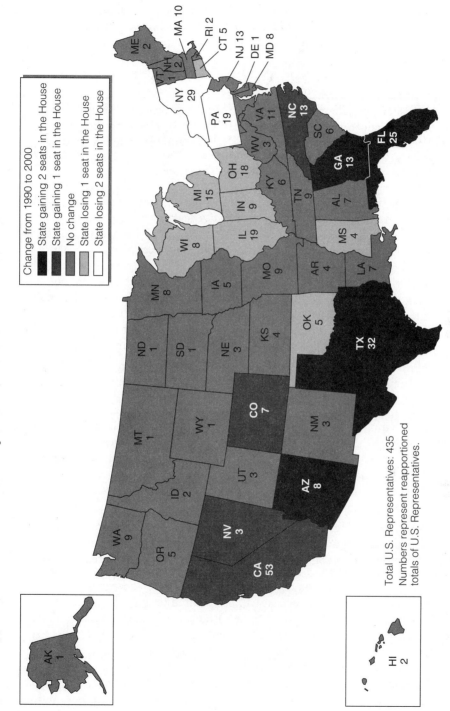

Change from 1990 to 2000

State gaining 2 seats in the House
State gaining 1 seat in the House
No change
State losing 1 seat in the House
State losing 2 seats in the House

ME 2

VT 1
NH 2
MA 10
RI 2
CT 5
NJ 13
DE 1
MD 8

NY 29
PA 19
WV 3
VA 11
NC 13
SC 6
GA 13
FL 25

OH 18
KY 6
TN 9
AL 7
MS 4

MI 15
IN 9
IL 19
WI 8

MN 8
IA 5
MO 9
AR 4
LA 7

ND 1
SD 1
NE 3
KS 4
OK 5
TX 32

MT 1
WY 1
CO 7
NM 3
AZ 8

WA 9
OR 5
ID 2
UT 3
NV 3
CA 53

AK 1

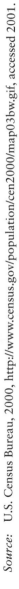

HI 2

Total U.S. Representatives: 435
Numbers represent reapportioned
totals of U.S. Representatives.

Source: U.S. Census Bureau, 2000, http://www.census.gov/population/cen2000/map03bw.gif, accessed 2001.

correctly predict "The Emerging Republican Majority" (Phillips 1969). In 2002, John Judis and Ruy Teixeira took a new look at the demographics of voters and decided that they now portend an emerging democratic majority (Judis and Teixeira 2002).

In truth, the strategies for "packaging and selling" a candidate are not unlike those used for any other product or service. Census and survey data are used to map the demographic characteristics of voters in a candidate's area and an appropriate plan of action is prepared. Remember, though, that to be elected, a candidate must appeal to *voters,* not necessarily to all people who reside in an area. Since not all residents may be eligible to vote (because they are too young, are not citizens, or are convicted felons) and because not all who are eligible will actually vote, these differences may be crucial. It has become increasingly clear that those candidates who pay attention to demographics have a greater chance of winning the opportunity to put their demographic literacy to work in effective legislation.

The United States is certainly not alone in the use of demographic data for political campaign work. Prior to the Mexican presidential election in 2000, Mexican expatriates living in the United States found themselves being courted by groups wanting them to drive to Mexico on election day to vote (absentee ballots are not allowed in Mexico) or at least to encourage their relatives in Mexico to vote (Millman 2000).

Even after election day, of course, demographics underlie many of the major social issues that confront national and state legislators, as I will discuss in virtually every chapter of this book. The demographics of the baby boom, for example, helped fuel inflation in the United States during the 1970s. Government policies in that period were oriented toward creating new jobs for the swelling numbers of labor force entrants, directly contributing to inflation through government expenditures and indirectly having an effect because Congress's attention was turned away from combating inflation. This same bulge in the young adult male population also contributed to the ability of the United States to get as involved as it did in the Vietnam War—the same "dogs of war" phenomenon mentioned earlier in connection with current issues in the Middle East.

The baby boom is still having an impact on legislative analysis in the new millennium. The question has become: How will the country finance the retirement and the health care needs of baby boomers as they age and retire? This set of questions loomed large in the debates in the U.S. presidential election in 2000, and many nations are facing similar issues, as declining fertility and increased longevity have contributed to the prospect of substantial increases in both the number and percentage of the older population. As the older cohorts begin to squeeze national systems of social insurance, legislative action will be required to make long-run changes in the financing and benefit structure of these systems if they are to survive. Changes will be made, of course, even if their exact shape is difficult to forecast. Increased self-reliance is one proposed solution: requiring people when younger to save for their own retirement, through mandatory contributions to mutual funds and other investment instruments. It may also be that, when the time comes, taxes will be raised on younger people in order to bail out those older people who, in fact, did not save enough for their retirement. Another solution could be to allow even more immigration in order to ensure a supply of young wage earners and taxpayers to support the retirement of the older population.

Demographics of Social Planning

Social agencies have also recognized the value of demographics for planning. School districts ask: How many students will there be next year? Or: Do we need to redraw school attendance areas? Police departments need to ask what the changes in population growth might portend for criminal activity in their areas. Health officials plan services by identifying specific high-risk populations (people who have an unusually high need for particular services) and attempting to meet their needs: Where is the best location for a new skilled-nursing facility? Or: Does the community hospital really need an additional 40 beds?

Demographics are widely used in American society to chart population movements and plan for social change. When such planning breaks down, as it often does in cities of third world nations where rapid population growth is occurring, the result can be a certain amount of chaos, social foment, and even political upheaval. However, in the United States, regional planning agencies have become commonplace in an effort to coordinate the wide range of public services that maintain a high quality of life. Laws (1994) has suggested that the overriding demographic changes to which social policy planners must respond in the United States include the transition in family and household structure, immigration, increased longevity, and the changing distribution of the population within metropolitan areas.

Perhaps most obvious in its need for demographic information is the educational system. Public elementary and secondary school districts cannot readily recruit students or market their services to new prospects; they rise and fall on demographic currents that determine enrollment. Of course, not every community experiences the national trend; there is variation among and between individual school districts and they have a need for precise information, because even within a district some geographic areas may be growing while others are diminishing in the number of school-aged children.

Do you close some schools and open new ones elsewhere? If so, what do you do with the old schools, and how do you pay for the new ones? Should children be bused from one area to another? If so, how is that to be organized and paid for? Can teachers be retrained to fill classes where demand is growing? These and a host of related questions push school district demographers to pore over questionnaires about the number of younger, pre-school-aged siblings reported by currently enrolled students; examine birth records to see where the families are that are having babies; and talk to realtors to assess the potential for home turnover and to evaluate the likely number of children among families moving in and out. From these data, they construct enrollment projections, which are a vital part of public education planning.

Educators must also bear in mind the important influence economic conditions have on job opportunities, and thus on the likelihood that people will be moving in or out. This means that school district demographers increasingly pay attention to regional economic models that are designed to forecast local economic trends on the theory that people follow jobs (and different kinds of people follow different kinds of jobs), so the creation of jobs within a region will influence local demographic trends (Exter 1996). Demographic conditions can also affect the school district in ways that go beyond the numbers. Immigrants to the United States from Latin America tend to have substantially lower levels of education than prevail in the

United States, so they may have relatively little experience with schools in their native country, much less in the U.S. When their children start attending school, they may be generally unable to help them with school work. Since parental involvement is a key ingredient in student successs, school districts are faced with the need to create new policies and programs to educate immigrant parents about what their children are experiencing in school.

Nor are colleges and universities immune to the effects of demographics (Fishlow 1997). The number of high school graduates in the United States declined each year between 1988 and 1992; then it began increasing again in the mid-1990s as the baby boomlet children reached college age, but demand for college has increased more rapidly than population growth because an increasing fraction of high school graduates are now attending college. In California, immigration since the 1960s has brought to the state a large number of young adults who then had children, and those kids began to move through high school and into colleges in the late 1990s. Between 1990 and 1995, the number of undergraduates in California's public universities and community colleges actually declined. But since then, the number has steadily increased and it is projected to continue climbing at least through 2011 (California Department of Finance 2002), contributing to what some educators have called Tidal Wave II (the first tidal wave, of course, was brought about by the baby boom).

The same age structure changes that influence the educational system also have an impact on the health care industry. Over the years, hospitals and other health care providers have learned that they have to "reposition" themselves in a classic marketing sense to meet the needs of a society that is changing demographically (Rives 1997). Part of that repositioning has been the shift from medical care (treating the medically needy) to a more comprehensive health care perspective (keeping people as healthy as possible, including treating them medically as needed) (Pol and Thomas 2001).

Most communities in North America, Europe, and Japan are facing the aging of their populations. This fact is significant when you consider that in the United States people aged 65 and older are three times more likely than those aged 18 to 44 to be admitted to a hospital, and they also spend an average of three more days each year in the hospital once admitted (National Center for Health Statistics 2002). As I discuss in Chapter 5, the ends of the age continuum are where the risks of death are highest, and the pattern of death rates is a good (albeit not perfect) index of demand for health services. In 1960, 59 percent of all deaths occurred to people 65 or older, and those aged 85 or older accounted for 11 percent of all deaths. By 2000, 75 percent of all deaths in the U.S. were people aged 65 and older, and 27 percent were people 85 and older (Minino and Smith 2001). As these percentages increase, the kind of physician training required obviously shifts to those diseases and conditions that affect older people.

At the local level, demographic detectives are usually available to spot the health care pitfalls and opportunities. As Kinstler and Pol (1990) have pointed out, local hospitals and physicians need to track local changes in age and sex structure (because of the implied shifts in demand from the population that is aging in place), and they must also keep an eye on in- and out-migration. As we will discuss in Chapter 7, migration can have a dramatic short-run impact on health care demand that may seem sudden if no one is paying attention to population trends in the area.

Unfortunately, broad health issues and problems often transcend the ability of local communities to find a remedy. Infant death resulting from low birth weight, for example, is a national issue, even though it must be dealt with at the local level. Likewise, the preference of older persons to remain at home even if they suffer from severe chronic illness is a national issue with repercussions at the community level. In general, the provision of health services is so costly, and the training of health personnel so lengthy and expensive, that legislative action is often required to permit the system of health services to respond adequately to demographic changes.

Crime, like health, is closely tied to the age and sex structure. Everybody knows that certain kinds of crimes—especially street crimes and drug-related crimes—are more likely to occur in neighborhoods with particular sociodemographic characteristics (Sampson and Morenoff 2003; Yablonsky 1997). Areas where the poverty level and unemployment are high are especially prone to these visible kinds of crime (as opposed to white collar crime, for example) (Short 1997). Crime occurs locally, but it may be influenced by demographic forces that extend well beyond the local region. For example, counties that are adjacent to the U.S.-Mexico border have discovered that their crime rates are elevated by crimes committed by people illegally crossing the border (Salant et al. 2001). The U.S. Department of Justice has responded with grants to these counties that cover at least part of the cost of incarcerating criminal illegal aliens.

"Build it and they will come" may apply to ballparks and it might have once applied to churches in America, but why leave it to chance? Religious organizations and private companies employ demographic researchers to keep track of household preferences in religion so that churches can respond most effectively to the needs of their members and can, in essence, "market" themselves to the demographically appropriate community. For example, the Presbyterian Church conducts a panel survey of its membership. In the late 1990s, the median age of members was 55; 61 percent were female; 98 percent described their race/ethnicity as white; and more than half lived within three miles of their church (The Presbyterian Panel 1998). This is the kind of information that can then be used to help locate new sites or expand local membership (Wendelken 1996). It has even been argued that demographic trends are encouraging a revival in religiosity in the United States. Roof (1999) has suggested that as baby boomers have aged they have become more religious (a not uncommon phenomenon among humans), but that being boomers they have been drawn to new types of Christian Protestant churches, rather than swelling the ranks of the more established denominations.

Demographics of Business

Making a profit in business requires having an edge on your competitors. This may mean that your product or service is better or at least different in some meaningful way, or it may mean that you have been able to find and keep customers that your competition has ignored (**marketing demographics** and **cluster marketing**), or it may mean that you have found a better location for your business than have your competitors based on an analysis of demographic data (**site selection demographics**). Profits are also maximized by investing in the businesses with the greatest opportunities for growth, based on changing demographics (**investment demographics**), as well by being able to employ and keep the best possible labor force (**demographics of human resource management**). In each instance, then, demographic knowledge can be one of the keys to generating higher profits: When it comes to understanding today's consumer marketplace, just about the only thing that's certain is that uncertainty reigns. The speed of technological change, the volatile global economy, the emergence of media-savvy and ever-more-demanding customers have all coalesced into the blur that characterizes business-as-usual at the beginning of this new century. Grappling with uncertainty in business planning requires more than guesswork, warns business guru Peter Drucker. It requires looking at "what has already happened that will create the future. The first place to look," says Drucker, "is in demographics" (quoted in Russell 1999).

Marketing Demographics   One of the most impressive (and profitable) ways in which demographics is used in business is for marketing. Demographics are employed to segment and target the market for a product, an approach so popular it has even been suggested that a television program's demographic base now determines its commercial success "far more than sheer audience numbers" (Lilly 2001:A13). **Segmentation** refers to the manufacturing and packaging of products or the provision of services that appeal to specific sociodemographically identifiable groups within the population. "Consumer markets are segmented on the basis of such demographic variables as geographic location, rate of product usage, income, age, sex, education, stage in the family life cycle, religion, race, and social class. Industrial markets are segmented demographically according to such variables as geographic location, kind of business, rate of product usage, and size of user" (Levy et al. 1994:177). Automobile manufacturers are most famous for segmenting the market and producing different cars to appeal to different categories of people. Ford did this very successfully with the original Mustang:

> Our market research showed that the youthful image of the new decade had a firm basis in demographic reality. Millions of teenagers, born in the postwar baby boom, were about to surge into the national marketplace. Here was a market in search of a car. Any car that would appeal to these young customers had to have three main features: great styling, strong performance and a low price (Iacocca 1984:64).

And, Ford continued that approach when it introduced its Focus as a car aimed specifically at "Echo Boomers (children of the baby boomers) and Gen Xers" (Kranhold 1999:B8).

Closely related to segmenting (sometimes indistinguishable in the literature) is the concept of **targeting.** It involves picking out particular sociodemographic characteristics of people who might purchase what you have to offer, then appealing to the consumer tastes and behavior reflected in those particular characteristics. Throughout this book, I will be emphasizing the ways in which sociodemographic characteristics can influence behavior. It is this relationship that is of interest to business. The kinds of products you buy and how much you will spend on those products is dictated, at least in part, by factors such as age, gender, education, income, socioethnic identity, household structure and living arrangements, and whether you live in a city and even where in the city you live.

People at different ages have different needs and tastes for products and differing amounts of money to spend. Companies catering to the youngest age group have to keep track of the number of births (their potential market) as well as the characteristics of the parents and grandparents (who spend the money on behalf of the babies). The baby market has seen some wild fluctuations in recent decades in the United States, as you can see in Figure 1.3. The number of babies being born each year plummeted during the 1960s and 1970s, rebounded in the 1980s, peaked in 1990, and slacked off in the 1990s, but is projected to rise steadily through the first part of the twenty-first century. It is a dangerous business, however, to be lulled into believing that every company dealing with baby products will necessarily live or die on the peaks and troughs of birth cycles. In 2000, there were 4.1 million births in the United States, compared to 4.3 million at the height of the baby boom in 1958. Yet, in 2000 there were 1.6 million *first* births, compared to only 1.1 million in 1958. If you have hung around new parents and new grandparents, you know that people react differently to first births than to others. In particular, they open their pocketbooks wider to pay for cribs, buggies, strollers, diapers, and every conceivable type of baby toy designed to stimulate and "improve the quality" of the baby's life. Businesses that cater to these needs, then, are as sensitive to birth order as they are to the absolute volume of births.

As the ups and downs in births work their way through the age structure, we can see that the age pattern of teenagers (ages 10 to 19) follows a wave that is a delayed reaction to the birth wave. Thus, as the baby market dried up in the 1970s (the babies born during this time are popularly known as Generation X), the young adult population was expanding, only to deflate again in the 1980s, but by then the baby market was growing again. Now, in the first part of the twenty-first century, the crop of babies born in the 1980s is moving through the teen and young adult years. Marketers, in fact, have labeled people born from 1977 through 1994 as Generation Y. In the 1990s, the parents of these children were spending money on their education and entertainment. Later, they will be competing for space in college and hunting for well-paying jobs, which will reinvigorate job-training and placement agencies. The number of people at each age is only part of the marketing equation, of course. The other part is how much money they have to spend, and it turns out that teenagers and young adults now have more money per person than their predecessors (Bounds 2000). This is due to a combination of factors, including a greater fraction of younger people who work part-time and an increase in the fraction of households with two wage earners but fewer children among whom discretionary income must be shared. The result has been an increase in the economic clout of young people—which especially benefits the food, entertainment, sports, and apparel markets.

The middle age group (represented in Figure 1.3 by people aged 40 to 49) was relatively unchanged in size and essentially unnoticed until the baby boom generation began moving into it in the 1980s. Since then, it has become a new "boom" industry. Laser eye surgery has surged, as have sales of walking shoes (running shoe sales slowed to a walk). When not walking, the middle-aged baby boomers were

Figure 1.3   Riding the Age Wave: Births and Selected Age Groups in the United States

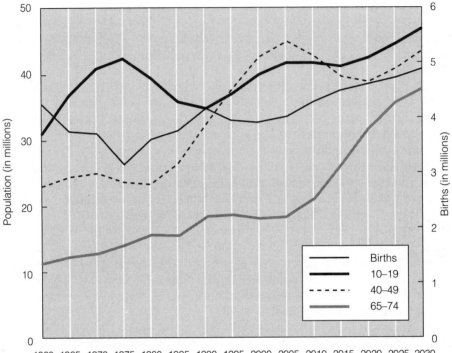

Sources: Data on births 1960–90 are from U.S. Census Bureau, 1996, Statistical Abstract of the United States (Washington, DC: Government Printing Office): Table 90; age data for 1960–90 are from the United Nations, 1997, The Sex and Age Distribution of the World Populations: The 1996 Revisions (New York: United Nations); birth and age data for 1995–2030 are from U.S. Census Bureau, 2000, Population Projections of the United States by Age, Sex, Race, Hispanic Origin, and Nativity: 1999 to 2100, Internet Release Date: January 13, 2000. (Note that these projections are not based on the results of the 2000 Census data. The numbers will almost certainly be somewhat different when the Census Bureau completes its revisions of the projections.)

driving their luxury or near-luxury sport utility vehicles (and are now eyeing the new hybrid cars). It was the baby boom reaching middle age that helped to fatten the nonfat market. People aged 45 to 49 are more likely than any other age to be concerned about the health effects of fat (Mogelonsky 1995), and as baby boomers began reaching that age, the market flourished for low-fat and nonfat foods. However, younger people, especially college students, have led the movement toward vegetarian meals (Walker 1995), and in fact vegetarians are disproportionately college-educated people with higher-than-average incomes (Fetto 2000). They will probably carry those food preferences into their middle ages at a time when the number of people 40 to 49 will dip as the baby boomers are replaced by the smaller cohort of Generation X.

The young-old population (ages 65 to 74) has been steadily increasing in numbers over time and, as you will learn in Chapter 9, has also become increasingly affluent. This segment of the population creates a market for a variety of things, from leisure travel, to appliances with larger print, to door handles that are shaped to be used more easily by arthritic fingers. Perhaps most importantly, the aging of the population in North America has spurred the marketing of health services and products aimed at that age group.

Johnson & Johnson provides a good example of a company that has kept its eye on the changing demographics not only of the United States, but of the world in general. The company got its start in the 1880s when Robert Wood Johnson began selling sterile bandages and surgical products—innovations built on Lister's germ theory that helped to lower death rates in hospitals. Later on, during the years of the baby boom, Johnson & Johnson flourished by selling baby products. As the baby boom waned, the company continued to diversify its product line in a demographically relevant way, including acquiring ownership of both Ortho Pharmaceuticals (the largest U.S. manufacturer of contraceptives—helping to keep the birth rate low—and a large manufacturer of drugs to treat chronic diseases associated with aging) and Tylenol (one of the world's most popular pain relievers).

The changing roles of women in society have also provided opportunities for businesses. Consider, for example, that more than half of all married women are now working, compared with one in four only a few decades ago. The consumer needs and interests of women have thus been dramatically altered and businesses have adjusted their targeting accordingly. As the status and roles of women have changed, there have been complementary shifts in the male population that require adjustments in marketing strategy. Increasing proportions of men are remaining single longer before marrying (and single after a divorce), with attendant problems of housekeeping and cooking, and married men are also increasingly likely to involve themselves in domestic affairs.

Cluster Marketing Just because you know the sociodemographic target groups for your business does not mean that you can easily locate and appeal to such people or households. However, the fact that "birds of a feather flock together" has meant that neighborhoods can be identified on the basis of a whole set of shared sociodemographic characteristics. This greatly facilitates the process of marketing to particular groups in a process known as cluster marketing. In the 1970s,

> . . . a computer scientist turned entrepreneur named Jonathan Robbin devised a wildly popular target-marketing system by matching zip codes with census data and consumer surveys. Christening his creation PRIZM (Potential Rating Index for Zip Markets), he programmed computers to sort the nation's 36,000 zips into forty "lifestyle clusters." Zip 85254 in Northeast Phoenix, Arizona, for instance, belongs to what he called the Furs and Station Wagons cluster, where surveys indicate that residents tend to buy lots of vermouth, belong to a country club, read Gourmet and vote the GOP ticket. In 02151, a Revere Beach, Massachusetts, zip designated Old Yankee Rows, tastes lean toward beer, fraternal clubs, Lakeland Boating and whoever the Democrats are supporting (Weiss 1988:xii).

The PRIZM system has made Robbin's company, Claritas Corporation, one of the largest and most successful demographics firms in the world. A core principle is that *where* you live is a good predictor of *how* you live (Weiss 2001). It combines demographic characteristics with lifestyle variables and permits a business to home in on the specific neighborhoods where its products can be most profitably marketed. In keeping with the changing demographics of America, Claritas added new neighborhood clusters to their system in the 1990s including "Latino America," concentrated in Miami, Chicago, and the Southwest and representing middle-class neighborhoods with a high percentage of families that are Hispanic, and "Family Scramble," which is a set of neighborhoods concentrated along the U.S.-Mexican border.

Site Selection Demographics   One of the most common uses to which demographic data are put is in helping businesses select a site (Pol and Thomas 1997). This is an extension of the concept of targeting because businesses that rely on walk-in traffic naturally prefer a location that is closest to the maximum number of people who are likely to want to buy goods or services from them. Site selection often involves the innovative use of geographic information systems. For example, in the mid-1990s American Honda Corporation wanted to locate its motorcycle dealerships in those neighborhoods closest to the people who would likely be buying them (a 30-something married male who attended college and earns about $33,200 per year; Hoerning 1996). Using geocoded census data, the likely neighborhoods were found, possible sites for the businesses were located, and these properties were evaluated according to the number of potential customers living within a given radius of each site. Demographics do not represent the sole measure of potential success—access to the site and visibility of the location are also important, and a firm has to be able to buy or lease the property for a price that allows it to make a profit, but the population component is the single most important factor for many retail businesses (Albert 1998; Fenker 1996).

Manufacturing and service firms also make use of demographics in site selection. Knowledge of the optimal sociodemographic characteristics for a firm's workers (for instance, males or females, preferred racial or ethnic groups, required level of education or background) can be crucial in helping a company decide where it should locate a new plant or facility. These decisions can then feed back to influence the community's long-term demographic structure if firms that need similar kinds of employees wind up clustering together in the same geographic region, such as the Silicon Valley in northern California. Of course, these are the same demographic decision tools that convinced some companies to move their manufacturing sites out of the United States into less developed nations.

Investment Demographics   To invest is to put your money to use for the purpose of securing a gain in income. Companies invest in themselves by plowing their profits back into the company in order to develop new products or expand or change markets; individuals (and companies also) invest by buying stocks or bonds of companies viewed as having a profitable future, or by purchasing real property whose value is expected to rise in the future. Clearly, a wide range of factors goes into the decision to invest in one product line, a particular company, or a specific piece of property, but one very important consideration is future demographic change. Companies that plan for demographic shifts have a better chance at success than those that do not. In fact, it has long been said that "Wall Street uses demographics to pinpoint opportunity" (Nordberg 1983:23).

Basically, making sound investment decisions (as opposed to lucky ones) involves peering into the future, forecasting likely scenarios, and then acting on the basis of what seems likely to happen. After reading this book, you should have a good feel for the shape of things to come demographically. Most people do not, but those who do have an edge in life. For example, in 1991, a veteran retail analyst in New York City looked at demographic trends and concluded that "two-worker families move less and end up putting more money into home furnishings and home improvement. A cohort of aging empty-nesters will do the same" (Oliver 1991:57). This led to the advice to buy stock in Home Depot, Pier 1 Imports, and Williams-Sonoma. All three have prospered since then, although of course we cannot be sure how much of the gain was due specifically to demographic trends.

What do the demographics suggest about investment opportunities? The fact that 90 percent of the world's population growth in the foreseeable future will occur in the less developed nations is, as already noted, an important reason for the globalization of business and the internationalization of investment. Developing nations have millions of babies, and each one needs some kind of diaper. Procter & Gamble, maker of Pampers disposable diapers, has found a huge market out there. Declining fertility in many developing countries is a very recent phenomenon, and in the meantime much of the world is experiencing a rapid increase in the number of young adults (trends that we will examine in detail throughout the book). From Malaysia to Argentina, young adults are buying cell phones, satellite dishes, color TVs, CD players, and the perennial favorites, Levi's and Coca-Cola. Companies selling in these markets are thus potential investment targets.

International investors have been particularly intrigued by the world's two most populous countries, China and India. General Motors, Chrysler, and Ford all have invested in car manufacturing in China, as have Volkswagen and Peugeot Citroën from Europe. The problem, of course, is that a huge population does not necessarily mean a huge market if most people are poor. Pizza Hut and McDonald's serve up fast food in China, but the average Chinese consumer cannot afford very many expensive goods. The same can be said for India, which is almost as populous, but less well-off than China. The so-called "consuming class" in India (those with at least some discretionary income) may number as many as 150 million (Baldauf 2001), but the middle class in the American or European sense in India represents only a tiny fraction of the total population. Nonetheless, it is growing and in that growth lies opportunity. Take toothpaste, for example. Colgate-Palmolive Company estimates that in a country of more than one billion, only 4 in 10 Indians use toothpaste (compared with 9 out of 10 in China; Jordan 1999), and Colgate's goal is to tap into the profits associated with cleaner teeth in India.

Demographics of Human Resource Management Whether or not a firm uses demographics in its marketing, site selection or investment decisions, nearly all firms must cope with the fact that their employees potentially are drawn from differing sociodemographic strata and geographic areas. A recognition of these differences may well enable managers to deal more humanely and successfully with their human resources.

Sensitivity to the needs of workers is viewed by most management professionals as a prerequisite for maximizing employee morale and productivity, and the underlying demographic makeup of a company influences what sorts of needs exist. A company that employs a large number of women may find itself dealing with the potential effects of maternity leaves, with the desirability of opening a child care center, and with protecting employees from sexual harassment and the higher crime risk that might be associated with working late and leaving for home after dark. A company that counts on immigrants or resident aliens for its labor force will find itself needing bilingual supervisors and a set of policies recognizing that different cultural backgrounds demand different kinds of inducements to productivity.

Effective managers also must be aware of the possible impact of demographic trends on their workforce. As the baby bust generation moved into the labor force in the 1980s and 1990s, these smaller-sized cohorts created greater competition among companies for the best young workers, keeping unemployment low (all other things staying the same) and leading some companies to think about ways to recruit younger workers from other states or other countries. In the meantime, the baby boomers were clogging up the middle management levels, significantly slowing the rate of promotion and leading to executive frustrations, and contributing to the popularity of "downsizing" American companies. A glance back at Figure 1.3 will be a quick reminder of how the U.S. labor force will be changing over time—changes with which many organizations will have to cope.

Summary and Conclusion

It is an often-repeated phrase that "demography is destiny," and the goal of this book is to help you to cope with the demographic part of your own destiny and that of your community, and to better understand the changes occurring all over the world. Demographic analysis helps you do this by seeking out both the causes and the consequences of population change. The absolute size of population change is very important, as is the rate of change, and of course, the direction (growth or decline). The past 200 years have witnessed almost nonstop growth in most places in the world, but the rate is slowing down, even though we are continuing to add nearly 10,000 people to the world's total every hour of every day.

You may prefer to bury your head in the sand and pretend that there is no world beyond your own circle of family and friends, but everything happening around you is influenced by demographic events. I refer not just to the big things like regional conflict, globalization, and massive migration movements, but even to little things that affect you directly, like the kinds of stores that operate in your neighborhood, the goods that are stocked on your local supermarket shelf, the availability of a hospital emergency room, and the kinds of jobs available to college graduates in your community. Influential decision makers in government agencies, social and health organizations, and business firms now routinely base their actions at least partly on their assessment of the changing demographics of an area. So, both locally and globally, demographic forces are at work to change and challenge your future. In the next chapter, I outline the basic facts of the global demographic picture.

Main Points

1. Demography is concerned with virtually everything that influences or can be influenced by population size, distribution, processes, structure, or characteristics.

2. The cornerstones of population studies are the processes of mortality (a deadly subject), fertility (a well-conceived topic), and migration (a moving experience).

3. Almost everything in your life has demographic underpinnings that you should understand.

4. Examples of global issues that have deep and important demographic components include terrorism and regional conflict, violence in sub-Saharan Africa, the backlash against immigrants, globalization, and the degradation of the environment.

5. There are also "local" uses for demographic information, which is usually labeled "demographics" and defined as the application of population theory and methods to the solution of practical problems.

6. Demographics is the central ingredient in congressional reapportionment and redistricting in the U.S.

7. Politicians also find demographics helpful in analyzing legislation and in developing their strategy for their own election to office.

8. Local agencies use demographics to plan for the adequate provision of services for their communities.

9. Education and criminal justice are areas of social planning in which demographics have been used extensively.

10. Health planners pay close attention to demographics because sickness and health are affected by sociodemographic characteristics and because the demand for services shifts with demographic trends.

11. A major use of demographics is to market products and services in the private sector.

12. Marketing demographics involve segmenting markets (tailoring products and services to a specific demographic group) and targeting (aiming the advertising of a product or service to a specific demographic group).

13. Cluster marketing relates demographic information about people to information about where they live.

14. Demographics are an important component of site selection for many types of businesses.

15. Investors use demographics to pinpoint areas of potential market growth, because population is a major factor behind social change (and thus opportunity).

16. Human resource managers use demographics to help increase their awareness of the special qualities and needs that exist among their present and prospective employees.

Suggested Readings

1. Paul Demeny and Geoffrey McNicoll, Editors, 2003, Encyclopedia of Population (2 Volumes). (New York: MacMillan Reference USA).

   An important reference source for any person (lay or professional) seeking an expert overview of virtually every major topic in the field of demography or population studies. This can be thought of as an update of both the Dictionary of Demography (William and Renee Petersen, 1985), and the International Encyclopedia of Population (edited by John Ross, 1982).

2. Dudley L. Poston Jr. and Michael Micklin, 2003, Handbook of Population. (New York: Plenum/Kluwer).

   This volume covers fewer topics, but deals with each in more detail than does the Encyclopedia of Population. It is thought of as an update of Hauser and Duncan's 1959 classic, The Study of Population.

3. The Economist, 2002, "Special Report: Demography and the West," The Economist, August 24th.

   An impressively concise summary of the different demographic paths that the United States and Europe are taking. The U.S. has buffered its falling fertility rates with high levels of immigration, whereas Europe has not yet embraced that demographic solution.

4. Lester R. Brown and associates, State of the World: A Worldwatch Institute Report on Progress Toward a Sustainable Society (New York: W. W. Norton), published annually.

   The goal of this annual set of research papers by scholars at the Worldwatch Institute in Washington is to analyze what the current trends in population growth (and related issues) mean for the future of the globe.

5. Michael J. Weiss, 2001, The Clustered World: How We Live, What We Buy, and What It All Means about Who We Are: The Clustering of America (New York: Little Brown).

   A journalistic but very detailed description of the PRIZM system of clustering zip codes according to demographic and lifestyle characteristics. Weiss describes the 62 lifestyle clusters and suggests that America is characterized more as a salad bowl than a melting pot.

Web Sites of Interest

Remember that Web sites are not as permanent as books and journals, so I cannot guarantee that each of the following websites still exists at the moment you are reading this. It is usually possible, however, to retrieve the contents of old websites at http://www.archive.org.

1. **http://www.census.gov**
   The Web site of the U.S. Census Bureau has several very useful features, including U.S. and world population information and U.S. and world population clocks (where you can check the latest estimate of the total U.S. and world populations).

2. **http://www.prb.org**
   The Population Reference Bureau (PRB) in Washington, D.C., is a world leader in developing and distributing population information. The site includes regularly updated information about PRB's own activities, as well as linkages to a wide range of other population-related Web sites all over the world.

3. **http://www.un.org/popin**

   The Population Division of the United Nations has created this site to link you not only to its research, but also to the other UN agencies that produce population-related data and reports.

4. **http://www.un.org/Pubs/CyberSchoolBus/infonation/e_infonation.htm**

   This United Nations Web site allows you to choose several countries at a time and look at demographic, economic, and social indicators for those nations.

5. **http://cluster1.claritas.com**

   The Web site for Claritas (a Latin term for clarity), the geodemographics firm that developed the PRIZM cluster marketing approach, allows you to put in a ZIP code of your choice (such as your own neighborhood) for free to see how that neighborhood would be characterized in demographic terms.

CHAPTER 2
Global Population Trends

Figure 2.1 Cartogram of Countries of the World by Population Size

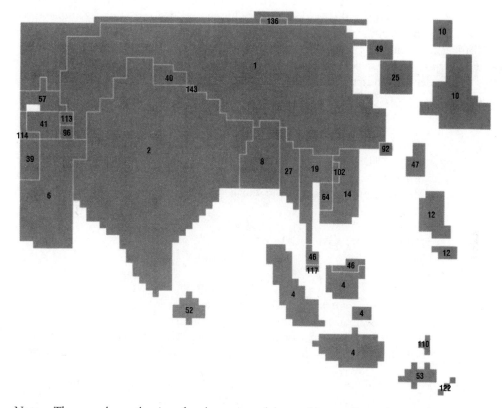

Note: The map shows the size of each country of the world according to its population as estimated for the year 2000 (based on data from the U.S. Bureau of the Census International Data Base). See Appendix for population data.

"The increase of population is the most revolutionary phenomenon of our times." With that, Spanish philosopher Ortega y Gasset neatly summarized the situation more than 50 years ago, and it is still difficult to quarrel with his assessment. Consider that at this moment you are sharing the planet, vying for resources, with more than six billion other people, and before the year 2030 yet another two billion souls will have joined you at the world's table. This in and of itself is pretty impressive, but it becomes truly alarming when you realize that this is a huge leap up from the "only" 2.5 billion in residence as recently as 1950. This phenomenon is obviously without precedent, so we are sailing in uncharted territory.

In order to deal intelligently with a future that will be shared with billions more people than today, we have to understand why the populations of so many countries are still growing (and why some are not) and what happens to societies as their patterns of birth, death, or migration change. In this chapter, I trace the history of population growth in the world to give you a clue as to how we got ourselves into the current situation. Then I will take you on a brief guided tour through each of the world's major regions, highlighting current patterns of population size and rates of growth, with a special emphasis on the world's 10 most populous nations.

World Population Growth

A Brief History

Human beings (*homo sapiens*) have been around for at least 200,000 years (Cann and Wilson 2003; Wilson and Cann 1992). For almost all of that time, their presence on the earth was scarcely noticeable. Humans were hunter-gatherers living a primitive existence marked by high fertility and high mortality, and only very slow population growth. Given the amount of space required by a hunting-gathering society, it seems unlikely that the earth could support more than several million people living like that (Biraben 1979; Coale 1974; McEvedy and Jones 1978), so it is no surprise that the population of the world on the eve of the **Agricultural Revolution** 10,000 years ago (-8,000) is estimated at about four million (see Table 2.1). Many people argue that the Agricultural Revolution occurred slowly but pervasively across the face of the earth precisely because the hunting-gathering populations were growing just enough to push the limit of the **carrying capacity** of that way of life (Boserup 1965; Cohen 1977; Harris and Ross 1987; Sanderson 1995). Carrying capacity refers to the number of people that can be supported in an area given the available physical resources and the way that people use those resources (Miller 2004). Hunting and gathering use resources *extensively* rather than *intensively,* so it was natural that over tens of thousands of years humans would move into the remote corners of the earth in search of sustenance. Eventually, people in most of those corners began to use the environment more intensively, leading to the more sedentary, agricultural way of life that has characterized most of human society for the past 10,000 years.

The population began to grow more noticeably after the Agricultural Revolution, as Table 2.1 shows. Between 8000 B.C., and 5000 B.C., about 372 people on average were being added to the world's total population each year, but by 500 B.C.,

Table 2.1   Population Growth Was Very Slow in the Earlier Years of Human Existence, but Has Accelerated in the Past 250 Years

Year	Population (in millions)	Average Annual Growth Rate	Doubling Time	Average Annual Population Increase
−8000	4			
−5000	5	0.01	9,277	372
−3000	14	0.05	1,340	7,207
−2000	27	0.07	1,051	17,733
−1000	50	0.06	1,120	30,809
−500	100	0.14	498	138,629
1	211	0.15	463	314,473
500	198	−0.01	−5,414	−25,233
1000	290	0.08	906	220,784
1100	311	0.07	1,000	214,223
1200	380	0.20	342	767,552
1300	396	0.04	1,673	163,322
1400	362	−0.09	−769	−324,967
1500	473	0.27	258	1,265,044
1600	562	0.17	400	968,924
1700	645	0.14	504	882,791
1750	764	0.34	203	2,599,021
1800	945	0.42	163	4,010,381
1850	1,234	0.53	129	6,585,656
1900	1,654	0.59	118	9,688,904
1910	1,750	0.57	122	9,915,685
1920	1,860	0.61	113	11,338,690
1930	2,070	1.07	65	22,143,229
1940	2,300	1.05	65	24,232,919
1950	2,556	1.47	47	37,573,200
1960	3,039	1.33	52	40,418,700
1970	3,707	2.07	33	76,734,900
1980	4,454	1.69	41	75,272,600
1990	5,278	1.56	44	82,336,800
2000	6,071	1.25	55	75,887,500
2010	6,822	1.17	59	80,007,822
2020	7,524	1.00	69	75,391,959
2030	8,065	0.80	86	64,343,206
2040	8,492	0.65	106	55,537,184
2050	8,919	0.51	135	45,502,212

Note: Negative doubling times indicate the number of years for the population to reach half its current size.

Sources: The population data from −8000 through 1990 are drawn from U.S. Census Bureau, International Programs Center, "Historical Estimates of the World Population," (http://www.census.gov/ipc/www/worldhis.html), accessed 2003. The numbers from −8000 through 1940 reflect the average of the estimates made by Biraben (1979), Durand (1967), and McEvedy and Jones (1978). Population figures for 2000 through 2050 are medium variant projections from the United Nations Population Division, 2003, World Population Prospects: The 2002 Revision (www.un.org/esa/population/publications/wpp2002/), accessed 2003.

as major civilizations were being established in China and Greece, the world was adding nearly 139,000 people each year to the total. By the time of Christ (the Roman Period, 1 A.D.) there may well have been more than 200 million people on the planet, increasing by more than 300,000 each year. There was some backsliding in the third through fifth centuries A.D., when increases in mortality led to declining population size in the Mediterranean area as the Roman Empire collapsed, and in China as the Han empire collapsed (McEvedy and Jones 1978). Population growth recovered its momentum only to be swatted down by the plague, or Black Death, that arrived in Europe in the middle of the fourteenth century and didn't leave until the middle of the seventeenth century (Cantor 2001). The rate of growth began clearly to increase after that, especially in Europe, and on the eve of the **Industrial Revolution** in the middle of the eighteenth century (about 1750), the population of the world was approaching one billion people and was increasing by 2.6 million every year.

It is quite likely that the Industrial Revolution occurred in part because of this population growth. It is theorized that the Europe of 300 or 400 years ago was reaching the carrying capacity of its agricultural society, so Europeans first spread out looking for more room and then began to invent more intensive uses of their resources to meet the needs of a growing population (Harrison 1993). The major resource was energy, which, along with the discovery of fossil fuels (first coal, then oil, and more recently natural gas), helped to light the fire under industrialization.

Since the beginning of the Industrial Revolution approximately 250 years ago, the size of the world's population has increased even more dramatically, as you can see graphically in Figure 2.2. For tens of thousands of years the population of the world grew slowly, and then within less than 300 years, the number of people mushroomed to six billion. There can be little question why the term **population explosion** was coined to describe recent demographic events. The world's population did not reach one billion until after the American Revolution—the United Nations fixes the year at 1804 (United Nations Population Division 1994)—but since then we have been adding each additional billion people at an accelerating pace. The two billion mark was hit in 1927, just before the Great Depression and 123 years after the first billion. In 1960, only 33 years later, came three billion; and four billion came along only 14 years after that, in 1974. We then hit five billion 13 years later, in 1987, and we passed the six-billion milestone 12 years later in 1999 (officially on October 12th of that year). Current projections suggest that each of the next few billions will come in quick succession, with the seventh billion expected in 2012, the eighth billion in 2027, and the ninth billion in 2052 (based on the United Nations Population Division 2002 Revisions). It is possible that we will eventually level off at around 10 billion (Lutz, Sanderson, and Scherbov 2001), although it is obviously difficult to say what will happen a half century or more from now. This pattern of growth over time can be seen in Figure 2.2, where the curve to the left of the vertical line is what has already happened, and the curve to the upper right is the projection made by the United Nations out to the middle of this century. Notice that as the curve approaches nine billion, the rate of population growth is expected to drop to about 0.5 percent per year (see Table 2.1)—a rate last seen in the world in the nineteenth century. The critical difference, of course, is that in the nineteenth century that rate was adding less than six million people per year to the world's total and in the twenty-first century it will be adding 45 million a year.

Figure 2.2   The World's Population Has Exploded in Size

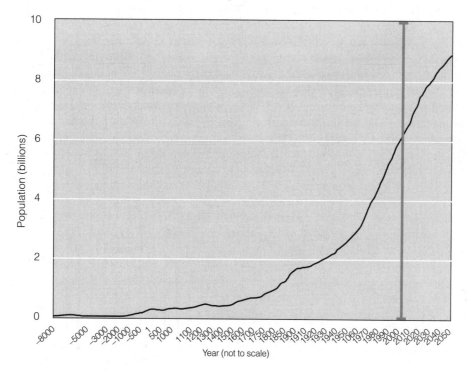

Source: Data in Table 2.1.

How Fast Is the World's Population Growing Now?

It is popular to think of population growth in terms of an explosion, and I have labeled Figure 2.2 with that term. However, let me suggest that population growth is,
in many respects, more like a protracted thunderstorm than an explosion. An explosion happens quickly, then it is over, and you pick up the broken pieces and move on
(if you have survived). A thunderstorm rolls in, stays with you for a while, deluging
you, causing you to change the way you do things. The aftermath of an explosion is
destruction, whereas after a storm you are not necessarily destroyed, but rather are
left with huge problems that have to be dealt with. Population growth is thus not
really like an explosion because the people do not go away; rather, the number increases and then stays large—creating huge problems that have to be dealt with.

   In Figure 2.3, I have graphed the data from Table 2.1 showing that the *rate* of
population growth for the world peaked in the early 1960s and has been declining
since then. This ought to be good news, but the deluge is measured in the number of
people, not just the rate of increase. So, you can see in Figure 2.3 that as we build on
an ever larger base of human beings, the lower rates of growth are still producing
very large absolute increases in the human population. When you build on a base
of 6.3 billion (the population in 2003), the seemingly slow rate of growth of about

Figure 2.3 More than 76 Million People Are Still Being Added to the World's Total Population Each Year, Despite the Drop in the Growth Rate

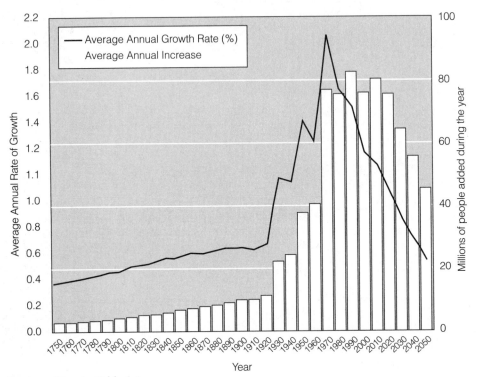

Source: Data in Table 2.1.

1.2 percent per year for the world still translates into the annual addition of more than 76 million people. Put another way, during the next 12 months, approximately 133 million babies will be born in the world, while 57 million of all ages will die, resulting in that net addition of more than 76 million people (see Table 2.2).

Back in 1950, as the world's population was really taking off (about when the term "population explosion" was coined), the rate of growth was higher than it is now (1.47 percent, as shown in Table 2.1), yet because the total population size was only 2.5 billion, we were adding "only" 38 million new people each year, as opposed to the current 76 million.

In Figure 2.3, you can see that the rate of growth in 1960 was lower than in 1950. This was due to a terrible famine in China in 1959–60, which was produced by Mao Zedong's Great Leap Forward program of 1958, in which the Chinese government "leapt forward" into industrialization by selling "surplus" grain to finance industrial growth. Unfortunately, the grain was not surplus, and the confiscation of food amounted to a self-imposed disaster that led to the deaths of 30 million Chinese in the following two years (1959 and 1960) (Becker 1997). Yet, even though the Chinese famine may have been the largest single disaster in human history, world population growth quickly rebounded, and the growth rate hit its record high shortly after that.

Table 2.2   In One Day the World's Population Increases by More than 200,000 People

Time Period	Births	Deaths	Natural Increase
Year	132,594,000	56,826,000	75,768,000
Day	363,271	155,688	207,583
Hour	15,136	6,487	8,649
Minute	252	108	144
Second	4	2	2

Source: Based on estimates for 2003 from the United Nations Population Division.

Even as the rate of population growth declines in the world, the volume of population growth is still astounding. You can see in Table 2.2 that more than 200,000 people are added to the total each day, the net result of 363,000 births less the 156,000 people who die every day (of whom about 20,000 are infants). Almost all of this increase is taking place in the less developed nations, as I will discuss later in this chapter.

The Power of Doubling—How Fast Can Populations Grow?

Human populations, like all living things, have the capacity for exponential increase. A common way of measuring the growth potential of any combination of birth and death rates is to calculate the doubling time, the time required for a population to double in number if the current rate of growth continues. You can calculate this easily for yourself by remembering the "rule of 69." The doubling time is approximately equal to 69 divided by the growth rate (in percent per year). So, if we estimate the world's rate of growth in the year 2003 to be 1.2 percent per year, we can calculate that the doubling time is 58 years. Where does the 69 come from in the doubling formula? Exponential growth is expressed mathematically by natural **logarithms.** Thus, to find out how long it would take a population to double in size, we first must find the natural logarithm (ln) of two. This turns out to be 0.69, which we multiply by 100 to get rid of the decimal point. Then dividing the rate of growth into 69 tells us how many years would be required for a population to double. Similarly, if we wanted to know how long it would take a population to triple in size, we would first find the natural logarithm of three, which is 1.10, or 110 when multiplied by 100. Dividing 110 by the rate of population growth then tells us how long it would take for the population to triple in size.

The incredible power of doubling can be illustrated by the tale of the Persian chessboard. The story is told that the clever inventor of the game of chess was called in by the King of Persia to be rewarded for this marvelous new game. When asked what he would like his reward to be, his answer was that he was a humble man and deserved only a humble reward. Gesturing to the board of 64 squares that he had devised for the game, he asked that he be given a single grain of wheat on the first square, twice that on the second square, twice *that* on the third square, and so on, until each square had its complement of wheat. The king protested that this was far too modest a prize, but the inventor persisted and the king finally relented. When the Overseer of the Royal Granary began counting out the wheat, it started out small enough: 1, 2, 4, 16, 32 . . . , but by the time the 64th square was reached, the number was staggering—nearly 18.5 quintillion grains of wheat (about 75 billion metric tons!) (Sagan 1989). This, of course, exceeded the "carrying capacity" of the royal granary in the same way that successive doublings of the human population threaten to exceed the carrying capacity of the planet.

Looking back at Table 2.1 you can see that it took several thousand years for the population to double from four million to eight million, a few thousand years to double from eight to 16 million, about a thousand to double from 16 to 32, and less than a thousand to double to 54 million. About 400 years elapsed between the European renaissance and the Industrial Revolution, and the world's population doubled in size during that time. But from 1750, it took only a little more than a 100 years to double again, and the next doubling occurred in less than 100 years. The most recent doubling (from three to six billion) took only about 40 years.

Will we double again in the future? Most projections suggest not (Bongaarts and Bulatao 2000; Lutz et al. 2001; United Nations Population Division 2003c), but we might come perilously close to it by the middle of this century, and—more importantly—we don't really know at this point how we will feed, clothe, educate, and find jobs for the 6.3 billion alive now, much less several billion more in the future. The social, economic, and political implications of this growth in human numbers are, to say the least, immense.

Once you realize how rapidly a population can grow, it is reasonable to wonder why early growth of the human population was so slow.

Why Was Early Growth So Slow?

The reason the population grew so slowly during the first 99 percent of human history was that death rates were very high, and at the same time very few populations have ever tried to maximize the number of children born (Abernethy 1979; Livi-Bacci 2001). During the hunting-gathering phase of human history, it is likely that life expectancy at birth averaged about 20 years (Livi-Bacci 2001; Petersen 1975) At this level of mortality, more than half of all children born will die before age five, and the average woman who survives through the reproductive years will have to bear nearly seven children in order to assure that two will survive to adulthood. This is still well below the biological limit of fertility, however, as I discuss in Chapter 6. Research in the twentieth century among the last of the hunting-gathering populations in sub-Saharan Africa suggests that a premodern woman might have deliberately limited the number of children born by spacing them a few years apart to make it easier to nurse and carry her youngest child and to permit her to do her work (Dumond 1975). She may have accomplished this by abstinence, abortion, or possibly even **infanticide** (Howell 1979; Lee 1972). Similarly, sick and infirm members of society were at risk of abandonment once they were no longer able to fend for themselves. For most groups of people, the balance between a large number of births and an almost equally large number of deaths led over time to only slight increases in population size. Of course, many groups may have become extinct because they were not able to maintain a positive balance between births and deaths.

It was once believed that the Agricultural Revolution increased growth rates as a result of people settling down in stable farming communities, where death rates were lowered. Sedentary life was thought to have improved living conditions because of the more reliable supply of food. The theory went that birth rates remained high but death rates declined slightly, and thus the population grew. However, archaeological evidence combined with studies of extant hunter-gatherer groups has offered another explanation for growth during this period (Spooner 1972). Possibly the sedentary life and the high-density living associated with farming actually raised death rates by creating sanitation problems and heightening exposure to communicable diseases. Growth rates probably went up because fertility rates rose as new diets improved the ability of women to conceive and bear children (see Chapter 6). Also, it became easier to wean children from the breast earlier because of the greater availability of soft foods, which are easily eaten by babies. This would have shortened the birth intervals, and the birth rate could have risen on that account alone.

It should be kept in mind, of course, that only a small difference between birth and death rates is required to account for the slow growth achieved after the Agricultural Revolution. Between 8000 B.C. and 1750 A.D., the world was adding an average of only 67,000 people each year to the population. At the beginning of the twenty-first century, that many people are being added every seven and a half hours.

Why Are More Recent Increases So Rapid?

The acceleration in population growth after 1750 was due almost entirely to the declines in the death rate that accompanied the Industrial Revolution. First in Europe and North America and more recently in less developed countries, death rates have decreased sooner and much more rapidly than have fertility rates. The result has been that many fewer people die than are born each year. In the more developed countries, declines in mortality at first were due to the effects of economic development and a rising standard of living—people were eating better, wearing warmer clothes, living in better houses, bathing more often, drinking cleaner water, and so on (McKeown 1976). These improvements in the human condition helped to lower exposure to disease and also to build up resistance to illness. Later, after 1900, much of the decline in mortality was due to improvements in medical technology, especially vaccination against infectious diseases.

Declines in the death rates first occurred in only those countries experiencing economic development. In each of these areas, primarily Europe and North America, fertility also began to decline within at least one or two generations after the death rate began its drop. However, since World War II, medical and public health technology has been available to virtually all countries of the world regardless of their level of economic development. In the less developed countries, although the risk of death has been lowered dramatically, birth rates have gone down less significantly, and the result is continued high levels of population growth. As you can see in Table 2.3, virtually all of the growth of the world's population is originating in

Table 2.3   Less Developed Regions Are the Sites of Future Population Growth (to the Year 2050)

		Area		
		More Developed Nations	Less Developed Nations	World
	Population in 2000 (in millions)	1,194	4,877	6,071
Medium fertility variant	Projection to year 2050 (in millions)	1,220	7,699	8,919
	Increase between 2000 and 2025 (in millions)	26	2,822	2,848
	Percent of world increase attributable to each area	1	99	
Low fertility variant	Projection to year 2050 (in millions)	1,084	6,325	7,409
	Increase between 2000 and 2025 (in millions)	−110	1,448	1,338
	Percent of world increase attributable to each area	0	100	
High fertility variant	Projection to year 2050 (in millions)	1,370	9,263	10,633
	Increase between 2000 and 2025 (in millions)	176	4,386	4,562
	Percent of world increase attributable to each area	4	94	

Source: Based on data in United Nations Population Division, 2003, World Population Prospects: The 2002 Revision (New York: United Nations), Table 1.

less developed nations. I say "originating" because some of that growth then spills into the more developed countries through migration, although the United Nations' projections do not reflect the volume of migration that may well occur over the next few decades. For this reason, the medium variant projection for 2050 (considered by the United Nations to be the most likely projection of demographic events) assumes a very small increase in population in the more developed nations between 2000 and 2050, but an actual decline in the size of the European population, where countries currently have very low fertility and also very restrictive immigration laws. As I will discuss later in the chapter, the aging of the European population and the threat of depopulation will likely result in a loosening of the immigration laws.

Are We Headed for a Population "Implosion"?

An implosion is something that collapses into itself—the opposite of an explosion. As the rate of population growth has slowed down over the past two decades, there

has been talk of a **population implosion** (Eberstadt 2001; Wattenberg 1997), implying that "the world is in for some rapid downsizing" (Singer 1999:22). As you can
see by looking at the data in Tables 2.1 through 2.3, this is either just rhetoric designed to gain attention or a misunderstanding of the actual demographic situation
of the world.

Most of the discussion about an implosion comes about in reaction to the
changing population projections made by demographers at the Population Division
of the United Nations. Every projection made for the year 2000 by the United Nations between 1958 and 1998 was higher than the 6.1 billion that was eventually estimated for that year. However, none was much over that—the highest being one
made in 1968 that projected 6.5 billion by 2000 (Bongaarts and Bulatao 2000).
Looking beyond 2000, the United Nations in 1993 projected that the world's population in 2025 could reach 8.5 billion (United Nations Population Division 1993),
but they more recently projected a lower figure of 7.9 billion for 2025 (United Nations Population Division, 2003c). The differences are due to the devastating effects
of AIDS mortality in sub-Saharan Africa, and lower than expected fertility in both
Europe and in China.

In commenting on their projections published in 2003, demographers at the
United Nations noted that, "As a result of these trends, the population of more developed regions, currently at 1.2 billion, is anticipated to change little during the
next 50 years. In addition, because fertility levels for most of the developed countries are expected to remain below replacement level during 2000–2050, the populations of 30 developed countries are projected to be smaller by mid-century than
today (e.g., 14 percent smaller in Japan; 22 percent smaller in Italy, and between 30
and 50 percent smaller in the cases of Bulgaria, Estonia, Georgia, Latvia, the Russian Federation and Ukraine)" (United Nations Population Division, 2003c:vi). This
is, to be sure, the warning bell of a population implosion for some European nations. However, this could be at least delayed by a drop in the age at which European women have children, even if they were not to increase that number (Lutz,
O'Neill, and Scherbov 2003). Furthermore, it is unlikely that these countries will resist the pressure to accept immigrants from less developed nations, where the population is projected to "rise steadily from 4.9 billion in 2000 to 7.7 billion in 2050
(medium variant). Particularly rapid growth is expected among the least developed
countries whose population is projected to

> . . . rise from 668 million to 1.7 billion despite the fact that their fertility is projected to
> decline markedly in the future (from 5.1 children per woman today to 2.5 children per
> woman in 2045–2050). With sustained annual growth rates higher than 2.5 per cent be-
> tween 2000 and 2050, the populations of Burkina Faso, Mali, Niger, Somalia, Uganda
> and Yemen are projected to quadruple, passing from 85 million to 369 million in total
> (United Nations Population Division, 2003c:vii).

The demographics of the world are shifting, to be sure, but there is no implosion in
sight. The rest of your life will almost certainly take place in a world in which there
will be more people tomorrow than there are today.

How Many People Have Ever Lived?

In a burst of poetic license, not to mention perhaps unprecedented ethnocentricity, William Matthews wrote in 1980 that "there are now more of us alive than ever have been dead" (Matthews 1980). It is certainly true that the population is increasing each year in numbers that strain credulity—could it be that today's 6.3 billion people exceed in number all the people who have ever been born and died over the entire history of human beings?

No, actually not. As a matter of fact, our current contribution to history's total represents only a relatively small fraction of all people who have ever lived. This question has intrigued scholars and has been examined with some regularity over the decades. In 1960, for example, William Deevey estimated that "a cumulative total of about 110 billion individuals seem to have passed their days, and left their bones, if not their marks, on this crowded market" (Deevey 1960:197).

Others have made similar types of calculations (Desmond, Wellemeyer, and Lorimer 1962; Haub 1995; Westing 1981), but the most analytical of the estimates has been made by Nathan Keyfitz (1966; 1985). I have used Keyfitz's formulas to estimate the number of people who have ever lived and as you can see in Table 2.4, the results of my calculations suggest that a total of 60.9 billion people have been born over the past 200,000 years, of whom the 6.3 billion alive in 2003 constitute 10.4 percent. Livi-Bacci (2001) has a lower figure—7.3 percent—but that is based on a longer period of time than the 200,000 years I have used.

Perhaps more importantly, Keyfitz (1985) has pointed out that the dramatic drop in infant and childhood mortality over the last two centuries means that babies are now far more likely than ever to grow up to be adults. Thus the adults alive today represent a considerable fraction of all people who have ever lived to adulthood. Furthermore, if we look at specific categories of people, such as engineers or college professors, then it is probable that the vast majority of such individuals who have ever lived on earth are still alive today.

The increase in size of the world's population is not the only important demographic change to occur in the past few hundred years. In addition, there has been a massive redistribution of population.

Table 2.4   How Many People Have Ever Lived?

Historical Period	Number of People Born During the Period (in billions)	Cumulative Total Ever Born (in billions)
200,000 BC to 8,000 BC	2.1	2.1
8,001 BC to AD 0	16.7	18.8
AD 1 to 1799	29.4	48.2
1800 to 1899	3.6	51.8
1900 to 1949	2.6	54.4
1950 to 2003	6.5	60.9

Source: Calculated by the author based on formulas from Keyfitz (1966) for estimates from 200,000 BC to 1949 and based on population size and birth rate estimates from the United Nations and the Population Reference Bureau for the period 1950–2003.

Redistribution of the World's Population through Migration

As populations have grown disproportionately in different areas of the world, the pressures or desires to migrate have also grown. This pattern is predictable enough that we label it the migration transition part of the overall demographic transition (which I will discuss more in the next chapter). Migration streams generally flow from rapidly growing areas into less rapidly growing ones, for example, from Latin America and Asia to the United States, from Asia to Canada, from Africa and Asia to Europe. In earlier decades, as population grew dense in a particular region, people were able to move to some other less populated area, much as high-pressure storm fronts move into low-pressure weather systems. This is the pattern of migration that characterized the expansion of European populations into other parts of the world, as European farmers were seeking farmland in less densely settled areas. This phenomenon of European expansion is especially notable because as Europeans moved around the world, they altered patterns of life, including their own, wherever they went.

European Expansion   Beginning in the fourteenth century, migration out of Europe started gaining momentum, and this expansion of the European population virtually revolutionized the entire human population. With their gun-laden sailboats, Europeans began to stake out the less developed areas of the world in the fifteenth and sixteenth centuries, and this was only the beginning. Migration of Europeans to other parts of the world on a massive scale took hold in the nineteenth century, when the European nations began to industrialize and swell in numbers. As Kingsley Davis has put it:

> Although the continent was already crowded, the death rate began to drop and the population began to expand rapidly. Simultaneous urbanization, new occupations, financial panics, and unrestrained competition gave rise to status instability on a scale never known before. Many a bruised or disappointed European was ready to seek his fortune abroad, particularly since the new lands, tamed by the pioneers, no longer seemed wild and remote but rather like paradises where one could own land and start a new life. The invention of the steamship (the first one crossed the Atlantic in 1827) made the decision less irrevocable (Davis 1974:98).

Before the great expansion of European people and culture, Europeans represented about 18 percent of the world's population, with almost 90 percent of these people of European origin living in Europe itself. By the 1930s, at the peak of European dominance in the world, people of European origin in Europe, North America, and Oceania accounted for 35 percent of the world's population (Durand 1967). By the beginning of the twenty-first century, the percentage has declined to 16, and it is projected to drop to 13 percent by the middle of this century (United Nations Population Division, 2003c). However, that may be a bit of an exaggeration, since the rate of growth in North American and European countries is increasingly influenced by immigrants and births to immigrants from developing nations.

"South" to "North" Migration   Since the 1930s, the outward expansion of Europeans has ceased. Until then, European populations had been growing more rapidly than the populations in Africa, Asia, and Latin America, but since World War II that trend has been reversed. The less developed areas now have by far the most rapidly growing populations. It has been said that "population growth used to be a reward for doing well; now it's a scourge for doing badly" (Blake 1979). This change in the pattern of population has resulted in a shift in the direction of migration. There is now far more migration from less developed (the "South") to developed areas (the "North") than the reverse. Furthermore, since migrants from less developed areas generally have higher levels of fertility than natives of the developed regions, their migration makes a disproportionate contribution over time to the overall population increase in the developed area to which they have migrated. As a result, the proportion of the population whose origin is one of the modern world's less developed nations tends to be on the rise in nearly every developed country. Within the United States, for example, data from Census 2000 show that non-Hispanic whites (the European origin population) no longer are the majority in the state of California, and it is projected that the Hispanic-origin population will represent the majority of Californians by the middle of this century (California Department of Finance 1998), a projection amply bolstered by the fact that in 2003 a majority of all births in California were to Hispanic mothers.

When Europeans migrated, they were generally filling up territory that had very few people, because they tended to be moving in on land used by hunter-gatherers who, as noted above, use land extensively rather than intensively. Those seemingly empty lands or frontiers have essentially disappeared today, and as a consequence migration into a country now results in more noticeable increases in population density. And, just as the migration of Europeans was typically greeted with violence from the indigenous population upon whose land they were encroaching, migrants today routinely meet prejudice, discrimination, and violence in the places to which they have moved. The difference is that today's migrants tend to be showing up in cities, not the countryside, and this is part of a broader aspect of the entire urban transition—an important offshoot of the migration transition.

The Urban Revolution   Until very recently in world history, almost everyone lived in basically rural areas. Large cities were few and far between. For example, Rome's population of 650,000 in 100 A.D. was probably the largest in the ancient world (Chandler and Fox 1974). It is estimated that as recently as 1800, less than 1 percent of the world's population lived in cities of 100,000 or more. More than one-third of all humans now live in cities of that size.

The redistribution of people from rural to urban areas is most marked in the industrialized nations. For example, in 1800 about 10 percent of the English population lived in urban areas, primarily London. By the year 2000, 90 percent of the British lived in cities. Similar patterns of urbanization have been experienced in other European countries, the United States, Canada, and Japan as they have industrialized. In the less developed areas of the world, urbanization was initially associated with a commercial response to industrialization in Europe, America, and Japan. In other words, in many areas where industrialization was not occurring, Europeans had established colonies or trade relationships. The principal economic activities in these areas were not industrial but commercial in nature, associated with buying and selling. The wealth acquired by people engaged in these activities naturally attracted attention, and urban centers sprang up all over the world as Europeans sought populations to whom they could sell their goods. Thus, urban populations began to grow in some countries even without industrialization, as places sprang up where goods were exchanged and where services were developed to meet the needs of merchants.

During the second half of the twentieth century, when the world began to urbanize in earnest, the underlying cause was the rapid growth of the rural population (I discuss this in more detail in Chapter 11). The rural population in every less developed nation outstripped the ability of the agricultural economy to absorb it. Paradoxically, in order to grow enough food for an increasing population, people have had to be replaced by machines in agriculture (as I will discuss in Chapter 12), and that has sent the redundant rural population off to the cities in search of work. Herein lie the roots of many of the problems confronting the world in the twenty-first century.

Geographic Distribution of the World's Population

The five largest countries in the world account for nearly half the world's population (49 percent as of the year 2003) but only 21 percent of the world's land surface. These countries include China, India, the United States, Indonesia, and Brazil, as you can see in Table 2.5. Rounding out the top 10 are Pakistan, Bangladesh, Russia, Japan, and Nigeria. Within these most populous 10 nations reside 60 percent of all people. It turns out that you have to visit only 15 countries in order to shake hands with two-thirds of all the people in the world. In doing so, you would travel across 37 percent of the earth's land surface. The rest of the population is spread out among 180 or so other countries, accounting for the remaining 63 percent of the earth's terrain.

If you set a goal to be as efficient as possible in maximizing the number of people you could visit while minimizing the distance you would travel, your best bet would be to schedule a trip to China and the Indian subcontinent. As of 2003, 42 percent of all the world's people were living in those two contiguous regions of Asia, and you can see how these areas stand out in the map of the world drawn with country size proportionate to population (see Figure 2.1, at the beginning of this chapter). Throughout this area, the average population density is a whopping 500 people per square mile, compared with the overall world average of 109 per square mile and the American average of 83 per square mile.

Population growth in Asia is not a new story. In Table 2.6 you can see that in 1500, as Europeans were venturing beyond their shores, China and India (or more technically the Indian subcontinent, including the modern nations of India, Pakistan, and Bangladesh) were already the most populous places on earth, and all of Asia accounted for 53 percent of the world's 461 million people. Five centuries later, the population in Asian countries accounts for 61 percent of all the people on earth, although it is projected to drop back to 58 percent by the year 2050. Sub-Saharan Africa, on the other hand, had nearly 100 million people in 1500, almost as many as in Europe. However, contact with Europeans tended to be deadly for Africans (as it was for Indians in North and South America) because of disease, violence, and

Table 2.5　The 20 Most Populous Countries in the World: 1950, 2003, and 2050

Rank	1950 Country	1950 Population (in millions)	1950 Area (000 sq. mi.)	2003 Country	2003 Population (in millions)	2003 Area (000 sq. mi.)	2050 Country	2050 Population (in millions)	2050 Area (000 sq .mi.)
1	China	563	3,601	China	1,304	3,601	India	1,531	1148
2	India	370	1,148	India	1,065	1,148	China	1,395	3601
3	Soviet Union	180	8,650	United States	294	3,536	United States	409	3536
4	United States	152	3,536	Indonesia	220	705	Pakistan	349	298
5	Japan	84	145	Brazil	178	3,265	Indonesia	294	705
6	Indonesia	83	705	Pakistan	154	298	Nigeria	258	352
7	Germany	68	135	Bangladesh	147	50	Bangladesh	255	50
8	Brazil	53	3,265	Russia	143	6,521	Brazil	233	3265
9	United Kingdom	50	93	Japan	128	145	Ethiopia	171	386
10	Italy	47	114	Nigeria	124	352	Democratic Republic of the Congo (former Zaire)	152	905
11	Bangladesh	46	50	Mexico	103	737	Mexico	140	737
12	France	42	212	Germany	82	135	Philippines	130	126
13	Pakistan	39	298	Vietnam	81	126	Egypt	127	384
14	Nigeria	32	352	Philippines	80	115	Vietnam	118	126
15	Mexico	28	737	Egypt	72	384	Japan	110	145
16	Spain	28	193	Ethiopia	71	386	Iran	105	632
17	Vietnam	26	126	Turkey	71	297	Uganda	103	93
18	Poland	25	118	Iran	69	632	Russia	101	6521
19	Ethiopia	22	386	Thailand	63	197	Turkey	98	297
20	Egypt	21	384	France	60	213	Yemen	84	204
Top 20		1,959	24,248		4,509	22,843		6,163	23,511
World		2,556	57,900		6,301	57,900		8,918	57,900
top 5		53	29		49	21		45	16
top 10		65	37		60	34		57	25
top 20		77	42		72	39		69	41

Source:　Based on data in United Nations Population Division, 2003, World Population Prospects: The 2002 Revision, Annex Tables 1 and 2.

Table 2.6 Geographic Distribution of the World's Population, from 400 BC to AD 2050

Region	Number of People (in millions) in the Region as of the Year:										
	-400	0	500	1000	1500	1750	1850	1900	1950	2000	2050
China	19	70	32	56	84	220	435	415	563	1,275	1,395
India/Pakistan/Bangladesh	30	46	33	40	95	165	216	290	455	1,298	2,135
Japan	1	2	5	4	10	26	30	45	84	127	110
Rest of Asia	45	52	49	52	56	89	109	153	296	980	1,582
Europe	32	43	41	43	84	146	288	422	547	728	632
North Africa	10	14	11	9	9	10	12	43	52	172	306
Sub-Saharan Africa	7	12	20	30	78	94	90	95	169	624	1,497
North America (U.S. and Canada)	1	2	2	2	3	3	25	90	172	316	448
Latin America	7	10	13	16	39	15	34	75	167	520	768
Oceania	1	1	1	1	3	3	2	6	13	31	46
	153	252	207	253	461	771	1,241	1,634	2,518	6,071	8,919
Percentage											
China	12.4	27.8	15.5	22.1	18.2	28.5	35.1	25.4	22.4	21.0	15.6
India/Pakistan/Bangladesh	19.6	18.3	15.9	15.8	20.6	21.4	17.4	17.7	18.1	21.4	23.9
Japan	0.7	0.8	2.4	1.6	2.2	3.4	2.4	2.8	3.3	2.1	1.2
Rest of Asia	29.4	20.6	23.7	20.6	12.1	11.5	8.8	9.4	11.8	16.1	17.7
Europe	20.9	17.1	19.8	17.0	18.2	18.9	23.2	25.8	21.7	12.0	7.1
North Africa	6.5	5.6	5.3	3.6	2.0	1.3	1.0	2.6	2.1	2.8	3.4
Sub-Saharan Africa	4.6	4.8	9.7	11.9	16.9	12.2	7.3	5.8	6.7	10.3	16.8
North America (U.S. and Canada)	0.7	0.8	1.0	0.8	0.7	0.4	2.0	5.5	6.8	5.2	5.0
Latin America	4.6	4.0	6.3	6.3	8.5	1.9	2.7	4.6	6.6	8.6	8.6
Oceania	0.7	0.4	0.5	0.4	0.7	0.4	0.2	0.4	0.5	0.5	0.5
	100.0	100.0	100.0	100.0	100.0	100.0	100.0	100.0	100.0	100.0	100.0

Sources: Adapted from Biraben, Jean-Noël, 1979, "Essai Sur L'Évolution Du Nombre Des Hommes." Population 34 (1):13–24, Table 1, and updated by the author to 2000 and 2050 using data from United Nations Population Division, 2003, World Population Prospects: The 2002 Revision, Table 1.

slavery. In the twentieth century, sub-Saharan Africa rebounded in population size, comprising 10 percent of the total world population in 2000. Of course, high mortality from HIV/AIDS is slowing down the rate of growth in sub-Saharan Africa, but the United Nations projects that because of the continued high fertility levels in the region, sub-Saharan Africa could account for 16.8 percent of the world's population in 2050—back to where it was in percentage terms in the year 1500, but with more than twice as many people as are projected to be in Europe in that year.

Differential rates of population growth help explain the shifting distribution of people around the globe, and they can also be the cause of very different outlooks on life. The world looks very different from the perspective of a low-mortality, low fertility rich nation than it does from a higher mortality, higher fertility, poor country, as I point out in the essay that accompanies this chapter.

Global Variation in Population Size and Growth

> World population is currently growing at a rate of 1.2 per cent annually, implying a net addition of 77 million people per year. Six countries account for half of that annual increment: India for 21 per cent; China for 12 per cent; Pakistan for 5 per cent; Bangladesh, Nigeria and the United States of America for 4 per cent each (United Nations Population Division 2003c:vi).

Three of the 10 most populous countries in the world—Nigeria, Pakistan, and Bangladesh—are growing even more rapidly than the world average. Among the other seven of the world's 10 largest countries, Russia is growing most slowly (indeed, Russia and much of Eastern Europe and the Baltic states appear to be depopulating), followed by Japan, China, the United States, and then Brazil. In Figure 2.4, you can see that the most rapidly growing regions in the world tend to be in the midlatitudes, and these are nations that are least developed economically, whereas the slowest growing are the richer nations, which tend to be more northerly and southerly. It has not always been that way, however.

Before the Great Depression of the 1930s, the populations of Europe and especially North America were the most rapidly growing in the world. During the decade of the 1930s, growth rates declined in those two areas to match approximately those of most of the rest of the world, which was about 0.75 percent per year—a doubling time of 93 years. Since the end of World War II, the situation has changed again, and now Europe and North America rank among the more slowly growing populations, with rapid growth in the less developed countries of Asia, Latin America, and Africa now responsible for most of the world's population increase. Let's examine these trends in more detail, focusing particular attention on the 10 most populous nations, with a few other countries included to help illustrate the variability of demographic situations in which countries find themselves.

North America

Canada and the United States—North America—have a combined population of 326 million as of 2003, representing 5 percent of the world's total. Canada's 31 mil-

Figure 2.4   Rates of Population Growth Are Highest in the Middle Latitudes

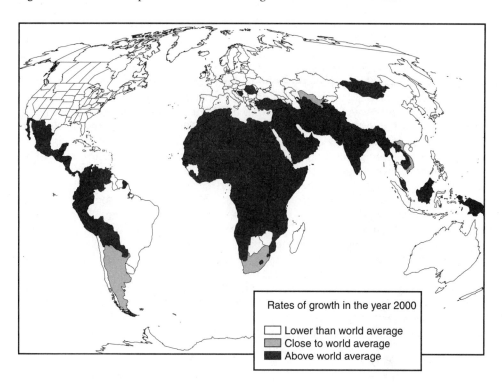

Rates of growth in the year 2000

☐ Lower than world average
▨ Close to world average
■ Above world average

lion people account for about 10 percent of North America's total, and the United States, with about 275 million, has the remaining 90 percent. The demographic histories of the two countries are intertwined but are not identical.

United States   It does not take a demographer to notice that the population of the United States has undergone a total transformation since John Cabot (an Italian hired by the British to search the new world) landed in Newfoundland in 1497 and claimed North America for the British. As was true throughout the western hemisphere, the European guns and diseases rather quickly decimated the native American Indian population, making it easier to establish a new culture. It is estimated that the population of North America in 1650 consisted of about 50,000 European colonists and two to three million native American Indians—the Europeans were outnumbered by as much as 60 to 1. By 1850, disease and warfare had reduced the Indian population to perhaps as few as 250,000, while the European population had increased to 25 million. Indeed, it was widely assumed that the Indian population was on the verge of disappearance (Snipp 1989).

Although early America was a model of rapid population growth (at least for the European-origin population), it was also a land of substantial demographic contrasts. Among the colonies existing in the seventeenth century, for example, those in New England seem to have been characterized by very high birth rates (women had an average of seven to nine children) yet relatively low mortality rates (infant

BOOMSTERS VERSUS DOOMSTERS: COULD POPULATION GROWTH BECOME A PONZI SCHEME?

There has been a genuine tension in the world for the past several decades in terms of whether population growth is good (this would make you a "boomster") or bad (in which case we label you a "doomster"). Throughout human history, people have seen a growing population as a sign of good times—economically and in other ways as well. The current lamentations in Europe about the threat of population decline underscore that kind of attitude. Yet, the idea that population growth might be bad because it can lead to environmental catastrophe is also an old notion, and history offers examples of human societies that have overshot their carrying capacity and so are no longer with us (the Mayan civilization being one of the more prominent examples). These contrasts in philosophy are interesting because the movement to lower the world's rate of population growth got started in richer countries, and was initially resisted by leaders of many less developed countries, who believed that Westerners promoting a slowdown in population growth were racist and/or genocidal. Yet, paradoxically, if you are a resident of the rich part of the world, you may actually stand to gain personally from population growth, whereas if you are in a poor country, it may only further victimize you.

The personal gains from population growth reflect the basic economic principle that wealth can be found in new markets, and so population growth is a good thing because it offers people an opportunity to make money by selling things to an ever expanding group of people. But how many people can actually profit from continued population growth? This is an important question, and it brings to mind the famous Ponzi scheme in the early part of the twentieth century.

In 1919, Charles Ponzi was an Italian immigrant living in Boston and aspiring to be a rich man. The story is told that he didn't intend to be dishonest, but he created one of the most famous pyramid schemes in history. In the process of trying to start up an export business, he discovered that an international reply postcard purchased for the equivalent of one cent US in Spain could be traded in at a U.S. Post Office for six cents worth of U.S. postage (Knutson 2000). The earnings potential inherent in this transaction seemed staggering to Ponzi and in December 1919 he filed for a business license in Boston to operate his Security Exchange Company. He promised investors a 50 percent return on their money within 90 days and, of course, he had a long list of takers. We may have outgrown our gulls in the evolutionary process, but we are still a gullible species. It turned out that regulatory difficulties and exchange-rate problems prevented Ponzi from being able to profit by exchanging the international reply postcards in the way he had originally imagined. But that did not prevent him or his early investors from making money. As long as new investors were pumping in new money, he had funds to pay the interest promised to earlier investors. The very act of paying investors excited new investors and kept the ball rolling for several months and made him a very rich man for a very short period of time. Eventually people caught on to the fact that he wasn't making money, he was just taking money, and he was arrested, the business was shut down, and Ponzi spent time in jail. The last investors in the scheme never saw their money again, but the first investors in the scheme got rich if they were clever or lucky enough to have withdrawn their principal before the collapse.

The Ponzi scheme was not a unique scam. Most recently, the dot.coms have been described as a "natural" Ponzi scheme by Robert Shiller (2000). Dot.coms were described to investors as representing a new type of economy in which high sales volume represented a raging bull market even if companies were losing money on those sales. Profits would eventually emerge, investors were told. A lot of instant millionaires were created, and then, when the bubble burst, a lot of people were left wondering where their money went. I suggest to you that population growth has emerged in this new millennium as potentially another type of "natural" Ponzi scheme. It comes packaged as none other than the touted savior of our global future—sustainable development.

Population growth in the world today has generated a huge group of people who are poor and need jobs in order to be lifted out of poverty.

Current United Nations and World Bank policies are oriented toward reducing poverty as the way to solve the problems of world environmental degradation. Poverty is to be reduced through the reduction of trade barriers, thus allowing multinational companies access to natural and human resources in these countries. This is supposed to create more wealth, which like a rising tide will lift people above poverty, at which point it will be easier to deal with the environmental problems that are created by more people consuming more resources. Put another way, the solution to our problems lies in using more resources, albeit more efficiently than in the past, and if we are willing to invest in these companies in order to allow them to accomplish their goal, they will repay us with a handsome rise in stock value that will allow us to share in the global production of more wealth.

Why is this like a Ponzi scheme? First, it assumes an unlimited amount of natural resources upon which companies can draw—just as a Ponzi scheme depends on an unlimited supply of new investors. And, secondly, it ignores the vastly greater number of people who will need a lift out of poverty a decade or two from now than there are right now. Consequently, companies should indeed be able to generate profits for the early investors and for a while it will seem as though we are making progress toward our goal. But the combination of a huge addition to the world's population at the same time that we are trying to raise everybody else's standard of living, without reducing our own, leads to a very high probability that we will exhaust our resource base long before we reach our goal of having a larger population sustained at a higher standard of living. If we continue to follow the economic model that currently drives the concept of sustainable development, it is likely that only a small fraction of people—the early investors, if you will—are going to reap the benefit of long-term high standards of living. The early investors are, of course, likely to live in richer countries. The last investors in the scheme—the ones who are assured of losing their shirts—are most apt to be in poorer countries.

How has this situation come about? There are three important reasons:

1. There exists a lack of appreciation for the dynamics of demographic transitions, and the way in which economic productivity can be enhanced or curtailed by the changing demographic structures of a country or region. These are issues I will explore with you in the remainder of the book.

2. Neoclassical economic theory ignores the real world of a closed ecosystem by treating things like resources and pollution as externalities or, when thinking of them in economic terms, concludes that they can be dealt with through a standard pricing mechanism. A growing literature explores the ins and outs of this topic, and I will touch on it later in Chapter 12.

3. People in the North (the rich countries) believe that the population explosion is over, and that population growth is no longer something the world has to worry about. But population growth has not ended. Even if we think we can see light at the end of the tunnel, we are still a long way from being out of the tunnel. I have tried in this chapter to illustrate the ways in which in our demographically divided world can lead to very different perceptions of the world population situation, depending on where you live.

If we recognize that the world's current plan for sustainable development is really a Ponzi scheme that cannot be sustained in the face of the likely patterns of future population growth, then we can try to use the age and urban transitions to our advantage, as I will discuss in subsequent chapters. This is most readily accomplished by rapid declines in fertility in the less wealthy nations. It is true that they use fewer resources and pollute less than the rich, but the reality is that rich nations are not going to willingly hand over their resources to less rich nations. Furthermore, people in those less rich nations have many of the same aspirations as do the people in rich nations for a high standard of living. At any level of technology, this is easier to accomplish with fewer rather than more people.

mortality rates in Plymouth colony may have been lower than in some of today's less developed nations, apparently a result of the fairly good health of Americans even during that era) (Demos 1965; Wells 1982). Demos notes that "the popular impression today that colonial families were extremely large finds the strongest possible confirmation in the case of Plymouth. A sample of some ninety families, about whom there is fairly reliable information, suggests that there was an average of seven to eight children per family who actually grew to adulthood" (1965:270). In the southern colonies during the same time period, however, life was apparently much harsher, probably because the environment was more amenable to the spread of disease. In the Chesapeake Bay colony of Charles Parish, higher mortality meant that few parents had more than two or three living children at the time of their death (Smith 1978).

Despite the regional diversity, the American population grew rather steadily during the seventeenth and eighteenth centuries, and though much of the increase in the number of Europeans in America was attributable to in-migration, the greater percentage actually was due to **natural increase** (the excess of births over deaths). The nation's first census, taken in 1790, shortly after the American Revolution, counted 3.9 million Americans, and although the population was increasing by nearly 120,000 a year, only about 3 percent of the increase was a result of immigration. With a crude birth rate of about 55 births per thousand population (comparable to the highest national birth rates in the world today) and a crude death rate of about 28 deaths per thousand, there were twice as many people being born each year as were dying. At this rate, the population was doubling in size every 25 years.

Although Americans may picture foreigners pouring in seeking freedom or fortune, it was not until the second third of the nineteenth century that migration became a substantial factor in American population growth. In fact, during the first half of the nineteenth century, in-migrants accounted for less than 5 percent of the population increase in each decade, whereas in every decade from the 1850s through the 1920s in-migrants accounted for at least 20 percent of the growth of population.

Throughout the late nineteenth and early twentieth centuries, the birth rate in the United States was falling. There is evidence that fertility among American Quakers began to be limited at about the time of the American Revolution (Wells 1971), and the rest of the nation was only a few decades behind their pace. By the 1930s, fertility actually dropped below the level required to replace the next generation (as I discuss more thoroughly in Chapter 7). Furthermore, restrictions on immigration had all but halted the influx of foreigners, and Americans were facing the prospect of potential depopulation.

The early post-World War II era upset forecasts of population decline; they were replaced by the realities of a population explosion. The period from the mid-1940s to the mid-1960s is generally known as the "baby boom" era, a time when the United States experienced a rapid rate of increase in population, accomplished almost entirely by increases in fertility. The baby boom, in turn, was followed in the late 1960s and early 1970s by a "baby bust." An echo of the baby boom was experienced as the "baby boomlet" of the 1980s, but fertility has since remained at or below the replacement level (Martin et al. 2002; Morgan 1996). Nonetheless, population growth, rather than population decline, has continued to be the order of the day, because in the 1960s and again in the 1990s a rewriting of the nation's immigration laws opened the nation's doors wider, and the result has been renewed high levels of migration into the United States. Indeed, the 900,000 immigrants added each year in the 1990s accounted for more than one-third of the total annual population increase. The long-term implication of this continued immigration is that variations in fertility levels in the United States are increasingly determined by fertility differences among the various racial and ethnic groups.

By current world standards, the United States today is one of the slower-growing countries, but its rate of growth is still one of the highest among the more developed nations. With a total rate of growth of about 0.9 percent per year, the U.S. population is growing at a rate that is similar to the sub-Saharan African nation of Zimbabwe and the Asian nation of Thailand. Fertility in the United States is low (2.1 children per woman), but not as low as in virtually all European countries. On the other hand, mortality is also very low (female life expectancy in the United States is now 80 years), but not as low as in the countries of Western Europe nor as low as Japan (which has the highest life expectancy in the world).

The only major demographic category in which the United States leads the world is in the number of immigrants received each year, but that has a lot to do with the demographic diversity that is so much a part of life in the United States. For example, the 35 million African-Americans in the United States in 2000 represented the best-educated, wealthiest, and fourth-largest (after Nigeria, Ethiopia, and South Africa) population of African-origin persons in the world. At the same time, there is little wonder that Hispanics have caught the attention of America when you consider that the number of Hispanics doubled between 1980 and 2000. The 31 million Hispanics represent more than 11 percent of the total U.S. population—nearly as much as the black population. And, of course, the Hispanic population is itself diverse, spanning three different major groups—Mexican-Americans (centered in Los Angeles), Puerto Ricans (primarily in New York City), and Cuban-Americans (concentrated in Miami). Asians represent a smaller, yet growing, population in the United States, especially along the Pacific Coast. Ethnically, it is an even more diverse group than Hispanics. Asian immigrants have also been attracted to Canada in large numbers.

Canada The French were the first Europeans to settle the area that has become Canada, and by 1760 the population of New France included about 70,000 people of European (largely French) origin, along with a diminishing population of indigenous peoples. In 1763, the French government ceded control of the region to the British. The French-origin population was thus "cut off from intercourse with the French, and they received no further demographic reinforcements" (Overbeek 1980:10). Twenty years later, the establishment of American independence led about 30,000 "loyalists" to leave the United States and migrate to Canada, the forerunner of a much more massive migration from England to Canada during the nineteenth century (Kalbach and McVey 1979). The conflict between the British-origin and French-origin populations helped to stimulate the British North America Act of 1867, which united all of the provinces of Canada into the Dominion of Canada, and every census since then has asked about "origins" as a way of keeping track of the numerical balance between these historically rival groups (Boyd, Goldmann, and

White 2000; Kralt 1990). In 1931, when Canada was granted independence from Great Britain, the Province of Québec, in which most French-speaking Canadians reside, accounted for 28 percent of the population of Canada. By 2001, that figure had dropped slightly to 24 percent.

In the seventeenth and eighteenth centuries, the high fertility of women in New France was legendary, and French speakers in Canada maintained higher-than-average levels of fertility until the 1960s. In the rest of Canada, fertility began to drop in the nineteenth century and reached very low levels in the 1930s, before rebounding after World War II in a baby boom that was similar in its impact on Canadian society to that experienced in the United States. This boom was similarly followed by a baby bust and then a small echo of the baby boom (Foot 1996). Canada now has a fertility level (1.5 children per woman) that is well below replacement, and women in Québec are bearing children at a rate even below the national level (1.4 children per women), making Francophone (French-speaking) Canada one of the lowest-fertility regions anywhere in the western hemisphere.

Just as fertility is lower in Canada than in the United States, so is mortality, making life expectancy in Canada about two years longer than in the United States. In both of these respects the demographic profile of Canada is more like Europe than it is the United States. However, when it comes to immigration, Canada reflects the Northern American history of being a receiving ground for people from other nations. Despite its lower fertility, Canada's overall rate of population growth exceeds that of the United States because it accepts more immigrants per person than the United States does. For example, in the 1990s Canada was admitting about 235,000 immigrants per year (about eight per thousand population) and the foreign-born population represented 18 percent of the Canadian total in 2001 (Statistics Canada 2003a). In the same period, there were 900,000 immigrants annually entering the United States (about three per thousand population), and in 2000 the foreign born accounted for 11 percent of the United States population (U.S. Census Bureau 2003a). Immigrants to Canada in recent years have come especially from Asia. Immigrants to the United States also come from Asia, but a larger fraction come from Latin America, especially Mexico.

Mexico and Central America

Mexico and the countries of Central America have been growing since the end of World War II as a result of rapidly dropping death rates and birth rates that have only more recently begun to drop but which remain above the world average. Mexico enumerated 97 million in its 2000 census and that represents three-fourths of the population of the region, with the remaining 25 percent distributed among (in order of size) Guatemala, El Salvador, Honduras, Nicaragua, Costa Rica, Panama, and Belize. The combined regional population of 144 million as estimated for the year 2003 is a little more than 2 percent of the world's total.

Mexico and Central America had developed more advanced agricultural societies than had North America at the time of European contact. The Aztec civilization in central Mexico and the remnants of the Mayan civilization farther south (centered near Guatemala) encompassed many millions more people than lived on the northern side of what is now the United States-Mexico border. This fact, combined with the Spanish goal of extracting resources (a polite term for plundering) from the New World rather than colonizing it, produced a very different demographic legacy from what we find in Canada and the United States.

Mexico was the site of a series of agricultural civilizations as far back as 2,500 years before the invasion by the Spanish in 1519. Within a relatively short time, the population of several million was cut by as much as 80 percent as a result of disease and violence. This population collapse (a true implosion) was precipitated by contact with European diseases, but it reflects the fact that mortality was already quite high before the arrival of the Europeans, and it did not take much to upset the demographic balance (Alchon 1997). By the beginning of the twentieth century, life expectancy was still very low in Mexico, less than 30 years (Morelos 1994), and fertility was very high in response to the high death rate. However, since the 1930s the death rate has dropped dramatically, and life expectancy in Mexico is now 78 years for women, nine years above the world average.

For several decades, this decline in mortality was not accompanied by a change in the birth rate, and the result was a massive explosion in the size of the Mexican population. In 1920, before the death rate began to drop, there were 14 million people in Mexico (Mier y Terán 1991). By 1950, that had nearly doubled to 26 million, and by 1970, it had nearly doubled again to 49 million. In the 1970s, the birth rate finally began to decline in Mexico. Mexican women had been bearing an average of six children each for many decades (if not centuries), but by the beginning of this century this figure had dropped to 2.7 children per woman. However, the massive build-up of young people has strained every ounce of the Mexican economy, encouraging out-migration, especially to the United States.

The other countries of Central America have experienced similar patterns of rapidly declining mortality, leading to population growth and its attendant pressures for migration to other countries where the opportunities might be better. Not every country in the region has experienced the same fertility decline as Mexico, however. In particular, Guatemala, Nicaragua, and Honduras are countries in which a high proportion of the population is indigenous (rather than being of mixed Indian and European origin) and they have birth rates that are close to or above four children per woman—much higher than the world average.

South America

The 358 million inhabitants of South America represent about 6 percent of the world's total population, with Brazil alone accounting for half of that. The modern history of Brazil began when Portuguese explorers found a hunter-gatherer indigenous population in that region and tried to enslave them to work on plantations. These attempts were unsuccessful, and the Portuguese wound up populating the colony largely with African slaves. The four million slaves taken to Brazil represent more than one-third of all slaves transported from Africa to the western hemisphere between the sixteenth and nineteenth centuries (Thomas 1997). The Napoleonic Wars in Europe in the early part of the nineteenth century allowed Brazil, like most Latin American countries, to gain independence from Europe, and the economic development that followed ultimately led to substantial migration into Brazil from Europe during the latter part of the nineteenth century. The result is a society that is now about half European-origin and half African-origin or mixed race.

Brazil boasts a land area nearly equal in size to the United States, upon which an estimated 177 million people currently reside. Since 1965, Brazil has experienced a reduction in fertility that has been described as "nothing short of spectacular" (Martine 1996:47). In 1960, the average Brazilian woman was giving birth to six children; this number dropped to 2.3 by the beginning of this century. This decline was largely unexpected, since the country was not experiencing dramatic economic improvement nor was there a big family planning campaign. For many years, the influence of the Catholic Church was strong enough to cause the government to forbid the dissemination of contraceptive information or devices (Martine 1996). For a variety of reasons that we will delve into in Chapter 6, Brazilian women have been using abortion and sterilization to limit family size.

The infant mortality rate of just under 33 deaths per 1,000 live births in Brazil is below the world average, while the female life expectancy of 73 years is above the world average. More than 90 percent of Brazilian baby girls can expect to still be around at age 50, but the odds are slimmer for boys, for whom life expectancy at birth (65 years) is just at the world average.

The other countries of South America that are predominantly European-origin, such as Argentina, Chile, and Uruguay, have fertility levels very similar to Brazil's but higher life expectancy. On the other hand, the countries with a large indigenous population—principally the descendants of the Incan civilization—such as Peru, Bolivia, and Ecuador, tend to have higher fertility, higher mortality, and higher rates of population growth.

Europe

The combined population of western, southern, northern, central, and eastern Europe is about 727 million, or about 12 percent of the world's total. Russia is the most populous, accounting for 20 percent of Europe's population, followed by Germany, France, the United Kingdom, and Italy.

Several European nations, especially the former Eastern Bloc nations, are experiencing negative population growth. Foremost among these is Russia, in which the breakup of the Soviet Union was foreshadowed by a rise in death rates (Feshbach and Friendly 1992; Shkolnikov, Meslé, and Vallin 1996) that has been followed by a baby bust that has lowered the average number of children being born per woman to 1.3. These trends in Russia mirror demographic events in East Germany just before and after the fall of the Berlin Wall, and in both cases the situation has been described as "a society convulsed by its stresses" (Eberstadt 1994b:149). At the same time, it appears that the low fertility in these countries is not just a temporary phenomenon caused by political and economic upheaval. Rather, it appears that the motivation to have large families has disappeared and has been replaced by a propensity to try to improve the family's standard of living by limiting the number of children (Avdeev and Monnier 1995).

The rest of Europe has experienced very low birth rates without the drop in life expectancy that has plagued Russia. Central and Eastern Europe (including Russia) are actually depopulating, both as a result of more deaths than births and also from out-migration. The rest of Europe is either not growing or is growing only very slowly. Where population growth is occurring, such as in France and the United Kingdom, it is largely attributable to the immigration of people from less developed nations (Hall and White 1995; Sardon 2002).

It should not be a surprise that fertility and mortality are both low in Europe, since that is the part of the world where mortality first began its worldwide decline approximately 250 years ago and where fertility began *its* worldwide decline about 150 years ago. What is surprising, however, is how low the birth rate has fallen. It is especially low in the Mediterranean countries of Italy and Spain, where fertility has dropped well below replacement level—in predominantly Catholic societies where fertility for most of history has been higher than in the rest of Europe. Sweden and the other Scandinavian countries have emerged with fertility rates that are now among the highest in Europe, although still below the replacement level, and Chesnais (1996) has offered the intriguing thesis that an improvement in the status of women may be required to push fertility levels in Europe back up to the replacement level. We return to this theme in Chapters 6 and 10.

I have already discussed the fact that a major consequence of the low birth rate is an aging of the population that has left Europe with too few young people to take jobs and pay taxes. Into this void have swept millions of immigrants, many of them illegal, and Europeans are very divided in their reaction to this phenomenon. Some see the immigrants as the necessary resource that will keep the economy running and pension checks flowing for aging Europeans. Others see the immigrants as a very real threat to the European way of life, coming as they mainly do from Africa and Asia.

Northern Africa and Western Asia

The area most often referred to as the Middle East includes Egypt and western Asia, while the remainder of northern Africa to the west of Egypt is usually referred to as the Maghrib. The entire region is characterized especially by the presence of Islam (with the obvious notable exception of Israel) and by the fact that it includes a globally disproportionate share of countries with the highest rates of population growth. Total population in the region is 392 million, representing 6 percent of the world's total. Egypt is the most populous of the countries in northern Africa and western Asia, followed by Turkey, and together they account for about one-third of the region's total population. There are 72 million Egyptians crowded into the narrow Nile Valley. With its rate of growth of 2.1 percent per year, Egypt's population could double in less than 35 years, and this rapid growth constantly hampers even the most ambitious strategies for economic growth and development. This explosion in numbers is due, of course, to the dramatic drop in mortality since the end of World War II. In 1937, the life expectancy at birth in Egypt was less than 40 years (Omran 1973; Omran and Roudi 1993), whereas by now it has risen to 70 for females. Even with such a drop, however, death rates are just at the world average, and so there is room for improvement.

As mortality declined, fertility remained almost intransigently high in Egypt until very recently, save for brief dips during World War II and again during the wars with Israel in the late 1960s and early 1970s. For as long as statistics had been kept, Egyptian women had been bearing an average of six children, until the late 1970s. Massive family planning efforts were initiated in the 1970s under President Sadat and then reinvigorated in the 1980s under President Mubarak. These programs, especially in combination with increasing levels of education among women (Fargues 1997; 2000a), have had an effect, and the estimated fertility level is now an average of 3.5 children per woman, much lower than it used to be but still very high. Because of the high fertility, a very high proportion (36 percent) of the population is under age 15.

It is the size and rate of increase in the youthful population that has been especially explosive throughout northern Africa and western Asia. The rapid drop in mortality after World War II, followed by a long delay in the start of fertility decline, produced a very large population of young people in need of jobs. They have spread throughout the region looking for work, and many have gone to Europe and North and South America. But the economies have not been able to keep up with the demand for jobs, and this has produced a generation of young people who, paradoxically, are better educated than their parents but who face an uncertain future in an increasingly crowded world. The declining birth rates hold out hope that the situation will ease in a few decades (Fargues 1994; 2000), but in the meantime the demographic situation has fueled discontent and has threatened the political stability of the region, as I discussed in the essay in Chapter 1.

Sub-Saharan Africa

Sub-Saharan Africa is the place from which all human life originated, according to most evidence (Wilson and Cann 1992), and the 711 million people living there now comprise 11 percent of the world's total. Nigeria accounts for about 1 in 5 of those 711 million, followed by Ethiopia and Zaire. If you do a few quick calculations in Table 2.6, you will discover that between 1000 A.D. and 1500 A.D., before extensive European contact with sub-Saharan Africa, that part of the world was increasing in population at a higher rate than anywhere else. However, Europeans brought their diseases to Africa and, much more significantly, they commercialized slavery to an extent never previously known in human history. More than 11 million Africans were enslaved and sent off to the western hemisphere between the sixteenth and nineteenth centuries (Thomas 1997), contributing to a decline in population size in sub-Saharan Africa between 1600 and 1850.

Until only a few decades ago, death rates in Africa were frequently as high as 40 deaths per 1,000 population and were associated with life expectancy in the range of 30 to 40 years, which is lower than the United States has experienced at any time since the American Revolution. Currently there are 27 countries in the world with life expectancy for females of less than 50 years; all but one (Afghanistan) are in sub-Saharan Africa. This area of the world has long had higher death rates than anywhere else, but since the 1980s the AIDS pandemic has lowered life expectancy, sometimes dramatically, in many countries in the region. However, do not be fooled into thinking that this signals a population implosion:

> Despite the devastating impact of the HIV/AIDS epidemic, the populations of the affected countries are generally expected to be larger by mid-century than today, mainly because most of them maintain high to moderate fertility levels. However, for the seven most affected countries in Southern Africa, where current HIV prevalence is above 20 percent, the population is projected to increase only slightly, from 74 million in 2000 to 78 million in 2050, and outright reductions in population are projected for Botswana, Lesotho, South Africa and Swaziland (United Nations Population Division 2003c:viii).

East Asia

There are 1.5 billion people living in East Asia, with the region dominated demographically by China (accounting for 85 percent of the population in this part of the world) and Japan (with 8 percent of the region's population). East Asia includes 25 percent of the world's total population, but its share will diminish a bit in the future as China continues to brake its population growth.

China   The People's Republic of China has a population of 1.3 billion people, maintaining for the time being its long-standing place as the most populous country in the world. With more than one-fifth of all human beings, China dominates the map of the world drawn to scale according to population size (see Figure 2.1). If we add in the Chinese in Taiwan (which the government of mainland China still claims as its own), Singapore, and the overseas Chinese elsewhere in the world (Poston, Mao, and Yu 1994), closer to one out of every four people is of Chinese origin. Nonetheless, China's share of the world's total population actually peaked in the middle of the nineteenth century, as you can see in Table 2.6. In 1850, more than one in three people were living in China, and that fraction has steadily declined over time.

For years the Chinese tried to ignore their demographic bulk, perhaps fearful that outside reports of population size and growth rates would not take sufficient account of the fact that the Communist revolution inherited a very large problem. In 1982, China took stock of the magnitude of its problem with its first national census since 1964 (which had been taken shortly after the terrible famine that I mentioned earlier). A total of just slightly more than one billion people were counted, and the results seemed to reinforce the government's belief that it was on the right track in vigorously pursuing a fairly coercive one child policy to cut the birthrate (which I will discuss in some detail in Chapter 6). The general government attitude was summed up in the mid-1990s as follows:

> Despite the outstanding achievements made in population and development, China still confronts a series of basic problems including a large population base, insufficient cultivated land, under-development, inadequate resources on a per capita basis and an uneven social and economic development among regions. . . . Too many people has impeded seriously the speed of social and economic development of the country and the rise of the

standard of living of the people. Many difficulties encountered in the course of social and economic development are directly attributable to population problems (Peng 1996:7).

Fertility decline actually began in China's cities in the 1960s and spread rapidly throughout the rest of the country in the 1970s, when the government introduced the family planning program known as *wan xi shao,* meaning "later" (marriage), "longer" (birth interval), "fewer" (children) (Banister 1987; Goldstein and Feng 1996). In 1979, this was transformed into the one-child policy, but fertility was already on its way down by that time. Between 1963 and 1983, China experienced what Blayo (1992) called a "breathtaking drop in TFRs [total fertility rates, see Chapter 6] from 7.5 to 2.5 children" (p. 213), and what Riley and Gardner (1996) suggest is "the most rapid sustained fertility decline ever seen in a population of any size" (p. 1).

China's fertility has continued to decline, although at a slower pace. China's birthrate has now dropped to 1.7 children per woman, which is below the replacement rate, although as I will discuss later in the book, that does not yet mean that the population has stopped growing. Furthermore, it is well above the government goal of one child per woman. Life expectancy at birth for females is 73 years, several years above the world average. Despite its low birth rate, the number of births each year in China is more than twice the number of deaths just because China is paying for its previous high birth rate. There are so many young women of reproductive age (women born 20 to 40 years ago when birth rates were much higher) that their babies vastly outnumber the people who are dying each year. As a result, the rate of natural increase in China is essentially the same as in the United States, despite the lower fertility rate. Of course, since China has more than four times as many inhabitants as the United States, it is adding 11 million people each year to its territory, whereas the United States is accumulating "only" an additional 2.5 million (and that includes immigrants). Thus, population growth remains a serious concern in China, but the concern is now turning from the young population to the rapidly increasing number and proportion of older Chinese—the inevitable consequence of a rapid decline in fertility in a nation where mortality is also low.

Japan   Let me also briefly mention Japan, which has the lowest level of mortality in the world, with a female life expectancy at birth of 85 years. Japan's health (accompanied by its wealth) translates demographically into very high probabilities of survival to old age—indeed, more than half of all Japanese born in the year 2000 will likely still be alive at age 80. This very low mortality rate is accompanied by very low fertility. Japanese women are bearing an average of 1.3 children each, of whom virtually all will live to adulthood. This is a level well below that required to replace each generation (2.1), and Japan is currently growing in number only because there are still enough women of reproductive age that, even at 1.3 each, they produce more babies than there are people dying each year. Japan's low mortality and low fertility have produced a population in which only 14 percent are under age 15, whereas 19 percent are 65 or older, with a forecasted rise to 26 percent aged 65 or older by the year 2025.

The resulting "graying" of Japan has caused the same concern as in Europe (and in China and Korea) about how to deal with the changing age structure. As the population ages, who will take over the jobs to keep the economy going, and who will care for the elderly? The United Nations (2000) has suggested that a policy of "replacement migration" may be needed if these countries are to maintain their demographic and economic viability. Japan, like other Asian countries, has an extremely restrictive immigration policy, so migrants currently contribute little to demographic change in Japan.

South and Southeast Asia

South and southeast Asia is home to about two billion people, one-third of the world's total. In contrast to Japan, this region of the world is dominated by youthful populations struggling with grinding poverty. The Indian subcontinent dominates this area demographically—India, Pakistan, and Bangladesh encompass two-thirds of the region's population. But Indonesia, the world's fourth most populous nation (and the one with the largest Muslim population in the world), is also part of southeast Asia, as are three other countries on the top 20 list—the Philippines, Vietnam, and Thailand.

India, Pakistan, and Bangladesh   Second in population size in the world is India, with the current population estimated to be just beyond one billion, but projected to be 1.5 billion (more populous than China) by the middle of this century (see Table 2.5). Mortality is somewhat higher in India than in China, and the birth rate is quite a bit higher than in China. Indian females have a life expectancy at birth of 64 years—below the world average, but a substantial improvement over the life expectancy of less than 27 years that prevailed back in the 1920s (Adlahka and Banister 1995). The infant mortality rate of 66 per 1,000 is above the world average, but it is also far lower than it was just a few decades ago. Women are bearing children at a rate of 3.1 each, and most children in India now survive to adulthood. With an annual growth rate of 1.7 percent, the Indian population is adding 18 million people to the world's total each year. The population of the Indian subcontinent is already more populous than mainland China (see Table 2.6), and that does not take into account the millions of people of Indian origin who are living elsewhere in the world.

It appears that fertility prior to 1961 was high and had remained virtually constant (Jain and Adlakha 1982), but since then it has declined, although not nearly as rapidly as in China. Furthermore, these declines have been nearly matched by declining mortality, thus maintaining the rate of population growth in India at a level consistently above the world's average. In fact, given its birth rate, if its death rate were to drop to a level equal to that of China, India would now be adding 21 million people a year, instead of its present 18 million.

India's population is culturally diverse, and this is reflected in rather dramatic geographic differences in fertility and rates of population growth within the country. For example, in the southern states of Kerala and Tamil Nadu, fertility had dropped below the replacement level by the mid-1990s. By comparison, in the four most populous states in the north (Bihar, Madhya Pradesh, Rajasthan, and Uttar Pradesh), where 40 percent of the Indian population lives, the average woman was still bearing five children (Adlahka and Banister 1995).

At the end of World War II, when India was granted its independence from British rule, the country was divided into predominantly Hindu India and predominantly Muslim Pakistan, with the latter having territory divided between West Pakistan and East Pakistan. In 1971, a civil war erupted between the two disconnected Pakistans, and, with the help of India, East Pakistan won the war and became Bangladesh. Although Pakistan and Bangladesh are both Muslim, Bangladesh has a demographic profile that now looks more like India than Pakistan. The average woman in Bangladesh now gives birth to 3.6 children (about half a child more than in India), whereas fertility in Pakistan has remained much higher (currently 4.8 children per woman). The overall rate of population growth in Bangladesh is 2.2 percent per year (a bit higher than in India), but it is 2.7 percent per year in Pakistan. Still, both Pakistan and Bangladesh have grown so much since independence after World War II that, were they still one country, they would be the third most populous nation in the world, and would of course surpass Indonesia as the world's most populous Muslim nation.

Indonesia   Indonesia is a string of islands in Southeast Asia and remains the world's most populous Muslim nation, with an estimated 220 million people. A former Dutch colony, it has experienced a substantial decline in fertility in recent years, and Indonesian women now bear an average of 2.6 children each—the lowest level of fertility in any Muslim nation outside of southern Europe (where predominantly Muslim Albania and Bosnia-Herzogovina have fertility levels at or below replacement level). Life expectancy for women in Indonesia at birth is 70 years (just a tick above the world average). For several decades, Indonesia has dealt with population growth through a program of transmigration, in which people have been sent from the more populous to the less populous islands. These largely forested outer islands have suffered environmentally from the human encroachment (Fearnside 1997), without necessarily dealing successfully with Indonesia's basic dilemma, which is how to raise its burgeoning young adult population out of poverty. This dilemma has contributed to increased political instability, as well as a rise in the level of Islamic fundamentalism and terrorism.

Oceania

Oceania is home to a wide range of indigenous populations, including Melanesian and Polynesian, but European influence has been very strong, and the region is generally thought of as being "overseas European." Its population of 31 million is about the same number as Canada's, less than 1 percent of the world's total. Australia accounts for two-thirds of the region's population, followed by Papua New Guinea and New Zealand. In a pattern repeated elsewhere in the world, the lowest birth rates and lowest death rates (and thus the lowest rates of population growth) are found in countries whose populations are largely European-origin (Australia and New Zealand, in this case), whereas the countries with a higher fraction of the population that is of indigenous origin have higher birth rates, higher mortality, and substantially higher rates of population growth (exemplified in Oceania by Papua New Guinea).

Global Demographic Contrasts

This whirlwind global tour should have put you in mind of the tremendous demographic contrasts that exist in the modern world. In the less developed nations, the population continues to grow quickly, especially in absolute terms. In sub-Saharan Africa this is happening even in the face of the AIDS pandemic. Yet, in the more developed countries population growth has slowed, stopped, or in some places even started to decline. As we look around the world, we see that the more rapidly growing countries tend to have high proportions of people who are young, poor, prone to disease, and susceptible to political instability. The countries that are growing slowly or not at all tend to have populations that are older, richer, and healthier, and these are the nations that are politically more stable and are calling the shots in the world right now. I do not want you to become a demographic determinist, because the world is more complicated than that, but keep in mind that there is almost certainly something to the idea that "demography is destiny"—a country cannot readily escape the demographic changes put into motion by the universally sought-after decline in mortality. Each country has to learn how to read its own demographic situation, and cope as well as it can with the inevitable changes that will take place as it evolves through all phases of the demographic transition.

Summary and Conclusion

High death rates kept the number of people in the world from growing rapidly until approximately the time of the Industrial Revolution. Then improved living conditions, public health measures, and, more recently, medical advances dramatically accelerated the pace of growth. As populations have grown, the pressure or desire to migrate has also increased. The vast European expansion into less developed areas of the world, which began in the fifteenth and sixteenth centuries, is a notable illustration of massive migration and population redistribution. Today migration patterns have shifted, and people are mainly moving from less developed to more developed nations. Closely associated with migration and population density is the urban revolution—that is, the movement from rural to urban areas.

The current world situation finds China and India as the most populous countries, followed by the United States, Indonesia, and Brazil. Everywhere population is growing we find that death rates have declined more rapidly than have birth rates, but there is considerable global and regional variability in both the birth- and death rates and thus in the rate of population growth. Dealing with the pressure of an expanding young population is the task of developing countries, whereas more developed countries, along with China, have aging populations, and are coping with the fact that the demand for labor in their economies may have to be met by immigrants from more rapidly growing countries.

Demographic dynamics represent the leading edge of social change in the modern world. It is a world of more than six billion people, heading to nine billion by the middle of this century and perhaps more beyond that, even assuming that there is no backsliding in the fertility declines currently under way in many less developed nations. In order to cope with the demographic underpinnings of our lives, we need to have a demographic perspective that allows us to sort out the causes and consequences of population change. We turn to that in the next chapter.

Main Points

1. During the first 90 percent of human existence, the population of the world had grown only to the size of today's New York City.

2. Between 1750 and 1950, the world's population mushroomed from 800 million to 2.5 billion, and since 1950 it has expanded to more than six billion.

3. Doubling time is a convenient way to summarize the rate of population growth. It is found by dividing the average annual rate of population growth into 69.

4. Early population growth was slow not because birthrates were low but because death rates were high.

5. On the other hand, continuing population increases are due to dramatic declines in mortality without a commensurate decline in fertility.

6. World population growth has been accompanied by migration from rapidly growing areas into less rapidly growing regions. Initially, that meant an outward expansion of the European population, but more recently it has meant migration from less developed to more developed nations.

7. Migration has also involved the shift of people from rural to urban areas, and urban regions on average are currently growing more rapidly than ever before in history.

8. Although migration is crucial to the demographic history of the United States and Canada, both countries have grown largely as a result of natural increase—the excess of births over deaths—after the migrants arrived.

9. At the time of the American Revolution, fertility levels in North America were among the highest in the world. Now they are among the lowest, even though there are still more babies born each year than there are people dying.

10. The world's 10 most populous countries are the People's Republic of China, India, the United States, Indonesia, Brazil, Russia, Bangladesh, Russia, Japan, and Nigeria. Together they account for 60 percent of the world's population.

11. Almost all of the population growth in the world today is occurring in the less developed nations, leading to an increase in the global demographic contrasts among countries.

Suggested Readings

1. Massimo Livi-Bacci, 2001, A Concise History of World Population, Third Edition (Malden, MA: Blackwell Publishers).
   This is a very thoughtful and well-documented account of population growth from ancient to modern times.

2. Jared Diamond, 1997, Guns, Germs, and Steel (New York: W. W. Norton).

   Diamond is a physiologist who has explored the ways in which disease and technology have aided the process by which the human population has evolved in the modern world.

3. K. Bruce Newbold, 2002, Six Billion Plus: Population Issues in the Twenty-First Century (Lanham, MD: Rowman & Littlefield).

   As implied by the title, the focus of the book is on the consequences for the world of the momentum of population growth carried over into the twenty-first century.

4. Robert Wright, 2000, Nonzero: The Logic of Human Destiny (New York: Pantheon Books).

   Wright is a nonacademic writer whose fascinating book puts population growth into its historical and future contexts.

5. George J. Demko, Grigory Ioffe, and Zhanna Zayonchkovskaya, Editors, 1999, Population Under Duress: The Geodemography of Post-Soviet Russia (Boulder, CO: Westview Press).

   The collapse of the Soviet Union had distinct demographic roots, as well as a fascinating set of consequences that provide important lessons about the linkage of demographic and social change. The authors of this volume explore these themes in depth.

🌐 Web Sites of Interest

Remember that Web sites are not as permanent as books and journals, so I cannot guarantee that each of the following Web sites still exists at the moment you are reading this. It is usually possible, however, to retrieve the contents of old Web sites at http://www.archive.org.

1. **http://www.popexpo.ined.fr/english.html**
   Would you like to know not only how many people there are at this moment in the world but also how many there were when you were born? This Web site offers you that bit of interesting information (no, not trivia—this is serious) and a lot more.

2. **http://www.censusindia.net**
   You don't have to take anybody else's word for what's happening demographically in India. This Indian census Web site is in English and has lots of data for the country and its regions.

3. **http://www.cpirc.org.cn/en/eindex.htm**
   There is a great deal of useful and regularly updated information on China at this Web site hosted by the China Population Information and Research Center in Beijing.

4. **http://sedac.ciesin.columbia.edu/plue/gpw**
   The Gridded Population of the World is a database created from censuses, surveys, satellite imagery, and other sources, producing a very realistic picture of population density and other characteristics at the global level. Regional maps and data are also available at this Web site.

5. **http://web.ornl.gov/sci/gist//landscan/index.html**
   LandScan is another globally gridded set of population data, designed at the Oak Ridge National Laboratory for the U.S. government as a way of evaluating the population anywhere in the world at risk of potential disasters.

CHAPTER 3
Demographic Perspectives

Figure 3.1   Population Density of the World, 2002

Persons per
square kilometer

☐ 0–2
▨ 3–99
▨ 100–4,999
■ 5,000–160,345

To get a handle on population problems and issues, you have to put the facts of population together with the "whys" and "wherefores." In other words, you need a **demographic perspective**—a way of relating basic information to theories about how the world operates demographically. A demographic perspective will guide you through the sometimes tangled relationships between population factors (such as size, distribution, age structure, and growth) and the rest of what is going on in society. As you develop your own demographic perspective, you will acquire a new awareness about your community or your job, for example, or national and world political or social problems. You will be able to ask yourself about the influences that demographic changes have had (or might have had), and you will consider the demographic consequences of events.

In this chapter, I discuss several theories of how population processes are entwined with general social processes. There are actually two levels of population theory. At the core of demographic analysis is the technical side of the field—the mathematical and biomedical theories that predict the kinds of changes taking place in the more biological components of demography: fertility, mortality, and the distribution of a population by age and sex. Demography, for example, has played a central role in the development of the fields of probability, statistics, and sampling (see Kreager 1993). This hard core is crucial to our understanding of human populations, but there is a "softer" (although no less important) outer wrapping of theory that relates demographic processes to the real events of the social world (Schofield and Coleman 1986). The linkage of the core with its outer wrapping is what produces your demographic perspective.

Two questions have to be answered before you will be able to develop your own perspective: (1) What are the causes of population growth (or, at least, population change)? and (2) What are the consequences of population growth or change? In this chapter, I discuss several perspectives that provide broad answers to these questions and that also introduce you to the major lines of demographic theory. The purpose of this review is to give you a start in developing your own demographic perspective by taking advantage of what others have learned and passed on to us.

I begin the chapter with a brief review of premodern thinking on the subject of population. Most of these ideas are what we call **doctrine,** as opposed to **theory.** Early thinkers were certain they had the answers and certain that their proclamations represented the truth about population growth and its implications for society. By contrast, the essence of modern scientific thought is to assume that you do not have the answer and to acknowledge that you are willing to consider evidence regardless of the conclusion to which it points. In the process of sorting out the evidence, we develop tentative explanations (theories) that help guide our thinking and our search for understanding. In demography, as in all of the sciences, theories replace doctrine when new systematically collected information (censuses and other sources discussed in the next chapter) becomes available, allowing people to question old ideas and formulate new ones (see, for example, Hirschman 1999). Table 3.1 summarizes the doctrines and theories discussed in the chapter.

Premodern Population Doctrines

Until about 2,500 years ago, human societies probably shared a common concern about population: they valued reproduction as a means of replacing people lost through universally high mortality. Ancient Judaism, for example, provided the prescription to "be fruitful and multiply" (Genesis 1:28). Indeed, reproductive power was often deified, as in ancient Greece, where it was the job of a variety of goddesses to help mortals successfully bring children into the world and raise those children to adulthood. In two of the more developed areas of the world 2,500 years ago, however, awareness of the potential for populations to grow beyond their resources prompted comment by well-known philosophers. In the fifth century B.C., the writings of the school of Confucius in China discussed the relationship between population and resources (Sauvy 1969), and it was suggested that the government should move people from overpopulated to underpopulated areas (an idea embraced in the twentieth century by the Indonesian government). Nonetheless, the idea of promoting population growth was clear in the doctrine of Confucius (Keyfitz 1973).

Plato, writing in *The Laws* in 360 B.C., emphasized the importance of population stability rather than growth. Specifically, Plato proposed keeping the ideal community of free citizens (much of the work was done by indentured laborers or slaves who had few civil rights) at a constant 5,040, which is handy because it can be divided by many different numbers. However, Charbit (2002:216) has suggested that "what inspired Plato in his choice of 5,040 is above all the fact that it is divisible by twelve, a number with a decisive sacred dimension," a legacy carried on in the 12 months of the year, among dozens (pun intended) of other aspects of modern life. Nonetheless, the number of people is still moderately small, and Plato felt that too many people led to anonymity, which would undermine democracy, whereas too few people would prevent an adequate division of labor and would not allow a community to be properly defended. Population size would be controlled by late marriage, infanticide, and migration (in or out as the situation demanded) (Plato 1960 [360 B.C.]). Plato was an early proponent of the doctrine that quality in humans is more important than quantity.

Plato's most famous student, Aristotle, was especially concerned that the population of a city-state not grow beyond the means of the families to support themselves. In *The Politics*, written in 340 B.C., he advocated that the number of children be limited by law, and that if a woman became pregnant after already having all the children that the law allowed, an abortion would be appropriate (Stangeland 1904). Echoes of this idea are found today in the one-child policy in China.

Table 3.1   Demographic Perspectives over Time

	Date	Demographic Perspective
Examples of Premodern Doctrines	~1300 B.C.	Book of Genesis—"Be fruitful and multiply."
	~500 B.C.	Confucius—Population growth is good, but governments should maintain a balance between population and resources.
	360 B.C.	Plato—population quality more important than quantity; emphasis on population stability.
	340 B.C.	Aristotle—population size should be limited and the use of abortion might be appropriate.
	~100 B.C.	Cicero—population growth necessary to maintain the Roman Empire.
	~400 A.D.	St. Augustine—abstinence is the preferred way to deal with human sexuality; the second best is to marry and procreate.
	~1280 A.D.	St. Thomas Aquinas—celibacy is *not* better than marriage and procreation.
	~1380 A.D.	Ibn Khaldun—population growth is inherently good because it increases occupational specialization and raises incomes.
	1500–1800	Mercantilism—increasing national wealth depends on a growing population that can stimulate export trade.
	1700–1800	Physiocrats—wealth of a nation is in land, not people; therefore population size depends on the wealth of the land, which is stimulated by free trade (laissez-faire).
Modern Theories	1798	Malthus—population grows exponentially while food supply grows arithmetically with misery (poverty) being the result in the absence of moral restraint.
	~1800	Neo-Malthusian—accepting the basic Malthusian premise that population growth tends to outstrip resources, but unlike Malthus believing that birth control measures are appropriate checks to population growth.
	~1844	Marxian—each society at each point in history has its own law of population that determines the consequences of population growth; poverty is not the natural result of population growth.
	1945	Demographic transition in its original formulation—the process whereby a country moves from high birth and death rates to low birth and death rates with an interstitial spurt in population growth. Explanations based originally on modernization theory.
	1962	Earliest studies suggesting the need to reformulate the demographic transition theory.
	1963	Theory of demographic change and response—the demographic response made by individuals to population pressures is determined by the means available to them to respond; causes and consequences of population change are intertwined.
	1968	Easterlin relative cohort size hypothesis—successively larger young cohorts put pressure on young men's relative wages, forcing them to make a tradeoff between family size and overall well-being.
	1971–present	Decomposition of the demographic transition into its separate transitions—mortality, fertility, age, migration, urbanization, and family and household.

In the Roman Empire, the reigns of Julius and Augustus Caesar were marked by clearly **pronatalist** doctrines—a necessity, given the very high mortality that characterized the Roman era (Frier 1999). In approximately 100 B.C., Cicero noted that population growth was seen by emperors as a necessary means of replacing war casualties and of ensuring enough people to help colonize the empire. Several scholars have speculated, however, that by the second century A.D., as the old pagan Roman Empire was waning in power, the birth rate in Rome may have been declining (Stangeland 1904; Veyne 1987). In a thoroughly modern sentiment, Pliny ("the younger") complained that " . . . in our time most people hold that an only son is already a heavy burden and that it is advantageous not to be overburdened with posterity" (quoted in Veyne 1987:13).

The Middle Ages in Europe, which followed the decline of Rome and its transformation from a pagan to a Christian society, were characterized by a combination of both pronatalist and **antinatalist** Christian doctrines. Christianity condemned polygamy, divorce, abortion, and infanticide, practices that had kept earlier Roman growth rates lower than they otherwise might have been. But the writings of Paul in the New Testament led the influential Christian leader, mystic, and writer Augustine (354–430) to argue that virgins were the highest form of human existence. Human sexuality was, in his view, a supernaturally good thing but also an important cause of sin (because most people are unable to control their desires) (O'Donnell 2001). He believed that abstinence was the best way to deal with sexuality, but the second-best state was marriage, which existed for the purpose of procreation. His Christian philosophy held that the world has a beginning, a middle, and an end, the end being eternal life in the city (community) of God. Thus, his doctrine of otherworldliness held that if all men would abstain from intercourse, then "so much more speedily would the City of God be fulfilled and the end of the world hastened" (quoted in Keyfitz 1972:43). This was an economically stagnant, fatalistic period of European history, and for centuries Europeans were content with the idea that population was a matter best regulated by God.

Intellectual stagnation in Europe was countered by the flowering of Islamic scholarship in the Middle East and North Africa. By the fourteenth century, one of the great Arab historians and philosophers, Ibn Khaldun, was in Tunis writing about the benefits of a growing population. In particular, he argued that population growth creates the need for specialization of occupations, which in turn leads to higher incomes, concentrated especially in cities: "Thus, the inhabitants of a more populous city are more prosperous than their counterparts in a less populous one. . . . The fundamental cause of this is the difference in the nature of the occupations carried on in the different places. For each town is a market for different kinds of labour, and each market absorbs a total expenditure proportionate to its size" (quoted in Issawi 1987:93). Ibn Khaldun was not a utopian. His philosophy was that societies evolved and were transformed as part of natural and normal processes. One of these processes was that "procreation is stimulated by high hopes and resulting heightening of animal energies" (quoted in Issawi 1987:268).

While Europe muddled through the Middle Ages, Islam (which had emerged in the seventh century A.D.) was expanding throughout the Mediterranean. Muslims took control of southern Italy and the Iberian peninsula and, under the Ottoman Empire, controlled the Balkans and the rest of southeastern Europe. Europe's reaction to this was the Crusades, a series of wars launched by Christians to wrestle control away from Muslims. These expeditions were largely unsuccessful from a military perspective, but they did put Europeans into contact with the Muslim world, which ultimately led to the Renaissance—the rebirth of Europe:

> The Islamic contribution to Europe is enormous, both of its own creations and of its borrowings—reworked and adapted—from the ancient civilizations of the eastern Mediterranean and from the remoter cultures of Asia. Greek science and philosophy, preserved and improved by the Muslims but forgotten in Europe; Indian numbers and Chinese paper; oranges and lemons, cotton and sugar, and a whole series of other plants along with the methods of cultivating them—all these are but a few of the many things that medieval Europe learned or acquired from the vastly more advanced and more sophisticated civilization of the Mediterranean Islamic world (Lewis 1995:274).

The cultural reawakening of Europe took place in the context of a growing population, as I noted in the previous chapter. Not surprisingly then, new murmurings were heard about the place of population growth in the human scheme of things. The Renaissance began with the Venetians, who had established trade with Muslims and others as the eastern Mediterranean ceased to be a Crusade war zone in the thirteenth century. In that century, an influential Dominican monk, Thomas Aquinas, argued that marriage and family building were not inferior to celibacy, thus implicitly promoting the idea that population growth is an inherently good thing.

By the end of the fourteenth century, the plague had receded from Europe; by the sixteenth-century, the Muslims (and Jews) had been expelled from southern Spain, and the Europeans had begun their discovery and exploitation of Africa, the Americas, and south Asia. Cities began to grow noticeably, and Giovanni Botero, a sixteenth century Italian statesman, wrote that " . . . the powers of generation are the same now as one thousand years ago, and, if they had no impediment, the propagation of man would grow without limit and the growth of cities would never stop" (quoted in Hutchinson 1967:111). The seventeenth and eighteenth centuries witnessed an historically unprecedented trade (the so-called **Columbian Exchange**) of food, manufactured goods, people, and disease between the Americas and most of the rest of the world (Crosby 1972), undertaken largely by European merchants, who had the best ships and the deadliest weapons in the world (Cipolla 1965).

This rise in trade, prompted at least in part by population growth, generated the doctrine of **mercantilism** among the new nation-states of Europe. Mercantilism maintained that a nation's wealth was determined by the amount of precious metals it had in its possession, which were acquired by exporting more goods than were imported, with the difference (the profit) being stored in precious metals. The catch here was that a nation had to have things to produce to sell to others, and the idea was that the more workers you had, the more you could produce. Furthermore, if you could populate the new colonies, you would have a ready-made market for your products. Thus population growth was seen as essential to an increase in national revenue, and mercantilist writers sought to encourage it by a number of means, including penalties for non-marriage, encouragements to get married, lessening penalties for illegitimate births, limiting out-migration, and promoting immigration of productive laborers. It is important to keep in mind that these doctrines were concerned with the wealth and welfare of a specific country, not all of human society. "The underlying doctrine was, either tacitly or explicitly, that the nation which became the strongest in material goods and in men would survive; the nations which lost in the economic struggle would have their populations reduced by want, or they would be forced to resort to war, in which their chances of success would be small" (Stangeland 1904:183).

Mercantilist doctrines were supported by the emerging demographic analyses of people like John Graunt, William Petty, and Edmund Halley (all English) in the seventeenth century and Johann Peter Süssmilch, an eighteenth-century chaplain in the army of Frederick the Great of Prussia (now Germany). In 1662, a Londoner named **John Graunt,** who is sometimes called the father of demography, analyzed the series of Bills of Mortality in the first known statistical analysis of demographic data (Sutherland 1963). Although he was a haberdasher by trade, Graunt used his spare moments to conduct studies that were truly remarkable for his time. He discovered that for every 100 people born in London, only 16 were still alive at age 36 and only 3 at age 66 (Dublin, Lotka, and Spiegleman 1949; Graunt 1662 [1939])—suggesting very high levels of mortality. With these data he uncovered the high incidence of infant mortality in London and found, somewhat to the amazement of people at the time, that there were regular patterns of death in different parts of London. Graunt "opened the way both for the later discovery of uniformities in many social or volitional phenomena like marriage, suicide and crime, and for a study of these uniformities, their nature and their limits; thus he, more than any other man, was the founder of statistics" (Willcox 1936:xiii).

One of Graunt's close friends (and possibly the person who coaxed him into this work) was William Petty, a member of the Royal Society in London. Petty circulated Graunt's work to the Society (which would not have otherwise paid much attention to a "tradesman"), and this brought it to the attention of the scientific world of seventh-century Europe. Several years later, in 1693, Edmund Halley (of Halley's comet fame) became the first scientist to elaborate on the probabilities of death. Although Halley, like Graunt, was a Londoner, he came across a list of births and deaths kept for the city of Breslau in Silesia (now Poland). From these data, Halley used the life-table technique (discussed in Chapter 5) to determine that the expectation of life in Breslau between 1687 and 1691 was 33.5 years (Dublin et al. 1949).

Then, in the eighteenth century, Süssmilch built on the work of Graunt and others and added his own analyses to the observation of the regular patterns of marriage, birth, and death in Prussia and believed that he saw in these the divine hand of God ruling human society (Hecht 1987; Keyfitz 1973). His view, widely disseminated throughout Europe, was that a larger population was always better than a smaller one, and, in direct contradistinction to Plato, he valued quantity over quality. He believed that indefinite improvements in agriculture and industry would postpone overpopulation so far into the future that it wouldn't matter.

The issue of population growth was more than idle speculation, because we know with a fair amount of certainty that the population of England, for example, doubled during the eighteenth century (Petersen 1979). More generally, during the period from about 1650 to 1850, Europe as a whole experienced rather dramatic population growth as a result of the disappearance of the plague, the introduction of the potato from the Americas, and evolutionary changes in agricultural practice—probably a response to the Little Ice Age (Fagan 2000)—that preceded (and almost certainly stimulated) the Industrial Revolution (Cohen 1995). The increasing interest in population encouraged the publication of two important essays on population size, one by David Hume (Hume 1752 [1963]) and the other by Robert Wallace (Wallace 1761 [1969]), which were then to influence Malthus, whom I discuss later. These essays sparked considerable debate and controversy, because there were big issues at stake: "Was a large and rapidly growing population a sure sign of a society's good health? On balance, were the growth of industry and cities, the movement of larger numbers from one social class to another—in short, all of what we now term 'modernization'—a boon to the people or the contrary? And in society's efforts to resolve such dilemmas, could it depend on the sum of individuals' self-interest or was considerable state control called for?" (Petersen 1979:139). These are questions that we are still dealing with 250 years later.

So the population had, in fact, increased during the Mercantilist era, although probably not as a result of any of the policies put forth by its adherents. However, it was less obvious that the population was better off. Rather, the Mercantilist period had become associated with a rising level of poverty (Keyfitz 1972). Mercantilism relied on a state-sponsored system of promoting foreign trade, while inhibiting imports and thus competition. This generated wealth for a small elite but not for most people. One of the more famous reactions against mercantilism was that mounted in the middle of the eighteenth century by François Quesnay, a physician in the court of King Louis XV of France (and an economist when not "on duty"). While mercantilists argued that wealth depends on the number of people, Quesnay turned that around and argued that the number of people depends on the means of subsistence (a general term for level of living). The essence of this view, called **physiocratic** thought, was that land, not people, is the real source of wealth of a nation. In other words, population went from being an independent variable, causing change in society, to a dependent variable, being altered by societal change. As you will see throughout this book, both perspectives have their merits.

Physiocrats also believed that free trade (rather than the import restrictions demanded by mercantilists) was essential to economic prosperity. This concept of "laissez-faire" was picked up by Adam Smith, one of the first modern economic theorists. Central to Smith's view of the world was the idea that, if left to their own devices, people acting in their own self-interest would produce what was best for the community as a whole. Smith differed slightly from the physiocrats, however, on the idea of what led to wealth in a society. Smith believed that wealth sprang from the labor applied to the land (we might now say the "value added" to the land by labor), rather than it being just in the land itself. From this idea sprang the belief that there is a natural harmony between economic growth and population growth, with the latter depending always on the former. Thus, he felt that population size is determined by the demand for labor, which is determined by the productivity of the land. These ideas are important to us because Smith's work served as an inspiration for the Malthusian theory of population, as Malthus himself acknowledges (see the preface to the sixth edition of Malthus 1872) and as I discuss later.

The Prelude to Malthus

The eighteenth century was the Age of Enlightenment in Europe, a time when the goodness of the common person was championed. This perspective that the rights of individuals superseded the demands of a monarchy inspired the American and French Revolutions and was generally very optimistic and utopian, characterized by a great deal of enthusiasm for life and a belief in the perfectibility of humans. In France, these ideas were well-expressed by Marie Jean Antoine Nicolas de Caritat, marquis de Condorcet, a member of the French aristocracy who forsook a military career to pursue a life devoted to mathematics and philosophy. His ideas helped to shape the French Revolution, but despite his sympathy with that cause, he died in prison at the hands of revolutionaries. In hiding before his arrest, Condorcet had written a *Sketch for an Historical Picture of the Progress of the Human Mind* (Condorcet 1795 [1955]). He was a visionary who "saw the outlines of liberal democracy more than a century in advance of his time: universal education; universal suffrage; equality before the law; freedom of thought and expression; the right to freedom and self-determination of colonial peoples; the redistribution of wealth; a system of national insurance and pensions; equal rights for women" (Hampshire 1955:x).

Condorcet's optimism was based on his belief that technological progress has no limits. "With all this progress in industry and welfare which establishes a happier proportion between men's talents and their needs, each successive generation will have larger possessions, either as a result of this progress or through the preservation of the products of industry, and so, *as a consequence of the physical constitution of the human race, the number of people will increase*" (Condorcet 1795 [1955]:188; emphasis added). He then asked whether it might not happen that eventually the happiness of the population would reach a limit. If that happens, Condorcet concluded, "we can assume that by then men will know that . . . their aim should be to promote the general welfare of the human race or of the society in which they live or of the family to which they belong, rather than foolishly to encumber the world with useless and wretched beings" (p. 189). Condorcet thus saw prosperity and population growth increasing hand in hand, and he felt that if the limits to growth were ever reached, the final solution would be birth control.

On the other side of the English Channel, similar ideas were being expressed by William Godwin (father of Mary Wollstonecraft Shelley, author of *Frankenstein*, and father-in-law of the poet Percy Bysshe Shelley). Godwin's *Enquiry Concerning Political Justice and Its Influences on Morals and Happiness* appeared in its first edition in 1793, revealing his ideas that scientific progress would enable the food supply to grow far beyond the levels of his day, and that such prosperity would not lead to overpopulation because people would deliberately limit their sexual expression and procreation. Furthermore, he believed that most of the problems of the poor were due not to overpopulation but to the inequities of the **social institutions,** especially greed and accumulation of property (Godwin 1793 [1946]).

Thomas Robert Malthus had recently graduated from Cambridge and was a country curate and a nonresident fellow of Cambridge as he read and contemplated the works of Godwin, Condorcet, and others who shared the utopian view of the perfectibility of human society. Although he wanted to be able to embrace such an openly optimistic philosophy of life, he felt that intellectually he had to reject it. In doing so, he unleashed a controversy about population growth and its consequences that rages to this very day.

The Malthusian Perspective

The **Malthusian** perspective derives from the writings of Thomas Robert Malthus, an English clergyman and college professor. His first *Essay on the Principle of Population as it affects the future improvement of society; with remarks on the speculations of Mr. Godwin, M. Condorcet, and other writers* was published anonymously in 1798. Malthus's original intention was not to carve out a career in demography, but only to show that the unbounded optimism of the physiocrats and philosophers was misplaced. He introduced his essay by commenting that "I have read some of the speculations on the perfectibility of man and society, with great pleasure. I have been warmed and delighted with the enchanting picture which they hold forth. I ardently wish for such happy improvements. But I see great, and, to my understanding, unconquerable difficulties in the way to them" (Malthus 1798 [1965]:7).

These "difficulties," of course, are the problems posed by his now famous **principle of population.** He derived his theory as follows:

> I think I may fairly make two postulata. First, that food is necessary to the existence of man. Secondly, that the passion between the sexes is necessary, and will remain nearly in its present state. . . . Assuming then, my postulata as granted, I say, that the power of population is indefinitely greater than the power in the earth to produce subsistence for man. Population, when unchecked, increases in a geometrical ratio. Subsistence increases only in an arithmetical ratio. . . . By the law of our nature which makes food necessary to the life of man, the effects of these two unequal powers must be kept equal. This implies a strong and constantly operating check on population from the difficulty of subsistence. This difficulty must fall somewhere; and must necessarily be severely felt by a large portion of mankind. . . . Consequently, if the premises are just, the argument is conclusive against the perfectibility of the mass of mankind (Malthus 1798 [1965]:11).

Malthus believed that he had demolished the utopian optimism by suggesting that the laws of nature, operating through the principle of population, essentially prescribed poverty for a certain segment of humanity. Malthus was a shy person by nature (James 1979; Petersen 1979), and he seemed ill prepared for the notoriety created by his essay. Nonetheless, after owning up to its authorship, he proceeded to document his population principles and to respond to critics by publishing a substantially revised version in 1803, slightly but importantly retitled to read *An Essay on the Principle of Population; or a view of its past and present effects on human happiness; with an inquiry into our prospects respecting the future removal or mitigation of the evils which It occasions.* In all, seven editions of Malthus's essay on population were published, and as a whole they have undoubtedly been the single most influential work relating population growth to its social consequences. Although Malthus relied heavily on earlier writers such as Hume and Wallace, he was the first to draw in a systematic way a picture that links the consequences of growth to its causes.

Causes of Population Growth

Malthus believed that human beings, like plants and nonrational animals, are "impelled" to increase the population of the species by what he called a powerful "instinct," the urge to reproduce. Further, if there were no checks on population growth, human beings would multiply to an "incalculable" number, filling "millions of worlds in a few thousand years" (Malthus 1872 [1971]:6). We humans, though, have not accomplished anything nearly that impressive. Why not? Because of the **checks to growth** that Malthus pointed out—factors that have kept population growth from reaching its biological potential for covering the earth with human bodies.

According to Malthus, the ultimate check to growth is lack of food (the **"means of subsistence"**). In turn, the means of subsistence are limited by the amount of land available, the "arts" or technology that could be applied to the land, and "social organization" or land ownership patterns. A cornerstone of his argument is that populations tend to grow more rapidly than the food supply does (a topic I return to in Chapter 12), since population has the potential for growing geometrically—two parents could have four children, sixteen grandchildren, and so on—while he believed that food production could be increased only arithmetically, by adding one acre at a time. He argued, then, that in the natural order, population growth will outstrip the food supply, and the lack of food will ultimately put a stop to the increase of people (see Figure 3.2).

Figure 3.2   Over Time, Geometric Growth Overtakes Arithmetic Growth

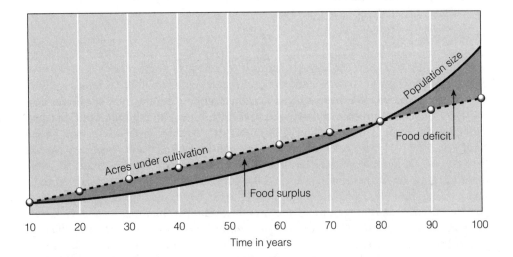

Note: If we start with 100 acres supporting a population of 100 people and then add 100 acres of cultivated land per decade (arithmetic growth) while the population is increasing by 3 percent per year (geometric growth), the result is a few decades of food surplus before population growth overtakes the increase in the acres under cultivation, producing a food deficit, or "misery," as Malthus called it.

Of course, Malthus was aware that starvation rarely operates directly to kill people, since something else usually intervenes to kill them before they actually die of starvation. This "something else" represents what Malthus calls **positive checks,** primarily those measures "whether of a moral or physical nature, which tend prematurely to weaken and destroy the human frame" (Malthus 1872 [1971]:12). Today we would call these the causes of mortality. There are also **preventive checks**—limits to birth. In theory, the preventive checks would include all possible means of birth control, including abstinence, contraception, and abortion. However, to Malthus the only acceptable means of preventing a birth was to exercise **moral restraint;** that is, to postpone marriage, remaining chaste in the meantime, until a man feels "secure that, should he have a large family, his utmost exertions can save them from rags and squalid poverty, and their consequent degradation in the community" (1872 [1971]:13). Any other means of birth control, including contraception (either before or after marriage), abortion, infanticide, or any "improper means," was viewed as a vice that would "lower, in a marked manner, the dignity of human nature." Moral restraint was a very important point with Malthus, because he believed that if people were allowed to prevent births by "improper means" (that is, prostitution, contraception, abortion, or sterilization), then they would expend their energies in ways that are, so to speak, not economically productive.

I should point out that as a scientific theory, the Malthusian perspective leaves much to be desired, since he constantly confuses moralistic and scientific thinking (see Davis 1955b). Despite its shortcomings, however, which were evident even in his time, Malthus's reasoning led him to draw some important conclusions about the consequences of population growth that are still relevant to us.

Consequences of Population Growth

Malthus believed that a natural consequence of population growth was poverty. This is the logical end result of his arguments that (1) people have a natural urge to reproduce, and (2) the increase in the food supply cannot keep up with population growth. In his analysis, Malthus turned the argument of Adam Smith upside down. Instead of population growth depending on the demand for labor, as Smith (and the physiocrats) argued, Malthus believed that the urge to reproduce always forces population pressure to precede the demand for labor. Thus, "overpopulation" (as measured by the level of unemployment) would force wages down to the point where people could not afford to marry and raise a family. At such low wages, with a surplus of labor and the need for each person to work harder just to earn a subsistence wage, cultivators could employ more labor, put more acres into production, and thus increase the means of subsistence. Malthus believed that this cycle of increased food resources leading to population growth leading to too many people for available resources leading then back to poverty was part of a natural law of population. Each increase in the food supply only meant that eventually more people could live in poverty.

As you can see, Malthus did not have an altogether high opinion of his fellow creatures. He figured that most of them were too "inert, sluggish, and averse from labor" (1798 [1965]:363) to try to harness the urge to reproduce and avoid the increase in numbers that would lead back to poverty whenever more resources were available. In this way, he essentially blamed poverty on the poor themselves. There remained only one improbable way to avoid this dreary situation.

Avoiding the Consequences

Borrowing from John Locke, Malthus argued that "the endeavor to avoid pain rather than the pursuit of pleasure is the great stimulus to action in life" (1798 [1965]:359). Pleasure will not stimulate activity until its absence is defined as being painful. Malthus suggested that the well-educated, rational person would perceive in advance the pain of having hungry children or being in debt and would postpone marriage and sexual intercourse until he was sure that he could avoid that pain. If that motivation existed and the preventive check was operating, then the miserable consequences of population growth could be avoided. You will recall that Condorcet had suggested the possibility of birth control as a preventive check, but Malthus objected to this solution: "To remove the difficulty in this way, will, surely in the opinion of most men, be to destroy that virtue, and purity of manners, which the advocates of equality, and of the perfectibility of man, profess to be the end and object of their views" (1798:154). So the only way to break the cycle is to change human nature. Malthus felt that if everyone shared middle-class values, the problem would solve itself. He saw that as impossible, though, since not everyone has the talent to be a virtuous, industrious, middle-class success story, but if most people at least tried, poverty would be reduced considerably.

To Malthus, material success is a consequence of the human ability to plan rationally—to be educated about future consequences of current behavior—and he was a man who practiced what he preached. He planned his family rationally, waiting to marry and have children until he was 39, shortly after getting a secure job in 1805 as a professor of history and political economy at East India College in Haileybury, England (north of London). Also, although Marx later thought that Malthus had taken the "monastic vows of celibacy" and other detractors attributed 11 children to him, he and his wife, 11 years his junior, had only three children (Nickerson 1975; Petersen 1979).

To summarize, the major consequence of population growth, according to Malthus, is poverty. Within that poverty, though, is the stimulus for action that can lift people out of misery. So, if people remain poor, it is their own fault for not trying to do something about it. For that reason, Malthus was opposed to the English Poor Laws (welfare benefits for the poor), because he felt they would actually serve to perpetuate misery. They permitted poor people to be supported by others and thus not feel that great pain, the avoidance of which might lead to birth prevention. Malthus argued that if every man had to provide for his own children, he would be more prudent about getting married and raising a family. In his own time, this particular conclusion of Malthus brought him the greatest notoriety, because the number of people on welfare had been increasing and English parliamentarians were trying to decide what to do about the problem. Although the Poor Laws were not abolished, they were reformed largely because Malthus had given legitimacy to public criticism of the entire concept of welfare payments (Himmelfarb 1984). The

Malthusian perspective that blamed the poor for their own poverty endures, contrasted with the equally enduring view of Godwin and Condorcet that poverty is the creation of unjust human institutions. Two hundred additional years of debate have only sharpened the edges of the controversy.

Critique of Malthus

The single most obvious measure of Malthus's importance is the number of books and articles that have attacked him, beginning virtually the moment his first essay appeared in 1798. The three most strongly criticized aspects of his theory have been (1) the assertion that food production could not keep up with population growth, (2) the conclusion that poverty was an inevitable result of population growth , and (3) the belief that moral restraint was the only acceptable preventive check. Malthus was not a firm believer in progress; rather, he accepted the notion that each society had a fixed set of institutions that established a stationary level of living. He was aware, of course, of the Industrial Revolution, but he was skeptical of its long-run value and agreed with the physiocrats that real wealth was in agricultural land. He was convinced that the increase in manufacturing wages that accompanied industrialization would promote population growth without increasing the agricultural production necessary to feed those additional mouths. Although it is clear that he was a voracious reader (Petersen 1979) and that he was a founder of the Statistical Society of London (Starr 1987), it is also clear that Malthus paid scant attention to the economic statistics that were available to him. "There is no sign that even at the end of his life he knew anything in detail about industrialization. His thesis was based on the life of an island agricultural nation, and so it remained long after the exports of manufacturers had begun to pay for the imports of large quantities of raw materials" (Eversley 1959:256). Thus, Malthus either failed to see or refused to acknowledge that technological progress was possible, and that its end result was a higher standard of living, not a lower one.

The crucial part of Malthus's ratio of population growth to food increase was that food (including both plants and nonhuman animals) would not grow exponentially. Yet when Charles Darwin acknowledged that his *Origin of the Species* was inspired by Malthus's essay, he implicitly rejected this central tenet of Malthus's argument. "Darwin described his own theory as 'the doctrine of Malthus applied with manifold force to the whole animal and vegetable kingdoms; for in this case there can be no artificial increase of food, and no prudential restraint from marriage.' Thus plants and animals, even more than men, would increase geometrically if unchecked" (Himmelfarb 1984:128).

Malthus's argument that poverty is an inevitable result of population growth is also open to scrutiny. For one thing, his writing reveals a certain circularity in logic. In Malthus's view, a laborer could achieve a higher standard of living only by being prudent and refraining from marriage until he could afford it, but Malthus also believed that you could not expect prudence from a laborer until he had attained a higher standard of living. Thus, our hypothetical laborer seems squarely enmeshed in a catch-22. Even if we were to ignore this logical inconsistency, there are problems with Malthus's belief that the Poor Laws contributed to the misery of the poor by discouraging them from exercising prudence. Historical evidence has revealed that between 1801 and 1835 those English parishes that administered Poor Law allowances did not have higher birth, marriage, or total population growth rates than those in which Poor Law assistance was not available (Huzel 1969; Huzel 1980; Huzel 1984). Clearly, problems with the logic of Malthus's argument seem to be compounded by his apparent inability to see the social world accurately. "The results of the 1831 Census were out before he died, yet he never came to interpret them. Statistics apart, the main charge against him must be that he was a bad observer of his fellow human beings" (Eversley 1959:256).

Neo-Malthusians

Those who criticize Malthus's insistence on the value of moral restraint, while accepting many of his other conclusions, are typically known as **neo-Malthusians.** Specifically, neo-Malthusians favor contraception rather than simple reliance on moral restraint. During his lifetime, Malthus was constantly defending moral restraint against critics (many of whom were his friends) who encouraged him to deal more favorably with other means of birth control. In the fifth edition of his *Essay,* he did discuss the concept of **prudential restraint,** which meant the delay of marriage until a family could be afforded without necessarily refraining from premarital sexual intercourse in the meantime. He never fully embraced the idea, however, nor did he ever bow to pressure to accept anything but moral restraint as a viable preventive check.

Ironically, the open controversy actually helped to spread knowledge of birth control among people in nineteenth-century England and America. This was aided materially by the trial and conviction (later overturned on a technicality) in 1877–78 of two neo-Malthusians, Charles Bradlaugh and Annie Besant, for publishing a birth control handbook (*Fruits of Philosophy: The Private Companion of Young Married People,* written by Charles Knowlton, a physician in Massachusetts, and originally published in 1832). The publicity surrounding the trial enabled the English public to become more widely knowledgeable about those techniques (Chandrasekhar 1979; Himes 1976). Eventually, the widespread adoption of birth control broke the connection between intercourse and fertility that had seemed so natural to Malthus: "In time fertility behavior came to be better epitomized by inverting the Malthusian assumption than by adhering to it. . . . The supplanting of the social control of fertility *within* marriage completely undermined the validity of the assumption that population growth rates would rise in step with prosperity" (Wrigley 1988:44; emphasis in original).

Criticisms of Malthus do not, however, diminish the importance of his work:

> There are good reasons for using Malthus as a point of departure in the discussion of population theory. These are the reasons that made his work influential in his day and make it influential now. But they have little to do with whether his views are right or wrong. . . . Malthus' theories are not now and never were empirically valid, but they nevertheless were theoretically significant (Davis 1955b:541).

As I noted earlier, part of Malthus's significance lies in the storm of controversy his theories stimulated. Particularly vigorous in their attacks on Malthus were Karl Marx and Friedrich Engels.

The Marxian Perspective

Karl Marx and Friedrich Engels were both teenagers in Germany when Malthus died in England in 1834, and by the time they had met and independently moved to England, Malthus's ideas already were politically influential in their native land. Several German states and Austria had responded to what they believed was overly rapid growth in the number of poor people by legislating against marriages in which the applicant could not guarantee that his family would not wind up on welfare (Glass 1953). As it turned out, that scheme backfired on the German states, because people continued to have children, but out of wedlock. Thus, the welfare rolls grew as the illegitimate children had to be cared for by the state (Knodel 1970). The laws were eventually repealed, but they had an impact on Marx and Engels, who saw the Malthusian point of view as an outrage against humanity. Their demographic perspective thus arose in reaction to Malthus.

Causes of Population Growth

Neither Marx nor Engels ever directly addressed the issue of why and how populations grew. They seem to have had little quarrel with Malthus on this point, although they were in favor of equal rights for men and women and saw no harm in preventing birth. Nonetheless, they were skeptical of the eternal or natural laws of nature as stated by Malthus (that population tends to outstrip resources), preferring instead to view human activity as the product of a particular social and economic environment. The basic **Marxian** perspective is that each society at each point in history has its own law of population that determines the consequences of population growth. For **capitalism,** the consequences are overpopulation and poverty, whereas for **socialism,** population growth is readily absorbed by the economy with no side effects. This line of reasoning led to a vehement rejection of Malthus, because if Malthus was right about his "pretended 'natural law of population'" (Marx 1890 [1906]:680), then Marx's theory would be wrong.

Consequences of Population Growth

Marx and Engels especially quarreled with the Malthusian idea that resources could not grow as rapidly as population, since they saw no reason to suspect that science and technology could not increase the availability of food and other goods at least as quickly as the population grew. Engels argued in 1865 that whatever population pressure existed in society was really pressure against the means of employment rather than against the means of subsistence (Meek 1971). Thus, they flatly rejected the notion that poverty can be blamed on the poor. Instead, they said, poverty is

WHO ARE THE NEO-MALTHUSIANS?

"Picture a tropical island with luscious breadfruits [a Polynesian plant similar to a fig tree] hanging from every branch, toasting in the sun. It is a small island, but there are only 400 of us on it so there are more breadfruits than we know what to do with. We're rich. Now picture 4,000 people on the same island, reaching for the same breadfruits: Number one, there are fewer to go around; number two, you've got to build ladders to reach most of them; number three, the island is becoming littered with breadfruit crumbs. Things get worse and worse as the population gradually expands to 40,000. Welcome to a poor, littered tropical paradise" (Tobias 1979:49). This scenario would probably have drawn a nod of understanding from Malthus himself, and it is typical of the modern neo-Malthusian view of the world.

One of the most influential neo-Malthusians is the biologist Garrett Hardin. In 1968, he published an article that raised the level of consciousness about population growth in the minds of professional scientists. Hardin's theme was simple and had been made by Kingsley Davis (1963) as he developed the theory of demographic change and response: Personal goals are not necessarily consistent with societal goals when it comes to population growth. Hardin's metaphor is "the tragedy of the commons." He asks us to imagine an open field, available as a common ground for herdsmen to graze their cattle. "As a rational being, each herdsman seeks to maximize his gain. Explicitly or implicitly, more or less consciously, he asks, 'What is the utility to me of adding one more animal to my herd?'" (Hardin 1968:1244). The benefit, of course, is the net proceeds from the eventual sale of each additional animal, whereas the cost lies in the chance that an additional animal may result in overgrazing of the common ground. Since the ground is shared by many people, the cost is spread out over all, so for the individual herdsman, the benefit of another animal exceeds its cost. "But," notes Hardin, "this is the conclusion reached by each and every rational herdsman sharing a commons. Therein is the tragedy. Each man is locked into a system that compels him to increase his herd without limit—in a world that is limited" (1968:1244). The moral, as Hardin puts it, is that "ruin is the destination toward which all men rush, each pursuing his own best interest in a society that believes in the freedom of the commons. Freedom in a commons brings ruin to all" (1968:1244).

Hardin reminds us that most societies are committed to a social welfare ideal. Families are not completely on their own. We share numerous things in common: education, public health, and police protection, and in the United States we are guaranteed a minimum amount of food and income at the public expense. This leads to a moral dilemma that is at the heart of Hardin's message: "To couple the concept of freedom to breed with the belief that everyone born has an equal right to the commons is to lock the world into a tragic course of action" (Hardin 1968:1246). He was referring, of course, to the ultimate Malthusian clash of population and resources, and Hardin was no more optimistic than Malthus about the likelihood of people voluntarily limiting their fertility before it is too late.

Meanwhile, in the 1960s the world was becoming keenly aware of the population crisis through the writings of the person who is arguably the most famous of all neo-Malthusians, Paul Ehrlich. Like Hardin, Ehrlich is a biologist (at Stanford University), not a professional demographer. His ***Population Bomb*** (Ehrlich 1968) was an immediate sensation when it came out in 1968 and continues to set the tone for public debate about population issues. In

the result of a poorly organized society, especially a capitalist society. Implicit in the writings of Marx and Engels is the idea that the normal consequence of population growth should be a significant increase in production. After all, each worker obviously was producing more than he or she required, or else how would all the dependents (including the wealthy manufacturers) survive? In a well-ordered society, if there were more people, there ought to be more wealth, not more poverty (Engels 1844 [1953]).

the second edition of his book (Ehrlich 1971), he phrased the situation in three parts: "too many people," "too little food," and, adding a wrinkle not foreseen directly by Malthus, "environmental degradation" (Ehrlich called Earth "a dying planet").

In 1990, Ehrlich, in collaboration with his wife Anne, followed with an update titled **The Population Explosion** (Ehrlich and Ehrlich 1990), reflecting their view that the bomb they worried about in 1968 had detonated in the meantime. The level of concern about the destruction of the environment has grown tremendously since 1968. Ehrlich's book had inspired the first Earth Day in the spring of 1970 (an annual event ever since in most communities across the United States), yet in their 1990 book Ehrlich and Ehrlich rightly question why, in the face of the serious environmental degradation that has concerned them for so long, have people regularly failed to grasp its primary cause as being rapid population growth? "Arresting population growth should be second in importance only to avoiding nuclear war on humanity's agenda. Overpopulation and rapid population growth are intimately connected with most aspects of the current human predicament, including rapid depletion of nonrenewable resources, deterioration of the environment (including rapid climate change), and increasing international tensions" (Ehrlich and Ehrlich 1990:18).

Ehrlich thus argues that Malthus was right—dead right. But the death struggle is more complicated than that foreseen by Malthus. To Ehrlich, the poor are dying of hunger, while rich and poor alike are dying from the by-products of affluence—pollution and ecological disaster. Indeed, this is part of the "commons" problem. A few benefit, all suffer. What does the future hold? Ehrlich suggests that there are only two solutions to the population problem: the birth rate solution (lowering the birth rate) and the death rate solution (a rise in the death rate). He views the death rate solution as being the most likely to happen, because, like Malthus, he has little faith in the ability of humankind to pull its act together. The only way to avoid that scenario is to bring the birth rate under control, perhaps even by force.

A major part of Ehrlich's contribution has been to encourage people to take some action themselves, to spread the word and practice what they preach. Like Hardin, Ehrlich feels that population growth is outstripping resources and ruining the environment. If we sit back and wait for people to react to this situation, disaster will occur. Therefore, we need to act swiftly to push people to bring fertility down to replacement level by whatever means possible.

Neo-Malthusians thus differ from Malthus because they reject moral restraint as the only acceptable means of birth control and because they face a situation in which they see population growth as leading not simply to poverty but also to widespread calamity. For neo-Malthusians, the "evil arising from the redundancy of population" (Malthus 1872 [1971]: preface to the fifth edition) has broadened in scope, and the remedies proposed are thus more dramatic.

Gloomy they may seem, but the messages of Ehrlich and Hardin are important and impressive and have brought population issues to the attention of the entire globe. One of the ironies of neo-Malthusianism is that if the world's population does avoid future calamity, people will likely claim that the neo-Malthusians were wrong. In fact, however, much of the stimulus to bring down birth rates (including emphasis on the reproductive rights of women as the alternative to coercive means) and to find new ways to feed people and protect the environment has come as a reaction to the concerns raised by neo-Malthusians.

Not only did Marx and Engels feel generally that poverty was not the end result of population growth, they argued specifically that even in England at that time there was enough wealth to eliminate poverty. Engels had himself managed a textile plant owned by his father's firm in Manchester, and he believed that in England more people had meant more wealth for the capitalists rather than for the workers because the capitalists were skimming off some of the workers' wages as profits for themselves. Marx argued that they did that by stripping the workers of their tools and then, in essence, charging the workers for being able to come to the factory to work. For example, if you do not have the tools to make a car but want a job making cars, you could get hired at the factory and work eight hours a day. But, according to Marx, you might get paid for only four hours, the capitalist (owner of the factory) keeping part of your wages as payment for the tools you were using. The more the capitalist keeps, of course, the lower your wages and the poorer you will be.

Furthermore, Marx argued that capitalism worked by using the labor of the working classes to earn profits to buy machines that would replace the laborers, which, in turn, would lead to unemployment and poverty. Thus, the poor were not poor because they overran the food supply, but only because capitalists had first taken away part of their wages and then taken away their very jobs and replaced them with machines. Thus, the consequences of population growth that Malthus discussed were really the consequences of capitalist society, not of population growth per se. Overpopulation in a capitalist society was thought to be a result of the capitalists' desire for an industrial reserve army that would keep wages low through competition for jobs and, at the same time, would force workers to be more productive in order to keep their jobs. To Marx, however, the logical extension of this was that the growing population would bear the seeds of destruction for capitalism, because unemployment would lead to disaffection and revolution. If society could be reorganized in a more equitable (that is, socialist) way, then the population problems would disappear.

It is noteworthy that Marx, like Malthus, practiced what he preached. Marx was adamantly opposed to the notion of moral restraint, and his life repudiated that concept. He married at the relatively young age (compared with Malthus) of 25, proceeded to father eight children, including one illegitimate son, and was on intimate terms with poverty for much of his life.

In its original formulation, the Marxian (as well as the Malthusian) perspective was somewhat provincial, in the sense that its primary concern was England in the nineteenth century. Marx was an intense scholar who focused especially on the historical analysis of economics as applied to England, which he considered to be the classic example of capitalism. However, as his writings have found favor in other places and times, revisions have been forced upon the Marxian view of population.

Critique of Marx

Not all who have adopted a Marxian world view fully share the original Marx-Engels demographic perspective. Socialist countries have had trouble because of the lack of political direction offered by the Marxian notion that different stages of social development produce different relationships between population growth and economic development. Indeed, much of the so-called Marxian thought on population is in fact attributable to Lenin, one of the most prolific interpreters of Marx. For Marx, the Malthusian principle operated under capitalism only, whereas under pure socialism there would be no population problem. Unfortunately, he offered no guidelines for the transition period. At best, Marx implied that the socialist law of population should be the antithesis of the capitalist law. If the birth rate were low under capitalism, then the assumption was that it should be high under socialism; if abortion seemed bad for a capitalist society, it must be good for a socialistic society. Thus, it was difficult for Russian demographers to reconcile the fact that demographic trends in the former Soviet Union were remarkably similar to trends in other developed nations. Furthermore, Soviet socialism was unable to alleviate one of the worst evils that Marx attributed to capitalism, higher death rates among people in the working class than among those in the higher classes (Brackett 1967). The flip side of that is that birth rates dropped to such low levels throughout pre-1990 Marxist Eastern Europe that it was no longer possible to claim (as Marx had done) that low birth rates were bourgeois.

In China, the empirical reality of having to deal with the world's largest national population has led to a radical departure from Marxian ideology. As early as 1953, the Chinese government organized efforts to control population by relaxing regulations concerning contraception and abortion. Ironically, after the terrible demographic disaster that followed the "Great Leap Forward" in 1958 (see Chapter 2), a Chinese official quoted Chairman Mao as having said, "A large population in China is a good thing. With a population increase of several fold we still have an adequate solution. The solution lies in production" (Ta-k'un 1960:704). Yet by 1979 production no longer seemed to be a panacea, and with the introduction of the one-child policy, the interpretation of Marx took an about-face as another Chinese official wrote that under Marxism the law of production "demands not only a planned production of natural goods, but also the planned reproduction of human beings" (Muhua 1979:724).

Thus, despite Marx's denial of a population problem, the Marxist government in China dealt with one by rejecting its Marxist-Leninist roots and embracing instead one of the most aggressive and coercive government programs ever launched to reduce fertility through restraints on marriage (the Malthusian solution), contraception (the neo-Malthusian solution), and abortion (a remnant of the Leninist approach) (Teitelbaum and Winter 1988). In a formulation such as this, Marxism was revised in the light of new scientific evidence about how people behave, in the same way that Malthusian thought has been revised. Bear in mind that although the Marxian and Malthusian perspectives are often seen as antithetical, they both originated in the midst of a particular milieu of economic, social, and demographic change in nineteenth-century Europe.

The Prelude to the Demographic Transition Theory

The population-growth controversy, initiated by Malthus and fueled by Marx, emerged into a series of nineteenth-century and early-twentieth-century reformulations that have led directly to prevailing theories in demography. In this section, I briefly discuss three individuals who made important contributions to those reformulations: John Stuart Mill, Arsène Dumont, and Émile Durkheim.

Mill

The English philosopher and economist John Stuart Mill was an extremely influential writer of the nineteenth century. Mill was not as quarrelsome about Malthus as Marx and Engels had been; his scientific insights were greater than those of Malthus at the same time that his politics were less radical than those of Marx and Engels. Mill accepted the Malthusian calculations about the potential for population growth to outstrip food production as being axiomatic (a self-truth), but he was more optimistic about human nature than Malthus was. Mill believed that although your character is formed by circumstances, one's own desires can do much to shape circumstances and modify future habits (Mill 1873 [1924]).

Mill's basic thesis was that the standard of living is a major determinant of fertility levels. "In proportion as mankind rises above the condition of the beast, population is restrained by the fear of want, rather than by want itself. Even where there is no question of starvation, many are similarly acted upon by the apprehension of losing what have come to be regarded as the decencies of their situation in life" (1848 [1929]:Book I, Chap. 10). The belief that people could be and should be free to pursue their own goals in life led him to reject the idea that poverty is inevitable (as Malthus implied) or that it is the creation of capitalist society (as Marx argued). One of Mill's most famous comments is that "the niggardliness of nature, not the injustice of society, is the cause of the penalty attached to overpopulation" (1848 [1929]:Book 1, Chap. 13). This is a point of view conditioned by Mill's reading of Malthus, but Mill denies the Malthusian inevitability of a population growing beyond its available resources. Mill believed that people do not "propagate like swine, but are capable, though in very unequal degrees, of being withheld by prudence, or by the social affections, from giving existence to beings born only to misery and premature death" (1848 [1929]:Book I, Chap. 7). In the event that population ever did overrun the food supply, however, Mill felt that it would likely be a temporary situation with at least two possible solutions: import food or export people.

The ideal state from Mill's point of view is that in which all members of a society are economically comfortable. At that point he felt (as Plato had centuries earlier) that the population would stabilize and people would try to progress culturally, morally, and socially instead of attempting continually to get ahead economically. It does sound good, but how do we get to that point? It was Mill's belief that before reaching the point at which both population and production are stable, there is essentially a race between the two. What is required to settle the issue is a dramatic improvement in the living conditions of the poor. If social and economic development is to occur, there must be a sudden increase in income, which could give rise to a new standard of living for a whole generation, thus allowing productivity to outdistance population growth. According to Mill, this was the situation in France after the Revolution:

> During the generation which the Revolution raised from the extremes of hopeless wretchedness to sudden abundance, a great increase of population took place. But a generation has grown up, which, having been born in improved circumstances, has not

learnt to be miserable; and upon them the spirit of thrift operates most conspicuously, in keeping the increase of population within the increase of national wealth (1848 [1929]:Book II, Chap. 7).

Mill was further convinced that an important ingredient in the transformation to a nongrowing population is that women do not want as many children as men do, and if they are allowed to voice their opinions, the birth rate will decline. Mill, like Marx, was a champion of equal rights for both sexes, and one of Mill's more notable essays, *On Liberty,* was co-authored with his wife. He reasoned further that a system of national education for poor children would provide them with the "common sense" (as Mill put it) to refrain from having too many children.

Overall, Mill's perspective on population growth was significant enough that we find his arguments surviving today in the writings of many of the twentieth-century demographers whose names appear in the pages that follow. However, before getting to those people and their ideas, it is important to acknowledge at least two other nineteenth-century individuals whose thinking has an amazingly modern sound: Arsène Dumont and Émile Durkheim.

Dumont

Arsène Dumont was a late-nineteenth-century French demographer who felt he had discovered a new principle of population that he called "social capillarity" (Dumont 1890). **Social capillarity** refers to the desire of people to rise on the social scale, to increase their individuality as well as their personal wealth. The concept is drawn from an analogy to a liquid rising into the narrow neck of a laboratory flask. The flask is like the hierarchical structure of most societies, broad at the bottom and narrowing as you near the top. To ascend the social hierarchy often requires that sacrifices be made, and Dumont argued that having few or no children was the price many people paid to get ahead. Dumont recognized that such ambitions were not possible in every society. In a highly stratified aristocracy, few people outside of the aristocracy could aspire to a career beyond subsistence. However, in a democracy (such as late-nineteenth-century France), opportunities to succeed existed at all social levels. Spengler (1979) has succinctly summarized Dumont's thesis: "The bulk of the population, therefore, not only strove to ascend politically, economically, socially, and intellectually, but experienced an imperative urge to climb and a palsying fear of descent. Consequently, since children impeded individual and familial ascension, their number was limited" (p. 158).

Notice that Dumont added an important ingredient to Mill's recipe for fertility control. Mill argued that it was fear of social slippage that motivated people to limit fertility below the level that Malthus had expected. Dumont went beyond that to suggest that social aspiration was the root cause of a slowdown in population growth. Dumont was not happy with this situation, by the way. He was upset by the low level of French fertility and used the concept of social capillarity to propose policies to undermine it. He believed that socialism would undercut the desire for upward social mobility and would thus stimulate the birth rate.

Durkheim

While Dumont was concerned primarily with the causes of population growth, another late-nineteenth-century French sociologist, Émile Durkheim, based an entire social theory on the consequences of population growth. In discussing the increasing complexity of modern societies, characterized particularly by increasing divisions of labor, Durkheim proposed that "the division of labor varies in direct ratio with the volume and density of societies, and, if it progresses in a continuous manner in the course of social development, it is because societies become regularly denser and more voluminous" (Durkheim 1893 [1933]:262). Durkheim proceeded to explain that population growth leads to greater societal specialization, because the struggle for existence is more acute when there are more people.

If you compare a primitive society with an industrialized society, the primitive society is not very specialized. By contrast, in industrialized societies there is a lot of differentiation; that is, there is an increasingly long list of occupations and social classes. Why is this? The answer is in the volume and density of the population. Growth creates competition for society's resources, and in order to improve their advantage in the struggle, people specialize. Durkheim's thesis that population growth leads to specialization was derived (he himself acknowledged) from Darwin's theory of evolution. In turn, Darwin acknowledged his own debt to Malthus. You will notice that Durkheim also clearly echoes the words of Ibn Khaldun, although it is uncertain whether Durkheim knew of Ibn Khaldun's work.

The critical theorizing of the nineteenth and early twentieth centuries set the stage for a more systematic collection of data to test aspects of those theories and to examine more carefully those that might be valid and those that should be discarded. As population studies became more quantitative in the twentieth century, a phenomenon called the demographic transition took shape and took the attention of demographers.

The Theory of the Demographic Transition

Although it has dominated recent demographic thinking, the **demographic transition** theory actually began as only a description of the demographic changes that had taken place in the advanced nations over time. In particular, it described the transition from high birth and death rates to low birth and death rates, with an interstitial spurt in growth rates leading to a larger population at the end of the transition than there had been at the start. The idea emerged as early as 1929, when Warren Thompson gathered data from "certain countries" for the period 1908–27 and showed that the countries fell into three main groups, according to their patterns of population growth:

> Group A (northern and western Europe and the United States): From the latter part of the nineteenth century to 1927, these countries had moved from having very high rates of natural increase to having very low rates of increase "and will shortly become stationary and start to decline in numbers" (Thompson 1929:968).

> Group B (Italy, Spain, and the "Slavic" peoples of central Europe): Thompson saw evidence of a decline in both birth rates and death rates but suggested that "it appears prob-

able that the death rate will decline as rapidly or even more rapidly than the birth rate for some time yet. The condition in these Group B countries is much the same as existed in the Group A countries thirty to fifty years ago" (1929:968).

Group C (the rest of the world): In the rest of the world, Thompson saw little evidence of control over either births or deaths.

As a consequence of this relative lack of voluntary control over births and deaths (a concept we will question later), Thompson felt that the Group C countries (which included about 70–75 percent of the population of the world at the time) would continue to have their growth "determined largely by the opportunities they have to increase their means of subsistence. Malthus described their processes of growth quite accurately when he said 'that population does invariably increase, where there are means of subsistence. . . .'" (Thompson, 1929:971).

Thompson's work, however, came at a time when there was little concern about overpopulation. The "Group C" countries had relatively low rates of growth because of high mortality and, at the same time, by 1936, birth rates in the United States and Europe were so low that Enid Charles published a widely read book called *The Twilight of Parenthood,* which was introduced with the comment that "in place of the Malthusian menace of overpopulation there is now real danger of underpopulation" (Charles 1936:v). Furthermore, Thompson's labels for his categories had little charisma (it is difficult to build a theory around categories called A, B, and C).

Sixteen years after Thompson's work, Frank Notestein (1945) picked up the threads of his thesis and provided labels for the three types of growth patterns that Thompson had simply called A, B, and C. Notestein called the Group A pattern **incipient decline,** the Group B pattern **transitional growth,** and the Group C pattern **high growth potential.** That same year, Kingsley Davis (1945) edited a volume of The Annals of the American Academy of Political and Social Sciences titled *World Population in Transition,* and in the lead article (titled "The Demographic Transition") he noted that "viewed in the long-run, earth's population has been like a long, thin powder fuse that burns slowly and haltingly until it finally reaches the charge and explodes" (Davis 1945:1). The term **population explosion,** alluded to by Davis, refers to the phase that Notestein called transitional growth. Thus was born the term demographic transition. It is that process of moving from high birth and death rates to low birth and death rates, from high growth potential to incipient decline (see Figure 3.3).

At this point in the 1940s, however, the demographic transition was merely a picture of demographic change, not a theory. But each new country studied fit into the picture, and it seemed as though some new universal law of population growth—an evolutionary scheme—was being developed. The apparent historical uniqueness of the demographic transition (all known cases have occurred within the last 200 years) has spawned a host of alternative names, such as the "vital revolution" and the "demographic revolution."

Between the mid-1940s and the late 1960s, rapid population growth became a worldwide concern, and demographers devoted a great deal of time to the demographic transition perspective. By 1964, George Stolnitz was able to report that "demographic transitions rank among the most sweeping and best-documented

Figure 3.3 The Demographic Transition

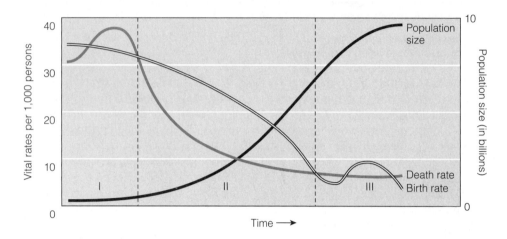

Note: The demographic transition is divided roughly into three stages. In the first stage, there is high growth potential because both birth and death rates are high. The second stage is the transition from high to low birth and death rates. During this stage, the growth potential is realized as the death rate drops before the birth rate drops, resulting in rapid population growth. Finally, the last stage is a time when death rates are as low as they are likely to go, while fertility may continue to decline to the point that the population might eventually decline in numbers.

trends of modern times . . . based upon hundreds of investigations, covering a host of specific places, periods and events" (Stolnitz 1964:20). As the pattern of change took shape, explanations were developed for why and how countries pass through the transition. These explanations tended to be cobbled together in a somewhat piecemeal fashion from the nineteenth- and early-twentieth-century writers I discussed earlier in this chapter, but overall they were derived from the concept of **modernization.**

The modernization theory is based on the idea that in premodern times human society was generally governed by "tradition," and that the massive economic changes wrought by industrialization forced societies to alter traditional institutions: "In traditional societies fertility and mortality are high. In modern societies fertility and mortality are low. In between, there is demographic transition" (Demeny 1968:502). In the process, behavior has changed and the world has been permanently transformed. It is a macro-level theory that sees human actors as being buffeted by changing social institutions. Thus, individuals did not deliberately lower their risk of death to precipitate the modern decline in mortality. Rather, societywide increases in income and improved public health infrastructure brought about this change. Similarly, people did not just decide to move from the farm to town to take a job in a factory. Economic changes took place that created those higher-wage urban jobs while eliminating many agricultural jobs. These same economic forces improved transportation and communication and made it possible for individuals to migrate in previously unheard of numbers.

Modernization theory provided the vehicle that allowed the demographic transition to move from a mere description of events to a demographic perspective. In its initial formulations, this perspective was perhaps best expressed by the sentiments "take care of the people and population will take care of itself" or "development is the best contraceptive" (Teitelbaum 1975). These were views that were derivable from Karl Marx, who was in fact one of the early exponents of the modernization theory (Inglehart and Baker 2000). The theory drew on the available data for most countries that had gone through the transition. Death rates declined as the standard of living improved, and birth rates almost always declined a few decades later, eventually dropping to low levels, although rarely as low as the death rate. It was argued that the decline in the birth rate typically lagged behind the decline in the death rate because it takes time for a population to adjust to the fact that mortality really is lower, and because the social and economic institutions that favored high fertility require time to adjust to new norms of lower fertility that are more consistent with the lower levels of mortality. Since most people value the prolongation of life, it is not hard to lower mortality, but the reduction of fertility is contrary to the established norms of societies that have required high birth rates to keep pace with high death rates. Such norms are not easily changed, even in the face of poverty.

Birth rates eventually declined, it was argued, as the importance of family life was diminished by industrial and urban life, thus weakening the pressure for large families. Large families are presumed to have been desired because they provided parents with a built-in labor pool, and because children provided old-age security for parents (Hohm 1975). The same economic development that lowered mortality is theorized to transform a society into an urban industrial state in which compulsory education lowers the value of children by removing them from the labor force, and people come to realize that lower infant mortality means that fewer children need to be born to achieve a certain number of surviving children (Easterlin 1978). Finally, as a consequence of the many alterations in social institutions, "the pressure from high fertility weakens and the idea of conscious control of fertility gradually gains strength" (Teitelbaum 1975:421).

Critique of the Demographic Transition Theory

It has been argued that the concept underlying the demographic transition is that population stability (sometimes called "homeostasis"—see, for example, Lee 1987) is the normal state of affairs in human societies and that change (the "transition") is what requires explanation (Kreager 1986). But that is only partially true. In its original formulation, the demographic transition theory explained high fertility as a reaction to high mortality. As mortality declines, the need for high fertility lessens, and so birth rates go down. There is a spurt of growth in that transition period, but presumably the consequences will not be serious if the decline in mortality was produced by a rise in the standard of living, which, in its turn, produces a motivation for smaller families. But what will be the consequences if mortality declines and fertility does not? That situation presumably is precluded by the theory of demographic transition, but the demographic transition theory has not been capable of predicting levels of mortality or fertility or the timing of the fertility decline. This is because the initial explanation for the demographic behavior during the transition tended to be **ethnocentric**. It relied almost exclusively on the sentiment that "what is good for the goose is good for the gander." In other words, if this is what happened to the developed countries, why should it not also happen to other countries that are not so advanced? One reason might be that the preconditions for the demographic transition are considerably different now from what they were when the industrialized countries began their transition.

For example, prior to undergoing the demographic transition, few of the currently industrialized countries had birth rates as high as those of most currently less developed countries, nor indeed were their levels of mortality so high. Yet when mortality did decline, it did so as a consequence of internal economic development, not as a result of a foreign country bringing in sophisticated techniques of disease prevention, as is the case today. A second reason might be that the factors leading to the demographic transition were actually different from what for years had been accepted as true. Perhaps it is not just change that requires explanation but also differences in the starting and ending points of the transition. Perhaps, then, the modernization theory, in and of itself, did not provide an appropriate picture of historical development. These problems with the original explanations of the demographic transition led to new research and a reformulation of the perspective.

Reformulation of the Demographic Transition Theory

One of the most important social scientific endeavors to cast doubt on the classic explanation was the European Fertility Project, directed by Ansley Coale at Princeton University. In the early 1960s, J. William Leasure, then a graduate student in economics at Princeton, was writing a doctoral dissertation on the fertility decline in Spain, using data for each of that nation's 49 provinces. Surprisingly, his thesis revealed that the history of fertility change in Spain was not explained by a simple version of the demographic transition theory. Fertility in Spain declined in contiguous areas that were culturally similar, even though the levels of urbanization and economic development might be different (Leasure 1962). At about the same time, other students began to uncover similarly puzzling historical patterns in European data (Coale 1986). A systematic review of the demographic histories of Europe was thus begun in order to establish exactly how and why the transition occurred. The focus was on the decline in fertility, because it is the most problematic aspect of the classic explanation. These new findings have been used to help revise the theory of the demographic transition.

One of the important clues that a revision was needed was the discovery that the decline of fertility in Europe occurred in the context of widely differing social, economic, and demographic conditions (van de Walle and Knodel 1967). Economic development emerges, then, as a sufficient cause of fertility decline, although not a necessary one (Coale 1973). For example, many provinces in Europe experienced a rapid drop in their birth rate even though they were not very urban, infant mortality rates were high, and a low percentage of the population was in industrial occupations. The data suggest that one of the more common similarities in those areas that have undergone fertility declines is the rapid spread of **secularization.** Secularization is an attitude of autonomy from otherworldly powers and a sense of responsibility for one's own well-being (Leasure 1982; Lesthaeghe 1977). It is difficult to know exactly why such attitudes arise when and where they do, but we do know that industrialization and economic development are virtually always accompanied by secularization. Secularization, however, can occur independently of industrialization. It might be thought of as a modernization of thought, distinct from a modernization of social institutions. Some theorists have suggested that secularization is part of the process of Westernization (see, for example, Caldwell 1982). In all events, when it pops up, secularization often spreads quickly, being diffused through social networks as people imitate the behavior of others to whom they look for clues to proper and appropriate conduct.

Education has been identified as one potential stimulant to such altered attitudes, especially mass education, which tends to emphasize modernization and secular concepts (Caldwell 1980). Education facilitates the rapid spread of new ideas and information, which would perhaps help explain another of the important findings from the Princeton European Fertility Project, that the onset of long-term fertility decline tended to be concentrated in a relatively short period of time (van de Walle and Knodel 1980). The data from Europe suggest that once marital fertility had dropped by as little as 10 percent in a region, the decline spread rapidly. This "tipping point" occurred whether or not infant mortality had already declined (Watkins 1986). Some areas of Europe that were similar with respect to socioeconomic development did not experience a fertility decline at the same time, whereas other provinces that were less similar socioeconomically experienced nearly identical drops in fertility. The data suggest that this riddle is solved by examining cultural factors rather than socioeconomic ones. Areas that share a similar culture (same language, common ethnic background, similar lifestyle) were more likely to share a decline in fertility than areas that were culturally less similar (Watkins 1991). The principal reason for this is that the idea of family planning seemed to spread quickly until it ran into a barrier to its communication. Language is one such barrier (Leasure 1962; Lesthaeghe 1977), and social and economic inequality in a region is another (Lengyel-Cook and Repetto 1982). Social distance between people turns out to effectively inhibit communication of new ideas and attitudes.

What kinds of ideas and attitudes might encourage people to rethink how many children they ought to have? To answer this kind of question we must shift our focus from the macro (societal) level to the micro (individual) level and ask how people actually respond to the social and economic changes taking place around them (see Kertzer and Hogan 1989 for an interesting Italian case study). A prevailing individual-level perspective is that of **rational choice theory** (sometimes referred to as RAT) (Coleman and Fararo 1992). The essence of rational choice theory is that human behavior is the result of individuals making calculated cost-benefit analyses about how to act and what to do. For example, Caldwell (1976; 1982) has suggested that "there is no ceiling in primitive and traditional societies to the number of children who would be economically beneficial" (1976:331). Children are a source of income and support for parents throughout life, and they produce far more than they cost in such societies. The **wealth flow,** as Caldwell calls it, is from children to parents. But the process of modernization eventually results in the tearing apart of large, extended family units into smaller, nuclear units that are economically and emotionally self-sufficient. As that happens, children begin to cost parents more (including the cost of educating them as demanded by a modernizing society), and the amount of support that parents get from children begins to decline (starting with the income lost because children are in school rather than working). As the wealth flow reverses and parents begin to spend their income on children, rather than deriving income from them, the economic value of children declines, and people no longer derive any economic advantage from children. Economic rationality would dictate having zero children, but in reality, of course, people continue having children for a variety of social reasons (that I detail in Chapter 6).

Overall, then, the principle ingredient in the reformulation of the demographic transition perspective is to add "ideational" factors to "demand" factors as the likely causes of fertility decline (Cleland and Wilson 1987). The original version of the theory suggested that modernization reduces the demand for children and so fertility falls—if people are rational economic creatures, then this is what should happen. But the real world is more complex, and the diffusion of ideas can shape fertility behavior along with, or even in the absence of, the usual signs of modernization. People may generally be rational, but Massey (2002) has reminded us that much of human behavior is still powered by emotional responses that supercede rationality. We are animals and while we may have vastly greater intellectual capacities than other species, we are still influenced by a variety of nonrational forces, including our hormones (Udry 1994; 2000).

This does not necessarily mean that Wallerstein was correct in 1976 when he declared that modernization theory was dead (Wallerstein 1976). On the contrary, there is evidence from around the world that "industrialization leads to occupational specialization, rising educational levels, rising income levels, and eventually brings unforeseen changes—changes in gender roles, attitudes toward authority and sexual norms; declining fertility rates; broader political participation; and less easily led publics" (Inglehart and Baker 2000:21). This is not a linear path, however. "Economic development tends to push societies in a common direction, but rather than converging, they seem to move on parallel trajectories shaped by their cultural heritages" (Inglehart and Baker 2000:49).

One strength of reformulating the demographic transition is that nearly all other perspectives can find a home here. Malthusians note with satisfaction that fertility first declined in Europe primarily as a result of a delay in marriage, much as Malthus would have preferred. Neo-Malthusians can take heart from the fact that rapid and sustained declines occurred simultaneously with the spread of knowledge about family planning practices. Marxists also find a place for themselves in the reformulated demographic transition perspective, because its basic tenet is that a change in the social structure (modernization) is necessary to bring about a decline in fertility. This is only a short step away from agreeing with Marx that there is no universal law of population, but rather that each stage of development and social organization has its own law, and that cultural patterns will influence the timing and tempo of the demographic transition—when it starts and how it progresses.

The Theory of Demographic Change and Response

The work of the European Fertility Project focused on explaining regional differences in fertility declines. This was a very important theoretical development, but not a comprehensive one because it only partially dealt with a central issue of the demographic transition theory: How (and under what conditions) can a mortality decline lead to a fertility decline? To answer that question, Kingsley Davis (1963) asked what happens to individuals when mortality declines. The answer is that more children survive through adulthood, putting greater pressure on family resources, and people have to reorganize their lives in an attempt to relieve that pressure; that is, people respond to the demographic change. But note that their response will be in terms of personal goals, not national goals. It rarely matters what a government wants. If individual members of a society do not stand to gain by behaving in a particular way, they probably will not behave that way. Indeed, that was a major argument made by the neo-Malthusians against moral restraint. Why advocate postponement of marriage and sexual gratification rather than contraception when you know that few people who postpone marriage are actually going to postpone sexual intercourse, too? In fact, Ludwig Brentano (1910) quite forthrightly suggested that Malthus was insane to think that abstinence was the cure for the poor.

In all events, Davis argued that the response that individuals make to the population pressure created by more members joining their ranks is determined by the means available to them. A first response, nondemographic in nature, is to try to increase resources by working harder—longer hours perhaps, a second job, and so on. If that is not sufficient or there are no such opportunities, then migration of some family members (typically unmarried sons or daughters) is the easiest demographic response. This is, of course, the option that people have been using forever (undoubtedly explaining in large part why human beings have spread out over the planet). In the early eighteenth century, Richard Cantillon, an Irish-French economist, was pointing out what happened in Europe when families grew too large (and this was even before mortality began markedly to decline):

> If all the labourers in a village breed up several sons to the same work, there will be too many labourers to cultivate the lands belonging to the village, and the surplus adults must go to seek a livelihood elsewhere, which they generally do in cities. . . . If a tailor makes all the clothes there and breeds up three sons to the same, yet there is work enough for but one successor to him, the two others must go to seek their livelihood elsewhere; if they do not find enough employment in the neighboring town they must go further afield or change their occupation to get a living . . . (Cantillon 1755 (1964):23).

But what will be the response of that second generation, the children who now have survived when previously they would not have, and who have thus put the pressure on resources? Davis argues that if there is in fact a chance for social or economic improvement, then people will try to take advantage of those opportunities by avoiding the large families that caused problems for their parents. Davis suggests that the most powerful motive for family limitation is not fear of poverty or avoidance of pain as Malthus argued; rather it is the prospect of rising prosperity that will most often motivate people to find the means to limit the number of children they have (I discuss these means in Chapter 6). Davis here echoes the themes of Mill and Dumont, but adds that at the very least, the desire to maintain one's relative status in society may lead to an active desire to prevent too many children from draining away one's resources. Of course, that assumes the individuals in question have already attained some status worth maintaining.

One of Davis's most important contributions to our demographic perspective is, as Cicourel put it, that he "seems to rely on an implicit model of the actor who makes everyday interpretations of perceived environmental changes" (Cicourel 1974:8). For example, people will respond to a decline in mortality only if they notice it, and then their response will be determined by the social situation in which they find themselves. Davis's analysis is important in reminding us of the crucial link between the everyday lives of individuals and the kinds of population changes that take place in society. Another demographer who extended the scope of the demographic transition with this kind of analysis is Richard Easterlin.

The Easterlin Relative Cohort Size Hypothesis

The **Easterlin relative cohort size hypothesis** (also sometimes known as the relative income hypothesis) is based on the idea that the birth rate does not necessarily respond to absolute levels of economic well-being but rather to levels that are relative to those to which one is accustomed (Easterlin 1978; 1968). Easterlin assumes that the standard of living you experience in late childhood is the base from which you evaluate your chances as an adult. If you can easily improve your income as an adult compared to your late childhood level, then you will be more likely to marry early and have several children. On the other hand, if you perceive that it is going to be rough sledding as an adult even to match the level of living you were accustomed to as a child, that fear will probably lead you to postpone marriage or at least postpone childbearing.

So far the theory of relative income is strikingly similar to Mill's writing 100 years earlier. But Easterlin goes on to ask what factors might cause you to be in a relatively advantageous or disadvantageous position as you begin adulthood. The answer lies in the relationship between economic conditions and the demographic responses to those conditions. The demographic response is a function of the age structure changes that occur in the context of declining mortality and fertility. Macunovich (2000) has summarized the scenario as follows:

> The increase in relative cohort size that occurs as a result of declining mortality during the demographic transition in part determines when the fertility portion of the transition begins. The increasing proportion of young adults generates a downward pressure on young men's relative wages; this in turn causes young adults to accept a tradeoff between family size and material wellbeing. This acceptance of a tradeoff could mark a turning point in a society's regulation of fertility, setting in motion a "cascade" or "snowball" effect in which total fertility rates tumble as social norms regarding individual control of fertility and acceptable family sizes begin to change (p. 257).

If young people are relatively scarce in society and business is good, they will be in relatively high demand. In nearly classic Malthusian fashion, they will be able to command high wages and thus be more likely to feel comfortable about getting married and starting a family. Of course, how comfortable they are will depend on how much those wages can buy compared to what they are accustomed to. If young people are in relatively abundant supply, then even if business is good, the competition for jobs will be stiff and it will be difficult for people to maintain their accustomed level of living, much less marry and start a family.

Easterlin's thesis presents a model of society in which demographic change and economic change are closely interrelated. Economic changes produce demographic changes, which in turn produce economic changes, and so on. The model, however, has a certain middle-class bias. What about the people at the bottom end of the economic ladder, for whom relative deprivation does not necessarily apply because they already have so little? Are they caught in a constant Malthusian cycle of overpopulation and poverty? Mill suggested in 1848 that they were, unless one entire generation could be catapulted into the middle class. To be more broadly applicable, the theory may have to take into account the importance of social class as an independent factor influencing demographic behavior (Lesthaeghe and Surkyn 1988).

The importance of the Easterlin hypothesis can be judged by the large number of studies that have attempted to test it empirically. Several studies provide evidence that, in general, changes in cohort size and the birth rate tend to move in directions predicted by Easterlin (Butz and Ward 1979; Crenshaw, Ameen, and Christenson 1997; Lee 1976; Pampel 1993). On the other hand, tests of specific aspects of the Easterlin hypothesis have often failed to support it (Behrman and Taubman 1989; Kahn and Mason 1987; Wright 1989) or have supported it only partially (Pampel 1996). Nonetheless, Macunovich (2000) makes the case that properly specified models will demonstrate the usefulness of the theory, especially in predicting the timing of the fertility transition. The Easterlin hypothesis is intuitively appealing; the idea of a demographic feedback cycle, which is at the core of Easterlin's thinking, is compelling, and relative cohort size is certainly a factor that will influence various kinds of social change. This kind of interaction between population change and societal change is, in fact, at the heart of the realization among demographers that the demographic transition is really a whole suite of transitions, rather than one big transition.

The Demographic Transition Is Really a Set of Transitions

The reformulation driven by the European Fertility Project, the theory of demographic change and response, the Easterlin hypothesis, and other research all generated the insight that the demographic transition is actually a set of interrelated transitions. Taken together, they help us understand not just the causes but the consequences of population change. Usually (but not always) the first transition to occur is the **mortality transition**—the shift from deaths at younger ages due to communicable disease to deaths at older ages due to degenerative diseases. This transition is followed by the **fertility transition**—the shift from natural (and high) to controlled (and low) fertility, typically in a delayed response to the mortality transition. The predictable changes in the age structure (the **age transition**) brought about by the mortality and fertility transitions produce social and economic reactions as societies adjust to constantly changing age distributions. The rapid growth of the population occasioned by the pattern of mortality declining sooner and more rapidly than fertility almost always leads to overpopulation of rural areas, producing the **migration transition,** especially toward urban areas, creating the **urban transition.** The **family and household**

Figure 3.4   The Transitions that Comprise the Demographic Transition's Impact on Society

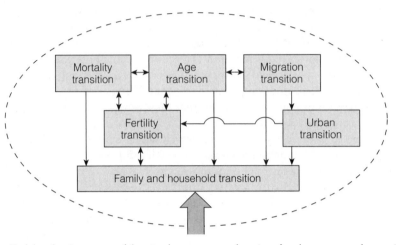

Note: Each box has its own set of theories that serve as explanations for phenomena under consideration.

transition is occasioned by the massive structural changes that accompany longer life, lower fertility, an older age structure, and urban instead of rural residence—all of which are part and parcel of the demographic transition. The interrelationships among these transitions are shown in Figure 3.4.

The Mortality Transition

The transition process almost always begins with a decline in mortality, which is brought about by changes in society that improve the health of people and thus their ability to resist disease, and by scientific advances that prevent premature death. However, death rates do not decline evenly by age; rather it is the very youngest and the very oldest—but especially the youngest—whose lives are most likely to be saved by improved life expectancy. Thus, the initial impact of the mortality transition is to increase the number of young people who are alive, ballooning the bottom end of the age structure in a manner that looks just like an increase in the birth rate. This sets all the other transitions in motion.

The Fertility Transition

The fertility transition can begin without a decline in mortality (as happened in France), but in most places it is the decline in mortality, leading to greater survival of children, that eventually motivates people to think about limiting the number of children they are having. Throughout most of human history, the average woman had two children who survived to adulthood. The decline in mortality, however, obviously increases that number and thereby threatens the very foundation of the household economy. At the community or societal level, the increasing number of young people creates all sorts of pressures to change, often leading to peer pressure to conform to new standards of behavior, including the deliberate control of reproduction. Another set of extremely important changes that occur in the context of mortality transition is that the scope of life expands for women as they, too, live longer. They are increasingly empowered to delay childbearing and to have fewer children because they begin to realize that most of their children will survive to adulthood and they, themselves, will survive beyond the reproductive ages, beyond their children's arrival into adulthood. This new freedom gives them vastly more opportunities than ever before in human history to do something with their lives besides bearing and raising children. This realization may be a genuine tipping point in the fertility transition, leading to an almost irreversible decline.

As fertility declines, the mortality transition is itself pushed along, because the survival of children is enhanced when a woman has fewer children among whom to share resources. Also affected by the fertility decline is the age structure, which begins to cave in at the younger ages as fewer children are being born, and as most are now surviving through childhood. In its turn, that shift in the age transition to an increasingly older age structure has the potential to divert societal resources away from dealing primarily with the impact of children, to dealing with broader social concerns, including raising the standard of living. A higher standard of living can then redound to the benefit of health levels throughout society, adding fuel to the fire of increasing life expectancy.

The Age Transition

In many respects, the age transition is the "master" transition in that the changing number of people at each age that occurs with the decline of mortality and then the decline in fertility presents the most obvious demographic pressure for social change. When both mortality and fertility are high, the age structure is quite young, but the decline in mortality makes it even younger by disproportionately increasing the number of young people. Then, as fertility declines, the youngest ages are obviously again affected first, since births occur only at age zero, so a fertility decline shows up first as simply fewer young children than before. However, as the bulge of young people born prior to the fertility decline push into the older ages while fertility begins to decline, the age structure moves into a stage that can be very beneficial to economic development in a society—a large fraction of the population is composed of young adults of working age who are having fewer children as dependents at the same time that the older population has not yet increased in size enough to create problems of dependency in old age. As we will see, this phase in the age transition is often associated with a golden age of advancement in the standard of living.

That golden age can be transitory, however, if a society has not planned for the next phase of the age transition, when the older population begins to increase more rapidly than the younger population. The baby bulge created by the initial declines in mortality reaches old age at a time when fertility has likely declined, and so the age structure has a much greater number and a higher fraction of older people than ever before. We are only now learning how societies will respond to this challenge of an increasingly older population.

The Migration Transition

Meanwhile, back at the very young age structure put into motion by declining mortality, the theory of demographic change and response suggests that in rural areas, where most of the population lived for most of human history, the growth in the number of young people will lead to an oversupply of young people looking for jobs, which will encourage people to go elsewhere in search of economic opportunity. The Europeans, who experienced the first wave of population growth because they experienced the mortality transition first, still lived in a world where there was "empty" land. Of course, it wasn't really empty, as the Americas and the islands of the South Pacific (largely Australia and New Zealand) were populated mainly by hunter-gatherers who used the land extensively, rather than intensively, as I mentioned in Chapter 2. This rotation of land use meant that as Europeans arrived in the Americas, they perceived land as being not used and so claimed it for themselves. We all know the consequences of this for the indigenous populations, as all of the land was eventually claimed for intensive human use. So, whereas the Europeans could initially spread out from rural Europe to rural areas in the Americas and the South Pacific, migrants from rural areas today no longer have that option.

Notice in Figure 3.4 that there is a double-arrow connecting the age and migration transitions. This is because migration takes people (mainly young adults) out of one area and puts them in another area, thus affecting the age structure in both places. As I have already mentioned, this difference in age structures between two places contributes to current migration patterns of people from younger societies filling in the "empty" places in the age structure of older societies.

The Urban Transition

With empty lands filling up, migrants from the countryside in the world today have no place to go but to cities, and cities have historically tended to flourish by absorbing labor from rural areas. A majority of people now live in cities, and by the end of the twenty-first century, almost all of us will be there. The urban transition thus begins with migration from rural to urban areas but then morphs into the urban "evolution" as most humans wind up being born in, living in, and dying in cities. The complexity of human existence is played out in cities, leading us to expect a constant dynamism of urban places for most of the rest of human history. Because urban places are historically associated with lower levels of fertility than rural areas, as the world's population becomes increasingly urban we can anticipate that this will be a major factor in bringing and keeping fertility levels down all over the world.

The Family and Household Transition

The dramatic changes taking place in family and household structure since World War II, especially in Europe, led van de Kaa (1987) to talk about the "second demographic transition." A demographic centerpiece of this change in the richer countries has been a fall in fertility to below-replacement levels, but van de Kaa suggested that the change was less about not having babies than it was about the personal freedom to do what one wanted, especially among women. So, rather than grow up, marry, and have children, this transition is associated with a postponement of marriage, a rise in single living, cohabitation, and prolonged residence in the parental household. There has also been an increasing lack of permanence in family relationships, leading to "rising divorce rates and high separation rates of cohabitants. And the forms of family reconstitution are shifting away from remarriage in favor of post-marital cohabitation as well" (Lesthaeghe 1998:5–6).

It is reasonable to think, however, that this transition in family and household structure is not so much a second transition as it is another set of transitions within the broader framework of the demographic transition. As I show in Figure 3.4, the family and household transition is influenced by all the previously mentioned transitions. The mortality transition is pivotal because it gives women (and men, too, of course) a dramatically greater number of years to live in general, and more specifically a greater number of years that do not need to be devoted to children. Low mortality reduces the pressure for a woman to marry early and start bearing children while she is young enough for her body to handle that stress. Furthermore, when mortality was high, marriages had a high probability of ending in widowhood when one of the partners was still reasonably young, and families routinely were reconstituted as widows and widowers remarried. But low mortality leads to a much longer time that married couples will be alive together before one partner dies and this alone is related to part of the increase in divorce rates.

The age transition plays a role at the societal level, as well, because over time the increasingly similar number of people at all ages—as opposed to a majority of people being very young—means that any society is bound to be composed of a greater array of family and household arrangements. Diversity in families and households is also encouraged by migration (which breaks up and reconstitutes families) and by the urban transition, especially since urban places tend to be more tolerant of diversity than are smaller rural communities.

The rest of this book is devoted to more detailed examinations of each of these transitions, and you will discover that there are a variety of theoretical approaches that have developed over time to explain the causes and consequences of each set of transitions. Your own demographic perspective will be honed by looking for similarities and patterns in the transitions that link them together and, more importantly, link them to their potential impact on society.

Impact on Society

The modern field of population studies came about largely to encourage and inspire deeper insight into the causes of changes in fertility, mortality, migration, age and sex structure, and population characteristics and distribution. Demographers spent most of the twentieth century doing that, but always with an eye toward new things that could be learned about what demographic change meant for human society. Unlike in Malthus's day, population growth is no longer viewed as being caused by one set of factors nor as having a simple prescribed set of consequences.

Perhaps the closest we can come at present to "big" theories are those that try to place demographic events and behavior in the context of other global change, especially political change, economic development, and Westernization. One of the more ambitious and influential of these theorists is Jack Goldstone (1986; 1991), whose work incorporates population growth as a precursor of change in the "early modern world" (defined by him as 1500 to 1800). He argues that population growth in the presence of rigid social structures produced dramatic political change in England and France, in the Ottoman Empire, and in China. Population growth led to increased government expenditures, which led to inflation, which led to fiscal crisis. In these societies with no real opportunities for social mobility, population growth (which initially increases the number of younger persons) led to disaffection and popular unrest and created a new cohort of young people receptive to new ideas. The result in the four case studies he analyzes was rebellion and revolution.

Stephen Sanderson (1995) promotes the idea that population growth has been an important stimulus to change throughout human history, but especially since the Agricultural Revolution. Thus, his scope is much broader than that of Goldstone. Sanderson largely synthesizes the work of others (such as Boserup 1965; Cohen 1977; and Johnson and Earle 1987). "Had Paleolithic hunter-gatherers been able to keep their populations from growing, the whole world would likely still be surviving entirely by hunting and gathering" (Sanderson 1995:49). Instead, population growth generated the Agricultural Revolution (an idea discussed in Chapter 2) and then the Industrial Revolution. The sedentary life associated with the Agricultural Revolution increased social complexity (a very Durkheimian idea), which led to the rise of civilization (cities) and the state (city-states and then nation-states).

Regional and cultural differences in population growth and age structures have been implicated in the patterns of instability and conflict that characterize the world in the early twenty-first century. The consequences of population growth figure prominently in Huntington's *Clash of Civilizations and the Remaking of World Order* (Huntington 1996) as they do in Robert Kaplan's *The Coming Anarchy: Shattering the Dreams of the Post Cold War* (Kaplan 2000). These and other writers have encouraged social scientists to realize that population change may seem largely imperceptible to us as it is occurring, so it requires an analytic observer to tell us what is going on. Thus, if we can look back and see that in the past there were momentous historical consequences of population growth, can we look forward into the future and project similar kinds of influences? Many people (including me) would say yes, and the rest of this book will show you why.

Summary and Conclusion

A lot of thinking about population issues has taken place over a very long period of time, and in this chapter I have traced for you the progression of demographic thinking from ancient doctrines to contemporary systematic perspectives. Malthus was not the first, but he was certainly the most influential of the early modern writers. Malthus believed that a biological urge to reproduce was the cause of population growth and that its natural consequence was poverty. Marx, on the other hand, did not openly argue with the Malthusian causes of growth, but he vehemently disagreed with the idea that poverty is the natural consequence of population growth. Marx denied that population growth was a problem per se—it only appeared that way in capitalist society. It may have seemed peculiar to you that I discuss a person who denied the importance of a demographic perspective in a chapter dedicated to that very importance. However, the Marxian point of view is sufficiently prevalent today among political leaders and intellectuals in enough countries that this attitude becomes in itself a demographic perspective of some significance. Furthermore, his perspective on the world finds its way into many aspects of current mainstream thinking, including modernization theory, that underly aspects of the demographic transition theory.

The perspective of Mill, who seems very contemporary in many of his ideas, was somewhere between that of Malthus and Marx. He believed that increased productivity could lead to a motivation for having smaller families, especially if the influence of women was allowed to be felt and if people were educated about the possible consequences of having a large family. Dumont took these kinds of individual motivations a step further and suggested in greater detail the reasons why prosperity and ambition, operating through the principle of social capillarity, generally lead to a decline in the birth rate. Durkheim's perspective emphasized the consequences more than the causes of population growth. He was convinced that the complexity of modern societies is due almost entirely to the social responses to population growth—more people lead to higher levels of innovation and specialization.

More recently developed demographic perspectives have implicitly assumed that the consequences of population growth are serious and problematic, and they move directly to explanations of the causes of population growth. The original theory of the demographic transition suggested that growth is an intermediate stage between the more stable conditions of high birth and death rates to a new balance of low birth and death rates. Reformulations of the demographic transition perspective have emphasized its evolutionary character and have shown that the demographic transition is not one, monolithic change, but rather that it encompasses several interrelated transitions: A decline in mortality will almost necessarily be followed by a decline in fertility, and by subsequent transitions in migration, urbanization, the age structure, and the family and household structure in society.

As I explore with you the causes and consequences of population growth and the uses to which such knowledge can be applied, you will need to know about the sources of demographic data. What is the empirical base of our understanding of the relationship between population and society? We turn to that topic in the next chapter.

Main Points

1. A demographic perspective is a way of relating basic population information to theories about how the world operates demographically.

2. Population doctrines and theories prior to Malthus vacillated between pronatalist and antinatalist and were often utopian.

3. Condorcet and Godwin were utopians whose books stimulated Malthus to write his *Essay on the Principle of Population,* which appeared in 1798 and

has been one of the most influential works ever written on population growth and its societal consequences.

4. According to Malthus, population growth is generated by the urge to reproduce, although growth is checked ultimately by the means of subsistence.

5. The natural consequences of population growth according to Malthus are misery and poverty because of the tendency for populations to grow faster than the food supply. Nonetheless, he believed that misery could be avoided if people practiced moral restraint—a simple formula of chastity before marriage and a delay in marriage until you can afford all the children that God might provide.

6. Karl Marx and Friedrich Engels strenuously objected to the Malthusian population perspective because it blamed poverty on the poor rather than on the evils of social organization.

7. Marx and Engels believed that overpopulation was a product of capitalism and that in a socialist society there would either be enough resources per person or else people would be motivated to keep families small.

8. Not surprisingly, very few people have bought the Malthusian idea of moral restraint, although there are many who agree that population growth has the potential to outstrip food production. Such people are usually called neo-Malthusians and believe in the use of birth control.

9. Revisions of Marxian ideology frequently include a more active government role in trying to influence birth limitation.

10. John Stuart Mill argued that the standard of living is a major determinant of fertility levels, but he also felt that people could influence their own demographic destinies.

11. Arsène Dumont argued that personal ambition generated a process of social capillarity that induced people to limit their number of children in order to get ahead socially and economically.

12. Émile Durkheim built an entire theory of social structure on his conception of the consequences of population growth.

13. The demographic transition theory is a perspective that emphasizes the importance of economic and social development, which leads first to a decline in mortality and then, after some time lag, to a commensurate decline in fertility. It is based on the experience of the developed nations, and is derived from the modernization theory.

14. Reformulations of the demographic transition theory suggest that secularization and the cultural diffusion of ideas may be important keys to declining fertility.

15. The theory of demographic change and response emphasizes that people must perceive a personal need to change behavior before a decline in fertility will take place, and that the kind of response they make will depend on what means are available to them.

16. The relative cohort size hypothesis views changes in the birth rate as being a response to levels of economic well-being related to those to which one is accustomed, particularly when combined with birth cohorts of unequal size.

17. The demographic transition is really a set of transitions, including the mortality, fertility, age, migration, urban, and family/household transitions.

Suggested Readings

1. United Nations, 1973, The Determinants and Consequences of Population Trends (New York: United Nations).

   This volume was written by well-known and highly expert demographers and is the best single source for a comprehensive review of major demographic doctrines and early theories.

2. Thomas Robert Malthus, 1798, An Essay on the Principles of Population, 1st Ed. (available from a variety of publishers).

   Despite the fact that numerous good summaries of Malthus have been written, you owe it to yourself to sample the real thing.

3. Ansley Coale and Susan Cotts Watkins, (eds.) 1986, The Decline of Fertility in Europe: The Revised Proceedings of a Conference on the Princeton European Fertility Project (Princeton: Princeton University Press).

   The most comprehensive statement in the literature of ways in which the Princeton European Fertility Project caused a rethinking of the theory of the demographic transition.

4. Simon Szreter, 1993, "The Idea of Demographic Transition and the Study of Fertility Change: A Critical Intellectual History," Population and Development Review 19(4):659–702.

   Szreter examines the intellectual roots of the demographic transition and suggests that demographers need to step back and reassess the adequacy of the theory for explaining fertility change in society.

5. Ron Lesthaeghe, 1998, "On Theory Development: Applications to the Study of Family Formation." *Population and Development Review* 24(1):1–14.

   Lesthaeghe offers an insider's view into theorizing about demographic issues, using the "second demographic transition" as an example.

🌐 Web Sites of Interest

Remember that Web sites are not as permanent as books and journals, so I cannot guarantee that each of the following Web sites still exists at the moment you are reading this. It is usually possible, however, to retrieve the contents of old Web sites at http://www.archive.org.

1. **http://www-groups.dcs.st-and.ac.uk/~history/**
   [Click on "Biographies Index," then on "C," then on "Condorcet"]
   The Marquis de Condorcet, who helped to inspire Malthus's essay, is the subject of this Web site, located at the University of St. Andrews in Scotland. It includes biographical information and a list of his publications.

2. **http://www.ecn.bris.ac.uk/het/malthus/popu.txt**
   The beauty of this Web site, located in the Department of Economics at the University of Bristol in England, is that it contains the full text of Malthus's first (1798) Essay on Population.

3. **http://www.ined.fr/englishversion/figures/indexE.htm**
   French demographers have played key roles in developing population studies, and the National Institute of Demographic Studies (INED) in Paris carries on that tradition. At this

part of the INED Web site, you can find a very useful glossary of demographic terms (in English), as well as data for most countries of the world.

4. **http://www.popcouncil.org**
The Population Research Council is a policy-oriented research center in New York City founded in 1952 by John D. Rockeller 3rd. It originated the journal *Population and Development Review* (now published by Blackwell), whose articles tend to focus on issues that directly or indirectly test demographic theories and perspectives. At this Web site you can, among other things, peruse abstracts of articles published in the journal to stay up to date on recent research.

5. **http://www.emagazine.com/november-december_1996/1196conv.html**
This interview with neo-Malthusians Paul and Anne Ehrlich may be several years old, but it is not outdated in terms of the Ehrlichs' style and overall point of view.

CHAPTER 4
Demographic Data

Figure 4.1   Population Center of the United States Based on Data from the Decennial Censuses

Source: U.S. Census Bureau: http://www.census.gov/geo/www/cenpop/cntpop2k.html, accessed 2003

SOURCES OF DEMOGRAPHIC DATA

POPULATION CENSUSES
    The Census of the United States
        Who is Included in the Census?
        Coverage Error

Measuring Coverage Error
Content Error
Sampling Error
Continuous Measurement—
    American Community Survey

I have offered you a variety of facts thus far in describing the history of population growth and in providing you with an overview of the world's population situation. I do not just make up these numbers, of course, and so in this chapter I discuss the various kinds of demographic data we draw on to know what is happening in the world. To analyze the demography of a particular society, we need to know how many people live there, how they are distributed geographically, how many are being born, how many are dying, how many are moving in, and how many are moving out. That, of course, is only the beginning. If we want to unravel the mysteries of why things are as they are and not just describe what they are, we have to know about the social, psychological, economic, and even physical characteristics of the people and places being studied. Furthermore, we need to know these things not just for the present, but for the past as well. Let me begin the discussion, however, with sources of basic information about the numbers of living people, births, deaths, and migrants.

Sources of Demographic Data

The primary source of data on population size and distribution, as well as on demographic structure and characteristics, is the **census of population,** and after an overview of the history of population censuses, I will take a closer look at census taking in North America—the United States, Canada, and Mexico. The major source of information on the population processes of births and deaths is the registration of **vital statistics,** although in a few countries this task is accomplished by **population registers.** These sources are often supplemented with data from **sample surveys** as well as **historical sources. Administrative data** provide much of the information about population changes at the local level and about geographic mobility and migration.

Population Censuses

For centuries, governments have wanted to know how many people were under their rule. Rarely has their curiosity been piqued by scientific concern, but rather governments wanted to know who the taxpayers were, or they wanted to identify potential laborers and soldiers. The most direct way to find out how many people there are is to count them, and when you do that you are conducting a population census. The United Nations defines a census of population more specifically as "the total process of collecting, compiling and publishing demographic, economic and social data pertaining, at a specified time or times, to all persons in a country or delimited territory" (United Nations 1958:3). In practice, this does not mean that every person actually is seen and interviewed by a census taker. In most countries, it means that one adult in a household answers questions about all the people living in that household. These answers may be written responses to a questionnaire sent by mail or verbal responses to questions asked in person by the census taker.

The term *census* comes from Latin for "assessing" or "taxing." For Romans, it meant a register of adult male citizens and their property for purposes of taxation, the distribution of military obligations, and the determination of political status (Starr 1987). Thus, in 119 A.D. a person named Horos from the village of Bacchias left behind a letter on papyrus in which he states: "I register myself and those of my household for the house-by-house census of the past second year of Hadrian Caesar our Lord. I am Horos, the aforesaid, a cultivator of state land, forty-eight years old, with a scar on my left eyebrow, and I register my wife Tapekusis, daughter of Horos, forty-five years old. . . . " (Winter 1936:187).

As far as we know, the earliest governments to undertake censuses of their populations were those in the ancient civilizations of Egypt, Babylonia, China, India, and Rome (Shryock, Siegel, and associates [condensed by E. Stockwell] 1976). For several hundreds of years, citizens of Rome were counted periodically for tax and military purposes, and this enumeration of Roman subjects was extended to the entire Empire, including Roman Egypt, in 5 B.C. The Bible records this event as follows: "In those days a decree went out from Caesar Augustus that all the world should be enrolled. This was the first enrollment, when Quirinius was governor of Syria. And all went to be enrolled, each to his own city" (Luke 2:1–3). You can, of course, imagine the deficiencies of a census that required people to show up at their birthplaces rather than paying census takers to go out and do the counting. Actually, all that was required was that the head of each household provide government officials with a list of every household member (Horsley 1987).

In the seventh century A.D., the Prophet Mohammed led his followers to Medina (in Saudi Arabia), and after establishing a city-state there, one of his first activities was to conduct a written census of the entire Muslim population in the city (the returns showed a total of 1,500) (Nu'Man 1992). William of Normandy used a similar strategy in 1086. After having conquered England in 1066, William ordered an enumeration of all the landed wealth in the newly acquired territory in order to determine how much revenue the landowners owed. Data were recorded in the Domesday Book, the term *domesday* being a corruption of the word *doomsday,* which is the day of final judgment. The census document was so named because it was the final proof of legal title to land. The Domesday Book was not really what we think of today as a census, because it was an enumeration of "hearths," or household heads and their wealth, rather than of people. In order to calculate the total population of England in 1086 from the Domesday Book, you would have to multiply the number of "hearths" by some estimate of household size. More than 300,000 households were included, so if they averaged five persons per household, the population of England at the time was approximately 1.5 million (Hinde 1995). The population actually was larger than that because, in fact, the Domesday Book does not cover London, Winchester, Northumberland, Durham, or much of northwest England, and the only parts of Wales included are certain border areas (U.K. Public Records Office 2001).

The European renaissance began in northern Italy in the fourteenth century, and the Venetians and then the Florentines were interested in counting the wealth of their region, as William had been after conquering England. They developed a *catasto* that combined a count of the hearth and individuals. Thus, unlike the Domesday Book, the Florentine catasto of 1427 recorded not only the wealth of households but also data about each member of the household. In fact, so much information was collected that most of it went unexamined until the modern advent of computers (Herlihy and Klapisch-Zuber 1985). The value of a census was well known to François de Salignac de La Mothe-Fénelon, who was a very influential French political philosopher of the late seventeenth and early eighteenth centuries. He was the tutor to the Duke of Burgundy and much of his writing was intended as a primer of government for the young Duke:

> Do you know the number of men who compose your nation? How many men, and how many women, how many farmers, how many artisans, how many lawyers, how many tradespeople, how many priests and monks, how many nobles and soldiers? What would you say of a shepherd who did not know the size of his flock? . . . A king not knowing all these things is only half a king (quoted in Jones 2002:110).

By that description, Louis XIV (the "Sun King") and his grandson Louis XV were only partial kings, because the demographic evidence now suggests that the French population was growing in the eighteenth century, rather than declining, as the royal advisors (including the physiocrat Quesnay) believed at the time. Fénelon's books and essays were widely read in the early eighteenth century, which ushered in the modern era of nation-states, in turn giving rise to a genuine quest for accurate population information (Hollingsworth 1969). Indeed the term *statistic* is derived from the German word meaning "facts about a state." Sweden was one of the first of the European nations to regularly keep track of its population, with the establishment in 1749 of a combined population register and census administered in each diocese by the local clergy (Statistika Centralbyran [Sweden] 1983). Denmark and several Italian states (before the uniting of Italy in the late nineteenth century) also conducted censuses during the eighteenth century (Carr-Saunders 1936), as did the United States (where the first census was conducted in 1790). England launched its first census in 1801.

By the latter part of the nineteenth century, the statistical approach to understanding business and government affairs had started to take root in the Western world (Cassedy 1969). The population census began to be viewed as a potential tool for finding out more than just how many people there were and where they lived. Governments began to ask questions about age, marital status, whether and how people were employed, literacy, and so forth.

Census data (in combination with other statistics) have become the "lenses through which we form images of our society." Frederick Jackson Turner announced this famous view on the significance of the closing of the frontier on the basis of data from the 1890 census. Our national self-image today is confirmed or challenged by numbers that tell of drastic changes in the family, the increase in ethnic diversity, and many other trends. Winston Churchill observed that "first we shape our buildings and then they shape us. The same may be said of our statistics" (Alonso and Starr 1982:30).

The potential power behind the numbers that censuses produce can be gauged by public reaction to a census. In Germany, the enumeration of 1983 was postponed to 1987 because of public concern that the census was prying unduly into private lives. Germany did not conduct another census until 2002, well after reunification, and even then it was a sample census, not a complete enumeration. In the past few decades, protests have occurred in England, Switzerland, and the Netherlands, as well. In the latter case, the census scheduled for the 1980s was actually canceled after a survey indicating that the majority of the urban population would not cooperate (Robey 1983), and no census has been taken since. However, the Netherlands maintains a population register, which I discuss later, so they do have good demographic information even in the absence of a census.

Since the end of World War II, the United Nations has encouraged countries to enumerate their populations in censuses, often providing financial as well as technical aid. Between 1953 and 1964, 78 percent of the world's population (including that of mainland China) was enumerated by census. In the most recent census round (1995 through 2004), 89 percent of the world's population has been or is scheduled to be enumerated, based on data provided by the United Nations Statistics Division (2003).

The world's two largest nations, China and India, each conducted a census at the beginning of the twenty-first century. In November 2000, the Chinese government undertook the most ambitious census in world history when it counted its 1.26 billion inhabitants, 12 percent more than 10 years earlier (BBC News 2001). The task itself involved 10 million volunteer and government enumerators (Chang 2000b), nearly equivalent to the entire population of Belgium. The year 2001 saw India well into its second 100 years of census taking, the first census having been taken in 1881 under the supervision of the British. The Indian census used "only" 2.2 million enumerators to count 1.027 billion Indians as of March 2001, a 21 percent increase over 10 years earlier (India Registrar General & Census Commissioner 2001).

In contrast to India's regular census-taking, another of England's former colonies, Nigeria (the world's 10th most populous nation), has had more trouble with these efforts. Nigeria's population is divided among three broad ethnic groups: the Hausa-Fulani in the north, who are predominately Muslim; the Yorubas in the southwest, who are of various religious faiths; and the largely Christian Ibos in the southeast. The 1952 census of Nigeria indicated that the Hausa-Fulani had the largest share of the population, and so they dominated the first post-colonial government set up after independence in 1960. The newly independent nation ordered a census to be taken in 1962, but the results showed that northerners accounted for only

30 percent of the population. A "recount" in 1963 led somewhat suspiciously to the north accounting for 67 percent of the population. This exacerbated underlying ethnic tensions, culminating in the Ibos declaring independence. The resulting Biafran war (1967–70) saw at least three million people lose their lives before the Ibo rejoined the rest of Nigeria. A census in 1973 was never accepted by the government, and it was not until 1991 that the nation felt stable enough to try its hand again at enumeration, after agreeing that there would be no questions about ethnic group, language, or religion, and that population numbers would not be used as a basis for government expenditures. The official census count was 88.5 million people, well below the 110 million that many population experts had been guessing in the absence of any real data (Okolo 1999). Nigeria has not scheduled another census since the one in 1991.

Lebanon has not been enumerated since 1932, when the country was under French colonial rule (Domschke and Goyer 1986). The nearly equal number of Christians and Muslims in the country and the political strife between those groups have combined to make a census a very sensitive political issue. Before the nation was literally torn apart by civil war in the 1980s, the Christians had held a slight majority with respect to political representation. But Muslims almost certainly now hold a demographic majority (Faour 1991), and a census could either bolster or deny a group's claim to power and thus create the potential for further divisiveness.

I should note that censuses historically have been unpopular in that part of the world. The Old Testament of the Bible tells us that in ancient times King David ordered a census of Israel in which his enumerators counted "one million, one hundred thousand men who drew the sword. . . . But God was displeased with this thing [the census], and he smote Israel. . . . So the Lord sent a pestilence upon Israel; and there fell seventy thousand men of Israel" (1 Chronicles 21). Fortunately, in modern times, the advantages of census taking seem more clearly to outweigh the disadvantages. This has been especially true in the United States, where records indicate that no census has been followed directly by a pestilence.

The Census of the United States

Population censuses were part of Colonial life prior to the creation of the United States. A census had been conducted in Virginia in the early 1600s, and most of the northern colonies had conducted a census prior to the Revolution (U.S. Census Bureau 1978b). A population census has been taken every 10 years since 1790 in the United States as part of the constitutional mandate that seats in the House of Representatives be apportioned on the bases of population size and distribution. Article 1 of the U.S. Constitution directs that "Representatives and direct taxes shall be apportioned among the several states which may be included within this union, according to their respective numbers. . . . The actual Enumeration shall be made within three years after the first meeting of the Congress of the United States, and within every subsequent term of ten years, in such manner as they shall by law direct."

Even in 1790 the government used the census to find out more than just how many people there were. The census asked for the names of head of family, free white males aged 16 years and older, free white females, slaves, and other persons

(Shryock et al. 1976). The census questions were reflections of the social importance of those categories. For the first 100 years of census taking in the United States, the population was enumerated by U.S. marshals. In 1880, special census agents were hired for the first time, and finally in 1902 the Census Bureau became a permanent part of the government bureaucracy (Francese 1979; Hobbs and Stoops 2002). Beyond a core of demographic and housing information, the questions asked on the census have fluctuated according to the concerns of the time. Interest in international migration, for example, rose in 1920 just before the passage of a restrictive immigration law, and the census in that year added a battery of questions about the foreign-born population. In 2000, a question was added about grandparents as caregivers, replacing a question on fertility, and providing insight into the shift in focus from how many children women were having to the issue of who is taking care of those children. Questions are added and deleted by the Census Bureau through a process of consultation with Congress, other government officials, and census statistics users.

One of the more controversial items for the Census 2000 questionnaire was the question about race and ethnicity. The growing racial and ethnic diversity of the United States has led to a larger number of interracial/interethnic marriages and relationships producing children of mixed origin (also called multiracial). Previous censuses had asked people to choose a single category of race to describe themselves, but there was a considerable public sentiment that people should be able to identify themselves as being of mixed or multiple origins if, in fact, they perceived themselves in that way. Late in 1997, the government accepted the recommendation from a federally appointed committee that people of mixed racial heritage be able to choose more than one race category when filling out the Census 2000 questionnaire. Thus, a person whose mother is white and whose father is African-American was able to check both "White" and "Black or African-American" whereas in the past the choice would have had to be made between the two. There was still a separate question on "Hispanic/Latino/Spanish Origin" identity. An even deeper controversy concerns the question of whether "race" is even an appropriate category to ask about. Zuberi (2001) argues that because race is a social construct rather than a biological fact it should be treated differently than it currently is in social statistics, and Hirschman, Alba, and Farley (2000) suggest replacing the racial category in the census with a question about "origins," which may be a more socially meaningful way of looking at the issue of minority status within a society.

The census is designed as a complete enumeration of the population, but in the United States only a few of the questions are actually asked of everyone. For reasons of economy, most items in the census questionnaire have been administered to a sample of households in the last several censuses. From 1790 through 1930, all questions were asked of all applicable persons, but as the American population grew and Congress kept adding new questions to the Census, the savings involved in sampling grew, and in 1940 the Census Bureau began its practice of asking only a small number of items of all households, and using a sample of households to gather more detailed data. Fortunately, there has been no significant loss in accuracy. In 2000, there were 281 million people counted in more than 100 million households in the United States. Thus, a sample of one out of every six households in 2000 still yielded detailed data for nearly 50 million people.

The items of information obtained from everyone are often called the short-form items and include basic demographic and housing characteristics. The questionnaire for Census 2000 is reproduced as Figure 4.2. The first page asks that everyone in the household be listed by name, which allows the Census Bureau to check for duplicate listings of people (such as college students away from home). The first person listed is supposed to be someone in the household who owns, is buying, or rents this housing unit. This person used to be known as the "head of household" (and there is still a tax category in the U.S. for such a person), but the Census Bureau now refers to him or her as the "householder." Information is then requested for each person in the household regarding his or her relationship to the first person listed (the householder), sex, racial and ethnic identification, age and year of birth, marital status, and Hispanic/Latino/Spanish origin. These and the questions relating to characteristics of the housing unit comprise the short-form.

Approximately five out of six households received the short form to fill out, whereas one in six (about 17 percent of households) were asked to complete the longer, more detailed questionnaire shown in Figure 4.2. Table 4.1 lists the items included on the U.S. Census 2000 questionnaire, compared with a list of the data obtained by the 2001 Census of Canada and the 2000 Census of Mexico. The table indicates which items are asked of every household and which are asked of a sample of households. As I will discuss later, there is a good chance that the census of 2010 will include only the short form, with the detailed data being collected through the ongoing **American Community Survey.**

In theory, a census obtains accurate information from everyone. But the first question that census workers have to ask is: Who is supposed to be included in the census? Are visitors to the country to be included? Are people who are absent from the country on census day to be excluded?

Who Is Included in the Census?   There are several ways to answer that question, and each produces a potentially different total number of people. At one extreme is the concept of the **de facto population,** which counts people who are in a given territory on the census day. At the other extreme is the **de jure population,** which represents people who legally "belong" to a given area in some way or another, regardless of whether they were there on the day of the census. For countries with few foreign workers and where working in another area is rare, the distinction makes little difference. But an increasing number of countries, including several in Europe and nearly all of the Gulf states in the Middle East, have large numbers of guest workers from other countries, and thus have a larger de facto than de jure population. On the other hand, a country such as Mexico, from which migrants regularly leave temporarily to go to the United States, has a de jure population that is larger than the de facto.

Most countries (including the United States, Canada, and Mexico) have now adopted a concept that lies somewhere between the extremes of de facto and de jure, and they include people in the census on the basis of **usual residence,** which is roughly defined as the place where a person usually sleeps. College students who live away from home, for example, are included at their college address rather than being counted in their parents' household. People with no usual residence (the homeless, including migratory workers, vagrants, and "street people") are counted

Figure 4.2 Questionnaire for United States Census 2000

(continued)

Figure 4.2 (continued)

118

Person 1

Your answers are important! Every person in the Census counts.

1. What is this person's name? *Print the name of Person 1 from page 2.*
Last Name
First Name   MI

2. What is this person's telephone number? *We may contact this person if we don't understand an answer.*
Area Code + Number

3. What is this person's sex? *Mark ☒ ONE box.*
☐ Male
☐ Female

4. What is this person's age and what is this person's date of birth?
Age on April 1, 2000
Print numbers in boxes.
Month   Day   Year of birth

→ 5. NOTE: Please answer BOTH Questions 5 and 6.
Is this person Spanish/Hispanic/Latino? *Mark ☒ the "No" box if not Spanish/Hispanic/Latino.*
☐ No, not Spanish/Hispanic/Latino
☐ Yes, Mexican, Mexican Am., Chicano
☐ Yes, Puerto Rican
☐ Yes, Cuban
☐ Yes, other Spanish/Hispanic/Latino – *Print group.* ↘

→ 6. What is this person's race? *Mark ☒ one or more races to indicate what this person considers himself/herself to be.*
☐ White
☐ Black, African Am., or Negro
☐ American Indian or Alaska Native – *Print name of enrolled or principal tribe.* ↘

☐ Asian Indian   ☐ Native Hawaiian
☐ Chinese   ☐ Guamanian or Chamorro
☐ Filipino   ☐ Samoan
☐ Japanese   ☐ Other Pacific Islander – *Print race.* ↗
☐ Korean
☐ Vietnamese
☐ Other Asian – *Print race.* ↘

☐ Some other race – *Print race.* ↘

7. What is this person's marital status?
☐ Now married
☐ Widowed
☐ Divorced
☐ Separated
☐ Never married

8. a. At any time since February 1, 2000, has this person attended regular school or college? *Include only nursery school or preschool, kindergarten, elementary school, and schooling which leads to a high school diploma or a college degree.*
☐ No, has not attended since February 1 → Skip to 9
☐ Yes, public school, public college
☐ Yes, private school, private college

☞ Question is asked of all persons on the short (100-percent) and long (sample) forms.

Person 1 (continued)

8. b. What grade or level was this person attending? *Mark ☒ ONE box.*
☐ Nursery school, preschool
☐ Kindergarten
☐ Grade 1 to grade 4
☐ Grade 5 to grade 8
☐ Grade 9 to grade 12
☐ College undergraduate years (freshman to senior)
☐ Graduate or professional school (for example: medical, dental, or law school)

9. What is the highest degree or level of school this person has COMPLETED? *Mark ☒ ONE box. If currently enrolled, mark the previous grade or highest degree received.*
☐ No schooling completed
☐ Nursery school to 4th grade
☐ 5th grade or 6th grade
☐ 7th grade or 8th grade
☐ 9th grade
☐ 10th grade
☐ 11th grade
☐ 12th grade, NO DIPLOMA
☐ HIGH SCHOOL GRADUATE – high school DIPLOMA or the equivalent (for example: GED)
☐ Some college credit, but less than 1 year
☐ 1 or more years of college, no degree
☐ Associate degree (for example: AA, AS)
☐ Bachelor's degree (for example: BA, AB, BS)
☐ Master's degree (for example: MA, MS, MEng, MEd, MSW, MBA)
☐ Professional degree (for example: MD, DDS, DVM, LLB, JD)
☐ Doctorate degree (for example: PhD, EdD)

10. What is this person's ancestry or ethnic origin?

(For example: Italian, Jamaican, African Am., Cambodian, Cape Verdean, Norwegian, Dominican, French Canadian, Haitian, Korean, Lebanese, Polish, Nigerian, Mexican, Taiwanese, Ukrainian, and so on.)

11. a. Does this person speak a language other than English at home?
☐ Yes
☐ No → Skip to 12
b. What is this language?

(For example: Korean, Italian, Spanish, Vietnamese)

c. How well does this person speak English?
☐ Very well
☐ Well
☐ Not well
☐ Not at all

12. Where was this person born?
☐ In the United States – *Print name of state.*

☐ Outside the United States – *Print name of foreign country, or Puerto Rico, Guam, etc.*

13. Is this person a CITIZEN of the United States?
☐ Yes, born in the United States → Skip to 15a
☐ Yes, born in Puerto Rico, Guam, the U.S. Virgin Islands, or Northern Marianas
☐ Yes, born abroad of American parent or parents
☐ Yes, a U.S. citizen by naturalization
☐ No, not a citizen of the United States

14. When did this person come to live in the United States? *Print numbers in boxes.*
Year

15. a. Did this person live in this house or apartment 5 years ago (on April 1, 1995)?
☐ Person is under 5 years old → Skip to 33
☐ Yes, this house → Skip to 16
☐ No, outside the United States – *Print name of foreign country, or Puerto Rico, Guam, etc., below; then skip to 16.*

☐ No, different house in the United States

Person 1 (continued)

15 b. Where did this person live 5 years ago?

Name of city, town, or post office
[_ _ _ _ _ _ _ _]

Did this person live inside the limits of the city or town?
☐ Yes
☐ No, outside the city/town limits

Name of county
[_ _ _ _ _ _ _ _]

Name of state
[_ _ _ _ _ _ _ _]

ZIP Code
[_ _ _ _ _]

16 Does this person have any of the following long-lasting conditions:

	Yes	No
a. Blindness, deafness, or a severe vision or hearing impairment?	☐	☐
b. A condition that substantially limits one or more basic physical activities such as walking, climbing stairs, reaching, lifting, or carrying?	☐	☐

17 Because of a physical, mental, or emotional condition lasting 6 months or more, does this person have any difficulty in doing any of the following activities:

	Yes	No
a. Learning, remembering, or concentrating?	☐	☐
b. Dressing, bathing, or getting around inside the home?	☐	☐
c. (Answer if this person is 16 YEARS OLD OR OVER.) Going outside the home alone to shop or visit a doctor's office?	☐	☐
d. (Answer if this person is 16 YEARS OLD OR OVER.) Working at a job or business?	☐	☐

18 Was this person under 15 years of age on April 1, 2000?
☐ Yes → Skip to 33
☐ No

2045

Form D-61B

5

19 a. Does this person have any of his/her own grandchildren under the age of 18 living in this house or apartment?
☐ Yes
☐ No → Skip to 20a

b. Is this grandparent currently responsible for most of the basic needs of any grandchild(ren) under the age of 18 who live(s) in this house or apartment?
☐ Yes
☐ No → Skip to 20a

c. How long has this grandparent been responsible for the(se) grandchild(ren)? *If the grandparent is financially responsible for more than one grandchild, answer the question for the grandchild for whom the grandparent has been responsible for the longest period of time.*
☐ Less than 6 months
☐ 6 to 11 months
☐ 1 or 2 years
☐ 3 or 4 years
☐ 5 years or more

20 a. Has this person ever served on active duty in the U.S. Armed Forces, military Reserves, or National Guard? *Active duty does not include training for the Reserves or National Guard, but DOES include activation, for example, for the Persian Gulf War.*
☐ Yes, now on active duty
☐ Yes, on active duty in past, but not now
☐ No, training for Reserves or National Guard only → Skip to 21
☐ No, never served in the military → Skip to 21

b. When did this person serve on active duty in the U.S. Armed Forces? *Mark [X] a box for EACH period in which this person served.*
☐ April 1995 or later
☐ August 1990 to March 1995 (including Persian Gulf War)
☐ September 1980 to July 1990
☐ May 1975 to August 1980
☐ Vietnam era (August 1964–April 1975)
☐ February 1955 to July 1964
☐ Korean conflict (June 1950–January 1955)
☐ World War II (September 1940–July 1947)
☐ Some other time

c. In total, how many years of active-duty military service has this person had?
☐ Less than 2 years
☐ 2 years or more

Person 1 (continued)

21 LAST WEEK, did this person do ANY work for either pay or profit? *Mark [X] the "Yes" box even if the person worked only 1 hour, or helped without pay in a family business or farm for 15 hours or more, or was on active duty in the Armed Forces.*
☐ Yes
☐ No → Skip to 25a

22 At what location did this person work LAST WEEK? *If this person worked at more than one location, print where he or she worked most last week.*

a. Address (Number and street name)
[_ _ _ _ _ _ _ _]

(If the exact address is not known, give a description of the location such as the building name or the nearest street or intersection.)

b. Name of city, town, or post office
[_ _ _ _ _ _ _ _]

c. Is the work location inside the limits of that city or town?
☐ Yes
☐ No, outside the city/town limits

d. Name of county
[_ _ _ _ _ _ _ _]

e. Name of U.S. state or foreign country
[_ _ _ _ _ _ _ _]

f. ZIP Code
[_ _ _ _ _]

23 a. How did this person usually get to work LAST WEEK? *If this person usually used more than one method of transportation during the trip, mark [X] the box of the one used for most of the distance.*
☐ Car, truck, or van
☐ Bus or trolley bus
☐ Streetcar or trolley car
☐ Subway or elevated
☐ Railroad
☐ Ferryboat
☐ Taxicab
☐ Motorcycle
☐ Bicycle
☐ Walked
☐ Worked at home → Skip to 27
☐ Other method

If "Car, truck, or van" is marked in 23a, go to 23b. Otherwise, skip to 24a.

b. How many people, including this person, usually rode to work in the car, truck, or van LAST WEEK?
☐ Drove alone
☐ 2 people
☐ 3 people
☐ 4 people
☐ 5 or 6 people
☐ 7 or more people

24 a. What time did this person usually leave home to go to work LAST WEEK?
☐ a.m. ☐ p.m.

b. How many minutes did it usually take this person to get from home to work LAST WEEK?
Minutes [_ _ _]

25 Answer questions 25-26 for persons who did not work for pay or profit last week. Others skip to 27.

a. LAST WEEK, was this person on layoff from a job?
☐ Yes → Skip to 25c
☐ No

b. LAST WEEK, was this person TEMPORARILY absent from a job or business?
☐ Yes, on vacation, temporary illness, labor dispute, etc. → Skip to 26
☐ No → Skip to 25d

c. Has this person been informed that he or she will be recalled to work within the next 6 months OR been given a date to return to work?
☐ Yes → Skip to 25e
☐ No

d. Has this person been looking for work during the last 4 weeks?
☐ Yes
☐ No → Skip to 26

e. LAST WEEK, could this person have started a job if offered one, or returned to work if recalled?
☐ Yes, could have gone to work
☐ No, because of own temporary illness
☐ No, because of all other reasons (in school, etc.)

26 When did this person last work, even for a few days?
☐ 1995 to 2000
☐ 1994 or earlier, or never worked → Skip to 31

Form D-61B

6

(continued)

120

Figure 4.2 (continued)

Person 1 (continued)

27 **Industry or Employer** – *Describe clearly this person's chief job activity or business last week. If this person had more than one job, describe the one at which this person worked the most hours. If this person had no job or business last week, give the information for his/her last job or business since 1995.*

a. **For whom did this person work?** *If now on active duty in the Armed Forces, mark $\boxtimes$ this box → ☐ and print the branch of the Armed Forces.*

Name of company, business, or other employer

b. **What kind of business or industry was this?** *Describe the activity at location where employed. (For example: hospital, newspaper publishing, mail order house, auto repair shop, bank)*

c. **Is this mainly** – Mark $\boxtimes$ *ONE box.*
☐ Manufacturing?
☐ Wholesale trade?
☐ Retail trade?
☐ Other (agriculture, construction, service, government, etc.)?

28 **Occupation**

a. **What kind of work was this person doing?** *(For example: registered nurse, personnel manager, supervisor of order department, auto mechanic, accountant)*

b. **What were this person's most important activities or duties?** *(For example: patient care, directing hiring policies, supervising order clerks, repairing automobiles, reconciling financial records)*

29 **Was this person** – Mark $\boxtimes$ *ONE box.*
☐ Employee of a PRIVATE-FOR-PROFIT company or business of an individual, for wages, salary, or commissions
☐ Employee of a PRIVATE NOT-FOR-PROFIT, tax-exempt, or charitable organization
☐ Local GOVERNMENT employee (city, county, etc.)
☐ State GOVERNMENT employee
☐ Federal GOVERNMENT employee
☐ SELF-EMPLOYED in own NOT INCORPORATED business, professional practice, or farm
☐ SELF-EMPLOYED in own INCORPORATED business, professional practice, or farm
☐ Working WITHOUT PAY in family business or farm

30 a. **LAST YEAR, 1999, did this person work at a job or business at any time?**
☐ Yes
☐ No → *Skip to 31*

b. **How many weeks did this person work in 1997?** *Count paid vacation, paid sick leave, and military service.*
Weeks

c. **During the weeks WORKED in 1999, how many hours did this person usually work each WEEK?**
Usual hours worked each WEEK

31 **INCOME IN 1999** – Mark $\boxtimes$ *the "Yes" box for each income source received during 1999 and enter the total amount received during 1999 to a maximum of $999,999. Mark $\boxtimes$ the "No" box if the income source was not received and mark $\boxtimes$ the "Loss" box next to the dollar amount.*

For income received jointly, report, if possible, the appropriate share for each person; otherwise, report the whole amount for only one person and mark $\boxtimes$ the "No" box for the other person. If exact amount is not known, please give best estimate.

a. **Wages, salary, commissions, bonuses, or tips from all jobs** – *Report amount before deductions for taxes, bonds, dues, or other items.*
☐ Yes
  Annual amount – *Dollars*
  $ _____ .00
☐ No

b. **Self-employment income from own nonfarm businesses or farm businesses, including proprietorships and partnerships** – *Report NET income after business expenses.*
☐ Yes
  Annual amount – *Dollars*
  $ _____ .00 ☐ Loss
☐ No

Person 1 (continued)

31 c. **Interest, dividends, net rental income, royalty income, or income from estates and trusts** – *Report even small amounts credited to an account.*
☐ Yes
  Annual amount – *Dollars*
  $ _____ .00 ☐ Loss
☐ No

d. **Social Security or Railroad Retirement**
☐ Yes
  Annual amount – *Dollars*
  $ _____ .00
☐ No

e. **Supplemental Security Income (SSI)**
☐ Yes
  Annual amount – *Dollars*
  $ _____ .00
☐ No

f. **Any public assistance or welfare payments from the state or local welfare office**
☐ Yes
  Annual amount – *Dollars*
  $ _____ .00
☐ No

g. **Retirement, survivor, or disability pensions** – *Do NOT include Social Security.*
☐ Yes
  Annual amount – *Dollars*
  $ _____ .00
☐ No

h. **Any other sources of income received regularly such as Veterans' (VA) payments, unemployment compensation, child support, or alimony** – *Do NOT include lump-sum payments such as money from an inheritance or sale of a home.*
☐ Yes
  Annual amount – *Dollars*
  $ _____ .00
☐ No

32 **What was this person's total income in 1997?** *Add entries in questions 31a–31h; subtract any losses. If net income was a loss, enter the amount and mark $\boxtimes$ the "Loss" box next to the dollar amount.*
☐ None   OR $ _____ .00 ☐ Loss
  Annual amount – *Dollars*

☞ *Question 32 is asked of all households on the short (100-percent) and long (sample) forms.*

HOUSING QUESTIONS

➜ **Now, please answer questions 33–53 about your household.**

33 **Is this house, apartment, or mobile home** –
☐ Owned by you or someone in this household with a mortgage or loan?
☐ Owned by you or someone in this household free and clear (without a mortgage or loan)?
☐ Rented for cash rent?
☐ Occupied without payment of cash rent?

34 **Which best describes this building?** *Include all apartments, flats, etc., even if vacant.*
☐ A mobile home
☐ A one-family house detached from any other house
☐ A one-family house attached to one or more houses
☐ A building with 2 apartments
☐ A building with 3 or 4 apartments
☐ A building with 5 to 9 apartments
☐ A building with 10 to 19 apartments
☐ A building with 20 to 49 apartments
☐ A building with 50 or more apartments
☐ Boat, RV, van, etc.

35 **About when was this building first built?**
☐ 1999 or 2000
☐ 1995 to 1998
☐ 1990 to 1994
☐ 1980 to 1989
☐ 1970 to 1979
☐ 1960 to 1969
☐ 1950 to 1959
☐ 1940 to 1949
☐ 1939 or earlier

36 **When did this person move into this house, apartment, or mobile home?**
☐ 1999 or 2000
☐ 1995 to 1998
☐ 1990 to 1994
☐ 1980 to 1989
☐ 1970 to 1979
☐ 1969 or earlier

37 **How many rooms do you have in this house, apartment, or mobile home?** *Do NOT count bathrooms, porches, balconies, foyers, halls, or half-rooms.*
☐ 1 room   ☐ 6 rooms
☐ 2 rooms   ☐ 7 rooms
☐ 3 rooms   ☐ 8 rooms
☐ 4 rooms   ☐ 9 or more rooms
☐ 5 rooms

38 How many bedrooms do you have; that is, how many bedrooms would you list if this house, apartment, or mobile home were on the market for sale or rent?

☐ No bedroom
☐ 1 bedroom
☐ 2 bedrooms
☐ 3 bedrooms
☐ 4 bedrooms
☐ 5 or more bedrooms

39 Do you have COMPLETE plumbing facilities in this house, apartment, or mobile home; that is, 1) hot and cold piped water, 2) a flush toilet, and 3) a bathtub or shower?

☐ Yes, have all three facilities
☐ No

40 Do you have COMPLETE kitchen facilities in this house, apartment, or mobile home; that is, 1) a sink with piped water, 2) a range or stove, and 3) a refrigerator?

☐ Yes, have all three facilities
☐ No

41 Is there telephone service available in this house, apartment, or mobile home from which you can both make and receive calls?

☐ Yes
☐ No

42 Which FUEL is used MOST for heating this house, apartment, or mobile home?

☐ Gas: from underground pipes serving the neighborhood
☐ Gas: bottled, tank, or LP
☐ Electricity
☐ Fuel oil, kerosene, etc.
☐ Coal or coke
☐ Wood
☐ Solar energy
☐ Other fuel
☐ No fuel used

43 How many automobiles, vans, and trucks of one-ton capacity or less are kept at home for use by members of your household?

☐ None
☐ 1
☐ 2
☐ 3
☐ 4
☐ 5
☐ 6 or more

44 Answer ONLY if this is a ONE-FAMILY HOUSE OR MOBILE HOME – All others skip to 45.

a. Is there a business (such as a store or barber shop) or a medical office on this property?

☐ Yes
☐ No

b. How many acres is this house or mobile home on?

☐ Less than 1 acre → Skip to 45
☐ 1 to 9.9 acres
☐ 10 or more acres

c. In 1999, what were the actual sales of all agricultural products from this property?

☐ None
☐ $1 to $999
☐ $1,000 to $2,459
☐ $2,500 to $4,999
☐ $5,000 to $9,999
☐ $10,000 or more

45 What are the annual costs of utilities and fuels for this house, apartment, or mobile home? If you have lived here less than 1 year, estimate the annual cost.

a. Electricity

Annual cost – Dollars
$ [ ] .00
OR
☐ Included in rent or in condominium fee
☐ No charge or electricity not used

b. Gas

Annual cost – Dollars
$ [ ] .00
OR
☐ Included in rent or in condominium fee
☐ No charge or gas not used

c. Water and sewer

Annual cost – Dollars
$ [ ] .00
OR
☐ Included in rent or in condominium fee
☐ No charge

d. Oil, coal, kerosene, wood, etc.

Annual cost – Dollars
$ [ ] .00
OR
☐ Included in rent or in condominium fee
☐ No charge or these fuels not used

2049

---

46 Answer ONLY if you PAY RENT for this house, apartment, or mobile home – All others skip to 47.

a. What is the monthly rent?

Monthly amount – Dollars
$ [ ] .00

b. Does the monthly rent include any meals?

☐ Yes
☐ No

47 Answer questions 47a–53 if you or someone in this household owns or is buying this house, apartment, or mobile home; otherwise, skip to questions for Person 2.

a. Do you have a mortgage, deed of trust, contract to purchase, or similar debt on THIS property?

☐ Yes, mortgage, deed of trust, or similar debt
☐ Yes, contract to purchase
☐ No → Skip to 48a

b. How much is your regular monthly mortgage payment on THIS property? Include payment only on first mortgage or contract to purchase.

Monthly amount – Dollars
$ [ ] .00
OR
☐ No regular payment required → Skip to 48a

c. Does your regular monthly mortgage payment include payments for real estate taxes on THIS property?

☐ Yes, taxes included in mortgage payment
☐ No, taxes paid separately or taxes not required

d. Does your regular monthly mortgage payment include payments for fire, hazard, or flood insurance on THIS property?

☐ Yes, insurance included in mortgage payment
☐ No, insurance paid separately or no insurance

48 a. Do you have a second mortgage or a home equity loan on THIS property? Mark [X] all boxes that apply.

☐ Yes, a second mortgage
☐ Yes, a home equity loan
☐ No → Skip to 49

b. How much is your regular monthly payment on all second or junior mortgages and all home equity loans on THIS property?

Monthly amount – Dollars
$ [ ] .00
OR
☐ No regular payment required

49 What were the real estate taxes on THIS property last year?

Yearly amount – Dollars
$ [ ] .00
OR
☐ None

50 What was the annual payment for fire, hazard, and flood insurance on THIS property?

Annual amount – Dollars
$ [ ] .00
OR
☐ None

51 What is the value of this property; that is, how much do you think this house and lot, apartment, or mobile home and lot would sell for if it were for sale?

☐ Less than $10,000
☐ $10,000 to $14,999
☐ $15,000 to $19,999
☐ $20,000 to $24,999
☐ $25,000 to $29,999
☐ $30,000 to $34,999
☐ $35,000 to $39,999
☐ $40,000 to $49,999
☐ $50,000 to $59,999
☐ $60,000 to $69,999
☐ $70,000 to $79,999
☐ $80,000 to $89,999
☐ $90,000 to $99,999
☐ $100,000 to $124,999
☐ $125,000 to $149,999
☐ $150,000 to $174,999
☐ $175,000 to $199,999
☐ $200,000 to $249,999
☐ $250,000 to $299,999
☐ $300,000 to $399,999
☐ $400,000 to $499,999
☐ $500,000 to $749,999
☐ $750,000 to $999,999
☐ $1,000,000 or more

52 Answer ONLY if this is a CONDOMINIUM –

What is the monthly condominium fee?

Monthly amount – Dollars
$ [ ] .00

53 Answer ONLY if this is a MOBILE HOME –

a. Do you have an installment loan or contract on THIS mobile home?

☐ Yes
☐ No

b. What was the total cost for installment loan payments, personal property taxes, site rent, registration fees, and license fees on THIS mobile home and its site last year? Exclude real estate taxes.

Yearly amount – Dollars
$ [ ] .00

⬆ Are there more people living here? If yes, continue with Person 2.

(continued)

121

Figure 4.2 (continued)

Table 4.1 Comparison of Items Included in the U.S. Census 2000 Questionnaire, the 2001 Census of Canada, and the 2000 Census of Mexico

Census Item	U.S. Census 2000	Canada 2001	Mexico 2000
Population Characteristics:			
Age	XX	XX	XX
Sex	XX	XX	XX
Relationship to householder (family structure)	XX	XX	XX
Race	XX	X	
Hispanic origin	XX		
Marital status	X	XX	XX
Fertility			XX
Income	X	X	XX
Sources of income	X	X	X
Health insurance			XX
Job benefits			X
Unpaid household activities		X	
Labor force status	X	X	XX
Industry, occupation, and class of worker	X	X	XX
Work status last year	X		
Veteran status	X		
Grandparents as caregivers	X		
Place of work and journey to work	X	X	XX
Journey to work	X	X	
Vehicles available	X		
Ancestry	X	X	
Place of birth	X	X	XX
Birthplace of parents		X	
Citizenship	X	X	
Year of entry if not born in this country	X	X	
Language spoken at home	X	X	XX
Language spoken at work		X	
Religion		X	XX
Educational attainment	X	X	XX
School enrollment	X	X	XX
Residence one year ago (migration)		X	
Residence five years ago (migration)	X	X	XX
International migration of family members			X
Disability (activities of daily living)	X	X	XX
Housing characteristics:			
Tenure (rent or own)	XX		XX
Type of housing	XX		XX
Material used for construction of walls			XX
Material used for construction of roof			XX

(continued)

Table 4.1   (continued)

Census Item	U.S. Census 2000	Canada 2001	Mexico 2000
Material used for construction of floors			XX
Repairs needed on structure		X	
Year structure built	X	X	X
Units in structure	X		
Rooms in unit	X	X	XX
Bedrooms	X	X	XX
Kitchen facilities	X		XX
Plumbing facilities	X		XX
Telephone in unit	X		X
Material possessions (TV, radio, etc.)			XX
House heating fuel	X		XX
Year moved into unit	X		
Farm residence	X		
Value	X		
Selected monthly owner costs	X	X	
Rent	X	X	

Note: XX = Included and asked of every household; X = Included but asked of only a sample of households. Questions asked on each census may be different; similar categories of questions asked do not necessarily mean strict comparability of data.

where they are found. On the other hand, visitors and tourists from other countries who "belong" somewhere else are not included, even though they may be in the country when the census is being conducted. At the same time, the concept of usual residence means that undocumented immigrants (who legally do not "belong" where they are found) will be included in the census along with everyone else.

Where you belong became a court issue following Census 2000 in the United States. The Census Bureau included in the census members of the military and the federal government who were stationed abroad. They were counted as belonging to the state in the U.S. that was their normal domicile, and this turned out especially to benefit North Carolina. However, Utah filed suit in federal court, objecting that Mormon missionaries from Utah who were serving abroad should also be counted as residents of Utah, rather than being excluded because they were living outside the United States. In 2001, the U.S. District Court ruled against that idea, and so North Carolina gained a seat in Congress on the basis of its "overseas residents" while Utah did not.

Knowing who should be included in the census does not, however, guarantee that they will all be found and accurately counted. There are several possible errors that can creep into the enumeration process. We can divide these into the two broad categories of **nonsampling error** (which includes **coverage error** and **content error**) and **sampling error.**

Coverage Error   The two most common sources of error in a census are coverage error and content error. A census is designed to count everyone, but there are always people who are missed, as well as a few who are counted more than once. The combination of the undercount and the overcount is called coverage error, or **net census undercount** (the difference between the undercount and the overcount). As I discuss below, there are several ways to measure and adjust for undercount, but it becomes more complicated (and political) when there is a **differential undercount,** when some groups are more likely to be underenumerated than other groups.

In the United States, the differential undercount has meant that racial/ethnic minority groups (especially African-Americans) have been less likely to be included in the census count than whites. Table 4.2 shows estimates of the net undercount in the last several censuses, along with the differential undercount of the black population. The overall undercount in the 1940 census was 5.6 percent, and you can see that it has been steadily declining since then as the Census Bureau institutes ever more sophisticated procedures. But you can also see that in 1940 more than 10 percent of African-Americans in the country were missed by the census. This was the year that the differential undercount was discovered as a result of a "somewhat serendipitous natural experiment" (Anderson and Fienberg 1999:29). Because of World War II, men were registering for the draft when that census was taken, providing demographers with a chance to compare census returns with counts of men registering for the draft. It turned out that 229,000 more black men signed up for the draft than would have been expected based on census data (Price 1947), signaling some real problems with the completeness of the census coverage in 1940. Since then, a great deal of time, effort, and controversy have gone into attempts to reduce both the overall undercount and the differential undercount. The numbers in Table 4.2 show that Census 2000 appears to have been more successful than any previous census in this regard, and I discuss this issue in more detail in the essay that accompanies this chapter.

Coverage is improved in the census by a variety of measures, such as having better address identification so that every household receives a questionnaire and having a high-profile advertising campaign designed to encourage a high response to the mail-out questionnaire. More than two-thirds (67 percent) of households responded to the mailed questionnaire in 2000, and the rest were contacted by the Census Bureau in the Non-Response Follow-Up (NRFU) phase of data collection. The Census Bureau sent staff members into the field to interview people who had not completed forms, and, in some cases, to find out about people whom they were unable to contact. When you combine this with the fact that one member of a household may have filled in the information for all household members, it is easy to see why so many people routinely think they have not been counted in the census—someone else answered for them.

Measuring Coverage Error   Right now you are probably asking yourself how the Census Bureau could ever begin to estimate the number of people missed in a census. This is not an easy task, and statisticians in the U.S. and other countries have experimented with a number of methods over the years. Two principal methods are typically used: (1) **demographic analysis (DA),** and (2) **dual-system estimation (DSE).**

The demographic analysis approach uses the **demographic balancing equation** to estimate what the population at the latest census should have been, and then compares that number to the actual count. The demographic balancing equation

Table 4.2   Net Undercount and Differential Undercount in U.S. Censuses

Year	Net undercount for total population (%)	Undercount of black population (%)	Undercount of white population (%)(*)	Differential undercount (percentage point difference between black and white undercount)
1940	5.6	10.3	5.1	5.2
1950	4.4	9.6	3.8	5.8
1960	3.3	8.3	2.7	5.6
1970	2.9	8.0	2.2	5.8
1980	1.4	5.9	0.7	5.2
1990	1.8	5.7	1.3	4.4
2000	1.1	2.1	1.0	1.1

Sources: Data for 1940 through 1980 are from Anderson and Fienberg (1999: Table 4.1); and data for 1990 and 2000 are from Robinson, West, and Adlakha (2002: Table 6). The undercount for 1940 through 1990 is based on demographic analysis, and the undercount for 2000 is based on the Accuracy Coverage and Evaluation Survey.
(*) The data for 1990 and 2000 represent the nonblack population rather than specifically the white population.

says that the population at time 2 is equal to the population at time 1 plus the births between time 1 and 2 minus the deaths between time 1 and 2 plus the in-migrants between time 1 and 2 minus the out-migrants between time 1 and 2. Thus, if we know the number of people from the previous census, we can add the number of births since then, subtract the number of deaths since then, add the number of in-migrants since then, and subtract the number of out-migrants since then to estimate what the total population count should have been. A comparison of this number with the actual census count provides a clue as to the accuracy of the census. Using these methods, the Census Bureau is able to piece together a composite rendering of what the population "should" look like. Differences from that picture and the one painted by the census can be used as estimates of under- or overenumeration. By making these calculations for all age, sex, and racial/ethnic groups, we can arrive at an estimate of the possible undercount among various groups in the population.

Of course, if we do not have an accurate count of births, deaths, and migrants, then our demographic-analysis estimate may itself be wrong, so this method requires careful attention to the quality of the noncensus data. And, you say, why should we even take a census if we think we can estimate the number of people more accurately without it? The answer is that the demographic-analysis approach usually only produces an estimate of the total number of people in any age, sex, racial/ethnic group, without providing a way of knowing the details of the population—which is what we obtain from the census questionnaire.

The dual-system estimation method involves comparing the census results with some other source of information about the people counted. After Census 2000 in the United States, the Census Bureau implemented its Accuracy and Coverage Evaluation (A.C.E.) Survey, which was similar to, albeit larger than, the postenumeration survey (PES) conducted after the 1990 census. This involved taking a carefully constructed sample survey right after the census was finished and then matching people in the sample survey with their responses in the census. This process can determine whether households and individuals within the households were counted in both the census and the survey (the ideal situation); in the census but not in the survey (possible but not likely); or in the survey but not in the census (the usual measure of underenumeration). Obviously, some people may be missed by both the census and the survey, but the logic underlying the method is analogous to the capture-recapture method used by biologists tracking wildlife (Choldin 1994). That strategy is to capture a sample of animals, mark them, and release them. Later, another sample is captured, and some of the marked animals will wind up being recaptured. The ratio of recaptured animals to all animals caught in the second sample is assumed to represent the ratio of the first group captured to the whole population, and on this basis the wildlife population can be estimated. Although some humans are certainly "wild," a few adjustments are required to apply the method to human populations.

The Census Bureau used both the DA and the DSE approaches in evaluating the accuracy of Census 2000 (Clark and Moul 2003; Robinson, West, and Adlakha 2002). Note that the dual-system estimation approach also provides a way of testing for content error by comparing people's responses on the census questionnaire with their answers to the postenumeration survey questionnaire.

Content Error   Although coverage error is a concern in any census, there can also be problems with the accuracy of the data obtained in the census (content error). Content error includes nonresponses to particular questions on the census or inaccurate responses if people do not understand the question. Errors can also occur if information is inaccurately recorded on the form or if there is some glitch in the processing (coding, data entry, or editing) of the census return. In comparison with other censuses, the United Nations rates the American census as highly accurate, especially with respect to recording age, one of the most important demographic characteristics (United Nations 1996a).

By and large, content error seems not to be a problem in the U.S. census, although the data are certainly not 100 percent accurate. There is always the potential for misunderstanding the meaning of a question and these problems appear to be greater for people with lower literacy skills (Iversen, Furstenberg, and Belzer 1999). In general, data from the United Nations suggest that the more highly developed a country is, the more accurate the content of its census data will be, and this is probably accounted for largely by higher levels of education.

Sampling Error   If any of the data in a census are collected on a sample basis (as is done in the United States, Canada, and Mexico), then sampling error is introduced into the results. With any sample, scientifically selected or not, differences are likely to exist between the characteristics of the sampled population and the larger group from which the sample was chosen. However, in a scientific sample, such as that used in most census operations, sampling error is readily measured based on the mathematics of probability. To a certain extent, sampling error can be controlled—samples can be designed to ensure comparable levels of error across groups or

TO ADJUST OR NOT TO ADJUST—THAT WAS THE QUESTION; OR WAS IT MUCH ADO ABOUT NOTHING?

The differential undercount, which I discuss in this chapter, matters for many reasons in the United States. In the first place, those communities that are less well counted will receive less legislative representation as a result of census-based apportionment and redistricting. They will also stand to lose out in the distribution of population-based funding from federal or state governments. Even more important, perhaps, is the fact that those who are typically missed in the U.S. censuses, including especially African-Americans and Hispanics, are more likely to vote for Democratic candidates than for Republican candidates. That puts the issue of the differential undercount very much in the political spotlight as each census rolls around, and Census 2000 was as politically charged as any census ever.

After the 1990 census, the Census Bureau had produced a set of very detailed adjusted numbers, reflecting the Bureau's estimates of undercount by age and race/ethnicity for each county in the United States. However, the U.S. Department of Commerce (the bureaucracy within which the Census Bureau is housed—and at that time headed by a Republican in the George H. W. Bush administration) had refused to make the adjusted numbers "official." So the adjusted numbers were there, but were not used for apportionment, redistricting, or the distribution of governmental resources. Had the adjustments been made in 1990, the states of Wisconsin and Pennsylvania would each have lost a seat in Congress and California and Arizona would each have gained a seat. The decision not to adjust sparked a flurry of lawsuits, of course,

and in 1996 the U.S. Supreme Court finally ruled unanimously that the Census Bureau was under no legal obligation to adjust the final figures, despite the acknowledged undercount of urban minority group members (Barrett 1996).

You might well think that a unanimous Supreme Court decision would settle an issue, but you would be wrong in this case, because in the meantime the Census Bureau had decided to take the advice offered by the National Academy of Sciences (Edmonston and Schultze 1995; National Research Council 1993; White and Rust 1997) and announced that for Census 2000 it intended to use statistical sampling techniques to estimate and include those people who were otherwise missed by the census enumeration. This is a process known as sampling for nonresponse follow-up (SNRFU for you lovers of acronyms). The basic idea was that rather than having a postenumeration survey to estimate how many people were missed by the census, the survey process would become part of the census and thus those who were estimated to have been missed would be included in the final count instead of being lamented and fought over later. The traditional methods of enumeration would be used to obtain the first 90 percent of responses, but the last 10 percent (the really hard-to-find people) would be estimated from a large (750,000 households) sample survey. This idea sparked a new flurry of Congressional mandates and lawsuits.

On one side of the debate were those members of Congress who wanted no sampling (i.e., no adjustment) whatsoever and who introduced bills in Congress to forbid the Census Bureau from

across geographic areas (U.S. Census Bureau 1997). However, nonsampling error and the biases it introduces in terms of some people being more likely to be included than others are present throughout the census process and can reduce the quality of results more than sampling error (Schneider 2003).

Continuous Measurement—American Community Survey   Almost all the detailed data about population characteristics obtained from the decennial censuses in the United States come from the "long form," which is administered to only about one in six households (see Table 4.1). The success of survey sampling in obtaining spending any money even on planning for sampling (Duff 1997). Some of these members of Congress were, in fact, proposing that the decennial census be reduced to a few questions that could fit on a postcard. On the other side of the debate was the Census Bureau, arguing that its Congressional mandate to make Census 2000 more accurate and yet less costly than the 1990 Census could only be met by the judicious use of sampling (Riche 1997).

Was the uproar over sampling genuinely an issue about how best to conduct a census? No, the real issue was politics: Those people missed by the census in the United States are perceived to be more sympathetic to the Democratic political party. It had been estimated that including all those people who have typically been missed by the census would cause Democrats to gain two dozen seats in the House of Representatives, at the expense, of course, of the Republican party (Spar 1998). Members of Congress took the issue to the Supreme Court and in January 1999 the Court ruled that any population counts that had been "statistically adjusted" to compensate for overcounts and undercounts could not be used for Congressional apportionment; only the "raw" numbers could be used for that purpose. However, that left open the question of whether adjustments should be made to the numbers for purposes other than apportionment.

The Census Bureau responded by establishing an Executive Steering Committee for A.C.E. Policy (ESCAP) charged with deciding whether or not statistical adjustments should be made to the Census 2000 numbers. The Clinton administration delegated the responsibility for making that decision to the Director of the Census Bureau, but when the Bush administration came into power just before the decision was to be made in 2001, the decision was put in the hands of the politically appointed Secretary of Commerce. The nation was poised for a battle over whether to adjust or not to adjust.

The drama deflated, however, when the ESCAP concluded in March 2001 that its review of the demographic-analysis approach and the dual-system estimation approach (the A.C.E. survey) showed that they were sufficiently inconsistent in their results that the Census Bureau could not recommend making an adjustment to the numbers (Census 2000 Initiative 2001). The Census 2000 numbers considerably exceeded the estimates based on the demographic-analysis approach, whereas the A.C.E. indicated that the Census 2000 data were of very high quality. The Secretary of Commerce thus made the announcement that no adjustments were going to be made. The city of Los Angeles and several other cities and counties immediately filed suit to force an adjustment, but the U.S. District Court wasted little time in dismissing that suit.

That did not end the story, however. A lawsuit was filed to force the Census Bureau to release the adjusted census figures—the numbers derived from the A.C.E. survey—even if the official numbers were not going to be adjusted. In 2003, the U.S. Court of Appeals for the Ninth Circuit ordered the data released and they are now available from a Web site at the University of California, Los Angeles: http://www.sscnet.ucla.edu/issr/da/Adjusted/adjust_web.html.

reliable demographic data has led the U.S. Census Bureau to prepare a process of "continuous measurement" that is expected to eliminate the need for the long form in subsequent decennial censuses. The vehicle for this is the monthly American Community Survey, which the Census Bureau implemented on an experimental basis in 1996 and has been expanding ever since (Torrieri and Jennifer 2000). Ultimately, this will be a "rolling survey" of three million American households each year, and it is designed to collect enough data over a 10-year period to provide detailed information down to the census block level. Data from the ACS are accessible on the Internet at http://www.census.gov/acs/www/.

The Census of Canada

The first census in Canada was taken in 1666 when the French colony of New France was counted on the order of King Louis XIV. This turned out to be a door-to-door enumeration of all 3,215 settlers in Canada at that time. A series of wars between England and France ended with France ceding Canada to England in 1763, and the British undertook censuses on an irregular but fairly consistent basis (Statistics Canada 1995). The several regions of Canada were united under the British North America Act of 1867, and that Act specified that censuses were to be taken regularly to establish the number of representatives that each province would send to the House of Commons. The first of these was taken in 1871, although similar censuses had been taken in 1851 and 1861. In 1905, the census bureau became a permanent government agency, now known as Statistics Canada.

Canada began using sampling in 1941, the year after the United States experimented with it. In 1956, Canada conducted its first quinquennial census (every five years, as opposed to every 10 years—the decennial census), and in 1971 Canada mandated that the census be conducted every five years. The U.S. Congress passed similar legislation in the 1970s but never funded those efforts, so the United States has stayed with the decennial census. Of course, the continuous measurement provided by the American Community Survey will effectively supplement the decennial censuses in the United States.

Two census forms are used in Canada, as in the United States—a short form with just a few key items (see Table 4.1) and a more detailed long form. In 2001, the long form went to a sample of 20 percent of Canadian households. Public opinion influences census activities in Canada, as it does in the United States, and so the 1996 Census of Canada included a set of questions on unpaid household activities because of a Saskatoon housewife's protest. She refused to fill out the 1991 Census form (and risked going to jail as a result) because the census form's definition of work did not include household work or child care. This helped to galvanize public opinion to include a set of questions on this type of activity in the 1996 census (De Santis 1996), and those questions were asked again in the 2001 census. On the other hand, the Canadian government decided that the number of children born to a woman might be too private a question to be asked any longer (it had been asked on every decennial census since 1941), and it was not included in the 1996 or 2001 censuses.

Canada's population is even more diverse than that of the United States and so the census asks several questions about language—indeed, the split between English and French speakers nearly tore the country apart in the 1990s. Detailed questions are also asked about race/ethnicity, place of birth, citizenship, and ancestry.

Statistics Canada estimates coverage error by comparing census results with population estimates (the demographic analysis approach), and by conducting a *Reverse Record Check* study to measure the undercoverage errors and also an *Overcoverage Study* designed to investigate overcoverage errors. The Reverse Record Check is the most important part of this, and involves taking a sample of records from other sources such as birth records and immigration records and then looking for those people in the census returns. An analysis of people not found who should have been there is a key component of estimating coverage error. The results of the Reverse Record Check and the Overcoverage Study are then combined to provide an estimate of net undercoverage As of this writing, only the demographic analysis results were available and they suggested a net undercount in the 2001 census of about 1 percent.

The Census of Mexico

Like Canada and the United States, Mexico has a long history of census taking. There are records of a census in the Valley of Mexico taken in the year 1116, and the subsequent Aztec Empire also kept count of the population for tax purposes. Spain conducted several censuses in Mexico during the colonial years, including a general census of New Spain (Nueva España as they knew it) in 1790. Mexico gained independence from Spain in 1821, but it was not until 1895 that the first of the modern series of national censuses was undertaken. A second enumeration was done in 1900, but since then censuses have been taken every 10 years (with the exception of the one in 1921, which was one year out of sequence). From 1895 through the 1970s, the census activities were carried out by the General Directorate of Statistics (Dirección General de Estadística), and there were no permanent census employees. However, the bureaucracy was reorganized for the 1980 census, and in 1983 the Instituto Nacional de Estadística, Geografía e Informática (INEGI) became the permanent government agency in charge of the census and other government data collection.

Fewer questions were asked in the 2000 Mexican census than in the U.S. or Canada, as you can see in Table 4.1. The 2000 Census was the first in Mexico to use a combination of a basic questionnaire administered to most households, plus a lengthier questionnaire administered to a sample of households. Furthermore, the sampling strategy was a bit different than in the United States and Canada. Most of the questions were asked of most households, and the sample involved asking 2.2 million households (about 10 percent of the total) to respond to a set of more detailed questions about topics included in the basic questionnaire. Especially noteworthy was a set of questions seeking information about family members who had been international migrants at any time during the previous five years.

Less income detail is obtained in Mexico than in Canada or the United States, and socioeconomic categories are more often derived from outward manifestations of income, such as housing quality, and material possessions owned by members of the household, about which there are several detailed questions. Since most Mexicans are "mestizos" (Spanish for mixed race, in this case mainly European and Amerindian), no questions are asked about race or ethnicity. The only allusion to diversity within Mexico on the basic questionnaire is found in the question about language, in which people are asked if they speak an Indian language. If so, they are also asked if they speak Spanish. On the long form administered to a sample of households, a question is also asked specifically about whether or not they belong to an indigenous group.

In Mexico, the evaluation of coverage error in the census has generally been made using the method of demographic analysis. On this basis, Corona (Corona Vásquez 1991) estimated that underenumeration in the 1990 Mexican census was somewhere between 2.3 and 7.3 percent. No analysis has yet been published of the 2000 census's accuracy.

Registration of Vital Events

When you were born, a birth certificate was filled out for you, probably by a clerk in the hospital where you were born. When you die, someone (again, typically a hospital clerk) will fill out a death certificate on your behalf. Standard birth and death certificates used in the United States are shown in Figure 4.3. Births and deaths, as well as marriages, divorces, and abortions, are known as vital events, and when they are recorded by the government and compiled for use they become vital statistics. These statistics are the major source of data on births and deaths in most countries, and they are most useful when combined with census data.

Registration of vital events in Europe actually began as a chore of the church. Priests often recorded baptisms, marriages, and deaths, and historical demographers have used the surviving records to try to reconstruct the demographic history of parts of Europe (Wall, Robin, and Laslett 1983; Wrigley 1974; Wrigley and Schofield 1981). Among the more demographically important tasks that befell the clergy was that of recording burials that occurred in England during the many years of the plague. In the early sixteenth century, the city of London ordered that the number of people dying be recorded in each parish, along with the number of christenings. Beginning in 1592, these records (or "bills") were printed and circulated on a weekly basis during particularly rough years, and so they were called the London Bills of Mortality (Laxton 1987; Lorimer 1959). Between 1603 and 1849, these records were published weekly (on Thursdays, with an annual summary on the Thursday before Christmas) in what amounts to one of the most important sets of vital statistics prior to the nineteenth-century establishment of official government bureaucracies to collect and analyze such data.

Initially, the information about deaths indicated only the cause (since one goal was to keep track of the deadly plague), but starting in the eighteenth century, the age of those dying was also noted. Yet despite the interest in these data created by the analyses of Graunt, Petty, Halley, and others mentioned in Chapter 3, people remained skeptical about the quality of the data and unsure of what could be done with them (Starr 1987). So it was not until the middle of the nineteenth century that civil registration of births and deaths became compulsory and an office of vital statistics was officially established by the English government, mirroring events in much of Europe and North America (Linder 1959). In fact, it was not until 1900 that birth and death certificates were standardized in the United States.

Today we find the most complete vital registration systems in the most highly developed countries and the least complete (often nonexistent) systems in the least developed countries. Such systems seem to be tied to literacy (there must be someone in each area to record events), adequate communication, and the cost of the bureaucracy required for such record keeping, all of which is associated with economic development. Among countries where systems of vital registration do exist, there is wide variation in the completeness with which events are recorded. Even in the United States, the registration of births is not yet 100 percent complete.

Although most nations have separate systems of birth and death registration, dozens of countries maintain **population registers.** These are lists of all people in the country. Alongside each name are recorded all vital events for that individual, typically birth, death, marriage, divorce, and change of residence. Such registers are kept

Figure 4.3   Standard Birth and Death Certificates Used in the United States

Note: Although each state in the United States may design its own birth and death certificates, these are the standard forms suggested by the National Center for Health Statistics of the U.S. Centers for Disease Control and Prevention.

Figure 4.3 (continued)

U.S. STANDARD CERTIFICATE OF DEATH

LOCAL FILE NO.   STATE FILE NO.

1. DECEDENT S LEGAL NAME (Include AKA s if any) (First, Middle, Last)	2. SEX	3. SOCIAL SECURITY NUMBER

4a. AGE-Last Birthday (Years)	4b. UNDER 1 YEAR		4c. UNDER 1 DAY		5. DATE OF BIRTH (Mo/Day/Yr)	6. BIRTHPLACE (City and State or Foreign Country)
	Months	Days	Hours	Minutes		

7a. RESIDENCE-STATE	7b. COUNTY	7c. CITY OR TOWN

7d. STREET AND NUMBER	7e. APT. NO.	7f. ZIP CODE	7g. INSIDE CITY LIMITS? ☐ Yes ☐ No

8. EVER IN US ARMED FORCES? ☐ Yes ☐ No	9. MARITAL STATUS AT TIME OF DEATH ☐ Married ☐ Married, but separated ☐ Widowed ☐ Divorced ☐ Never Married ☐ Unknown	10. SURVIVING SPOUSE S NAME (If wife, give name prior to first marriage)

11. FATHER S NAME (First, Middle, Last)	12. MOTHER S NAME PRIOR TO FIRST MARRIAGE (First, Middle, Last)

13a. INFORMANT S NAME	13b. RELATIONSHIP TO DECEDENT	13c. MAILING ADDRESS (Street and Number, City, State, Zip Code)

14. PLACE OF DEATH (Check only one: see instructions)

IF DEATH OCCURRED IN A HOSPITAL: ☐ Inpatient ☐ Emergency Room/Outpatient ☐ Dead on Arrival	IF DEATH OCCURRED SOMEWHERE OTHER THAN A HOSPITAL: ☐ Hospice facility ☐ Nursing home/Long term care facility ☐ Decedent s home ☐ Other (Specify):

15. FACILITY NAME (If not institution, give street & number)	16. CITY OR TOWN, STATE, AND ZIP CODE	17. COUNTY OF DEATH

18. METHOD OF DISPOSITION: ☐ Burial ☐ Cremation ☐ Donation ☐ Entombment ☐ Removal from State ☐ Other (Specify):	19. PLACE OF DISPOSITION (Name of cemetery, crematory, other place)

20. LOCATION-CITY, TOWN, AND STATE	21. NAME AND COMPLETE ADDRESS OF FUNERAL FACILITY

22. SIGNATURE OF FUNERAL SERVICE LICENSEE OR OTHER AGENT	23. LICENSE NUMBER (Of Licensee)

ITEMS 24-28 MUST BE COMPLETED BY PERSON WHO PRONOUNCES OR CERTIFIES DEATH	24. DATE PRONOUNCED DEAD (Mo/Day/Yr)	25. TIME PRONOUNCED DEAD

26. SIGNATURE OF PERSON PRONOUNCING DEATH (Only when applicable)	27. LICENSE NUMBER	28. DATE SIGNED (Mo/Day/Yr)

29. ACTUAL OR PRESUMED DATE OF DEATH (Mo/Day/Yr) (Spell Month)	30. ACTUAL OR PRESUMED TIME OF DEATH	31. WAS MEDICAL EXAMINER OR CORONER CONTACTED? ☐ Yes ☐ No

CAUSE OF DEATH (See instructions and examples)

Approximate interval: Onset to death

32. PART I. Enter the chain of events—diseases, injuries, or complications—that directly caused the death. DO NOT enter terminal events such as cardiac arrest, respiratory arrest, or ventricular fibrillation without showing the etiology. DO NOT ABBREVIATE. Enter only one cause on a line. Add additional lines if necessary.

IMMEDIATE CAUSE (Final disease or condition ------> resulting in death) a._____
Due to (or as a consequence of):

Sequentially list conditions, if any, leading to the cause listed on line a. Enter the UNDERLYING CAUSE (disease or injury that initiated the events resulting in death) LAST

b._____
Due to (or as a consequence of):

c._____
Due to (or as a consequence of):

d._____

PART II. Enter other significant conditions contributing to death but not resulting in the underlying cause given in PART I.

33. WAS AN AUTOPSY PERFORMED? ☐ Yes ☐ No
34. WERE AUTOPSY FINDINGS AVAILABLE TO COMPLETE THE CAUSE OF DEATH? ☐ Yes ☐ No

35. DID TOBACCO USE CONTRIBUTE TO DEATH? ☐ Yes ☐ Probably ☐ No ☐ Unknown	36. IF FEMALE: ☐ Not pregnant within past year ☐ Pregnant at time of death ☐ Not pregnant, but pregnant within 42 days of death ☐ Not pregnant, but pregnant 43 days to 1 year before death ☐ Unknown if pregnant within the past year	37. MANNER OF DEATH ☐ Natural ☐ Homicide ☐ Accident ☐ Pending Investigation ☐ Suicide ☐ Could not be determined

38. DATE OF INJURY (Mo/Day/Yr) (Spell Month)	39. TIME OF INJURY	40. PLACE OF INJURY (e.g., Decedent s home; construction site; restaurant; wooded area)	41. INJURY AT WORK? ☐ Yes ☐ No

42. LOCATION OF INJURY: State: City or Town:

Street & Number: Apartment No.:   Zip Code:

43. DESCRIBE HOW INJURY OCCURRED:	44. IF TRANSPORTATION INJURY, SPECIFY: ☐ Driver/Operator ☐ Passenger ☐ Pedestrian ☐ Other (Specify)

45. CERTIFIER (Check only one):

☐ Certifying physician-To the best of my knowledge, death occurred due to the cause(s) and manner stated.
☐ Pronouncing & Certifying physician-To the best of my knowledge, death occurred at the time, date, and place, and due to the cause(s) and manner stated.
☐ Medical Examiner/Coroner-On the basis of examination, and/or investigation, in my opinion, death occurred at the time, date, and place, and due to the cause(s) and manner stated.

Signature of certifier:

46. NAME, ADDRESS, AND ZIP CODE OF PERSON COMPLETING CAUSE OF DEATH (Item 32)

47. TITLE OF CERTIFIER	48. LICENSE NUMBER	49. DATE CERTIFIED (Mo/Day/Yr)	50. **FOR REGISTRAR ONLY**- DATE FILED (Mo/Day/Yr)

51. DECEDENT S EDUCATION-Check the box that best describes the highest degree or level of school completed at the time of death. ☐ 8th grade or less ☐ 9th - 12th grade; no diploma ☐ High school graduate or GED completed ☐ Some college credit, but no degree ☐ Associate degree (e.g., AA, AS) ☐ Bachelor s degree (e.g., BA, AB, BS) ☐ Master s degree (e.g., MA, MS, MEng, MEd, MSW, MBA) ☐ Doctorate (e.g., PhD, EdD) or Professional degree (e.g., MD, DDS, DVM, LLB, JD)	52. DECEDENT OF HISPANIC ORIGIN? Check the box that best describes whether the decedent is Spanish/Hispanic/Latino. Check the No box if decedent is not Spanish/Hispanic/Latino. ☐ No, not Spanish/Hispanic/Latino ☐ Yes, Mexican, Mexican American, Chicano ☐ Yes, Puerto Rican ☐ Yes, Cuban ☐ Yes, other Spanish/Hispanic/Latino (Specify) _____	53. DECEDENT S RACE (Check one or more races to indicate what the decedent considered himself or herself to be) ☐ White ☐ Black or African American ☐ American Indian or Alaska Native (Name of the enrolled or principal tribe) _____ ☐ Asian Indian ☐ Chinese ☐ Filipino ☐ Japanese ☐ Korean ☐ Vietnamese ☐ Other Asian (Specify)_____ ☐ Native Hawaiian ☐ Guamanian or Chamorro ☐ Samoan ☐ Other Pacific Islander (Specify)_____ ☐ Other (Specify)_____

54. DECEDENT S USUAL OCCUPATION (Indicate type of work done during most of working life. DO NOT USE RETIRED).

55. KIND OF BUSINESS/INDUSTRY

REV. 11/2003

NAME OF DECEDENT — For use by physician or institution

To Be Completed/Verified By: FUNERAL DIRECTOR

To Be Completed By: MEDICAL CERTIFIER

To Be Completed By: FUNERAL DIRECTOR

primarily for administrative (that is, social control) purposes, such as legal identification of people, election rolls, and calls for military service, but they are also extremely valuable for demographic purposes, since they provide a demographic life history for each individual. However, population registers vary widely in quality. The most complete and demographically usable ones are probably those found in the Netherlands (where they now serve as a substitute for the census), Denmark, Norway, Sweden, Finland, and Japan. Registers are expensive to maintain, but many countries that could afford them, such as the United States, tend to avoid them because of the perceived threat to personal freedom that can be inherent in a system that compiles and centralizes data.

Combining the Census and Vital Statistics

Although recording vital events provides information about the number of births and deaths (along with other events) according to such characteristics as age and sex, we also need to know how many people are at risk of these events. Thus, vital statistics data are typically teamed up with census data, which do include that information. For example, you may know from the vital statistics that there were 4,035,000 births in the United States in 2003, but that figure tells you nothing about whether the birth rate was high or low. In order to draw any conclusion, you must relate those births to the 292,400,000 people residing in the U.S. as of mid-2003, and only then do you discover a low birth rate of 13.8 births per 1,000 population, down from 16.7 in 1990.

Since no census has been taken since 2000, you may wonder how an estimate of the population can be produced for an **intercensal** year such as 2003. The answer is that once again census data are combined with vital statistics data (and migration estimates) using the demographic balancing equation (which I discussed earlier in the chapter)—the population in 2003 is equal to the population as of the 2000 census plus the births minus the deaths plus the in-migrants minus the out-migrants between 2000 and 2003. Naturally, deficiencies in any of these data sources will lead to inaccuracies in the estimate of the number of people alive at any time, and this is exactly what happened in the United States in 2000. The Census counted 281,421,906 people as of April 1, 2000, but the demographic balancing equation had produced an estimate for that day of only 274,520,000. The difference was probably accounted for largely by undocumented immigration between 1990 and 2000.

Administrative Data

Knowing that censuses and the collection of vital statistics were not originally designed to provide data for demographic analysis has alerted demographers everywhere to keep their collective eyes open for any data source that might yield important information. For example, an important source of information about immigration to the United States is the compilation of **administrative records** filled out for each person entering the country from abroad. These forms are collected and tabulated by the U.S. Citizenship and Immigration Service (USCIS) within the U.S.

Department of Homeland Security, which used to be the Immigration and Naturalization Service (INS) within the U.S. Department of Justice. Of course, we need other means to estimate the number of people who enter without documents and avoid detection by the government, and I discuss that more in Chapter 8.

Data are not routinely gathered on people who permanently leave the United States, but the administrative records of the U.S. Social Security Administration provide some clues about the number and destination of such individuals. One source of data on migration within the United States used by the Census Bureau is a set of data provided to them by the Internal Revenue Service (IRS). Although no personal information is ever divulged, the IRS can match Social Security numbers of taxpayers each year and see if their address has changed, thus providing a clue about geographic mobility from that type of administrative record.

At the local level, a variety of administrative data can be tapped to determine demographic patterns. School enrollment data provide clues to patterns of population growth and migration. Utility data on connections and disconnections can also be used to discern local population trends, as can the number of people signing up for government-sponsored health programs (Medicaid and Medicare) and income assistance (various forms of welfare).

Sample Surveys

There are two major difficulties with using data collected in the census, by the vital statistics registration system, or derived from administrative records: (1) They are usually collected for purposes other than demographic analysis (this is true even for the censuses in most countries) and thus do not necessarily reflect the theoretical concerns of demography, and (2) They are collected by many different people using many different methods and may be prone to numerous kinds of error. For these two reasons, in addition to the cost of big data-collection schemes, sample surveys are frequently used to gather demographic data. Sample surveys may provide the social, psychological, economic, and even physical data I referred to earlier as being necessary to an understanding of why things are as they are. Their principal limitation is that they provide less extensive geographic coverage than a census or system of vital registration.

By using a carefully selected sample of even a few thousand people, demographers have been able to ask questions about births, deaths, migration, and other subjects that reveal aspects of the "why" of demographic events rather than just the "what." In some poor or remote areas of the world, sample surveys can also provide good estimates of the levels of fertility, mortality, and migration in the absence of census or vital registration data.

Demographic Surveys in the United States

I have already mentioned the American Community Survey (ACS), which is becoming an important adjunct of the census itself. For several decades now, one of the most important sample surveys has been the Current Population Survey (CPS) conducted monthly by the U.S. Census Bureau in collaboration with the Bureau of Labor Statistics. Since 1943, thousands of households (currently more than 50,000) have been queried each month about a variety of things, although a major thrust of the survey is to gather information on the labor force. Each March, detailed demographic questions are also asked about fertility and migration and such characteristics as education, income, marital status, and living arrangements. These data have been a major source of demographic information about the American population, filling in the gap between censuses.

Since 1983, the Census Bureau has also been conducting the Survey on Income and Program Participation (SIPP), which is a companion to the Current Population Survey. Using a rotating panel of about 20,000 households, the SIPP gathers detailed data on sources of income and wealth, disability, and the extent to which household members participate in government assistance programs. The Census Bureau also regularly conducts the American Housing Survey for the U.S. Department of Housing and Urban Development, and this survey generates important data on mobility and migration patterns in the United States. The National Center for Health Statistics (NCHS) within the U.S. Centers for Disease Control and Prevention generates data about fertility and reproductive health in the National Survey of Family Growth (NSFG), which it conducts every five years or so, and data on health and disability from the regular National Health Interview Survey (NHIS).

Canadian Surveys

Canada has a monthly Labor Force Survey (LFS), initiated in 1947 to track employment trends after the end of World War II. It, too, is a rotating panel of more than 50,000 households, and although its major purpose is to produce data on the labor force (hence the name), it gathers data on most of the core sociodemographic characteristics of people in each sampled household, so it provides a continuous measure of population trends in Canada, just as the CPS and the ACS do in the United States. Since 1985, Statistics Canada has also conducted an annual General Social Survey, a sample of about 25,000 respondents designed to elicit detailed data about various aspects of life in Canada, such as health and social support, families, time use, and related topics.

Mexican Surveys

Mexico conducts a survey three times a year that is comparable in many ways to the CPS and the LFS. The National Survey of Employment (Encuesta Nacional de Empleo [ENE]) is a large national sample of households undertaken by INEGI that is designed to be representative of the entire country. As with the CPS, the goal is to provide a way of continuously measuring and monitoring the sociodemographic and economic characteristics of the population. Many of the population questions asked in the census (see Table 4.1) are also asked in the ENE, along with a detailed set of questions about the labor force activity of everyone in the household who is 12 years of age or older.

Demographic and Health Surveys

One of the largest social scientific projects ever undertaken was the World Fertility Survey, conducted under the auspices of the International Statistical Institute in the Netherlands. Between 1972 and 1982, a total of nearly 350,000 women of childbearing age were interviewed, encompassing 42 developing nations and 20 developed countries (Lightbourne, Singh, and Green 1982). These data contributed substantially to our knowledge of how and why people in different parts of the world were (or were not) controlling their fertility at that time. Concurrent with the World Fertility Survey was a series of Contraceptive Prevalence Surveys, conducted in Latin America, Asia, and Africa with funding from the U.S. Agency for International Development (USAID).

In 1984, the work of the World Fertility Survey and the Contraceptive Prevalence Surveys was combined into one large project called the Demographic and Health Surveys, which are now carried out by a private company, ORC Macro, with funding from USAID. The focus is on fertility, reproductive health, and child health and nutrition, but the data provide national estimates of basic demographic processes, structure, and characteristics since a few questions are asked about all members of each household in the sample. More than 170 surveys have been conducted in more than 70 developing countries in Africa, western Asia, and Latin America. This is a rich source of information, as you will see in subsequent chapters.

Demographic Surveillance Systems

In Africa, many people are born, live, and die without a single written record of their existence because of the poor coverage of censuses and vital registration systems (de Savigny 2003). To provide a way of tracking the lives of people in specific "sentinel" areas of sub-Saharan Africa (and to a lesser extent south Asia) an INDEPTH network has been created that works with individual countries to select one or two defined geographic regions that are representative of a larger population. A census is conducted in that region and then subsequent demographic changes are continuously measured by keeping track of all births, deaths, migration, and related characteristics of the population. Currently, this permits the estimation of fertility and death rates and their changes over time for 31 sites in 17 different countries of Africa and Asia (Korenromp et al. 2003).

European Surveys

Declining fertility in Europe has generated a renewed interest in the continent's demography and there are now several surveys in Europe that capture useful demographic information. In particular, the Population Activities Unit of the United Nations Economic Commission for Europe funded the Family and Fertility Surveys (FFS) in most European nations during the 1990s. Since 2000, they have funded a new round of surveys called "Generations and Gender: Research into their Behaviour and Quality of Life," which has a slightly different focus, but incorporates many of the questions asked in the FFS.

Historical Sources

The flip side of the rather detailed gathering of demographic data outlined in this chapter is that our modern understanding of population processes is shaped not only by our perception of current trends but also by our understanding of historical events. Historical demography requires that we almost literally dig up information about the patterns of mortality, fertility, and migration in past generations—to reconstruct "the world we have lost," as Peter Laslett (1971) has called it. You may prefer to whistle past the graveyard, but researchers at the Cambridge Group for the History of Population and Social Structure in the Department of Geography at Cambridge University (U.K.) have spent the past several decades developing ways to recreate history by reading dates on tombstones and organizing information contained in parish church registers and other local documents (Reher and Schofield 1993; Wrigley and Schofield 1981), extending methods developed especially by the great French historical demographer Louis Henry (Rosental 2003).

Historical sources of demographic information include censuses and vital statistics, but the general lack of good historical vital statistics is what typically necessitates special detective work to locate birth records in church registers and death records in graveyards. Even in the absence of a census, a complete set of good local records for a small village may allow a researcher to reconstruct the demographic profile of families by matching entries of births, marriages, and deaths in the community over a period of several years. Yet another source of such information is family genealogies, the compilation of which has become increasingly common in recent years.

The results of these labors can be of considerable importance in testing our notions about how the world used to work. For example, through historical demographic research we now know that the conjugal family (parents and their children) is not a product of industrialization and urbanization, as was once thought (Wrigley 1974). In fact, such small family units were quite common throughout Europe for several centuries before the Industrial Revolution and may actually have contributed to the process of industrialization by allowing the family more flexibility to meet the needs of the changing economy. In the United States, however, extended families may have been more common prior to the nineteenth century than has generally been thought (Ruggles 1994b). Conclusions such as this come from the Integrated Public Use Microdata Series (IPUMS) of the Minnesota Population Center at the University of Minnesota. Steven Ruggles and his associates have built an innovative (and accessible) database by selecting samples of household data from a long series of censuses (1850 to 2000) in the United States, and they are in the process of collecting similar census data sets from elsewhere in the world.

The African Census Analysis Project has also been organized at the University of Pennsylvania to archive and analyze microlevel census data from African nations. In subsequent chapters, we will also have numerous occasions to draw on the results of the Princeton European Fertility Project, which gathered and analyzed data on marriage and reproduction throughout nineteenth- and early-twentieth-century Europe.

By quantifying (and thereby clarifying) our knowledge of past patterns of demographic events, we are also better able to interpret historical events in a meaningful fashion. Wells (1985) has reminded us that the history of the struggle of American colonists to survive, marry, and bear children may tell us more about the determination to forge a union of states than a detailed recounting of the actions of British officials.

Demographic Uses of Geographic Information Systems

Demographers have been using maps as a tool for analysis for a long time, and some of the earliest analyses of disease and death relied heavily on maps that showed, for example, where people were dying from particular causes (Cliff and Haggett 1996). In the middle of the nineteenth century, London physician John Snow used maps to trace a local cholera epidemic. He was able to show that cholera occurred much more frequently among customers of a water company that drew its water from the lower Thames River (downstream), where it had become contaminated with London sewage, whereas another company was associated with far fewer cases of cholera because it obtained water from the upper Thames—prior to passing through London, where sewage was dumped in the river (Snow 1936). Today a far more sophisticated version of this same idea is available to demographers through **geographic information systems** (GIS). The advent of powerful desktop computers has created a "GIS revolution" (Longley and Batty 1996) that encourages demographers (and others, of course) to bring maps together with data in innovative ways (Ricketts et al. 1997; Rindfuss and Stern 1998; Weeks 2003).

A GIS is a computer-based system that allows us to combine maps with data that refer to particular places on those maps and then to analyze those data and display the results as thematic maps or some other graphic format. The computer allows us to transform a map into a set of areas (such as a country, state, or census tract), lines (such as streets, highways, or rivers), and points (such as a house, school, or a health clinic). Our demographic data must then be **geo-referenced** (associated with some geographic identification such as precise latitude-longitude coordinates, a street address, ZIP code, census tract, county, state, or country) so the computer will link them to the correct area, line, or point. If the computer knows that a particular set of latitude and longitude coordinates represents the map of Mexico, then our data for Mexico must be "referenced" to that particular location. Or, if we have a survey of households, then each variable for the household would be geo-referenced to the specific address (the point) of that household or, more probably, to the general neighborhood where the household is located (in order to preserve privacy and confidentiality). Demographic data are virtually always geographically referenced (and have been in the United States since 1850), and the Geography Division of the U.S. Census Bureau works closely with the Population Division to make sure that data are identified for appropriate levels of "census geography," as shown in Table 4.3.

The "revolutionary" aspect of GIS is that geo-referencing data to places on the map means we can combine different types of data (such as census and survey data) for the same place, and we can do it for more than one time (such as data for 1990 and 2000). This greatly increases our ability to visualize and analyze the kinds of demographic changes taking place over time and space. Since 1997, most of the Demographic and Health Surveys in less-developed countries have used global positioning

Table 4.3   The U.S. Census Provides Geographically Referenced Data for a Wide Range of Geographic Areas

---

Basic Geographic Hierarchy of Census Data:

United States
  Region (4)
    Division (9)
      State (50, plus the outlying areas of American Samoa, Guam, northern Mariana Islands, Palau, Puerto Rico, and the Virgin Islands)
       County (the basic administrative and legal subdivision of states)
        County Subdivision (also known as minor civil divisions, including towns and townships)
         Place (incorporated cities and unincorporated census-designated places)
         Census Tract/Block Numbering Area (tracts are small, relatively permanent subdivisions of a county, delineated for all metropolitan areas and other densely settled areas; block numbering areas are similar to census tracts except that they are set up for nonmetropolitan areas that have not been tracted)
          Block Group (a cluster of blocks within census tracts, usually containing about 400 housing units)
           Census Block (the smallest geographic unit, usually bounded on all sides by readily identifiable features such as streets, railroad tracks, or bodies of water)

Source: U.S. Census Bureau, 2000.

system (GPS) devices to record the location of each household in the sample in order to allow for more sophisticated spatial analysis of the survey data (ORC Macro 2000).

GIS and the U.S. Census Bureau

The computer—the revolutionary element that has created GIS (the maps and the demographic data have been around for a long time)—has vastly expanded our capacity to process and analyze data. It is no coincidence that census data are so readily amenable to being "crunched" by the computer; the histories of the computer and the U.S. Census Bureau go back a long way together. Prior to the 1890 census, the U.S. government held a contest to see who could come up with the best machine for counting the data from that census. The winner was Herman Hollerith, who had worked on the 1880 census right after graduating from Columbia University. His method of feeding a punched card through a tabulating machine proved to be very successful, and in 1886 he organized the Tabulating Machine Company, which in 1911 was merged with two other companies and became the International Business Machines (IBM) Corporation (Kaplan and Van Valey 1980). Then, after World War II, the Census Bureau sponsored the development of the first computer designed for mass data processing—the UNIVAC-I—which was used to help with the 1950 census and led the world into the computer age. Photo-optical scanning, which we now heavily rely on for entering data from printed documents into the computer, was also a by-product of the Census Bureau's need for a device to tabulate data from census forms. FOSIDC (film optical sensing device for input to computers) was first used for the 1960 census (U.S. Census Bureau 1999).

Another useful innovation was the creation for the 1980 U.S. Census of the DIME (Dual Independent Map Encoding) files. This was the first step toward computer mapping, in which each piece of data could be coded in a way that could be matched electronically to a place on a map. In the 1980s, several private firms latched onto this technology, improved it, and made it available to other companies for their own business uses.

By the early 1990s, the pieces of the puzzle had come together. The data from the 1990 U.S. Census were made available for the first time on CD-ROM and at prices affordable to a wide range of users. Furthermore, the Census Bureau reconfigured its geographic coding of data, creating what it calls the TIGER (Topologically Integrated Geographic Encoding and Referencing) system, a "digital (computer-readable) geographic database that automates the mapping and related geographic activities required to support the Census Bureau's census and survey programs" (U.S. Census Bureau 1993:A11). At the same time, and certainly in response to increased demand, personal computers came along that were powerful enough and had enough memory to be able to store and manipulate huge census files, including both the geographic database and the actual population and housing data. Not far behind was the software to run those computers, and a growing number of firms now make software for desktop computers that allow interactive spatial analysis of census and other kinds of data and then the production of high-quality color maps of the analysis results. Two of these firms—Environmental Systems Research Institute (ESRI—http://www.esri.com) and Geographic Data Technology (GDT—http://www.geographic.com)—were awarded contracts in 1997 to help the U.S. Census Bureau update its computerized Master Address File (the information used to continuously update the TIGER files) in order to improve accuracy in Census 2000. In a very real sense, the census and the TIGER files more specifically helped to spawn the now-booming GIS industry.

Knowledge and understanding are based on information, and our information base grows by being able to tap more deeply into rich data sources such as censuses and surveys. GIS is an effective tool for doing this, and you will see numerous examples of GIS at work in the remaining chapters. You can also see it at work on the Internet. Virtually all of the data from Census 2000 are being made available on the U.S. Census Bureau's Web site at http://factfinder.census.gov and the Web site allows you to create thematic maps "on the fly."

Where Can You Go for Information?

The power of the Internet and the World Wide Web have greatly expanded the scope of demographic information available to people all over the world. I could be camped out in a remote valley in the Andes Mountains in Peru, but if I have my laptop computer with a wireless connection to a cellular service provider, I can log onto the Internet and discover the latest demographic information for almost any country in the world, using some of the Web sites listed at the end of each chapter in this book. Nonetheless, there is still a lot of population information, especially the details and historical data, that is available only on paper, although that may well change in your lifetime. The U.S. Census Bureau is rapidly moving toward the day when virtually all of its data will be accessed through the Internet, rather than through traditional printing on paper. Statistics Canada (www.statcan.ca) and INEGI (http://www.inegi.gob.mx) also have much of their data on the Internet.

A widely available printed source of international data is the *Demographic Yearbook,* published by the United Nations every year since 1948, and the United Nations has many Internet resources available at http://www.un.org/popin. The Population Reference Bureau regularly publishes information on the world's population in its *Population Bulletin,* and it also publishes an annual *World Population Data Sheet* that is a handy reference for up-to-date estimates of basic demographic facts and can be accessed at http://www.prb.org. Many countries publish summaries of demographic, social, and economic data, and these are often available at libraries and increasingly on the Internet—a Google search on the Internet should track them down for you.

Many periodicals concentrate on issues specific to population studies, including *Demography, Population Studies, Population and Development Review, Population Research and Policy Review, Studies in Family Planning, Family Planning Perspectives, International Family Planning Perspectives, Population and Environment, Genus, European Journal of Population (Revue européenne de démographie), Espaces Populations et Sociétés, Population* (also French, but they publish an English edition), *Mathematical Population Studies, Journal of Population Economics, Social Biology,* and *International Migration Review.* The journal *Demographic Research* is available only on the Internet at http://www.demographic-research.org/. Many important articles dealing with demographic issues are published in nondemographic journals, such as the *American Sociological Review, American Journal of Sociology, Social Forces, Annals of the Association of American Geographers, Professional - Geographer, Science, Nature,* and *Environment and Planning A.* Although all of the above journals publish articles that may refer to any geographic area in the world, demographic studies of Canada may also be found in *Canadian Studies in Population* and *Cahiers Québécois de Démographie,* and research about Mexico is often published in *Estudios Demográficos y Urbanos.* Finally, at the more popular level is *American Demographics,* which is explicitly aimed at interpreting population statistics for nonacademic audiences. You should be able to find it on your local bookseller's magazine rack, as well as on the Internet at http://www.american demographics.com.

Summary and Conclusion

The working bases of any science are facts and theory. In this chapter, I have discussed the major sources of demographic information, the wells from which population data are drawn. Censuses are the most widely known and used sources of data on populations, and humans have been counting themselves in this way for a long time. However, the modern series of more scientific censuses dates only from the late eighteenth and early nineteenth centuries. The high cost of censuses, combined with the increasing knowledge we have about the value of surveys, has meant that even so-called complete enumerations often include some kind of sampling. That is certainly true in North America, as the United States, Canada, and Mexico all use sampling techniques in their censuses. Even vital statistics can be estimated using sample surveys, especially in developing countries, although the usual pattern is for births and deaths (and often marriages, divorces, and abortions) to be registered with the civil authorities. Some countries take this a step farther and maintain a complete register of life events for everybody.

Knowledge can also be gleaned from administrative data gathered for nondemographic purposes. These are particularly important in helping us measure migration. It is not just the present that we attempt to measure; historical sources of information can add much to our understanding of current trends in population growth and change. Our understanding is increasingly enhanced by incorporating our demographic data into a geographic information system, permitting us to ask questions that were not really answerable before the advent of the computer.

In this and the preceding three chapters, I have laid out for you the basic elements of a demographic perspective. With this in hand (and hopefully in your head as well), we are now ready to probe more deeply into the analysis of population processes, to come to an appreciation of how important the decline in the death rate is, yet why it is still so much higher in some places than in others, why birth rates are high in some places and low in others, and why some people move and others do not.

Main Points

1. In order to study population processes and change, you need to know how many people are alive, how many are being born, how many are dying, how many are moving in and out, and why these things are happening.

2. A basic source of demographic information is the population census, in which information is obtained about all people in a given area at a specific time.

3. Not all countries regularly conduct censuses, but between 1995 and 2004 it is expected that 89 percent of the world's population will have been enumerated in a census.

4. In the United States, censuses have been taken every 10 years since 1790.

5. In Canada, censuses have been taken every 10 years since 1851 and every five years since 1951.

6. In Mexico, censuses have been taken every 10 years since 1900, and every five years since 1990.

7. Errors in the census typically come about as a result of nonsampling errors (the most important source of error, including coverage error and content error) or sampling errors.

8. Coverage error is typically measured by demographic analysis and/or a dual system of estimation based on a postcensus survey or record check.

9. It has been said that censuses are important because if you aren't counted, you don't count.

10. Information about births and deaths usually comes from vital registration records—data recorded and compiled by government agencies. The most complete vital registration systems are found in the most highly developed nations, while they are often nonexistent in less developed areas.

11. Most of the estimates of the magnitude of population growth and change are derived by combining census data with vital registration data (as well as administrative data), using the demographic balancing equation.

12. Sample surveys are sources of information for places in which census or vital registration data do not exist or where reliable information can be obtained less expensively by sampling than by conducting a census.

13. Parish records and old census data are important sources of historical information about population changes in the past.

14. Geographic information systems combine maps and demographic data into a computer system for innovative types of analysis.

Suggested Readings

1. David A. Swanson and Jacob S. Siegel, 2003, Methods and Materials of Demography, Second Edition (San Diego: Academic Press).

   The first edition of this book, by Henry Shryock and Jacob Siegel, was published in 1976 and immediately became a classic in the field of population studies. It has now been revised and updated to reflect the important changes occurring over the past 25 years in the sources of data and techniques for analyzing them.

2. Margo J. Anderson and Stephen E. Fienberg, 1999, Who Counts? The Politics of Census-Taking in Contemporary America (New York: Russell Sage Foundation).

   The authors are both veterans of discussions over the years about how to ensure the most accurate possible census, and in this book they describe the increasing politicization of that issue in the United States.

3. Dominique Arel, 2002, "Demography and Politics in the First Post-Soviet Censuses: Mistrusted State, Contested Identities," Population (English Edition) 57(6):801–828.

   The breaking up of the Soviet Union meant the restructuring of the census in the new Russian Federation, and this process has offered important insights into the social and political issues that lie behind censuses everywhere in the world.

4. Monica Boyd, Gustave Goldmann, and Pamela White, 2000, "Race in the Canadian Census," in Leo Driedger and Shiva S. Halli, editors, Race and Racism: Canada's Challenge (Montreal: McGill/Queen's University Press).

   Canada has been a country marked by demographic diversity from its earliest national roots, and these authors trace the way in which the census has reflected the country's angst over how best to measure the ethnic and cultural diversity of the population.

5. Christine E. Bose, 2001, Women in 1900: Gateway to the Political Economy of the 20th Century (Philadelphia: Temple University Press).

   This is a fascinating example of the social scientist as detective using historical census data to allow women to "speak" to us from the past.

✇ Web Sites of Interest

Remember that Web sites are not as permanent as books and journals, so I cannot guarantee that each of the following Web sites still exists at the moment you are reading this. It is usually possible, however, to retrieve the contents of old Web sites at http://www.archive.org.

1. **http://www.census.gov**
   The home page of the U.S. Census Bureau. From here you can locate an amazing variety of information, including the latest releases of data from Census 2000 and all of the surveys conducted by the Census Bureau. This is one of the most accessed Web sites in the world.

2. **http://www.statcan.ca**
   The home page of Statistics Canada, which conducts the censuses in Canada. From here you can obtain data from the 1996 and 2001 Censuses and track other demographically related information about Canada, including vital statistics and survey data.

3. **http://inegi.gob.mx**
   The home page of INEGI (Instituto Nacional de Estadística, Geografía, y Informática), which is the government agency in Mexico that conducts the censuses and related demographic surveys, as well as compiling the vital statistics for Mexico. You can obtain the latest information from the 2000 Mexican census from this site.

4. **http://www.cdc.gov/nchswww/nchshome.htm**
   In many countries (including Canada and Mexico), a central statistical agency conducts censuses and also collects vital statistics. Not so in the United States, where these functions were divided up in the 1940s. The vital statistics data are collected from each state, tabulated, analyzed, and disseminated by the National Center for Health Statistics (NCHS), which is part of the U.S. Centers for Disease Control and Prevention (CDC).

5. **http://www.library.uu.nl/wesp/populstat/populhome.html**
   Jan Lahmeyer of the University of Utrecht in the Netherlands has compiled census and other population statistics from all over the world, and this Web site provides data on the "growth of the population per country in a historical perspective, including the administrative divisions and principal towns." It is a very useful research resource.

PART TWO
Population Processes

CHAPTER 5
The Mortality Transition

Figure 5.1 Global Variability in Maternal Mortality Rates

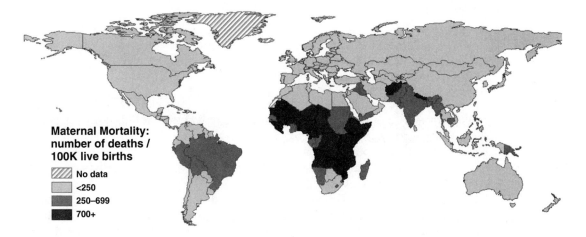

Maternal Mortality: number of deaths / 100K live births

- No data
- <250
- 250–699
- 700+

Source: Adapted by the author from data in UN Statistical Division. *The World's Women 2000*. Trends and Statistics.

149

It isn't that people now breed like rabbits, rather that we no longer die like flies—declining mortality, not rising fertility, is the root cause of the revolutionary increase in the world's population size and growth. Only within the last century has mortality been brought under control to the point that most of us are now able to take a long life pretty much for granted. Human triumph over disease and early death surely represents one of the most significant improvements ever made in the condition of human life, and we can rightly be proud of ourselves. Nevertheless, one by-product of our triumph is current world growth and the problems that have grown along with population size. Furthermore, the problems will continue to grow, for although mortality is increasingly under control, there are still wide variations in life expectancy in different parts of the world, as well as among different groups of people within countries. These differences represent a potential reservoir of population growth, since further declines in death rates will trigger even higher rates of growth unless we are able to curtail fertility at the same rate.

It was once widely believed that differences in death rates were genetic or biological in nature and thus difficult to change, but we now know that most variations are due to social, not biological, causes. I will pursue this point by beginning the chapter with a discussion of the differences between the biological and social components of mortality and then show how these two come together in the nearly universal patterns of death by age and sex. The pattern of death by age is central to the mortality transition and important, therefore, to the measurement of mortality. After reviewing the causes of death, I turn to an overview of the ways in which mortality is measured. In the remainder of the chapter, we will explore the mortality transition (also known as the epidemiological transition), including changes over time, differences between countries, and differences within societies.

Life Span and Longevity

Mortality is an inherently biological phenomenon. One aspect of this is **life span,** which refers to the oldest age to which human beings can survive. The second aspect, **longevity,** refers to the ability to remain alive from one year to the next—the ability to resist death. Life span seems to be almost entirely a biological phenomenon, whereas longevity has both biological and social components.

Life Span

What is the human life span? We do not yet have a good theory about aging to help us to predict how long humans *could* live, so we must be content to assume that the oldest age to which a human actually *has* lived (a figure that may change from day to day) is the oldest age to which it is possible to live. Claims of long human life span are widespread, but confirmation of those claims is more difficult to find. The oldest authenticated age to which a human has ever lived is 122 years and 164 days, an age achieved by a French woman, Jeanne Louise Calment, who died in August 1997. Her authenticated birth date was February 21, 1875, and on her 120th birthday in 1995, she was asked what kind of future she expected. "A very short one," she replied (Wallis 1995:85).

You have probably heard reports of people who claim to have lived longer than that, such as María do Carmo Jeronimo, who was reportedly 129 years old when she died in Brazil in 2000. But she had been born into slavery in Brazil and could not prove her age because no birth certificate existed for her—a common situation for people born prior to the twentieth century. A well-studied case is that of another ex-slave named Charlie Smith who claimed to have been born in Liberia in 1842 and brought to the United States in 1854, and who died at a reported age of 137. In the United States, the usual way to check on the reported age of an older person for whom no birth certificate exists is to look back through the census returns for information about those individuals when they were younger (census returns are made public 72 years after a census is taken and are available on microfilm at the National Archives in Washington, D.C., or at its 13 regional locations—http://www.archives. gov). Unfortunately, Charlie Smith is such a common name, and he had moved so often, that he was impossible to trace (Meyers 1978). However, a marriage record in Florida that appears to refer to him has led investigators to believe that his age was exaggerated by at least 33 years, and it is possible that he wasn't even quite 100 years old when he died (McWhirter 1983).

Age exaggeration is often suspected among very old people. In a piece of fascinating detective work, Meyers (1978) was able to dig through old census data and show that a Pennsylvania man who died at a reported age of 112 in 1866 had, in fact, "aged" 27 years in the 10-year period between the 1850 and 1860 censuses and had "aged" eight years in the six-year period from the 1860 census to his death. Clearly, there was some exaggeration going on. Other evidence of age exaggeration among American centenarians has been discovered by Rosenwaike (1979), who also found significant discrepancies between ages recorded on death certificates and ages recorded in the census. In the year 2000 in the United States, 85 people died whose age was recorded as 110 or older. Of those, 37 (44 percent) were black, and research suggests that age exaggeration is more common among African-Americans than among whites (Hill, Preston, and Rosenwaike 2000), but it is not a problem peculiar to the United States, since examples of age exaggeration have been found in Denmark (Jeune and Skytthe 2001), and almost certainly exist in every society.

So we know that humans can live to age 122 and maybe even beyond, yet very few people come close to achieving that age. Most, in fact, can expect to live scarcely more than half that long (life expectancy at birth for the world as a whole is estimated to be about 67 years). It is this concept, the age to which people *actually* survive, their demonstrated ability to stay alive as opposed to their potential, that we refer to as longevity.

Longevity

Biological Factors   Longevity is usually measured by **life expectancy,** the statistically average length of life (or average expected age at death), and is greatly influenced by the society in which we live (as I discuss later in connection with the mortality/epidemiological transition) and by the genetic characteristics with which we are born. The strength of vital organs, predisposition to particular diseases, metabolism rate, and so on, are biological factors over which we presently have little control. Many of the most severe biological weaknesses, though, tend to display themselves rather soon after birth, and, as a result, mortality is considerably higher in the first year or so of life than it is in the remainder of childhood and early adulthood. This is true for all animal populations, not just humans. Furthermore, as I discuss below, there is a nearly universal pattern in all animal populations that death rates are highest at the youngest and oldest ages. There is bound to be a strong biological component to this pattern. And, among humans it is also true that females tend to have higher probabilities of survival at every age than do males, and this also seems to have a strong biological component.

There is little doubt that part of our ability to stay alive is influenced by our genetic inheritance. Studies of identical twins separated at birth and raised in different environments show that their average age at death is more similar than nonidentical twins, and that both groups of twins have life expectancies that are more similar than you would expect by chance (Herskind et al. 1996). Nonetheless, the available evidence suggests that no more than 35 percent of the variability in longevity is due to inherited characteristics (Carey and Judge 2001). The remaining differences in mortality are due to social (including economic) factors that influence when and why death occurs.

Social Factors   The social world influences the risk of death in a variety of ways that can be reasonably reduced to two broad categories: (1) the social and economic infrastructure (how much control we exercise over nature) and (2) lifestyle (how much control we exercise over ourselves). The infrastructure of society refers to the way in which wealth is generated and distributed, reflecting the extent to which water and milk are purified, diseases are vaccinated against, rodents and other pests controlled, waste eliminated, and food, shelter, clothing, and acute care and long-term care medical assistance are made available to members of society. Most of the rest of this chapter will examine this theme in detail. Within any particular social setting, however, death rates may also be influenced by lifestyle. An increasing body of evidence has implicated smoking, drug use, excessive alcohol use, fatty food, and too little exercise as lifestyle factors that may shorten longevity.

Although one key to a long life may be your "choice" of long-lived parents, prescriptions for a long life are most often a brew of lifestyle choices. A typical list of ways to maximize longevity includes regular exercise, daily breakfast, normal weight, no smoking, only moderate drinking, seven to eight hours of sleep daily, regular meal-taking, and an optimistic outlook on life. These suggestions, by the way, are not unique to the Western world, nor are they particularly modern. A group of medical workers studying older people in southern China concluded that the important factors for long life are fresh air, moderate drinking and eating, regular exercise, and an optimistic attitude (Associated Press 1980). Similarly, note the words of a Dr. Weber, who was 83 in 1904 when he published an article in the *British Medical Journal* outlining his prescriptions for a long life:

> Be moderate in food and drink and in all physical pleasures; take exercise daily, regardless of the weather; go to bed early, rise early, sleep for no more than 6-7 hours; bathe daily; work and occupy yourself mentally on a regular basis—stimulate the enjoyment of life so that the mind may be tranquil and full of hope; control the passions; be resolute about preserving health; and avoid alcohol, narcotics, and soothing drugs (quoted in Metchnikoff 1908).

Other examples of the way in which social and psychological processes can apparently influence death have been given by David Phillips. In the first of a series of studies, Phillips (1974) found that mortality from suicide tends to increase right after a famous person commits a well-publicized suicide (see also Stack 1987; Wasserman 1984). Thus, some people seem to "follow the leader" when it comes to dying. Further, people do so in more invidious ways than just simple suicide. In follow-up studies, Phillips has found that the number of fatal automobile crashes (especially single-car, single-person crashes) goes up after publicized suicides (Phillips 1977) and (incredibly enough) that private airplane accident fatalities also increase just after newspaper stories about a murder-suicide. It appears "that murder-suicide stories trigger subsequent murder-suicides, some of which are disguised as airplane accidents" (Phillips 1978:748). In another study, Phillips demonstrated that mass-media violence can also trigger homicides. He discovered that in the United States between 1973 and 1978, homicides regularly increased right after championship prize fights. Furthermore, the more heavily publicized the fight, the greater the rise in homicides (Phillips 1983).

The mind is a wondrous thing, illustrated by the intriguing finding that Chinese-Americans who are born in a year that Chinese astrology considers ill-fated and who have a disease that Chinese medicine considers to be ill-fated, have significantly lower life expectancy than normal (Phillips 1993). It may be, of course, that Chinese astrology and medicine are correct about the fates, but Phillips thinks it is more likely that people succumb to an earlier death because of psychosomatic processes— their *belief* that they are fated to die. Even your name might affect your longevity. Christenfeld, Phillips and Glynn (1999) examined 27 years of death certificates in California and concluded that males with "positive" initials such as A.C.E. or W.I.N. or G.O.D. lived 4.5 years longer on average than a control group with neutral or ambiguous initials (such as J.R.W. or D.J.H.). Conversely, men with "negative" initials just as P.I.G. or D.U.D. or B.U.M. died an average of 2.8 years earlier than the control group. Effects were less strong for females, though still present, and the findings could not be explained by race, socioeconomic status, or any other variables available on the death certificate. There is no clear explanation for this except that somehow the impact of your initials on health must be psychosocial in nature, although as one critic put it, we have to remember that this is only an "initial" study.

It is easier, of course, to die than to resist death, adding interest to another angle of Phillips's research. He has found that there is a tendency for people who are near death to postpone dying until after a special event, especially a birthday. The story is often told that Thomas Jefferson lingered on his deathbed late on the evening of July 3, 1826, until his physician assured him that it was past midnight and was now the fourth of July, whereupon Jefferson died. Phillips and Smith (1990) found in two large samples of nearly three million people that women especially are indeed more likely to die in the week right after their birthday than in any other week of the year, suggesting the deliberate prolongation of life. Men, on the other hand, show a peak mortality just before their birthday, suggesting a "deadline" for death.

Age and Sex Patterns of Mortality

Age Differentials in Mortality

Humans are like most other animals with respect to the general pattern of death by age—the very young and the old are most vulnerable, whereas young adults are least likely to die. In Figure 5.2, you can see that the pattern of death by age is similar whether the actual death rates are high or low. After the initial year of life, there is a period of time, usually lasting at least until middle age, when risks of death are relatively low. Beyond middle age, mortality increases, although at a decelerating rate (Manton, Stallard, and Corder 1997). As populations age, those groups with higher death rates die out in greater numbers than those with lower death rates, transforming the older population into one consisting mostly of people with low death rates (Carey and Judge 2001).

The genetic or biological aspects of longevity have led many theorists over time to believe that the age patterns of longevity shown in Figure 5.2 could be explained by a simple mathematical formula similar perhaps to the law of gravity and other laws of nature. The most famous of these was put forward in 1825 by Benjamin Gompertz and describes a simple geometric relationship between age and death rates from the point of sexual maturity to the extreme old ages (Olshansky and Carnes 1997). These mathematical models are interesting, but they have so far not been able to capture the actual variability in the human experience with death. Part of the problem, as we are reminded by Carey and Judge (2001), is that we may know what *kills* us, but we are less certain about what it is that allows us to *survive*. So, even if we were able to rid ourselves of all diseases, we do not know how long we might live. We do know, however, that since the youngest and oldest ages are so critical to longevity, it is worth looking at those rates in a bit more detail.

Infant Mortality   There are few things in the world more frightening and awesome than the responsibility for a newborn child, fragile and completely dependent on others for survival. In many societies, the fragility and dependency are translated into high **infant mortality rates** (the number of deaths during the first year of life per 1,000 live births). In some of the less developed nations, especially in equatorial

Figure 5.2   The Very Young and the Old Have the Highest Death Rates

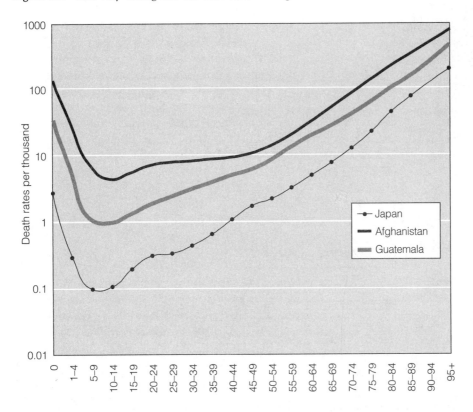

Note: Afghanistan has among the highest death rates in the world, Japan is among the lowest, and Guatemala is at the world average. Yet all three countries exhibit the virtually universal age pattern of mortality—high at both ends and lowest in the middle. Data are for females, but the pattern is the same for males.

Source: Adapted from data sets in United Nations, 2003, *World Population Prospects,* the 2002 Revision.

Africa, infant death rates are as high as 155 deaths per 1,000 live births. That figure is for Sierra Leone early in the twenty-first century, plagued by drought and famine and in a region of the world that has some of the highest mortality rates ever recorded for a human population (McDaniel 1992).

Infant death rates are closely correlated with life expectancy, and Figure 5.3 shows the infant mortality rates for a sample of countries around the world. Japan has had the world's lowest rate of infant mortality among larger countries for a number of years and its rate is now at three deaths per 1,000 live births. The world's previous leader, Sweden, is now second, with a rate of four, which is the same as in France and much of Europe. Canada's rate is five per 1,000, whereas the United States has a slightly higher rate of seven per 1,000. Mexico's level of 25 per 1,000 is obviously higher, but it is still well below the world average of 55 per 1,000.

Figure 5.3   Variations in Infant Mortality Around the World

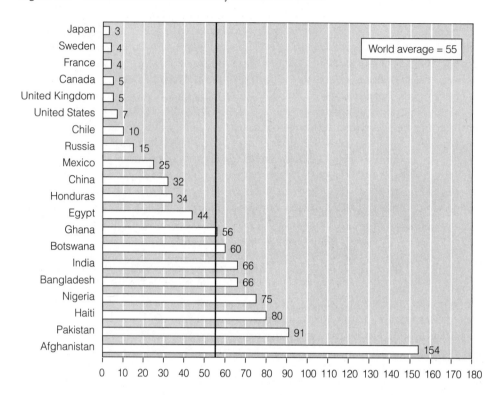

World average = 55

Country	Value
Japan	3
Sweden	4
France	4
Canada	5
United Kingdom	5
United States	7
Chile	10
Russia	15
Mexico	25
China	32
Honduras	34
Egypt	44
Ghana	56
Botswana	60
India	66
Bangladesh	66
Nigeria	75
Haiti	80
Pakistan	91
Afghanistan	154

Source: Data from Population Reference Bureau, *2003 World Population Data Sheet.*

Why are babies so vulnerable? One of the most important causes of death among infants is dehydration, which can be caused by almost any disease or dietary imbalance, with polluted water being a common source of trouble for babies. How can dehydration and other causes of death among infants be avoided? The answer is perhaps best summed up by mentioning the two characteristics common to people in places where infant death rates are low—high levels of education and income. These are key ingredients at both the societal level and the individual level. In general, those countries with the highest levels of income and education are those with enough money to provide the population with clean water, adequate sanitation, food and shelter, and, very importantly, access to health care services.

At the individual level, education here can refer simply to knowledge of a few basic rules that would avoid unnecessary infant death. Afghanistan has one of the highest infant death rates and the highest maternal mortality rate in the world. An important reason is that the lack of medical facilities in that country means that most births are attended by a traditional midwife who may be "dangerously igno-rant. Some refuse to tie umbilical cords, so that 'dirty' blood can flow from the baby. Others seek to cure infections by placing dead mice—and other treats—in the vagina. Many insist babies be born in a bowl of dirt. Tetanus contracted by new-borns from the dirt is dismissed as demonic possession" (*The Economist* 2003a:42).

Income is important in order to provide babies with a nutritious, sanitary diet that prevents diarrhea, an important cause of death among infants. Nursing mothers can best provide this service if their diet is adequate in amount and quality. Income is also frequently associated with the ability of a nation to provide, or an individual to buy, adequate medical protection from disease. In places where infant death rates are high, communicable diseases are a major cause of death, and most of those deaths could be prevented with medical assistance. For example, between 1861 and 1960, the infant death rate in England and Wales dropped from 160 to 20, with more than two-thirds of that decline due to the control of communicable diseases.

Throughout the world, infant health has been aided especially by the fact that the World Health Organization of the United Nations has promoted the use of **oral rehydration therapy (ORT),** which involves administering an inexpensive glucose and electrolyte solution to replenish bodily fluids. Oral rehydration therapy was first shown to be effective in clinical trials in Bangladesh in 1968 (Nalin, Cash, and Islam 1968), and it has proven to be very effective in controlling diarrhea among infants (and adults as well) all over the world (Centers for Disease Control and Prevention 1992).

When infant mortality drops to low levels, such as in advanced nations like the United States and Canada, prematurity becomes the single most important reason for deaths among infants, and in many cases prematurity results from lack of proper care of the mother during pregnancy. Pregnant women who do not maintain an ade-quate diet, smoke, take drugs, or in general do not care for themselves have an elevated chance of giving birth prematurely, thus putting their baby at a distinct dis-advantage in terms of survival after birth.

Throughout the world, the infant mortality rate is a fairly sensitive indicator of societal development because as the standard of living goes up, so does the average level of health in a population, and the health of babies typically improves earlier and faster than of people at other ages. This greater ability to resist death generally holds up throughout the reproductive years (with the exception of maternal mortal-ity, which I discuss below), but beyond that death rates start to increase.

Mortality at Older Ages   It has been said that in the past parents buried their chil-dren; now, children bury their parents. This describes the mortality transition in a nutshell. The postponement of death until the older ages means that the number of deaths among friends and relatives in your own age group is small in the early years and then accumulates more rapidly in the later decades of life. But even at the older ages, there are revolutionary changes taking place in death rates. In the more devel-oped countries of the world, the risk of death has been steadily going down even at the very oldest ages. We have not yet unlocked the key to living beyond age 122, but we are pushing toward the day when a large fraction of people will approach that age before they die.

As life expectancy has increased and people survive in greater proportions to older ages, societies experience less variability in the ages at which people die. In-stead of people being likely to die at almost any age, death gets compressed into a narrow range of ages. Wilmoth and Horiuchi (1999) have calculated, for example,

Figure 5.4 The "Rectangularization" of Mortality in the United States

Sources: Data for 1901–10 and 1950–60 are from the Berkeley Mortality Database (http://demog. berkeley.edu/wilmoth/mortality), based on life tables prepared by the Office of the Chief Actuary of the Social Security Administration. Data for 1998 are based on death data from the National Center for Health Statistics (Murphy 2000). The "Alive at 100" data were generated by the author.

that the variability in ages at death in Sweden the 1950s was only about one-fourth of what it had been 100 years before that. The result of this compression of death into a narrow range at the older ages is called **rectangularization** (Comfort 1979; Fries 1980). This means that the curve of the proportion of people surviving to any given age begins to square off, rather than dropping off smoothly. Figure 5.4 gives you an example of this using data for females in the United States.

Death rates by age for 1901–10 show that the proportion surviving drops off fairly quickly at the younger ages because of high infant and child mortality and then it drops off fairly smoothly after that until everybody has died off at around 100. By the middle of the century, in 1951–60, the proportions alive at each successive age are noticeably larger than they were at the beginning of that century, and the trend continued on to the end of the century (data for 1998). An almost totally rectangular situation is shown as the extreme case in Figure 5.4. If everyone survived to age 100, and then died quickly after that, mortality would be compressed into a very short time period, and the survival curve would be squared off at the oldest ages, as you can see. In general, the limited data available seem to support the idea that compression and rectangularization are occurring (see, for example, Kannisto 2001; Wilmoth and Horiuchi 1999), although not everybody agrees (Lynch and Brown 2001). The principal argument against it is that it assumes a fixed human life span of around 120 years. If we are somehow able to crack that barrier, then people could live to increasingly older ages, which would "decompress" mortality and smooth out the mortality curve at the older ages (Caselli and Vallin 2001). These things may not matter much to you now, but as you approach old age, you'll start to give them more thought—in fact, you'll give them more thought in Chapter 9, where I discuss the consequences for society of the aging process.

Sex and Gender Differentials in Mortality

Although the age pattern of death is the most obvious way in which biology affects our lives, it is also true that at every age there are differences between males and females in the likelihood of death. Some of these differentials seem to be strictly biological in origin (the "sex" differences) whereas others are induced by society (the "gender" differences), although it is not always easy to tell the difference between the biological and social influences.

The most basic health difference between males and females is that women generally live longer than men do, and the gap had been widening until recently. In 1900, women could expect to live an average of two years longer than men in the United States, and by 1975, the difference had peaked at 7.8 years. Since then, the difference has dropped to 5.4 years as of 2000, but the survival advantage of women is nearly universal among the nations of the world. Canada has followed a similar pattern, with a 1.8-year advantage in 1920 expanding to a 7.2-year advantage in 1980 before dropping back slightly to a 7.0-year advantage in 2001 (U.S. Census Bureau 2001a). Mexico is still in the phase of a widening gender gap. In 1960, for example, women in Mexico had a life expectancy that was about the same as in Canada and the United States in 1920, and their advantage over men was 3.2 years. By 2001, female life expectancy in Mexico had reached the level of Canada and the United States in 1970, and the advantage for women had grown to 6.2 years.

This phenomenon has attracted curiosity for a long time, and it has been suggested facetiously that the early death of men is nature's way of repaying those women who have spent a lifetime with demanding, difficult husbands. However, the situation has been more thoroughly investigated by a variety of researchers. It is possible, perhaps even probable, that a real biological superiority exists for women in the form of an immune function, perhaps imparted by the hormone estrogen (Waldron 1983; 1986), but it is very difficult to measure this biological advantage (Vallin 1999).

Even if this innate tendency exists for women to better resist the force of mortality, the gender difference is more complicated than that. For example, there is some evidence that a widening gender gap is a normal accompaniment to the epidemiological transition because, for reasons not yet well understood, women are more likely than men to experience a decline in degenerative disease, which takes over as the main reason for death in the later stages of the transition (Gage 1994; Lopez 1983). This argues for a biological interpretation of the difference, and Retherford (1975) pointed out that a large number of studies have shown that throughout the animal kingdom females survive longer than males, which could indicate a basic biological superiority in the ability of females to survive relative to males (biologists refer to this as an aspect of *sexual dimorphism*). In human populations, the survival advantage of women is widespread, but it is not quite universal. This implies that there are also social factors operating. One such factor is the status of women. It is in those countries where women are most dominated by men (especially in western and southern Asia) that women have been least likely to outlive men.

In Bangladesh, the preference of parents for sons means that the health of girls may be neglected compared to that of boys (Muhuri and Prestion 1991), repeating a pattern found in India (Langford and Storey 1993). Still, this does not mean that girls are universally discriminated against (Basu 1993). In the rice-growing regions of India, where the work of women is more highly valued, life expectancy for women is higher than in the wheat-growing areas, where men do more of the work (Kishor 1993).

In North America, as in Europe and other parts of the world, the trend toward greater independence of women has been accompanied by a noticeable gender gap in life expectancy, perhaps because women have been gaining in their knowledge of preventive health measures (such as proper nutrition and exercise) and because greater independence enhances the ability to seek proper medical care. On the other hand, as the empowerment of women encourages them to engage in higher-risk behavior such as smoking, the gap can be expected to close somewhat. Since 1900, males have smoked cigarettes much more than have females, and this has helped to elevate male risks of death from cancer, degenerative lung diseases (such as chronic bronchitis and emphysema), and cardiovascular diseases (Preston 1970). However, cigarette smoking by women increased after World War II and in 2001 women were nearly as likely as men to be smokers (the smoking version of gender equality), but fortunately the percentage of people who smoke has been steadily declining since the 1960s (National Center for Health Statistics 2003). Of course, deaths associated with smoking tend to occur many years after smoking begins, and so the result of women's post-World War II smoking habits has been a predictable recent rise in death rates from lung cancer.

Using data for 1986, Rogers and Powell-Griner (1991) calculated that males and females who smoke heavily have similar (and lower-than-average) life expectancies, but nonsmoking males still have lower life expectancies than nonsmoking females (although still higher than smokers). Overall, Rogers and his colleagues (2000) estimate that smoking accounts for 25 percent of the gender gap.

Birth can be a traumatic and dangerous time not only for the infant, as discussed above, but for the mother as well. More than half a million women die each year from maternal causes—the equivalent of three jumbo jets crashing each day and killing all passengers. These deaths leave a trail of tragedy throughout the world, but disproportionately in less developed nations, where pregnancies are more frequent among women and health care systems are less adequate. There are three factors, in particular, that increase a woman's risk of death when she becomes pregnant: (1) lack of prenatal care that might otherwise identify problems with the pregnancy before the problems become too risky; (2) delivering the baby somewhere besides a hospital, where problems can be dealt with immediately; and (3) seeking an unsafe abortion because the pregnancy is not wanted (see, for example, Reed, Koblinsky, and Mosley 2000; Salter, Johnston, and Hengen 1997; Shiffman 2000).

Women are only at risk of a maternal death if they become pregnant, so maternal death rates need to take that risk into account. Thus, the **maternal mortality ratio** measures the number of maternal deaths per 100,000 live births. At the end of the twentieth century, the world average was estimated by the World Health Organization to be 400 per 100,000. In the United States, however, this figure is 12 per 100,000 and it is even lower (6) in Canada, while it is higher in Mexico (110). The highest levels of maternal mortality are found in sub-Saharan Africa, led (if that is the right word) by Sierra Leone, with a ratio of 1,800 maternal deaths per 100,000 live births. Canada is tied with several European nations for the lowest level.

Another way of measuring maternal mortality that takes into account the number of pregnancies a woman will have is to estimate a woman's lifetime risk of maternal death. As a woman begins her reproductive career by engaging in intercourse, the question asked is: What is the probability that she will die from complications of pregnancy and childbirth. This risk represents a combination of how many times she will get pregnant and the health risks she faces with each pregnancy, which are influenced largely by where she lives. For the average woman in the world, that probability is .013, or one chance in 75, but for women in sub-Saharan Africa, the risk is one in 11 (World Health Organization 2001a). Looked at another way, if you line up 11 young African women, one of them will wind up dying of maternal-related causes. In Sierra Leone (as in Somalia and Afghanistan), the risk is 1 in 7. By contrast, the risk is only 1 in 7,700 in Canada; 1 in 3,500 in the United States; and 1 in 220 in Mexico.

Causes of Death

In general, there are three major reasons why people die: (1) They are killed by diseases that can be transmitted from one person to another (infectious and parasitic diseases), (2) They degenerate, or (3) They are killed by products of the social and economic environment.

Infectious and Parasitic Diseases

For much of human history, infectious diseases have been the major cause of death, killing people before they had a chance to die of something else. These diseases are parasites that survive by feasting on the host. They are spread in different ways (by different vectors), and have varying degrees of severity. Malaria, for example, is typically spread by mosquitoes first biting an infected person. Then the blood from the malarial person spends a week or more in the mosquito's stomach, where the malarial spores develop and enter the mosquito's salivary gland. The disease is passed along with the mosquito's next bite to a human. The probability of dying from untreated malaria is more than 10 percent (Chin 1999).

Infectious diseases are usually either a virus (a noncellular organism) or a bacterium (a single-celled organism). Measles is an example of an acute viral disease that is severe in infancy and adulthood but less so in childhood. It is usually spread by droplets passed through the air when an infected person coughs or sneezes. If left untreated in an infant or adult, the chance of death is 5–10 percent. Vaccinations now protect most people in the developed world from measles, and the United Nations has been working to increase immunization elsewhere. In India in 1985, only 1 percent of children had been immunized against measles, but by 2002 the figure had jumped to 67 percent (World Health Organization 2004b). By contrast, only 40 percent of children in Nigeria were immunized in 2002.

The plague is perhaps the most famous infectious disease. Plague is caused by the bacterium *Yersinia pestis,* which lives mainly in flea-bearing rodents, especially wild ground squirrels, but also rats. That is important because rats live in closer proximity to humans, who are thus at risk of being bitten by a flea living on a plague-ridden rat. The disease can either infect the lymph nodes and cause bubonic plague, or it can attack the lungs and cause pneumonic plague. The latter is especially problematic because the disease then can be spread by coughing, an easier method of transmission than a flea bite. The "Black Death" in medieval Europe was probably bubonic plague, although Wood and his colleagues (2003) think that it may have been a slightly different strain than the plague that still pops up in various parts of Asia, especially India, not to mention the rural southwest of the United States. As recently as 2003, ground squirrels carrying the bubonic plague were discovered in a campground near the Palomar Observatory in San Diego County, California (*San Diego Union-Tribune* 2003). Untreated bubonic plague has a case fatality rate of about 50 percent, whereas untreated pneumonic plague is nearly always fatal—if you get it, you die if you do not seek medical attention. Fortunately, tetracycline and other antibiotics are effective against the disease as long as treatment is started immediately after the infection is discovered.

A major victory against infectious disease was scored in 1979 when smallpox was eliminated from the world. Two years had passed without a reported case (and the World Health Organization (WHO) had offered a $1,000 reward to ferret out any cases), so WHO officials declared the disease eradicated (except for sample vials contained in laboratories in the United States and Russia).

Infectious and parasitic diseases are far less important as a cause of death in the United States, Canada, and other developed societies now than they were in the past, as I discuss later in more detail in describing the mortality/epidemiological transition. As seen in Table 5.1, the top two killers in the United States, Canada, and Mexico are heart disease and cancer and in fact the top four causes of death in all three countries are degenerative, even though they are not identical in disease categories. All three countries have accidents (social causes) on the top 10 list and all three also have pneumonia and influenza (infectious diseases) among the leading causes of death. HIV/AIDS literally exploded on the scene in the early 1980s to become a worldwide pandemic, but it has just recently dropped off the top 10 lists in the United States and Canada, and has not yet made it to the top 10 list in Mexico. However, its global impact is severe enough that it warrants additional discussion.

Table 5.1   The Leading Causes of Death in the United States Compared to Canada and Mexico

Cause of Death	Major Category	Percent of Deaths from this Cause in:		
		USA (2001)	Canada (1997)	Mexico (2001)
Diseases of the heart	Degenerative	29.0	26.6	15.9
Malignant neoplasms	Degenerative	22.9	27.2	12.7
Cerebrovascular diseases	Degenerative	6.8	7.4	5.8
Chronic lower respiratory disease	Degenerative	5.1	4.5	2.5
Accidents	Social	4.2	4.0	8.0
Diabetes mellitus	Degenerative	3.0	2.6	11.3
Influenza and pneumonia	Infectious	2.6	3.7	2.6
Alzheimer's disease	Degenerative	2.2	n.a.	n.a.
Nephritis, nephrotic syndrome, and nephrosis	Degenerative	1.6	1.2	n.a.
Septicemia	Infectious	1.3	n.a	4.1
All other causes		21.3	22.8	37.1

Sources: United States data are from Anderson, 2003, #2433:Table C; Canada data are from Statistics Canada (2003); Mexico data are from INEGI (2003).

HIV/AIDS   At about the same time that the world was apparently rid of smallpox, another deadly viral disease came along to remind us of our tenuous control over mortality. I refer to acquired immunodeficiency syndrome (AIDS), which turned out to be worse than the one eliminated. Classic smallpox (variola major) had a case fatality rate as high as 40 percent (Chin 1999), which is certainly terrible, but AIDS appears to have the potential to kill virtually everyone who develops its symptoms unless an infected person is treated with newly developed antiretroviral drugs that slow down the progression of HIV to AIDS among infected persons. The spread of HIV can be prevented, especially by using condoms during intercourse and by not sharing needles to inject drugs. These relatively simple control measures have been very effective in North America and Europe, but they have been slow to catch on in sub-Saharan Africa, where prevalence rates and new infection rates are by far the highest in the world, as can be seen in Table 5.2. Although HIV prevalence rates are still fairly low in Asia and North Africa, they are nonetheless rising in those areas of the world because governments have been slow to recognize and respond to local risks of transmission among injection drug users, prostitutes, and men having sex with men.

In North America and Western Europe, the openness with which AIDS has been discussed has helped to slow down the spread of the disease by encouraging the use of condoms or even abstinence and by increasing the chance that someone infected with HIV will seek treatment that may forestall or perhaps even prevent HIV from progressing to AIDS and premature death. But in much of the rest of world, including Eastern Europe, Asia, and especially Africa, the situation is dramatically different because the vast majority of people who are infected with HIV either do not know of or do not acknowledge their infection, and so they continue to spread the disease to others.

Table 5.2   Regional HIV/AIDS Statistics and Features, End of 2003

Region	Adults and children living with HIV/AIDS	Adults and children newly infected with HIV	Adult prevalence (%)*	Adult & child deaths due to AIDS
Sub-Saharan Africa	25.0–28.2 million	3.0–3.4 million	7.5–8.5	2.2–2.4 million
North Africa & Middle East	470,000–730,000	43,000–67,000	0.2–0.4	35,000–50,000
South & Southeast Asia	4.6 – 8.2 million	610,000–1.1 million	0.4–0.8	330,000–590,000
East Asia & Pacific	700,000–1.3 million	150,000–270,000	0.1–0.1	32,000–58,000
Latin America	1.3–1.9 million	120,000–180,000	0.5–0.7	49,000–70,000
Caribbean	350,000–590,000	45,000–80,000	1.9–3.1	30,000–50,000
Eastern Europe & Central Asia	1.2–1.8 million	180,000–280,000	0.5–0.9	23,000–37,000
Western Europe	520,000–680,000	30,000–40,000	0.3–0.3	2,600–3,400
North America	790,000–1.2 million	36,000–54,000	0.5–0.7	12,000–18,000
Australia & New Zealand	12,000–18,000	700–1,000	0.1–0.1	<100
TOTAL	40 million (34–46 million)	5 million (4.2–5.8 million)	1.1% (0.9–1.3%)	3 million (2.5–3.5 million)

*The proportion of adults (15 to 49 years of age) living with HIV/AIDS in 2003, using 2003 population numbers.

Note: The ranges around the estimates in this table define the boundaries within which the actual numbers lie, based on the best available information. These ranges are more precise than those of previous years, and work is under way to increase even further the precision of the estimates that will be published mid-2004.

Source: UNAIDS (2004: Introduction Table 1).

The incredibly high rate of infection in sub-Saharan Africa jumps out from Table 5.2, which shows that more than half of all people estimated to be infected with HIV at the end of 2003 were living in that region of the world and that approximately 8 percent of all adults in sub-Saharan Africa are probably infected. In sub-Saharan Africa, women are as likely as men to be infected, since the disease is spread largely through heterosexual intercourse. The difference between sub-Saharan Africa and much of the rest of the world appears to be related to the fundamentally different pattern of sexual networking prevalent throughout the region. Premarital and extramarital sexual intercourse have long been fairly common in sub-Saharan Africa. Factors that seem to encourage these sexual patterns include long periods of postpartum abstinence on the part of women, which may encourage males to seek sexual encounters elsewhere; polygyny, which institutionalizes male sexual adventurousness; and the low status of women, which forces them to be sexually submissive, all compounded by high rates of male migration for employment, which separates husbands and wives for long periods of time and encourages men, in particular, to have intercourse with other partners (Caldwell 2000). In such an environment, any sexually related disease has increased opportunities for transmission through heterosexual channels. Other sexually transmitted diseases are already much more prevalent in sub-Saharan Africa than elsewhere in the world, and the open sores associated with those diseases increase the risk of HIV transmission.

Probably the most disturbing aspect of AIDS in Africa has been the widespread denial of the disease's existence, and the general lack of political support for putting prevention programs into place. In particular, condom use in Africa has been slowed down by suspicions that condoms themselves carried the disease, and by cultural norms that associate the use of condoms with prostitution, thus limiting their use within marriage, even though one or both partners in a marriage may have been at risk for HIV due to their own extramarital sexual activity (Messersmith et al. 2000). Added to this was the belief of President Mbeki of South Africa that AIDS was not caused by HIV. This slowed down the government response in that country until 2003, when a report by former president Nelson Mandela finally forced Mbeki to change his position and respond to the serious problem of HIV/AIDS in that country. The impact of AIDS on sub-Saharan Africa is astounding. Since people of reproductive age are the principal victims, the disease has created millions of orphans in Africa. When the disease first hit Africa, it was called the "slim" disease, because sufferers lost so much weight as a result of the infection. Over time, it has been transformed into the "grandmother's disease" because it falls upon the older women to care for these orphans, as well for the sick and dying.

The above-average prevalence of HIV/AIDS in the Caribbean is accounted for especially by the high rate in Haiti, where the sexual networking patterns are similar to those in sub-Saharan Africa. The rates for south and Southeast Asia are influenced by Thailand, where the frequent use of commercial sex workers without frequent use of condoms led to the spread of the disease within the heterosexual population (VanLandingham et al. 1993). However, the government responded to this with a policy of 100 percent condom use among commercial sex workers, and the result has been a slowdown in the rate of new HIV infections in Thailand (UNAIDS/WHO 2000).

AIDS is the most severe of several manifestations of infection with the human immunodeficiency virus (HIV), and technically AIDS is not a single disease but rather a complex of diseases and symptoms whose underlying causes are still not fully understood. AIDS kills by attacking the immune system and rendering the victim susceptible to various kinds of infections and malignancies that then lead to death in a degenerative, disabling fashion.

HIV/AIDS probably entered the United States in the 1970s (Quinn 1987), but the first scientific awareness of it was a report by physicians in Los Angeles published in 1981 in the *Mortality and Morbidity Weekly Report* (MMWR) of the U.S. Centers for Disease Control (Gottlieb et al. 1981). They told of five young homosexual males who had died of a type of pneumonia that suggested a cellular-immune dysfunction. Later that year, a headline in *The Wall Street Journal* announced that a "Mysterious Ailment Plagues Drug Users, Homosexual Males." Only six years later the United Nations declared AIDS to be a global emergency, and by the end of 2003 the WHO reported that there were more than 40 million people in the world living with HIV, in addition to the 21 million people who have died since the epidemic began (UNAIDS 2004). The United Nations estimates that in 2003 there were three million deaths from AIDS in the world, and an additional five million new HIV infections. It is now the leading cause of death in Africa and the fourth leading cause of death in the world (Centers for Disease Control and Prevention 2001).

We know with a fair degree of certainty that the disease has been around for decades, perhaps even centuries, in chimpanzees (de Groot et al. 2002; Ewald 1994), especially in the sooty mangabey chimpanzee in West Africa (Gao et al. 1999). The transfer from chimpanzee to humans may have come from humans killing infected animals for food or possibly through the use of animal blood for some magical potion (Gould 1993). Regardless of the mechanism, it must have happened as early as the 1950s, since an Englishman who had been to Africa died in 1959 of what we now know was AIDS (Gould 1993), and a reexamination of a 1959 blood sample (the so-called ZR59 sample) from a man in the Congo tested positive for HIV (Zhu et al. 1998).

About a year after ZR59 was collected, the Congo erupted into one of the bloodiest and most disruptive civil wars in African history. War, and the refugees and starvation that result from it, provide ideal circumstances for any disease to spread, and the activities of armies composed largely of young men are particularly likely to give sexually transmitted infection a boost. Without the Congolese war, HIV-1 might, like its cousin HIV-2, still be confined to a small area of Africa (*The Economist* 1998b:82).

The progression from HIV infection to AIDS is fairly slow in the first five years after initial infection, but within 10 years after initial infection, it is estimated that 50 percent of HIV-infected persons will have developed AIDS symptoms (Heymann, Chin, and Mann 1990; Schinaia et al. 1991). It is this slow progression that has made AIDS so deadly, because infected people have many years in which to spread the disease. Many new viruses, like Ebola, kill their hosts so quickly that the epidemic quite literally dies out after a short period of time. On the other hand, HIV has a fairly low probability of being transmitted in any single sexual encounter, but the host lives long enough to have multiple encounters, thereby increasing the odds of passing the infection along (Goudsmit 1997).

Pneumonia and Influenza   The impact of AIDS has overshadowed the quiet deadliness of influenza, which has killed millions of people throughout history. Influenza is much like a cold but more severe. It is an acute viral infection of the respiratory tract that produces a fever, headache, sore throat, cough, and sometimes nausea, vomiting, and diarrhea. In August 1918, as World War I was ending, a particularly virulent form of the flu apparently mutated almost spontaneously in West Africa (Sierra Leone), although it was later called "Spanish Influenza." For the next year, it spread quickly around the world, killing more than 20 million people in its path, including more than 500,000 in the United States and Canada—disproportionately healthy young adults (Crosby 1989).

Pneumonia is a family of bacterial respiratory infections and is lumped together with influenza in government summaries of causes of death. It is often called "opportunistic" (like a vulture), because it is a disease that preys especially on physically weak individuals (infants, older people, and others with suppressed immune systems). For a reasonably healthy person, the pneumonia case fatality rate is about 5–10 percent, even with treatment, but that can rise to 20–40 percent for people who have pneumonia on top of another underlying disease (Chin 1999). Twice as many people die of pneumonia and influenza as die of AIDS in the United States, and the figure is four times as many in Canada, and five times as many in Mexico.

Emerging Infectious Diseases   In the 1980s, HIV/AIDS emerged as a terrible new infectious disease, but this did not occur as an isolated threat to the health of the world. Just as malaria may have emerged when humans cleared forests and settled into agricultural villages (Pennisi 2001), so it is that the desire of humans to add more animal protein to their diet may be creating opportunities for coronaviruses (animal viruses) to be spread to humans. As I noted above, it is believed that HIV crossed to humans from chimpanzees, and the evidence suggests that SARS (severe acute respiratory syndrome) may have come from animals captured for food in China (Lingappa JR 2004). SARS emerged in China in November 2002, by March 2003 it had become a global health threat, and by the end of 2003 it had already killed nearly 1,000 people worldwide (WHO 2003).

Birds can also be sources of disease. Although West Nile virus has existed in Africa and the Middle East for decades, it was brought to New York City in 1999, apparently by an imported infected bird that was bitten by a mosquito which then bit a human who became sick and died. It has since spread to other communities all over the country. Then, late in 2003, a new H5N1 Avian influenza (popularly called "bird flu") was reported in Asia and has thus far killed dozens of people (Crispin 2004; WHO 2004a). This is a disease discovered in Asian poultry farms that has apparently now crossed over to humans.

Degeneration from Chronic Diseases

Degeneration refers to the biological deterioration of a body. In American medicine, "[C]hronic diseases have been referred to as chronic illnesses, noncommunicable diseases, and degenerative diseases. They are generally characterized by uncertain etiology, multiple risk factors, a long latency period, a prolonged course of illness, noncontagious origin, functional impairment or disability, and incurability" (Taylor et al. 1993). The major **chronic diseases** associated with degeneration, in order of importance as a cause of death in the United States, Canada (with one exception), and Mexico (with one exception), are heart disease, malignant neoplasms (cancer), and cerebrovascular disease (stroke), (see Table 5.1). The exceptions are that in Canada cancer is just slightly ahead of heart disease as the leading cause of death, and in Mexico diabetes is a more important cause of death than stroke.

Deaths from coronary heart disease occur as a result of a reduced blood supply to the heart muscle, most often caused by a narrowing of the coronary arteries, which can be a consequence of atherosclerosis, "a slowly progressing condition in which the inner layer of the artery walls become thick and irregular because of plaque—deposits of fat, cholesterol, and other substances. As the plaque builds up, the arteries narrow, the blood flow is decreased, and the likelihood of a blood clot increases" (Smith and Pratt 1993).

The second leading cause of death is cancer, a group of diseases that kill by generating uncontrolled growth and spread of abnormal cells. These cells, if untreated, may then metastasize (invade neighboring tissue and organs) and cause dysfunction and death by replacing the normal tissue in vital organs (Brownson et al. 1993). In the United States, lung cancer is responsible for more cancer deaths than any other type (28 percent), followed by colon and rectal cancer (11 percent), breast cancer (9 percent), and prostate cancer (7 percent).

Cerebrovascular disease (stroke) is the third leading cause of death. It is actually part of the family of **cardiovascular diseases,** along with coronary heart disease, but whereas heart disease produces death by the failure to get enough blood to the heart muscle, stroke is the result of the rupture or clogging of an artery in the brain. This causes a loss of blood supply to nerve cells in the affected part of the brain, and these cells die within minutes (Smith and Pratt 1993).

Fourth on the list of most important causes of death in both the United States and Canada (but much further down the list in Mexico) is chronic lower respiratory disease. This also is a family of problems, including bronchitis, emphysema, and asthma. The underlying functional problem is difficulty breathing, symptomatic of inadequate oxygen delivery. Diabetes mellitus is a disease that inhibits the body's production of insulin, a hormone needed to convert glucose into energy (Bishop et al. 1993). Like most of the other degenerative diseases, diabetes is part of a group of related diseases, all of which can lead to further health complications such as heart disease, blindness, and renal failure.

Products of the Social and Economic Environment

Despite the widespread desire of humans to live as long as possible, we have devised myriad ways to put ourselves at risk of **accidental death** as a result of the way in which we organize our lives and deal with products of our technology. Furthermore, we are the only known species of animal that routinely kills other members of the same species (homicide) for reasons beyond pure survival, and we seem to be unique in being able to kill ourselves intentionally (suicide). These uniquely human causes of death are included among the top killers in North America. Of course, other aspects of our lifestyle also contribute to our risk of infectious and chronic disease, as discussed in the accompanying essay.

Accidents   Every time a person is killed in an automobile accident, slips and is killed in the bathtub, puts a gun to the temple and pulls the trigger, or is murdered by a mugger, that death is attributable to the social and economic environment. Accidental deaths generally occur in the absence of degeneration or transmissible disease, although an otherwise sick person may be more susceptible to having an accident. The only types of accidents not directly attributable to the social and economic environment are those due to natural phenomena such as floods, tornadoes, avalanches, or earthquakes. However, in most of these instances deaths can be attributed to human risk taking. For instance, people die in an earthquake not from the quake itself, but from the building that falls on top of them because it was unable to withstand the shaking. Nevertheless, buildings continue to be built and occupied very near, and in some cases right on top of, known fault lines. Likewise, floods generally kill those who unwisely build homes in flood plains or who venture away from secure areas. Tornadoes, too, tend to kill those who do not take precautions and, for whatever reason, have not found a basement, ravine, or ditch to hide in. Also intriguing is the finding that tornadoes are more likely to touch down in urban counties than in rural counties, raising the possibility that human urban settlements may themselves increase the odds of being victimized by a tornado (Aguirre et al. 1993).

About half of all accidental deaths in the United States are attributable to motor vehicles, compared with one-third in Canada and less than one-fifth in Mexico. Yet, despite the differences in the reasons for accidental deaths, in all three countries they are in the top five causes of death. Each year there are tens of thousands of lives lost in traffic accidents in Canada, the United States, and Mexico; tens of thousands more are injured and face permanent disability. These victims are disproportionately young males, and alcohol is involved in a high, although declining, fraction of cases. Alcohol also plays a role in other accidents, including bicycle fatalities. A study in Baltimore, for example, found that nearly one-half of the males aged 25 to 34 who had died in bike crashes between 1987 and 1991 had alcohol in their blood (Englehart 1994).

Suicide   The WHO estimates that about one million people each year commit suicide and that as many as 20 times that number attempt suicide but are not successful (World Health Organization 2002). Suicide rates are highest by far in Eastern Europe, especially in Russia and several of the countries that were part of the former Soviet Union. In general, Europe has the highest rates, whereas the United States and Canada are right at the world average and Mexico is well below the world average.

In general, suicide rates rise through the teen years (a phenomenon that has always received considerable publicity), peak in the young adult ages, plateau in the middle years, and then rise in the older ages. Almost universally among human societies, the suicide rate is considerably higher for males than for females (WHO 2002), and so around the world it is older men who are most prone to suicide. Beyond these general patterns, however, the actual difference from one country to another in the suicide rate seems to be a cultural phenomenon (Cutright and Fernquist 2000; Pampel 1998).

Homicide   Men are not only more successful at killing themselves, they are also more likely to be killed by someone else. Homicide rates are highest for young adult males in virtually every country for which data are available (WHO 2001b). In the United States in 1998, males were three times more likely than females to be homicide victims, but the homicide rate peaks at ages 15 to 24. For white males at this age, the homicide rate in 1998 was 12 per 100,000 compared with a staggering 97 deaths per 100,000 African-American males aged 15 to 24 (Murphy 2000).

Homicide death rates in the United States are higher than for any other industrialized nation except Russia (Anderson, Kochanek, and Murphy 1997), possibly reflecting the cultural acceptance of violence as a response to conflict (Straus 1983) combined with the ready availability of guns (which are used in two-thirds of homicides in the United States). The remarkable contrast between African-American and white homicide rates within the United States has existed for decades and appears to be most readily explained by the proposition that "economic stress resulting from the inadequate or unequal distribution of resources is a major contribution to high rates of interpersonal violence" (Gartner 1990:95). Put another way, a "subculture of exasperation" (Harvey 1986) promotes a "masculine way of violence" (Staples 1986).

By combining homicide, suicide, and vehicle fatalities into one index, we have an overall rate of violent mortality. Using data from WHO, Chesnais (2000) has calculated that Russia is the most violent of the more industrial societies with a rate of 209 violent deaths per 100,000 population. France is second (because of its high automobile fatality rate) at 73. The U.S. has a rate of 56, whereas Canada is 45—low, but not as low as the United Kingdom's 28.

The "Real" Causes of Death

The causes of death discussed above reflect those items listed on a person's death certificate. The World Health Organization has worked diligently over the years to try to standardize those causes under a set of guidelines called the International Classification of Diseases (ICD), so that the pathological conditions leading to death will be identified consistently from one person to the next and from one country to the next. This enhances comparability, but it ignores the actual things going on that contribute to that death. Thus, when public concern first arose over the role of alcohol in traffic fatalities, there were no data available to suggest whether a person who died in an accident was a victim of his or her own alcohol use or the alcohol use of someone else. Similarly, a person who dies of lung cancer or heart disease may really be dying of smoking, no matter what the actual pathological condition that led immediately to death.

There is a vast amount of literature in the health sciences tracing the etiology (origins) of the diseases listed on death certificates, and in 1993 two physicians (McGinnis and Foege) culled those studies in order to estimate the "real" or "actual" causes of death in the United States in 1990, in comparison with the 10 leading causes of death as shown in Table 5.1. The actual causes of death, as traced by McGinnis and Foege (1993), offer a different picture. Of the 2,148,000 people who died in the United States in 1990, 400,000 (19 percent) died as a result of tobacco use. Tobacco has been traced to cancer deaths (especially cancers of the lung, esophagus, oral cavity, pancreas, kidney, and bladder), cardiovascular deaths (coronary heart disease, stroke, and high blood pressure), chronic lung disease, low birth weight, and other problems of infancy as a result of mothers who smoke, and to accidental deaths from burning cigarettes. Smoking has emerged as an increasingly important real cause of death throughout the world, and with a slowdown in tobacco use in developed countries, the burden of smoking has shifted to developing, countries, where more than half of the world's smoking-related deaths now occur (Ezzati and Lopez 2003).

The second most important "real" cause of death is diet and activity patterns of the United States population, which account for 300,000 deaths annually (14 percent of the total in 1990). The major dietary abuses identified in the literature include high consumption of cholesterol, sodium, and animal fat, while the principal activity pattern of concern is lack of activity—a sedentary lifestyle. Poor diet and inactivity contribute to heart disease and stroke, cancers (especially colon, breast, and prostate), and diabetes mellitus.

Alcohol misuse is the third (albeit a distant third) real cause of death in the United States, although McGinnis and Foege note that the consequences of alcohol misuse extend well beyond death and include the ruination of lives due to alcohol dependency. Alcohol contributes to death from cirrhosis, vehicle accidents, injuries in the home, drowning, fire fatalities, job injuries, and some cancers. Health care practitioners have been more cautious in their assessment of the positive effects of limited alcohol consumption. A great deal of publicity has been given to the beneficial effects of red wine in keeping heart disease rates lower in France than in the United States, but closer inspection of death rates suggests that red wine has not protected the French against other causes of death, and the French have death rates that are almost identical to those in the United States. Klatsky and Friedman (1995) have noted "a J-shaped alcohol-mortality curve, with the lowest risk among drinkers who take less than three drinks daily" (p. 16).

Less equivocal is the danger from microbial agents—infectious diseases (beyond HIV or infections associated with tobacco, alcohol, or drug use). These represent the fourth leading cause of death and are, in theory at least, largely preventable through appropriate vaccination and sanitation. Toxic agents are responsible for an estimated 60,000 deaths annually in the United States. These agents include occupational hazards, environmental pollutants, contaminants of food and water supplies, and components of commercial products. Toxins are known to contribute to cancer and to diseases of the heart, lungs, liver, kidneys, bladder, and the neurological system.

Firearms contribute to an estimated 36,000 deaths annually, including 16,000 homicides, 19,000 suicides, and 1,400 unintentional deaths. The United States is unique in the world in the number of deaths from firearms, and the most pressing problem is that of guns in the hands of teenagers and young adults, who disproportionately use the weapons to kill themselves and/or someone else. In 1990, there were another 30,000 deaths calculated as being the result of sexual behavior, including 21,000 from sexually acquired HIV infection, 5,000 from infant deaths resulting from unintended pregnancies, 4,000 from cervical cancer, and 1,600 from sexually acquired hepatitis B infection. Illicit use of drugs is estimated to cause 20,000 deaths annually by contributing to infant deaths (through the mother's use of drugs), as well as to drug overdose, suicide, homicide, motor vehicle deaths, HIV infection, pneumonia, hepatitis, and heart disease.

Researchers have explored the real causes of death in other countries as well. For example, Nizard and Muñoz-Pérez (1994) found that 16 percent of deaths in France in 1986 were attributable to tobacco use (a little lower than in the United States), and that 10 percent of deaths in France are traceable to alcohol consumption (higher than in the United States).

Thus far I have discussed life expectancy and death rates in some detail, but I have not actually defined those rates for you. So, before proceeding to an analysis of differences and trends in mortality levels around the world, it will be useful to consider the methods by which mortality is usually measured.

Measuring Mortality

In measuring mortality, we are attempting to estimate the **force of mortality,** the extent to which people are unable to live to their biological maximum. The ability to measure accurately varies according to the amount of information available, and as a consequence, the measures of mortality differ considerably in their level of sophistication. The least sophisticated and most often quoted measure of mortality is the crude death rate.

Crude Death Rate

The **crude death rate (CDR)** is the total number of deaths in a year divided by the average total population. In general form:

$$CDR = \frac{d}{p} \times 1,000$$

where d represents the total number of deaths occurring in a population during any given year, and p is the total average (midyear) population in that year. It is called crude because it does not take into account the differences by age and sex in the likelihood of death. Nonetheless, it is frequently used because it requires only two pieces of information, total deaths and total population, which often can be estimated with reasonable accuracy, especially in developing countries where the cost of censuses and vital registration systems may limit their availability.

Differences in the CDR between two countries could be due entirely to differences in the distribution of the population by age, even though the force of mortality is actually the same. Thus, if one population has a high proportion of old people, its crude death rate will be higher than that of a population with a high proportion of young adults, even though at each age the probabilities of death are identical. For example, in 2000, Mexico had a crude death rate of 4 per 1,000, scarcely one-third of the 11 per 1,000 in Lithuania in that year. Nonetheless, the two countries actually had an identical life expectancy at birth of 72 years. The difference in crude death rates was accounted for by the fact that only 5 percent of Mexico's population was aged 65 and older, whereas the elderly accounted for 13 percent of Lithuania's population. Mexico's crude death rate was also lower than the level in the United States (9 per 1,000 in 2000). Yet in Mexico a baby at birth could expect to live five years less than a baby in the United States. The younger age structure in Mexico puts a smaller fraction of the population at risk of dying each year, even though the actual probability of death at each age is higher in Mexico than in the United States.

In order to account for the differences in dying by age (and sex), we must calculate age/sex-specific death rates.

Age/Sex-Specific Death Rates

To measure mortality at each age and for each sex we must have a vital registration system (or a large survey) in which deaths by age and sex are reported, along with census or other data that provide estimates of the number of people in each age and sex category. The age/sex-specific death rate ($_nM_x$ or ASDR) is measured as follows:

$$_nM_x = \frac{_nd_x}{_nP_x} \times 100{,}000$$

where $_nd_x$ is the number of deaths in a year of people of a particular age group in the interval x to $x + n$ (typically a five-year age group, where x will be lower limit of the age interval and n represents the width of the interval in years of age) divided by the average number of people of that age, $_nP_x$, in the population (again, usually defined as the midyear population). It is typically multiplied by 100,000 to get rid of the decimal point.

In the United States in 2001, the ASDR for males aged 65 to 69 was 2,316 per 100,000, while for females it was 1,486. In 1900, the ASDR for males aged 65 to 69 was 5,000 per 100,000, and for females 5,500. Thus, we can see that over the course of the twentieth century, the death rates for males aged 65 to 69 dropped by 54 percent, while for females the decline was 73 percent. To be sure, in 1900, the death rate for females was actually a bit higher than for males, whereas by 1990 it was well below that for males.

Age-Adjusted Death Rates

It is possible to compare crude death rates for different years or different regions, but it is analytically more informative if the data are adjusted for differences in the age structure of the populations prior to making those comparisons. The usual method is to calculate age-specific death rates for two different populations and then apply those rates to a standard population. For this reason, this method is also known as **standardization.** The formula for the age-adjusted death rate (AADR) is as follows:

$$AADR = \sum_i {_nws_x} \times {_nM_x}$$

where $_nws_x$ is the standard weight representing this age group's proportion in the total population and $_nm_x$ is the age-specific death rate as calculated in the previous section. We can apply this methodology to compare the crude death rate in Egypt in 1996 (6.2 deaths per 1,000 population) with the United States in that same year (8.7 deaths per 1,000 population). Could it be that mortality is really higher in the United States than in Egypt? We use the United States population as the standard weights and apply the age-specific death rates for Egypt to the United States age-sex structure in 1996 to see what the crude death rate would be in Egypt if its age-sex structure were identical to the U.S. The result is that the age-adjusted death rate for Egypt in 1996 was 15.1 deaths per thousand—nearly twice that of the United States.

Life Tables

Although the age-adjusted death rate takes the age differences in mortality into account, it does not provide an intuitively appealing measure of the overall mortality experience of a population. We would like to have a single index that sums that up, and so we turn to a frequently used index called **expectation of life at birth,** or more generally **life expectancy.** This measure is derived from a **life table,** which is a fairly complicated statistical device, but used so widely that I have included a brief discussion of it here for you.

The life table is a statistical device for summarizing the mortality experience of a population and, as I mentioned in Chapter 3, it was first used in 1662 by John Graunt to uncover the patterns of mortality in London. Life expectancy can be summarized as the average age at death for a hypothetical group of people born in a particular year and being subjected to the risks of death experienced by people of all ages in that year. The expectation of life at birth for U.S. females in 2001 of 79.8 years (see Table 5.3) does not mean that the average age at death in that year for females was 79.8. What it does mean is that if all the females born in the United States in the year 2001 had the same risks of dying throughout their lives as those indicated by the age-specific death rates in 2001, then their average age at death would be 79.8. Of course, some of them would have died in infancy while others might live to be 120, but the age-specific death rates for females in 2001 implied an average of 79.8.

Note that life expectancy is based on a hypothetical population, so the *actual* longevity of a population would be measured by the average age at death. Since it is undesirable to have to wait decades to find out how long people are actually going to live, the hypothetical situation set up by life expectancy provides a useful, quick comparison between populations.

One of the limitations of basing the life table on rates for a given year, however, is that in most instances the death rates of older people in that year will almost certainly be higher than will be experienced by today's babies when they reach that age. This will especially be true for a country that is in the midst of a rapid decline in mortality, but even in the United States in the twenty-first century, current life tables are assumed to underestimate the actual life expectancy of people at all ages (Bongaarts and Feeney 2003; Murray and Lopez 1996).

Life Table Calculations   Life table calculations, as shown in Table 5.3 for U.S. females for 2001 and in Table 5.4 for U.S. males in 2001, begin with a set of age/sex-specific death rates, and the first step is to find the probabilities of dying during any given age interval. Table 5.3 is called an abridged life table because it groups ages into five-year categories, rather than using single years of age. The probability of dying ($_nq_x$) between ages x and $x + n$ is obtained by converting age/sex-specific death rates ($_nM_x$) to probabilities. A probability of death relates the number of deaths during any given number of years (that is, between any given exact ages) to the number of people who started out being alive and at risk of dying. For most age groups, except the very youngest (less than five) and oldest (85 and older), for which special adjustments are made, death rates ($_nM_x$) for a given sex for ages x to $x + n$ may be converted to probabilities of dying according to the following formula:

$$_nq_x = \frac{(n)(_nM_x)}{1 + (a)(n)(_nM_x)} .$$

This formula is only an estimate of the actual probability of death, because the researcher rarely has the data that would permit an exact calculation, but the difference

Table 5.3 Life Table for Females in the United States, 2001

(1)	(2)	(3)	(4)	(5)	(6)	(7)	(8)	(9)	(10)
					Of 100,000 hypothetical people born alive:		Number of years lived		Expectation of life
Age interval	Number of females in the population	Number of deaths in the population	Age-specific death rates in the interval	Probabilities of death (proportion of persons alive at beginning who die during interval)	Number alive at beginning of interval	Number dying during age interval	In the age interval	In this and all subsequent age intervals	Average number of years of life remaining at beginning of age interval
x to x + n	nPx	nDx	nMx	nqx	lx	ndx	nLx	Tx	ex
Under 1	1,969,539	12,091	0.006139	0.006107	100,000	611	99,481	7,977,418	79.8
1–4	7,484,746	2,208	0.000295	0.001179	99,389	117	397,276	7,877,937	79.3
5–9	9,827,338	1,366	0.000139	0.000695	99,272	69	496,188	7,480,661	75.4
10–14	10,202,614	1,561	0.000153	0.000765	99,203	76	495,826	6,984,473	70.4
15–19	9,841,558	3,789	0.000385	0.001923	99,127	191	495,160	6,488,647	65.5
20–24	9,615,385	4,500	0.000468	0.002337	98,937	231	494,105	5,993,487	60.6
25–29	9,327,957	5,205	0.000558	0.002786	98,705	275	492,839	5,499,382	55.7
30–34	10,267,287	7,721	0.000752	0.003753	98,430	369	491,228	5,006,543	50.9
35–39	11,140,440	13,168	0.001182	0.005893	98,061	578	488,860	4,515,315	46.0
40–44	11,479,684	20,342	0.001772	0.008821	97,483	860	485,266	4,026,454	41.3
45–49	10,543,157	27,117	0.002572	0.012778	96,623	1,235	480,030	3,541,188	36.6
50–54	9,408,392	36,100	0.003837	0.019003	95,389	1,813	472,411	3,061,159	32.1
55–60	7,324,845	43,722	0.005969	0.029406	93,576	2,752	461,001	2,588,747	27.7
60–64	5,829,812	55,459	0.009513	0.046460	90,824	4,220	443,572	2,127,747	23.4
65–69	5,122,956	76,122	0.014859	0.071634	86,605	6,204	417,513	1,684,175	19.4
70–74	4,892,734	113,257	0.023148	0.109409	80,401	8,797	380,012	1,266,662	15.8
75–79	4,367,495	162,366	0.037176	0.170073	71,604	12,178	327,576	886,649	12.4
80–84	3,219,721	198,821	0.061751	0.267464	59,426	15,894	257,395	559,073	9.4
85+	3,104,651	447,998	0.144299	1.000000	43,532	43,532	301,678	301,678	6.9

Source: Death data are from U.S. Centers for Disease Control and Prevention, *National Vital Statistics Reports*, Vol. 52, No. 3, September 18, 2003: Table 3.

Table 5.4 Life Table for Males in the United States, 2001

(1)	(2)	(3)	(4)	(5)	(6)	(7)	(8)	(9)	(10)
					Of 100,000 hypothetical people born alive:		Number of years lived		Expectation of life
Age interval	Number of males in the population	Number of deaths in the population	Age-specific death rates in the interval	Probabilities of death (proportion of persons alive at beginning who die during interval)	Number alive at beginning of interal	Number dying during age interval	In the age interval	In this and all subsequent age intervals	Average number of years of life remaining at beginning of age interval
x to x + n	nP_x	nD_x	nM_x	nq_x	l_x	nd_x	nL_x	T_x	e_x
Under 1	2,064,150	15,477	0.007498	0.007451	100,000	745	99,367	7,449,792	74.5
1–4	7,837,838	2,900	0.000370	0.001479	99,255	147	396,668	7,350,425	74.1
5–9	10,335,329	1,726	0.000167	0.000835	99,108	83	495,334	6,953,758	70.2
10–14	10,706,140	2,441	0.000228	0.001139	99,025	113	494,845	6,458,424	65.2
15–19	10,422,625	9,766	0.000937	0.004674	98,913	462	493,407	5,963,578	60.3
20–24	10,061,658	14,197	0.001411	0.007030	98,450	692	490,521	5,470,171	55.6
25–29	9,591,445	13,006	0.001356	0.006757	97,758	661	487,140	4,979,650	50.9
30–34	10,417,328	15,751	0.001512	0.007532	97,098	731	483,660	4,492,510	46.3
35–39	11,102,948	23,727	0.002137	0.010628	96,366	1,024	479,271	4,008,850	41.6
40–44	11,298,228	34,437	0.003048	0.015125	95,342	1,442	473,106	3,529,579	37.0
45–49	10,225,516	46,567	0.004554	0.022514	93,900	2,114	464,215	3,056,473	32.6
50–54	9,010,668	58,281	0.006468	0.031825	91,786	2,921	451,628	2,592,258	28.2
55–60	6,865,714	65,801	0.009584	0.046799	88,865	4,159	433,928	2,140,631	24.1
60–64	5,288,415	79,157	0.014968	0.072141	84,706	6,111	408,254	1,706,703	20.1
65–69	4,409,656	102,110	0.023156	0.109444	78,595	8,602	371,473	1,298,449	16.5
70–74	3,887,802	139,471	0.035874	0.164607	69,994	11,521	321,164	926,976	13.2
75–79	3,057,378	171,311	0.056032	0.245737	58,472	14,369	256,439	605,812	10.4
80–84	1,929,321	169,431	0.087819	0.360047	44,103	15,879	180,819	349,373	7.9
85+	1,299,108	217,533	0.167448	1.000000	28,224	28,224	168,554	168,554	6.0

Source: Death data are from U.S. Centers for Disease Control and Prevention, *National Vital Statistics Reports*, Vol. 52, No. 3, September 18, 2003: Table 3.

between the estimation and the "true" number will seldom be significant. The principal difference between reality and estimation is the fraction a, where a is usually 0.5. This fraction implies that deaths are distributed evenly over an age interval, and thus the average death occurs halfway through that interval. This is a good estimate for every age between 5 and 84, regardless of race or sex (Anderson 1999; Chiang 1984). At the younger ages, however, death tends to occur earlier in the age interval. The more appropriate fraction for ages zero to one is 0.85 and for ages one to four is 0.60. Another special case is the oldest age group. Deaths tend to occur later, rather than earlier, in the interval at the oldest ages. However, Table 5.3 uses 85+ as the oldest age group and so avoids the specific age detail at the very oldest ages. Note that since the interval 85+ is open-ended, going to the highest age at which people might die, the probability of death in this interval is 1.0000—death is certain.

In Table 5.3, the age-specific death rates for females in 2001 in the United States are given in column (4). In column (5), they have been converted to probabilities of death from exact age x to exact age $x + n$. Once the probabilities of death have been calculated, the number of deaths that would occur to the hypothetical life table population is calculated. The life table assumes an initial population of 100,000 live births, which are then subjected to the specific mortality schedule. These 100,000 babies represent what is called the **radix** (l_0). During the first year, the number of babies dying is equal to the radix (100,000) times the probability of death. Subtracting the babies who died ($_1d_0$) gives the number of people still alive at the beginning of the next age interval (l_1). These calculations are shown in columns (7) and (6) of Table 5.3. In general:

$$_nd_x = (_nq_x)(l_x)$$

and

$$l_x + n = l_x - {_nd_x}$$

The final two columns that lead to the calculation of life expecancy are related to the concept of number of years lived. During the five-year period, for example, between the fifth and the 10th birthdays, each person lives five years. If there were 99,203 people sharing their 10th birthdays, then they all would have lived a total of $5 \times 99,203 = 496,015$ years between their fifth and tenth birthdays. Of course, if a person died after the fifth but before the tenth birthday, then only those years that were lived prior to dying would be added in. The lower the death rates, the more people there are who will survive through an entire age interval and thus the greater the number of years lived will be. The number of years lived ($_nL_x$) can be estimated as follows:

$$_nL_x = n(l_x - a \, {_nd_x})$$

The fraction a is 0.50 for all age groups except zero to one (for which 0.85 is often used) and one to five (for which 0.60 is often used). Furthermore, this formula will not work for the oldest, open-age interval (85+ in Table 5.3), since there are no survivors at the end of that age interval and the table provides no information about how many years each person will live before finally dying. The number of years lived in this group is estimated by dividing the number of survivors to that oldest age (l_{85}) by the death rate at the oldest age (M_{85}):

$$L_{85+} = \frac{l_{85}}{M_{85}}.$$

The results of these calculations are shown in column (8) of Table 5.3. The years lived are then added up, cumulating from the oldest to the youngest ages. These calculations are shown in column (9) and represent T_x, the total number of years lived in a given age interval and all older age intervals. At the oldest age (85+), T_x is just equal to $_nL_x$. But at each successively younger age (e.g., 80 to 85), T_x is equal to T_x at all older ages (e.g., 85+, which is T_{85}), plus the number of person-years lived between ages x and $x + n$ (e.g., between ages 80 and 85, which is $_5L_{80}$). Thus at any given age:

$$T_x = T_x + n + {_nL_x}.$$

The final calculation is the expectation of life (e_x), or average remaining lifetime. It is the total years remaining to be lived at exact age x and is found by dividing T_x by the number of people alive at that exact age (l_x):

$$e_x = \frac{T_x}{l_x}.$$

Thus, for U.S. females in 2001, the expectation of life at birth (e_0) was 7,977,418 / 100,000 = 79.8, while at age 55 a female could expect to live an additional 27.7 years. For males (Table 5.4), the comparable numbers are a life expectancy at birth of 74.5 and at age 55 of an additional 24.1 years.

Although it has required some work, we now have a sophisticated single index that summarizes the level of mortality prevailing in a given population at a particular time. I should warn you that the formulas I have provided to generate the data in Tables 5.3 and 5.4 are very close, but not identical, to those produced by the U.S. National Center for Health Statistics (NCHS) (see Arias 2002), because the NCHS uses slightly more complex formulas for its calculations at the youngest and oldest ages. The resulting differences in life table values are, however, very small.

Other Applications of the Life Table   The life table as a measure of mortality is actually a measure of duration—of the length of human life. We can apply the same techniques to measure the duration of anything else. Life table techniques have been used to study marriage patterns (see Schoen and Weinick 1993), migration (see Long 1988), school dropouts, labor force participation, fertility, family planning, and other isues (see Chiang 1984; Preston, Heuveline, and Guillot 2001; Smith 1992). Life tables are the backbone of the insurance industry, since actuaries use life tables to estimate the likelihood of occurrence of almost any event for which an insurance company might want to issue a policy. For example, the Retirement Protection Act of 1994 in the United States mandated that standard life tables be established for use in calculating pension plans, charging the Secretary of the Treasury with the task of generating those tables in order to protect consumers (Society of Actuaries 2000). Life table analysis has even been used by General Motors to estimate the cost of maintaining warranties given the probability of a particular car part failing ("dying") from a specific model-year "population" (Merrick and Tordella 1988).

By adding to the mathematical complexity of the table, we can gain additional insights into some of these phenomena by separating out specific events or by combining events. In the former instance we produce a *multiple-decrement* table and in the latter case we get a *multistate* table (Smith 1992), also known as an increment-decrement life table (Palloni 2001). A multiple-decrement table can, for example, isolate the impact of a specific cause of death on overall mortality levels. In a classic study, Preston, Keyfitz, and Schoen (1972) found that in 1964 the expectation of life for all females in the United States was 73.8 years. Deleting heart disease as a cause of death would have raised life expectancy by 17.1 years, to 90.9; whereas deleting cancer as a cause of death would have produced a gain of 2.6 years in life expectancy, to 76.4. In a different type of analysis, we might use a multiple-decrement table to ask what are the probabilities at each age of leaving the labor force before death? Of dying before your spouse? Or of discontinuing a contraceptive method and becoming pregnant?

Multistate tables move us to the level of sophistication that allows for a succession of possible contingencies—several migrations; entering and leaving the labor force, including periods of unemployment and retirement; school enrollment, with dropouts and reentries; moving through different marital statuses; and moving from one birth parity to another. Elaborate discussion of these techniques is, of course, beyond the scope of this book, but you should be aware that such methods do, in fact, exist to help untangle the mysteries of the social world (see, for example, Keyfitz 1985; Palloni 2001; Siegel 2002; Smith 1992).

The Mortality/Epidemiological Transition

Having discussed the more prominent causes of death and the usual methods for measuring mortality levels, we are ready to analyze the trends and levels in mortality. We must first explain how and why mortality has declined over time across different regions of the globe. This phenomenon is best described by the mortality (also known as the epidemiological) transition, discussed in this section. Then, after looking at the mortality transition, we explore why it is that everywhere we go in the world there are some fairly predictable differences within each society in the risks of death—differences between urban and rural places, differences by social characteristics, and differences by age and sex.

Mortality Changes Over Time

Most of us take our long life expectancy for granted. Yet scarcely a century ago, and for virtually all of human history before that, death rates were very high and early death was commonplace. Within the past 200 years, and especially during the twentieth century, country after country has experienced an epidemiological transition

(Olshansky and Ault 1986; Omran 1971; Omran 1977)—a long-term shift in health and disease patterns that has brought death rates down from very high levels in which people die young, primarily from communicable disease, to low levels, with deaths concentrated among the elderly, who die from degenerative diseases. In this process, the variability by age in mortality is reduced or *compressed*, leading to an increased rectangularization of mortality, meaning that most people survive to advanced ages, and then die quickly once reaching that advanced age (as I discussed earlier in the chapter). These are mortality's contributions to the overall demographic transition. We can perhaps best appreciate the epidemiological transition by following it through time, from premodern societies to the present.

Premodern Mortality   In much of the world and for most of human history, life expectancy probably fluctuated between 20 and 30 years (United Nations 1973; Weiss 1973). At this level of mortality, only about two-thirds of babies survived to their first birthday, and only about one-half were still alive at age five, as seen in Table 5.5. This means that one-half of all deaths occurred before age five. At the other end of the age continuum, around 10 percent of people made it to age 65 in a premodern society (the consequences of that are discussed in Chapter 9). So in the premodern world, about one-half the deaths were to children under age five and only about one in 10 were to a person aged 65 or older.

Table 5.5   The Meaning of Improvements in Life Expectancy

Period	Life Expectancy (Females)	Percentage Surviving to Age				Percentage of Deaths		Number of Births Required for ZPG
		1	5	25	65	>5	65+	
Premodern	20	63	47	34	8	53	8	6.1
	30	74	61	50	17	39	17	4.2
Lowest in world today (comparable to Europe/U.S. in nineteenth century)	40	82	73	63	29	27	29	3.3
World average circa 2000	68	97	96	94	74	4	74	2.1
Mexico	76	98	98	98	83	2	83	2.1
United States	80	99	99	98	86	1	86	2.1
Canada	83	99	99	98	88	1	88	2.1
Japan (highest in the world)	83	99	99	98	89	1	89	2.1

Source: Based on stable population models in Ansley Coale and Paul Demeny, 1966, Regional Model Life Tables and Stable Populations (Princeton, NJ: Princeton University Press), updated with the use of software provided by Eduardo E. Arriaga, 1994, Population Analysis with Microcomputers (Washington, D.C.: U.S. Bureau of the Census).

In hunter-gatherer societies, it is likely that the principal cause of death was poor nutrition—people literally starving to death—combined perhaps with selective infanticide and geronticide (the killing of older people) (McKeown 1988), although there is too little evidence to do more than speculate about this (Livi-Bacci 1991). As humans gained more control over the environment by domesticating plants and animals (the Agricultural Revolution), both birth and death rates probably went up, as I mentioned in Chapter 2. It was perhaps in the sedentary, more densely settled villages common after the Agricultural Revolution that infectious diseases became a more prevalent cause of death. People were almost certainly better fed, but closer contact with one another, with animals, and with human and animal waste would encourage the spread of disease, a situation that prevailed for thousands of years.

The Roman Era to the Industrial Revolution   Life expectancy in the Roman era is estimated to have been 22 years (Petersen 1975), but by the Middle Ages nutrition in Europe had probably improved enough to raise life expectancy to more than 30 years. The plague, or Black Death, hit Europe in the fourteenth century, having spread west from Asia (Cantor 2001). It is estimated that one-third of the population of Europe may have perished from the disease between 1346 and 1350. The plague then made a home for itself in Europe and, as Cipolla says, "[F]or more than three centuries epidemics of plague kept flaring up in one area after another. The recurrent outbreaks of the disease deeply affected European life at all levels—the demographic as well as the economic, the social as well as the political, the artistic as well as the religious" (Cipolla 1981:3).

I mentioned in Chapter 2 that Europe's increasing dominance in oceanic shipping and weapons gave it an unrivaled ability not only to trade goods with the rest of the world, but to trade diseases as well. The most famous of these disease transfers is the so-called **Columbian Exchange,** involving the diseases that Columbus and other European explorers took to the Americas (and a few that they took back to Europe). Their relative immunity to the diseases they brought with them, at least in comparison with the devastation those diseases wrought on the indigenous populations, is one explanation for the relative ease with which Spain was able to dominate much of Latin America after arriving there around 1500. The populations in Middle America at the time of European conquest were already living under conditions of "severe nutritional stress and extremely high mortality" (McCaa 1994:7), but contact with the Spaniards turned a bad situation into what McCaa (1994) has called a "demographic hell" with high rates of orphanhood and with life expectancy probably dipping below 20. Spain itself was hit by at least three major plague outbreaks between 1596 and 1685, and McNeill (1976) suggests that this may have been a significant factor in Spain's decline as an economic and political power.

Industrial Revolution to the Twentieth Century   The plague had been more prevalent in the Mediterranean area (where it is too warm for the fleas to die during the winter) than farther north or east, and the last major sighting of the plague in Europe was in the south of France, in Marseilles, in 1720. It is no coincidence that this was the eve of the Industrial Revolution. The plague retreated (rather than disappeared) probably as a result of "changes in housing, shipping, sanitary practices,

and similar factors affecting the way rats, fleas, and humans encountered one another" (McNeill 1976:174).

By the early nineteenth century, after the plague had receded and as increasing income improved nutrition, housing, and sanitation, life expectancy in Europe and the
United States was approximately 40 years. As Table 5.5 shows, this was a transitional stage at which there were just about as many deaths to children under age five
as there were deaths at age 65 and over. Infectious diseases (including influenza, acute
respiratory infections, pneumonia, enteric fever, malaria, cholera, and smallpox)
were still the dominant reasons for death, but their ability to kill was diminishing.

Although death rates began to decline in the middle of the nineteenth century,
improvements were at first fairly slow to develop for various reasons. Famines were
frequent in Europe as late as the middle of the nineteenth century—the Irish potato
famine of the late 1840s and Swedish harvest failures of the early 1860s are prominent examples. These crop failures were widespread, and it was common for local
regions to suffer greatly from the effects of a poor harvest, because poor transportation made relief very difficult. Epidemics and pandemics of infectious diseases, including the 1918 influenza pandemic, helped to keep death rates high even into this
century.

Until recently, then, increases in longevity primarily were due to environmental
improvements—better nutrition and increasing standards of living—not to better
medical care:

> Soap production seems to have increased considerably in England, and the availability of
> cheap cotton goods brought more frequent change of clothing within the economic feasi-
> bility of ordinary people. Better communication within and between European countries
> promoted dissemination of knowledge, including knowledge of disease and the ways to
> avoid it, and may help to explain the decline of mortality in areas which had neither an
> industrial nor an agricultural revolution at the time (Boserup 1981:124–125).

McKeown and Record (1962), who did the pioneering research in this area, argue
that the factors most responsible for nineteenth century mortality declines were improved diet and hygienic changes, with medical improvements largely restricted to
smallpox vaccinations. Preston and Haines (1991), on the other hand, concluded
that public health knowledge was more important in controlling infectious disease
than was nutrition.

> In 1900, the United States was the richest country in the world (Cole and Deane
> 1965:Table IV). Its population was also highly literate and exceptionally well-fed. On the
> scale of per capita income, literacy, and food consumption, it would rank in the top
> quarter of countries were it somehow transplanted to the present. Yet 18 percent of its
> children were dying before age 5, a figure that would rank in the bottom quarter of the
> contemporary countries. Why couldn't the United States translate its economic and so-
> cial advantages into better levels of child survival? Our explanation is that infectious dis-
> ease processes . . . were still poorly understood . . . (Preston and Haines 1991:208).

Of course, the epidemiological transition was already under way in North
America and Europe, but genuine public health knowledge emerged only in the late

nineteenth and early twentieth centuries and then began to be widely diffused. Along with this came a new assumption that child health was a public responsibility, not just a private family matter.

The experience of the North American population has been generally comparable to that of Europeans, although there is some evidence that during the eighteenth and nineteenth centuries, mortality was slightly lower in the United States than in Europe. Nonetheless, by 1900 the expectations of life at birth were nearly identical in England and the United States, and they have remained that way since.

Life expectancy has increased enormously in the United States since the mid-nineteenth century, as seen in Figure 5.5. In 1850, the numbers were 40.5 years for females and 38.3 years for males. Referring to Table 5.5, this meant that about 72 babies out of 100 would survive to age five and about 30 percent of people born would still be alive at age 65. Figure 5.5 also shows that life expectancy began to increase more rapidly as we moved into the twentieth century. The data for Canada from 1920 on show that life expectancy was generally a bit higher than in the United States, whereas Mexico began the century with a life expectancy that was

Figure 5.5   Life Expectancy Has Improved Substantially in the United States, Canada, and Mexico

	1850	1860	1870	1880	1890	1900	1910	1920	1930	1940	1950	1960	1970	1980	1990	2000
■								60.6	62.1	66.3	70.9	74.3	76.5	79.1	80.1	82.9
□	40.5	40.9	42.0	43.5	44.5	48.3	51.8	54.6	61.6	65.2	71.1	73.1	74.8	77.4	78.8	79.8
▣						29.8	25.4	31.5	36.7	45.7	53.7	59.5	62.7	70.3	71.5	75.6

Sources: Data for the United States 1850 through 1970 are from the U.S. Bureau of the Census, 1975, Historical Statistics of the United States, Colonial Times to the 1970 Bicentennial Edition, Part I (Washington, D.C.: Government Printing Office); Tables B107–115 and B126–135 (data for 1850 through 1880 refer only to Massachusetts); Data for Mexico from 1900 to 1950 are from Martha Mier y Terán, 1991, "El Gran Cambio Demográfico," Demos 5:4–5; Data for Canada from 1920 to 1970 are from Statistics Canada, Catalogue no. 82-221-XDE; all other data are from U.S. Bureau of the Census, International Data Base. All data for Canada and the U.S. are for females, data from 1950 to 2000 are for females for Mexico, but 1900 through 1940 refers to both sexes combined.

not quite 30 years, which has now improved so rapidly that its levels that are now very close to those of Canada and the United States.

Mortality in Europe was changing in a pattern similar to that in North America. For example, an analysis of data from Italy concluded that in the 100 years between 1887 and 1986, life expectancy had gone from 38 years to 79 years (for females), and that 65 percent of that gain was due to the fact that mortality for infectious diseases had been nearly wiped out (Caselli and Egidi 1991).

Prior to the Spanish invasion in the sixteenth century, the area now called Latin America was dotted with primitive civilizations in which medicine was practiced as a magic, religious, and healing art. In an interesting reconstruction of history, Ortiz de Montellano (1975) conducted chemical tests on herbs used and claimed to have particular healing powers by the Aztecs in Mexico. He found that a majority of the remedies he was able to replicate were, in fact, effective. Most of the remedies were for problems very similar to those for which Americans spend millions of dollars a year on over-the-counter drugs: coughs, sores, nausea, and diarrhea. Unfortunately, these remedies were not sufficient to combat most diseases and mortality remained very high in Mexico until the 1920s, when it started to decline at an accelerating rate. Since the 1920s, death rates have been declining so rapidly that Mexico has reduced mortality to the level that the United States achieved in the early 1970s.

On the other hand, Eastern Europe is one area of the world where mortality has experienced a backslide. Data for Russia, Hungary, and several other eastern European nations all exhibit similar downward trends (Carlson and Watson 1990; U.S. Census Bureau 2001a). To the extent that death rates reflect the underlying social and economic situation of a nation, the rise in death rates in Eastern Europe and the former Soviet Union perhaps foreshadowed the massive political and economic upheaval in the early 1990s. In Russia, life expectancy had been falling behind other Western nations since at least the 1970s (Vishnevsky and Shkolnikov 1999), and it has been suggested that Russia's health system was unable to move beyond the control of communicable disease to the control of the degenerative diseases that kill people in the later stages of the epidemiological transition. With the collapse of the Soviet Union in 1989, the health of Russians undoubtedly was further threatened by "political instability and human turmoil" (Chen, Wittgenstein, and McKeon 1996:526).

World War II as a Modern Turning Point   As mortality has declined throughout the world, the control of communicable diseases has been the major reason, although improved control of degenerative disease has also played a part (Gage 1994). This is true for the less developed nations of the world today, just as it was for Europe and North America before them. However, there is a big difference between the more developed and less developed countries in what precipitated the drop in death rates. Whereas socioeconomic development was a precursor to improving health in the developed societies, the less developed nations have been the lucky recipients of the transfer of public health knowledge and medical technology from the developed world. Much of this has taken place since World War II.

World War II conjures up images of German bombing raids on London, desert battles in Egypt, D-Day, and the nuclear explosion in Hiroshima. It was a devastating war costing more lives than any previous combat in history. Yet it was also the staging ground for the most amazing resurgence in human numbers ever witnessed. To keep their own soldiers alive, each side in that war spent millions of dollars figuring out how to control disease, clean up water supplies, and deal with human waste. All of the knowledge and technology was transferred to the rest of the world at the war's end, leading immediately to significant declines in the death rate. Not every area of the world has benefited equally, however, so there is still considerable unevenness in the pattern of mortality decline throughout the underdeveloped areas of the world. Among less developed nations, mortality tends to be lowest in Latin America, followed by Asia, with most African nations trailing behind in terms of their success at battling disease, with the HIV/AIDS epidemic being a major setback in that part of the world.

It took only half a century in Latin America for mortality to fall to a point that had taken at least five centuries in European countries. Countries no longer have to develop economically to improve their health levels if public health facilities can be emulated and medical care imported from European countries. As Arriaga (1970) noted: "Because public health programs in backward countries depend largely on other countries, we can expect that the later in historical time a massive public health program is applied in an underdeveloped country previously lacking public health programs, the higher the rate of mortality decline will be."

Included among those techniques that can be used to lower mortality even in the absence of economic development are eradication of disease-carrying insects and rodents, chlorination of drinking water, good sewage systems, vaccinations, dietary supplements, use of new drugs, better personal hygiene, and oral rehydration therapy to control diarrhea in infants.

The decade of the 1950s was the period in which mortality declined most rapidly in the less developed nations. More recently, however, that improvement has slowed:

> This should not be surprising, because the pace of mortality decline during the 1950s was rather unique. It would be hard to maintain such a rapid rate of mortality change over a long period of time. The 1950s were the years of the application of low cost and massive health programs which produced a large decline in mortality, as the simpler causes of death were reduced. For continuation of rapid mortality decline, more expensive programs, including implementation of additional public health programs, would have been required (Arriaga 1982:9).

This is another way of noting that communicable diseases generally are easier to bring under control than are other causes of death, and this is one of the reasons why mortality differentials still persist in country after country.

What are the "routes" to low mortality? One genuine difficulty in drawing conclusions about the exact reasons for mortality decline is that we know less than you might think about the factors that promote or retard the decline of mortality. As Schofield and Reher (1991) noted in a review of the European mortality decline, "there is no simple or unilateral road to low mortality, but rather a combination of many different elements ranging from improved nutrition to improved education" (p. 17). Nonetheless, Caldwell (1986) has pointed out that while a high level of national income is nearly always associated with higher life expectancies, the bigger question is whether it is possible for poorer countries to lower their mortality levels. The answer is at least a qualified "yes," and there are several ways to do it. Geography makes some difference. Caldwell notes that islands and other countries with very limited territories seem to be able to lower their mortality more readily than territorially larger nations. Being "in the path" of European expansion has also been fortuitously beneficial to some countries (such as Costa Rica and Sri Lanka) because it increases the opportunities for the transfer of death control technology.

The implementation of a basic primary health care system is probably the single most important element in lowering mortality, but it is not enough to provide a health care system. It must be used appropriately, and for this reason much of the improvement in health is associated with education. Since children account for a large share of the population at risk of death in a poorer country, the education of the caregivers—generally, the women in the population—will have an important bearing on the societal death rate.

The education of women is a cultural variable more than an economic one and is closely related to the status of women in society. Caldwell points out that the poorer countries of the world that have higher-than-expected life expectancies all share in common an egalitarian attitude toward women, equal access for women to education, and political openness (his "star" examples are Sri Lanka, Costa Rica, and the Indian state of Kerala).

The Nutrition Transition   We know how to keep ourselves alive in ways that would have been nearly unimaginable to people a century or more ago. But the epidemiological transition has not come without its own potential costs to health. Popkin (1993; 2002) has argued that there is evidence of a **nutritional transition** that accompanies the epidemiological transition. Hunting and gathering populations had a pattern of collecting food; early agriculturalists had a pattern related to avoiding famine; then as time progressed and the economy improved, the pattern evolved to what Popkin and his associates call "receding famine." More recently and especially in Western nations, the diet associated with industrialization has shifted to one that is high in fat, cholesterol, and sugar (and often accompanied by a sedentary lifestyle) called the "degenerative disease pattern."

There is ample evidence that people living in the wealthier societies of the world are larger in size than ever before in history (see the essay accompanying this chapter). More importantly, modern society is increasingly associated with obesity and with less active lifestyles than ever before, and these factors threaten to limit our ability to push life expectancy to higher levels. The next stage of the nutrition transition is to recognize that our incomes have increased at a faster pace than the human body's adaptation to not having to live on the edge of famine (the human condition until recently). In the past, our body's ability to store fat is what saved us when the food ran out for a while. We no longer face such periodic shortages and so that fat isn't used up unless we consciously engage in regular exercise, and we have become less likely to do that because our incomes allow us to buy the means to avoid such exercise. If we are going to reduce the burden of degenerative disease and prolong our health, we are almost certainly going to have to restructure our everyday life to reduce the consumption of fat and sugar while increasing our level of exercise.

Regional Differences in Mortality

Japan has the highest life expectancy in the world (taking over a few years ago from Sweden, which had been in first place for a long time). This probably is due to diet and culturally based health practices, in combination with societal wealth, which engenders access to clean water, good sewage systems, a high standard of living, and excellent medical care. These characteristics are not peculiar to Japan. East Asians, in general, have lower-than-average death rates, and Asians in the United States have the highest life expectancy of any group.

A comparison of life expectancies in different regions of the world can be seen in Figure 5.6. In this graph, each box represents the middle 50 percent of all countries in the region, and the line through the box shows the median life expectancy

Figure 5.6   There Is Considerable Regional Variability in Life Expectancy in the World

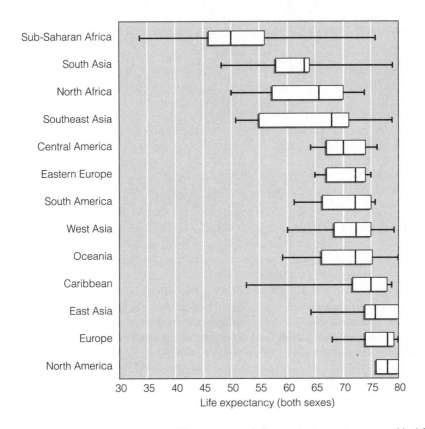

Note: In this graph each box shows the middle 50 percent of all countries in a region arranged by life expectancy for men and women combined. The line through the box represents the median life expectancy among countries in the region and the lines to the right and to the left of the box stretch to the highest and lowest life expectancies for countries in the region.

Source: Data refer to the year 2000 and are drawn from the World Population Data Sheet of the Population Reference Bureau.

MORTALITY CONTROL AND THE ENVIRONMENT

"Live long and prosper." Every newborn child should have such a toast offered on her or his behalf, but it is not easy to accomplish. For most of human history, children could look forward neither to a long life nor a prosperous one, and the achievement of both has required that we bring the environment under our control. We do this, for example, by growing a greater abundance of nutritious food than nature would otherwise provide, by killing the bacteria in our water supply, by protecting ourselves from disease-carrying insects and rodents, by using herbs and chemicals to concoct medications that kill the parasites that attack our bodies, and other devices and concoctions that help to repair or replace failing body parts. We control nature by draining swamps, clearing forests, plowing land, building roads and bridges, constructing water and sewage systems, building dams and levies, and so forth. Then we protect ourselves from nature by building houses that keep predators at bay, keep the rain and snow outside, and climatize the indoor atmosphere so we don't get too hot or too cold, using large amounts of generated energy in the process.

Controlling nature and protecting ourselves from its ravages is not done without a cost, of course. In the process we rearrange our relationship to nature, and risk degrading the environment to the point of unsustainability. This is all because of the natural linkage between living long and prospering—the two go together, as you can see in the accompanying diagram. The obvious impact of a declining death rate on the environment is that more people mean more resources used. This is the classical Malthusian view—population growth means that more people are trying to live at the same standard of living. That model simply says that we live long, but don't prosper. The prosperity comes from our greater per-person productivity as we become more clever and efficient in using environmental resources for our personal and collective improvement. Two important components to this are linked to the decline in mortality. The lower death rate is a result of our being healthier, and healthy people work harder and longer and thus are more productive. But there is also a psycho-social aspect, which I have labeled an increase in the "scope" of life. The prospect of a long life, unburdened by the threat of imminent death, means that we can think about life in the long term, making plans and implementing changes and improvements that would have been unthinkable and unimaginable in the days when life was so uncertain. This greater scope means that each of us has the potential to be more creative and productive in each year that we live.

So, on the one hand, declining mortality increases the size of the population and increases the demand for resources. But, at the same time, the demand for resources is increasing more quickly than the growth of population, because a healthier, longer-living population has the potential to prosper—to improve per-person productivity, which increases per-person use of resources. One of the most elemental of these impacts is on the food supply—itself an important ingredient in the long-term improvement of human health. As we have become healthier, we have become bigger people, and so the demand for food is increasing at a faster pace than the population is growing (see, for example, Fogel 1994). Consider this—in 1969, the 3.6 billion people alive consumed an average of 2,386 calories of food per day; but by 1999, the nearly six billion people alive were consuming food at the rate of 2,808 calories per day. Population had increased by 67 percent, but the total amount of food demanded had increased by 96 percent!

Can we go on this way? I will consider that question in more detail in Chapter 12, but let me give you an example from Brazil, which in that same period from 1969 to 1999 increased from 93 million to 171 million people. This was a result of a death rate that was declining much more quickly than the birth rate—Brazilians added six years to their life expectancy during that 30-year period. Brazilians, like many others in the world, were healthier because they were eating better, increasing their per-day calorie consumption from 2,440 in 1969 to 3,012 in 1999. Thus, Brazil experienced an 84 percent increase in population, and a 127 percent increase in food consumed. How did they manage this? Partly by encouraging population growth and farming in the Amazon region, with results that have threatened the fragile ecosystem of that area, as well as putting people squarely in the path of higher death rates from the malaria that is endemic to the Amazon (Caldas de Castro and Singer 2000).

Our greater productivity—that which makes us prosper—depends heavily on the use of energy, and the by-products of energy use include the emission of greenhouse gases, which have contributed to global warming. Mosquitoes and other disease-carrying insects flourish in warmer weather, and there is a real concern that the reemergence of infectious diseases will be among the many undesirable consequences of global warming (Epstein 2000). If that happens, of course, we run the risk that by living longer and prospering, we have created a heap of other troubles with which we will have to cope.

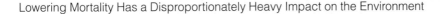

Lowering Mortality Has a Disproportionately Heavy Impact on the Environment

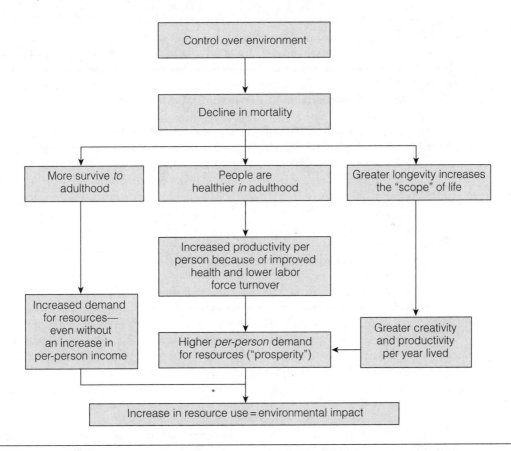

in the region. The lines ("whiskers") extending to the right and left of the box show, respectively, the highest and lowest life expectancies found in the region. Sub-Saharan Africa is clearly set off from other regions of the world in terms of life expectancy, with half of all countries in the region falling between 46 and 56 years and median life expectancy in the year 2000 of only 50, a level equivalent to the United States in about 1910. Africa remains the farthest behind of the world's regions in terms of the mortality transition. It has been estimated that between 1985 and 1990, 59 percent of all deaths on the African continent were to people under the age of 15, and this was before the real impact of AIDS had hit the region (Heligman, Chen, and Babakol 1993). The gains in life expectancy experienced in Africa are being eroded by AIDS and that will continue for the foreseeable future in that part of the world (Bongaarts 1996). "About 100 million additional deaths are expected in these African countries by 2025 because of the toll of the HIV epidemic. By 2025, these countries will have 14 per cent fewer inhabitants than they would have had in the absence of AIDS" (United Nations Population Division 2003c:xii).

The region with the next-to-lowest average level of life expectancy is South Asia, where half of the countries in the region have a life expectancy of between 58 and 64, with the median being 63, three years below the world average. In that region, the lowest life expectancy is found in Afghanistan (which had been the highest-mortality country in the world for some time until the AIDS pandemic spread through Africa), whereas the highest is found in Sri Lanka (as noted above). North Africa's median life expectancy is right at the world average, with the extremes in the region being represented by Tunisia at the high end and Western Sahara at the low end. Southeast Asia is the only other region in which a large fraction of countries are below the world average. The median of 68 years is just above the average, but the middle half of countries fall between 55 and 71 years. The lowest life expectancy is found in Cambodia, while the highest is found in Singapore. You can probably deduce for yourself that the very lowest life expectancy in the Caribbean region belongs to Haiti, another victim of the AIDS pandemic.

Mortality Differences within Countries

The regional differences in mortality are clear reminders that similar cultural and economic features of societies can have a major impact on human well-being. Our health is very dependent on massive infrastructure developments such as piped cleaned water, piped sewerage, transportation and communications systems that deliver food and other goods, and a health care system that is affordable and available. Regions in the world vary considerably in their access to these resources, and within the same regions and countries some people are more advantaged than others in these respects.

Urban and Rural Differentials   Until a few decades ago, cities were deadly places to live. Mortality levels were invariably higher there than in surrounding areas, since the crowding of people into small spaces, along with poor sanitation and contacts with travelers who might be carrying disease, helped maintain fairly high levels of communicable diseases. For example, life expectancy in 1841 was 40 years for native English males and 42 years for females, but in London it was five years less than that (see, for example, Landers 1993). In Liverpool, the port city for the burgeoning coal regions of Manchester, life expectancy was only 25 years for males and 27 years for females. In probability terms, a female child born in the city of Liverpool in 1841 had less than a 25 percent chance of living to her 55th birthday, while a rural female had nearly a 50 percent chance of surviving to age 55. Sanitation in Liverpool at that time was atrociously bad. Pumphrey notes that "pits and deep open channels, from which solid material (human wastes) had to be cleared periodically, often ran the whole length of streets. From June to October, cesspools were never emptied, for it was found that any disturbance was inevitably followed by an outbreak of disease" (Pumphrey 1940:141).

In general, we can conclude that the early differences in urban and rural mortality were due less to favorable conditions in the countryside than to decidedly unfavorable conditions in the cities. Over time, however, medical advances and environmental improvements have benefited the urban population more than the rural, leading to the current situation of better mortality conditions in urban areas. As the world continues to urbanize (see Chapter 11), a greater fraction of the population in each country will be in closer contact with systems of prevention and cure. At the same time, the sprawling slums of many third world cities blur the distinction between urban and rural and may produce their own unhealthy environments (Montgomery et al. 2003).

Occupational Differentials in Mortality   Differences in mortality by social status are among the most pervasive inequalities in modern society, and they are especially apt to show up in cities. The connection between income and health has been obvious for centuries, and whatever characteristic lowers your status in society may therefore put you at greater risk of death. Marx attributed the higher death rate in the working classes to the evils of capitalism and argued that mortality differentials would disappear in a socialist society. That may have been overly optimistic, but data do clearly suggest that by nearly every index of status, the higher your position in society, the longer you are likely to live.

I should note here that good data are not always easy to come by, since death certificates rarely contain information on occupation and virtually never on income or education, and when they do have occupation data, many indicate "retired" or "housewife," giving no further clues to social status. Thus, more circumlocutious means must be devised to obtain data on the likelihood of death among members of different social strata. An important method used is record linkage, such as in the study by Kitagawa and Hauser (1973) for the United States, in which death certificates for individuals in a census year were linked with census information obtained for that individual prior to death. Age and cause of death were ascertained from the death certificate, while the census data provided information on occupation, education, income, marital status, and race.

Among white American men aged 25 to 64 when they died in 1960, mortality rates for laborers were 19 percent above the average, while those for professional men were 20 percent below the average (Kitagawa and Hauser 1973). Similar kinds of results have been reported more recently for all Americans (not just males) followed in the National Longitude Mortality Study, in which the risk of death was clearly lowest for the highest occupational positions and highest for those in the lowest occupational strata (Gregorio, Walsh, and Paturzo 1997).

In England, researchers followed a group of 12,000 civil servants in London who were first interviewed in 1967–69 when they were aged 40 to 64. They were tracked for the next 10 years, and it was clear that even after adjusting for age and sex, the higher the pay grade, the lower the death rate. Furthermore, within each pay grade, those who owned a car (a more significant index of status in England than in the United States) had lower death rates than those without a car (Smith, Shipley, and Rose 1990). The importance of this study is that it relates to a country that has a highly egalitarian national health service. Even so, equal access to health services does not necessarily lead to equal health outcomes.

Nearly half of all work-related fatalities in the United States involve vehicles, but some occupations are inherently more dangerous than others, as you can see in Figure 5.7. Among all occupations in 2002, people involved in timber cutting had the highest rate of fatal work injuries per 100,000 workers, caused largely by being hit by a felled tree. People in fishing were especially at risk of drowning; airplane pilots were at risk of air crashes. Drivers and sales workers were at risk of being murdered in their vehicles by robbers. With the exception of pilots and navigators, high-risk occupations also tend to be lower on the social status scale than occupations whose fatality rate is much lower and therefore are not on the list.

By the way, socialism did not generally reduce occupational differences in mortality, no matter what Marx may have hoped for. Carlson and Tsvetarsky (1992) found that manual laborers in Bulgaria under socialism not only had higher death rates than nonmanual workers, but that the difference had been growing over time. This mirrored trends in other Eastern European nations. Indeed, it has been said that "communism could seriously damage your health; but post-communism is proving even worse" (*The Economist* 1995a:41).

It has to be admitted that "no one knows precisely why people with high status are more healthy" (Shweder 1997:E). Nonetheless, the most important aspects of occupation and social class that relate to mortality are undoubtedly income and education (Johnson, Sorlie, and Backlund 1999)—income to buy protection against and cures for diseases, and education to know the means whereby disease and occupational risks can be minimized.

Income and Education   There is a striking relationship between income and mortality in the United States. Kitagawa and Hauser's data for 1960 showed clearly that as income went up, mortality went down, and more recent studies have confirmed that conclusion. Menchik (1993) used data from the National Longitudinal Survey of Older Men to suggest that, at least for older men in the United States, income is more important than anything else in explaining social status differences in mortality. McDonough and her associates (1997) used data from the Panel Study of Income Dynamics to show that among people under 65, regardless of race or sex, income was a strong predictor of mortality levels.

As with income, there is a marked decline in the risk of death as education increases. A white male in 1960 with an eighth grade education had a 6 percent chance of dying between the ages of 25 and 45, whereas for a college graduate the

Figure 5.7   The 10 Deadliest Occupations* in the United States, 2002

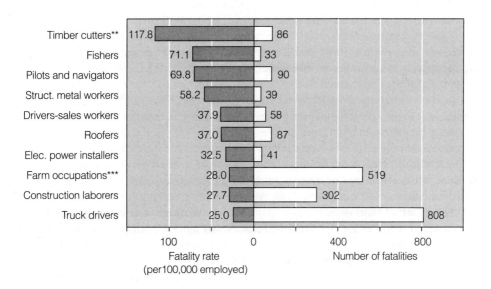

*Selected occupations had a minimum of 30 fatalities in 2002 and 45,000 employed.

**Timber cutters include the following: Timber cutting and logging occupations; supervisors, forestry, and logging workers.

***Farm occupations include the following: Nonhorticultural farmers, nonhorticultural farm managers, farm workers, and farm worker supervisors.

Rate = (fatal work injuries/employment) x 100,000 employed. Employment data extracted from the 2002 Current Population Survey (CPS). The fatality rates were calculated using employment as the denominator; employment-based rates measure the risk for those employed during a given period of time, regardless of exposure hours.

Note: The average fatality rate for all occupations is 4.0.

Source: U.S. Department of Labor, Bureau of Labor Statistics, *Census of Fatal Occupational Injuries,* 2002. http://www.bls.gov/iif/oshwc/cfoi/cfch0001.pdf, accessed 2004.

probability was only half as high (Kitagawa and Hauser 1973). For women, education makes an even bigger difference, especially at the extremes. A white female with a college education could expect, at age 25, to live 10 years longer than a woman who had only four or fewer years of schooling.

For virtually every major cause of death, white males with at least one year of college had lower risks of death than those with less education. The differences appear to be least for the degenerative chronic diseases and greatest for accidental deaths. This is consistent with the way you might theorize that education would affect mortality, since it should enhance an individual's ability to avoid dangerous, high-risk situations.

Kitagawa and Hauser's work was partially replicated by Duleep (1989) using 1973 Current Population Survey data matched to 1973–78 Social Security records. This study showed a persistence of the pattern of mortality differentials by education. Overall, death rates were 45 percent higher for white males aged 25 to 64 who had less than a high school education than those who had at least some college education. Pappas and his associates used the 1986 National Mortality Followback Survey and the 1986 National Health Interview Survey to further replicate Kitagawa and Hauser's analysis. They found that "despite an overall decline in death rates in the United States since 1960, poor and poorly educated people still die at higher rates than those with higher incomes and better education, and the disparity increased between 1960 and 1986" (Pappas et al. 1993:103). Data from Canada (Choiniére 1993), as well as from the Netherlands (Doornbus and Kromhout 1991), confirm that these patterns are not unique to the United States.

Race and Ethnicity   In most societies in which more than one racial or ethnic group exists, one group tends to dominate the others. This generally leads to social and economic disadvantages for the subordinate groups, and such disadvantages frequently result in lower life expectancies for the racial or ethnic minority group members. Some of the disadvantages are the obvious ones in which prejudice and discrimination lead to lower levels of education, occupation, and income and thus to higher death rates. But a large body of evidence suggests that there is a psychosocial component to health and mortality, causing marginalized peoples in societies to have lower life expectancies than you might otherwise expect (see, for example, Kunitz 1994; Ross and Wu 1995). In the United States, African-Americans and Native Americans have been particularly marginalized and have historically experienced higher-than-average death rates. In Canada and Mexico, the indigenous populations are similarly disadvantaged with respect to the rest of society.

United States data for 2001 from the National Center for Health Statistics show that at every age up to 70, African-American mortality rates are nearly double the rates for the white population (Murphy 2000). Between ages 20 and 59, death rates among blacks are more than double those for whites. In 1900, African-Americans in the United States had a life expectancy that was 15.6 years less than for whites, but that differential had been reduced to 5.2 years by 2001. Yet that is a larger gap than exists, for example, between the United States as a whole and the population of Mexico.

African-Americans have higher risks of death from almost every major cause of death than do whites (Rogers et al. 2000). There are three causes of death, in particular, that help to account for the overall difference. Most important is the higher rate of heart disease for African-Americans of both sexes (Keith and Smith 1988; Murphy 2000; Poednak 1989), which may be explained partly by the higher rates of unemployment among African-Americans (Guest, Almgren, and Hussey 1998; Potter 1991). Cancer and homicide rates are also important factors in the African-American mortality differential (Keith and Smith 1988). Especially disturbing is the fact that life expectancy for African-Americans actually declined between 1984 and 1988 and again between 1990 and 1991, and by 1995 it was only back to where it had been in 1984. This was particularly noticeable for males and was apparently caused by an increase in deaths from HIV/AIDS and by a rise in the already high homicide rates. For example, in 1990 the life expectancy for white males in the United States was 72.7 years. In that same year, in the Central Harlem section of New York City, an African-American male had a life expectancy of 53.8 years (about the same as Bangladesh at the time) (Findley and Ford 1993). At ages 35 to

44, African-American males in Harlem were more than 10 times as likely to die as the average male in the United States. This especially was due to the effect of AIDS, compounded by homicide, suicide, and the spread of infectious diseases in a poverty- and crime-ridden area.

The gap in mortality in the United States between whites and African-Americans could be closed completely with increased standards of living and improved lifestyle, according to a study by Rogers (1992). His analysis of data from the 1986 National Health Interview Survey suggests that sociodemographic factors such as age, sex, marital status, family size, and income account for most of the racial differences in life expectancy. In general, the data suggest that socioeconomic status is more important than lifestyle factors in explaining the racial difference in mortality (Hayward et al. 2000). This is generally consistent with the pattern for other racial/ethnic groups. For example, over time, the income and social status gap has narrowed between "Anglos" (non-Hispanic whites) and the Mexican-origin population in the United States. As this happened, differences in life expectancy between the groups disappeared (Bradshaw and Frisbie 1992; Rosenwaike 1991) and more recently have crossed over, as you can see in Table 5.6. Life expectancy in 2001 was higher for Hispanics in the United States than for the white, non-Hispanic population, and it is likely that at least some of this difference is due to the lower levels of mortality among immigrants to the United States (Rogers et al. 2000).

Asians have the highest life expectancy in the United States, as you can see in Table 5.6. Some of this is also due to the high proportion of immigrants among the Asian population, but some of it is almost certainly due to cultural and lifestyle differences, since East Asians, in particular, generally have the highest levels of life expectancy in the world, as I discussed above.

The data in Table 5.6 suggest that life expectancy is now higher for Native Americans than for non-Hispanic whites. It is not clear whether this is a "real" finding, or whether it is caused by differences in the definition of Native Americans in the death data (the numerator of the death rates used in the life table calculations) compared to the census data (the denominator data) (Hollman, Mulder, and Kallan 2000). Health conditions for American Indians in North America have been much more dismal for those living on reservations than for those who live outside of the reservations, but the death data are not reported on the basis of whether a person

Table 5.6   Life Expectancy by Race/Ethnicity in the United States, 2001

Race/ethnicity	Females	Males
Total population	74.5	79.8
White, non-Hispanic	75.0	80.1
Black, non-Hispanic	68.4	75.3
Hispanic	77.6	83.4
Asian/Pacific Islander	82.0	87.2
American Indian	77.0	82.2

Source: Calculated by the author from data in National Vital Statistics Reports, Vol. 52, No. 3, September 18, 2003, Tables 3 and 4.

resided on or off a reservation. As recently as World War II, the Inuit in Canada had a life expectancy of only 29 years (Young 1994), and given the health consequences for many tribes of having been herded onto reservations in the second half of the nineteenth and early twentieth centuries, the Native American population has experienced a rather rapid epidemiological transition (Trafzer 1997).

As immigrants have once again become more numerous in the United States, where you were born has reemerged as a characteristic of importance. Hummer and his associates (1999) used data from the National Health Interview Survey to calculate the probabilities of survival for different groups. They found that U.S.-born young adult blacks had the highest odds of dying when controlling for socioeconomic status, whereas older foreign-born blacks and Asians had the lowest likelihood of dying in comparison to other groups—Americanization isn't necessarily good for your health!

The United States is one of the few countries for which reliable data are available on ethnic differences in mortality, but earlier studies revealed that in India in 1931, members of the Parsis religious community had (and probably still have) an expectation of life at birth that was 20 years longer than that of the total population of India. The difference apparently is due to the relatively high economic position of the Parsis (Axelrod 1990). Similarly, in the 1940s, before Palestine was partitioned to create the state of Israel, the death rate among Muslims was two to three times that of the Jewish population, probably due to the higher economic status of the Jews (United Nations 1953). The mortality gap between Jews and Muslims within the modern state of Israel has narrowed considerably, but Jews continue to have a slightly higher life expectancy (Israel Central Bureau of Statistics 2001).

Marital Status   It has long been observed that married people tend to live longer than unmarried people. This is true not only in the United States but in other countries as well (Hu and Goldman 1990). A long-standing explanation for this phenomenon is that marriage is selective of healthy people; that is, people who are physically handicapped or in ill health may have both a lower chance of marrying and a higher risk of death. At least some of the difference in mortality by marital status certainly is due to this (Kisker and Goldman 1987).

Another explanation is that marriage is good for your health: protective, not just selective. In 1973, Gove examined this issue, looking at cause-of-death data for the United States in 1959–61. His analysis indicates that the differences in mortality between married and unmarried people "are particularly marked among those types of mortality where one's psychological state would appear to affect one's life chances" (Gove 1973:65). As examples, Gove noted that among men aged 25 to 64, suicide rates for single men were double those of married men. For women, the differences were in the same direction, but they were not as large. Unmarried males and females also had higher death rates from what Gove called "the use of socially approved narcotics" such as alcohol and cigarettes. Finally, Gove noted that for mortality associated with diseases requiring "prolonged and methodical care," unmarried people are also at a disadvantage. In these cases, the most extreme rates were among divorced men, who had death rates nine times higher than married men.

In general, Gove's analysis suggested that married people, especially men, have lower levels of mortality than unmarried people because their levels of social and psychological adjustment are higher. Other researchers have found that economic factors may also play a protective role. Married women are healthier than unmarried women partly because they have higher incomes (Hahn 1993), and unmarried men are especially more likely to die than married men if they are living below the poverty line (Smith and Waitzman 1994).

Does this mean that if you are currently single you should jump into a marriage just because you think it might prolong your life? Not necessarily, but Lillard and Panis (1996) found evidence in the Panel Study of Income Dynamics in the United States that relatively unhealthy men tend to marry early and to remain married longer, presumably using marriage as a protective mechanism. The flip side of that is that international comparisons of data suggest that it is the ending of a marriage that especially elevates the risk of death. Divorced males have a noticeably higher death rate than people in any other marital category (Fenwick and Barresi 1981; Hu and Goldman 1990).

Summary and Conclusion

The control of mortality has vastly improved the human condition but has produced worldwide population growth in its wake. There are, however, wide variations between and within nations with respect to both the probabilities of dying and the causes of death. The differences between nations exist because countries are at different stages of the epidemiological transition, the shift from high mortality (largely from infectious diseases, with most deaths occurring at young ages) to low mortality (with most deaths occurring at older ages and largely caused by degenerative diseases). On the other hand, differences within a society tend to be due to social status inequalities. Differences in mortality by social status are among the most pervasive inequalities in modern society. As status and prestige (indexed especially by education and income) go up, death rates go down. The social and economic disadvantages felt by minority groups often lead to lower life expectancies. Marital status is also an important variable, with married people tending to live longer than unmarried people. Females have a survival advantage over males at every age in most of the world, and a gender gap in mortality seems to be a feature of the epidemiological transition.

In general, infancy—birth to the first birthday—is the riskiest age until you are well into the retirement years. The vast majority of infant deaths in North America are due to prematurity and low birth weight, whereas infectious diseases kill many babies in developing countries. Mothers also put themselves at risk of death by getting pregnant. This risk is very high in the least developed nations of the world where women have several pregnancies in an environment of less-than-adequate health care, whereas the risk is very low in more developed nations. Declining death rates continue to revolutionize the older ages (as I will discuss in more detail in Chapter 9).

Although mortality rates are low in the more developed nations and are declining in most less developed nations, diseases that can kill us still exist if we relax our vigilance. The explosion of HIV/AIDS onto the world stage was a reminder of that, as is the emergence of SARS and Avian influenza. Worldwide efforts have been put into malaria control because that deadly disease has also been making a resurgence in many tropical regions of the world, reminding us that death control cannot be achieved and then taken for granted, for as Zinsser so aptly put it:

> However secure and well-regulated civilized life may become, bacteria, protozoa, viruses, infected fleas, lice, ticks, mosquitoes, and bedbugs will always lurk in the shadows ready to pounce when neglect, poverty, famine, or war lets down the defenses. And even in normal times they prey on the weak, the very young, and the very old, living along with us, in mysterious obscurity waiting their opportunities (Zinsser 1935:13).

If the thought of those lurking diseases scares you to death, then I suppose that too is part of the mortality transition. Also lurking around the corner is Chapter 6, in which we examine fertility concepts, measurements, and trends.

Main Points

1. Life span refers to the oldest age to which members of a species can survive, whereas longevity is the ability to resist death from year to year.

2. Although biological factors affect each individual's chance of survival, social factors are important overall determinants of longevity.

3. The most important biological determinants of death are age and sex, with the very young and the very old being at greatest risk, and with males generally having higher death rates than females.

4. Most deaths can be broadly classified as a result of infectious disease, degeneration, or the social and economic environment.

5. Infectious diseases, the control of which accounts for the worldwide drop in mortality, were historically the most important causes of death.

6. Although deaths from HIV/AIDS have slowed in the Americas and Western Europe, the disease is still spreading rapidly in the world, especially in sub-Saharan Africa, but also in eastern Europe and Asia.

7. Degeneration is associated with chronic diseases, which account for the majority of all deaths in developed nations. Heart disease is the most important of the chronic diseases as a cause of death.

8. Accidents, homicides, and suicides are the most important causes of death associated with living in the social environment.

9. The most important "real" cause of death in the United States (and increasingly in the world as a whole) is the use of tobacco.

10. Mortality is measured with the crude death rate, the age-specific death rate, and life expectancy.

11. The changes over time in death rates and life expectancy are captured by the perspective of the mortality or epidemiological transition.

12. Significant widespread improvements in the probability of survival date back only to the nineteenth century and have been especially impressive since the end of World War II. The drop in mortality, of course, precipitated the massive growth in the size of the human population.

13. The role played by public health preventive measures in bringing down death rates is exemplified by the saying a century ago that the amount of soap used could be taken as an index of the degree of civilization of a people.

14. Living in a city used to verge on being a form of latent suicide, but now cities tend to have lower death rates than rural areas.

15. Rich people live longer than poor people on average, and married people tend to live longer than unmarried people (they also have sex more often—is there a connection?)

16. The decline of mortality is a product of our control over the environment, but it has in turn put increasing pressure on that very same environment.

Suggested Readings

1. Abdel Omran, 1977, "Epidemiological Transition in the United States," Population Bulletin 32(2).

   Omran was the first to develop the concept of the epidemiological transition, and it is well summarized in this report.

2. Christopher J. L. Murray and Alan D. Lopez, Editors, 1996, The Global Burden of Disease: A Comprehensive Assessment of Mortality and Disability From Diseases, Injuries, and Risk Factors in 1990 and Projected to 2020 (Cambridge: Harvard University Press).

   This volume is part of a massive project undertaken by the Harvard School of Public Health in collaboration with the World Health Organization and the World Bank to assess the societal impact of disease and death on human societies. Although it is not an "easy read," there is a wealth of information packed into the book.

3. Norman F. Cantor, 2001, In the Wake of the Plague: The Black Death and the World it Made (New York: Free Press).

   Europe, and by historical extension the rest of the world, was remade as a result of having to cope with the huge population losses brought about periodically by the plague. This book, which was a New York Times bestseller, lays out the role of mortality in our lives.

4. Richard G. Rogers, Robert A. Hummer, and Charles B. Nam, 2000, Living and Dying in the USA: Behavioral, Health, and Social Differentials of Adult Mortality (San Diego: Academic Press).

   This book provides a comprehensive review of the literature on mortality differences in the United States and then offers important new insights into those relationships through a careful analysis of survey data.

5. Benjamin Caballero and Barry M. Popkin, Editors, 2002, The Nutrition Transition: Diet and Disease in the Developing World (San Diego: Academic Press).

   Barry Popkin and his associates have pioneered the concept that a nutrition transition accompanies the epidemiological transition, posing a new set of challenges to human health even as we bring some of the bigger causes of mortality under control.

Web Sites of Interest

Remember that Web sites are not as permanent as books and journals, so I cannot guarantee that each of the following Web sites still exists at the moment you are reading this. It is usually possible, however, to retrieve the contents of old Web sites at http://www.archive.org.

1. **http://www.nmfn.com/tn/learnctr—lifeevents—longevity**
   How long can you expect to live? Play the Longevity Game and calculate your life expectancy. This innovative "game" from Northwestern Mutual Life Insurance Company lets you calculate your own expected age at death, given the data that you enter about yourself, your health, and your lifestyle characteristics. When you are finished with that, you can go to **http://www.realage.com** and take the "Real Age Test" and record your real age; then go to **http://www.deathclock.com** and record your personal day of death; and then go to **http://channels.netscape.com/ns/atplay/lifeline.jsp** and do your lifeline calculation and write down your expected age at death.

2. Diseases are often spread by contact with "bodily fluids," especially fecal matter—which is why washing up is so important. Visit the following two web sites and consider your reaction to these findings: **http://www.biochemist.org/news/page.htm?item=8607** and
   http://www.washup.org/Cc-hand.ppt

3. **http://www.who.int/en**
   The World Health Organization of the United Nations, located in Geneva, Switzerland, is the first place to go for updates on disease outbreaks, as well as for global statistics of mortality and morbidity.

4. **http://www.mara.org.za/**
   The Mapping Malaria in Africa (MARA) project is an Africa-based international collaborative effort to map the incidence of the world's single biggest killer in the area (sub-Saharan Africa) where 90 percent of the world's deaths from malaria occur. You can download maps and data from this site.

5. **http://www.amnh.org/exhibitions/epidemic/index.html**
   In 1999, the American Museum of Natural History in Washington, D.C., organized an exhibit to show how epidemics emerge and diffuse. The exhibit is gone, but the Web site remains, offering you a useful glimpse into the way in which death can lurk around the corner.

CHAPTER 6
The Fertility Transition

Figure 6.1   Total Fertility Rates by Country

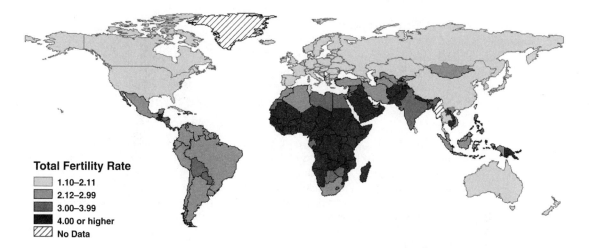

Total Fertility Rate
- 1.10–2.11
- 2.12–2.99
- 3.00–3.99
- 4.00 or higher
- No Data

Source: Adapted by the author from medium variant projections in United Nations, World Population Prospects 2002.

WHAT IS FERTILITY?
  The Biological Component
  The Social Component

MEASURING FERTILITY
  Period Measures of Fertility
  Cohort Measures of Fertility
  Fertility Intentions

THE PRECONDITIONS FOR A DECLINE
  IN FERTILITY

HOW CAN FERTILITY BE CONTROLLED?

PROXIMATE DETERMINANTS OF FERTILITY
  Limiting Exposure to Intercourse
  Lowering the Risk of Conception
    Premodern or "Traditional"
      Methods
    Breast-feeding
    Contraception
    Surgical Contraception

How Effective Are Different
  Contraceptive Techniques?
Ending a Pregnancy Before Birth
The Relative Importance of the Proximate
  Determinants

EXPLANATIONS FOR HIGH FERTILITY
  Need to Replenish Society
  Children as Security and Labor
  Desire for Sons

MOTIVATIONS FOR LOWER FERTILITY
  LEVELS
  The Supply-Demand Framework
    Income, Wealth, Prestige, and Fertility
    Education and Fertility
  The Innovation/Diffusion and "Cultural"
    Perspective
  The Role of Public Policy

> I am inclined to think that the most important of Western values is the habit of a low birth-rate. If this can be spread throughout the world, the rest of what is good in Western life can also be spread. There can be not only prosperity, but peace. But if the West continues to monopolize the benefits of low birth-rate(s), war, pestilence, and famine must continue, and our brief emergence from those ancient evils must be swallowed in a new flood of ignorance, destitution and war.

When Bertrand Russell, whose words these are (1951:49), died in 1970 at age 89, he had witnessed almost the entire demographic transition in the Western nations; birth rates in many less developed areas, however, seemed stubbornly high and few people would have predicted much of a drop. Yet drop they did, and early in the twenty-first century we are witnessing a fertility transition that represents a truly incredible revolution in the control of human reproduction, comparable in many ways to the amazing progress in postponing death, which I discussed in the previous chapter.

The **fertility transition** is the shift from high fertility, in which there is only minimal individual deliberate control, to low—perhaps very low—fertility, which is entirely under a woman's (or more generally a couple's) control. It has been summarized by Lloyd and Ivanov (1988) as the shift from "family building by fate" to "family building by design." The transition almost always involves a delay in childbearing to older ages (at least beyond the teen years) and also an earlier end to childbearing. This process helps to free women and men alike from the bondage of unwanted parenthood and helps to time and space those children who are desired, which has the desirable side effect of improving the health of mothers and their children. To control fertility does not necessarily mean to limit it, yet almost everywhere you go in the world, the two concepts are nearly synonymous. This suggests, of course, that as mortality declines and the survival of children and their parents is assured, people generally want relatively small families, and the wider the range of means available to accomplish that goal, the greater the chance of success.

The central questions of the fertility transition are why, when, and how does fertility decline from high to low levels? This means that we have to start with an understanding of why fertility has been high for all of the previous millennia of human history, and then proceed to explanations for its decline. Before we get to that, however, we need to define some concepts related to human reproduction, and we need to review the ways in which we measure it, so that we can talk more intelligently about trends in fertility over time and across the globe.

What Is Fertility?

When demographers speak of **fertility,** we are referring to the number of children born to women. This can be confusing, because physicians and agriculturalists routinely use the term to refer to reproductive *potential*—how fertile a woman is, or how fertile the soil might be, is assumed to refer to the capacity to grow things (humans in the first instance, crops in the second). But, in population studies fertility means the actual reproductive performance of women or men—how many children have they parented? Note that although our concern lies primarily with the total impact of childbearing on a society, we have to recognize that the birth rate is the accumulation of millions of individual decisions to have or not have children. Thus, when we refer to a "high-fertility society," we are referring to a population in which most women have several children, whereas a "low-fertility society" is one in which most women have few children. Naturally, some women in high-fertility societies have few children, and vice versa.

Fertility, like mortality, is composed of two parts, one biological and one social. The biological component refers to the capacity to reproduce, and while obviously a necessary condition for parenthood, it is not sufficient alone. Whether children will actually be born and, if so, how many, given the capacity to reproduce, is largely a result of the social environment in which people live.

The Biological Component

The physical ability to reproduce is usually called **fecundity** by demographers. A fecund person can produce children; an infecund (sterile) person cannot. However, since people are rarely tested in the laboratory to determine their level of fecundity, most estimates of fecundity are actually based on levels of fertility (by which we mean the number of children a person has had). Couples who have tried unsuccessfully for at least 12 months to conceive a child are usually called "infertile" by physicians (demographers would say "infecund"). The 1995 National Survey of Family Growth (NSFG) showed that 7 percent of American couples (where the wife was aged 15 to 44) are infecund/infertile by that criterion (Abma et al. 1997; Stephen and Chandra 2000)—a decline from 11 percent in 1965 (Mosher and Pratt 1990). A more general concept is the idea of **impaired fecundity** (also know as *subfecundity*), measured by a woman's response to survey questions about her fecundity status. A woman is classified as having impaired fecundity if she believes it is impossible for her to have a baby, if a physician has told her not to become pregnant because the pregnancy would pose a health risk for her or her baby, or if she has been continuously married for at least 36 months, has not used contraception, and yet has not gotten pregnant. In 1995, 10 percent of American women fell into that category—an increase from 8 percent in 1982 (Chandra and Stephen 1998).

For most people, fecundity is not an all-or-none proposition and varies according to age. Among women it tends to increase from **menarche** (the onset of menstruation, which usually occurs in the early teens), peak in the twenties, and then decline to **menopause** (the end of menstruation) (Kline, Stein, and Susser 1989). Male fecundity increases from puberty to young adulthood, and then gradually declines, but men are generally fecund to a much older age than are women.

At the individual level, very young girls occasionally become mothers (indeed, in 2002, there were more than 7,000 babies born to mothers under 15 years of age in the United States), whereas at the other end of the age continuum, there were 263 mothers in that year whose age was listed as 50–54 (Martin et al. 2003). Until the mid-1990s, the oldest verified mother in the world had been an American named Ruth Kistler, who gave birth to a child in Los Angeles, California, in 1956 at the age of 57 years, 129 days (McFarlan et al. 1991). Now, however, hormone treatment of postmenopausal women suggests that a woman of virtually any age might be able to bear a child by implanting in a woman an embryo created from a donated egg impregnated with sperm, and this has been done successfully for several women over the age of 50, including a 63-year-old Filipino-American woman, also in Los Angeles, in 1996 (Paulson et al. 2002).

The world's verified most prolific mother was a Russian woman in the eighteenth century who gave birth to 69 children. She actually had "only" 27 pregnancies, but experienced several multiple births (Mathews et al. 1995). But let us put the extremes of individual variation aside. Assuming that most couples are normally fecund, how many babies could be born to women in a population that uses no method of fertility control? If we assume that an average woman can bear a child during a 35-year span between the ages of 15 and 49, that each pregnancy lasts a little less than nine months (accounting for some pregnancy losses such as miscarriages), and that, in the absence of fertility limitation, there would be an average of about 18 months between the end of one pregnancy and the beginning of the next, then the average woman could bear a child every 2.2 years for a potential total of 16 children per woman (see Bongaarts 1978). This can be thought of as the maximum level of reproduction for an entire group of people.

No known society has ever averaged as many as 16 births per woman, and there are biological reasons why such high fertility is unlikely. For one thing, pregnancy is dangerous (in the previous chapter, I noted the high rates of maternal mortality in many parts of the world), and many women in the real world would die before (if not while) delivering their 16th child, assuming they had not died from other disease in a high-mortality society. Another problem with the above calculation is the assumption that all couples are "normally" fecund. The principal control a woman has over her fecundity is to provide herself with a good diet and physical care. Without such good care, of course, the result may well be lower fertility. In some sub-Saharan African countries, such as Cameroon and Nigeria, more than one-third of the women of reproductive age are infecund/infertile (Larsen 1995). Studies of fertility rates of U.S. blacks also suggest that the drop in fertility between the late 1880s and early 1930s was due in part to the deteriorating health conditions of black women, and that part of the post-World War II baby boom among blacks was due to improved health conditions, especially the eradication of venereal disease and tuberculosis (Farley 1970; McFalls and McFall 1984; Tolnay 1989).

Naturally, disease is not the only factor that can lower the level of fecundity in a population. Nutrition also plays a role, and Frisch (1978; 2002) was among the first to suggest that a certain amount of fat must be stored as energy before menstruation and ovulation can occur on a regular basis. Thus, if a woman's level of nutrition is too low to permit fat accumulation, she may experience **amenorrhea** (a temporary absence or suppression of menstruation) and/or **anovulatory** cycles, in which no egg is released. For younger women, the onset of puberty may be delayed until an undernourished girl reaches a certain critical weight (Komlos 1989). Conversely, improved nutrition has been linked to girls beginning menstruation at earlier ages than their less well-fed counterparts. Such findings have been reported for Bangladesh (Haq 1984), as well as for the United States (Kapolowitz and Oberfield 1999). Increased levels of fat among girls in the United States appear to have stimulated hormonal change and induced puberty in an increasing proportion of preteenagers. Frisch (2002) reports that in 1800 the average age at menarche in the United States was 16.0, dropping to 14.7 by 1880, and to 12.7 in the post-World War II era. The nutrition transition, discussed in the previous chapter, thus has the effect of making it easier for younger women to conceive, even though it is paradoxically associated with the lower mortality that would reduce the need for women to bear children at a young age.

Since the maximum level of fertility described above would require modern levels of health and nutrition, it is not the level that we would expect to find in premodern societies. A slightly different concept, **natural fertility,** has historically been defined as the level of reproduction that exists in the absence of deliberate (or at least modern) fertility control (Coale and Trussell 1974; Henry 1961). This is clearly lower than the maximum possible level of fertility, and it may be that the secret of human success lies in the very fact that as a species we have not actually been content to let nature take its course (Potts and Short 1999); that rather than there being some "natural" level of fertility, humans have always tried to exercise some control over reproduction:

> The genius of the species has not been to rely on a birth rate so high that it can overcome almost any death rate, no matter how high. The genius of the species is rather to have few offspring and to invest heavily in their care and training, so that the advantages of a cultural adaptation can be realized. Throughout 99 percent of hominid history, then, fertility was kept as low as it could be, given the current mortality (Davis 1986:52).

The clear implication of this idea is that the social component of human reproductive behavior is more important than the biological capacity to reproduce. Rodgers and his associates (2001) have used data from Denmark on twins to conclude that "slightly more than one-quarter of the variance in completed fertility is attributable to genetic influence" (p. 39). Perhaps it is only a coincidence that this is very nearly the same proportion that has been assigned to the biological component of mortality (as I discussed in Chapter 5).

The Social Component

Opportunities and motivations for childbearing vary considerably from one social environment to another, and the result is great variability in the average number of children born to women. As I mentioned in Chapter 2, hunter-gatherer societies were probably motivated to space children several years apart, thus keeping fertility lower than it might otherwise have been. It would be difficult to be pregnant and have other small children and be on the move. It may also have been, of course, that women in hunter-gatherer societies had sufficiently little body fat that their risk of conception was low enough to provide adequate spacing between children. Agricultural societies provide an environment in which more children may be advantageous, and where improved nutrition might well have improved a woman's chances of becoming pregnant more often. The low mortality and high standard of living of urban industrial and postindustrial societies reduce the demand for children well below anything previously imagined in human existence, yet paradoxically the biological capacity to reproduce is probably the highest it has ever been because people are healthier than they have ever been in human history.

It was the combination of modern medical science and a prosperous agricultural community that produced the world's most famous high-fertility group. The Hutterites are an Anabaptist (opposed to childhood baptism) religious group who live in agrarian communes in the northern plains states of the United States and the western provinces of Canada. In the late nineteenth century, about 400 Hutterites migrated to the United States from Russia, having originally fled there from eastern Europe (Kephart 1982), and in the span of about 100 years they have doubled their population more than six times to a current total of more than 30,000. In the 1930s, the average Hutterite woman who survived her reproductive years could expect to give birth to at least 11 children (Eaton and Mayer 1954).

The secret to high Hutterite fertility has been a fairly early age at marriage, a good diet, good medical care, and a passion to follow the biblical prescription to "be fruitful and multiply." Also, of course, they engage regularly in sexual intercourse without using contraception or abortion, believing as they do that any form of birth control is a sin. In the past few decades, however, population growth has slowed among the Hutterites. Each Hutterite farming colony typically grows to a size of about 130 people. Then, in a manner reminiscent of Plato's Republic, the division of labor becomes unwieldy and the colony branches off to form a new group. Branching requires that additional land be purchased to establish the new colony, and the gobbling up of vacant land by Hutterites has caused considerable alarm, especially in Canada, where a large number of Hutterites now live. Laws have been passed in Canada restricting the Hutterites' ability to buy land and, at the same time, new technological changes in farming methods (which the Hutterites tend to keep up with) have changed the pattern of work in the colonies. These social dynamics have apparently had the effect of raising Hutterite women's average age at marriage by as many as four or five years. Furthermore, access to modern health care has led women at the other end of their reproductive years to agree to sterilization for "health reasons" (Peter 1987). Overall fertility levels among Hutterites in 1980 were only half what they had been in 1930 (Nonaka, Miura, and Peter 1994; Peter 1987), and this trend has probably continued into the twenty-first century as

Hutterites confront the scarcity of new land for the expansion of their colonies (White 2002).

Lest you think that the decline of fertility among Hutterites ends the story of extremely high fertility in the modern world, I offer you the case of the Shipibo Indians of the eastern Amazon region of Peru. This population was studied intensively by Hern (1977; 1992), whose analysis of data collected in 1964–69 and again in 1982–84 revealed a level of reproduction even higher than that of the Hutterites. This was accomplished especially by an average age at marriage of only 14 years and an average age at delivery of the first child of only 15.6 years. This is, by the way, a pattern similar to that found in the United States among Hmong refugees from the highlands of Laos. Upon their arrival in the United States in the early 1980s, the Hmong continued their pattern of early and frequent motherhood, achieving levels of fertility that, at least temporarily, also exceeded the Hutterites (Weeks et al. 1989).

In Figure 6.2, I have contrasted for you the fertility rates by age that produced the Hutterites' 11 children per woman only a few decades ago. For comparison, the

Figure 6.2   Hutterite Fertility Compared with Contemporary Mexico, United States, and Canada Fertility Levels

	15–19	20–24	25–29	30–34	35–39	40–44	45–49
Hutterite "Natural"	250	460	431	396	321	167	24
Mexico 2001	44	117	147	120	66	26	4
US 2001	59	113	113	86	35	7	0
Canada 2001	22	69	108	87	30	5	0

Sources: Data for Hutterites were drawn from Warren Robinson, 1986, "Another Look at the Hutterites and Natural Fertility," Social Biology, 33:65–76; Table 6, data for Mexico, United States, and Canada are from the International Data Base of the U.S. Bureau of the Census.

figure also offers you the age pattern of fertility for women as estimated for 2001 in Mexico (2.6 children per woman), the United States (2.0 children per woman), and Canada (1.6 children per woman). Notice that Hutterite fertility levels were dramatically higher at every age than contemporary American levels, but especially at ages 20–24. Mexico, on the other hand, has a pattern almost identical to the U.S., except just a bit higher at each age. The only real difference between the U.S. and Canada is that Canadian women wait longer than U.S. women to begin childbearing, but after that their pattern is very similar to the U.S.

You can easily visualize in Figure 6.2 that fertility levels in modern societies such as Mexico, the United States, and Canada are vastly lower than among the Hutterites, who themselves had levels of reproduction below the potential biological maximum for a group. But in order to understand how the fertility transition occurs, we must first consider how fertility is measured, to know how to assess the evidence for differences from place to place and changes over time.

Measuring Fertility

The measures of fertility used by demographers attempt generally to gauge the rate at which women of reproductive age are bearing live children. Since poor health can lead to lower levels of conception and higher rates of pregnancy "wastage" (spontaneous abortions and stillbirths), improved health associated with declining mortality can actually increase fertility rates by increasing the likelihood that a woman who has intercourse will eventually have a live birth. Most rates are based on **period** data, which refer to a particular calendar year and represent a cross section of the population at one specific time. **Cohort** measures of fertility, on the other hand, follow the reproductive behavior of specific birth-year groups (cohorts) of women as they proceed through the childbearing years. Some calculations are based on a **synthetic cohort** that treats period data as though they referred to a cohort. Thus, data for women aged 20–24 and 25–29 in the year 2000 represent the period data for two different cohorts. If it is assumed that the women who are now 20–24 will have just the same experience five years from now as the women who are currently 25–29, then a synthetic cohort has been constructed from the period data.

Period Measures of Fertility

A number of period measures of fertility are commonly used in population studies, including the crude birth rate, the general fertility rate, the child-woman ratio, the age-specific fertility rate, the total fertility rate, the gross reproduction rate, and the net reproduction rate. Each one tells a little different story because each is based on a slightly different set of data. I will discuss the one that requires the least data first. Then each successive measure requires a bit more information (or at least harder-to-get information) for its calculation.

The **crude birth rate** (CBR) is the number of live births (b) in a year divided by the total midyear population (p). It is usually multiplied by 1,000 to reduce the number of decimals:

$$CBR = \frac{b}{p} \times 1,000$$

The CBR is "crude" because (1) it does not take into account which people in the population were actually at risk of having the births and (2) it ignores the age structure of the population, which can greatly affect how many live births can be expected in a given year. Thus, the CBR (which is sometimes called simply "the birth rate") can mask significant differences in actual reproductive behavior between two populations and, on the other hand, can imply differences that do not really exist. For example, if a population of 1,000 people contained 300 women who were of childbearing age and 10 percent of them (30) had a baby in a particular year, the CBR would be (30 births/1,000 total people) = 30 births per 1,000 population. However, in another population, 10 percent of all women may also have had a child that year. Yet, if out of 1,000 people there were only 150 women of childbearing age, then only 15 babies would be born, and the CBR would be 15 per 1,000. Crude birth rates in the world at the start of the twenty-first century ranged from a low of eight per 1,000 (in Bulgaria and Latvia) to a high of 51 per 1,000 in Niger. The CBR in Canada was 11, compared with 14 in the United States, and 23 in Mexico.

Despite its shortcomings, the CBR is often used because it requires only two pieces of information: the number of births in a year and the total population size. If, in addition, a distribution of the population by age and sex is available, usually obtained from a census (but also obtainable from a large survey, especially in less developed nations), then more sophisticated rates can be calculated.

The **general fertility rate** (GFR) uses information about the age and sex structure of a population to be more specific about who actually has been at risk of having the births recorded in a given year. As you can see in Table 6.1, the GFR (which is sometimes simply called "the fertility rate") is the total number of births in a year (b) divided by the number of women in the childbearing ages ($_{30}F_{15}$ —denoting females starting at age 15 with an interval width of 30, i.e., women aged 15–44):

$$GFR = \frac{b}{_{30}F_{15}} \times 1,000$$

Smith (1992) has noted that the GFR tends to be equal to about 4.5 times the CBR. Thus, in 2002, the GFR in the United States of 64.9 was just slightly more than 4.5 times the CBR of 13.9 that year.

If vital statistics data are not available, it is still possible to estimate fertility levels from the age and sex data in a census or large survey. The **child–woman ratio** (CWR) provides an index of fertility that is conceptually similar to the GFR but relies solely on census data. The CWR is measured by the ratio of young children (aged zero to four) enumerated in the census to the number of women of childbearing ages (15 to 49):

$$CWR = \frac{^{4}p^{0}}{_{35}F_{15}} \times 1,000$$

Notice that an older upper limit on the age of women is typically used with the CWR than with the GFR, because some of the children aged 0 to 4 will have been born up to five years prior to the census date. Census 2000 counted 19,176,000

Table 6.1 Calculation of Fertility Rates, United States 2002

(1) Age group	(2) Midpoint of interval	(3) F Number of women in age group	(4) b Number of births to women in age group	(5) ASFR Age-specific birth rate (per 1,000)	(6) ASFR_f Number of female births to women in age group	(7) Female births per 1,000 women	(8) Proportion of female babies surviving to midpoint of age interval	(9) Surviving daughters per woman during 5-year interval	(10) Column (2) × Column (9)
10–14	12.5	10,450,000	7,315	0.7	3,573	0.3	0.9917	0.00170	0.02119
15–19	17.5	9,895,186	425,493	43.0	207,853	21.0	0.9903	0.10401	1.82019
20–24	22.5	9,865,888	1,022,106	103.6	499,299	50.6	0.9882	0.25006	5.62634
25–29	27.5	9,334,428	1,060,391	113.6	518,001	55.5	0.9857	0.27349	7.52109
30–34	32.5	10,395,836	951,219	91.5	464,670	44.7	0.9825	0.21957	7.13596
35–39	37.5	10,964,420	453,927	41.4	221,743	20.2	0.9777	0.09887	3.70750
40–44	42.5	11,540,723	95,788	8.3	46,792	4.1	0.9705	0.01968	0.83620
45–49	47.5	10,448,000	5,224	0.5	2,552	0.2	0.9705	0.00119	0.05630
TOTALS		82,894,481	4,021,463	402.6		196.7		0.96856	26.72478

Column (4): × sum of ages 15–44 in column (3) x 1000 = GFR: 64.9

Column (5): × 5 = TFR: 2,013.0

Column (7): × 5 and divided by 1,000 = GRR: 0.983

Column (9): divided by NRR = 27.6 = mean length of generation

Column (10): divided by NRR = 27.6 = mean length of generation

Source: Birth data are from Joyce A. Martin, Brady E. Hamilton, Paul D. Sutton, Stephanie J. Ventura, Fay Menacker, and Martha Uunson, 2003, "Births: Final Data for 2002," *National Vital Statistics Reports* 52(10); Death data are from Table 5.3.

children aged 0 to 4 in the United States and 61,577,000 women aged 15 to 49; thus, the CWR was 311 children aged 0 to 4 per 1,000 women of childbearing age. By contrast, the 2000 census in Mexico counted 10,635,000 children aged zero to four and 23,929,000 women aged 15–49, for a CWR of 444.

The CWR can be affected by the underenumeration of infants, by infant and childhood mortality (some of the children born will have died before being counted), and by the age distribution of women within the childbearing years, and researchers have devised various ways to adjust for each of these potential deficiencies (Smith 1992; Weeks et al. 2004). Just as the GFR is roughly 4.5 times the CBR, the CWR is approximately 4.5 times the GFR. The CWR for the United States in 2000, as noted above, was 311, which was slightly more than 4.5 times the GFR of 67.5.

As part of the Princeton European Fertility Project, a fertility index was produced that has been useful in making historical comparisons of fertility levels. The overall **index of fertility** (I_f) is the product of the proportion of the female population that is married (I_m) and the index of marital fertility (I_g). Thus:

$$I_f = I_m \times I_g$$

Marital fertility (I_g) is calculated as the ratio of marital fertility (live births per 1,000 married women) in a particular population to the marital fertility rates of the Hutterites in the 1930s. Since they were presumed to have had the highest overall level of "natural" fertility, any other group might come close to, but not likely exceed, that level. Thus, the Hutterites represent a good benchmark for the upper limit of fertility. An I_g of 1.0 would mean that a population's marital fertility was equal to that of the Hutterites, whereas an I_g of 0.5 would represent a level of childbearing only half that. Calculating marital fertility as a proportion, rather than as a rate, allows the researcher to readily estimate how much of a change in fertility over time is due to the proportion of women who are married and how much is due to a shift in reproduction within marriage.

One of the more precise ways of measuring fertility is the **age-specific fertility rate** (ASFR). This requires a rather complete set of data: births according to the age of the mother and a distribution of the total population by age and sex. The ASFR is the number of births (b) occurring in a year to mothers aged x to $x+n$ ($_nb_x$) per 1,000 women (p^f) of that age (usually given in five-year age groups):

$$ASFR = \frac{_nb_x}{_nF_x} \times 1,000$$

As you can see in Table 6.1, in 2002, in the United States, in 2002 there were 104 births per 1,000 women aged 20 to 24. However, in 1955 in the United States, childbearing activity for women aged 20 to 24 had been more than twice that, reflected in an ASFR of 242. In 2002, the ASFR for women aged 25 to 29 was 114, compared with 191 in 1955. Thus, we can conclude that between 1955 and 2002 fertility dropped more for women aged 20 to 24 (a 57 percent decline) than for women aged 25 to 29 (a 40-percent drop).

ASFRs require that comparisons of fertility be done on an age-by-age basis. Demographers have also devised a method for combining ASFRs into a single fertility index covering all ages. This is called the **total fertility rate** (TFR). The TFR uses the synthetic cohort approach and approximates knowing how many children women have had when they are all through with childbearing by using the age-specific fertility rates at a particular date to project what could happen in the future if all women went through their lives bearing children at the same rate that women of different ages were at that date. For example, as noted above, in 2002, American women aged 25 to 29 were bearing children at a rate of 114 births per 1,000 women per year. Thus, over a five-year span (from ages 25 through 29), for every 1,000 women we could expect 570 (= 5 x 114) births among every 1,000 women if everything else remained the same. By applying that logic to all ages, we can calculate the TFR as the sum of the ASFRs over all ages:

$$TFR = \Sigma ASFR \times 5$$

As shown in Table 6.1, the ASFR for each age group is multiplied by five only if the ages are grouped into five-year intervals. If data by single year of age are available, that adjustment is not required. The TFR can be readily compared from one population to another because it takes into account the differences in age structure, and its interpretation is simple and straightforward. The TFR is an estimate of the average number of children born to each woman, assuming that current birth rates remain constant and that none of the women dies before reaching the end of the childbearing years. In 2002, the TFR in the United States was 2,013 children per 1,000 woman, or 2.01 per woman, which was well below the 1955 figure of 3.60 children per woman. A rough estimate of the TFR (measured per 1,000 women) can be obtained by multiplying the GFR by 30 or, or by multiplying the CBR by 4.5 and then again by 30. Thus, in the United States in 2002, the TFR of 2,013 per 1,000 women was slightly more than, but still close to, 30 times the GFR of 64.9.

A further refinement of the TFR is to look at female births only (since it is only the female babies who eventually bear children), producing a measure called the **gross reproduction rate** (GRR). The most precise way to do this is to calculate age-specific birth rates using only female babies, and then the calculation of the TFR for females represents the GRR, as shown in Table 6.1:

$$GRR = (\Sigma ASBR_f \times 5)/1,000$$

The GRR is interpreted as the number of female children that a female just born may expect to have during her lifetime, assuming that birth rates stay the same and ignoring her chances of survival through her reproductive years. A value of one indicates that women will just replace themselves, while a number less than one, such as the 0.983 for the United States in 2002, indicates that women will not quite replace themselves, and a value greater than one indicates that the next generation of women will be more numerous than the present one.

The GRR is called "gross" because it assumes that a woman will survive through all her reproductive years. Actually, some women will die before reaching the oldest age at which they might bear children. The risk of dying is taken into account by the **net reproduction rate** (NRR). The NRR represents the number of female children that a female child just born can expect to bear, taking into account her risk of dying before the end of her reproductive years. It is calculated as follows:

$$NRR = \sum\left(\frac{_nb^f_x}{_np^f_x} \times \frac{l_x}{500,000}\right)$$

where $_nb^f_x$ represents the number of female children born to women between the ages of x and $x + n$, which is divided by the total number of women between the ages of x and $x + n$ ($_np^f_x$). This is the age-sex-specific birth rate, which, in this example assumes a five-year age grouping of women. Each age-sex-specific birth rate is then multiplied by the probability that a woman will survive to the midpoint of the age interval, which is found from the life table by dividing $_nL_x$ (the number of women surviving to the age interval x to $x+n$) by 500,000 (which is the radix multiplied by 500,000). Note that if a single year of age data were used, the denominator would be 100,000, rather than 500,000.

The NRR is always less than the GRR, since some women always die before the end of their reproductive periods. How much before, of course, depends on death rates. In a low-mortality society such as the United States, the NRR is only slightly less than the GRR—the GRR of 0.983 is associated with a NRR of 0.969, whereas in a high-mortality society, the GRR may be considerably higher than the NRR. As an index of **generational replacement,** an NRR of one indicates that each generation of females has the potential to just replace itself. This indicates a population that will eventually stop growing if fertility and mortality do not change. A value less than one indicates a potential decline in numbers, and a value greater than one indicates the potential for growth, unless fertility and mortality change. It must be emphasized that the NRR is not equivalent to the rate of population growth in most societies. For example, in the United States, the NRR in 2002 was less than one, yet the population was still increasing by more than 2.8 million people each year. The NRR represents the future potential for growth inherent in a population's fertility and mortality regimes. However, peculiarities in the age structure (such as large numbers of women of childbearing age), as well as migration, affect the actual rate of growth at any point in time.

By adding one more column to Table 6.1, we are able to provide another useful index called the **mean length of generation,** or the average age at childbearing (typically labeled T) (Palmore and Gardner 1983). Column 10 illustrates the calculation. You multiply the midpoint of each age interval (column 2) by the surviving daughters per woman for that age interval (column 9) and then you divide the sum of those calculations by the net reproduction rate (the sum of column 9), yielding a figure for 2002 of 27.6 years.

Cohort Measures of Fertility

Cohort data follow people through time as they age, rather than taking snapshots of different people at regular intervals, which is what period data do. Thus, the basic measure of cohort fertility is births to date, measured as the **cumulated cohort** fertility rate (CCFR), or the total number of **children ever born** (CEB) to women. For example, women born in 1915 began their childbearing during the Depression. By the time those women had reached age 25 in 1940, they had given birth to 890 babies per 1,000 women (Heuser 1976). By age 44 (in 1959), those women had finished their childbearing in the baby boom years with a completed fertility rate of 2,429 births per 1,000 women. We can compare those women with another cohort of women who were raised during the Depression and began their childbearing right after World War II. The cohort born in 1930 had borne a total of 1,415 children per 1,000 women by the time they were age 25 in 1955. This level is 60 percent greater than the 1915 cohort. By age 44 (in 1974), the 1930 cohort had borne 3,153 children per 1,000 women, 30 percent higher than the 1915 cohort. Indeed, examining cohort data for the United States, it turns out that the women born in 1933 were the most fertile of any group of American women since the cohort born in 1881.

Frejka and Kingkade (2001) have tracked more recent trends in U.S. fertility with cohort data by calculating the average number of children born to women by exact age 27 (the average age at childbearing, as you recall from Table 6.1) among successive birth cohorts of women. Women born in 1950–51 had given birth to an average of 1.201 children by age 27 (which for them was 1977–78), whereas women born in 1960–61 had given birth to an average of 1.057 children by age 27 (which was 1987–88), but the cohort of women born in 1970–71 had given birth to 1.072 children by age 27 (1998–99). This is indicative of a recent rise in fertility in the United States, which is probably a consequence of increasing immigration to the U.S., as I will discuss in more detail later.

Cohort information is very illuminating; keep in mind that we cannot always wait for women to go through their childbearing years to estimate their level of fertility, which is why we typically use the synthetic cohort approach to calculate the TFR.

Fertility Intentions

Our understanding of the fertility transition in the world has been quietly, but importantly, influenced by research on fertility intentions. These are data on what the women who are presently of childbearing age say they intend do in the future in terms of having children. The idea for collecting this kind of information was inspired by demographers who had failed to forecast the baby boom that occurred after World War II. They realized after the fact that many couples had been postponing having babies before the war (because of the Depression) and during the war (because of the uncertainty and disruptions caused by the war), but they still intended to have more babies. The eventual fulfillment of these intentions helped to fuel the baby boom, as I will discuss later.

Demographers thus try to estimate future fertility by asking women about the number of births they expect in the future. For example, in June 1971, the Current Population Survey asked women how many births they expected to have had by 1976. Women aged 18 to 24 (the birth cohorts of 1946 to 1953) responded with an average of 1.99 births. Overall, they expected to have 2.38 births on average during their lifetime. In 1976, when they were reinterviewed, it appeared that their expectations had exceeded reality; they were averaging only 1.76 births per woman, and it seemed unlikely that they would reach their lifetime expectations. But, by 1981, when interviewed a third time (now aged 28 to 34), their behavior had nearly caught up with their expectations of a decade earlier, and they were averaging 2.21 children each. What had happened? Women who were single in 1971 failed to anticipate their average delay in getting married and, within marriage, a postponement of children. Thus, their short-run expectations missed the mark (O'Connell and Rogers 1983). Yet ultimately they made up for lost time, and by 1986 they had reached their expected number of lifetime births.

Period rates are prone to this problem of being influenced by the *timing* of births, which may distort the underlying *quantum* of births (Bongaarts and Feeney 1998), but data on lifetime births expected by women can provide a clue to the number of births that will eventually be produced, even if the timing cannot be well predicted. For example, between 1976 and 1998, the TFR increased from 1.7 to 2.1 births per woman in the United States, seeming to show that fertility was on the rise during that 22-year period of time. However, data on lifetime births expected by women who were ages 30 to 34 actually went down from 2.4 in 1976 to 2.1 in 1998, while the number of lifetime births expected by women ages 18 to 24 and 25 to 29 remained virtually the same, at 2.1 and 2.0 respectively. This tells us that the TFR in 1976 was lower than it should have been because, as mentioned above, women were just postponing births to a later date. Although the timing had changed, young women in 1998 had exactly the same family size in mind (an average of 2.1 children) as did young women back in 1976.

The usefulness of data on fertility intentions has been the subject of considerable debate for a number of years (see, for example, Demeny 1988; Masnick 1981; Westoff and Ryder 1977). Westoff (1990) used data from 134 different fertility surveys conducted in 84 different countries to conclude that "the proportion of women reporting that they want no more children has high predictive validity and is therefore a useful tool for short-term fertility forecasting" (p. 84). A key issue is that fertility intentions assume that people have a fixed target in mind with respect to family size, whereas in reality people have children one at a time (except for the occasional multiple birth), and each new birth may change the calculation about how many additional children are wanted, especially since circumstances may change in the lives of individuals over time (Morgan 2001). Nonetheless, after an exhaustive review of the literature, Morgan (2001:160) concluded that "the evidence is clear: intentions strongly predict subsequent behavior." This then raises the obvious point—what is it that eventually convinces people that they should want to have a small family, given that people seemed to want large families until very recently in human history?

The Preconditions for a Decline in Fertility

In 1973, in response to the findings emerging from the Princeton European Fertility Project (which I discussed in Chapter 3), Ansley Coale tried to deduce how an individual would have to perceive the world on a daily basis if fertility were to be consciously limited as happens in the process of the fertility transition. In this revised approach to the demographic transition, he states that there are three **preconditions**

for a substantial fertility decline: (1) the acceptance of calculated choice as a valid element in marital fertility, (2) the perception of advantages from reduced fertility, and (3) knowledge and mastery of effective techniques of control (Coale 1973). Although the societal changes that produced mortality declines may also induce fertility change, they will do so, Coale argues, only if the three preconditions exist. The preconditions, though, may exist even in the absence of mortality decline. The important causal factors determining whether these preconditions will exist include "unmeasured traditions and habits of mind" (Burch 1975), especially secularization, as I discussed in Chapter 3 and will mention again later.

Coale's first precondition thus goes to the very philosophical foundation of individual and group life: Who is in control? If a supernatural power is believed to control reproduction, then it is unlikely that people will run the risk of offending that deity by impudently trying to limit fertility—even when they know that means of fertility control exist. On the other hand, the more secular people are (even if still religious), the more likely it is that they will believe that they and other humans have the ability to control important aspects of life, including reproduction. Control need not be in the hands of a god for a person not to be empowered. If a woman's life is controlled by her husband or other family members, then she is not going to run the risk of insult or injury by doing things that she knows are disapproved of by those who dominate her (Bledsoe and Hill 1998). The status of women, not just secularization, is an important part of this first, basic precondition for a decline in fertility.

The second precondition recognizes that more is required than just the belief that you can control your reproduction. You must have some reason, in particular, to want to limit fertility. Otherwise, the natural attraction between males and females will lead to unprotected intercourse and, eventually, to numerous children. What kinds of changes in society might motivate people to want fewer children? Davis (1963; 1967) suggested that people will be motivated to delay marriage and limit births within marriage if economic and social opportunities make it advantageous for them to do so. He argued further that having children per se rarely satisfies an end in itself but is generally a means to some other end, such as expanding the ego, proving gender roles, maintaining a descent group, continuing the inheritance of property, securing economic welfare in old age, assuring future family labor, or satisfying the believed demand of a deity. If the important goals change, or if the means available to achieve those goals change (for example, money instead of children), then the desire to have children may change.

Coale's third precondition involves the knowledge and mastery of effective means of fertility control. Specific methods of fertility control may be thought of as technological innovations, the spread of which is an example of diffusion. Women in the U.S. and Canada now typically use the pill to space children and then use surgical contraception to end reproduction after the desired number of children are born. This is a different set of techniques than prevailed 30 years ago, and 30 years from now the mix will doubtless be different still. At the same time that knowledge of methods is being diffused, part of the decision about what method of fertility regulation to use is based on the individual's cost-benefit calculation about the "costs of fertility regulation." The economic and psychosocial costs of various methods may well change over time and cause people to change their fertility behavior accordingly, keeping in mind that not all possible avenues of fertility control are open to all people. Abortion is a good example. Although it is a legal method of birth control in the United States, it is not "available" to people who object to it for religious or personal reasons.

The remainder of the chapter uses these three preconditions for a fertility decline as an organizing framework to understand the fertility transition. We first have to know how it is that humans are able to control fertility, recognizing that most methods of fertility limitation are very recent in human history. Then we need to explore the changes in social structure that may be associated with the way in which humans view their role in reproduction (the first precondition), and that also influence the motivations to use techniques of fertility control to limit fertility. During this discussion you should keep two points in mind: (1) The decline in mortality and the incredible increase in child survival precipitated societywide examinations of how to limit fertility, and (2) The three preconditions do not necessarily operate in a strictly linear fashion; it is possible, for example, that the availability of a particular type of contraceptive method (e.g., sterilization becoming available to Hutterite women) can encourage people to think in different ways about their control over reproduction and to reassess the number of children they want or intend to have.

How Can Fertility Be Controlled?

Earlier in the chapter, I discussed the fact that "natural" fertility is rarely as high as the maximum level that would be possible. In most societies, families are trying (or at least hoping) to have the number of surviving children that will be most beneficial to them. But people for most of human history have lived close to the subsistence level and have lived in the shadow of high death rates. Thus, it is not surprising that in such circumstances, couples are unlikely to have a preference for a specific number of children (van de Walle 1992). The vagaries of both child mortality and the food supply were apt to cause people to "play things by ear" rather than plan in advance the number of children desired. With high mortality, how many children are born is less important than how many survive—which is how the net reproduction rate is defined, as you will recall.

When we realize that it is net reproduction (surviving children, not just children ever born) that is of importance, we can see that human beings have been very clever at dealing with family size by controlling the *family,* rather than by controlling *fertility.* These societal interconnections are diagrammed in Figure 6.3. For example, higher-than-desired fertility in terms of live births can be responded to after a child is born by what Skinner (1997) has called "**child control,**" or by what Mason (1997) has labeled "postnatal control." There are at least three ways of dealing with a child who is not wanted or cannot be cared for by his or her parents after birth: (1) infanticide (known to have been practiced in much of Asia), or a less dramatic general neglect of or selective inattention to an unwanted child that leads to early death because of not being properly nourished or cared for (Cherfas 1980; Scrimshaw 1978; Simmons et al. 1982); (2) **fosterage** (sending, or even selling, an "excess" child to another family that needs or can afford it—a relatively common practice in sub-Saharan Africa and parts of Asia, and not uncommon in pretransition Europe); and (3) **orphanage,** which involves abandoning a child in such a way that she or he

Figure 6.3   The Interconnecting Influences on the Fertility Transition

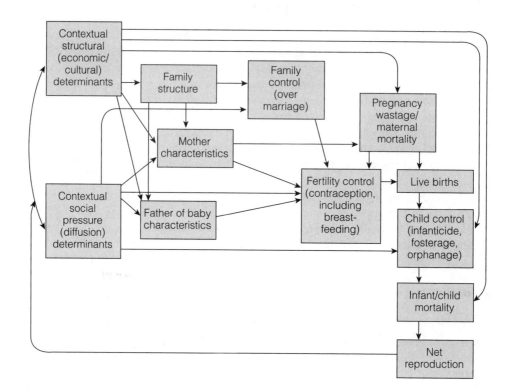

Source: Prepared by author.

is likely to be found and cared for by strangers. This was a pretransition practice in much of Europe, where children were abandoned on the steps of a church (Kertzer 1993).

Family control does not end at that point, of course. As more children survive through childhood, the child-control options become more difficult because the number of children who can no longer be afforded may stretch the limits of what families can get away with in terms of infanticide, fosterage, and orphanage. This hearkens back to the theory of demographic change and response, which I discussed in Chapter 3. As declining infant and child mortality impinges on familial resources, the family's reaction may be to work harder, especially if "child control" is not available to them. Then, as children become teenagers (sometimes even before that), they may be sent elsewhere in search of work, either to reduce the current burden on the family or more specifically to earn wages to send back to the parents. These decisions are typically made by the family, not by the individual young person, and they are made in direct, albeit belated, response to the drop in the death rate that permitted these "excess" children to survive to adulthood.

Families have also exercised important control over fertility by determining the age at which their daughters will be allowed to marry, and by heavily supervising the premarital activities of young female family members to ensure that they are not exposed to the risk of pregnancy prior to marriage. The practice of chaperoning teenagers and other unmarried young adults was very common in much of the world until fairly recently, but it requires that there be a substantial system of surveillance in place within a household, and that is generally inconsistent with a world in which both parents work and where the older generation (the grandparents) may reside separately from the parents and their children. These family and household changes have gone hand in hand with major advances in the variety and effectiveness of methods of controlling fertility by directly influencing its proximate determinants.

Proximate Determinants of Fertility

The means for regulating fertility have been popularly labeled (popular, at least, in population studies) the **intermediate variables** (Davis and Blake 1955). These represent 11 variables through which any social factor influencing the level of fertility will operate. Davis and Blake point out that there are actually three phases to fertility: *intercourse, conception,* and *gestation.* Intercourse is required if conception is to occur; if conception occurs, successful gestation is required if a baby is to be born alive. Table 6.2 lists the 11 intermediate variables according to whether they influence the likelihood of intercourse, conception, or gestation.

Although each of the 11 intermediate variables plays a role in determining the overall level of fertility in a society, the relative importance of each varies considerably. Bongaarts (1978; 1982) has been instrumental in refining our understanding of fertility control first by calling these variables the **proximate determinants of fertility** instead of intermediate variables, and secondly by suggesting that differences in fertility from one population to the next are largely accounted for by only four of those variables: proportion married, use of contraceptives, incidence of abortion, and involuntary **infecundity** (especially **postpartum** infecundity as affected by breast-feeding practices). These variables are noted with a checkmark in Table 6.2. Although it gets ahead of our story a bit, it is worth noting that the transition from high to low fertility is usually accomplished with a combination of declines in breast-feeding (a seemingly contradictory pattern that I discuss below), later marriage, increased use of **contraception,** and increased incidence of abortion.

Bongaarts does not mean to imply, however, that the other intermediate or proximate determinants are irrelevant to our understanding of fertility among humans, only that they are relatively less important, and for this reason I will focus attention on the most important determinants—those checked in Table 6.2. As we review these determinants, you will discover, by the way, that there is heavier emphasis on the behavior of women than of men. That is simply because if a woman never has intercourse, she will never have a child, whereas a man will never have a child no matter what he does. Of course, avoiding conception by sterilization or contraceptives can be done by either sex, but if conception occurs, it is only the woman who bears the burden and the risk of either pregnancy or abortion.

Table 6.2   The Proximate Determinants of Fertility—
The Intermediate Variables through which Social Factors Influence Fertility

Most Important of the Proximate Determinants	Proximate Determinants or Intermediate Variables
	I. Factors affecting exposure to intercourse ("intercourse variables").
	A. Those governing the formation and dissolution of unions in the reproductive period.
✔	1. Age of entry into sexual unions (legitimate and illegitimate).
	2. Permanent celibacy: proportion of women never entering sexual unions.
	3. Amount of reproductive period spent after or between unions.
	a. When unions are broken by divorce, separation, or desertion.
	b. When unions are broken by death of husband.
	B. Those governing the exposure to intercourse within unions.
	4. Voluntary abstinence.
	5. Involuntary abstinence (from impotence, illness, unavoidable but temporary separations).
	6. Coital frequency (excluding periods of abstinence).
	II. Factors affecting exposure to conception ("conception variables").
✔	7. Fecundity or infecundity, as affected by involuntary causes, but including breast-feeding.
✔	8. Use or nonuse of contraception.
	a. By mechanical and chemical means.
	b. By other means.
	9. Fecundity or infecundity, as affected by voluntary causes (sterilization, medical treatment, and so on).
	III. Factors affecting gestation and successful parturition ("gestation variables").
	10. Fetal mortality from involuntary causes (miscarriage).
✔	11. Fetal mortality from voluntary causes (induced abortion).

Sources: Kingsley Davis and Judith Blake, 1955, "Social Structure and Fertility: An Analytic Framework," Economic Development and Cultural Change 4(3). Used by permission; modified using information from John Bongaarts, 1982, "The Fertility-Inhibiting Effects of the Intermediate Fertility Variables," Studies in Family Planning 13:179–189.

Limiting Exposure to Intercourse

Permanent virginity is obviously very rare, but the longer past puberty a woman waits to begin engaging in sexual unions, the fewer children she will probably have because of the shorter time she will be at risk of bearing children (Variable 1 in Table 6.2). Perhaps due to lower surveillance by the family, the percentage of teenage girls aged 15–17 in the United States who had ever engaged in sexual intercourse rose to between one-third and one-half of girls by the late 1990s, depending on which survey results are used (Santelli et al. 2000). At the same time, however, the average age at first marriage has been rising in the United States, suggesting that women are sexually active for an increasingly longer time prior to marriage (helping to account for the increase in out-of-wedlock births, as I will discuss in Chapter 10). The percentage of teenage boys who have ever had intercourse seems to be on the decline, and in the late 1990s the percentage (50 percent or less) was closer to that of girls than it had previously been (Santelli et al. 2000). Probably in response to sex education programs, including the fear of HIV/AIDS, teenagers seem to be substituting various forms of nonvaginal sexual behavior for vaginal intercourse (Gates and Sonenstein 2000). The condom is the most popular type of contraception among U.S. teenagers, but scarcely half reported using it during their most recent sexual intercourse, according to a survey by Abma and Sonenstein (2001). Furthermore, there is evidence that many teenagers are clueless about the proper use of the condom and so may be less protected than they think (Crosby and Yarber 2001).

Modern contraception has altered the relationship between intercourse and having a baby, but it remains true that an effective way to postpone childbearing is to postpone engaging in sexual activity, particularly on the regular basis implied in marriage. Historically, those societies with a later age at marriage have been the ones in which fertility was lower, and it is still true that more traditional populations in sub-Saharan Africa and south Asia are characterized by early marriage for women and consequent higher-than-average levels of fertility (Singh and Samara 1996). In Niger (sub-Saharan Africa) and Bangladesh (South Asia), half of all women already have had at least one child by the time they reach 18 (Alan Guttmacher Institute 1998).

Permanent **celibacy** (Variable 2) refers to those women who never marry. It will not surprise you to learn that the greater the proportion of women in a society who never marry, the lower the level of fertility tends to be. The idea of celibacy has not caught on, though, and the proportion of women who never marry is consistently low, never reaching even 10 percent outside of Europe. The highest level of permanent celibacy occurs in northwestern Europe, which has historically also had among the lowest levels of fertility in the world. For decades, Ireland had the highest rate of non-marriage in the world, reaching a level of 18 percent of women aged 40–44 never having been married in 1971 (Ireland Central Statistics Office 1978). However, by the late 1990s, the rate had dropped to 12 percent (U.S. Census Bureau 2001a) and in all events remaining single no longer means that a woman will necessarily remain childless.

Under normal circumstances, once a woman is involved in a sexual union such as marriage, sexual intercourse will be more or less regularly engaged in. However, not all sexual unions are permanent, since some are broken up by divorce, separation, or desertion (Variable 3a), while some are broken up by the death of a spouse (Variable 3b). During that fecund period of life when a woman has had a sexual union broken and has not entered into a new one, months or years of pregnancy risk are lost, never to be recouped. Data for the United States and nearly every other country of the world reveal that regardless of the age at which a woman first married, if her marriage has broken up, her fertility will be lower than if she had remained continuously married. While I sincerely doubt that people deliberately break up a marriage just to keep from having children, the result is the same.

Voluntary abstinence (Variable 4) will clearly eliminate the risk of pregnancy within a marriage. This is not a popular option, but historically it was probably an important way in which fertility was brought down in England and other European countries (Szreter 1996). Abstinence seems now to be uncommon among couples in developed nations except shortly after the birth of a child, when abstinence does not have much effect on fertility anyway, since childbirth is normally followed by a period of one or two months of temporary sterility (called *postpartum amenorrhea,* and referring to the fact that a woman does not ovulate during this time). In less developed societies, postpartum (that is, after-childbirth) taboos on intercourse occasionally extend to several months or even a few years in societies in which intercourse is forbidden while a mother is nursing a child. The rationale is based generally on superstitions that intercourse will somehow be harmful to either mother or child (Davis and Blake 1955) or on the generally correct notion that a pregnancy will be harmful to the nursing child because it will reduce the quantity of the mother's milk. Voluntary abstinence after childbirth may have helped to keep birth rates below the potential maximum in many premodern societies by lengthening the interval between pregnancies, but abstinence on the part of the woman can also have the side-effect of encouraging her male partner to seek other sexual outlets, which then contributes to the spread of sexually transmitted disease.

Abstinence has been an effective means of fertility limitation for highly motivated people. For example, in 1970, Brandes (1975) studied a small Spanish peasant village in which there was considerable pressure to limit family size to two children. He relates the story of a 35-year-old man with five young children who "is constantly greeted with the refrain, '¡Atate al catre!'—Tie yourself to the bedpost!" Brandes goes on to explain:

> Through a combination of coitus interruptus and abstention, most villagers manage to keep births down. Though they openly complain about the frustrations which these methods entail, they are willing to make the sacrifice in order to assure their ability to provide their children the best possible opportunities. Those husbands who cannot exercise control are seen as somewhat animalistic (1975:43).

Thus, if the motivation exists to limit family size, people will do so, turning to the time-honored methods of abstinence and coitus interruptus (discussed later in the chapter). As recently as the 1980s, an American woman wrote to "Dear Abby" explaining that "the most effective method yet found for birth control is a large dog sleeping in the middle of the bed on top of the covers. Try it. It works. That's what we've been doing for 17 years" (Universal Press Syndicate 1982).

Involuntary abstinence (Variable 5) could be a result of either impotence or involuntary separation. It is not easy to tell how important the role played by impotence is in involuntary birth control. On the other hand, we do know that temporary separations are fairly common in many occupations, such as sales and transportation, and for various other reasons—separate vacations, hospitalization, military duty, and so on. In less developed nations, where migratory labor is more common than in the United States, we might also expect to find involuntary abstinence to lower fertility somewhat, if husbands are separated from their wives for considerable periods of time. However, there is also evidence that males return home often enough to impregnate their wives, so this effect on fertility is probably very small.

Within marriage (or any sexual union) the regularity of sexual intercourse (Variable 6) will influence the likelihood of a pregnancy, particularly if no contraception is used. In general, the more often a couple has intercourse, the more likely it is that the woman will get pregnant, primarily because it increases the chance that intercourse will occur during the fertile period of the menstrual cycle.

Lowering the Risk of Conception

Premodern or "Traditional" Methods   It is probable that at least some people in most societies throughout human history have pondered ways to prevent conception (Himes 1976). Abstinence, withdrawal, and the douche are the most ancient of such premodern means, but there is some historical evidence that various plants were used in earlier centuries to produce "oral contraceptives" and early-stage abortifacients (Riddle 1992). We do not know much about the actual effectiveness of such methods, but they were almost certainly far less effective in preventing conception or birth and far riskier for a woman's health than are modern methods. The Shipibo Indians in the Amazon basin of Peru, whose extraordinarily high fertility was noted earlier, nonetheless (or maybe because of this) know of an herb that is believed to have contraceptive powers. It is made from the root of the cyperus plant (a bog plant, not the same as the cypress), and they believe that it induced temporary sterility (Maxwell 1977). Its real level of effectiveness, however, appears to be nil (Hern 1976). To be sure, the lack of effectiveness of the premodern methods meant that a badly unwanted pregnancy was more likely to end in an attempted abortion or the woman trying to conceal her pregnancy and then abandoning the baby (probably leading to the infant's death) after a secret delivery (see, for example, van de Walle 2000).

There have been references to douching throughout recorded history, stretching back to ancient Egypt (Baird et al. 1996; Himes 1976), and it is one of the principal means of contraception mentioned by Charles Knowlton in his famous *Fruits of Philosophy* early in the nineteenth century (which I mentioned in Chapter 3). Over time, it has been recommended as a means of treating specific gynecological conditions and also as a contraceptive, on the theory that washing sperm out of the vagina right after intercourse (the "dash for the douche") might prevent conception. Unfortunately, for the one doing the douching, the sperm take only about 15 seconds to travel through the vagina into the cervical canal, so the effectiveness of douching is very limited. There is also some evidence that douching may increase the risk of pelvic inflammatory disease (Hatcher et al. 1994).

Withdrawal is an essentially (although not exclusively) male method of birth control. It has a long history (it is, in fact, referenced in the Bible). It is actually a form of incomplete intercourse (thus its formal name "coitus interruptus") because it requires the male to withdraw his erect penis from his partner's vagina just before ejaculation. The method leaves little room for error, especially since there may be an emission of semen just before ejaculation, but it is one of the more popular methods historically for trying to control fertility.

Breast-feeding   Even if a woman is regularly sexually active, she will not necessarily conceive a child if she may be involuntarily infecund, as I mentioned previously (Variable 7), and one of the major ways in which human societies differ in this regard is with respect to **breast-feeding** practices. Breast-feeding prolongs the period of postpartum amenorrhea and suppresses ovulation, thus producing in most women the effect of temporarily impaired fecundity. In fact, nature provides the average new mother with a brief respite from the risk of conception after the birth of a baby whether she breast-feeds or not; however, the period of infecundity is typically only about two months among women who do not nurse their babies, compared with 10–18 months among lactating mothers (Konner and Worthman 1980). Research suggests that stimulation of the nipple during nursing sets up a neuroendocrine reflex that reduces the secretion of the luteinizing hormone (LH) and thus suppresses ovulation (Jones 1988). A study in Indonesia found that women whose babies nurse intensively (several nursing bouts per day of fairly long duration) delay the return of menses by an average of 21 months, nearly twice the delay of those women who breast-feed with low intensity (Jones 1988). On the other hand, the cessation of lactation signals a prompt return of menstruation and the concomitant risk of conception in most women (Prema and Ravindranath 1982).

Although breast-feeding is a natural contraceptive and keeps fertility lower than it would otherwise be, the percentage of women who breast-feed their babies has tended to decline with modernization, and so we find the seemingly contradictory fact that breast-feeding declines as fertility declines. In the earlier decades of modernization, bottle-feeding was seen as preferable to breast-feeding because it gave the mother more flexibility to work and have someone else care for her child, and there was a certain cachet attached to being able to bottle-feed rather than engaging in the more "peasant" activity of breast-feeding. In Paris in the nineteenth century, there was an increase in the percentage of women working outside the home (especially among middle-class artisans and shopkeepers). Before bottle-feeding, a woman with a baby who wished to continue working had to hire a wet nurse. Most wet nurses were peasant women who lived in the countryside, and a mother would have to give up the child for several months if she wished to keep working. Infant mortality was as high as 250 deaths per 1,000 infants among those placed with wet nurses (Sussman 1977), yet in the early 1800s nearly one-fourth of all babies born in Paris were placed with wet nurses. Infants were a bother, and the risk was worth taking.

Over the past few decades, there has been concern that women in less developed nations were abandoning breast-feeding in favor of bottle-feeding. In the absence of some method of contraception, a decline in breast-feeding would, of course, increase fertility by spacing children closer to one another. Closer spacing poses a threat to the health of both mother and child, and bottle-feeding also is likely to raise the infant death rate, since in less developed nations bottle-feeding often is accompanied by watered-down formulas that are less nutritious than a mother's milk. Bacteria growing in unsterilized bottles and reused plastic liners can also lead to disease, especially diarrhea, which is often fatal to infants.

In response to these concerns, UNICEF and the World Health Organization approved a voluntary code in 1981 to regulate the advertising and marketing of infant formula. However, baby food companies lobbied strongly against the code, the

United States opposed it, and a year later only 6 percent of the countries who voted for it had actually adopted it (Adelman 1982). The pressure continued to mount, however, and in 1991 the Swiss firm Nestlé finally decided to limit its supply of free baby formula to third world hospitals (Freedman 1991), and other baby formula suppliers quickly followed suit. Of some interest is the publication in that same year of a study suggesting that the marketing practices of the infant food industry may have a smaller impact on a mother's decision to breast- or bottle-feed than is com-monly assumed (Stewart et al. 1991).

The women most likely to lead the movement back to breast-feeding in any given country are, somewhat ironically, the better educated (Akin et al. 1981; Hirschman and Butler 1981), the very same women whose fertility is apt to be kept low by deliberate use of other means of contraception rather than by the influence of lactation. In the United States in 1995, 81 percent of women with a college degree breast-fed their most recent baby, compared with less than half of the mothers with only a high school education or less (Abma et al. 1997).

In the 1980s, breast-feeding was elevated to the status of a natural contraceptive, especially through the issuance of the so-called Bellagio Consensus. In 1988, a group of experts met in Bellagio, Italy, to review the evidence regarding the use of breast-feeding as a deliberate method of preventing conception. "The consensus of the group was that the maximum birth spacing effect of breastfeeding is achieved when a mother 'fully' or nearly fully breastfeeds and remains amenorrheic. When these two conditions are fulfilled, breast feeding provides more than 98 percent protection from pregnancy in the first six months" (Kennedy, Rivera, and McNeilly 1989). Other analyses have confirmed the important fertility-reducing impact of lactation for those mothers who are willing to breast-feed on demand, whenever or wherever that might be (Guz and Hobcraft 1991).

Contraception   No matter how effective breast-feeding might be, the first thing we usually think of when we consider ways to control fertility is a **contraceptive device** (Variable 8), such as the pill. There are so many different types of contraceptives, not to mention new research being conducted on contraceptives, that it is useful to categorize methods before discussing them. In Table 6.3, I have divided contraceptive methods into three main categories, according to whether they are primarily female, male, or couple methods. Under each of those categories, I have listed methods according to whether their action is primarily barrier, chemical, natural, or surgical. Since individuals or couples often combine methods, this classification scheme is not mutually exclusive. **Surgical contraception** is actually captured by the category of voluntary infecundity (Variable 9), but it is now more commonly included among contraceptive methods.

Barrier methods include the diaphragm; cervical cap; a disposable vaginal sponge; the intrauterine device (IUD); the female condom; and vaginal spermicidal foams, jellies, suppositories, and film. The **diaphragm** is a rubber disk that is inserted deep into the vagina and over the mouth of the uterus sometime before sexual intercourse. Used alone, it is a fairly ineffective physical barrier. As with all barrier methods, it is normally used in conjunction with a **spermicide** that is spread in and around the diaphragm before insertion into the vagina. The spermicide acts to kill the sperm, and the diaphragm operates to keep the spermicide in place, preventing

Table 6.3   Contraceptive Methods

Primary Characteristic	Primary Use		
	Female	Male	Couple
Barrier	Diaphragm	Condom	
	Cervical cap		
	Female condom		
	Vaginal spermicides		
	IUD		
Chemical			
Precoital	Pill		
	Mini-pill		
	Implants		
	Injectables		
	Vaginal ring		
Postcoital	Emergency contraceptive pills		
Natural	Breast-feeding	Withdrawal	Abstinence
			Fertility awareness (NFP)
			Oral/anal sex and other forms of nonvaginal intercourse
Surgical	Tubal ligation	Vasectomy	

the passage of sperm into the uterus and the fallopian tubes, where fertilization occurs. Since sperm can live in the vagina for several hours, a woman must leave the diaphragm in place for a while after intercourse. Used in this way, the diaphragm can be very effective, and until the 1960s when oral contraceptives were placed on the market, the diaphragm (invented in 1883) was a fairly common method of birth control for married women. Because it requires a great deal of forethought, it is obviously not useful for those occasions, especially outside of marriage, when intercourse occurs spontaneously. Variations on the diaphragm theme have appeared over the years, and there really is not space in this chapter to review them all (there are Web sites listed at the end of the chapter where you can find detailed information on these and other contraceptive techniques).

The intrauterine device (IUD) was first designed in 1909, but it was not widely manufactured and distributed until the 1960s. At that time, it was the contraceptive technique that many family planners believed would bring an end to the world's population explosion. Its success however, has been much less spectacular than was hoped, and although it is still used by millions of women around the world, it has not enjoyed the wholesale acceptance that many had predicted. No one knows for certain how an IUD works to prevent a birth, but it is generally believed that, since it is a foreign object, it sets up a chemical reaction inside the uterus that may destroy the fertilized egg and/or prevent the fertilized egg from being implanted (Sivin 1989). About 2–3 percent of the women who have an IUD expel it, but for the remainder, the IUD theoretically can stay in place for long periods of time and is very effective in preventing pregnancy. In less developed areas of the world, however, high proportions of women who have an IUD inserted choose to have it removed even if they do not want to get pregnant. The principal reasons are infections and excessive bleeding caused by the IUD, although fear and superstition may also be causes. The IUD is one of the most popular contraceptive methods in China—much more widely used than the pill (Short, Linmao, and Wentao 2000)—and it is possible that more than two-thirds of all IUD users in the world are in China (Poston 1986).

The oral contraceptive, or "the pill" as it is popularly known, has revolutionized birth prevention for millions of women all over the world. The pill is a compound of synthetic hormones that suppress ovulation by keeping the *estrogen* level high in a female. This prevents the pituitary gland from sending a signal to the ovaries to release an egg. In addition, the *progestin* content of the pill makes the cervical mucus hostile to implantation of the egg if it is indeed released and may block the passage of sperm as well (similar to the action of hormones used in the IUDs). The combination pill is virtually 100-percent effective when taken as directed. It has estrogen and progestin in each pill and typically is taken for 21 days, followed by seven days when it is not taken and the woman has a menstrual flow. The sequential, or triphasic, pill varies the dosages of estrogen and progestin over the menstrual cycle (see Figure 5.2), trying to mimic the "natural" change in a woman's hormones while still protecting against pregnancy (Hatcher et al. 1994). The mini-pill contains only a progestin, is taken every day, thickens the cervical mucus, thins the cervical lining (making fetal implantation more difficult), and may even retard ovulation. It is not quite as effective as the combination pill, but has the advantage of having one less hormone for a woman's body to deal with. Mini-pills are especially useful for breast-feeding women, because they offer additional protection against pregnancy without affecting lactation.

In the 1960s and 1970s, the pill was the method of choice for women of all ages, and it continues to be the most popular nonsurgical method of birth control in the United States (see Table 6.4). All things considered, the pill would be just about the perfect solution except for the problem of side effects. When first taking the pill, women often report nausea, breast tenderness, and spotting, although these problems usually do not persist. More important by far are the risks of death that might be associated with the pill. Although the evidence is far from conclusive, it appears that oral contraceptives can aggravate problems such as high blood pressure, blood clotting, diabetes, and migraine headaches. Women with histories of these problems are normally advised not to use the pill. Much of the early scare about increased risk of cancer, however, has subsided as researchers have been able to lower the dosages of chemicals in the pills and as they have learned that oral contraceptives do not, in fact, increase the risk of breast cancer, even after prolonged use (Hatcher 1998; Wharton and Blackburn 1988). More encouraging are the studies suggesting that using the pill may actually reduce a woman's risk of anemia and certain cancers (Blackburn, Cunkelman, and Zlidar 2000).

Table 6.4 Sterilization Is the Most Commonly Used
Method of Fertility Control among Women in the United States

	15–44	15–19	20–24	25–29	30–34	35–39	40–44
Not currently using contraception	35.8	70.2	36.6	30.7	27.3	27.1	28.5
Among those using contraception							
Surgical methods:							
Female sterilized	27.7	0.3	3.9	17.0	29.4	40.9	49.8
Male sterilized	10.9	0.0	1.1	4.5	10.5	18.7	20.3
Nonsurgical methods							
Pill	26.9	43.6	52.2	39.0	28.5	11.1	5.9
Implant	1.4	2.7	3.8	2.0	0.7	0.3	0.1
Injectable	3.0	9.7	6.2	4.2	1.8	1.1	0.3
IUD	0.8	0.0	0.3	0.7	0.8	0.1	1.3
Diaphragm	1.9	0.0	0.6	0.9	2.3	0.3	2.7
Condom	20.4	36.6	26.3	24.2	18.4	16.9	12.3
Female condom	0.0	0.0	0.2	0.0	0.0	0.0	0.0
Periodic abstinence	2.3	1.3	0.9	1.7	3.2	2.9	2.5
NFP	0.3	0.0	0.2	0.3	0.4	0.5	0.3
Withdrawal	3.1	0.4	3.3	3.8	2.9	3.2	2.0
Other methods	1.6	0.1	1.4	1.7	1.8	1.2	2.5

Note: Although the pill is the most popular form of fertility control for younger women, from ages 30 on, sterilization (both male and female) is the most widely used method.

Source: J. Abma, A. Chandra, W. Mosher, L. Peterson, L. Piccinino, 1997, "Fertility, Family Planning, and Women's Health: New Data From the 1995 National Survey of Family Growth," National Center for Health Statistics, Vital Health Statistics 23(19): Table 41.

The pill has been embroiled in a storm of controversy about its potentially harmful side effects ever since its introduction. It is the most widely studied drug ever invented, and its principal inventor, Carl Djerassi of Stanford University, is understandably defensive. In 1981, he wrote:

> The point that these opponents of the Pill really miss is that the Pill as well as all other methods of fertility control should be available for any woman who is willing and able to use them given her particular circumstances. . . . The reality is that for many women throughout the world, the Pill is the best contraceptive method currently available (1981:47).

Another way of introducing synthetic hormones into the body is through implants. Norplant is a progestin-only implant developed by the Population Council in New York. It is a one-eighth-inch-long silicone rubber capsule containing a small amount of levonorgestrel, a synthetic progestin. Capsules are implanted just beneath the skin of a woman's upper arm and release a tiny amount of progestin every day for up to five years before needing to be replaced. Another variation on the theme is for a woman to receive a periodic injection of chemicals in a birth control shot called Depo-Provera. This is a highly effective contraceptive that requires a woman to be injected with a massive dose of synthetic progestin every three months. Although it had been tested in the United States since 1968 and had been approved for contraceptive use in at least 80 countries, the U.S. Food and Drug Administration delayed its approval for use in the United States until 1992 and it came on the market in 1993. Since that time it has become increasingly popular among teenage girls (Abma and Sonenstein 2001), apparently because it provides protection without anybody necessarily knowing about it, since no pills are being taken (Freedman 1998).

Yet another category of chemical contraceptive encompasses those methods that essentially back up all of the previously discussed techniques. **Emergency contraceptive pills (ECP)** (or *postcoital contraception*) are designed to avert pregnancy within 72 hours after intercourse (Ellertson 1996; Hatcher 1998). These are actually large doses of ordinary birth control pills taken over period of one or two days. The actual dosage depends on which birth control pills you use.

Men have very few options available to them in contraceptives. In fact, they are basically limited to the **condom.** This is a rubber or latex sheath inserted over the erect penis just prior to intercourse. During ejaculation, the sperm are trapped inside the condom, which is then removed immediately after intercourse while the penis is still erect, to avoid spillage. The condom is very effective, and when used properly in conjunction with a spermicidal foam, it is virtually 100-percent effective. The condom is of course also useful in preventing the spread of sexually transmitted diseases, including venereal disease and, more popularly, HIV/AIDS. Although use of the condom dropped off considerably during the 1960s and 1970s, by 1995 (the most recent year for which data are available), it had regained its place just after the pill in popularity among younger Americans (Piccinino and Mosher 1998) (see Table 6.4).

Until recently, the principal method of contraception that required couple cooperation was rhythm. Users of this method, however, are jokingly referred to as parents. Recent developments have altered this view somewhat, with the development of more sophisticated ways of trying to determine when in the menstrual cycle ovulation is occurring so that intercourse can be avoided during that time. One of the difficulties is that Mother Nature seems to have thought of that ahead of time, since sperm can live for several days in a woman's vaginal area, and so it turns out that intercourse a few days prior to ovulation can still lead to conception.

Other methods of nonvaginal sexual activity that explicitly require cooperation between both partners are mutual masturbation and oral-genital sex. In Davis and Blake's 1955 article, these forms of incomplete intercourse were listed as "perversions," but in more recent years, with a significant change in openness about sex and with the introduction of best-selling "how-to" manuals, they have become more openly acceptable techniques for engaging in sexual activity.

Surgical Contraception   The last of the conception variables that can be manipulated to achieve low fertility is voluntary infecundity or **sterilization** (Variable 9). These methods are now commonly known as surgical contraception, and represent the most popular means of preventing births in the United States (see Table 6.4), as well as in China (Short et al. 2000). For females, voluntary sterilization procedures largely involve **tubal ligation,** although there are other, more extreme surgical techniques such as a hysterectomy (the removal of the uterus) that are normally done for health reasons, rather than contraceptive reasons (although the result is the same).

As is true for females, there are drastic as well as simple means of sterilization for males. The drastic means is castration, which is removal or destruction of the testes. This generally eliminates sexual responsiveness in the male, causing him to be impotent (incapable of having an erection). Eunuchs (males who have been castrated) have an interesting place in history, but castration is practiced now mainly in the case of life-threatening disease. **Vasectomy** is the male surgical contraceptive, and like a tubal ligation, it does not alter a person's sexual response. A vasectomy involves cutting and tying off the vasa deferens, which are the tubes leading from each testicle to the penis. The male continues to generate sperm, but they are unable to leave the testicle and are absorbed into the body. Vasectomy is quite popular in the United States, as you can see in Table 6.4. In fact, from ages 35 to 44 it is the second most common method of fertility control among American couples, trailing only female sterilization. As early as 1982, sterilization of either sex had stolen the lead away from the pill as the contraceptive method of choice among married women of reproductive age. By 1995, nearly two-thirds of women aged 35 to 44 who used some method of contraception relied on surgical methods (for themselves or their partner).

How Effective Are Different Contraceptive Techniques?   Contraceptive effectiveness refers to how well a contraceptive works to prevent a pregnancy. There are two principal ways to measure this. The first is to look at **theoretical effectiveness,** meaning the percentage of the time that a method should prevent pregnancy after sexual intercourse if everything else is perfect. Thus, if we say that oral contraceptives are 100 percent effective, we mean that if a woman is in normal health and if she takes the pills exactly as directed, she will never get pregnant. Theoretical effectiveness is very difficult to assess and, in addition, it ignores the human failure that can occur with any method.

The more usual measure of a contraceptive's effectiveness is called **use-effectiveness,** which measures the actual pregnancy performance associated with a particular method. There are several specific ways in which use-effectiveness can be measured. The most common is called the Pearl method, or the pregnancy-failure rate, the number of contraceptive failures per 100 woman-years of contraceptive exposure. For example, if 200 women used the pill for a year, they have had a combined total of 200 woman-years of exposure. If four of the women became pregnant while using the pill, the pregnancy rate is two. In other words, there were two failures (accidental pregnancies) for every 100 woman-years of exposure. The lower the failure rate, the more effective the contraceptive. Use-effectiveness combines two kinds of failures: (1) method failure, such as when a woman gets pregnant even though she has an IUD in place, and (2) user failure, such as when a woman on the pill goes on a weekend trip and leaves her pills at home accidentally, or when a man is not sufficiently motivated to always use a condom.

The most precise way to measure use-effectiveness is to study couples who are using a specific method of contraception for some period of time (such as a year) and see what proportion of the couples wind up with an accidental pregnancy. Such data are presented in Table 6.5. If a couple were to use no method at all and rely

Table 6.5   Theoretical and Use-Effectiveness for Selected Contraceptive Techniques

Method	Theoretical Effectiveness	Use-Effectiveness (Percentage of Couples Experiencing an Accidental Pregnancy During the First Year of Use) Typical Findings
None	Chance	85.0
Implants (Norplant)	Virtually no failures	0.09
Injectables (Depo-Provera)	Virtually no failures	0.3
IUD	Virtually no failures	2.0
Pill	Virtually no failures	3.0
Male condom	Very few failures	12.0
Diaphragm/cap	Some failures	18.0
Withdrawal	Some failures	19.0
Fertility awareness	Some failures	20.0
Female condom	Some failures	21.0
Spermicides	Some failures	21.0
Vaginal sponge	Some failures	24.0

Source: Use-effectiveness data from R. Hatcher et al., 1994, Contraceptive Technology, sixteenth revised edition (New York: Irvington), Table 27-1.

simply on chance to avoid a pregnancy, there is an 85 percent chance of a pregnancy during the year. Implants and injectables improve on those odds considerably, with failure rates of .09 percent and 0.03 percent, respectively. The data in Table 6.5 show very clearly that, regardless of the particular method chosen, even the least effective contraceptive dramatically reduces the odds of a pregnancy compared with not using any method at all. Data from a variety of surveys also have shown clearly that the more highly motivated couples (that is, those who do not want any more pregnancies, as opposed to those merely spacing their children) have higher use-effectiveness rates than less motivated couples, regardless of the method chosen.

Ending a Pregnancy Before Birth

Assuming that conception has occurred, a live birth may still be prevented. This could happen as a result of involuntary fetal mortality (Variable 10), which is either a spontaneous abortion (miscarriage) or a stillbirth. More important for our discussion, though, is voluntary fetal mortality, or induced **abortion** (Variable 11). Induced abortions became legal in Canada in 1969 and in the United States in 1973, and they are legal in all three of the world's most populous nations (China, India, and the United States), as well as in Japan and virtually all of Europe, except Ireland (although Irish women may travel to England for an abortion). Worldwide, the demand for abortion is dropping, but it is still high, even in places where it is not legal, and it has been estimated that one of four pregnancies in the world may end in abortion (Henshaw, Singh, and Haas 1999), with China and Russia leading the list in total numbers of abortions per year, followed by Vietnam, Brazil (despite the fact that abortion is very restricted), and the United States. Back in the 1970s, abortion in many countries "changed from a largely disreputable practice into an accepted medical one, from a subject of gossip into an openly debated public issue" (Tietze and Lewit 1977:21). Abortion rates are highest in Central and Eastern Europe, followed by East Asia, and lowest in Western Europe. Table 6.6 provides illustrative data.

Abortion is probably the single most often-used form of birth control in the world and is resorted to when methods of contraception are otherwise not available or have failed. Abortions have played a major role in fertility declines around the world, and they are an important reason for the continued low birth rate in many countries, including the United States and Canada. The number of legally induced abortions reported in the United States increased steadily from 1973 to 1990, but the number has been declining since then (Ventura et al. 2000). Abortions have played a role in Mexico's fertility decline, as well, despite the fact that elective abortion is not legally available to women in Mexico.

There are two measures of abortion: (1) the abortion rate, which is the number of abortions per 1,000 women aged 15–44, and (2) the abortion ratio, which is the number of abortions per 100 known pregnancies. The latter is the best measure, but also the most difficult to estimate. In the United States, the abortion ratio rose steadily to a peak of about 36 abortions per 100 pregnancies in the mid-1980s and it has been declining since then, dropping to 31 in 1997 (Koonin et al. 2000). Abortion ratios are higher for unmarried women than for married, higher for African-Americans than for other racial/ethnic groups, and higher for teenagers than for older women. This profile has not changed much over time. The abortion ratio has remained fairly constant in Canada since abortion laws were liberalized. Canadian women are less likely than women in the United States to use abortion, and since the fertility rate in Canada is slightly lower than in the United States, the implication is that Canadian women are more efficient users of other methods of contraception than are women in the United States. In Mexico, the public appears to favor the broader legalization of abortion (Reider and Pick 1992), but abortions remain clandestine. Despite that fact, the abortion rate in Mexico is higher than that in Canada, although the abortion ratio is lower, as you can see in Table 6.6.

There are now two major different types of abortions: (1) medical and (2) surgical. A medical abortion refers to the use of the drug Mifepristone (formerly known as RU-486) to terminate a pregnancy after conception has occurred. This drug was developed in France and approved for use in the United States in 2000, sold under the name of Mifeprex. The drug blocks the action of progesterone, which is needed for a pregnancy to continue. It is then followed by another drug, misoprostol, and the combination of the two ends the pregnancy (American Health Consultants 2001). This approach to abortion can occur from the time a women knows she is pregnant up to 49 days after the beginning of her last menstrual period.

Surgical abortion is the "standard" method of induced abortion. In the United States, nearly all (97 percent) abortions are done during the first trimester (the first three months) of a pregnancy with a procedure called *vacuum aspiration,* in which a

Table 6.6   The Abortion Rate and Abortion Ratio Varies Widely thoughout the World

Country	Abortion Rate	Abortion Ratio
Russia	68	63
Belarus	68	62
Cuba	78	59
Ukraine	57	58
Vietnam	83	44
Kazakhstan	44	41
Chile	50	35
Brazil	41	30
China	26	27
United States	23	26
Sweden	19	25
Korea (South)	20	25
Canada	16	22
Japan	13	22
Italy	11	21
England and Wales	16	20
New Zealand	16	19
France	12	18
Mexico	25	17
Egypt	23	16
Philippines	25	16
Germany	8	14
Israel	14	13
Spain	6	13
Nigeria	25	12
Belgium	7	11
Netherlands	6	11
Ireland	6	9
Tunisia	9	8
India	3	2

Source: Stanley K. Henshaw, Susheela Singh, and Taylor Haas, 1999, "The Incidence of Abortion Worldwide," International Family Planning Perspectives 25(Supplement):S30–S38, Tables 2 and 3. Data refer to mid-1990s. Abortion rate is abortions per 1,000 women aged 15 to 44; abortion ratio is abortions per 100 pregnancies.

tube the size of a pencil is placed into the uterine cavity and then suction is applied to remove the fetus (Koonin et al. 2000). Later-term abortions may also be conducted surgically, but they are normally undertaken to preserve the health of the mother, rather than for contraceptive purposes.

The Relative Importance of the Proximate Determinants

Of all the proximate determinants of fertility, contraception is by far the most important. Without modern contraception, fertility can be maintained at levels below the biological maximum, perhaps even as low as three children per woman (largely through abstinence), but it is very difficult to achieve low levels of fertility without a substantial fraction of reproductive-age couples using some form of modern fertility control. The exact nature of the relationship between fertility and contraception has been measured with increasing accuracy and frequency over the past few decades, beginning with the Knowledge, Attitude, and Practice (KAP) surveys of the 1960s and 1970s and followed by the World Fertility Surveys and Contraceptive Prevalence Surveys of the 1970s and 1980s and then by the Demographic and Health Surveys of the 1980s to the present. In North America, these kinds of data have come to us through the National Surveys of Family Growth in the United States, through the General Social Survey in Canada, and through the National Fertility Studies in Mexico. In Europe, such data were derived in the 1990s from the Fertility and Family Dynamics Surveys.

Contraceptive use is usually measured by calculating the rate of **contraceptive prevalence,** which is the percentage of "at risk" women of reproductive age (15 to 44 or 15 to 49) who are using a method of contraception. Being at risk of a pregnancy means that a woman is in a sexual union and is fecund, but is not currently pregnant. There are approximately one billion women in the world right now who are at risk of pregnancy, and about 60 percent of them are using some contraceptive method (Population Reference Bureau, 2003). Contraceptive methods tend to be lumped into two broad categories: modern and traditional. The modern methods include the pill, IUD, implants, injectables, sponge and cervical cap, diaphragm, condom (male and female), and voluntary sterilization. Thus, the definition of contraception in most surveys encompasses two of the intermediate variables listed in Table 6.2: contraception (Variable 8) and voluntary infecundity (Variable 9). The traditional methods include total abstinence, periodic abstinence, withdrawal, and other methods. Periodic abstinence is not really a "traditional" method. In fact, the timing of ovulation in the menstrual cycle (which led to the development of the rhythm method of periodic abstinence) was unknown until the 1930s when Kyusako Ogino and Herman Knaus independently discovered the fact that peak fecundity in women occurs at the approximate midpoint between menses and that, despite the variability in the amount of time between the onset of menses and ovulation, the interval between ovulation and the next menses is fairly constant at about 14 days (Population Information Program 1985). Nonetheless, the high failure rate of the rhythm method (in contrast to newer methods of periodic abstinence) removes it from the modern category.

As of the year 2001, data on contraceptive prevalence were available for 167 countries. The data come from surveys conducted principally in the 1990s, although the dates are obviously not the same for all countries. In Figure 6.4, I have charted the percentage of married women using contraception (including both modern and traditional types) against the total fertility rate (TFR, an index similar to the total number of children born to women, as I discuss in this chapter). Each triangle on the graph represents one of those 167 countries in terms of its total fertility rate and contraceptive prevalence. Thus, Ethiopia, in the upper left of the graph, has a very

Figure 6.4   Higher Levels of Contraceptive Use Lead to Lower Levels of Fertility

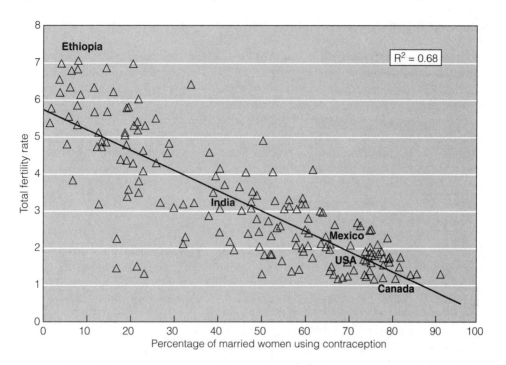

Source: Adapted from data in U.S. Census Bureau International Database, accessed 2001; TFR data are estimates for 2001; contraceptive data are from the 1990s.

high fertility rate (7.0 children per woman) and a very low 4 percent of married women using any method of contraception. In the middle of the graph is India, with a TFR of 3.0 and a middling 41 percent of women using contraception. Mexico is farther down the line with a TFR of 2.6 and a fairly high 67 percent of women using contraception. Toward the lower right-hand side of the graph are the United States and Canada, both with a low TFR (2.0 in the United States and 1.6 in Canada) and a high percentage of women using contraception (71 and 75, respectively).

You do not need a degree in statistics to see that there is a close relationship between these two variables, and the R^2 of .68 confirms that fact for those of you familiar with correlation coefficients. This is not a perfect relationship, however, and you can see in the graph that not every country is exactly on the line (which would represent a perfect relationship between fertility and contraceptive prevalence). Some countries have very high fertility, even though they appear to have reasonably high levels of contraceptive prevalence. In general, the explanation is that in these countries couples are using relatively less effective methods of contraception, although in some cases it is also true that couples have not been using contraception long enough for it to yet have a huge impact on the birth rate.

On the other hand, other countries have quite low levels of fertility, even though the rate of contraceptive prevalence seems quite modest. In these cases, induced legal abortion rather than contraception causes the low fertility rate. Most central and eastern European nations fall into this category, as do several east Asian nations, as I have mentioned elsewhere in this chapter. In several Latin American countries, the evidence suggests that illegal induced abortions also play an important auxiliary role in limiting fertility (Frejka and Atkin 1996). Thus, if contraceptive effectiveness were increased generally, and in particular if effective contraceptive use replaced abortion, the fit between fertility and contraceptive prevalence would approach a perfect relationship (Bongaarts and Westoff 2000).

Knowing as we now do how fertility can be controlled, we are almost ready to ask why it is that couples make the conscious decision to limit fertility—the decision that sets the fertility transition in motion. First, though, it helps to be reminded of why fertility was high in the first place.

Explanations for High Fertility

You will recall that for the first 99 percent of human history, mortality was very high. Only those societies with sufficiently high fertility managed to survive over the years. There may have been lower-fertility societies in the past, but we know nothing about them because they did not reproduce themselves. Societies that did survive did not take for granted that people would have enough children to keep the population going. They instituted multiple inducements—pronatalist pressures—to encourage the appropriate level of reproduction. It is the undoing of those, and their being replaced by different kinds of pressures to keep fertility low, that we have to understand if we are to explain the fertility transition. So, let me discuss the general idea of the need to replenish society and then review the major inducements used by high-mortality societies, including the value of children as security and labor and the desire for sons.

Need to Replenish Society

A crucial aspect of high mortality is that a baby's chances of surviving to adulthood are not very good. Yet if a society is going to replace itself, an average of at least two children for every woman must survive long enough to be able to produce more children. So, under adverse conditions, any person who limited fertility might threaten the very existence of society. In this light, it is not surprising that societies have generally been unwilling to leave it strictly up to the individual or to chance to have the required number of children. Societies everywhere have developed social institutions to encourage childbearing and reward parenthood in various ways. For example, among the Kgatla people in South Africa, mortality was very high during the 1930s when they were being studied by Schapera. He discovered that "to them it is inconceivable that a married couple should for economic or personal reasons deliberately seek to restrict the number of its offspring" (Schapera 1941:213). Several social factors encouraged the Kgatla to desire children:

> A woman with many children is honored. Married couples acquire new dignity after the birth of their first child. Since the Kgatla have a patrilineal descent system (inheritance

passes through the sons), the birth of a son makes the father the founder of a line that will perpetuate his name and memory . . . [and the mother's] kin are pleased because the birth saves them from shame (Nag 1962:29).

In a 1973 study of another African society, the Yoruba of western Nigeria, families with fewer than four children were (and still are) looked upon with horror. Ware reports that "even if it could be guaranteed that two children would survive to adulthood, Yoruban parents would find such a family very lonely, for many of the features of the large family which have come to be negatively valued in the West, such as noise and bustle, are positively valued by them" (Ware 1975:284). The 1989 Kenya Demographic and Health Survey (DHS) discovered that in the coastal province of that eastern African nation, the average ideal family size of 5.6 children was higher than the actual level of reproduction. Women wanted larger families than they were having, but they were slowed down by involuntary infecundity (Cross, Obungu, and Kizito 1991). The 1998 Kenya DHS showed that the average ideal family size had dropped to below the number of children that women were having—but fertility was still very high.

Social encouragements to fertility have been discussed in a more general form by Davis:

> We often find for example that the permissive enjoyment of sexual intercourse, the ownership of land, the admission to certain offices, the claim to respect, and the attainment of blessedness are made contingent upon marriage. Marriage accomplished, the more specific encouragements to fertility apply. In familistic societies where kinship forms the chief basis of social organization, reproduction is a necessary means to nearly every major goal in life. The salvation of the soul, the security of old age, the production of goods, the protection of the hearth, and the assurance of affection may depend upon the presence, help, and comfort of progeny. . . . [T]his articulation of the parental status with the rest of one's statuses is the supreme encouragement to fertility (Davis 1949:561).

You may notice from these examples that social pressures are not actually defined in terms of the need to replace society, and an individual would likely not recognize them for what they are. By and large, the social institutions and norms that encourage high fertility are so taken for granted by the members of society that anyone who consciously said, "I am having a baby in order to continue the existence of my society," would be viewed as a bit weird. Further, if people really acted solely on the basis that they had to replace society, then higher-fertility societies would now actually have much lower levels of fertility, since in all such countries the birth rates exceed the death rates by a substantial margin.

The *societal* disconnection between infant and child mortality and reproductive behavior probably explains why at the *individual* level there is not much evidence of a relationship between infant deaths in a family and the number of children born to those parents (Montgomery and Cohen 1998; Preston 1978). Thus, the more culturally oriented perspectives on fertility begin with the assumption that fertility is maintained at a high level in the premodern setting by institutional arrangements that bear no logical relationship to mortality. Premodern groups accepted high mortality, especially among children, as a given, and so they devised various ways to ensure that fertility would be high enough to ensure group survival. Pronatalist pressures encourage family members to bring power and prestige to themselves and to their group by having children, and this may have no particular relationship to the level of mortality within a family.

Children as Security and Labor

In a premodern society, human beings were the principal economic resource. Even as youngsters they were helpful in many tasks, and as they matured they provided the bulk of the labor force that supported those, such as the aged, who were no longer able to support themselves. More broadly speaking, children can be viewed as a form of insurance that rural parents, in particular, have against a variety of risks, such as a drought or a poor harvest (Cain 1981). Though at first blush it may seem as though children would be a burden under such adverse conditions, many parents view a large family as providing a safety net—at least one or two of the adult children may be able to bail them out of a bad situation. One important way that this may happen is that one or more children may migrate elsewhere and send money home. Although children may clearly provide a source of income for parents until they themselves become adults (and parents), it is less certain that children will actually provide for parents in their old age. Despite an almost worldwide norm that children should care for parents in old age, there is very little empirical evidence to suggest a positive relationship between fertility and the perceived need for old-age security (Dharmalingam 1994; Vlassoff 1990). For now, it is most noteworthy that in a premodern setting, the *quantity* of children may matter more than the *quality,* and the nature of parenting is more to *bear* children than to *rear* them (Gillis, Tilly, and Levine 1992). Still, the noneconomic, nonrational part of society (the sexist part) intrudes by often suggesting that male children are more desirable than female children.

Desire for Sons

Although it is obvious that in many less-developed countries the status of women is steadily improving, it is nonetheless still true that in many societies around the world, desired social goals can be achieved only by the birth and survival of a son— indeed, most known societies throughout human history have been dominated by men. Since in most societies males have been valued more highly than females, it is easy to understand why many families would continue to have children until they have at least one son. Furthermore, if babies are likely to die, a family may have at least two sons in order to increase the likelihood that one of them will survive to adulthood (an "heir and a spare"). For example, given the level of mortality in Pakistan, the total fertility rate of 4.8 children per woman in 2003 was just slightly higher than the level of fertility required to ensure an average of two surviving sons per woman. In Angola, where mortality is much higher, the total fertility rate of 6.8 in 2003 was also very close to, albeit still above, the level at which the average woman would have two sons surviving to adulthood.

India is a country where the desire for a surviving son is strong, since the Hindu religion requires that parents be buried by their son (Mandelbaum 1974). Malthus was very aware of this stimulus to fertility in India and, in his *Essay on Population,* quoted an Indian legislator who wrote that under Hindu law a male heir is "an object of the first importance. 'By a son a man obtains victory over all people; by a son's son he enjoys immortality; and afterwards by the son of that grandson he reaches the solar abode'" (Malthus 1872 [1971]:116). Such beliefs, of course, also serve to ensure that society will be replaced in the face of high mortality.

Yet, as Arnold (Arnold 1988; Arnold, Choe, and Roy 1998) reminds us, the desire for sons cannot alone account for high birth rates. In Korea and China, two Asian societies with very strong preferences (Park and Cho 1995), the drop in fertility has been rapid, and Vietnam, another nation with a marked preference for sons (Goodkind 1995), is also in the midst of a rapid fertility decline. The major impact of son preference in the midst of a fertility decline is to increase the chance that a female fetus may be aborted, leading to the phenomenon of the "missing females" in China (Coale and Banister 1994)—fewer girls enumerated in censuses at younger ages than you would expect, given the number of boys enumerated.

In the 1980s, there was a great deal of speculation and concern that the missing females were victims of female infanticide. Since female infanticide had been fairly common during the precommunist era, the probability seemed great that the one-child policy in China would lead a couple to kill or abandon a newborn female infant, reserving their one-child quota for the birth of a boy (Mosher 1983). But further analyses of data in both China and Korea (where a similar pattern of fertility decline has occurred without a coercive one-child policy) suggest that sex-selective abortion, combined with the nonregistration of some female births, accounts for almost all of the "missing" females, and that the role of infanticide probably was exaggerated (Coale and Banister 1994; Park and Cho 1995). Infanticide in most cases is probably a result of abandonment of children, and there is evidence of a large number of orphans in China who are foundlings, suggesting that at least some children who are abandoned are found and do survive (Johnson 1996).

Male preference was also an indelible part of European patterns of primogeniture, which were designed to maintain a family's wealth by passing it on only to the oldest son. Over time, however, the preference for sons in North America and Europe has abated as a consequence of increasing gender equity, and data from 17 European countries suggest a strong tendency for a mixed-sex composition, with some countries exhibiting a slight preference for males and others a slight preference for females (Hank and Kohler 2000).

Encouragements to high fertility often persist even after mortality declines. Childbearing is rarely an end in itself, but rather a means to achieve other goals, so if the attainment of the other goals is perceived as being more important than limiting fertility, a woman may continue to risk pregnancy because she is **ambivalent** about having a child—caught between competing pressures and thus unsure of how she feels. Smelser (1997) suggests that ambivalence can be thought of as a rational-choice situation in which a person is rational, but has no choice. A woman may rationally prefer a smaller family, yet not be in a position to exercise that choice in her life. Some of the factors that may enhance feelings of ambivalence at various points in a woman's life include a woman's role being exclusively associated with reproduction (that is, where males do not help with child rearing), lack of women participating in work outside of the immediate family, low levels of education, lack of communication between husband and wife, lack of potential for social mobility, and an extended family system in which couples need not be economically independent to afford children (Davis 1955a). Most of these factors are related to the domination of women by men and go to the heart of women's empowerment (Dixon-Mueller 2001).

Despite publicity in more developed countries about women needing fertility drugs and other help to get pregnant, remember that for most women in most of human history it was easier to have several children than to limit the number to one or two, regardless of the level of motivation. For the average person, a high level of desire and access to the means of fertility control are required to keep families small. Thus, the fertility transition was by no means automatically assured just because mortality declined.

Motivations for Lower Fertility Levels

The fertility transition is most often explained as some combination of rational factors embodied in the supply-demand framework, and sociocultural influences captured by the innovation/diffusion perspective. These complementary perspectives come together in the preconditions for a fertility decline that emerged especially from the historical study of fertility transitions in Europe. Let me discuss each of these in turn.

The Supply-Demand Framework

The demographic transition envisions a world in which the normal state of affairs is a balance between births and deaths. Mortality is assumed to decline for reasons that are often beyond the control of the average person (**exogenous factors**), but a person's reproductive behavior is dominated by a rational calculation of the costs and benefits to himself or herself (**endogenous factors**) of maintaining high fertility in the face of declining mortality. The idea is that people will eventually perceive that lower mortality has produced a situation in which more children are going to survive than can be afforded and, at that point, fertility will decline.

The economist Richard Easterlin (mentioned earlier in Chapter 3) is especially notable for his work in this regard, and the resulting perspective is somewhat clumsily called "the theory of supply, demand, and the costs of regulation," or, in shorthand, "the supply-demand framework." It is also known as "the new household economics" because the household, rather than the individual or the couple, is often taken as the unit of analysis. High fertility, for example, may help households avoid risk in the context of low economic development and weak institutional stability, especially when children generate a positive net flow of income to the parents. Under those conditions, it is rational to want to produce a large number of children.

The supply-demand framework draws its concepts largely from the field of neo-classical economics, which assumes that people make rational choices about what they want and how to go about getting it. The essence of the theory is that the level of fertility in a society is determined by the choices made by individual couples within their cultural (and household) context (Bongaarts 1993; Bulatao and Lee 1983; Easterlin 1978; Easterlin and Crimmins 1985; McDonald 1993; Robinson 1997). Couples strive to maintain a balance between the potential supply of children (which is essentially a biological phenomenon determined especially by fecundity) and the demand for children (which refers to a couple's ideal number of surviving children). If mortality is high, the number of surviving children may be small, and the supply may essentially equal the demand. In such a situation, there is no need for fertility regulation. However, if the supply begins to exceed the demand, either because infant and child survival has increased, or because the **opportunity costs** of children are rising, then couples may adjust the situation by using some method of fertility regulation. The decision to regulate fertility will be based on the couple's perception of the costs of doing so, which include the financial costs of the method and the social costs (such as stigmas attached to the use of methods of fertility control).

What are the opportunity costs of children? Let us assume that people are rational and that they make choices based on what they perceive to be their best self-interest. One set of choices that most people make in life revolves around children. The idea that children might be thought of as "commodities" was introduced in 1960 by University of Chicago economist Gary Becker, whose work on the economic analysis of households and fertility earned him a Nobel prize in 1992. Becker's theory treated children as though they were consumer goods that require both time and money for parents to acquire. Then he drew on classic microeconomic theory to argue that for each individual a utility function could be found that would express the relationship between a couple's desire for children and all other goods or activities that compete with children for time and money (Becker 1960). It is important to note that time as well as money is being considered, for if money were the only criterion, then one would expect (in a society where there are social pressures to have children) that the more money a person had, the more children he or she would want to have. Yet we know that in virtually every more-developed nation, those who are less well-off financially tend to have more children than do those who are more well-off.

With the introduction of time into the calculations, along with an implicit recognition that social class determines a person's tastes and lifestyle, Becker's economic theory turns into a trade-off between quantity and quality of children. For the less well-off, the expectations that exist for children are presumed to be low and thus the cost is at its minimum. In the higher economic strata, the expectations for children are presumed to be greater, both in terms of money and especially in terms of time spent on each child. The theory asserts that parents in the higher strata also are exposed to a greater number of opportunities to buy goods and engage in time-consuming activities. Thus, to produce the kind of child desired, the number must be limited.

When an advanced education, a prestigious career, and a good income were not generally available to women, the lack of such things was not perceived as a cost of having children. But when those advantages are available, reducing or foregoing them for the sake of raising a family may be perceived as a sacrifice. Again, the reflexive nature of the fertility-women's status connection is apparent. As fertility has gone down, more time has become available for women to pursue alternate lifestyles; and as the alternatives grow in number and attractiveness, the costs of having children have gone up. The benefits of having children are less tangible, but no less important than the costs. They include psychological satisfaction and proof of adulthood, not to mention being more integrated into the family and community. The latter reasons reflect a broad category of reward that society offers for parenthood—social approval. In addition, most people are most comfortable in families with children, since by definition everyone was raised in a situation involving at least one child. Thus, having children may allow you vicariously to relive (and perhaps revamp) your own childhood, helping to recreate the past and take the sting out of any failures you may experience as an adult. In a more instrumental way, children tend to provide a means for establishing a network of social relationships in a community, through school, organized sports, and activity groups. The rewards of childbearing thus are greatest in terms of the personal and social satisfaction derived from them, since in more-developed nations there is certainly little, if any, economic advantage derived from having children.

The fertility transition is not something that occurs in a vacuum—it occurs because other changes are taking place in society to which individual couples or people within households are responding. The improved health and lower mortality that are part of the epidemiological transition actually increase the potential supply of children because healthier women are better able to conceive and bear a child successfully to birth. But the changes in society that are generating that improved health are assumed by the supply-demand framework to be largely economic in nature, and it is the changing economic circumstances, not the decline in mortality per se, that couples are believed to respond to as they choose to lower their demand for children. An improved economy generates other things in life that compete with children. Other sources of income are more productive than what children can produce, and there are other ways to spend one's time and money besides just on children. Thus, if the means for fertility regulation are sufficiently effective, the lowered demand for children will lower fertility.

Income, Wealth, Prestige, and Fertility   Historically, the most persistent socioeconomic factors related to fertility are income, wealth, and prestige (all of which are further related to power). A study of primitive societies has shown that population control was related to a competition for power and prestige (Douglas 1966), which often led to an emphasis on high fertility (there may be power in numbers) rather than low fertility. However, as Benedict (1972) has noted, in urban industrial societies prestige and wealth tend to be linked with low fertility rather than high fertility. This reversal may seem puzzling at first glance. After all, it would seem on an a priori basis that as people acquired wealth and prestige they would acquire more children, since presumably they could afford more rather than fewer. But, of course, the supply-demand framework suggests that the key is in the availability of resources. In most countries, wealth and prestige (which are both closely related to income) are scarce economic and social commodities, and it may require sacrifices of one kind or another if an individual is to beat the competition. One sacrifice is the large family. In 1938, an Englishman put it rather succinctly: ". . . in our existing economic system, apart from luck, there are two ways of rising in the economic system; one is by ability, and the other by infertility. It is clear that of two equally able men—the one with a single child, and the other with eight children—the one with a single child will be more likely to rise in the social scale" (Daly 1971:33).

In 1942, Joseph Schumpeter discussed the emerging attitudes of middle-class families toward childbearing and rearing. Heralding yuppie ideas, but 40 years ahead of his time, he posed the question "that is so clearly in many potential parents' minds: 'why should we stunt our ambitions and impoverish our lives in order to be insulted and looked down upon in our old age?'" (Schumpeter 1942:157). This is a very different view of the world from that in which family members work primarily for the mutual survival of all other family members or the view from the bottom of the social heap in which the only thing that matters is day-to-day survival.

Increasing income and acquiring wealth may require that a family be kept small, whereas already having wealth may permit, and even encourage, the growth of families. Often people who have kept families small in order to improve their incomes and acquire wealth and prestige are past their reproductive years when they reach their goal (if they reach it), or they may have grown comfortable with a small family and decide not to have more children even though they finally may be able to afford them. This difference in the timing of fertility and wealth is one important point to keep in mind in thinking about the relationship between supply and demand—women have a limited range of years within which to supply the desired number of children.

Education and Fertility   Although wealth and prestige are characteristics of premodern as well as contemporary societies, mass education is a new dimension that has been added to the mix by the Renaissance and the Industrial Revolution it engendered. As I mentioned in Chapter 3, education (and its closely related cousin, secularization) is the single best clue to a person's attitude toward reproduction, and thus a clue to how responsive a person will be to changes that could lead to a lower demand for children (see, for example, Axinn and Barber 2001). An increase in education (probably to a level beyond primary) is strongly associated with the kind of rational decision making implied in the supply-demand framework. Furthermore, the better-educated members of society are most likely to be the agents of change who will encourage the diffusion of an innovation such as fertility limitation.

I have never seen a set of data from anywhere in the world that did not show that more educated women had lower fertility than less educated women in that society. It is nearly axiomatic, and it is the identification of this kind of **fertility differential** that helps to build our understanding of reproductive dynamics in human societies, because it causes us to ask what it is about education that makes reproduction so sensitive to it. In general terms, the answer is that education offers to people (men and women) a view of the world that expands their horizon beyond the boundaries of traditional society, and causes them to reassess the value of children and reevaluate the role of women in society. Education also increases the opportunity for social mobility, which, in turn, sharpens the likelihood that people will be in the path of innovative behavior, such as fertility limitation, that they may try themselves.

In Figure 6.5, I have summarized data from Demographic Health Surveys in less-developed nations, along with data from Mexico, Canada, and the United States, showing

Figure 6.5   The Education of Women Is an Important Part of the Fertility Transition

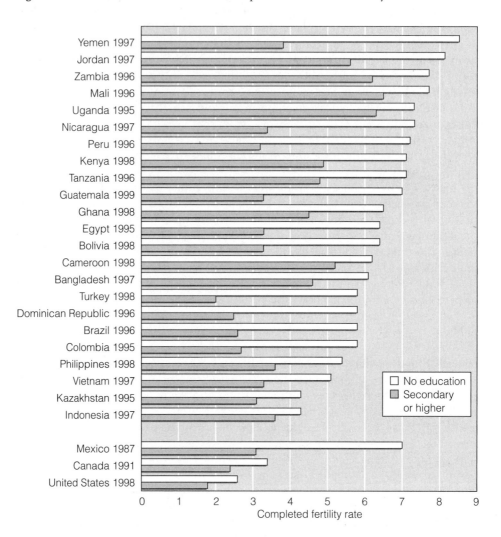

Sources: Data for most countries are from the Demographic Health Surveys, obtained through the Macro International STAT compiler (http://www.measuredhs.com; accessed 2001). Data for Canada are from T. R. Balakrishnan, Evelyne LaPierre-Adamcyk, and Karol J. Krótke, 1993, Family and Childbearing in Canada: A Demographic Analysis (Toronto: University of Toronto Press), Table 3.5; data for Mexico are from the Dirección General de Planificación Familiar, México: Encuesta Nacional Sobre Fecudidad y Salud, 1987 (Columbia, MD: IRD/Macro International), 1989, Table 6.4; and data for the United States are from Amara bachu and Martin O'Connell, 2000, "Fertility of American Women: June 1998," Current Population Reports P20-526, Appendix Table 3 (data refer to women aged 35–44, and the lowest level of education reported is for women who are not high school graduates).

the relationship between fertility and education. The data are for women aged 40–44, so the number of children ever born approximates the completed fertility rate—the total number of children these women will ever have. The countries are arranged by the fertility level of women with no education, representing the situation in virtually all human societies prior to the fertility transition. Below that is the bar showing the completed fertility rate for women with secondary education (the equivalent of high school) or higher—representing the kind of structural change in a society that is likely to change the relationship between the supply and demand for children. The differences in fertility by education tend to be largest in Latin America and lowest in Africa. Part of this is probably due to cultural factors. In Latin America, women with no education are apt to be indigenous, while the more educated population is more likely to be *mestizo*. In Africa, there are less likely to be these kinds of cultural differences, and so the education difference is not compounded by this sort of cultural variability. Educated women are also more likely to live in urban places where fertility tends to be lower, but the differential by education exists in both rural and urban areas.

You can see in Figure 6.5 that education always leads to fertility differentials within a population, but similar levels of education do not always lead to similar levels of fertility in different populations. Part of the reason for this difference may be economic: Education will buy you more in some societies than in others, and the more it will buy you, the higher the opportunity costs of children may be. Another reason may be cultural—education is not necessarily comparable from one country to another. On the other hand, education is such a critical determinant of fertility that Bongaarts (2003) has suggested that low levels of schooling in developing countries may well cause the fertility transitions in those countries to stall.

One must also recognize that motivations for low fertility do not appear magically just because one aspires to wealth, or has received a college education, or is an only child and also prefers a small family. Motivations for low fertility arise out of our communication with other people and other ideas. Fertility behavior, like all behavior, is in large part determined by the information we receive, process, and then act on. The people with whom, and the ideas with which, we interact in our everyday lives shape our existence as social creatures.

The Innovation/Diffusion and "Cultural" Perspective

Not all social scientists agree that human behavior is described by rational neoclassical economic theory. In particular, sociologists, anthropologists, and cultural geographers have often been drawn to the idea that many changes in society are the result of the diffusion of innovations (Brown 1981; Rogers 1995). We know, for example, that much of human behavior is driven by fads and fashions. Last year's style of clothing will go unworn by some people this year, even though the clothes may be in very good repair, just because that is not what "people" are wearing. These "people" are important agents of change in society—those who, for reasons that may have nothing to do with money or economic factors, are able to set trends. You may call it charisma, or karma, or just plain influence, but some people set trends and others do not. We see it happen many times in our lives. Often these change agents are members of the upper strata of society. They may not be the inventors of the innovation, but when they adopt it, others follow suit. Notice, too, that the innovation may be technological, such as the cellular telephone, or it may be attitudinal and behavioral, such as deciding that two children is the ideal family size and then using the most popular means to achieve that number of children.

In Chapter 3, I mentioned that the fertility history of Europe suggests a pattern of geographic diffusion of the innovation of fertility limitation within marriage. The practice seemed to spread quickly across regions that shared a common language and ethnic origin, despite varying levels of mortality and economic development (Watkins 1991). This finding led to speculation that fertility decline could be induced in a society, even in the absence of major structural changes such as economic development, if the innovation could be properly packaged and adopted by the appropriately influential change agents. However, this is where the concept of "culture" comes into play, because some societies are more prone to accept innovations than are others (Pollack and Watkins 1993).

To accept an innovation and to change your behavior accordingly, you must be "empowered" (to use an increasingly overworked term) to believe that it is within your control to alter your behavior. Not all members of all societies necessarily feel this way. In many premodern and "traditional" societies, people accept the idea that their behavior is governed by God, or multiple gods, or more generally by "fate," or more concretely by their older family members (dead or alive). In such a society, an innovation is likely to be seen as an evil intrusion and is not apt to be tolerated.

You can perhaps appreciate, then, that the diffusion of an innovation requires that people believe that they have some control over their life, which is the essence of the rational-choice model that underlies the economic approach to the fertility transition, and is essentially the first of Coale's preconditions for a fertility decline. In other words, the supply-demand model and the innovation-diffusion model tend to be complementary to one another, not opposed to one another. Both approaches can be helpful in explaining why fertility declines.

In any social situation in which influential couples are able to improve their own or their children's economic and social success by concentrating resources on a relatively smaller number of children, other parents may feel called upon to follow suit if they and their offspring are to be socially competitive (Caldwell 1982; Handwerker 1986; Turke 1989). The importance of "influential couples" is sometimes ignored by North Americans who prefer the ideal of a classless society. European demographers offer the reminder that two enduring theories of social stratification have strong implications for fertility behavior: (1) Cultural innovation typically takes place in higher social strata as a result of privilege, education, and concentration of resources, whereas lower social strata adopt new preferences through imitation and (2) Rigid social stratification or closure of class or caste inhibits such downward cultural mobility (Lesthaeghe and Surkyn 1988). Thus, the innovative behavior of influential people will be diffused downward through the social structure, as long as there are effective means of communication among and between social strata. From our perspective, the innovation of importance is the preference for smaller families, implemented through the innovations of delayed marriage/sexual partnership and/or fertility limitation.

The Role of Public Policy

It is in the modern world of nation-states that the fertility transition has taken place, and so we cannot ignore the role that public policy may play in making it relatively more or less attractive to have children, and relatively easier or harder to control fertility. In Chapter 3, I alluded to the laws that existed in England and the United States limiting a couple's access to birth control information, so that only the most highly motivated couples were able to effectively limit their fertility. You will also recall that Marx was adamantly opposed to government intervention in the reproductive lives of people. So, in 1920s Russia, after the Communist revolution, Lenin repealed anti-abortion laws and abolished the restrictions on divorce in order to free women; the result was a fairly rapid decline in the birth rate (although it turned out to be too rapid for the government's taste and in the 1930s abortions were again made illegal). On the other hand, the Cuban response to a Marxist government was exactly the opposite. Shortly after the Cuban revolution in 1959, the crude birth rate soared from 27 births per 1,000 population in 1958 to 37 per 1,000 in 1962. A Cuban demographer, Juan Perez de la Riva, has explained that after the revolution, rural unemployment disappeared, new opportunities arose in towns, and an exuberant optimism led to a lowering of the age at marriage and an abandonment of family planning (Stycos 1971)—an ironically "Malthusian" response to a Marxist reorganization of society. The birth rate in Cuba later did reestablish its prerevolutionary decline, facilitated by eased restrictions on abortion and increasing availability of contraceptives (Diaz-Briquets and Perez 1981; Hollerbach 1980). However, low fertility has been maintained over time by improved female education and increased participation of women in the labor force (Catasús Cervera and Fraga 1996).

In Egypt, there has been long and sustained governmental support for family planning programs, but Fargues (1997) has argued that fluctuations in the birth rate over the years have been much more responsive to state policies affecting the economy than they have been to family planning programs. In particular, the government's policy of liberalizing the economy through market reforms has probably increased the motivation of couples to limit fertility because it has created opportunities to increase levels of income and wealth, which in turn compete with children for parents' attention. Conversely, Fargues (2000b) suggests that decades of conflict between Israelis and Arabs have kept fertility rates in both Israel and Palestine higher than they would otherwise be, in particular preventing a fertility transition from occurring within the Palestinian population, especially among those living in the Gaza Strip (Khawaja 2000).

How is the Fertility Transition Accomplished?

There is no single straight path that a population is likely to take to get from high fertility to low fertility, but there are some patterns that show up more regularly than others. In particular, birth rates are apt to drop noticeably at the two age extremes of a woman's reproductive career. One of the first signs of a fertility decline in a population is an increase in the age at which a woman has her first birth. You might call this the transition from children having children to women having children, and it is part and parcel of the rising status of women in society (which I discuss further in the essay that accompanies this chapter). This may be accomplished more through abstinence than anything else, but in societies where girls are sexually active prior to marriage this will obviously require either an effective contraceptive or the availability of abortion.

REPRODUCTIVE RIGHTS, REPRODUCTIVE HEALTH, AND THE FERTILITY TRANSITION

The ability of women to control their own reproduction, and the overall level of their reproductive health, are closely related to the changes that occur in the context of the demographic transition. By now you are familiar with the fact that pronatalist pressures have always been strong in societies characterized by high mortality and high fertility, especially agricultural societies. In those areas, several children must be born just to ensure that enough will survive to replace the adult membership. Thus, one component of the social status of women is that with a regime of high mortality, women are busy with pregnancy, nursing, and child care, and men, who are biologically removed from the first two of these activities, are able to manipulate and exploit women by tying the status of women to their performance in reproduction and the rearing of children. Furthermore, high mortality means that childbearing must begin at an early age, because the risk of death even as an adult may be high enough that those younger, prime reproductive years cannot afford to be "wasted" on activities other than family building. In a premodern society with a life expectancy of about 30 years, fully one-third of women age 20 died before reaching age 45, making it imperative that childbearing begin as soon as possible. Of course, the catch in all of this, as you will recall, is that pregnancy and childbirth are major causes of death for women between the ages of 20 and 45. This means that reducing fertility is a major cause of the improvement in women's reproductive health (Velkoff and Adlakha 1998).

Women who marry young and begin having children may be "twice cursed"—having more years to be burdened with children and also being in a more vulnerable position to be dominated by a husband. Men need not marry as young as women since they are not the childbearers, and they also remain fecund longer. The older and more socially experienced a husband is compared to his wife, the easier it may be for him to dominate her. It is no coincidence that in Africa, western Asia, and southern Asia, where women are probably less free than anywhere in the world (and where fertility is higher than anywhere else), men are consistently several years older than their wives. In the Gambia,

husbands are nine years older than their wives, in Afghanistan they are eight years older, and in Egypt they are almost six years older (United Nations Population Division 2000). By contrast, in most European countries the difference averages two to three years, and in the United States since 1970, the age gap at first marriage has been two years (Clarke 1995). In general, women who marry young and are in a position to be dominated by an older husband are likely to have poorer levels of health than would otherwise be the case.

Among the uglier aspects of traditional approaches to reproductive "health" is the practice of **female genital mutilation (FGM),** sometimes known as *female circumcision,* which involves practices that are technically *clitoridectomy* and *infibulation.* Clitoridectomy accounts for about 80 percent of such mutilation practices (Chelala 1998) and involves the total removal of the clitoris, whereas infibulation involves cutting the clitoris, the labia minora, and adjacent parts of the labia majora and then stitching up the two sides of the vulva. These are useless and dangerous practices to which an estimated two million girls and women in at least 28 countries in northern and sub-Saharan Africa and parts of Asia are subjected annually. The effect is to dramatically lower a woman's enjoyment of sexual intercourse, but as was true for foot binding in an earlier era in China, the real purpose is to "control access to females and ensure female chastity and fidelity" (Mackie 1996).

The migration of refugees from African countries such as Somalia, where the practice is common, to North America and Europe helped to ignite worldwide knowledge of and concern about FGM and there are now international movements in place bringing pressure on governments to make it illegal. A significant event occurred in 1997 when the highest court in Egypt upheld the government's ban on FGM. There is evidence as well that educated women are less willing to submit to FGM, so the trend in Africa toward increasing female education bodes well for the eventual elimination of this form of male domination (Islam and Uddin 2001).

Female genital mutilation appears not to adversely affect a woman's ability to bear children (Larsen and Yan 2000), but the same cannot be said for violence against women, which is also common in many less-developed nations. Women in India seem especially likely to be beaten by their husbands and there is evidence that this kind of brutality is associated with elevated levels of pregnancy loss and infant mortality (Jejeebhoy 1998).

Three demographic processes—a decline in mortality, a drop in fertility, and increasing urbanization—have importantly influenced the ability of women to expand their social roles and improve their life chances. A major factor influencing the rise in the status of women has been the more general liberation of people from early death. The decline in mortality does not mean that pressures to have children have evaporated. That is far from the case, but there is a greater chance that the pressures will be less; indeed, remaining single and/or childless is more acceptable for a woman now than at any time in history.

In North America, Europe, and much of Latin America and Asia, most of a woman's adult life is now spent doing something besides bearing and raising children, because she is having fewer children than in previous generations and she is also living longer. An average American woman bearing two children in her late 20s and early 30s would, at most, spend about 20 years bearing and rearing them. That many years is far fewer, of course, than she will actually have of relative (indeed increasing) independence from child-rearing obligations, since if her two children are spaced two years apart and the first child is born when she is 30, by age 38 her youngest child will be in school all day, and she will still have 44 more years of expected life. Is it any wonder, then, that women have searched for alternatives to family building and have been recruited for their labor by the formal economic sector? In a sociocultural setting in which the reproduction of children consumes a great deal of societal energy, the domestic labor of women is integral to the functioning of the economy (see, for example, Cosio-Zavala 1997). With a slackening in demand for that type of activity, it is only natural that a woman's time and energy would be employed elsewhere, and elsewhere is increasingly likely to be in a city.

Cities are more likely than rural areas to provide occupational pursuits for both women and men that encourage a delay in marriage (thus potentially lowering fertility) and lead to a smaller desired number of children within marriage. Furthermore, since urbanization involves migration from rural to urban areas, this has meant that women, as they migrate, are distanced a bit from the pressures to marry and have children that may have existed for them while living in their parents' homes. Thus migration may lead to a greater ability to respond independently to the social environment of urban areas, which tend to value children less than rural areas. Young adults are especially prone to migration and every adult who moves may well be leaving a mother behind. This means the migrant will have more time on her hands to look for alternatives and question the social norms that prescribe greater submissiveness, a lower status, and fewer out-of-the-home opportunities for women than for men.

Of course, it isn't quite that simple. For one thing, the process of urbanization in the Western world initially led to an increase in the dependency of women before promoting increased gender equity (Nielsen 1978). Urbanization is typically associated with a transfer of the workplace from the home to an outside location—a severing of the household economy and the establishment of what Kingsley Davis (1984b) has called the "breadwinner system," in which a member of the family (usually the male) leaves home each day to earn income to be shared with other family members. In premodern societies, women generally made a substantial contribution to the family economy through agricultural work and the marketing of produce (Boserup 1970), but the city changed all that. Men were expected to be breadwinners (a task that women had previously shared), while women were charged with domestic responsibility (tasks that men had previously shared). From our vantage point in history, the breadwinner system seems "traditional," but from a longer historical view, it is really an anomaly. Thus, the idea of men and women sharing economic responsibility for the family is a return to the way in which most human societies have been organized for most of human history.

(continued)

REPRODUCTIVE RIGHTS, REPRODUCTIVE HEALTH, AND THE FERTILITY TRANSITION (CONTINUED)

The Demographic Linkage Between the Fertility Transition and Reproductive Health

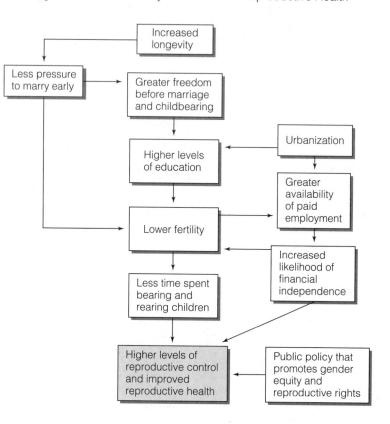

As the life expectancy of the urban woman increases and as her childbearing activity declines, the lack of alternative activities is bound to create pressures for change. In the figure accompanying this essay, I have diagrammed the major paths by which mortality, fertility, and urbanization influence the status of women and lead to more egalitarian gender roles and improved reproductive health. Increased longevity eventually lessens the pressure for high fertility and lessens the pressure to marry early. These changes permit a woman greater freedom for alternative activities before marrying and having children, as well as providing more years of life beyond childbearing. Women are left to search for the alternatives, which are importantly wrapped up in higher levels of education. Society is then of-

fered a "new" resource—nondomestic female labor. This can create new opportunities for a woman's economic independence, which is key to controlling her life, including her reproduction.

Having greater control over her own life, enhanced by lower fertility, also improves a woman's health in the process, even without government programs designed to increase reproductive health. Keep in mind, however, that public policy supporting gender equity, reproductive rights, and reproductive health can go a long way toward accelerating these changes in society. For its part, the fertility transition can be viewed as a key element in the broader pattern of changes involved in the demographic transition associated with women being able to take control of their lives and their bodies.

At older ages, women may decide not to have an additional child, and so completed family size becomes five children born instead of six, or four instead of five. Research suggests that even when couples have an ideal family size in mind, they make decisions about children one at a time. This information is captured by what demographers call **parity progression ratios,** which represent the proportion of women with a given number of children (parity refers to how many children have already been born) who "progress" to having another child. Thus, as fertility declines, parity progression ratios will decline at the higher parities. The consequence is a "compression" of reproduction (to borrow the term I used in Chapter 5 in discussing mortality) into a shorter number of years. Instead of women routinely having babies between the ages of 15 and 44, most women will have all of their children in their 20s and early 30s. The decline in parity progression ratios at the older ages may be especially accomplished with surgical contraception, although as I indicated earlier, the time-honored method was abstinence. In between the younger and older ages, the use of effective temporary contraception or the availability of abortion or emergency contraception provides couples with the opportunity to space children as they see most appropriate. In that process, they may decide, for example, not to have a third child now that they have two—making their own personal contribution to the fertility transition.

In Figure 6.6, I provide some examples of how age-specific fertility rates change over time in the context of the fertility transition, using illustrations from Latin America (Mexico), South Asia (India), West Asia (Jordan), and Africa (Kenya). You can see that each of these four countries had a variation on the same theme—the percentage decline in the age-specific fertility rates since the 1970s has been highest at the older ages, next highest among adolescents, and least dramatic among women aged 20–34. Keep in mind, however, that the majority of babies are born to those women in the 20–34 age group, so it is their absolute decline in fertility that makes the difference in the long run.

All of the perspectives on the fertility transition discussed above assume that fertility will not decline until people see limiting fertility as being in their interest. The supply-demand framework assumes that people are making rational economic choices between the quantity and quality of children, whereas the innovation-diffusion perspective argues that social pressure is the motivation, regardless of the underlying economic circumstances. Coale's three preconditions do not specify what the motivating factors might be, leaving open the possibility of some mix of economic and social motivations, sometimes stimulated or retarded by public policy decisions that make it easier or harder for people interested in controlling their reproduction to actually do so. In the real world, a combination of forces seems to produce the observed fertility transitions and we can see these combinations at work in the following country-specific illustrations.

Regional Differences in the Fertility Transition

It is fair to say that by the early part of the twenty-first century there is no region of the world that has not experienced at least the early stages of the fertility transition. Nearly all areas of the world have experienced a decline in fertility over the past few

Figure 6.6   Changes in ASFRs in the Context of the Fertility Transition

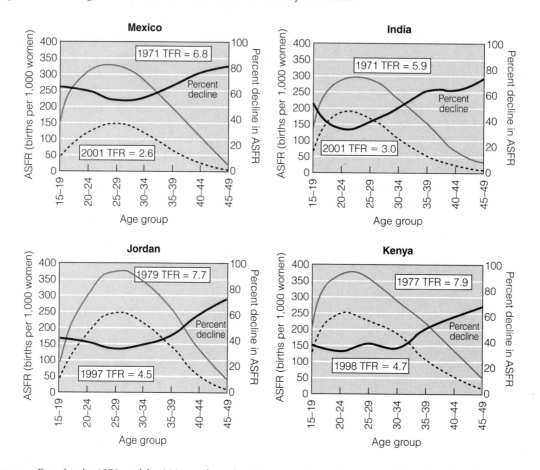

Source: Data for the 1970s and for 2001 are from the U.S. Census Bureau International Data Base (http://www.census. gov/ipc/www), accessed 2001; Data for Jordan for 1997 and Kenya for 1998 are from the Demographic and Health Survey *STATcompiler* (http://www.measuredhs.com), accessed 2001.

decades, as you can see in Table 6.7. Even in 1950, Europe had the lowest fertility levels in the world, and it has now dropped to below replacement throughout that region. The major question being asked is whether or not it will ever rise, or indeed if it might even drop to a lower level. You can see that the United Nations' medium projections (typically the ones they think are most likely) suggest that fertility might go up again in Europe. That is pure speculation. In the meantime, there is evidence that within Eastern Europe, in particular, the one-child family is becoming the norm (Sobotka 2003).

The most dramatic drop in fertility has been in East Asia, where the average total fertility rate in 1950 was 5.68, but it has now dropped to 1.78, although it may rise a bit over the course of the next half century. The remainder of Asia continues to

Table 6.7 Regional Differences in the Fertility Transition, 1950–2050

	Total Fertility Rates		
	1950–1955	2000–2005	2045–2050
WORLD	5.02	2.69	**2.02**
More-developed regions	2.84	**1.56**	1.85
Less-developed regions	6.16	2.92	**2.04**
Least-developed countries	6.64	5.13	2.47
AFRICA	6.74	4.91	2.40
Eastern Africa	6.97	5.61	2.48
Middle Africa	5.91	6.28	2.59
Northern Africa	6.82	3.21	**1.90**
Southern Africa	6.46	2.79	**1.87**
Western Africa	6.85	5.56	2.48
ASIA	5.89	2.55	**1.91**
Eastern Asia	5.68	**1.78**	**1.85**
Southcentral Asia	6.08	3.25	**1.91**
Southeastern Asia	5.95	2.55	**1.86**
Western Asia	6.46	3.45	2.19
EUROPE	2.66	**1.38**	**1.84**
Eastern Europe	2.91	**1.18**	**1.85**
Northern Europe	2.32	**1.61**	**1.85**
Southern Europe	2.65	**1.32**	**1.85**
Western Europe	2.39	**1.58**	**1.85**
LATIN AMERICA AND THE CARIBBEAN	5.89	2.53	**1.86**
Caribbean	5.22	2.39	**1.90**
Central America	6.87	2.76	**1.88**
South America	5.69	2.45	**1.85**
NORTHERN AMERICA	3.47	**2.05**	**1.85**
Canada	3.73	**1.48**	**1.85**
United States of America	3.45	2.11	**1.85**
OCEANIA	3.90	2.34	**1.92**
Australia/New Zealand	3.27	**1.75**	**1.85**
Melanesia	6.29	3.91	**2.01**
Micronesia	6.41	3.40	**1.91**
Polynesia	6.89	3.16	**1.95**

Source: Adapted from data in United Nations, 2003, World Population Prospects: The 2002 Revision; numbers in bold-faced type are total fertility rates that are below replacement level; data for 2045–2050 are based on the medium projections.

have fertility well above replacement level, although the United Nations thinks that all areas but western Asia will drop to below replacement by the middle of this century.

Sub-Saharan Africa had very high fertility in 1950 and has experienced the slowest decline. Indeed, in the middle portion of Africa, fertility has actually gone up! The United Nations expects that at the middle of this century, sub-Saharan Africa will still have by far the highest fertility levels in the world, but significantly lower than current levels. However, northern Africa, where fertility still remains well above replacement level, is expected to continue its decline over this century, and the same is expected of southern Africa.

In Latin America, there is a distinction between countries that are heavily composed of indigenous populations (much of Central America and northern South America) and the remainder of the region. Latin America had very high fertility in 1950, and it has dropped the most in the Caribbean and South America, although it is still well above replacement. Nonetheless, the United Nations thinks that the entire region could be below replacement level by the middle of this century. In North America, there is a divergence between the United States, where fertility is just at replacement level, and Canada, where it has dropped to below replacement. Australia and New Zealand, like Canada, have taken a path more similar to Europe than to the U.S.

These regional trends give us a feel for what is happening, but not necessarily for why it's happening. It helps to take a look at a few countries in more detail to better understand the fertility transition.

Case Studies In The Fertility Transition

There are as many interesting stories of the fertility transition as there are countries and even subregions within countries, but for illustrative purposes I am going to first choose a country that has had low fertility for a long time and is now well below replacement level—the United Kingdom (along with comments about other European nations). Then I will review the situation in two of the world's most populous nations—China and the United States—each of which has a different perspective on the fertility transition. Finally, I will offer the example of Jordan—a country that still has a way to go toward replacement-level fertility, even though it has already moved a considerable distance in that direction.

United Kingdom and Other European Nations

Historical Background   In England and other parts of Europe, the beginnings of a potential fertility decline may well have existed even before the Industrial Revolution touched off the dramatic rise in the standard of living. In English parishes, there is evidence that withdrawal (coitus interruptus) was used to reduce marital fertility during the late seventeenth and early eighteenth centuries, and it was apparently also a major reason for a steady decline in marital fertility in France during the late eighteenth and early nineteenth centuries. Abortion was quite probably also fairly common (Wrigley 1974). Furthermore, preindustrial birth rates were much higher in the European colonies of America than in Europe points to the fact that fertility limitation in Europe was widely accepted and practiced, especially through the mechanism of deliberately delayed marriage, as well as other, more direct means. The evidence for the use of some means of fertility control as far back as the eighteenth century in Europe is circumstantial, to be sure, but powerful nonetheless. Consider the comment by the great Scottish economist Adam Smith, writing in 1776, that "barrenness, so frequent among women of fashion, is very rare among those of inferior station. Luxury in the fair sex, while it enflames perhaps the passion for enjoyment, seems always to weaken, and frequently to destroy altogether, the powers of generation" (Smith 1776:I.viii.37).

The enormous economic and social upheaval of industrialization took place earlier in England than anywhere else, and by the first part of the nineteenth century, England was well into the Machine Age. For the average worker, however, it was not until the latter half of the nineteenth century that sustained increases in real wages actually occurred. During the first part of that century, the Napoleonic Wars were tripling the national debt in England, increasing prices by as much as 90 percent without an increase in production. Thus, during most of Malthus's professional life, his country was experiencing substantial inflation and job insecurity. These relatively adverse conditions undoubtedly contributed to a general decline in the birth rate during the first half of the nineteenth century brought about largely by delayed marriage (Wrigley 1987).

After about 1850, economic conditions improved considerably, and the first response was a rise in the birth rate (as the marriage rate increased), followed by a long-run decline. This was a period in which all of Ansley Coale's preconditions for a fertility decline existed: (1) People had apparently accepted calculated choice as a valid element, (2) People perceived advantages from lowered fertility, and (3) People were aware of at least reasonably effective means of birth control. As I already mentioned, the British were accustomed to thinking in terms of family limitation, and delayed marriage, abstinence, and coitus interruptus within marriage were known to be effective means to reduce fertility. In the second half of the nineteenth century, then, motivation to limit family size came in the form of larger numbers of surviving children combined with aspiration for higher standards of living.

It is important to remember that the restriction of fertility was in many ways a return to preindustrial patterns, in which an average of about two children survived to adulthood in each generation. Thus, as we discussed earlier in relation to the theories of the fertility transition, mortality declines produced changes in the lives of people to which they had to respond. The English reacted in ways consistent with the theory of demographic change and response. They responded to population growth by migrating and by delaying marriage and then, only when those options were played out, did marital fertility clearly decline (Friedlander 1983). In a world now characterized by very open sexuality, it is sometimes hard to believe that large numbers of people were willing to suppress their sexuality in order to achieve some measure of economic success, but Szreter and Garrett (2000) remind us that:

> The available historical evidence suggests that the late marriage regime of early modern Britain entailed systematic sexual restraint among young adults up to their mid-20s, the point at which marriage could be realistically anticipated. A sexual culture of this sort might lend itself to restraint *after* marriage, as well, if and when the need arose (p. 70).

The best-known explanation of the fertility decline in England in the latter half of the nineteenth century is that offered by J. A. Banks (1954) in *Prosperity and Parenthood* and its sequel, *Feminism and Family Planning in Victorian England* (Banks and Banks 1964). Banks's thesis is by now a familiar one: that the rising standard of living in England, especially among the middle classes, gave rise to a declining fertility by (1) raising expectations of upward social mobility, (2) creating fears of social slippage (you had to "keep up with the Joneses"), and (3) redefining the roles of women from housewife to a fragile luxury of a middle-class man.

While it is true that birth rates dropped more quickly in the upper social strata of English society than in the lower social strata, by 1880 all segments of English society were experiencing fertility declines. From about 1880 to 1910, England shared with much of continental Europe in what Knodel and van de Walle (1979) have called "the momentous revolution of family limitation." Fertility has continued on a slow downward trend since then, interrupted by a postwar baby boom, which peaked in the mid-1960s. Since about 1964, however, birth rates have resumed their downward trend—England's baby bust (Hobcraft 1996)—and England has settled into a consistent pattern of below-replacement-level fertility, with a TFR for the year 2003 of only 1.6 children per woman, accomplished by widespread use of contraceptives (especially the pill), delayed marriage, delayed childbearing, and the use of surgical contraception when the desired family size has been achieved.

Figure 6.7 shows the change in fertility that has taken place in England over the course of the past 200 years. As dramatic as the change has been, you can see that it follows the same pattern as the more recent declines in less-developed countries, as shown in Figure 6.6. In 1800, the TFR in England is estimated to have been 5.6 children per woman (Livi-Bacci 2000). The transition down to a TFR of 1.6 means that contraception is used by 82 percent of sexually active couples (United Nations Population Division 2001), and that, along with abortion, means that women do not have to delay sexual activity even though they delay marriage and/or having children. And surgical contraception means that when they decide to stop having children, it is a permanent decision.

Below-Replacement Fertility   England mirrors other European nations in starting the twenty-first century with below-replacement-level fertility. Only Albania has above-replacement-level fertility, and then only barely so (a TFR of 2.3 according to U.N. estimates). There has been a good deal of hand-wringing over this in Europe because the low fertility, especially when combined with increasing life expectancy, is producing an increasingly older population with a shrinking base in the younger ages. As I mentioned in Chapter 2, governments are worried about how old-age pensions will be funded and how economic growth will be maintained if there are too few young people compared to the older population. However, this low fertility should be not too surprising because it can be explained in terms of both the supply-demand framework and the diffusion of innovations.

Europeans grow up knowing that they and any children they have will almost certainly survive to a rather old age. They also know that highly effective contraceptives make it possible to engage in sexual activity without fear of pregnancy (and of course the use of condoms will protect them as well from sexually transmitted diseases, including HIV). Should the contraceptive fail, either emergency contraception

Figure 6.7   The Fertility Transition in England

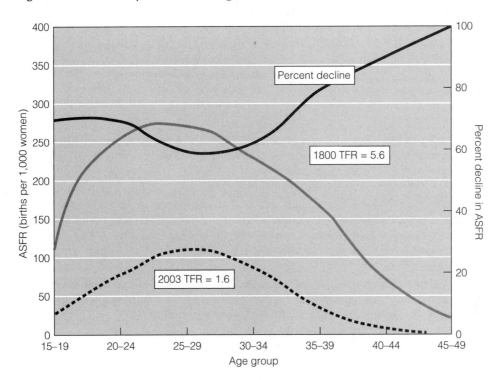

Source: Data for all countries for 2001 are from the U.S. Census Bureau International Data Base (http://www.census.gov/ipc/www), accessed 2001. Data for England in 1800 are based on fertility and mortality estimates from Livi Bacci (2000) applied to spreadsheets developed by the U.S. Census Bureau (Arriaga 1994).

or abortion is available. Thus, European women are in almost total control over the supply of children. The important question is how large is the demand? Surveys throughout Europe suggest that Europeans would prefer two children, but at the same time women are investing heavily in their own education and are seeking jobs and careers comparable to those of men. While gender equity may be approachable on those two fronts, it is has been less obvious within marriage. Women are still expected to be the principal providers of care to children and to their husbands and this extra burden placed on women has created a climate of caution about getting married and having those two children. Women know that they will bear a disproportionate share of the burden of child rearing and household labor, while suffering substantial opportunity costs from delayed careers and lost wages. The movement toward gender equity in education and in the labor force has thus not been matched by gender equity in domestic relationships (McDonald 2000), and so the demand for children has dropped to what we might think of as female replacement—a woman having a child (whether male or female) that allows her to experience reproduction and replacement, but not much more than that. This mismatch seems to be greatest in southern Europe, most notably Italy and Spain, where attitudes about the woman's role in the family are more traditional than in the rest of Europe (Dalla Zuanna 2001; de Sandre 2000). Chesnais (1996) and McDonald (2000) have offered the suggestion that in highly developed countries, a *rise* in the status of women within family-oriented institutions may be necessary to bring fertility back up to replacement level. Japan is very much in that same position.

China

At the time of the communist revolution in 1949, the average woman in China was bearing 6.2 children and the population, which was already more than half a billion, was growing rapidly. The government of the People's Republic of China realized decades ago that the population problem was enormous, and so it implemented the largest, most ambitious, most significant, and certainly the most controversial policy to slow population growth ever undertaken in the world. The 1978 constitution of the People's Republic of China declared that "the state advocates and encourages birth planning" and the reasons for this were spelled out in 1979 by Vice Premier Chen Muhua (Chen 1979). She argued that there are three major explanations for the fact that in China population control was now "dictated and demanded by the socialist mode of production" (1979:94). These were: (1) Too rapid an increase in population is detrimental to the acceleration of capital accumulation, (2) Rapid population increase hinders the efforts to quickly raise the scientific and cultural level of the whole nation, and (3) Rapid population growth is detrimental to the improvement of the standard of living.

The goal of the Chinese government at that time was, incredibly enough, to achieve zero population growth (ZPG) by the year 2000, with the population stabilizing at 1.2 billion people. To accomplish this meant that at least one generation of Chinese parents had to limit their fertility to only one child, because the youthful age structure in China in the 1970s meant a high proportion of people were in their childbearing years. If all of those women had two children, the population would still be growing too fast. How did they go about trying to achieve this goal? The first step was to convince women not to have a third child (third or higher-order births accounted for 30 percent of all births in 1979; Chen 1979). The second step was to promote the one-child family. These goals have been accomplished partly by increased social pressure (propaganda, party worker activism, and probably coercion as well) and partly by the increased manufacture and distribution of contraceptives. The heart of the policy, though, has been a carefully drawn system of economic incentives (rewards) for one-child families and disincentives (punishments) for larger families.

The one-child policy was directed by the central government, but implemented at the local level in ways that varied from place in place. In general, the policy was that in cities couples with only one child who pledged that they would have no more children (and were using some form of fertility limitation) could apply for a one-child certificate. The certificate entitled the couple to a monthly allowance to help with the cost of child rearing until the child reached age 14. Furthermore, one-child couples were to receive preference over others in obtaining housing and were to be allotted the same amount of space as a two-child family; their child was given preference in school admissions and job applications; and the parents with a one-child certificate were promised a larger-than-average pension when they retired.

In the countryside, the incentives tended to be a bit different. One-child rural families received additional monthly work points (which determined the rural payments in cash and in kind) until the child reached age 14. These one-child families could also get the same grain ration as a two-child family. In addition, all rural families received the same-sized plot for private cultivation, regardless of family size, thus indirectly rewarding the small family. Each province in China has been encouraged by the central government to tailor specific policies to meet the particular needs of its residents, and some of the more widely implemented policies have included an increasingly heavy tax on each child after the second, and the expectation that for each child after the second, parents will pay full maternity costs as well as full medical and educational costs.

Figure 6.8 reveals that the total fertility rate was already declining in China at the time the government implemented the one-child policy, and it may well be argued that the government policy merely reinforced changes in reproduction that were already well under way. Furthermore, fertility fell not only in urban areas where the

Figure 6.8   The Fertility Transition in China

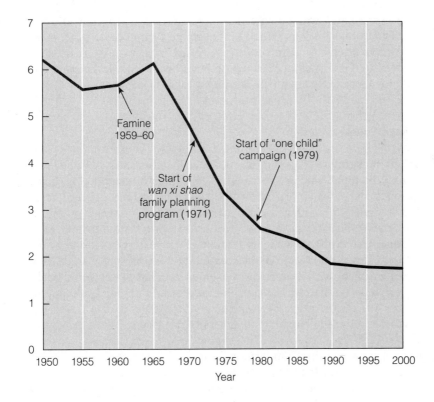

Source: TFR data are from United Nations Population Division, 2003, World Population Prospects: The 2002 Revision (New York: United Nations).

motivation for small families might be greatest, but in rural areas as well (Attané 2002).

In 1971, the government instituted the *wan xi shao* (later, longer, fewer) campaign and that helped to accelerate fertility decline in China. Thus fertility was already on its way down in many parts of China when the one-child policy was implemented in 1979. The one-child policy in China was never intended to be anything more than an interim measure that would finally put the brake on population growth in that country (Greenhalgh 1986). From the beginning, the plan had been to ease back to a two-child family after hitting the 1.2 billion level of population size. Even a quick glance at Figure 6.8 calls into question whether the one-child policy was ever as potent an instrument for fertility decline as is often thought or whether it mainly helped to publicize and maintain a fertility decline that was already well under way. Several researchers have concluded that socioeconomic development in China has been at least as important a factor as government policy in motivating couples to want fewer children (see Attané 2001; Poston and Gu 1987; Tu 2000), although the government's attitude has certainly facilitated and enhanced those motivations (Greenhalgh, Chuzhu, and Nan 1994). It is also intriguing to note the cultural similarities between mainland China and Taiwan (which, as you know, is actually claimed by China as part of its territory) and the coincidence of rapid fertility declines in the two countries—one with a clearly defined set of incentives and disincentives, and the other with a more normal voluntary family planning program, but both with rapidly expanding economies during the latter part of the twentieth century. The United Nations estimates that 83 percent of married women of reproductive age are using a modern contraceptive in China, with the IUD being the most popular "temporary" method, backed up by abortion. Surgical contraception is the method of choice once a couple has completed childbearing.

The United States

Historical Background   Around 1800, when Malthus was writing his *Essay on Population,* he found the growth rate in America to be remarkably high and commented on the large frontier families about which he had read. Indeed, it is estimated that the average number of children born per woman in colonial America was about eight. It is probably no exaggeration to say that early in the history of the United States, American fertility was higher than any European population had ever experienced. Early data are not very reliable, but in 1963 Ansley Coale and Melvin Zelnick made new estimates of crude birth rates in the United States going back as far as 1800; these indicate that the crude birth rate of nearly 55 per 1,000 population was higher than the rate in any less-developed country today (Coale and Zelnick 1963). Even in 1855, the crude birth rate in America was 43 per 1,000, comparable to current levels in sub-Saharan Africa. The American Civil War seems to have been a turning point in marital fertility and fertility declined virtually unabated from about 1870 until the Great Depression of the 1930s, during which time it bottomed out at a low level only recently reapproached (Hacker 2003). Why the precipitous drop?

As I discussed in Chapter 2, almost all voluntary migrants to the North American continent were Europeans. The people who made up much of the population of the early United States came from a social environment in which fertility limitation was known and practiced. Despite the frontier movement westward, America in the century after the Revolution was urbanizing and commercializing rapidly. Furthermore, the United States was experiencing the process of secularization, and people's lives were increasingly loosened from the control of church and state. Malthus had commented that, with respect to all aspects of life, including reproduction, "despotism, ignorance, and oppression produced irresponsibility; civil and political liberty and an informed public gave grounds for expecting prudence and restraint" (quoted by Wrigley 1988:39). Bolton and Leasure (1979) have shown that throughout Europe, the early decline in fertility occurred near the time of revolution, democratic reform, or the growth of a nationalist movement. Analogously, Leasure (1989) found that the decline in fertility in the United States in the nineteenth century was closely associated with a rise in what he calls the "spirit of autonomy," measured early in the century by the proportion of the population in an area belonging to the more tolerant Protestant denominations (Congregational, Presbyterian, Quaker, Unitarian, and Universalist) and measured later in the century by educational level.

Lower fertility was accomplished by a rise in the average age at marriage and by various means of birth control within marriage, including coitus interruptus, abortion (even though it was illegal), and extended breast-feeding (Sanderson 1995). Nineteenth-century America also witnessed the secret spread of knowledge about douching and periodic abstinence, neither of which is necessarily very effective on the face of it, but the fact that women were searching for ways to prevent pregnancy is clear evidence of the motivation that women had to limit fertility (Brodie 1994).

New immigrants arriving in the United States in the late nineteenth and early twentieth centuries came especially from southern and eastern Europe, where ideas of contraception were less well known. As a nurse in New York City working amongst immigrants in the first decade of the twentieth century working, Margaret Sanger witnessed firsthand the tragic health consequences for mothers and their babies of having too many children. Sanger herself decided to have only three children after watching her own Irish Catholic mother die prematurely after having 11 children. Her patients kept asking her about the secrets that middle-class women must know that allowed them to keep their families small. But the secrets she did have at that time, including coitus interruptus and abstinence, were dismissed by women as being impossible without a husband's cooperation, which they knew they would never get. In 1912, after helplessly watching a young mother with several children die from a botched self-inflicted abortion, she "resolved to seek out the root of evil, to do something to change the destiny of mothers whose miseries were vast as the sky" (Sanger 1938:92).

Sanger immersed herself in finding out all she could about contraception, and began to write on the subject, landing herself in continuing legal difficulty for publishing "pornographic" material. In 1915, she was introduced to a newly designed diaphragm developed in the Netherlands. It required a health professional to fit a woman for the right size, but it was far more effective than anything else that existed at that time and it was the most effective contraceptive in the world until the pill came along in the 1960s (Douglas 1970). The next year, in 1916, Sanger opened her notorious birth control clinic in Brooklyn. She spent the remainder of her life trying to legalize the publication of information about family planning, and to legalize the distribution and use of contraceptives themselves. Until 1965, it was technically illegal in the United States for even a married couple to use any method of birth control. In that year the U.S. Supreme Court ruled in *Griswold v. Connecticut* that married couples had a right to privacy that extended to the use of contraceptives. The Court granted that same right to unmarried couples in 1972, and the following year extended the same argument to the legalization of abortion. Sanger played a critical role in the contraceptive revolution; she was the founder in 1939 of the Planned Parenthood Federation of America and in 1952 helped to create the International Planned Parenthood Federation.

After World War I, the use of condoms became widespread in the United States (and in Europe as well) and, along with withdrawal and abstinence (and the clandestine use of the diaphragm), contributed to very low levels of fertility during the Depression (Himes 1976). The condom was available for sale not because it was a contraceptive (that use was illegal), but rather because it was a "prophylactic" that prevented the spread of sexually transmitted disease. During the Depression, fertility fell to levels below generational replacement. The United States was not unique in this respect, but that bottoming out did cap the most sustained drop in fertility the world had seen until very recently. It was undoubtedly a response to the economic insecurity of the period, especially since that insecurity had come about as a quick reversal of increasing prosperity. Fear of social slippage was thus a very likely motive for keeping families small. The American demographic response was for many couples to defer marriage and to postpone having children, hoping to marry later on and have a larger family. Gallup polls starting in 1936 indicate that the average ideal family size was three children, and that most people felt that somewhere between two and four was what they would like. Thus, people were apparently having fewer children than they would have liked to have under ideal circumstances.

In 1933, the birth rate hit rock bottom because women of all ages, regardless of how many children they already had, lowered their level of reproduction, but this was mainly a matter of timing. From 1934 on, the birth rates for first and second children rose steadily (reflecting people getting married and having small families), while birth rates for third and later children continued to decline (reflecting the postponement of larger families) until about 1940 (Grabill, Kiser, and Whelpton 1958). Just as the United States was entering World War II in late 1941 and 1942, there was a momentary rise in the birth rate as husbands went off to war, followed by a lull during the war. However, the end of World War II signaled one of the most dramatic demographic phenomena in North American history—the **baby boom.**

The Baby Boom   Most of you can probably appreciate that immediately after the end of a war, families and lovers are reunited and the birth rate goes up temporarily as people make up for lost time. This occurred in the United States as well as in England and Canada and several of the other countries actively involved in World War II. Surprisingly, these baby booms lasted not for one or two years, but for several years after the war. Birth rates in the United States continued to rise through the 1950s, as the total fertility rate went from 2.19 in 1940 to 3.58 in 1957, an increase of nearly 1.5 children per woman. Note that in the United States the term "baby boomers" is usually applied to people born between 1946 and 1964, but the "boom" peaked in 1957—12 years after the war ended.

An important contribution to the baby boom was the fact that, after the war, women started marrying earlier and having their children sooner after marriage. For example, in 1940, the average first child was born when the mother was 23.2 years old, whereas by 1960, the average age had dropped to 21.8. This had the effect of bunching up the births of babies, which in earlier times would have been more spread out. Further, not only were young women having children at younger ages, but older women were having babies at older-than-usual ages, due at least in part to their having postponed births during the Depression and the war. After the war, many women stopped postponing and added to the crop of babies each year. These somewhat mechanical aspects of a "catching up" process explain only the early part of the baby boom. What accounts for its prolongation?

We do not have a definitive answer to this question (and remember that it occurred in other countries besides just the U.S.), but a widely discussed explanation is offered by Easterlin (1968; 1978), which I mentioned in Chapter 3 as the relative-cohort size hypothesis—a spin-off of the supply-demand perspective. Easterlin begins his analysis by noting that the long-term decline in the birth rate in the United States was uneven, sometimes happening more rapidly than at other times. In particular, the birth rate declined less rapidly during times of greater economic growth. If a young man could easily find a well-paying job, he could get married and have children; if job hunting was more difficult, marriage (and children within marriage) would be postponed. Thus, it was natural that the postwar baby boom occurred, because the economy was growing rapidly during that time. What was unusual was that economic growth was more rapid than in previous decades, and the resultant demand for labor was less easily met by large numbers of immigrants, because in the 1920s the United States had passed very restrictive immigration laws. Furthermore, the number of young people looking for work was rather small because of the low birth rates in the 1920s and 1930s. Finally, the demand for labor was not easily met by females, since there was a distinct bias against married women working in the United States, particularly a woman who had any children. Some states passed legislation that actually restricted married women from working in certain occupations. To be sure, women did work, especially single women, but their opportunities were limited. Thus, economic expansion, restricted immigration, a small labor force, and discrimination against women in the labor force meant that young men looking for jobs could find relatively well-paying positions, marry early, and have children. Indeed, income was rising so rapidly after the war and on into the 1950s that it was relatively easy for couples to achieve the lifestyle to which they were accustomed, or even to which they might moderately aspire, and still have enough money left over to have several children.

In 1958, the crude birth rate and the general fertility rate in the United States registered clear declines—a downward change that carried into the late 1970s—as you can see in Figure 6.9. At first, the decline was due merely to the fact that the trend of earlier marriage and closer spacing of children had bottomed out. The number of women in their childbearing years had also declined as the relatively small number of babies born during the Depression reached childbearing age. At this point in the early 1960s, there was no discernible trend toward smaller families. The ideal family size among Americans had remained quite stable between 1952 and 1966, ranging only between 3.3 and 3.6 children. But in 1967, Blake discovered in a national sample taken the year before that "young women (those under age 30) gave

Figure 6.9   The Baby Boom, the Baby Bust, and the Baby Boomlet in the United States

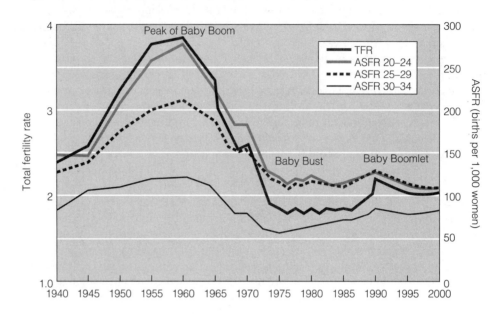

Note: The total fertility rate shows the overall pattern of reproductive behavior in the United States, while the age-specific birth rates show the contribution of individual age groups to that overall pattern.

Source: Data for 1940 through 1959 are from National Center for Health Statistics, 1990, Vital Statistics of the United States, 1988, Vol. I, Natality (Hyattsville, MD: NOHS), Table 1–6; data from 1960 through 2000 are from the U.S. Census Bureau International Data Base (http://www.census.gov/ipc/www), accessed 2001.

'two' children as their ideal more frequently than they had in any surveys since the early nineteen-fifties" (Blake 1967:20). This was the first solid evidence that the desired family size might be on the way down—that the baby bust period had arrived.

The Baby Bust, Baby Boomlet, and Beyond   Social and economic factors in the late 1960s suggested that fertility might continue to decline for awhile. The rate of economic growth had slackened off, and there was no longer a labor shortage. As Norman Ryder very presciently noted:

> In the United States today the cohorts entering adulthood are much larger than their predecessors. In consequence they were raised in crowded housing, crammed together in schools, and are now threatening to be a glut on the labor market. Perhaps they will have to delay marriage, because of too few jobs or houses, and have fewer children. It is not entirely coincidental that the American cohorts whose fertility levels appear to be the highest in this century were those with the smallest number (Ryder 1965:845).

This was, of course, the kernel of Easterlin's relative cohort hypothesis. As it turned out, between the early 1960s and the mid-1970s, the average age at marriage among American women went up slightly and the number of children born to women aged

20–24 went steadily down, declining by 27 percent between 1960 and 1971. The period 1967 to 1971 was especially dramatic. Between those two dates, the total number of lifetime births expected for currently married American women aged 18–24 dropped from 2.85 to 2.38. By 1977, the birth expectations had continued to drop at a slower pace to 2.14 births (U.S. Census Bureau 1979b). Expectations for older women also dropped steeply between 1967 and 1977, but the decline came later than for the younger women. It was, of course, younger couples who were most influenced by the cohort squeeze as outlined by Easterlin.

Almost all of the fertility decline was due to a drop in marital fertility (Gibson 1976), mainly as a result of more efficient use of contraception, and to a rise in the use of abortion. As fertility dropped, family size ideals dropped as well. Gallup surveys reported by Blake (1974) indicated, for example, that the proportion of white women under age 30 saying that two children was an ideal number rose dramatically from a low of 16 percent in 1957 to 57 percent in 1971. The decline in fertility following the baby boom peak thus seemed to signal a major shift in the norms surrounding parenthood in American society. "Motherhood is becoming a legitimate question of *preferences*. Women are now entitled to seek rewards from the pursuit of activities other than childrearing" (Ryder 1990:477, emphasis added). I have already commented on Caldwell's theory of the reversal of income flow between parents and children. The baby boomers have been trapped in the final turn of that transition—sandwiched between parents who feel that children owe them something and children who feel that parents owe them something.

The baby bust troughed in the mid-1970s and was followed by a baby boomlet, as you can see in Figure 6.9. In much publicized fashion, the boomlet was sparked by women in their early 30s whose birth rate had risen almost uninterruptedly since 1975. Birth rates also rose for women in their late 30s and early 40s, but I have not shown those data in Figure 6.8 in order to keep the graph more readable and, besides, almost 80 percent of all births in the United States are to women under the age of 35.

These general trends hide a great deal of complexity in American fertility patterns, and two trends, in particular, are worth noting at this point: (1) a rise in out-of-wedlock births and (2) an increase in childlessness. Both of these trends have consequences for family and household structure and I will discuss them in more detail in Chapter 10. I have already indicated that the general explanation for the rise in out-of-wedlock births seems to be that women are engaging in sexual activity at younger ages, but marrying at older ages. Thus, the sexually active time before marriage is increasing, and it would be unusual under these circumstances if out-of-wedlock births did not go up.

One of the more important aspects of childlessness is that it illustrates the control that now exists over fertility in the United States. Brodie (1994) suggests that a substantial fraction of women even in nineteenth century America intended not to have children, but were thwarted, so to speak, by their inability to prevent pregnancy. The vastly improved fertility control technology that is now available makes it much easier for a woman to implement such a desire. It can also be said that, given the set of high monetary and opportunity costs of children in advanced societies, it would be very surprising, indeed, if there were not a noticeable fraction of women who chose not to have children.

No matter how important the diffusion of innovative ideas and technologies may have been in initiating the fertility transition, most of the evidence seems clearly to suggest that some variant of the rational-choice model now dictates the reproductive decisions made by most Americans (see, for example, Rindfuss, Morgan, and Offutt 1996; Yamaguchi and Ferguson 1995). That raises the question again, of course, of why people have children at all in a society like the United States. Why doesn't everyone opt for childlessness? The answer, as I alluded to earlier in the chapter, is that children do provide benefits to their parents, even though the return to the parents is rarely economic. Rather, children provide a source of "social capital" for parents, as Schoen and his associates (Schoen et al. 1997) have suggested, building on rational-choice theory. Children establish for their parents a network with other relatives, as well as with other people in the community, and they do that not just while they are young, but potentially for the rest of the parents' lives. Humans are social creatures, and through children we assure ourselves of a social network.

The United States is the most populous wealthy country in the world and it also has the highest fertility rate among developed nations. This seems to be due especially to the influx of immigrants since the 1980s (especially immigrants from Mexico) who have brought with them views of the world that place a higher value on the number of children in a family than is generally true of the native-born non-Hispanic population (Bean, Swicegood, and Berg 2000). Figure 6.10 shows the changes in total fertility rates by major ethnic group in the United States between 1990 and 2002. You can see that white non-Hispanics had below-replacement fertility already by 1990 and it didn't change much between then and 2002, whereas fertility did decline for every other group, enough so that every group but Hispanics had at or below-replacement fertility by 2002. Among Hispanics, the Mexican-origin population is the largest group and you can see that their fertility declined somewhat between 1990 and 2002, but it is still well above replacement. Fertility has been declining steadily in Mexico since the 1970s, but it is still higher than in the United States and so immigrants continue to come from a culture with expectations for a larger family size. In fact, the TFR for the Mexican-origin population in the United States as shown in Figure 6.10 is virtually the same (2.9 children per woman) as the Population Reference Bureau's estimate of the current TFR in Mexico (Population Reference Bureau 2003). No one is yet sure whether or when Hispanics in the United States will lower their fertility to replacement level, but that will be a decisive factor in future fertility levels because it is already the case that in populous Southwestern states such as California and Texas a majority of births are to women of Mexican origin.

Jordan

Jordan is a small Middle Eastern nation sandwiched between Israel to the west and Saudi Arabia, Iraq, and Syria on its other sides, and it has the distinction of being home to the world's largest number of Palestinians. Like so many countries in the region, Jordan's modern boundaries were set somewhat arbitrarily at the end of World War I when the remnants of the Ottoman Empire were put under the "protectorate," or "mandate," of various European powers—England, in the case of Jordan. In 1946, Jordan was granted independence, and its area at that time included

Figure 6.10   Changes in Fertility by Ethnic Group, United States, 1990–2002

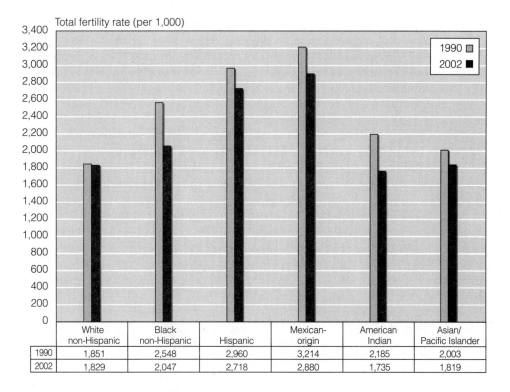

Total fertility rate (per 1,000)

	White non-Hispanic	Black non-Hispanic	Hispanic	Mexican-origin	American Indian	Asian/ Pacific Islander
1990	1,851	2,548	2,960	3,214	2,185	2,003
2002	1,829	2,047	2,718	2,880	1,735	1,819

Source: Adapted from Martin, 2003 #2453: Tables 4 and 9.

territory on both sides of the Jordan River. However, Jordan was part of the Arab coalition that attacked Israel in 1967, and Israel has militarily occupied the West Bank since that date. Many Arabs who had fled to the West Bank from Palestine in 1948, when it was partitioned to create the modern state of Israel, fled farther east into Jordan as a result of the 1967 war.

Fertility in Jordan is high, but declining. The estimate of the total fertility rate for 1997 was 4.5 children per woman (see Figure 6.6) and is based on the 1997 Demographic and Health Survey. The United Nations estimates the TFR in the 2000–2005 period to be 3.6 and, given the declining mortality rate in Jordan, that TFR translates into a net reproduction rate of 1.7, indicating that the next generation could be almost twice the size of the current generation. Still, the fertility transition is under way and part of the decline is due to an increase in the age at marriage among women, but part is due as well to an increase in the percentage of married women who are using methods of fertility control. In 1976, 17 percent of women were estimated to have been using a modern method of contraception—that had increased to 26 percent by 1990 and to 36 percent by 1997, according to data from the DHS, with the IUD being the most popular method, followed distantly by the pill.

Attitudinal changes are taking place especially with regard to the acceptability of family planning. In 1985, a survey of husbands in Jordan revealed that more than half of that nation's married men believed that family-size decisions were up to God, and more than 40 percent explicitly did not believe in practicing contraception (Warren et al. 1990). We do not have more recent data from husbands themselves, but data from the 1997 DHS show that 86 percent of women responded that both they and their husbands approved of family planning. Percentages of approval were lower in rural than in urban places, and were lowest for those with the least education and highest for those with the most education.

There are now nearly six million people in Jordan, and the population is currently growing at a rate that will double in less than 30 years. Only since the mid-1990s has the government moved from not having a position on population growth to believing that the rate of growth is too high. Policy or not, the government has consistently made contraceptives available through government-run clinics. Since most Jordanians are Muslim, it is tempting to think that Islam is responsible for high fertility. As I have argued elsewhere (Weeks 1988), it is unlikely that religion per se is the motivating force, and a 1997 survey of religious leaders in Jordan revealed that they, as well as the general public, believed that family planning was consistent with the beliefs of Islam (Underwood 2000). Rather than religion, low levels of education have contributed to the maintenance of traditional beliefs (which tend to support high fertility regardless of the specific religious persuasion), and the flames of high fertility are fanned by the Palestine Liberation Organization (PLO), which views a high birth rate as a "secret weapon" in its fight to regain the land conquered by Israel (Waldman 1991). Jordan is one of the few Arab nations that has granted citizenship to Palestinians, and a large fraction of the Jordanian population is of Palestinian origin.

The counterpoint to Jordan's high birth rate is the area's limited supply of water, and there is a serious question as to how long the region's ecological resources can sustain the current rate of population growth. To avoid further disaster, it seems necessary not simply for the birth rate to decline, but for it to decline at an even more rapid rate than it is currently doing.

Summary and Conclusion

Fertility has both a biological and a social component. The capacity to reproduce is biological (although it can certainly be influenced by the environment), but we have to look to the social environment to find out why women are having a particular number of children. Understanding the fertility transition also requires that we understand how to measure fertility and in this chapter I outlined for you the major techniques for doing so.

Ansley Coale's three preconditions for a fertility decline offer a useful framework for conceptualizing the kinds of changes that must occur in a society if reproduction is going to drop to significantly lower levels. These include the acceptance of calculated choice about reproductive behavior, a motivation to limit fertility, and the availability of means by which fertility can be limited. With respect to the latter, fertility control can be accomplished by a variety of means, but in general they include ways by which intercourse can be prevented, conception can be prevented if intercourse takes place, and successful gestation can be prevented if conception occurs. An understanding of these proximate determinants or intermediate variables is an important preliminary to a good analysis of fertility trends and levels.

The fertility transition is viewed by many as having an essentially economic interpretation, emphasizing the relationship between the supply of children (which is driven by biological factors) and the demand for children (based on a couple's calculations about the costs and benefits of children), given the costs (monetary and psychosocial) of fertility regulation. This is the supply-demand framework. It is complemented by those who argue that fertility limitation is an innovation that is diffused through societies across social strata and over distances in ways that may be independent of economic factors. These perspectives on the fertility transition can be used to help understand the fertility situation in the modern world, ranging from those countries where fertility remains high, to those where it has dropped to or below the replacement level. People expect (and receive) pressure from others to have children, even in lower fertility countries such as the United States. This is why it is often difficult for fertility to decline when mortality declines, especially if no other economic or social changes have preceded the mortality decline. If marriage and childbearing are prestigious when mortality is high, why should it be any different when mortality is low? Social norms may change slowly in response to the new environment, because the link between mortality levels and fertility levels is not usually consciously recognized.

Theories of the fertility transition emphasize the role of wealth and economic development in lowering levels of fertility, although it is clear that these are not sufficient reasons for fertility to decline. You must also assess the overall social environment in which change is occurring. When there are desired and scarce resources, wealth, prestige, status, education, and other related factors often help to lower fertility because they change the way people perceive and think about the social world and their place in it. Human beings are amazingly adaptable when they want to be. When people believe that having no children or only a few children is in their best interest, they behave in that way. Sophisticated contraceptive techniques make it easier, but they are not necessary, as the histories of fertility decline in places like the United Kingdom and the United States illustrate.

One of the most important ways in which societies change in the modern world is through migration. Migrants bring not only themselves but also their ideas with them when they move, and as communication and transportation get increasingly easier, they are more apt to diffuse ideas and innovations back to their place of origin. In the next chapter, we turn our attention, then, to this next aspect of the demographic transition: the migration transition.

Main Points

1. Fertility refers to the number of children born to women (or fathered by men).

2. Fecundity, on the other hand, refers to the biological capacity to produce children.

3. The group of people who are always pointed out when high fertility is discussed is the Hutterites, who live in the northern plains of the United States and in Canada; in the 1930s, each married Hutterite woman averaged 11 children.

4. The principal measures of fertility used by demographers are the crude birth rate, general fertility rate, child-woman ratio, age-specific fertility rate, total fertility rate, gross reproduction rate, net reproduction rate, and birth expectations. Don't let people tell you what the birth rate is without finding out what rate they are referring to and whether the rate refers to period or cohort data.

5. Ansley Coale's three preconditions for a fertility decline include: (1) acceptance of calculated choice in reproductive decision making; (2) motivations to limit fertility; and (3) the availability of means by which fertility can be regulated.

6. The fertility transition represents the shift from "natural fertility" to more deliberate fertility limitation, and is associated with a drop in fertility at all ages, but especially at the older ages (beyond the 30s) and younger ages (under 20).

7. Some women wonder if they should have a baby after 35, but the answer is that, no, 35 is probably enough.

8. In the era before modern contraception, people controlled family size *after* the birth of children, since they were unable to prevent a birth.

9. Fertility can be controlled by a combination of the proximate determinants of fertility, which affect the exposure to intercourse, the risk of conception given intercourse, and the likelihood of a live birth, given conception.

10. For most of human history, high mortality meant that societies were more concerned with maintaining reasonably high levels, rather than contemplating a decline in fertility—surviving children, not children ever born, was the goal.

11. The need for children as sources of labor and security against risk, as well as the desire for sons, are some of the structural features of a society that may contribute to high fertility.

12. The supply-demand perspective on the fertility transition suggests that couples strive to maintain a balance between the potential supply of children and the demand (desired number of surviving children), given the cost of fertility regulation.

13. An important concept in the supply-demand framework is the tradeoff between the quantity and quality of children, expressed typically as the opportunity costs of children.

14. Increased education is a very important factor that raises the opportunity costs of children and encourages women to limit fertility.

15. The innovation-diffusion model of fertility draws on sociological and anthropological evidence that much of human behavior is driven by the diffusion of new innovations—both technological and attitudinal—that may have little to do with a rational calculus of costs and benefits.

16. Government public policy may influence the course of the fertility transition by stimulating or retarding the motivations for limiting fertility and/or the availability of means of fertility control.

17. The fertility transition is typically accomplished through a later age at marriage, through older women deciding not to have that additional child, and through

women in their prime reproductive years using effective means of fertility control.

18. Virtually all wealthy societies except the United States now have below-replacement fertility levels, whereas in almost all less-developed nations in the world today are genuine stirrings of a fertility decline, as high-fertility norms and behavior give way to low-fertility preferences.

19. The baby boom is one of the most famous demographic events in United States history (and a baby boom was part of the demographic landscape of England and Canada, as well), but it was really a readily explainable detour on a long road of declining fertility punctuated by the baby bust and the baby boomlet.

20. In the world today, a woman gives birth to a child more than four times each second. (We've got to find this woman and stop her!)

Suggested Readings

1. Rodolfo A. Bulatao and John B. Casterline, Editors, 2001, Global Fertility Transition, A Supplement to Volume 27 of Population and Development Review (New York: Population Council).

   This collection of research papers and commentaries reviews the rather dramatic changes that emerged in world fertility levels at the end of the twentieth century, and which are setting the stage for demographic change in the twenty-first century.

2. John B. Casterline, Editor, 2001, Diffusion Processes and Fertility Transition (Washington, DC: National Academy Press).

   The decline of fertility even in relatively low-income countries has spurred research into the processes of diffusion that can help us understand why and when the fertility transition occurs, and this book looks at these issues from a variety of angles.

3. C. Y. Cyrus Chu and Ronald Lee, Editors, 2000, Population and Economic Change in East Asia (A Supplement to Volume 26 of Population and Development Review) (New York: Population Council).

   This book offers detailed insight into the fertility transitions that have taken place rather suddenly in East Asia over the past few decades. Much of the economic miracle of Asia is related to its rapidly changing demographic structure, fueled in part by a rapid decline in fertility, and this book discusses these issues in depth.

4. V. M. Zlidar, R. Gardner, S. O. Rutsetin, L. Morris, H. Goldberg, and K. Johnson, 2003, "New Survey Findings: The Reproductive Revolution Continues." Population Reports, Series M, No. 17. (Baltimore: Johns Hopkins Bloomberg School of Public Health, the INFO Project), http://www.populationreports.org/m17.

   This report is packed with data and analyses summarizing recent research on fertility trends, contraceptive use, fertility preferences, and the relationship between the fertility transition and other aspects of reproductive health.

5. United Nations Population Division, 2000, Below Replacement Fertility, Special Issue Nos. 40/41 of the Population Bulletin of the United Nations (New York: United Nations).

   As an increasing number of countries even outside Europe reach below-replacement fertility levels, an issue that was almost unthinkable 30 years ago has now become an important and widely debated topic, as reflected in the articles in this special volume.

❖ Web Sites of Interest

Remember that Web sites are not as permanent as books and journals, so I cannot guarantee that each of the following Web sites still exists at the moment you are reading this. It is usually possible, however, to retrieve the contents of old Web sites at http://www.archive.org.

1. http://www.rhgateway.org
   This Web site is actually a portal or gateway to other sites on its server and on the Internet covering nearly all topics related to reproductive health, including contraceptive methods, and almost every topic covered in this chapter. It provides a good set of navigation and search tools so you can find the information you are looking for.

2. http://www.familyplanet.org
   The David and Lucille Packard Foundation and the William and Flora Hewlett Foundation have combined the resources of their respective population programs to support this Web site, which has as one of its goals promoting awareness of the importance of international family planning.

3. http://www.nyu.edu/projects/sanger
   Margaret Sanger spent much of her adult life in New York City, in and around Greenwich Village, near New York University (NYU) and the History Department at NYU hosts this Web site, which includes background on her life and access to her writings about the struggle for reproductive freedom for Americans.

4. http://www.measuredhs.com
   Most of the information we have about fertility and reproductive health in developing countries comes from the Demographic and Health Surveys (DHS) conducted by Macro International as part of the Measure project. At this Web site, you can produce your own summary of results by using the *DHS STATcompiler*.

5. http://www.unfpa.org/
   The UNFPA is the United Nations Population Fund—the population "outreach" arm of the UN. They publish an informative annual report that is available at this site. Recent annual reports have focused on issues of reproductive rights and gender equity and their relationship (implied or explicit) to the fertility transition.

CHAPTER 7
The Migration Transition

Figure 7.1 Net Migrant Receiving and Sending Countries of the World

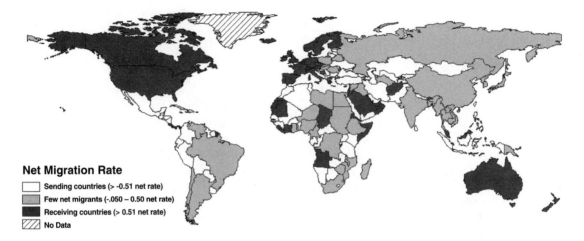

Net Migration Rate

☐ Sending countries (> -0.51 net rate)
▦ Few net migrants (-.050 – 0.50 net rate)
■ Receiving countries (> 0.51 net rate)
▨ No Data

Source: Adapted by the author from medium variant projections in United Nations, World Population Prospects 2002.

"The sole cause of man's unhappiness," quipped Pascal in the seventeenth century, "is that he does not know to stay quietly in his room." If this is so, unhappiness is enjoying unprecedented popularity as people are choosing to leave their rooms, so to speak, in record numbers. Sometimes they are fleeing from unhappiness; sometimes they are producing it. Always they are responding to and, in their turn, creating change. Because migration brings together people who have probably grown up with quite different views of the world, ways of approaching life, attitudes, and behavior patterns, it contributes to many of the tensions that confront the world, leading Kingsley Davis to comment that "so dubious are the advantages of immigration that one wonders why the governments of industrial nations favor it" (1974:105). Even if a country tried to slam its doors to immigrants, however, will that stop people from migrating? Nearly 80 million people are being added to the world's population each year, and the young adult population in many less-developed countries is booming to the point of nearly bursting at the seams. What are these people to do? As it becomes ever harder for a person to find a niche in the world economy, a would-be worker is often compelled to move.

An old Mexican saying goes, "Don't ask God to give it to you, ask Him to put you where it is." "Where it is" for many in Mexico is the United States. Today's pilgrims are from places such as Jalisco, Sinaloa, and Michoacán. They look to their northern neighbor for economic advancement in life and, although entrants from Mexico often cross the border with every intention of returning home, many never do. Similar tales are told of Algerians migrating to France, Moroccans to Spain, and Turks to Germany.

This vast transnational migration is part of the **migration transition** that is a component of the broader demographic transition, as I discussed earlier. Population growth changes the ratio of people to resources and this forces various kinds of local adjustment. One such adjustment is to move somewhere else—to "get out of town" or to "head west, young person." Humans have been migrating throughout history (or else we would not be found in every nook and cranny of the globe), but the advent of relatively inexpensive and quick ground, water, and air transportation has given migration a new dimension at the very same time (and for many of the same reasons) that rapidly declining mortality has accelerated population growth. Migration may occur within the same country (**internal migration**) or between countries (**international migration**). In either case, migration within the context of the demographic transition often involves people moving from rural to urban areas and that is an important enough topic on its own that I have devoted the entirety of Chapter 11 to the *urban transition*.

Migration has no biological components in the way that mortality and fertility do. As far as we know, humans as a species are by nature neither sedentary (prone to staying in one place) nor inherently mobile (prone to moving from place to place). At the same time, the fact that we study migration rather than immobility means that we assume that people prefer not to move and so it is the moving that needs to be explained. The relationship of migration to the demographic transition arises out of the fact that control of mortality and fertility has historically occurred within the context of urban places and then been diffused to rural areas. The population growth resulting from the decline in mortality in rural areas creates the paradoxical situation in which many of the people working in agriculture need to be replaced by machines so that enough food can be grown for a burgeoning population. Thus, people become less useful in agriculture as the population grows. Fortunately, the same forces creating this situation typically are creating employment opportunities in cities and so these changes in both rural and urban economies help to spur the movement of the population from rural to urban places.

However, even if there were no demographic transition—and even when all demographic transitions are completed—migration still has the potential to profoundly alter a community or an entire country within a short time. In-migration and out-migration can increase or decrease population size, respectively, far more quickly than either mortality and fertility. And even if the number of in-migrants just equals the number of out-migrants, the flow of people in and out will affect the social and economic structure of a community.

Migration, then, is a huge topic of tremendous importance for all human societies, regardless of levels of fertility and mortality. At the same time, however, migration is influenced by relative levels of mortality and fertility and migration then has its own influence on mortality and fertility, not just in the places to which migrants go, but also in the places from which they came. The topic of migration is far too huge for one chapter in a book, and so my relatively limited purpose in this chapter is provide you with an overview of the ways in which we define, measure, and conceptualize the migration process and understand the migration transition, looking separately at the situation within and between countries. I also examine migration that occurs forcibly, including slavery and refugee movements.

Defining Migration

Migration is defined as any permanent change in residence. It involves the "detachment from the organization of activities at one place and the movement of the total round of activities to another" (Goldscheider 1971:64). Thus the most important aspect of migration is that it is spatial by definition. You cannot be a migrant unless you "leave your room." However, just because you leave your room, you are not necessarily a migrant. You may be a traveler or perhaps a daily commuter from your home to work. These activities represent **mobility,** but not *migration*. You might be a temporary resident elsewhere (such as a construction worker on a job away from home for a few weeks or even months), or a seasonal worker (returning regularly to a permanent home), or a **sojourner** (typically an international migrant seeking temporary paid employment in another country). Again, such people are mobile, but they are not migrants because they have not changed their residence permanently. Of course, even when you change your permanent residence, if your new home is only a short distance away and you do not have to alter your round of activities (you still go to the same school, have the same job, shop at the same stores), then you are a **mover** (and maybe even a shaker), but not a migrant. All migrants are movers, but not all movers are migrants.

Less clear conceptually are *transients* and *nomads*. Technically, you could say that because they are constantly changing their residence and round of activities, then they should be thought of as migrants. However, the lack of a permanent residence creates a problem in defining them as migrants. Most demographers deal with this by simply ignoring transients and nomads, and I do much the same, since they represent a very small part of most contemporary human populations.

Internal And International Migrants

Internal migration, that which occurs within a country, typically is "free" or voluntary in the sense that people are choosing to migrate or not, often basing that decision on economic factors. This is not to say that, within a country, people are never forced to move. Witness the migration of hundreds of thousands of Chinese who have been forced to relocate so that the Three Gorges Dam can be built, or the massive transmigration that Indonesia periodically attempts—moving people from the crowded island of Java to other, less populous islands. But such internal migration, although forced, is usually planned. People's needs are anticipated in advance and, presumably, the migration is expected to improve the lives of the people involved.

Migration across international boundaries is usually voluntary, but it typically means that a person has met fairly stringent entrance requirements, is entering without documents (which carries a load of stress with it), or is being granted refugee status, fleeing from a political, social, or military conflict. You can easily imagine that any kind of international migration is apt to be more stressful than internal migration, and on top of that is heaped the burden of accommodating to a new culture and often a new language, being dominated perhaps by a different religion, being provided different types and levels of government services, and adjusting to different sets of social expectations and obligations.

With reference to your area of origin (the place you left behind), you are an **out-migrant,** whereas you become an **in-migrant** with respect to your destination. If you move from one country to another, you become an international migrant—an **emigrant** in terms of the area of origin and an **immigrant** in terms of the area of destination. Because commuters and sojourners also may cross international boundaries, the United Nations has tried to tighten the definition of an international migrant by developing the concept of a **long-term immigrant,** which includes all persons who arrive in a country during a year and whose length of stay in the country of arrival is more than one year (Kraly and Warren 1992).

International migration can be further differentiated between **legal immigrants, illegal (or undocumented) immigrants, refugees,** and **asylees.** Legal immigrants are those who have governmental permission to live in the place to which they are migrating, whereas illegal or undocumented migrants do not. A refugee is defined by the United Nations (and by most countries of the world) as "any person who is outside his or her country of nationality and is unable or unwilling to return to that country because of persecution or a well-founded fear of persecution. Claims of persecution may be based on race, religion, nationality, membership in a particular social group, or public opinion" (United States Citizenship and Immigration Services 2004). An asylee is a refugee—with a geographic twist. He or she already is in the country to which they are applying for admission, whereas a refugee is outside the country at the time of application.

You can see that the definition of migration is confounded by the fact that migration is an activity (changing residence) carried out by people (the migrants) under varying legal and sociopolitical circumstances. If we have this much trouble defining migration, you can be sure that it is hard to measure.

Measuring Migration

Defining migration as a permanent change of residence still leaves several important questions open that have to be answered before we can measure the phenomenon and know who is a migrant and who is not. For example, how far does a person have to move to be considered a migrant instead of just a mover? That is fairly straightforward in the case of international migrants, but not so easily determined for internal migrants. As a rule of thumb, people moving within a country are classified as migrants if they move across administrative boundaries. For example, the U.S. Census Bureau usually defines a migrant as a person who has moved to a different county within the U.S. Note, however, that from the standpoint of a city within a county, a migrant would be anyone moving into or out of the city limits. From the standpoint of a local school district, a migrant would be anyone moving into or out of the school district's boundaries. This issue of geographic scale is one that must be dealt with in all migration research. A birth is a birth, and a death is a death, but whether or not you are considered a migrant when you move varies according to who is asking the question.

Another question that has to be asked is: What do we mean by permanent? Most people who move tend to move more than once in their life, so we have to decide how long you must stay at the new place before your move is considered permanent. The United Nations has somewhat arbitrarily decided that anyone who spends at least one year in the new locale within another geographic region of the same country is a migrant, and all who move permanently to another country can be defined unambiguously as migrants, regardless of how long they stay. Nonetheless, it is sometimes the case that the data you are dealing with will determine what your definition of permanent will be. The decennial census in the United States has routinely asked people a question about where they lived five years prior to the census. So, a migrant is defined from the census data as someone who lived in a different county (or a different state or a different country) five years ago, regardless of how many times they may have moved in between. Of course, such a definition fails to capture the migration of someone who moved out and then back to the point of origin within that five-year time period. To tangle the situation further, migration may involve more than a single individual—a family or even an entire village may migrate together. A ghost town, it has been suggested, does not necessarily signal the end of a community, only its relocation.

Stocks versus Flows

One of the more important things to keep in mind as we discuss the migration transition is that it involves both a process and a transformation. The process is that people move from one place to another and this represents the **migration flow.** The transformation is that the **migrant stock** changes as people move into and out of a given place. The fact that more than one million legal immigrants were admitted to the United States in 2002 represents a measure of the flow of people from other places into the country. Those people were added to the stock of immigrants (the foreign-born population) already residing in the United States at the time. As of the 2000 Census, there were 31 million residents of the U.S. (11 percent of the total) who were born in another country. The difference between the stock and the flow is, of course, that if one million had left the country just as one million were entering, then the in- and out-flows would have been pretty large, but the stock itself would not have changed numerically. On the other hand, the composition of the stock of migrants could have been changed by the flows, even if the stock did not change numerically.

Information about the migrant stock comes largely from censuses and surveys where everybody is asked a question about whether or not they used to live in a different place. If they did (and they meet the criteria for a migrant), then they are part of the stock of migrants. If one additional question is asked about where they came from, then we can infer something about migration flows, as well, since we can then measure the number of people in the stock who flowed from point A to point B.

Migration flows can come from a variety of other sources. The U.S. Census Bureau also regularly conducts the American Housing Survey, which tracks changes in the nation's housing stock and thereby generates data on **residential mobility** (changing residence regardless of how long or short the distance) within the country. The Census Bureau also has an arrangement with the Internal Revenue Service that allows the Bureau to periodically examine the address changes reported by people on tax returns and that provides yet another way of tracking migration flows within the country. Data on flows into the U.S. are kept by the Citizenship and Immigration Services branch of the Department of Homeland Security (formerly the Immigration and Naturalization Service branch of the Department of Justice). Flows of undocumented immigrants have to be estimated from arrests by the Border Patrol and by various surveys that attempt to assess the legal status of immigrants.

We also do not have a good source of data on migration flows out of the United States and so we must rely on outside estimates, such as the number of Social Security checks sent to people living outside the country. The same can be said for Canada, where Citizenship and Immigration Canada (CIC) records the arrival of immigrants, but has no data on people who leave the country. This stands in contrast to at least a few European nations that, as I mentioned in Chapter 4, maintain population registers and thus have a pretty accurate fix on the extent of both internal and international migration. Most countries, however, have little available information, and we either do not know what is happening or have to rely on sample surveys or other indirect evidence (such as the stock of foreign-born people counted in a census) to infer patterns of migration flow. Confused? Don't worry, you're in good company, but remember the definitions and distinctions we have just discussed as we sort through the major ways by which we index migration.

Migration Indices

When data are available, migration is measured with rates similar to those constructed for fertility and mortality. These rates can be used to measure internal or international migration, depending on the focus of your analysis and the data you have at your disposal.

Gross or total out-migration represents the flow of all people who leave a particular region during a given time period (usually a year), and so the **crude or gross rate of out-migration** (*OMigR*) relates those people to the total midyear population (*p*) in the region (and then we multiply by 1,000):

$$OMigR = \frac{OM}{p} \times 1,000$$

Similarly, the **crude or gross rate of in-migration** (*IMigR*) is the ratio of all people who moved into the region (the flow) during a given year relative to the total midyear population in that region:

$$IMigR = \frac{IM}{p} \times 1,000$$

The gross rate of in-migration is a little misleading because the midyear population refers to the people living in the area of destination, which is not the group of people at risk of moving in (indeed, they are precisely the people who are *not* at risk of moving in because they are already there). Nonetheless, the in-migration rate does provide a sense of the impact that in-migration has on the region in question, so it is useful for that reason alone. The difference between those who move in and those who move out is called net migration. If these numbers are the same, then the net migration is zero, even if there has been a lot of migration activity. If there are more in-migrants than out-migrants, the net migration is positive; and if the out-migrants exceed the in-migrants, the net migration is negative. The **crude net migration rate** (*CNMigR*) is thus the net number of migrants in a year per 1,000 people in a population, and it is the difference between the net in- and out- migration rates, calculated as follows:

$$CNMigR = IMigR - OMigR$$

The total volume (flow) of migration also may be of interest to the researcher because it can have a substantial impact on a community even if the net rate is low. This is measured as the **total or gross migration rate** (*TmigR*)—also sometimes called the **migration turnover rate**—which is the sum of in-migrants and out-migrants divided by the midyear population, or more simply the in-migration rate plus the out-migration rate:

$$TMigR = IMigR + OMigR$$

Migration effectiveness (E) measures how "effective" the total volume of migration is in redistributing the population (Manson and Groop 2000; Plane and Rogerson 1994). For example, if there were a total of 10 migrants in a region in a year and all 10 were in-migrants, the "effectiveness" of migration would be 10/10, or 100 percent; whereas if four were in-migrants and six were out-migrants, the effectiveness would be much lower: (4 − 6)/10, or −20 percent. In general, the rate of effectiveness is as follows:

$$E = \frac{CNMigR}{TMigR} \times 100$$

There is no universally agreed-upon measure of migration that summarizes the overall levels in the same way that the total fertility rate summarizes fertility and life expectancy summarizes mortality. However, one way of measuring the contribution that migration makes to population growth is to calculate the ratio of net migration to natural increase (the difference between births and deaths); this is called the **migration ratio** (*MigRatio*):

$$MigRatio = \frac{NetMig}{b-d} \times 1,000$$

For example, between 2001 and 2002, there were an estimated 231,440 net international immigrants into the state of California, along with an estimated 76,200 net internal (domestic) migrants, for a total of 307,640 net migrants into the state (Office of Immigration Statistics 2003). During that same one-year period, there were 528,151 births in California and 232,790 deaths, so the natural increase was 295,361 people. The ratio of the net migrants to natural increase was thus 307,640 to 295,361 or 1.04. You can see that migration was a slightly more important component of growth in California than was natural increase, contributing 51 percent of the growth in population during that year.

Because we often do not have complete sets of data on the number of in- and out-migrants, we can "back into" the migration rate by solving the demographic balancing equation (which I discussed in Chapter 4) for migration. This is known as the **components of change (or residual) method of estimating migration.** The demographic equation states that population growth between two dates is a result of the addition of births, the subtraction of deaths, and the net effect of migration (the number of in-migrants minus the number of out-migrants). If we know the amount of population growth between two dates, and we also know the number of births and deaths, then by subtraction we can estimate the amount of net migration. Let me give you an example. Based on the 1990 census of the United States, we know that on April 1, 1990, there were 248,709,873 residents counted in the country. Between that date and April 1, 2000, there were 39,865,670 births and 22,715,464 deaths in the country. Thus on April 1, 2000, we should have expected the census to find 265,860,079 residents if no migration had occurred. However, the 2000 census counted 281,421,906 people. That difference of 15,561,827 people we estimate to be the result of migration (note that a small fraction of the difference could also be the result of differences in coverage error between the two censuses, as discussed in Chapter 4).

In an analogous way, we can also calculate intercensal net migration rates for specific age groups by gender. If we know the number of people at ages 15–24 in 1990, for example, and if we can estimate how many of them died between 1990 and 2000, then we know how many people aged 25–34 there should have been in 2000 in the absence of migration. Any difference between the observed and the expected number in 2000 can be attributed to migration. Typically, we use a life table (see Chapter 5) to calculate the proportion of people who will die between two different ages, and we call this whole procedure the **forward survival (or residual) method of migration estimation.** For example, in 1990 in the United States, there were 17,996,033 females aged 15–24. Life table values (see Table 5.3) suggest that 99.5 percent of those women (or 17,906,052) should still have been alive at ages 25–34 in 2000 (the forward survival). In fact, Census 2000 counted 20,083,131 women in that age group, or 2,177,079 more than expected. We assume, then, that those "extra" people (the residual) were immigrants, so this is a measure of net migration. This assumption of course ignores any part of that difference that may have been due to differences in the coverage error in the two censuses. If one census undercounted the population more or less than the next census, this could account for at least part of the residual that is otherwise being attributed to migration. Conversely, if we knew the number of migrants, then the difference between the actual and expected number of migrants would tell us something about the accuracy of the census.

Now, having worn you out trying to measure the nearly unmeasurable, let us move on to yet another difficult (but inherently more interesting) task: explaining the migration transition. We will do this first in the context of migration within countries, and then we will turn our attention to international migration.

The Migration Transition within Countries

In the premodern world, rates of migration typically were fairly low, just as birth and death rates were generally high. The demographic transition helped to unleash migration, and it is now evident that a **migration transition** has occurred virtually everywhere in concert with the fertility and mortality transitions discussed in earlier chapters (Zelinsky 1971). The theory of demographic change and response (see Chapter 3) suggested that migration is a ready adaptation that humans (or other animals, for that matter) can make to the pressure on local resources generated by population increase. However, people do not generally move at random—they tend to go where they believe opportunity exists. Because the demographic transition occurred historically in the context of economic development, which involves the centralization of economic functions in cities, migrants have been drawn to cities, and the urban transition is a central part of the migration transition, as discussed more fully in Chapter 11. Yet the mere existence of a migration transition does not explain why people move, who moves, and where they go. We need to dig deeper for those explanations.

Why Do People Migrate?

Over time, the most frequently heard explanation for migration has been the so-called **push-pull theory,** which echoes common sense by saying that some people move because they are pushed out of their former location, whereas others move because they have been pulled or attracted to someplace else. This idea was first put forward by Ravenstein (1889), who analyzed migration in England using data from the 1881 census of England and Wales. He concluded that pull factors were more important than push factors: "Bad or oppressive laws, heavy taxation, an unattractive climate, uncongenial social surroundings, and even compulsion (slave trade, transportation), all have produced and are still producing currents of migration, but none of these currents can compare in volume with that which arises from the desire inherent in most men to 'better' themselves in material respects." Thus Ravenstein is saying that it is the desire to get ahead more than the desire to escape an unpleasant situation that is most responsible for the voluntary migration of people, at least in late-nineteenth-century England. This theme should sound familiar to you. Is it not the same point made by Davis (1963) in discussing personal motivation for having small families (see Chapter 6)? Remember, Davis argued that it is the pursuit of pleasure or the fear of social slippage, not the desire to escape from poverty, that motivates people to limit their fertility.

In everyday language, we could label the factors that might push a person to migrate as stress or strain. However, it is probably rare for people to respond to stress by voluntarily migrating unless they feel there is some reasonably attractive alternative, which we could call a pull factor. The social science model conjures up an image of the rational decision maker computing a cost-benefit analysis of the situation. The potential migrant weighs the push and pull factors and moves if the benefits of doing so exceed the costs. For example, if you lost your job, it could benefit you to move if there are no other jobs available where you live now, unemployment compensation and welfare benefits have expired, and there is a possibility of a job at another location. Or, to be more sanguine about your employability, the process may start, for example, when you are offered an excellent executive spot in a large firm in another city. Will the added income and prestige exceed the costs of uprooting the family and leaving the familiar house, community, and friends behind? In truth, whether or not you migrate will likely depend on a more complicated set of circumstances than this simple example might suggest. The decision to move usually develops over a fairly long period of time, proceeding from a desire to move, to the expectation of moving, to the actual fact of migrating (Rossi 1955). In Rossi's longitudinal sample of families in the 1950s, half of those interviewed expressed a desire to move, but only about 20 percent of them actually did so. Sell and DeJong (1983) produced a set of longitudinal data for the 1970s that reinforced Rossi's findings— migration rates reflect a whole spectrum of attitudes, ranging from people who are "entrenched nonmovers" (who have no desire to move and no expectation of moving and who do not migrate) to "consistent decision-maker movers" (who desire to, expect to, and do migrate).

Between the desire to move and the actual decision to do so there also may be **intervening obstacles** (Lee 1966). The distance of the expected destination, the cost of getting there, poor health, and other such factors may inhibit migration. These obstacles are hard to predict on any wide scale, however, and so we tend to lump them together with the overall "costs" of moving and concentrate our attention on explaining the desire to move. Economic variables dominate most explanations of why people migrate. Beginning in 1998, a question was added to the Current Population Survey in the U.S. asking people who had moved during the previous year why they had moved. The results show that, as expected, people mentioned a work-related reason more often than any other for migrating (defined as moving between counties). Shorter-distance moves (within the same county) are most likely to be housing related (Schachter 2001b).

Migration associated with career advancement, as happens so often in the military, academics, and large companies, illustrates the hypothesis that migration decisions "arise from a system of strategies adopted by the individual in the course of passing through the life cycle" (Stone 1975:97). If it is assumed that people spend much of their life pursuing various goals, then migration may be seen as a possible means—an **implementing strategy**—whereby a goal (such as more education, a better job, a nicer house, a more pleasant environment, and so on) might be attained. Although this is not a startling new hypothesis (it is little more than a modern restatement of Ravenstein's nineteenth-century conclusions), it is nonetheless a very reasonable one. Indeed, Lee (1966) has observed that two of the more enduring generalizations that can be made about migration are:

1. Migration is selective (that is, not everyone migrates, only a selected portion of the population).

2. The heightened propensity to migrate at certain stages of the life cycle is important in the selection of migrants.

One particular stage of life disproportionately associated with migration is that of reaching maturity. This is the age at which the demand or desire for obtaining more education tends to peak, along with the process of finding a job or a career, and getting married.

Agreement is nearly universal on the general ideas about migration I have outlined above, but "the devil is in the details." There are so many different aspects to the process of migration that no one has yet produced an all-encompassing theory of migration. Figure 7.2 provides an overview of the major aspects of the migration process that require explanation, adapted from a conceptual model devised by De Jong and Fawcett (1981) and revised by De Jong (2000). These are analogous in certain respects to the three preconditions for a fertility decline. The migration process can be thought of as having three major stages including: (1) the propensity to migrate in general, (2) the motivation to migrate to a specific location, and (3) the actual decision to migrate.

Figure 7.2   A Conceptual Model of Migration Decision Making

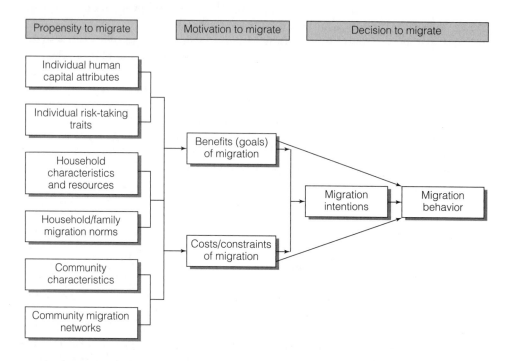

Source: Adapted by the author from Gordon De Jong and James Fawcett, 1981, "Motivations for Migration: An Assessment and a Value-Expectancy Research Model," in G. DeJong and R. Gardner (eds.), Migration Decision Making (New York: Pergamon Press), Figure 2.2; and Gordon De Jong, 2000, "Expectations, Gender, and Norms in Migration Decision-Making," Population Studies 54:307–319, Figure 1.

The migration process begins with individuals and household members in the context of a given culture and society, represented by the community in which they live. The decision about who will migrate and when and to where may often be part of a household strategy for improving the group's quality of life—consistent with the perspective of demographic change and response, as you will recall from Chapter 3. Furthermore, the household decision is not made in a vacuum; it is influenced by the sociocultural environment in which the household members live. Individual and household characteristics are important because of the selectivity of migrants—households with no young adults are less likely to contemplate migration. Social and cultural norms are important because they provide the context in which people might think consciously of migration as a necessary or desirable thing to do. Social norms can play a role in discouraging migration by emphasizing the importance of place and community or, on the other hand, political and economic instability may cause people to rethink their commitment to an area.

Personal traits are important because some people are greater risk takers than others. Rajulton (1991) has suggested that people can be scaled along a dimension of *migrability,* reflecting the probability that, all things being equal, they would risk a migration. In the United States, the "average" person will move 11 times in his or her lifetime (Hansen 1994), but that figure hides a lot of variability. Some people account for a disproportionate amount of migration by migrating frequently, while others never move. Part of the explanation for the propensity to move may be cultural, of course. Long (1991) examined rates of residential mobility (which are highly correlated to migration rates) for a number of developed nations and found that the "overseas European" nations, including the United States, Canada, Australia, and New Zealand—populated by migrants who supplanted the indigenous population—are the countries with the highest rates of mobility.

Demographic characteristics combine with societal and cultural norms about migration to shape the values people hold with respect to migration—the benefits they hope to gain by migrating. Such benefits represent clusters of motivations to move, including the desires for wealth, status, comfort (better living or working conditions), stimulation (including entertainment and recreation), autonomy (personal freedom), affiliation (joining family or friends), and morality (especially religious beliefs). At the same time, personal traits (such as being a risk-taking person) combine with the opportunity structure within the household and within the community to affect the costs and constraints that might keep a person from migrating. All of these personal and social environmental factors combine to affect a person's expectation of actually achieving the goals they have in mind that might be facilitated by migration. The amount of information a person has about the comparative advantages of moving also contributes to the expectation of attaining migration values or goals—that balancing of the likelihood of obtaining the benefits with the perceived costs involved in migrating. If the benefits appear to outweigh the costs, then the person may decide to migrate.

Given the intention to move, a person may discover that, by making adjustments in his or her current situation, personal or family goals can be achieved without having to move. "Such adjustments might include a change in occupation, alterations to the physical structure of the house, a change in daily and friendship patterns, or lifestyle changes" (De Jong and Fawcett 1981:56). Finally, the intention to move (or to stay) leads ultimately to the act of moving (or staying) itself, although unanticipated events may affect that decision. Indeed, events may overtake a person in something akin to the diffusion model, so that even though there was no original intention to migrate, the person migrates anyway. We have all been swept up in situations and done things that we wouldn't necessarily have done had we given them more thought.

Who Migrates?

Selectivity by Age   In virtually every human society, young adults are far more likely to migrate than people at any other age. This is about as close as we are likely to get to a "law" of migration (Tobler 1995). The data in Figure 7.3 show the age pattern of intercounty migrants in the United States between 2001 and 2002, using data from the Current Population Survey. As you can see, young adults were much more mobile than people of other ages, and although these data are for the United States for 2001–2002, the same pattern has existed in the United States in the past and holds true in other countries as well (Long 1988).

Figure 7.3   Young Adults Are Most Likely to Be Migrants

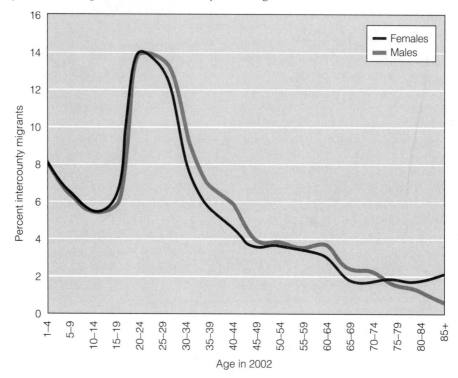

Source: Adapted from U.S. Census Bureau, Current Population Survey, March 2002, Geographical Mobility: March 2001 to March 2002, Table 1, http://www.census.gov/population/www/socdemo/migrate/cps2002.html, accessed 2004.

The young adult ages, 20–29, are clearly those at which migration predominates. One in seven American females aged 20–24 was an intercounty migrant during the year, whereas migration peaked a little later for males—25–29—at which ages one in seven was a migrant during 2001–2202. From there, the percentage of people who migrate drops off steeply, with a few little bumps around the time of retirement. At ages younger than 20, children typically are just following their parents around, so it is not surprising that younger children (who have the youngest parents) move more than older children. We can see then that age is an important determinant of migration because it is related to life-cycle changes that affect most humans in most societies.

Selectivity by Life Cycle   In most contemporary societies, it is expected that young adults will leave their parents' home, establish an independent household, get a job, marry, and have children. Some people choose to ignore one or more of these expectations, of course, but they exist nonetheless as behavioral guidelines. They also influence migration because each of these several phases of young adulthood may precipitate a move. For example, those who are not currently married (single, divorced, separated, and widowed) have the highest migration rates (Schachter 2001a), and longitudinal data from the Census Bureau's Survey of Income and Program Participation (SIPP) have shown that people whose marital status changed between two consecutive interviews were far more likely to have moved than those whose marital status remained unaltered (DeAre 1990).

The incidence of migration also varies according to the number and ages of children. Among young couples, the smaller the family and the younger the children, the greater the probability of migration, but once a child is old enough to start school, the temptation to move seems to go down. Migration, in its turn, may temporarily disrupt family-building activity. The period just before migration, as people plan for their move, has been found to be a time of lower-than-expected fertility, although it may be followed by a "catch-up" time after migration is completed (Bean and Swicegood 1985; Goldstein and Goldstein 1981). Of course, as fertility has fallen, the impact of children on migration decisions has almost certainly declined.

Selectivity by Gender   In the United States, women have virtually the same rates of migration as do men (see Figure 7.3), reflecting increasing gender equity. Women are increasingly apt to migrate on their own, rather than to move only because they are trailing a husband. In so-called traditional societies, the role of women is to be at home caring for children and other family members and under such circumstances migration is dominated more by males than by females. In fact, many models of household decision making about migration take for granted that it is the male of the household who will be making the migration decision (Cerrutti and Massey 2001). Therefore, it should not surprise you that men are more likely to outnumber women among migrants in those areas of the world where the status of women is lowest—Africa and Asia—whereas women are as likely or even more likely to be migrants in Europe, North America, Latin America, and the Caribbean (Chant and Radcliffe 1992; Clarke 2000).

Any generalization about gender and migration is likely to be misleading, however, because, like all aspects of migration, this one is complicated by varying regional patterns. Findley (1999) has shown that women represent a slight majority of migrants throughout Latin America, but this pattern is only loosely connected to any rising status of women. Most women are following their husband or other family members, and the occupational opportunities for women who migrate may be increasing, but they tend to be in gender-segregated jobs such as domestic work. A more varied picture emerges for Asia, where two countries in particular—Thailand and the Philippines—account for a large fraction of the women who are migrating in that region (Hugo 1999). Domestic service and commercial sex and entertainment are key inducements (for lack of a better word) to the migration of women within and from these countries.

Global patterns of migration show a rise in the percentage of all migrants who are women (Campani 1995) and, especially for single women, migration can be seen as a form of emancipation from traditional gender roles. But women on average are still disadvantaged compared to men when it comes to migration. They are still less likely than men to have made the decision to move, even when they are involved in the move, and their own situation may actually suffer as a result of being a trailing spouse. Data consistently show that when families move, employment opportunities for women are apt to be less favorable than they were prior to the move (Boyle et al. 2001; Jacobsen and Levin 1997).

Where Do People Migrate within Their Own Country?

The decision about whether you are going to move cannot be divorced from where you might be going. Data from the American Housing Surveys and the more recent Current Population Surveys suggest that about one in five internal migrants in the United States is involved in a job transfer, implying that the choice of destination may have been in someone else's hands. An additional one in 10 migrants was moving closer to relatives, and obviously their location fixes the migrant's destination, but the fact that they had to move to be near relatives is itself a function of somebody having previously moved—otherwise the relatives would still all be together.

The migration transition within countries is essentially the story of population growth in rural areas leading to a redundancy of that population, so that people look elsewhere for jobs and a livelihood. The city is almost always where the jobs are, and so the migration transition is, as I have already noted, largely an urban transition. Initially, then, the movement of people within a country is from rural to urban areas. Eventually, of course, the size of the rural population stabilizes at the relatively low fraction of the total population that is required to grow the food and harvest any other natural resources required by the nation's economy. At this point, it may no longer be appropriate to speak of a migration transition, but rather to refer to a **migration evolution,** which implies that the population is largely urban-based, and people are moving between and within urban places.

We can expect that every country in the world will experience a migration transition as part of the overall demographic transition, but some countries are apt to experience a larger migration evolution than others. The United States, for example, is quite literally a nation on the move, and it always has been. Data from the 2002 Current Population Survey provide the estimate that 41 million Americans (15 percent of the population) aged one and older in 2002 were living in a different house than in the year before. Some of these people had undoubtedly moved more than once during that period, so that represents a probable minimum of mobility. Of those movers, 16 million crossed county lines and would thus be considered migrants. An additional 1.6 million persons moved in from outside the United States during the previous year, including U.S. citizens returning from abroad (about one-third of that total) and immigrants. At the same time, we can note that the number of migrants has kept pace with the population in the United States, and the actual rate of migration (measured as the *percentage* of people moving each year) has remained remarkably steady since the end of World War II and, if anything, may have declined, not risen (Gober 1993; Schachter 2001a).

As I will discuss in more detail in Chapter 11, the migration transition in the United States is essentially complete, but population movements are part and parcel of American life, and so the country continues to evolve demographically as a consequence of emerging patterns of migration. There were several decades when migration was in the direction of the industrializing centers in the northeastern and north central states and to the rich farmland and industry in the Midwest. The strongest of these movements was the one westward (Shryock 1964). At first this meant that the mountain valley areas west of the Atlantic seacoast were migration destinations; then the plains states were settled; and, especially since the end of World War II, the Pacific Coast states have been popular destinations.

Until about 1950, migrants had also been heading out of the southern states and into the northeastern and north central states. This generally represented rural-to-urban migration out of the economically depressed South into the industrialized cities of the North. In the 1950s, this pattern of net out-migration from the South reversed itself and the northeastern and north central states found themselves increasingly to be migration origins rather than destinations. In the 1950s, migrants began heading not only west, but south as well. This pattern has continued into the twenty-first century. Table 7.1 shows that the northeastern states (the "Rust Belt" or "Snow Belt") have had more out-migrants than in-migrants, and people have been moving west and especially south to the Sun Belt. In the 1960s and 1970s, the movement still was more strongly to the west than to the south, but since then the South has been gaining more through net migration than the western states have.

In the 1990s, the nearly unthinkable pattern unfolded of net out-migration from California, at least in terms of internal migrants. As a result of the defense cutbacks associated with the end of the cold war, on top of a general economic recession, California's economy slowed dramatically in the early 1990s, and in 1993 demographers estimated that 252,000 more persons left California for other states than moved to California from elsewhere in the United States (California Department of Finance 1994). Most of the people leaving California were going to the nearby states of Arizona, Nevada, Utah, Idaho, and Washington (Gober 1993). In spite of the net domestic out-migration, California's population continued to increase numerically, not

Table 7.1   Americans at the Beginning of the Twenty-First Century Were Heading South

Destination in 2002:	Origin in 2001				
	Northeast	Midwest	South	West	Totals
Northeast	*−220*	81	235	129	445
Midwest	84	*−225*	304	142	530
South	487	401	*392*	394	1,282
West	84	272	351	*52*	717
Totals	665	754	890	665	2,974

Note: Between 2001 and 2002, there were 2,974,000 Americans who migrated between the four major geographic regions of the country. This table shows the number of migrants (in thousands) according to their origin (where they lived in 2001) and their destination (where they lived in 2002). The diagonal (highlighted) shows the net number of migrants for that region. Thus, there were 890,000 people who moved out of the South between 2001 and 2002, but they were replaced by 1,282,000 in-migrants, for the largest regional net gain of 392,000 migrants.

Source: Adapted from U.S. Census Bureau, Current Population Survey, March 2002, Geographical Mobility: March 2001 to March 2002, Tables 18 and 20, http://www.census.gov/population/socdemo/migration/cps2002/tab18-1.xls, accessed 2004.

only through an excess of births over deaths, but also through the continuation of net international migration from Latin American and Asian countries. By the late 1990s, the flow of migrants had turned around again and was heading back to California, and between 2001 and 2002 there were 76,000 more in-migrants to the state than there were out-migrants (Demographic Research Unit 2003).

Despite the fluctuations in the flow of internal migrants into California, the influx of international migrants remained quite constant and this flow has confounded the pattern of internal migration to a certain extent. In the first place, immigrants tend to congregate in relatively few states (especially California, Texas, New York, and Florida), and within those states they tend to congregate in specific metropolitan areas, such as Los Angeles, Houston, New York City, and Miami, respectively (Frey and Liaw 1998). Furthermore, when they move internally in the United States, they may be more apt to move in order to be close to relatives and others of the same nationality than they are to move for purely economic reasons (Kritz and Nogle 1994). Still, Gober (1999) reminds us that there are now so many different immigrant and ethnic groups in the United States that we make such generalizations at our own peril.

Internal migration patterns among immigrants represent part of an increasingly complex picture of overall migration patterns in the United States, encouraged by the economic restructuring that has been taking place since the 1980s (Pandit and Withers 1999). As William Frey (1995:333) has noted:

> The spatial demographic shifts that characterize the 1980s and 1990s are a far cry from the 1970s days of Snowbelt urban declines, Texas oil booms, and California dreaming. Nor is there much talk of a back-to-nature rural renaissance. The new, post-1980 urban revival is an uneven one—rewarding corporate nodes, information centers, and other tie-ins to the global economy. Areas specializing in high-tech manufacturing, recreation, or retirement have also grown. And while these kinds of areas can be found in most parts of the country, they are now especially prominent in newly developing regions—the South Atlantic coastal states, and states around California.

Frey elaborates five trends that he believes capture the complexity of contemporary patterns of population movement in the United States: (1) an uneven urban revival—a select few metropolitan areas (those with diverse economies that can withstand industrial restructuring) are gaining migrants at the expense of others; (2) regional racial division—the influx of immigrants from Asia and Latin America has dramatically diversified the racial and ethnic composition of the major receiving states (California, Texas, and New York), and has separated those states from the rest of the country, which remains predominantly non-Hispanic white; (3) regional divisions by skill level and poverty—the "hourglass" economy that has come to characterize the highly industrialized nations has widened the income gap between college graduates and those with less education, and the geographic redistribution of knowledge-based industries creates in its wake a migration of those with higher education; (4) baby boom and elderly realignments—the early baby boomers fueled the movement west and south, whereas the later baby boomers are having to look elsewhere for opportunities; in the meantime, the elderly continue to move to the Sun

Belt; and (5) suburban dominance and city isolation—suburban areas "captured the bulk of employment and residential growth in 1980s. The modal commuter now both lives and works in the suburbs" (Frey 1995:275).

The current trend, then, is a multiplicity of migration trends, reminding us yet again of the complexity of migration. Economic motives may dominate individual decisions to migrate, but even in the United States, there are broader influences on where economic opportunities are likely to be, and who is likely to be attracted to them.

Referring back to Figure 7.2, we can make the general statement that internal migration is more strongly influenced by individual characteristics of people, whereas international migrants are more apt to be influenced by the social and political climate and by the opportunity structure (especially the lack of barriers to migration that influence the costs and constraints of migration). The kinds of migration goals that internal migrants have are also likely to differ somewhat from those of international migrants. Because international migration is a more drastic change in life than internal migration, and because it has important implications for social, political, and economic policy, it has received considerable attention in the literature. In particular, Massey and his associates (Massey et al. 1993; Massey et al. 1994; Massey and Espinosa 1997; Massey, Durand, and Malone 2002) have reviewed and evaluated various theories that try to explain contemporary patterns of international migration.

Migration between Countries

Although the migration transition originally referred to the mobility of people *within* a nation, it has expanded its scope to become an international global phenomenon of movement *between* countries. Of the factors laid out in Figure 7.2, the most important elements in explaining migration in the modern world appear to be: (1) the creation of new opportunity structures for migration, which raise the benefit of migrating (pull factors) partly by undermining existing local relationships between people and resources (push factors); while (2) cheaper and quicker transportation and communication can (a) increase the information that people have about a potential new location (lowering the risk of migration by closing the gap between the anticipated benefits and the perceived likelihood of attaining those goals) and (b) make it easier to migrate and to return home if things do not work out (also lowering the risk of migration by making the decision to migrate more reversible).

Step migration and **chain migration,** two migration strategies that have stood out over time, help determine where migrants go. Step migration is a process whereby migrants attempt to reduce the risk of their decision by sort of inching away from home. The rural resident may go to a nearby city, and from there to a larger city, and perhaps eventually to a huge megalopolis. Chain migration reduces risk because it involves migrants in an established flow from a common origin to a predetermined destination where earlier migrants have already scoped out the situation and laid the groundwork for the new arrivals. This is a pattern that especially characterizes migration from Mexico to the United States and it reminds us that the choice of *where* to move is a large component of the decision to migrate internationally.

Why Do People Migrate Internationally?

The major theories that exist to help explain various aspects of international migration, as outlined by Massey and his associates (1993; 1994), include first those that focus on the initiation of migration patterns: (1) neoclassical economics; (2) the new household economics of migration; (3) dual labor market theory; and (4) world systems theory. Then there are three perspectives that help to explain the perpetuation of migration, once started: (1) network theory; (2) institutional theory; and (3) cumulative causation. All of these perspectives are aimed at explaining the *flow* of migrants between countries.

The Neoclassical Economics Approach   By applying the classic supply-and-demand paradigm to migration, this theory argues that migration is a process of labor adjustment caused by geographic differences in the supply of and demand for labor. Countries with a growing economy and a scarce labor force have higher wages than a region with a less developed economy and a larger labor force. The differential in wages causes people to move from the lower-wage to the higher-wage region. This continues until the gap in wages is reduced merely to the costs of migration (both monetary and psychosocial). At the individual level, migration is viewed as an investment in **human capital** (investments in individuals that can improve their economic productivity and thus their overall standard of living). People choose to migrate to places where the greatest opportunities exist. This may not be where the wages are currently the highest, but rather where the individual migrant believes that, in the long run, his or her skills will earn the greatest income. These skills include education, experience, training, and language capabilities.

This approach has been used to explain internal as well as international migration. It is also the principle that underlies Ravenstein's conceptualization of push factors (especially low wages in the region of origin) and pull factors (especially high wages in the destination region).

The New Household Economics of Migration   The neoclassical approach assumes that the individual is the appropriate unit of analysis, but the new household economics of migration approach argues that decisions about migration are often made in the context of what is best for an entire family or household. This approach accepts the idea that people act collectively not only to maximize their expected income, but also to minimize risk. Thus migration is not just a way to get rid of people; it is also a way to diversify the family's sources of income. Migrating members of the household have their journey subsidized and then remit portions of their earnings back home. This cushions households against the risk inherent in societies with weak institutions. If there is no unemployment insurance, no welfare, no bank from which to borrow money or even to invest money safely, then the remittances from migrant family members can be cornerstones of a household's economic well-being. I will return later to this topic.

Dual Labor Market Theory   This theory offers a reason for the creation of opportunities for migration. It suggests that in developed regions of the world there are essentially two kinds of job markets—the primary sector, which employs well-educated people, pays them well, and offers them security and benefits; and the secondary labor market, characterized by low wages, unstable working conditions, and lack of reasonable prospects for advancement. It is easy enough to recruit people into the primary sector, but the secondary sector is not so attractive. Historically, women, teenagers, and racial and ethnic minorities were recruited into these jobs, but in the past few decades women and racial and ethnic minority groups have succeeded in moving increasingly into the primary sector, at the same time that the low birth rate has diminished the supply of teenagers available to work. Yet the lower echelon of jobs still needs to be filled, and so immigrants from developing countries are recruited—either actively (as in the case of agricultural workers) or passively (the diffusion of information that such jobs are available).

World Systems Theory   This theory offers a different perspective on the emerging opportunity structure for migration in the contemporary world. The argument is that, since the sixteenth century (and as part of the Industrial Revolution in Europe), the world market has been developing and expanding into a set of core nations (those with capital and other forms of material wealth) and a set of peripheral countries (in essence, the rest of the world) that have become dependent on the core, as the core countries have entered the peripheral countries in search of land, raw materials, labor, and new consumer markets.

   According to world systems theory, migration is a natural outgrowth of disruptions and dislocations that inevitably occur in the process of capitalist development. As capitalism has expanded outward from its core in Western Europe, North America, Oceania, and Japan, ever-larger portions of the globe and growing shares of the human population have been incorporated into the world market economy. As land, raw material, and labor within peripheral regions come under the influence and control of markets, migration flows are inevitably generated (Massey et al. 1993:445).

   Migration flows do not tend to be random, however. In particular, peripheral countries are most likely to send migrants (including refugees and asylees) to those core nations with which they have had the greatest contact, whether economic, political, or military (Rumbaut 1991).

Network Theory   Once migration has begun, it may well take on a life of its own, quite separate from the forces that got it going in the first place. Network theory argues that migrants establish interpersonal ties that "connect migrants, former migrants, and non-migrants in origin and destination areas through ties of kinship, friendship, and shared community origin. They increase the likelihood of international movement because they lower the costs and risks of movement and increase .the expected net returns to migration" (Massey et al. 1993:449). Once started, migration sustains itself through the process of diffusion until everyone who wishes to migrate can do so. In developing countries, such migration eventually may become a rite of passage into adulthood for community members, having little to do with economic supply and demand.

Institutional Theory   Once started, migration also may be perpetuated by institutions that develop precisely to facilitate (and profit from) the continued flow of immigrants. These organizations may provide a range of services, from humanitarian protection of exploited persons to more illicit operations such as smuggling people across borders and providing counterfeit documents, and might include more benign services such as arranging for lodging or credit in the receiving country. These organizations help perpetuate migration in the face of government attempts to limit the flow of migrants.

Cumulative Causation This perspective recognizes that each act of migration changes the likelihood of subsequent decisions about migration because migration has an impact on the social environments in both the sending and receiving regions. In the sending countries, the sending back of remittances increases the income levels of migrants' families relative to others in the community, and in this way may contribute to an increase in the motivation of other households to send migrants. Migrants themselves may become part of a culture of migration and be more likely to move again, increasing the overall volume of migration. In the receiving country, the entry of immigrants into certain occupational sectors may label them as "immigrant" jobs, which reinforces the demand for immigrants to fill those jobs continually.

Which Theories Are Best? Massey and his associates (1994) attempted to evaluate the adequacy of each of the just-discussed theories in explaining contemporary patterns of international migration. Their conclusion was that each of the theories is supported in some way or another by the available evidence and, in particular, none of the theories is specifically refuted. This serves only to underscore the point made at the beginning of the chapter that migration is a very complex process. No single theory seems able to capture all of its nuances, but all of the previous perspectives add something to our understanding of migration. Recognizing now that the reasons for migrating are numerous and complex, we also must bear in mind that when people migrate, the impact is felt deeply at both individual and societal levels.

Who Migrates Internationally?

You can appreciate from the theories of international migration that it is dominated by economic incentives. People migrate for job-related reasons and then, very often, their family members follow them in a pattern of chain migration that involves family reunification. The flow of labor ought to be explained by a simple supply-and-demand model, with people moving from places where there aren't enough jobs to places where there are jobs. And, to a certain extent, that model does shape the "big picture" in the world right now. Improved communication and transportation technology have greatly facilitated a time-honored way of solving short-term labor shortages—the importation of workers from elsewhere. Population growth in less-developed nations has put incredible pressure on resources in those nations, while the declining rate of population growth in the more-developed nations has, in many instances, heightened the demand for lower-cost workers from less-developed nations. Labor shortages in northern and western Europe, in the United States and Canada, in Japan, and Australia—the more-developed societies with aging populations—have created opportunities for workers from Africa, western and southern

Asia, and Latin America. At the same time, the regional ebb and flow of economies stimulate movements of large numbers of people among the less-developed nations.

The problem for the receiving societies is that it is not a simple process to import labor. If you call the plumber in your neighborhood to come work on your house, he or she will come over and do the work, get paid, and go back home. But if you call a plumber from another country, they are apt to bring not just themselves, but also family members, and they are likely to have children, and the arrival of these people will create both a new supply of labor and a rising demand for goods and services, both public and private. This is not necessarily a bad thing, but it raises a bigger and broader set of issues with which not just you, but the entire community, must deal.

Guest labor programs sparked the migration from Mexico to the United States, as I discuss in the essay accompanying this chapter. They also flourished in post-World War II Europe, and they have been very popular in the Middle East and Africa (Castles and Miller 1998; Gould and Findlay 1994). The Arab oil-producing nations became centers of rapid immigration in the late 1970s and early 1980s as they benefited economically from the rise in oil prices. Libya and Saudi Arabia, in particular, became attractive destinations for Egyptian and Jordanian migrants, along with substantial numbers of Indians and Pakistanis. But the idea behind guest labor programs has been that people would come to the host country, work under contract for a certain period of time, and then go back to their country of origin. However, like the proverbial brother-in-law sleeping on your living room couch, they do not necessarily leave when their time is up.

A common way to deal with guest laborers who become permanent residents is a variation on the segmented assimilation model. Many countries (including virtually all of the oil-producing Gulf states, for example) deny citizenship to those people who were not born in the country, and then deny it to their children because the children were not born to citizens. Thus, the workers and their offspring may have the legal right to live and work in their adopted country, but they will never be fully participating citizens. This has also been accomplished informally in several European countries, as immigrant communities evolve into ethnic minority groups who then suffer discrimination as a result of that labeling. This kind of discrimination probably has the unintended consequence of encouraging the immigrants and their children to maintain close ties with relatives in the country of origin. Close ties may offer emotional strength, but often they also include an important monetary component, as the residents in the economically more robust countries send money back home.

Migration Origins and Destinations

Global Patterns of Migration The massive waves of international migration that characterized the nineteenth and early twentieth centuries have already been described in Chapter 2. They primarily represented the voluntary movement of people out of Europe into the "new" worlds of North and South America and Oceania. Restrictive immigration laws throughout the world (not just in the United States) and the worldwide economic depression between World Wars I and II severely limited

IS MIGRATION A CRIME? ILLEGAL IMMIGRATION IN GLOBAL CONTEXT

Illegal or undocumented immigration is not a problem peculiar to the United States. All over the globe, people without papers are moving in the millions. Is that a crime? Actually, that question could get us into a huge philosophical discussion, since different cultures define crime in different ways, but in general if a government passes a law and you violate that law, you have committed a crime. If a government says that only people with official permission may enter a country and reside permanently with full legal rights, and you enter that country without permission, then you have violated the law and, in that sense, you have committed a crime. On the other hand, we could as easily call this a semantic problem—it's a matter of words. If you show up without documents, then you are simply undocumented (and that term is increasingly preferred to the word "illegal") instead of being a criminal. After all, the usual penalty for being caught as an undocumented immigrant is to be returned to your country of origin, rather than serving time in jail. On the other hand, the penalty for being an illegal immigrant if you are *not* caught is that you may be exploited because unscrupulous employers know they can turn you in to authorities for deportation if you dare to complain. This can lead to people being forced to work in sweatshops, under near-slavery conditions, with little dignity and few legal rights (Appleyard and Taran 2000).

The reason for the rise in undocumented immigration throughout the world is straightforward—there are more people who want to get into the more-developed countries than the governments of more-developed countries are willing to let in; so people sneak in. Some people want to get a better job; some want to join family members who are already there (with or without documents); some are fleeing for their lives from an awful situation and are seeking asylum. There were an estimated nine million undocumented immigrants (the U.S. government uses "undocumented," "unauthorized," and "illegal" to describe this population) in the United States as of the year 2004 (Passel, Capps, and Fix 2004).

It is likely that more than 50 percent of all undocumented immigrants in the United States are from Mexico and given the geographic proximity and the long-standing economic disparities between the two countries, it would be truly remarkable if there were not a good deal of interaction and migration taking place. That interaction includes the apprehension of more than one million people trying to enter the U.S. from Mexico each year. When the Border Patrol (now part of the Customs and Border Protection agency) apprehends "UDAs" (undocumented aliens), most are returned to Mexico and of course most of them will keep trying to cross until they are successful. People are motivated not only by the lure of a job, but also because they may already have a job and without papers the only way they can go between their home in Mexico and their job in the U.S. is by crossing the border someplace other than a legal checkpoint. Border Patrol statistics on apprehensions, for example, always show a seasonal pattern that drops off throughout the year to a low point just before Christmas, and then spikes to a high after Christmas (U.S. Border Patrol 2003). The clear implication is that people working illegally in the U.S. have gone home to celebrate the holidays in Mexico, but then attempt to return after the holidays are over.

I should point out that there are two primary ways by which a person becomes an undocumented immigrant. A person may enter the country with a tourist or other visa and then overstay the visa. For example, you may arrive with a student visa, attend college in the United States, but then decide not to return home, at which point you become an illegal immigrant. The more usual and dramatic way is to cross the border without papers. People may try to do this on their own, but often they pay a smuggler to assist them. Most of the crossings into the U.S. are on land, and by building massive fences along the border at the two historically most popular entry places—San Diego and El Paso—the U.S. has made it increasingly more challenging for people to cross illegally. Unfortunately, the fences have largely served just to divert migrants to cross the border in other areas that are either rugged mountains or barren deserts, and in both cases the risk of death from dehydration, frost, or injury is not inconsequential (Nevins 2002). In 2001, the U.S. Border Patrol responded to an increase in the number of injuries and deaths of people crossing the border by launching a publicity campaign in Mexico aimed at describing the potential dangers of trying to cross into the U.S. The U.S. and Mexican governments also began discussions about proposed measures designed to improve the situation for those attempting to cross the border without documentation.

People trying to enter the European Union (EU) without papers will often use the Mediterranean or the Adriatic Seas to do so. Only eight miles separate Africa from Europe at the Strait of Gibraltar—the western end of the Mediterranean where it meets the Atlantic—and Africans are willing to pay $600 or more for a seat in a small boat for a rough nighttime journey from Morocco to Spain, across what is sometimes called the "Sea of Death" (Simons 2000; *The Economist* 2000a). If they make it, the payoff for these largely unskilled workers is a job in agriculture or construction in Portugal and Spain, whose below-replacement fertility levels are creating vacancies in the labor market (Mendoza 1999). Not all European countries are members of the European Union and since legal migration into EU countries is very limited, undocumented immigration has increased over time. A study in Romania (which is not yet a member of the EU) uncovered a pattern in which young Romanians will "do a season" in Paris, sneaking illegally into France to work as a rite of passage, and sending money back to Romania to ease life in the rural villages from which these undocumented immigrants came (Diminescu 1996).

The stories of undocumented immigrants into Europe mirror those of immigrants from Mexico to the United States. There are many villages in west central Mexico in which Massey and his associates estimate that nearly all males have migrated at least once to the United States by age 30 (Massey and Espinosa 1997). Many remit money home, but do not themselves return, preferring instead to remain in the United States. Although the principal jobs available to immigrants from Mexico are low-wage, low-skill jobs (as predicted by the segmented dual labor market theory), experience and English-language skills offer opportunities for better-paying jobs. Furthermore, as the number of Mexicans living

in the United States has increased and the network of immigrant families has widened, the costs of arriving in the new country have dropped substantially, making it less risky for villagers from Mexico to "try out" the U.S. labor market.

Migration from Mexico to the United States began in earnest early in this century as a reaction to the Mexican Revolution, which started in 1910 and ended in the creation of the modern United Mexican States (the official name of the Republic of Mexico). The migration northward from Mexico began to increase in the 1920s and 1930s. That flow was then halted by a combination of the Great Depression, which raised unemployment levels in the United States, and the concomitant discrimination against immigrants that surfaced during this period, leading to massive deportations of many immigrants (legal and otherwise), including many from Mexico.

Labor shortages in agriculture during World War II, however, led to a renewed invitation for Mexican workers to migrate to the United States. In 1942, the United States signed a treaty with the government of Mexico to create a system of contract labor whereby Mexican laborers ("braceros"—literally those who work with their arms, which are *brazos* in Spanish) would enter the United States for a specified period of time to work. After the war ended, the *bracero* program remained in place, but it was not until the early 1950s that the number of Mexican contract workers began to increase noticeably (Garcia y Griego, Weeks, and Ham-Chande 1990). By the mid-1950s, the contract workers had been joined by many undocumented immigrants from Mexico, and the United States reacted in 1954 by deporting more than one million Mexicans (some later found to be U.S. citizens) in what was called "Operation Wetback."

The bracero program was ended formally in 1964, and in 1965 the new immigration act, which ended the national origins quota system, also put a numerical limit on the number of legal immigrants to the United States from countries in the Western Hemisphere. Neither action noticeably slowed the migration from Mexico, which was by then well entrenched because there was and still is a constant demand for immigrant labor (Krissman

(continued)

IS MIGRATION A CRIME? ILLEGAL IMMIGRATION IN GLOBAL CONTEXT (CONTINUED)

2000). However, both of these government actions did jointly conspire to increase the number of immigrants classified as illegal or undocumented (U.S. Immigration and Naturalization Service 1991). This is a critical point with respect to undocumented immigration from Mexico: The flow from Mexico to the U.S. began because workers were needed and the government agreed to provide them with temporary work visas. More recently, however, the labor supply is still needed, but the government has been less willing to provide documentation for the immigrants, so they "enter without inspection" (EWI—another way of describing people in this category).

Undocumented immigration represents a true conundrum because it is a problem with no easy solution that everyone can agree to. There are those in the United States and Europe who argue that the solution is simple—fortify the borders and do not let anyone in who has not been pre-approved, imme-

diately deport anyone who is found inside the country without papers, and do a better job of punishing those who knowingly hire undocumented immigrants (Simcox 1999). The latter issue was supposed to have been taken care of by the 1986 Immigration Reform and Control Act, but over time the continued demand for labor in the United States has not been met by legal immigrants and in seeming recognition of this, the U.S. government has devoted very few resources to tracking down and fining these employers. This changing emphasis was reflected in the 1996 passage of the Illegal Immigration Reform and Immigrant Responsibility Act (IIRIRA), the main thrust of which was to focus Border Patrol attention on deterring, finding, and deporting illegal immigrants.

So, the reality is that people continue to enter without papers and this "crime" begets other genuinely serious crimes related to exploitation of the

international migration in the 1920s and 1930s. However, World War II unleashed a new cycle of European and Asian migration—this time a forced push of people out of war-torn countries as boundaries were realigned and ethnic groups were transferred between countries (United Nations 1979). Shortly thereafter, the 1947 partition of the Indian subcontinent into India and Pakistan led to the transfer of more than 15 million people—Muslims into Pakistan and Hindus into India. Meanwhile, in the Middle East, the partitioning of Palestine to create a new state of Israel produced 700,000 Palestinian out-migrants and an influx of a large proportion of the North African and Middle Eastern Jewish population into that area. Substantial migration into Israel from Europe, the Soviet Union, and other areas continued well into the 1960s (Wolffsohn 1987). In the 1980s, the flow of migrants into Israel began to dry up, replaced by a small but steady stream of out-migrants, but this trend was quickly turned around in the early 1990s following the Soviet Union's decision to allow Soviet Jews to emigrate. Those choosing to leave headed primarily for Israel and the United States.

Another unexpected political event in Eastern Europe in the late 1980s and early 1990s was the collapse of the Berlin Wall and the amazingly rapid reunification of Germany. Stimulated by Gorbachev's policy of openness in the Soviet Union (and by the former USSR's economic inability to continue subsidizing other communist nations), the reunification of Germany was both the cause and effect of migration from East Germany into West Germany (Heilig, Büttner, and Lutz 1990). Between 1950 and 1988, more than three million East Germans had fled to the West, but most of those had done so before the Berlin Wall went up in 1961. Then, in 1989, East Gerimmigrants. The Mexican government employs troops to patrol its northern border with the U.S., not to prevent people from crossing, but to keep them from being robbed by their countrymen who know that these people have money to pay a smuggler. The smugglers then take advantage of the immigrants by failing to warn them of the dangers of crossing the desert without water, or of going into the mountains in winter without a jacket, and of course the smugglers will leave the immigrants to fend for themselves at the first sign of trouble. Once in the U.S., the immigrants are, as I mentioned earlier, susceptible to many kinds of abuses from employers and others who know that the immigrants have few resources, legal or otherwise, to fight back. This set of circumstances leads many people to believe that illegal immigrants need to have amnesty granted as soon as possible in order to protect their human rights. Immigrants understand this idea, of course, and the most common route to amnesty is to become legal by marrying a U.S. citizen (Jasso et al. 2000).

At the opposite end of the spectrum from completely shutting down the border is the idea of an open border, which would allow people to come and go in quick response to changes in demand for their services. This idea generally appeals more to prospective employers, who would like to hire cheaper labor, while it has relatively little appeal to the currently employed, who do not relish the thought of increased competition for their jobs. This idea is not likely to be implemented in the near future because there is still a fair amount of anti-immigrant sentiment in the United States and in Europe. This idea may seem more feasible in two or three decades when the number of young adults needing work begins to subside in Mexico and other countries in response to the declining birth rate.

many relaxed its visa policies, allowing East Germans to visit West Germany, and Hungary relaxed the patrol of its border with Austria, allowing vacationing East Germans to escape to the West. Within weeks, migration from east to west was transformed from a trickle into a flash flood.

This east-west migration in Europe is in many respects a continuation of a pattern that has evolved over centuries. Between 1850 and 1913 (the start of World War I), more than 40 million Europeans moved from east to west to populate North America (Hatton and Williamson 1994), but as that was occurring, Polish, Slavic, and Ukrainian workers were migrating west to Germany and France, and Italians were migrating west to France to find work. The cold war, which cut off much of that flow, was simply a temporary aberration in a long-term trend.

To this trend has been added a mixture of other patterns: (1) south-north migration (particularly of migrant laborers from developing countries of the "south" to developed countries of the "north"); (2) a flow of migrant laborers from some of the poorer developing countries to some of the "emerging" economies, especially in south and southeast Asia; (3) a flow of workers in the Persian Gulf region from the non-oil-producing to the oil-producing nations; and (4) a flow of refugees, especially in Africa and western Asia. We have moved into what Castles and Miller (1998) have called "the age of migration." "The hallmark of the age of migration is the global character of international migration: the way it affects more and more countries and regions, and its linkages with complex processes affecting the entire world" (Castles and Miller 1998:283). Sassen (2001) has argued that the globalization of migration is, in part at least, a consequence of economic globalization—economic connections lead inevitably to flows of people. Income differences between regions attract migrants from the lower- to higher-income places, and globalization has helped to both increase the income differences and increase the chance that people will migrate in hopes of improving their economic situation.

Figure 7.4 shows the countries of the world that sent or received the greatest absolute number of net international migrants (including refugees) in the last half decade of the twentieth century, according to United Nations estimates. You can see that the United States is the top receiving country, followed by Rwanda (which was accepting refugees), the Russian Federation (which was affected by ethnic Russians leaving the former Republics of the Soviet Union), Germany, and Canada. At the other extreme, China is estimated to have lost the greatest number of people each year (contributing to a phenomenon known as the *Chinese Diaspora*), but that is just a tiny fraction of its total population (see, for example, Ma and Cartier 2003). Mexico is second on the list, followed by Kazakhstan (which sent many migrants to Russia after the breakup of the Soviet Union), the Congo (an outflow of refugees from a prolonged civil war), and India (which, like China, has experienced a substantial diaspora).

The global picture in terms of net migration rates is shown in Figure 7.1 at the beginning of the chapter. The principal receiving countries include the United States, Canada, Australia, most of the nations of Western Europe and Scandinavia, and Russia (which was receiving immigrants—especially repatriated Russians—from its former republics during the 1990s). The sending countries tend to be in Asia and Latin America (especially Mexico).

Figure 7.4 The United States Takes in More Immigrants (Net) than Any Other Country, and China Sends More Emigrants than Any Other Country

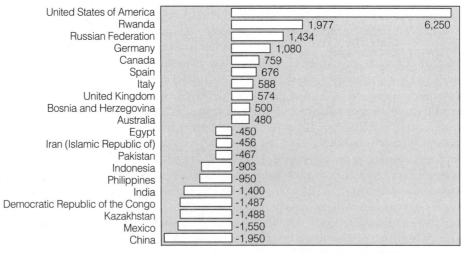

Net migrants (000's) 1995–2000

Source: Adapted from data in United Nations Population Division, 2003, World Population Prospects: The 2002 Revision (New York: United Nations), Table 10.

Figure 7.5   Countries with the Largest International Migrant Stock, 2000

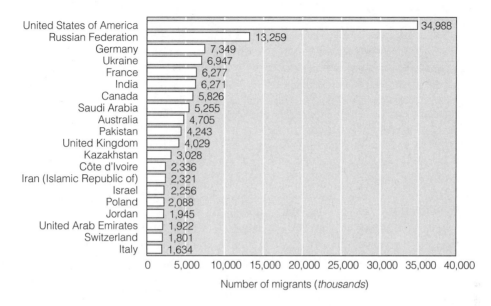

Source: United Nations Population Division, 2002. International Migration Report 2002 (New York: United Nations), Figure 1. The material was originally published by the United Nations.

Net migration rates tell us about flows of migrants, but over time those flows can vary from place to place. The longer-term effect of those flows is revealed by the stock of migrants within each country. Figure 7.5 shows the United Nations' estimates of stocks, in which the United States shows up clearly as the most migrant-friendly country in the world, followed by Russia, Germany, the Ukraine, and France.

Migration into the United States   Prior to World War I, there were few restrictions on migration into the United States and Canada, so the number of immigrants was determined more by the desire of people to move than anything else. Particularly important as a stimulus to migration, of course, was the drop in the death rate in Europe during the nineteenth century, which launched a long period of population growth, with its attendant pressure on Europe's economic resources. Economic opportunities in America looked very attractive to young Europeans who were competing with increasing numbers of young people for jobs. Voluntary migration from Europe to the temperate zones of the world—especially the United States—represents one of the significant movements of people across international boundaries in history. The social, cultural, economic, and demographic impacts of this migration have been truly enormous (Davis 1988).

Immigration to the United States in most of the nineteenth century was dominated by people arriving from northern and western Europe (beginning with England, Ireland, Scotland, and Sweden, then stretching to the south and east to draw immigrants from Germany, especially), as seen in Table 7.2. There was a lull during

Table 7.2 The Geographic Origin of Immigrants to the United States Has Changed Dramatically over the Decades

Years	Total	Asia	Latin America	S/E Europe	N/W Europe	Other	% Foreign-born
1821–1830	143,439	30	9,287	3,142	97,932	33,048	
1831–1840	599,125	55	19,800	5,902	503,403	69,965	
1841–1850	1,713,251	141	20,746	5,301	1,633,864	53,199	9.7
1851–1860	2,598,214	41,538	15,411	21,236	2,490,650	29,379	13.1
1861–1870	2,314,824	64,759	12,729	25,691	2,193,328	18,317	14.0
1871–1880	2,812,191	124,160	20,404	127,582	2,527,983	12,062	13.3
1881–1890	5,246,613	69,942	33,393	602,450	4,526,338	14,490	14.7
1891–1900	3,687,564	73,751	36,876	1,910,158	1,648,341	18,438	13.6
1901–1910	8,795,386	325,429	184,703	6,147,975	2,093,302	43,977	14.7
1911–1920	5,735,811	246,640	401,507	3,326,771	1,737,951	22,942	13.2
1921–1930	4,107,209	110,895	591,438	1,178,769	2,209,679	16,428	11.6
1931–1940	528,431	16,381	51,258	149,546	306,490	4,756	8.8
1941–1950	1,035,039	37,261	154,221	132,485	660,355	50,717	6.9
1951–1960	2,515,479	153,444	558,436	402,477	1,303,018	98,104	5.5
1961–1970	3,321,677	428,496	1,282,167	541,433	996,503	73,078	4.7
1970–1981	4,493,314	1,586,140	1,810,806	512,238	458,318	125,812	6.2
1981–1990	7,338,062	2,737,097	3,456,227	388,917	528,340	227,481	7.9
1991–2000	9,095,417	2,892,179	4,319,191	1,078,236	233,142	572,669	11.1

Sources: Immigration data are from U.S. Office of Immigration Statistics, 2003, 2002 Yearbook of Immigration Statistics, http://uscis.gov/graphics/shared/aboutus/statistics/ybpage.htm, Table 3, accessed 2004; data on foreign-born are from Lisa Lollock, 2001, "The Foreign-Born Population in the United States: March 2000," *Current Population* Reports P20-534, and the 2000 Census of Population.

the American Civil War, but after the war the pace of immigration resumed. By the late nineteenth and early twentieth centuries, the immigrants from northern and western Europe were being augmented by people from southern and eastern Europe (Spain and Italy, then stretching farther east to Poland and Russia). Immigration to the United States and Canada reached a peak in the first decade of the twentieth century, when 1.6 million entered Canada and nearly nine million entered the United States, accounting for more than one in 10 of all Americans at that time. "They came thinking the streets were paved with gold, but found that the streets weren't paved at all and that they were expected to do the paving" (Leroux 1984).

This represents one of the most massive population shifts in history and virtually all of the theories of international migration discussed earlier have something to offer by way of explanation (Hatton and Williamson 1994; Massey et al. 1994; Moch 1992). Compared to the United States, European wages were low and unemployment rates high. Capital markets were beginning to have a disruptive effect in some of the less-developed areas of southern and eastern Europe. Eastern Europe was also undergoing tremendous social and political instability, and the Russian pogrom against Jews caused many people to flee the region.

Nor was all this emigration from Europe aimed at North America. Millions of Italians and Spaniards, as well as Austrians, Germans, and other Europeans, settled throughout Latin America during this period. Most notable among the destinations were Brazil and Argentina (Sanchez-Albornoz 1988). The migration peaked at about the time of World War I, and Europe has never since experienced emigration of this magnitude, partly because wages rose in Europe, helping to keep people there (and encouraging some return migration from the Americas), compounded by the Great Depression of the 1930s (which also encouraged return migration to one's "roots"). In the latter part of the nineteenth century, the percentage of the population enumerated in the U.S. Census as being foreign born hovered near 14 percent, as seen in Table 7.4. Never since has it reached that level. I mention that particularly because the rapid increase in the foreign-born population at the end of the twentieth century in the United States and Canada, which I will discuss below, is something that both countries have seen before.

The Great Depression, coupled with more restrictive immigration laws (discussed in Chapter 13), followed by World War II, effectively slowed migration to the United States and Canada to a trickle during the 1930s and 1940s. In the 1950s, there was a brief post-World War II upsurge in migration from northern and western Europe, but as seen in Figure 7.6, the 1950s represented a transition time to a new set of "origins and destinies" for immigrants to North America (Rumbaut 1994). Since the 1960s, European immigrants to the United States have been replaced almost totally by those from Latin America (especially Mexico, but also Nicaragua and El Salvador) and Asia (especially the Philippines, China, and India).

Although only one person is recorded as having migrated from Mexico to the United States in 1820, the number of Mexican and other Latin American immigrants has increased so tremendously since then that the legal and undocumented immigrants from Mexico probably account for one in four of all foreigners now moving into the United States. Between 1820 and 2002, more than six million Mexicans migrated legally into the United States, with more than half of them arriving just since 1980 (Office of Immigration Statistics 2003). Migration between Mexico and the United States is the largest sustained flow of migrant workers in the contemporary world and this flow of migrants is closely connected with the population increase in Mexico. As Mexico's population has grown, so has the difficulty of finding adequate employment for the burgeoning number of young adults. That "push," accompanied by the "pull" of available jobs with higher wages in the United States, has stimulated a tremendous migration stream. The flow in this stream can build quickly as the number of social ties between sending and receiving areas grows, "creating a social network that progressively reduces the cost of international movement. . . . The range of social contacts in the network expands with the entry of each migrant, thus encouraging still more migration and ultimately leading to the emergence of international migration as a mass phenomenon" (Massey et al. 1987:5). Data from Massey's *Mexican Migration Project* suggest that migration streams are much easier to start than to stop, because migration cumulatively begets more migration as community members in the sending area derive a real benefit (human capital) from migration and as expanded networks (**social capital**) make it increasingly easier to migrate (Massey and Espinosa 1997). In this context, human capital refers to the benefits derived by the migrants who have been to the United States, whereas social capital refers to the links with friends and relatives already abroad.

Immigrants from Latin America are arriving from countries in which the mandatory level of education is typically six to eight years of schooling, and they represent a relatively poorly educated, lower-skilled group. The principal exception to this generalization has been Cuban immigrants, but even recent immigrants from Cuba are less well educated and have fewer skills than the group of exiles who initially fled Castro's takeover of the island. The low levels of education of Mexican migrants means that they are not very conversant with the educational system of either their country of origin or destination, and it is then harder for them to support their children in the context of the American educational system. This puts the children of Mexican immigrants at a disadvantage compared to many of the Asian immigrant groups, and increases the probability that their experience in the U.S. will be a type of segmented assimilation (Portes and Rumbaut 2001).

A large portion of Mexican immigrants to the United States expect eventually to return to Mexico (even if they eventually do not). This has dampened their enthusiasm for becoming U.S. citizens, and the U.S., unlike many other countries, technically does not allow dual citizenship. This is one reason why Mexican immigrants have been somewhat politically underrepresented. The Mexican government has worried about this over time, and in 1998, the Mexican Nationality Act went into effect allowing Mexicans to retain their Mexican nationality even if they become a citizen of another country such as the United States. This is not the same as dual citizenship, however, because dual nationals are not allowed to run for public office, vote, or serve in the armed forces in Mexico. However, the dual-nationality status means that Mexican nationals who become U.S. citizens will have the right to buy and sell land in Mexico free of the restrictions imposed on foreigners, to receive better treatment under investment and inheritance laws in Mexico, to attend public schools and universities as Mexicans, and to access other Mexican government services and jobs.

Asian immigration to the United States began in the middle of the nineteenth century, but a series of increasingly restrictive immigration laws (which I will discuss in Chapter 13) kept the number relatively low until the late 1960s, when the laws were substantially liberalized. In the entire decade from 1841–50, scarcely more than 100 Asians immigrated (see Table 7.2), but in the following decade the demand for labor on the West Coast produced a 10-year total of 41,000 Asian immigrants, virtually all of whom were from China. Between 1871 and 1880, the level of immigration from China climbed to 124,000, and that volume led to an unfortunate series of restrictive laws limiting the ability of Chinese to enter the country. As migration from China abated, migration from Japan increased; in the decade from 1901–10, only 21,000 Chinese immigrated, whereas 130,000 came from Japan.

The National Origins Quota system, which went into effect in the 1920s (see Chapter 13), dramatically reduced migration from all parts of Asia. Thus, of the 528,000 immigrants to the United States between 1931 and 1940, only 16,000 were from Asia. However, following the 1965 revisions of the immigration laws in the United States, the picture changed. Since 1965, Asians have accounted for more than four in 10 new arrivals into the United States, and in the period from 1991–2000 the list was led by the Philippines, followed by China (including the People's Republic of China, Taiwan, and Hong Kong), India, and Vietnam.

In 1898, the United States gained control of the Philippines from Spain as a result of the Spanish-American War. From that date until 1936, immigrants from the Philippines were routinely excluded from the United States, as the Chinese and Japanese had been (Tyner 1999). Since Filipinos were technically U.S. nationals, they were not subject to the usual immigration laws, but their entry was generally restricted, and by 1930, only 45,000 Filipinos lived in the United States, almost all of them male. The majority of these men came without families, and either returned home or stayed and married non-Filipino women. The United States granted the Philippines its independence in 1936 and that immediately brought them under federal immigration law, which effectively excluded them altogether from entering the country until after the end of World War II. At the time, there was an influx of Filipino veterans who had served in the U.S. military during the war. However, the biggest boon to Filipino immigration was the change in the law in the 1960s that gave highest immigration preference to family members. Americans of Filipino origin began to sponsor the immigration of their relatives (especially parents and siblings). Thus, since 1980, more than three-fourths of the 40,000 to 60,000 immigrants each year from the Philippines have been relatives of U.S. citizens (U.S. Immigration and Naturalization Service 2000).

Migration out of the United States   The United States Immigration and Naturalization Service does not keep track of people who emigrate from the country (U.S. Immigration and Naturalization Service 2002), but that doesn't mean that people are not leaving. Using data from the Current Population Surveys and other sources, Woodrow-Lafield (1996) estimated that there were approximately 220,000 emigrants from the United States each year as of the mid-1990s, of whom a majority are foreign-born persons (probably returning to their country of origin). The U.S. Census Bureau has projected a steady increase in out-migration, exceeding 300,000 per year by 2003 (Hollman, Mulder, and Kallan 2000).

The evidence suggests that there are two main patterns to return migration—which accounts for most emigration from the United States (Reagan and Olsen 2000; Rumbaut 1997). The first pattern is for some people who migrate into the country to quickly become discouraged or disenchanted, or perhaps to reach a short-term goal, and then to return home, having stayed in the United States for only a short while. The other pattern is for people to migrate to the United States when young, and then retire back to their country of origin. We know, for example, that each month the Social Security Administration sends out nearly a half million checks to foreign addresses (U.S. Social Security Administration 2001a), and it is likely that most of these are being sent to people who migrated to the United States to work, and then when they reached retirement age, went back from whence they came—where their retirement check in dollars is likely to go farther than in the United States. Estimates for both the United States and Canada suggest that there are typically about five immigrants for each emigrant (U.S. Immigration and Naturalization Service 2002; Warren and Kraly 1985).

Migration into Canada Canada's immigration experience has been similar, but not identical, to the pattern in the United States. Although Canada has historically had a high level of immigration, it also experienced considerable emigration (people leaving to return home, or entering the United States from Canada) until after World War II, as shown in Figure 7.6. Net migration jumped in the late 1980s and early 1990s as a result of immigration policy changes, but has slackened some since then. However, it is still at historically high levels.

The 2001 Census data for Canada show that one in 10 Canadians speaks a language other than English or French at home, and the number of people reporting a mother tongue other than English or French totals nearly 5 million of Canada's 30 million residents. Chinese is the most common language spoken at home other than English or French, replacing German, Italian, and Ukrainian, the leading nonofficial languages as reported in the 1971 census (*Migration News* 1998). As in the United States, recent immigrants to Canada are far less likely to have come from Europe than was true in the past. In the 1990s, more than half of all immigrants to Canada were from somewhere in Asia (Boyd and Vickers 2000).

Figure 7.6 Immigration into Canada

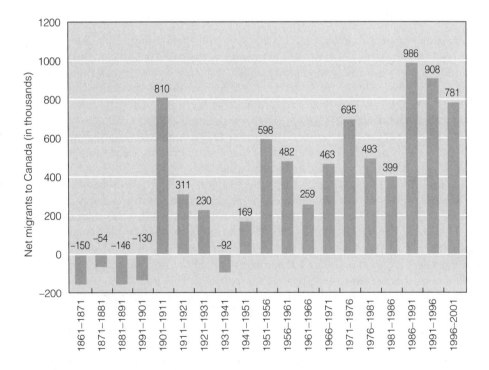

Source: Adapted from Statistics Canada, 2003: http://www.statcan.ca/english/Pgdb/demo03.htm; accessed 2004.

Consequences of Migration

Why does migration matter? It has the potential to significantly affect the lives of the people who move, the people they leave behind, and the new people with whom they interact after making the move. And, depending on the size of the migration flow over time, migration has the potential to dramatically alter the demographic structure of a community within a very short period of time, as well as to subtly, but importantly, alter the social structure over the long term as the immigrant stock adds its children (part of the foreign stock) and the stock changes its flavor, so to speak.

Consequences for Migrants   Although migrants typically move in order to improve their lives, we must recognize that, for the individual migrant, relocation may nonetheless produce anxiety and stress as a new social environment has to be negotiated. Part of that negotiation process may be to deal with discrimination (possibly including violence) that is often the result of **xenophobia,** fear and mistrust of strangers on the part of people already residing in the place to which the migrant has moved. One of the ways in which migrants cope with a new environment is to seek out others who share their cultural and geographic backgrounds. This is often aided or even forced by the existence of an **enclave** (a place in a larger community within which members of a particular subgroup tend to concentrate) of recent and former migrants from the same or similar areas. In fact, the development of an enclave may facilitate migration, since a potential migrant need not be too fearful of the unknown. The enclave in the host area has guides to the new environment—former migrants who have made the adaptation and stand ready to aid in the social adjustment and integration of new migrants. "For subsequent generations, preservation of the ethnic community, even if more dispersed, can also have significant advantages. Among the entrepreneurially inclined, ethnic ties translate into access to working capital, protected markets, and pools of labor" (Portes and Rumbaut 1990:54).

Although finding people of similar background may ease the coping burden for a new migrant, there is some evidence to suggest that the long-run social consequences of "flocking together" (especially among relatives) will retard the migrant's adjustment to and assimilation into the new setting. John F. Kennedy's comment that "the way out of the ghetto lies not with muscle, but with the mastery of English," is more than a facile phrase. It points to a key to educational success and labor force entry (Portes and Rumbaut 2001). In general, it seems that immigrants, even those living in ethnic enclaves, will have weaker social ties in the new country than they probably would have had in the old country (Wierzbicki 2003).

On arrival in the host area, an immigrant may go through a brief period of euphoria and hopefulness—a sort of honeymoon. However, that may be followed by a period of shock and depression, especially for refugees. Rumbaut (1995b:260) quotes a Cambodian refugee to the United States he interviewed:

> I was feeling great the first few months. But then, after that, I started to face all kinds of worries and sadness. I started to see the real thing of the United States, and I missed home more and more. I missed everything about our country: people, family, relatives and friends, way of life, everything. Then, my spirit started to go down; I lost sleep; my

physical health weakened; and there started the stressful and depressing times. By now [almost 3 years after arrival] I feel kind of better, a lot better! Knowing my sons are in school as their father would have wanted [she was widowed], and doing well, makes me feel more secure.

Immigrants undergo a process of **adaptation** or adjustment to the new environment, in which they adjust to the new physical and social environment and learn how best to negotiate everyday life. Some immigrants never go beyond this, but most proceed to some level of **acculturation,** in which they adopt the host language, bring their diet more in line with the host culture, listen to the music and read the newspapers, magazines, and books of the host culture, and make friends outside of their immigrant group. This may be more likely to happen if the immigrant has children, because children often are exposed to the new culture more intensively than are adults. Language use frequently is employed as an indicator of acculturation. Many migrants never go beyond this, but some migrants (and especially their children raised in the host culture) **assimilate,** in which they take on not just the outer trappings of the host culture, but also assume the behaviors and attitudes of members of the host culture (Alba and Nee 1997; Rumbaut 1997). Intermarriage with a member of the host society is often used as an index of assimilation.

These individual adjustments to a receiving society assume an open society and assume that immigrants are considered on an individual basis. In fact, nations rarely are open with regard to immigrants, and because immigration occurs regularly in clumps (with groups of refugees, or new guest workers arriving nearly *en masse*), immigrants often are treated categorically. Although assimilation is one model by which a society might incorporate immigrants into its midst, there are at least three other types of incorporation: **integration** (mutual accommodation); **exclusion** (in which immigrants are kept separate from most members of the host society and are maintained in separate enclaves or ghettos); and **multiculturalism** or *pluralism* (in which immigrants retain their ethnic communities but share the same legal rights as other members of the host society) (Zlotnick 1994).

Multiculturalism, in particular, is enhanced by a new class of migrant—the **transnational migrant**—who sets roots in the host society while still maintaining strong linkages to the donor society (Schiller, Basch, and Blanc 1995). Such individuals have also been called "skilled transients"—relatively skilled workers moving internationally on assignment and, in the process, having an impact on the area of destination while always intending to return to the area of origin (Findlay 1995). In parts of sub-Saharan Africa, a type of transnationalism has been institutionalized into the social structure among migratory laborers (Guilmoto 1998), and there is evidence that elsewhere in the world less skilled workers are adopting this strategy of living dual lives—working in one environment, but maintaining familial ties in another (Levitt, DeWind, and Vertovec 2003; Smith and Guarnizo 1998).

Children of Immigrants   Most of the explanations about how immigrants deal with their new society focus on adult immigrants who were raised in one country and now are adapting and adjusting to another. The children born to them in the new country (the second generation) will have the task of growing up mainly (or only) knowing the new country, but having to deal with immigrant parents. Note, however, that the youngest immigrants (those who are prepubescent, approximately 12 or younger) are actually in that same situation—too young for the country of origin to have had a strong impact on their own development, and growing up in the new country almost as though they had been born there. This is sometimes known as the 1.5 generation—halfway between the first generation (the immigrants), and the second generation (the children of immigrants) (Rumbaut 1997).

The path that receiving societies often have in mind for the children of immigrants is a straight-line process of assimilation from the country of their parents' origin into the country of their own birth. You are already primed for the idea, however, that when it comes to migration, life is never going to be so simple. In Europe and increasingly in the Americas, the second generation may fall into one of two possible patterns of **segmented assimilation,** in which the children of immigrants (1) adopt the host language and behavior, but find themselves identified with a racial/ethnic minority group that effectively limits their full participation in society, or (2) assimilate economically into the new society, but retain strong attachments to the ethnic group of origin (Portes 1995). These variations in the path of assimilation are occasioned especially by the fact that immigrants to Europe and to North America are likely to come from different ethnic and cultural backgrounds than prevail in the host country. In Canada, these are known as *visible minorities* and children of immigrants who are visible minorities are more likely to experience some form of segmented assimilation rather than assimilating directly into the mainstream of Canadian society (Boyd and Grieco 1998).

Societal Consequences   Although the consequences of migration for the individual are of considerable interest (especially to the one uprooted), a more pervasive aspect of the social consequences of migration is the impact on the demographic composition and social structure of both the donor and host areas. The demographic composition is influenced by the selective nature of migration, particularly the selectivity by age. The **donor area** typically loses people from its young adult population, those people then being added to the **host area.** Further, because it is at those ages that the bulk of reproduction occurs, the host area has its level of natural increase augmented at the expense of the donor area. This natural-increase effect of migration is further enhanced by the relatively low probability of death of young adults compared with the higher probability in the older portion of a population.

The selective nature of migration, when combined with its high volume, such as in the United States and Canada, helps to alter the patterns of social relationships and social organization in both the host and donor communities. Extended kinship relations are weakened, although not destroyed, and local economic, political, and educational institutions have to adjust to shifts in the number of people serviced by each.

Remittances as a Factor in International Migration   Hidden from the view of most eyes is the tremendous economic benefit that less-developed sending countries receive when their citizens go off to work in the more-developed nations. Many, if not most, such workers send part of their pay back home, thereby raising the standard of living of the family members who stayed behind, and encouraging a certain dependence on the income being earned by the migrating family members (Stark and Lucas 1988).

"Migradollars," as they are often called, can finance the purchase of homes and consumer durables in the country of origin, provide startup money for new businesses, and even provide enough resources in poor communities of Mexico, for example, to lower the infant death rate (Kanaiaupuni and Donato 1999). In fact, the United Nations estimates that remittances may account for more than 10 percent of the total economy in several Central American and Central European countries (United Nations Population Division 2002b). Remittances are not received without a cost, however, which is the possibility that family members in the sending countries will become dependent on the remittances as a source of income, and will be vulnerable to economic changes not just in their own country, but also in the country to which their remitting relatives have moved. This is not something that people are going to think about in their initial cost-benefit analysis, but it is something they may have to deal with later on.

Overall, then, migration has the greatest short-run impact on society of any of the three demographic processes. It is a selective process that always requires changes and adjustments on the part of the individual migrant. More importantly, when migration occurs with any appreciable volume, it may have a significant impact on the social, cultural, and economic structure of both donor and host regions. Because of their potential impact, patterns of migration are harbingers of social change in a society. Of course, the impact may be different if people are forced to migrate, rather than doing so of their own volition.

Forced Migration

Sitting on the sidelines of the migration transition are people who may not have originally intended to move, but wound up being forced to do so anyway. A **forced migrant** is, as you might expect, someone who has been forced to leave his or her home because of a real or perceived threat to life and well-being (Reed, Haaga, and Keely 1998). There may be at least 50 million people alive at this moment who have been forced to migrate. Most are internally displaced—still in their original country but not in their original residence. Many more have been forced into another country and so are refugees. A third category, and an historically important one, is the worst one of all, which is slavery.

Refugees

There were more than 25 million "uprooted" people in the world as of 2003, of whom about 14 million were refugees, with the remainder being asylum seekers and internally displaced people, according to data compiled by the United Nations High Commissioner for Refugees and the U.S. Committee for Refugees. Refugees from Palestine actually represent the largest group, although the majority of these individuals have been born into refugee status and so the United Nations does not always include them in their statistics. Jordan is the only one of the Middle Eastern countries that typically has allowed any Palestinian to become a citizen of the country of asylum, and so Palestinians have remained as refugees for nearly a half century in other countries. Afghanistan has produced the largest group of "current" refugees (all of them within the last two decades), as a result first of the Soviet Union becoming embroiled in a civil war there, followed by the imposition of Taliban rule in the country, followed by the U.S.-led attack to drive the Taliban out of Afghanistan. The majority of these refugees went to neighboring Pakistan and Iran, and a large number have since been repatriated, but remain in displaced person camps.

There are essentially three solutions to the problem of refugee populations, including: (1) repatriation to the country of origin; (2) resettlement in the country to which they initially fled; and (3) resettlement in a third country. None of these is easy to accomplish and a comparison of the countries producing refugees with those who have taken in refugees shows you why, as you can see in Table 7.3. Afghanistan's refugees are disproportionately next door on one side in Pakistan and next door on the other side in Iran. Palestinian refugees are also close to their area of origin—congregating in the West Bank, Jordan, and the Gaza Strip. Sudan has taken in almost as many refugees from Eritrea as it has generated (and those Sudanese refugees are in Tanzania, Kenya, and Ethiopia). Since birth rates are high among virtually all of these groups, many of these refugees are children who have been born outside their parents' country of origin. There are no easy solutions to the problems

Table 7.3 Origin of Major Refugee Population in 2002[1]

Country of Origin[2]	Main Countries of Asylum	Total
Afghanistan	Pakistan/Iran	2,481,000[3]
Burundi	Tanzania/D.R. Congo	574,000
Sudan	Uganda/Ethiopia/D.R. Congo/Kenya/ Central African Rep.	505,200
Angola	Zambia/D.R. Congo/Namibia/Congo	433,000
Somalia	Kenya/Yemen/Ethiopia/United Kingdom/ USA/Djibouti	429,000
Democratic Rep. Congo	Tanzania/Congo/Zambia/Burundi/Rwanda	415,000
Iraq	Iran/Germany/Netherlands/Sweden	401,000
Bosnia-Herzegovina	Serbia-Montenegro/USA/Sweden/Demark/ Netherlands	372,000
Viet Nam	China/USA	348,000
Eritrea	Sudan/Ethiopia	316,000

[1]More than four million Palestinians who are covered by a separate mandate of the U.N. Relief and Works Agency for Palestine Refugees in the Near East (UNRWA) are not included in this report. However, Palestinians outside the UNRWA area of operations such as those in Iraq or Libya, are of concern to UNHCR. At year end, their number was 428,710.

[2]This table includes UNHCR estimates for nationalities in industrialized countries on the basis of recent refugee arrivals and asylum seeker recognition.

[3]This is a UNHCR estimate, although government figures say there are 2.0 million refugees in Iran and 1.8 million refugees in Pakistan.

Source: United Nations High Commissioner for Refugees, Refugees by Numbers 2003; http://www. unhcr.ch/cgi-bin/texis/vtx/home/opendoc.pdf?tbl=STATISTICS&id=3d0754&page=statistics, Table 3, accessed 2004. The material was originally published by the United Nations.

faced by these individuals. To some, refugees represent a pool of potential human capital; to others, a trough of potential trouble.

Slavery

There can be no doubt that the most hideous of migratory movements are those endured by slaves. Slavery has existed within various human societies for millennia, and even in the 1990s it was reported that slave-trading was still taking place in the Sudan in sub-Saharan Africa (Davis 1998). McDaniel (1995:11) has summarized the early historical situation as follows:

> The international slave trade in Africans began with the Arab conquests in northern and eastern Africa and the Mediterranean coast in the seventh century. From the seventh to the eleventh century, Arabs and Africans brought large numbers of European slaves into the North African ports of Tangier, Algiers, Tunis, Tripoli, and Fez. In fact, most of the slaves traded throughout the Mediterranean before the fall of Constantinople were European. Between the thirteenth and fifteenth centuries Africans, along with Turks, Russians, Bulgarians, and Greeks, were slaves on the plantations of Cyprus.

However, the most massive migration of slaves was that of the Atlantic slave trade, which transported an estimated 11 million African slaves to the western hemisphere between the end of the fifteenth century and the middle of the nineteenth century (Thomas 1997). The slaves came largely from the west coast of sub-Saharan Africa, from countries that now comprise Senegal, Sierra Leone, Ivory Coast, Dahomey, Benin, Cameroon, Gabon, Nigeria, and the Congo. "The preponderance of Africans who were sold into slavery were taken by force. Some were taken directly by Arab or European slave traders, but most were sold into slavery by the elite Africans who had captured them in warfare or who were holding them either for their own use as slaves or to be traded as slaves later" (McDaniel 1995:14).

The destinations were largely the sugar and coffee plantations of the Caribbean and Brazil, but hundreds of thousands were also sold in the United States to serve as laborers on cotton and tobacco plantations. The slave traders themselves were initially Portuguese and Spanish, but the French, Dutch, and especially the British were active later on. It was eventually the British, however, who pushed for a worldwide abolition of slavery. Slavery was abolished throughout the British Empire, including Canada, in 1833, although Canada had never tolerated slavery, and the impact of slavery on Canada was primarily that it was a place of refuge for those seeking to escape slavery. Similarly in Mexico, there is little of history of slavery and in 1827 Mexico declared that no person could be born a slave in Mexico. In the United States, it was not until 1865 (in the context of the Civil War) that the Thirteenth Amendment to the Constitution finally abolished slavery.

Summary and Conclusion

Migration is any permanent change of residence. It is the most complex of the three population processes because we have to account for the wide variety in the number of times people may move, the vast array of places migrants may go, and the incredible diversity of reasons there may be for who goes where, and when. Of importance to demographers is the fact that the migration transition is an inevitable consequence of the mortality transition, which lowers the death rate prior to the initiation of the fertility transition. This unleashes population growth in rural areas that cannot be accommodated by agricultural economies, and so people (usually young adults) migrate to where the jobs are—usually cities within their own country, but sometimes to other countries.

For decades, migration theory advanced little beyond the basic idea of push and pull factors operating in the context of migration selectivity. More recently, conceptual models have developed that are very reminiscent of the explanations of the fertility transition. To begin with, we need a model of how the migration decision is arrived at (not unlike the first precondition for a fertility decline—the acceptance of the idea that you are empowered to act). Then, most importantly for the study of migration, we need to understand what might motivate a person to migrate. A variety of theoretical perspectives has been offered, including the neoclassical approach, the new household economics of migration approach, the theory of the dual labor market, and world systems theory. Finally, the means available to migrate represent those things that "grease the skids," making it easy to migrate once a person is motivated to do so. These include not only improved transportation and communication but also the development of social networks and institutions that facilitate the migration process.

Migration has dynamic consequences for the migrants themselves, for the areas from which they came, and for the places to which they go. Some of these consequences are fairly predictable if we know the characteristics of the migrants. If immigrants are well-educated young adults, for example, they will be looking for well-paying jobs, they may add to the economic prosperity of an area, and they will probably be establishing families, which will further add to the area's population and increase the demand for services.

Throughout the world, population growth has induced an increase in the volume of migration, both legal and undocumented. "Temporary" labor migration has also increased throughout the world as jobs have become more available in the aging, more-developed societies for the burgeoning number of younger workers from less-developed countries. Understandably, guest workers are often reluctant to leave higher-income countries, even when the economies in those places slow down and pressure builds for foreigners to go home. Such people are only a few steps away from the unhappily large fraction of the world's migrants who are refugees, seeking residence in other countries after being forced out of their own. Of the more than 14 million refugees scattered throughout the globe at the beginning of the twenty-first century, disproportionate shares are found in western Asia and sub-Saharan Africa.

Although it is not always apparent, the quality of our everyday life is greatly affected by the process of migration, for even if we ourselves never move, we will spend a good part of our lifetime adjusting to people who have migrated into our lives and to the loss of people who have moved away. Each new person coming into our life greatly expands our social capital by increasing the potential size of our social network, especially since many people who move away do so physically but not symbolically; that is, we remain in communication.

In the next chapter, I put the dynamics of migration together with those of mortality and fertility to show you how they collectively influence the age/sex structure. That may not sound very interesting, but in fact it is through the changing age structure that population growth makes itself felt, and so the age structure is one of the most important drivers of social change in the world. For that reason, it is like a stealth bomber—without a really good radar system (in this case, the tools of demography), you may not see it until it's too late to do anything about it.

Main Points

1. Migration is the process of changing residence and moving your whole round of social activities from one place to another.

2. International migrants move between countries, whereas internal migrants do their moving within national boundaries.

3. Migration can be assessed in terms of flows (the movement of people) or stocks (the characteristics of people according to their migration status).

4. Migration can be measured by a variety of indices designed especially to capture aspects of the flow of migrants into and out of a place.

5. The migration transition initially referred to internal migration occurring as a result of population growth, which created a redundant rural population.

6. Explanations of why people move typically begin with the push-pull theory, first formulated in the late nineteenth century.

7. Migration is selective and is associated especially with age, gender, and different stages in the life cycle, giving rise to the idea that migration is an implementing strategy—a means to a desired end.

8. Migration within countries tends to be for economic reasons, but housing-related moves are especially common in the United States.

9. Ross Baker once suggested that the "First Law of Demographic Directionality" is that a body that has headed west remains at west, but that is no longer so true in the United States.

10. Major theories offered to explain international migration include neoclassical economics, the new household economics of migration, the dual labor market theory, world systems theory, network theory, institutional theory, and cumulative causation.

11. Global patterns of migration include the east-west movement of Europeans, the south-north movement of labor, and the circulation of labor from less-developed to emerging economies.

12. In the nineteenth and early twentieth centuries, Europeans were the principal migrants to the United States and Canada; in the 1930s and 1940s, more migrants left than came; and since the 1960s, migration has largely been from Latin America and Asia.

13. An estimated 200,000–300,000 people leave the United States each year, probably to return to their place of origin.

14. Migration forces adjustment to a new environment on the part of the migrant, and forces a societal response to the immigrant on the part of the receiving society.

15. Major concepts in the adjustment of immigrants include adaptation, acculturation, assimilation, integration, incorporation, exclusion, multiculturalism, and transnationalism.

16. Remittances from guest laborers represent an important economic resource for the countries that send the migrants.

17. There are more than 14 million refugees in the world, concentrated in Africa and western Asia.

Suggested Readings

1. Charles Hirschman, Philip Kasinitz, and Josh De Wind, Editors, 1999, The Handbook of International Migration: The American Experience (New York: Russell Sage Foundation).

   The chapters in this important volume are written by the top researchers in the field of international migration as it relates to the United States. Although the focus is only on the U.S., the concepts and theories are applicable on a global scale.

2. Douglas Massey, Joaquín Arango, Graeme Hugo, Ali Kouaouci, Adela Pellegrino, and J. Edward Taylor, 1993, "Theories of International Migration: A Review and Appraisal," Population and Development Review 19(3):431–466; and 1994, "An Evaluation of International Migration Theory: The North American Case," Population and Development Review 20(4):699–752.

   These two articles, representing a matched set, will have a long-lasting influence on the field of international migration analysis because they bring together a long-needed review of theoretical perspectives.

3. Stephen Castles and Mark J. Miller, 2003, The Age of Migration, Third Edition (New York: Guilford Press).

   There is no better source than this book for a good global perspective on the volume and direction of international migration.

4. Philip Martin and Jonas Widgren, 2002, "International Migration: Facing the Challenge," Population Bulletin 57(1); and Philip Martin and Elizabeth Midgley, 2003, "Immigration: Shaping and Reshaping America," Population Bulletin 58(2).

   The first of these two reports from the Population Reference Bureau provides an excellent summary of the global migration situation, with discussions of regional patterns as well. The second report focuses on the impact of immigration on the United States.

5. Alejandro Portes and Rubén G. Rumbaut, 2001, Legacies: The Story of the Immigrant Second Generation (Berkeley and Los Angeles: University of California Press).

   The United States, Canada, and most European countries face a future in which a large fraction of the population will be children of immigrants. Portes and Rumbaut have been studying the phenomenon in the United States and report their findings in this volume.

⊕ Web Sites of Interest

Remember that Web sites are not as permanent as books and journals, so I cannot guarantee that each of the following Web sites still exists at the moment you are reading this. It is usually possible, however, to retrieve the contents of old Web sites at http://www.archive.org.

1. http://uscis.gov/graphics/index.htm
   This is the Web site of the U.S. Citizenship and Immigration Services (USCIS), which is part of the U.S. Department of Homeland Security. The site includes information about and for immigrants and refugees.

2. http://www.cic.gc.ca/english/
   In Canada, the immigration agency is known as Citizenship and Immigration Canada (CIC) and its Web site includes information on government policy, as well as links to research on immigration.

3. http://migration.ucdavis.edu
   This Web site was developed by Philip Martin at the University of California, Davis, and is a tremendous resource, especially because it includes full issues of the newsletter *Migration News*, in which newspaper, magazine, and journal articles and books about migration are summarized and commented on.

4. http://www.migrationinformation.org
   The Migration Information Source is a project of the Migration Policy Institute in Washington, DC, and offers a wide range of news and resources useful to policy makers and people in academia.

5. http://www.refugees.org
   The United States Committee for Refugees (USCR) is a private, nonprofit organization working to help refugees throughout the world. This site includes information about refugees and asylees for almost every country of the world. Because the estimation of refugees is as much an art as it is a science, you should also visit the site of the United Nations High Commissioner for Refugees (UNHCR) at http://www.unhcr.ch/cgi-bin/texis/vtx/home for comparisons.

PART THREE
Population Structure and Characteristics

CHAPTER 8
The Age Transition

Figure 8.1 Percentage of the Population that Is Under Age 15

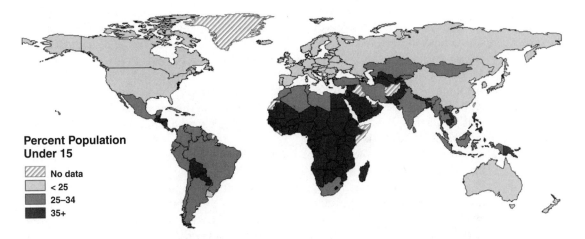

Percent Population Under 15
- No data
- < 25
- 25–34
- 35+

Source: Adapted by the author from data in United Nations Development Program 2003. Human Development Report 2003: http://www.uudp.org/hdr2003/pdf/hdr03_HDI.pdf.

Discussions of population growth often make it seem as though population increases the same way your bathtub fills up with water—evenly all over. But you know that in the previous chapters the issues of age and sex have come up over and over again. Mortality differs by age and sex; fertility differs by age and sex; migration differs by age and sex. It is time, then, to deal more explicitly with the concepts of age and sex, because population growth takes place age by age and somewhat differently for males than females. It matters not just how fast the population is growing, but which age and sex groups are growing and which ones are growing faster or slower than others. In particular, predictable changes occur in the distribution of a population by age and sex as a society goes through the demographic transition. This **age transition** represents a shift from a very young population in which there are slightly more males than females to an older population in which there are more females than males. In between, bumps and dents in the age and sex structure represent powerful forces for social, economic, and political change—and a source of demographic change, as well.

In general, it is the interaction of fertility, mortality, and migration that produces the **age and sex structure,** which can be viewed as a key to the life of a social group—a record of past history and a portent of the future. Population processes not only produce the age and sex structure but are, in turn, affected by it; another example of the complexity of the world when seen through your demographic "eye." It would not be exaggerating too much to say that changes in the age and sex structure affect virtually all social institutions. In this chapter, I escort you through that complexity by first defining the concepts of age and sex as they relate to population dynamics, and then examining how we measure changes in a population's distribution by age and sex. Then I look at the impact of each of the population processes on the age transition—the change in the age and sex distribution in the context of the demographic transition. I conclude the chapter with illustrations of how the age transition is influencing the world around you.

The Concepts of Age and Age Cohorts

Age and sex influence the working of society in important ways because society assigns social roles and frequently organizes people into groups on the basis of their age and gender (the social component of sex). Age is a biological characteristic, but it is constantly changing, whereas sex is biological in nature, but does not change (except by human intervention in rare cases). It is the changing nature of age that creates such a dynamic process because younger people are treated differently from older people, and different kinds of behavior are expected of people as they move through different ages. At the same time, biological changes inherent in the aging process influence what societies expect of people and influence how people behave. At very young and very old ages, people are more dependent on others for survival, and so the proportions of people at these ages will influence how society works. In the teen years, reproductive hormones influence behavior and every society has to deal with this phenomenon as well.

The idea that societies have separate sets of expected roles and obligations for people of different ages is captured by the concept of **age stratification,** whereas

cohort flow captures the notion that at each age we are influenced by the historical circumstances that similarly affect other people who are the same age. The perspective of age stratification and cohort flow was first put forward as a cohesive package by Riley (1976a; 1976b) and Foner (1975). The notion of age status has been around for some time; Kingsley Davis noted in 1949 that "all societies recognize age as a basis of status, but some of them emphasize it more than others" (Davis 1949: 104). Likewise, I mentioned in Chapter 6 that Ryder (1964) had already signaled the importance of cohorts in analyzing social change when he commented that "social change occurs to the extent that successive cohorts do something other than merely repeat the patterns of behavior of their predecessors" (p. 461). What the perspective of age stratification and cohorts does is to integrate these concepts in a way that helps explain social change, and thus helps us understand how and why the age transition is a predictable component of the demographic transition.

The age stratification theory begins with the proposition that age is a basis of social differentiation in a manner analogous to stratification by social class (Foner 1975). The term *stratification* implies a set of inequalities, and in this case it refers to the fact that societies distribute resources unequally by age. These resources include not only economic goods but also such crucial intangibles as social approval, acceptance, and respect. This theory is not a mere description of status, however; it introduces a dynamic element by recognizing that aging is a process of social mobility. Foner (1975) notes that "as the individual ages, he too moves within a social hierarchy. He goes from one set of age-related social roles to another and at each level receives greater or lesser rewards than before" (p. 156). Contrasted to other forms of social mobility, however, that may rely on merit, luck, or accident of birth, social mobility in the age hierarchy is "inevitable, universal and unidirectional in that the individual can never grow younger" (p. 156).

What aspects of life are influenced by age (and in some instances by sex or gender as well)? In Table 8.1 have listed just a few of the important things that vary by age in most human societies. As the number and percentage of people at each age and sex change, the distribution of these characteristics will therefore also change, and this is the force for social, economic, and political change. For example, a very young population will have a relatively small fraction of its population in the labor force unless, as happens in poorer countries, children are put to work at a young age. However, in such a society, those younger people in the labor force will have lower-status occupations. It typically takes time and experience (including education and other training) to reach the higher occupational strata. Since income is closely related to occupation, it is the older adults who tend to have the highest incomes, and it is the maintenance of higher incomes for several years that increases the chance that people will accumulate wealth. Thus, all other things being equal, we would expect that a population with a high proportion of young to young-old adults would have more people in the labor force with higher incomes and more wealth than a population with a high proportion of children.

Age strata, though identifiable, are not viewed as fixed and unchanging. The assumption is that the number of age strata, and the prestige and power associated with each, are influenced by the needs of society and by characteristics of people at each age (their numbers and sociodemographic characteristics). European society of a few hundred years ago seems to have been characterized by three age strata—infancy,

Table 8.1   Aspects of Human Society That Vary by Age and Sex (or Gender)

Category	Characteristic or Activity
Demographic	Being sick and having restricted activities of daily living
	Dying
	Being sexually active
	Having a baby
	Moving or migrating
Social	Getting married/divorced
	Being involved in religious organizations and activity
	Being involved in political organizations and activity
	School enrollment
	Level of educational attainment
	Being involved in criminal or other socially disapproved behavior
Economic	Being in the labor force
	Occupation within the labor force
	Current income
	Level of accumulated wealth

adulthood, and old age (Aries 1962); and power (highest status) seems to have been concentrated in the hands of older people (Simmons 1960). Modern Western societies appear to have at least seven strata—infancy, childhood, adolescence, young adulthood, middle age, young-old, and old-old, with power typically concentrated in the hands of the middle-aged and the young-old.

As we age from birth to death, we are allocated to **social statuses** (your relative position or standing in society) and **social roles** (the set of obligations and expectations that characterize your particular position in society) considered appropriate to our age. Thus, children and adolescents are typically allocated to appropriate educational statuses, adults to appropriate positions of power and prestige, and the older population to positions of retirement. We all learn the roles that society deems appropriate to our age, and we reward each other for fulfilling those roles and tend to cast disapproval on those who do not fulfill the societal expectation. But neither the allocation process nor the overall **socialization** process (learning the behavior appropriate to particular social roles) is static. They are in constant flux as changing cohorts alter social conditions and as social conditions, in turn, alter the characteristics of cohorts. This leads us to the concept of cohort flow.

In population studies, a cohort refers to a group of people born during the same time period. As Riley (1976b:194–195) points out:

Each cohort starts out with a given size which, save for additions from immigration, is the maximum size it can ever attain. Over the life course of the cohort, some portion of its members survive, while others move away or die until the entire cohort is destroyed. Each cohort starts out also with a given composition; it consists of members born with certain characteristics and dispositions. Over the life course of the individual, some of these characteristics are relatively stable (a person's sex, color, genetic makeup, country of birth, or—at entry into adulthood in our society—the level of educational attainment are unlikely to change). . . . When successive cohorts are compared, they resemble each other in certain respects, but differ markedly in other respects: in initial size and composition, in age-specific patterns of survival (or longevity), and in the period of history covered by their respective life span.

At any given moment, a cross section of all cohorts defines the current age strata in a society. Figure 8.2 displays a Lexis diagram, a tool often used in population studies to help us discern the difference between period data (the cross-sectional snapshot of all ages at one time) and cohorts (of which there are many at any point in time). The diagram is named for a German demographer, Wilhelm Lexis (1875), who helped develop it in the nineteenth century as an aid for analyzing life table data (Vandeschrick 2001). Age is shown on the vertical axis, and time (the moment of birth in units of years) is shown along the horizontal axis. The cohort of people born in 2001 (starting at 2001 and ending just at 2002) "advance through life along a 45° line" (Preston, Heuveline, and Guillot 2001:31), which is represented by the area (C) in Figure 8.2.

Figure 8.2 Lexis Diagram Visualizes Relationship between Cohorts and Period Data

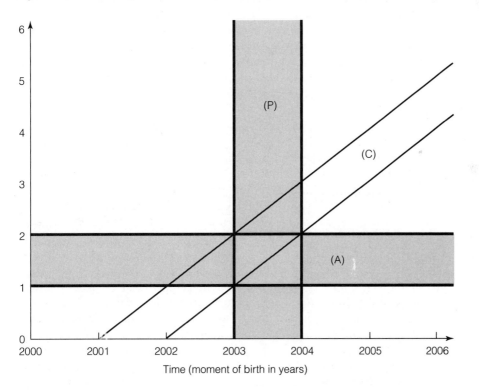

Source: Adapted from Samuel H. Preston, Patrick Heuveline, and Michel Guillot, 2001, Demography: Measuring and Modeling Population Processes (Oxford: Blackwell Publishers), Figure 2.1; and Christophe Vandeschrick, 2001, "The Lexis Diagram, a Misnomer," Demographic Research 4(3), Figure 10.

The period data, crosscutting many cohorts, is illustrated by the shaded area (*P*), whereas comparing people at the same age across many cohorts over time would be done with data shaded as (*A*) in Figure 8.2. Researchers use the Lexis diagram to calculate age-period-cohort (*APC*) rates that disentangle the combined influences of things specific to a particular age (the age effect), things unique to a time in history (the period effect), and things tunique to specific birth cohorts (the cohort flow effect) (Knorr-Held and Rainer 2001; O'Brien 2000; Robinson and Jackson 2001). Lung cancer, for example, is more likely to kill people at older ages (the age effect), but death rates from it will depend partly on when a person was diagnosed (the period effect influenced by the timing of new treatments for the disease), and partly on cohort effects (cohorts born from the 1920s through the 1940s were more heavily into cigarette smoking than earlier or later cohorts).

As cohorts flow through time, their respective sizes and characteristics may alter the allocation of status and thus their socialization into various age-related roles because members of each cohort are moving through history together, whereas each separate cohort moves through moments in history at a different age (and thus with a different potential effect) than every other cohort. Figure 8.3 tries to illustrate this idea graphically. The cohort born in 1900 had pretty much died out by 2000, but the lifetime experience of its members was very different from the cohort of 1920, which in turn was very different from the cohort of 1940. As cohorts move through time, their characteristics may change in response to changing social and economic conditions (such as wars, famines, and economic prosperity), and those changing conditions will influence the formation of new cohorts. This continual feedback between the dynamics of successive cohorts and the dynamics of other changes in society produces a constant shifting in the status and meaning attached to each age stratum, providing an evolutionary link between the **age structure** and the social structure (Gordon and Longino 2000).

Cohorts of different sizes and backgrounds, with different facets of social history having affected them, shape each age somewhat differently. One of the more widely discussed analyses of the impact of changing cohort size on a population is the "Easterlin relative cohort size hypothesis," which I first discussed in Chapter 3. You already know that the U.S. baby boom generation has, through sheer cohort size, vastly influenced the social world of the United States, and boomers in Canada have had a similarly dramatic impact on that nation. Neoclassical economic theory would suggest that, all things being equal, a larger cohort would increase the supply of labor relative to demand and thus depress wages, leaving the baby boom generation less well off than its parents. Has this happened?

Easterlin and his colleagues (Easterlin, Schaeffer, and Macunovich 1993) examined this issue and concluded that baby boomers are not less well off (at least economically) than their parents. The reason: Things were not equal. Early on, baby boomers anticipated and responded to the threat of depressed wages implied by their larger numbers. They adjusted. They altered their life-cycle decisions dramatically by delaying marriage, postponing children, combining both spouses' income, and combining mother's work outside the home with childbearing. These are now familiar attributes of the baby boom, and they represent the underlying causes of much of the social change the baby boom has produced in American society. In the simplest formulation, these changes can be seen as the consequences of the way in

Figure 8.3   Age Cohorts and Age Strata Are Closely Intertwined

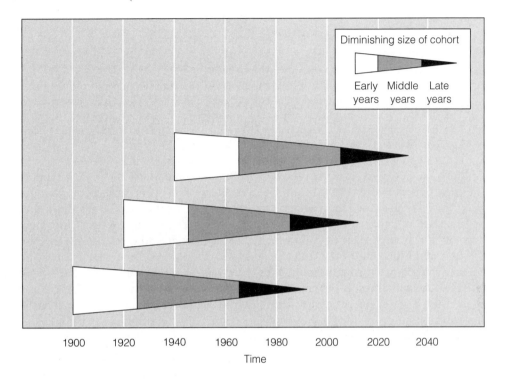

Note: Cohorts flow through time, gradually diminishing in size and impinging on society as they move through time. At any given date the number and characteristics of people in each life stage will influence the system of age strata.

Source: Adapted from M. Riley, M. Johnson, and A. Foner, Aging and Society: A Sociology of Age Stratification, Vol. III (New York: Russell Sage Foundation, 1972), p. 10. Used by permission.

which baby boomers have reacted to the age distribution pressures that have confronted them all of their lives.

The downside of the baby boom adjustments may be that the penalties attached to a large cohort have just been postponed to an older age. Easterlin and associates (Easterlin, Schaeffer, and Macunovich 1993) also note that baby boomers have not provided so well for their retirement years. They have neither the financial nor the familial resources for retirement that their parents had. Another example of the influence of the age structure on social phenomena is the change in the American suicide rate. This is another variation on the Easterlin theme that changing cohort size alters the fortunes of each cohort. Besides leading to delayed marriage, delayed parenthood, and slower social and economic mobility, changing cohort size may have contributed to a rise in the suicide rate. The Easterlin approach suggests that:

> When relative cohort size (the ratio of younger to older males) increases, it is expected to lead to an increase in suicide rates for young males (as their labor market position relative to their income aspirations deteriorates) and a decrease in suicide rates for older

males (as their labor market position relative to their income aspiration improves) (Ahlburg and Schapiro 1984:99).

The economic effect so crucial to the Easterlin perspective may be augmented by a psychosocial effect:

> The large baby boom cohort, because of imperfect socialization arising from the rising child/adult ratio, was "saddled with expectations and beliefs which were less realistic, less suited to particular individual lives, less flexible in the face of adversity" (Carlson 1980). Such imperfect socialization may be expected to result in an increase in the suicide rate (Ahlburg and Schapiro 1984:100).

Ahlburg and Schapiro show that the data from the United States do, in fact, support the idea that between 1948 and 1976 the suicide rate went up among younger people as (1) the size of the younger cohorts increased compared to the older, and (2) the ratio of children to adults increased. Thus the age structure itself may help account for at least part of the dramatic rise in teenage suicides during the late 1960s and 1970s, and portends a rise in suicide rates for older males as the baby boom cohort flows through time. So far there has been no evidence of that, so we will have to wait and see. In general, suicide rates have been going down at all ages in the United States since the early 1990s, which suggests a period effect (such as the implementation of suicide hotlines and support groups for emotionally troubled individuals) that may overcome the cohort effect. Nonetheless, there is a persistent difference in suicide rates (and many other things) between males and females.

The Concepts of Sex and Gender

I use the term *sex* when referring to the biological differences between males and females, reserving the term *gender* for the social aspects of behavior. There is enough overlap between biology and the social world, however, that this distinction is often pretty fuzzy. Sorting out which is which is still important, though, because we may not be able to do very much about biological differences, but we can do something about the fact that women are treated differently from men in most societies and different kinds of behavior are often expected from each. Women have been what Simone de Beauvoir (1953) called "The Second Sex" in a book that helped to spark feminism in the twentieth century:

> One is not born, but rather becomes a woman. No biological, psychological, or economic fate determines the figure that the human female presents in society; it is civilization as a whole that determines this creature (quoted by Clarke 2000:v)

The last half century since de Beauvoir's comments has seen a considerable amount of research that somewhat tempers her claim that civilization is the sole cause of gendered behavior. I mentioned sex dimorphism in Chapter 5 in connection with the apparent superiority of females with respect to longevity. The same concept has been applied to various aspects of human and other primate behavior in which hormones do seem to underlie certain kinds of behavior (for a review, see Udry 2000). The social environment is clearly the strongest influence on gendered behavior, but the biological differences between males and females cannot be ignored.

It is a common assumption that there are the same numbers of males and females at each age—actually, this is rarely the case. Migration, mortality, and fertility operate differently to create inequalities in the ratio of males to females (known as the **sex ratio**):

$$\text{Sex ratio} = \frac{\text{Number of males}}{\text{Number of females}} \times 100$$

A sex ratio greater than 100 thus means that there are more males than females, while a value of less than 100 indicates there are more females than males. The ratio can obviously be calculated for the entire population or for specific age groups.

Fertility has the most predictable impact on the sex ratio because in virtually every known human society more boys are born than girls. Sex ratios at birth are typically between 104 and 110. The United States tends to be on the low end of that range and Asian societies tend to be on the high end, but the data do not permit many generalizations. Despite a great deal of research into the question of why the sex ratio is not simply 100, no one really knows (Clarke 2000). This is perhaps a biological adaptation to compensate partially for higher male death rates (or vice versa, since we also are not sure why death rates are higher for males, as I mentioned in Chapter 5). In fact, data on miscarriages and fetal deaths suggest that more males are conceived than females, and that death rates are higher for males from the very moment of conception. Thus, some of the variability in the sex ratio at birth could be due to differences in fetal mortality.

Ever since it became possible to identify the sex of a child *in utero,* there has been concern that deeply rooted preferences for sons would lead to an increase in the sex ratio at birth as parents used sex-selective abortion to accomplish their reproductive goals. There is some evidence that this might be happening in Asian countries, but it is not clear that this is a widespread phenomenon throughout the world. More than 30 years ago, Westoff and Rindfuss (1974) argued that if these methods ever enjoyed widespread acceptance, there would be a short-run rise in the sex ratio at birth, since a preference for sons as first children (and for more total sons than daughters) is fairly common throughout the world, as discussed in Chapter 6. This is what appears to have happened, for example, in Korea (Clarke 2000). However, Westoff and Rindfuss also concluded that after an initial transition period, the sex ratio at birth would probably revert to the natural level of about 105 males per 100 females, because the disadvantage of too many or too few of either sex would be controlled by a shift to the other sex.

Mortality creates sex inequalities, because at every age males have higher death rates than females, as I discussed in Chapter 5. The cumulative effect of this is to produce a situation in which, as mortality declines, there are increasingly fewer males at older ages than females. Western nations, having experienced the earliest declines in mortality, have thus become increasingly characterized by having substantially more older females than males. This seems to be an inevitable shift inherent in the demographic transition. On the other hand, migration offers a way to alter the biologically driven component of the sex ratio. In some instances, females are more likely to migrate (and thus to be added to or subtracted from a specific age group). In other situations, males are more likely to be the migrants and thus also to produce changes in the sex ratio that trump those created by the combination of the sex ratio at birth and the mortality differential by sex.

Overall, there is a fairly predictable pattern in the sex ratio by age as a population moves through the demographic transition. These patterns are observable in Figure 8.4, where I have plotted the sex ratios at each age group in 2004 for three countries representing different cultural patterns with respect to the sex ratio: Ghana, the United States, and China). All three countries show the general pattern that the sex ratio declines with increasing age, but in Ghana the sex ratio is lower at birth and the younger ages and then declines a little more slowly than the other countries. China has the highest sex ratio at birth among these three countries, but

Figure 8.4   Sex Ratios by Age in Selected Countries, 2004

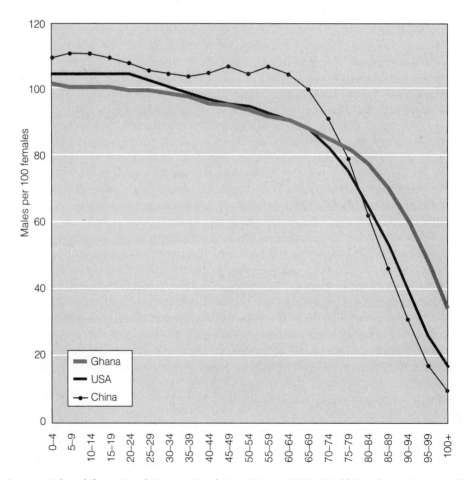

Source: Adapted from United Nations Population Division, 2003. World Population Prospect: The 2002 Revision (New York: United Nations): Tables DB4F2 and DB4F3.

then after about age 60 it drops much more quickly than in either Ghana or the U.S. The U.S. is in the middle in terms of sex ratio at birth, and in terms of the steepness with which the sex ratio drops in the older ages.

Now that you have a feel for the essential elements of age and sex as demographic concepts, let us examine how we visualize and measure age structures and the dynamics of the age transition.

Measuring the Dynamics of the Age Transition

A population is considered old or young depending on the proportion of people at different ages. In general, a population with about 35 percent or more of its people under age 15 is "young," and a population with about 12 percent or more of its people aged 65 or older can be considered "old." Further, as the proportion of young people increases relative to the total, we speak of the population as growing younger. Conversely, an aging population is one in which the proportion of older people is increasing relative to the total. We can graphically visualize the age structure by constructing an **age pyramid.** We quantify the age structure in three major ways, including calculating the **average age of a population,** the **dependency ratio,** and **growth rates by age.** Although we can compare these measures over two or more time periods to get a sense of trends, demographers have worked out more detailed ways of modeling the dynamics of age structure changes, including **stable** and **stationary populations,** and by computing **population projections.**

Population Pyramids

A population pyramid (or age pyramid) is a graphic representation of the distribution of a population by age and sex. It can graph either the total number of people at each age, or the percentage of people at each age. It is called a pyramid because the "classic" picture is of a high-fertility, high-mortality society (which characterized most of the world until only several decades ago) with a broad base built of numerous births, rapidly tapering to the top (the older ages) because of high death rates in combination with the high birth rate. Nigeria's age and sex structure as of 2004 still reflected the classic look of the population pyramids, as you can see in Figure 8.5. Until very recently, Mexico also looked like that, but the decline in fertility in Mexico since the 1970s has begun to narrow the base of the pyramid rather noticeably. Developed countries such as the United States and Canada have age and sex distributions that are more rectangular (in the case of the United States, where fertility is still at replacement level) or barrel-shaped (in the case of Canada, where fertility is below replacement level. No matter the shape, we still call the graph a population pyramid.

Average Age and Dependency Ratio

Although a picture may be worth a thousand words, there are times when we like to summarize an age structure in only a few short words or, even better, in a few

Figure 8.5 Illustrative Population Pyramids

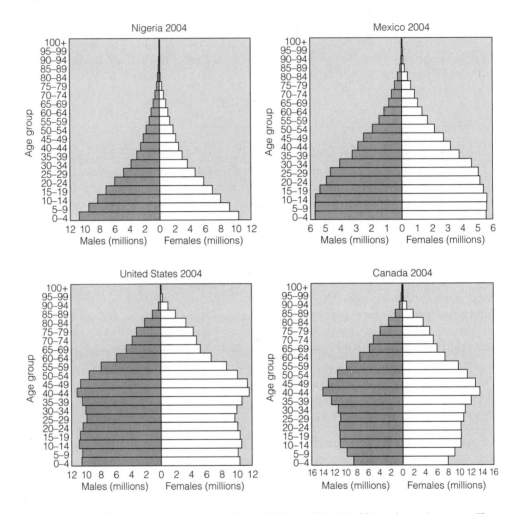

Source: Adapted from United Nations Population Division, 2003. World Population Prospects: The 2002 Revision (New York: United Nations): Tables DB4F2 and DB4F3.

numbers. The average age and dependency ratio are two measures to help us do just that. The average age in a population is generally calculated as the median, which measures the age above which is found half of the population and below which is the other half. In Figure 8.5, the population pyramids of Nigeria, Mexico, the United States, and Canada reflect median ages of 17.6, 24.3, 35.7, and 38.5, respectively. Thus, the obvious differences in the shapes of the age distributions for less-developed and more-developed countries are reflected in the clear differences in median ages.

An index commonly used to measure the social and economic impact of different age structures is the dependency ratio—the ratio of the dependent-age population (the young and the old) to the working-age population. The higher this ratio is, the more people each potential worker has to support; conversely, the lower it is, the fewer the people dependent on each worker. For example: Suppose that a population of 100 people had 45 members under age 15, three people 65 or older, and the rest of approximately economically active ages (15–64). This is the situation in Nigeria, one of the highest-fertility nations in the world. Forty-eight percent of the population is of dependent age (0–14 and 65+), compared with the remaining 52 percent who are of working age. Thus the dependency ratio is 48/52 or 0.92, which means there are 0.92 dependents per working-age person—nearly one dependent for each person of working age. This is a fairly heavy load, especially since in most societies you will not find everyone of working age actually working.

We can compare this dependency ratio of 0.92 with that for a population of 100 people in which 18 are under 15, 13 are 65 or over, and the rest (69) are of working age. This is the situation in Canada and it produces a dependency ratio of 31/69 or 0.45, which means that one person of working age would be supporting about one-half as many dependents as are supported in Nigeria. When you consider that it is more difficult for a parent (usually the mother) to be as economically productive when she has to tend to the needs of children, you can appreciate that the actual difference in dependency between Nigeria and Canada is even greater than these numbers suggest.

The dependency ratio does not capture all the intricacies of the age structure, but it is a useful indicator of the burden (or lack thereof) that some age structures place on a population. For individuals with large families, the impact of a youthful age structure, for example, will be immediately apparent. But even for the childless or those with only a few children, the effect may be higher taxes to pay for schools, health facilities, and subsidized housing. For those in business (whether government or private), an age structure that includes numerous dependents may mean that workers are able to save less, having to spend it on families, while government taxes must go toward subsidizing food, housing, and education rather than toward financing industry or economic infrastructure, such as roads, railways, and power and communications systems.

Growth Rates by Age

I have stressed that population growth occurs by age, not really as an overall number. Increases in the average age tell us generally that a population is aging, and changes in the dependency ratio tell us that the age structure is going through alterations, but we cannot tell for sure where those changes are occurring. This is important because the age-graded aspects of society shown in Table 8.1 differ from age to age. This tells you that we should examine the growth rates by age if we are going to fully understand what is happening in a society.

In Table 8.2, I provide an illustration of these kinds of comparisons, contrasting the changes in age structure in China and India between 1980 and 2000. During that 20-year interval, the average income in China went from $168 per year to $727 per year—a 333-percent increase; whereas in India it went from 311 to 440—a 41-percent increase (World Bank 2000). Income changes such as this are normally an indicator of increases in per-person productivity in the labor force (Lindh and Malmberg 1999), and this is enhanced by favorable changes in the age structure. In Table 8.2, you can see that between 1980 and 2000, there was an actual decline in numbers of people at all ages under 20 in China, whereas the rate of growth and the absolute increase of people was high in the ages from 30–39—prime ages for economic productivity. In India, by contrast, the rates of growth were positive at all ages and despite the fact that the overall rate of growth in India (1.93 percent per year) was higher during this 20-year period than in China (1.23 percent per year), there was no age group in India, except the very oldest, with as high a rate of growth as the adult population in China. It was true in India that the rate of growth was lower at the younger ages than at the older ages, reflecting the slow decline in fertility, but there was no "demographic windfall" (as I will discuss later in the chapter) as there was in China from the delayed effect of that country's rapid drop in fertility during the 1960s and 1970s. In both countries, you can see that the highest rates of growth are occurring at ages 80 and above. This is the result of the improved survival in increasingly larger younger cohorts (so that the ratio between a younger co-

Table 8.2   Changes in Number and Rate of Growth at Specific Ages: Comparison of India and China, 1980 to 2000.

	China		India	
Age Group	Growth rate 1980–2000	Absolute change 1980–2000	Growth rate 1980–2000	Absolute change 1980–2000
0–4	−0.15	−3,004	0.84	17,606
5–9	−0.97	−21,938	1.26	25,339
10–14	−0.50	−12,434	1.55	29,112
15–19	−0.35	−7,358	1.79	30,747
20–24	0.63	11,590	1.78	26,618
25–29	1.52	31,810	2.36	31,516
30–34	3.25	60,631	2.81	32,939
35–39	3.53	53,163	2.93	30,165
40–44	2.71	35,142	2.74	24,902
45–49	3.11	39,765	2.36	18,338
50–54	2.27	22,796	2.12	13,572
55–59	1.69	13,395	2.15	11,304
60–64	2.26	15,068	2.36	10,077
65–69	2.78	14,957	2.56	8,375
70–74	2.75	10,615	2.74	6,121
75–79	3.09	7,278	3.20	4,113
80+	4.90	7,203	4.93	3,963
Total	1.23	278,681	1.93	324,806

Source: Calculated from data in United Nations, Population Division, 2000, World Population Projections to 2150: The 1998 Revisions (New York: United Nations).

hort and its next older cohort is very high) and is the signal that increased old-age dependency will be an issue for both countries over the next few decades. China, with its improving income and declining number of children, may be in the better position to deal with that.

Stable and Stationary Populations

The long-run dynamics of an age structure are conveniently laid out by formal demographic models called stable and stationary populations. A stable population is one in which neither the age-specific birth rates nor the age-specific death rates have changed for a long time. Thus, a stable population is stable in the sense that the percentages of people at each age and sex do not change over time. However, a stable population could be growing at a constant rate (that is, the birth rate is higher than the death rate), it could be declining at a constant rate (the birth rate is lower than the death rate), or it could be unchanging (the birth rate equals the death rate). This is the case of **zero population growth (ZPG),** and if this prevails, we call it a stationary population. Thus, a stationary population is a special case of a stable population—all stationary populations are stable, but not all stable populations are stationary. The life table (discussed in Chapter 5) is one type of stationary population model.

For analytical purposes, a stable population is usually assumed to be closed to migration. Since 1760, when Leonhard Euler first devised the idea of a stable population, demographers have used the concept to explore the exact influence of differing levels of mortality and fertility on the age/sex structure. Such analyses are possible using a stable population model because it smoothes out the dents and bumps in the age structure created by migration and by shifts in the death rate or the birth rate. Thus, if demographers were forced to study only real populations, we would be unable to ferret out all of the kinds of relationships we are interested in. Demographers look at real populations and then apply stable population models to the real setting to understand the underlying demographic processes that influence the structure of a population by age and sex.

I will use stable population models later in the chapter to show you how mortality, fertility, and migration each influence the age and sex distribution of a population. Stable populations are also invaluable aids in estimating demographic measures. Let us suppose, for example, that we knew what the mortality and fertility rates were for a population, but no census had been conducted so we were unsure about the age and sex structure of the population. As I mentioned in Chapter 4, it is not uncommon for less-developed nations to have such incomplete information. We do not have to give up or sit on our thumbs waiting for the United Nations to finance a census. We can use a **model stable population**—the age and sex structure implied by an unchanging set of mortality and fertility rates (Coale and Demeny 1983). A country that experiences little migration and has had relatively little change in mortality and fertility will have an age and sex structure similar to that of a stable population. Thus, if we know (or can reasonably estimate) two important facts, such as age-specific death rates (so that we can calculate life expectancy) and the rate of population increase, we can determine what the age and sex structure would almost certainly have to look like. For example, in 1991, the results were just being released from the 1987 census in the eastern African nation of Zimbabwe. With a life expectancy in 1987 estimated to be 57 years, an annual rate of growth of 3.5 percent per year, and only a very modest decline in fertility up to that point, it was possible to use the stable population models to infer that the median age in Zimbabwe should have been 16.3, with 47 percent of the population under 15, and only 3 percent aged 65 or older. Subsequently, census data showed that these inferences were quite close to reality. The 1987 census revealed that the actual median age was 16.7, and that 45 percent of the population was under age 15, while 3 percent was aged 65 or older (United Nations 1991).

Population Projections

A population projection is the calculation of the number of persons we can expect to be alive at a future date given the number now alive and given reasonable assumptions about age-specific mortality and fertility rates (Keyfitz 1968). Population projections are among the more useful tools available for demographic analysis. By enabling us to see what the future size and composition of the population might be under varying assumptions about the trends in mortality and fertility, we can intelligently evaluate what the likely course of events will be many years from now. Also, by projecting the population forward through time from some point in history, we are able to determine the sources of change in the population over time. A word of caution, however, is in order. Population projections are always based on a conditional future—this is what will happen if a certain set of conditions are met.

Demographic theory is not now, nor is it likely ever to be, sophisticated enough to be able to predict future shifts in demographic processes, especially fertility and migration, over which we as individuals exercise considerable control. Thus we must distinguish projections from forecasts. A **population forecast** is a statement about what you *expect* the future population to be. This is different from a projection, which is a statement about what the future population *could be under a given set of assumptions*. As Keyfitz has observed, "Forecasts of weather and earthquakes, where the next few hours are the subject of interest, and of unemployment, where the next year or two is what counts, are difficult enough. Population forecasts, where one peers a generation or two ahead, are even more difficult" (Keyfitz 1982:746). Population projections are rarely right on the money, but in comparing past projections with subsequent censuses, the Panel on Population Projections of the National Research Council concluded that our ability to correctly guess the future is better over the short term than the long term, better for larger than for smaller countries, and better for more-developed than for less-developed nations (Bongaarts and Bulatao 2000). The U.S. Census Bureau has concluded that its projections are limited mainly by demographers' (and almost everybody else's) inability to predict turning points—those events, such as the baby boom, or the massive increase in immigration—that are nearly impossible to foresee but which have long-term impacts on population growth and change (Mulder 2001).

There are several ways in which a demographer might project the population. These include: (1) extrapolation methods, (2) the components-of-growth method, and (3) the cohort component method. In addition, some of these methods can be used to project backward, not just into the future. As you will see, given the importance of the age structure, the best method is the cohort component approach, which follows age cohorts through time, but it is important to have a quick overview of all the usual methods of projection in order to see why we usually prefer to use a method that incorporates information about the age distribution.

Extrapolation   The easiest way to project a population is to extrapolate past trends into the future. This can be done using either a linear (straight-line) extrapolation, or a logarithmic (curved-line) method. Both methods assume that we have total population counts or estimates at two different dates. If we know the rate of growth between two past dates, and if we assume that rate will continue into the future, then we can project what the population size will be at a future date. For example, in 1990, 248,709,873 people were enumerated in the U.S. Census, and in 2000 the census counted 281,421,906. The average annual linear rate of growth between those two dates was 0.013153 or approximately 13.2 per thousand per year, which we can calculate using the following formula:

$$r_{lin} = \left(\frac{\text{population at Time 2} - \text{population at Time 1}}{\text{population at Time 1}} \right) / n$$

In this example, the population at Time 1 is the census count in 1990; the population at Time 2 is the census count in 2000, and n is the number of years (10) between the two censuses. You can plug in the above numbers to see that the average annual linear rate of growth turns out to be 0.013153. Now, we use that rate of growth to extrapolate the population forward, for example, from our **base year** (the beginning year of a population projection) of 2000 to a **target year** (the year to which we project a population forward in time) of 2050, using the following formula:

Population in target year = Population in base year x $[1 + (r_{lin} \, x \, n)]$

In this formula, r_{lin} is the average annual linear rate of growth just calculated (0.013153) and n is the number of years (in this case 50) between the base year (2000) and the target year (2050). So, plugging in the numbers from above, the projected population in the year 2050 is

$$[281,421,906 \, x \, (1.6576)] = 466,494,629$$

You may recall from Chapter 2 that populations typically are thought to grow exponentially, not in a straight-line fashion, and this hearkens back to the formula that we used in that chapter to derive an estimate of the doubling time of the population. In that problem, our concern was estimating the likely time it would take a population to double in size when we knew the rate of population growth. By algebraically rearranging that equation, we can produce a formula that expresses the logarithmic growth of a population, assuming a constant rate of growth. This formula is as follows:

Population at Time 2 = Population at Time 1 $\times e^{rn}$

In this case, r represents the geometric or exponential rate of increase (r_{exp}) and is calculated with the following formula:

$$r_{exp} = [ln \text{ (Population at Time 2 / Population at Time 1)}] / n$$

The term *ln* represents the natural logarithm of the ratio of the population at Time 2 to the population at Time 1. It is one of the function buttons on most handheld calculators. Once again, *n* is the time between censuses. So, to calculate the exponential average annual rate of population growth between 1990 and 2000, first find that the ratio of the population at those two dates (281,421,906/248,709,873) is 1.13153 (which is what we found when calculating the linear rate of growth), then find that the natural logarithm of that number is 0.12357, which is then divided by 10 to find that the rate of increase is 0.012357 (or 12.3 per thousand per year). Next, plug this rate of growth (0.012357) back into the formula for exponential or logarithmic growth (above) in order to project the population forward from the base year of 2000 to the target year of 2050. The answer is:

$$281,421,916 \times e^{(.012357 \times 50)} = 522,014,164$$

This is of course a higher number than we found with the linear method of extrapolation, reminding us of the power of geometric growth (the power of doubling as discussed in Chapter 2). But is this really what the future holds for the United States? Is the United States really expecting a huge increase in population between 2000 and 2050? Notice that these extrapolation methods of population projection do not take into account births, deaths, or especially migration. If we have a way of projecting such information, then we can project the population to a target year using the **components-of-growth** method.

Components of Growth   This is an adaptation of the demographic balancing equation mentioned earlier in the book. The population of the United States in the year 2050 will be equal to the population in 2000 plus all the births between 2000 and 2050, less the deaths, plus the net migration between those two dates. But how will we figure out the number of births, deaths, and migrants that we might expect over the next 50 years? We know all of these population processes differ by age and sex, and we know that as the population grows we cannot expect that the number of births and deaths will remain constant over time. What is needed is a more sophisticated approach to figuring out what the components of growth are likely to be. This means a method that takes the age structure into account, which we call the cohort component method.

Cohort Component Method   To make a population projection using the **cohort component method,** we begin with a distribution of the population by age and sex (in absolute frequencies, not percentages) for a specific base year. Usually a base year is a year for which we have the most complete and accurate data—typically a census year. Besides the age and sex distributions, you need to have base-year age-specific mortality rates (that is, a base-year life table), base-year age-specific fertility rates, and, if possible, age-specific rates of in- and out-migration. Cohorts are usually arranged in five-year groups, such as ages 0 to 4, 5 to 9, 10 to 14, and so on, which facilitates projecting a population forward in time in five-year intervals. For example, if we are making projections from a base year of 2000 to a target year of 2050, we would make intermediate projections for 2005, 2010, 2015, and so forth.

With the base-year data in hand and a target year in mind, we must next make some assumptions about the future course of each component of population growth between the base year and the target year. Will mortality continue to drop? If so, which age will be most affected and how big will the changes be? Will fertility decline, remain stable, or possibly rise at some ages while dropping at others? If there is an expected change, how big will it be? Can we expect rates of in- and out-migration to change? Note that if our population is an entire country, our concern will be with international migration only, whereas if we are projecting the population of an area such as a state, county, or city, we will have to consider both internal and international migration.

The actual process of projecting a population involves several steps and is carried out for each five-year cohort between the base and target years. First, the age-specific mortality data are applied to each five-year age group in the base-year population to estimate the number of survivors in each cohort five years into the future. Since there were 9,303,809 females aged 20–24 in the United States in 2000 and the probability of a female surviving from age 20–24 to age 25–29 (derived from the life table, as discussed in Chapter 5) is 0.9977, then in 2005 we expect there to be 9,282,410 women surviving to age 25–29, before we make any adjustment for migration. This process of "surviving" a population forward through time is carried out for all age groups in the base-year population. The probabilities of migration (assuming that such data are available) are applied in the same way as are mortality data.

Fertility estimation is complicated by the fact that only women are at risk of having children, and then all those children are added into the population at age zero. The tasks include (1) calculating the number of children likely to be born during the five-year intervals, and (2) calculating how many of those born will also die during those intervals. The number to be born is estimated by multiplying the appropriate age-specific fertility rate by the number of women in each of the childbearing ages. Then we add up the total number of children and apply to that number the probability of survival from birth to the end of the five-year interval. Experience suggests that fertility behavior often changes more rapidly (both up and down) than demographers may expect, so population projectionists hedge their bets by producing a range of estimates from high to low, with a middle or medium projection that incorporates what the demographer thinks is the most likely scenario. The highest estimate reflects the demographer's estimate of the highest fertility trend possible in the future, along with the highest decline likely in mortality, and the maximum net immigration likely. Conversely, the lowest projection incorporates the most rapid decline in fertility, the least rapid drop in mortality, and the lowest level of net immigration probable.

Differences in fertility often account for the biggest chunk of variation between high and low projections. For example, in the projection of the population of Mexico made by the U.S. Census Bureau in the late 1970s (U.S. Census Bureau 1979a), the high projection for 1990 was 102,349,000, whereas the low projection was 88,103,000. Virtually all the difference was found in the ages zero to 14. The 1990 census in Mexico, however, counted "only" 86 million people. Part of the difference was almost certainly due to an acknowledged undercount in the Mexican census, but most of the difference was a result of the success of Mexico's campaign to lower the birth rate.

Calculating a cohort component projection is somewhat complicated and other sources contain more details (Murdock and Ellis 1991; Smith 1992; Smith, Tayman, and Swanson 2001). The U.S. Census Bureau uses these methods in preparing their series of population projections for the United States. Notice that they keep revising their projections because new information comes along that alters some of the previous assumptions. The cohort component method projections that they completed in 2000 included a large set of projections, allowing you to pick the combination of fertility, mortality, and migration that you think makes most sense down the road (Hollman, Mulder, and Kallan 2000). They use single years of age cohorts, rather than five-year age groups, in order to produce as much detailed information as possible. They also do the projections separately by race/ethnicity since different groups have somewhat different patterns of fertility, mortality, and migration. For example, under the "middle" scenario, the total fertility rates for Hispanics and Asians are projected to decline, while other groups will remain reasonably unchanged, but that still leaves the U.S. with a higher total fertility rate at the middle of the twenty-first century than at the beginning.

The U.S. Census Bureau had not at the time of this writing completed the revision of its population projections building on the base year of Census 2000. However, the United Nations had done such a projection, and their middle projection assumes a drop in the total fertility rate from 2.11 in the 2000–2005 period to 1.85 by 2050, an increase in life expectancy (both sexes combined) from 77.1 to 81.6, and maintenance of the current pattern of about 1.1 million net immigrants each year. These assumptions, calculated in detail by age and sex group, produce a total projected population in the year 2050 of 408,695,000. As high as that is, it is still lower than the projections based on simple extrapolations of past trends. As you look at these alternative projections, you will do well to keep in mind the old Chinese proverb: "Prediction is very difficult—especially with regard to the future."

We are particularly interested in the age-structure changes implied by these projections because it is the age structure that will allow us to read the tea leaves for the future. We'll return to that later in the chapter.

Backward or Inverse Projection   Population data are projected into the future in order to estimate what could happen down the road. Similar methods can be used to work backward to try to understand what happened in the past. The basic idea is to begin with census data that provide a reasonably accurate age and sex distribution for a given base year. Then, making various assumptions about the historical trends in fertility, mortality, and migration, you work back through time to "project" what earlier populations must have been like in terms of the number of people by age and sex. Wrigley and Schofield (1981) used this method to work backward from the 1871 census of England and Wales to reconstruct that region's population history. Whitmore (1992) used a complex backward projection model to show how it was possible for European contact to have led to a 90-percent depopulation of the indigenous peoples living in the Basin of Mexico at the time that Cortés arrived. Details of these methods are contained in other sources (Oeppen 1993; Smith 1992).

Cohort component methods of population projection give clues about the future (and even about the past) by letting us view alternative age and sex distributions. From this demographic information, we draw inferences about the impact on society by calling on our knowledge of how prior changes in the age structure have altered the fabric of social life. These kinds of interrelationships are what make the age transition such a critical part of the demographic transition.

The Age Transition

From the previous chapters, you know that a premodern society will have high fertility, high mortality, and probably very little in- or out-migration. This is associated with a young population, whereas the "end" of the transition is associated with an older population characterized by low fertility, low mortality, and probably a moderate amount of migration. This transition from a younger to an older population progresses through some predictable stages, with reasonably predictable consequences for the fabric of social life. In Figure 8.6, I have mapped the countries of the world in 2000 according to whether we would consider them young (defined as 35 percent or more of the population being under 15), old (defined as 12 percent or more of the population being 65 or older), or neither old nor young. Not surprisingly, this is the

Figure 8.6   Global Distribution of Younger and Older Populations

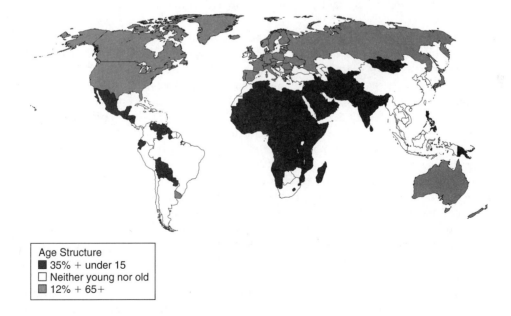

Age Structure
- ■ 35% + under 15
- □ Neither young nor old
- ▨ 12% + 65+

Source: Adapted by the author from data in the World Population Data Sheet 2000 of the Population Reference Bureau.

same picture you have seen before of where fertility is highest (the younger populations), lower (the older populations), and somewhere in between (the countries that are neither young nor old). Figure 8.6 includes information for 205 counties, of which 102 (50 percent) were young, 42 (20 percent) were old, and 61 (30 percent) were neither young nor old.

The Dynamics of the Age Transition

The pattern shown in Figure 8.6 does not happen by chance. Each of the three population processes has predictable impacts on the age structure. I have alluded to them before, but let me be more systematic here in the discussion, looking at the impact of each population process in the order I introduced them to you in the previous chapters.

The Impact of Declining Mortality   Mortality declines affect all ages and both sexes, but in virtually all societies the very youngest and the very oldest ages are most susceptible to death, and in modern societies (where maternal mortality is fairly low), males are more likely than females to die at any given age. As I discussed in Chapter 5, however, the early stages of the epidemiological transition are characterized by bringing communicable diseases under control. This tends to affect infants and children more than any other age. As a result, the early drop in mortality increases the proportion of children who survive and thus serves to increase the fraction of the population at the younger ages. Furthermore, as I discussed previously in the chapter, the decline in mortality is also apt to affect male children slightly more than females, thus slightly increasing the sex ratio at the younger ages (Mayer 1999).

Let me once again use data from stable population models to show you the impact on an age structure as life expectancy increases from 20 years for females (the premodern era) to 50 years for females (near the lower levels in the world today), if we assume that fertility does not change (remaining at the TFR of 6.2 required for replacement with a life expectancy of 20 years) and that there is no migration. These data are shown in Figure 8.7, which has pyramids drawn as line graphs rather than bar graphs to better visualize the overlap of the age patterns.

As mortality declines without any change in fertility, you can see in Figure 8.7-A that the age structure actually becomes more pyramidal than it was before, as a result of the greater survivability of children, which means that the younger population is growing faster than the population at the older ages. In fact, as mortality declines, the average age of females in this population drops from 25.5 to 21.8, and the dependency ratio increases from 0.71 to 0.99. The percentage of the population under age 15 increases from 36 to 45, whereas the percentage aged 65 and older increases from just under 3 percent to just over 3 percent. Because of the decline in mortality, this population is growing fast in overall terms. At a female life expectancy of 20, the crude birth rate and crude death rate were both 50 (remember that since they were equal, this was a stationary population), whereas the increase in life expectancy drops the death rate to 16, while the birth rate drops slightly to 46 (a function of the age structure effect on the crude birth rate, as I mentioned in

Figure 8.7   Impact of Decline in Mortality on Age Structure Early in the Epidemiological Transition

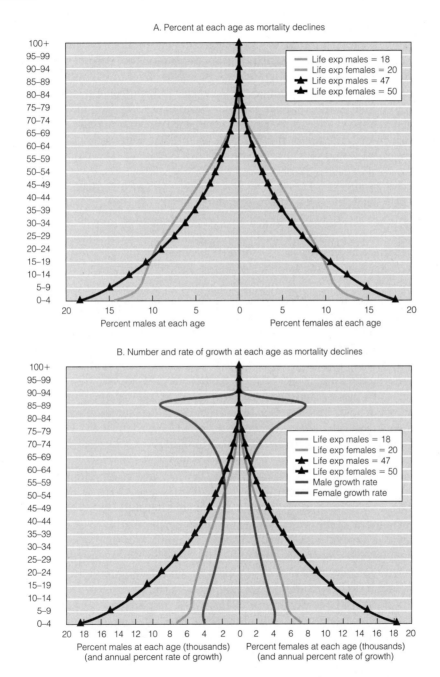

A. Percent at each age as mortality declines

Life exp males = 18
Life exp females = 20
Life exp males = 47
Life exp females = 50

Percent males at each age   Percent females at each age

B. Number and rate of growth at each age as mortality declines

Life exp males = 18
Life exp females = 20
Life exp males = 47
Life exp females = 50
Male growth rate
Female growth rate

Percent males at each age (thousands)
(and annual percent rate of growth)   Percent females at each age (thousands)
(and annual percent rate of growth)

Source: Adapted from Ansley Coale and Paul Demeny, 1983, Regional Model Life Tables and Stable Populations, 2nd edition, (New York: Academic Press). Data from "West" model stable populations at mortality levels 1 and 13.

Chapter 5). Thus, the control of mortality, unmatched by a drop in fertility, causes the rate of population growth to jump from 0 to 3 percent per year (which by the "rule of 69" implies a population doubling every 23 years).

The age pyramids in Figure 8.7-A do not really capture the impact of the decline in mortality, because careful study of that picture shows that the change in the percentage of the population at each age is not very dramatic, despite the tremendous difference that it makes to have a life expectancy of 50, compared to 20. This is because when mortality levels change, all ages tend to be affected in some way or another, even though some, especially the very young, are affected more than others. It is not so much the change in the percentage at each age that makes the drop in mortality important; rather it is the differential numbers and rates of growth by age caused by the absolute increase in population. This is what causes the population "thunderstorm" I referred to in Chapter 2.

Panel B of Figure 8.7 shows the age structure in terms of the absolute number of people at each age when life expectancy is 20, compared to what it is when life expectancy is 50, without any change in fertility and migration. I have assumed that the population began with 100,000 people and then doubled to 200,000, as implied by the rate of growth, as mortality declines. In this view, it is more obvious that there is a huge increase at the younger ages. This is what families and communities have to deal with as mortality declines—a huge increase in the number of young people. It looks like a baby boom, but in reality it's a survival boom. The broader gray lines show you the average annual rate of increase, based on the overall growth rate of 3 percent per year. Even though the population as a whole may be increasing at a rate of 3 percent, there is considerable variability, as you can see in Figure 8.7–B. The rate is highest at the youngest and oldest ages and lowest in the adult ages. It is no coincidence, of course, that the pattern of growth rates by age looks exactly like the pattern of age-specific death rates. In this example, the drop in death rates is the only thing influencing population change, so you can see graphically what that means from one age to the next. The high growth rates at the very oldest ages are, of course, building on a very small base of older people.

Over the long run, changes in mortality by themselves do not appreciably affect the percent of the population at each age in a society—but it obviously has a huge impact on the number of people at each age. In populations with medium to high mortality, a decline in mortality makes the population slightly younger—a seemingly paradoxical result, since it seems as though lower mortality should age the population by allowing people to live longer. This more expected pattern does occur when mortality is lower, but in higher-mortality populations, the disproportionately larger decline in infant and childhood mortality as overall levels of mortality decline tends to have the opposite effect. This impact of a mortality decline on the age structure is very noticeable in the short run if it is not accompanied by a decline in fertility.

In the short run, then, a decrease in mortality levels in all but low-mortality societies can substantially increase the number of young people, and one of the best studies of this effect is by Arriaga (1970) in his analysis of Latin American countries. Arriaga examined data from 11 countries for which information was available on the mortality decline from 1930 to the 1960s. He discovered that "of the 27 million people alive in all the eleven countries in the 1960s who would not have been alive if there had not been a mortality decline since the 1930s, 16 million—59 percent—were under 15" (1970:103). In relative terms, a lowering of mortality in Latin America noticeably raised the proportion of people at young ages, slightly elevated the proportion at old ages, and lowered the proportion at the middle ages (14–64). However, in absolute terms, the number of people at all ages increased. Declining mortality had an impact similar to that of a rise in fertility, while also making a contribution to higher fertility.

The appearance of higher fertility is, of course, produced by the greater proportion of children surviving through each age of childhood. It is as though women were bearing more children, thereby broadening the base of the age structure. The actual contribution to higher fertility is generated by the higher probabilities of women (and their spouses) surviving through the reproductive ages, since under conditions of high mortality, a certain percentage of women will die before giving birth to as many children as they might have. When death rates go down, a higher percentage of women live to give birth to more children, assuming that social changes are not producing motivations for limiting fertility.

Note that the only time a change in mortality generates a change in the percentage of the population at each age is when the mortality shifts are different at different ages. If there is a change in the probability of survival from one age to the next that is exactly equal for all ages and for both sexes, then the age and sex structure will remain unchanged. On the other hand, in a low-mortality society such as the United States, where nearly three-fourths of all deaths occur at ages 65 or older, a drop in mortality will age the population largely because the death rates are now so low at the younger ages that it is very hard to improve on them. So, at the later stages of the epidemiological transition, continued declines in mortality will age the population, assuming there is no change in fertility or migration. Figure 8.8 illustrates this point. Let us assume that a country had reached replacement-level fertility at a female life expectancy of 65 (just slightly lower than the world average in 2003), and that fertility does not change as life expectancy rises to 80 for females (very close to the United States level in 2003). You can see that the proportion of the population that is older increases, while the population that is younger decreases. In other words, the rate of growth is positive at ages over 55, and negative at every younger age. At a life expectancy of 65 with replacement-level fertility, the average age is 38; the average age rises to 41 as life expectancy improves to 80. The percentage of the population aged 65 and older increases from 15 percent to 24 percent, while the percentage under 15 drops from 21 percent to 19 percent. Once again, however, you can appreciate from Figure 8.8 that a drop in life expectancy does not have a spectacular impact on the proportion at each age. Fertility has a much more dramatic impact.

The Impact of Declining Fertility   Both migration and mortality can affect all ages and differentially affect each sex. The impact of fertility is not quite the same, and it is for this reason that changes in fertility tend to have the most dramatic long-term impact on the age structure (migration has the biggest short-term impact, as I discuss

Figure 8.8   Impact of Decline in Mortality on Age Structure Later in Epidemiological Transition

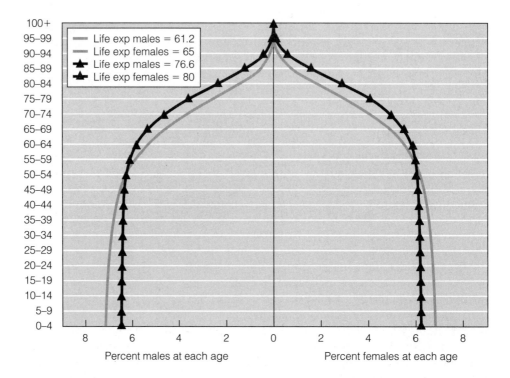

Source: Adapted from Ansley Coale and Paul Demeny, 1983, Regional Model Life Tables and Stable Populations, 2nd edition, (New York: Academic Press). Data from "West" model stable populations at mortality levels 21 and 25.

below). Fertility obviously adds people only at age zero to begin with, but that effect stays with the population age after age. Thus if the birth rate were to drop suddenly in one year, then as those people get older, there will still always be fewer of them than there are people of surrounding ages. For example, in Japan in 1966 the birth rate made a sudden one-year dip—this was the Year of "Hinoeuma"—the Fiery Horse. According to a widely held Japanese superstition, girls born in the Year of the Fiery Horse (which occurs every 60 years) will have troublesome characters, such as a propensity to murder their husbands. Thus, girls born in that year are hard to marry off, and many couples avoided having children in 1966, creating a permanent dent in the Japanese age structure. If fertility goes up, then there will be more people in each younger age group. Both of these situations—rising and declining fertility—have had strong influences on the age/sex structure of the United States, as I will discuss in a moment.

In Figure 8.9, I have used stable population models to show how different fertility levels can affect the shape of the age structure if everything else is held constant. Panel A of Figure 8.9 assumes (as do most stable population models) that no migration is occurring. Then it assumes that mortality is constant, with a female life

Figure 8.9   Using Stable Population Models Shows that Different Levels of Fertility Have Dramatically Different Effects on Age Structure

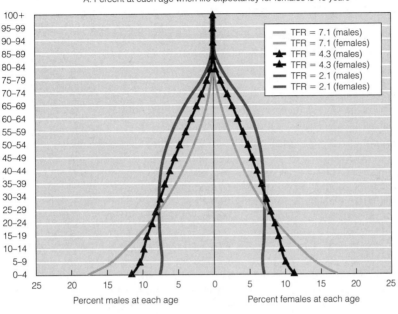

A. Percent at each age when life expectancy for females is 40 years

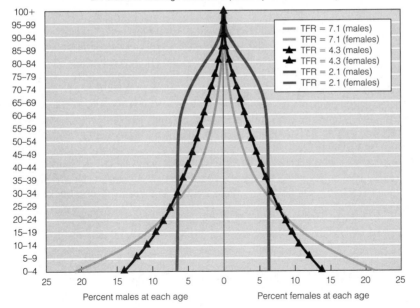

B. Percent at each age when life expectancy for females is 80 years

Source: Adapted from Ansley Coale and Paul Demeny, 1983, Regional Model Life Tables and Stable Populations, 2nd edition, (New York: Academic Press). Data from "West" model stable populations at mortality levels 21 and 25.

BABY BOOM, OIL BOOM, AND SOCIAL CHANGE IN SAUDI ARABIA

At the end of World War II, Saudi Arabia was home to scarcely more than three million people, many of whom were still nomadic. Vast underground reserves of petroleum and natural gas had been discovered in the 1930s after the establishment of the current kingdom under King Ibn Saud, but it was not until after World War II that they were aggressively exploited. In 1973, the creation of the OPEC oil cartel produced a boom in oil revenues that has catapulted Saudi Arabia into the ranks of the wealthier countries of the world. In the early days of the boom, the kingdom could not meet the labor demands with its own population and so foreign workers were brought in, especially from other Arab countries. For a while, it seemed to some as though Saudi Arabia might be wealthy enough to defy the demographic odds and maintain very high fertility even in the face of rapidly declining mortality. But by the year 2004, the total fertility rate of 5.7 was matched with a life expectancy of 72 years. The population of 25 million represented eight times more people than 50 years earlier and the very youthful age structure was putting severe strains on the traditional structure of Saudi society.

In the accompanying graph you can see the increase in the number of people at each age in Saudi Arabia between 1950 and 2004. Notice that the age structure was essentially the same in 1975 as it had been in 1950, except that there were more people. During that 25-year period, the total fertility rate remained at 7.2 children per woman, but life expectancy for women increased from 41 to 56. The percentage of the population under age 15 increased slightly from 42 percent to 44 percent, while the percentage of people 65 and older dropped slightly from just over 3 percent to just under 3 percent.

The age and sex structure for 2004 was particularly intriguing, however, and in "reading" it, we can see the signs of future social change. In the first place, you can see the huge surge in numbers of both males and females between 1975 and 2004. This seems truly like a population explosion. Because of continuing high fertility, the shape of the age structure was essentially the same for females

as in the earlier years, but you can see that there is a bulge of males in the adult ages—nearly two million more men than women in a population of 25 million, and all of the "excess" men are in the adult ages. Many of these men are foreigners hired by Saudi families to drive the women wherever they need to go. In fact, the United Nations estimates that more than one in four persons living in the country is a migrant (United Nations Population Division 2002a).

Saudi Arabia is the home to Islam and virtually all Saudis are Sunni Muslims (the major group within Islam), but since the mid-eighteenth century the country has adhered to a very traditional reformist version of Islam based on the teachings of Muhammad Abd Al-Wahhab. Among the traditions that have evolved over time is that women are not permitted to drive. There is no specific law that prohibits it, but the custom is designed to protect the honor and integrity of women, who might otherwise be put in a compromising position, and it is so strong a prohibition that few would dare disobey it. Instead, women must be driven either by a family member or by a person employed by the family for that purpose. Thus, the million or so foreign males hired to drive (Liesman 1999).

The problem for families, and for Saudi society in general, is that the population has been growing at about 3.5 percent per year, but national revenue from petroleum and natural gas production has been increasing by only 1 percent per year (Pope 2001). Per-person income, in fact, declined in the 1990s (World Bank 2000). Unemployment was estimated to be 30 percent in 2003 (http://www.arabicnews.com 2003), and the economy has been unable to generate enough jobs, even for university-educated Saudis. At the same time, the Saudi government's efforts to improve women's education levels means there is also an increasing number of professional women looking for jobs (Yamani 2000) and clamoring quietly for the chance to drive.

It is not clear yet exactly what shape the changes will take, but Saudi Arabia is a very different country now from what it was in 1950 and it will have to change dramatically in order to deal with a population that is projected to reach 55 million by

2050—more than twice the current population, and almost 20 times the population at the end of World War II. That projection, made by the United Nations Population Division (2003c), even assumes a drop to below-replacement-level fertility by 2050. Thus, you can see that there is a huge momentum built into the very youthful age structure of Saudi Arabia. A society in which the average couple is currently having almost six children, nearly all of whom will live to old age, is a society that is going to face agonizing pressures. The United Nations' assumption of declining fertility masks the kinds of changes this will involve. As women limit their fertility, they will be doing so in an economic context that will increasingly value their economic productivity, and this will almost certainly create pressures for full societal rights for women. At the same time, the youthful age structure also suggests that we can expect a Saudi diaspora as young people leave the Persian Gulf region in search of employment elsewhere—an ironic reversal of the situation only a few decades ago. As the events of 9/11 tell us, we can also anticipate that a rapidly growing population of unemployed young men will provide recruits for religious fundamentalist groups seeking change in Saudi Arabia and elsewhere in the world.

Saudi Arabia's Age Structure Is Influenced by High Fertility, Declining Mortality, and Immigration of Males

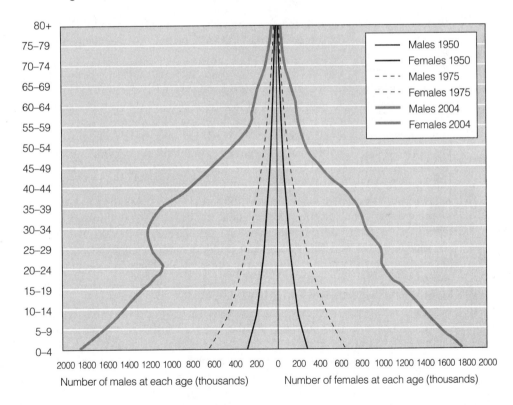

Source: Adapted by the author from data in United Nations, 2003, World Population Prospects: The 2002 Revision (New York: United Nations).

expectancy of 40 years, representing a typical population in the world until about 100 years ago, on the verge of the demographic transition. The high fertility level is equivalent to a total fertility rate of 7.1 (similar to Somalia at the start of the twenty-first century). All other things being equal, a high-fertility, high-mortality society will have a very youthful age distribution. Indeed, the average age in the population is 22.0 years, with 44.5 percent of the population under age 15, and 2.5 percent 65 or older; the dependency ratio is 0.95. The middle fertility level is equivalent to a total fertility rate of 4.3 children per woman (similar to Honduras at the beginning of the twenty-first century). This level of fertility still produces a youthful age structure, although less so than was associated with the higher fertility level. This population has an average age of 29.5 years and with 30.7 percent of the population under age 15 and 6.3 percent 65 or older; the dependency ratio is 0.68. At the low fertility level, we are talking about a total fertility rate of 2.1, which is exact replacement when mortality is low, but at this high mortality level it means that women are not reproducing themselves and the population will soon implode. This population has an average age of 36.2 years, and only 20.9 percent of the population is under age 15, while 11.5 percent is 65 or older; the dependency ratio is 0.61. You can see then that at high fertility levels, the age pyramid is a pyramid; at middle levels, it is still a pyramid, but not so dramatically; and at low fertility levels, the age structure takes on a barrel shape.

Now examine Figure 8.9–B, in which the age pyramids represent the same three levels of fertility, but this time they are paired up with a high life expectancy of 80 years for females, representing a population nearing the end of the demographic transition. You can see that despite a doubling of female life expectancy, the age structures at each of the three fertility levels are very nearly identical, regardless of mortality level. The only real difference is that at the higher life expectancy, the extremes are more noticeable. At the fertility level of 7.1 children, the average age is 19.9 and 50.7 percent of the population is under age 15, while only 2.6 percent is 65 or older; the dependency ratio is a whopping 1.20. This is the kind of age structure that is bound to cause problems for a country, as I discuss with respect to Saudi Arabia in the essay that accompanies this chapter. At the lower level of 4.3 children, the average age is 27.5 with 36.4 percent under 15 and 7.2 percent 65 or older, and a dependency ratio of 0.87. Finally, replacement-level fertility produces an average age of 40.9 with 18.7 percent of the population under 15 and 20.0 percent aged 65 or older; the dependency ratio is 0.80.

In general, the impact of fertility levels is so important that with exactly the same level of mortality, just altering the level of fertility can produce age structures that run the gamut from those that might characterize primitive to highly developed populations. The data I have shown you come from stable population models rather than the real world, but they help us know what to look for. For example, let us suppose we are looking at two countries with high female life expectancies of 76 years (such as the United Arab Emirates and Latvia). However, one country (United Arab Emirates) still had high fertility—a total fertility rate (TFR) of 4.9 in 2000, whereas the other had very low fertility (a TFR of only 1.2 in 2000). The respective age distributions of these two nations are already very different, because the United Arab Emirates (UAR) has had high fertility for a long time, whereas Latvia, part of the former Soviet Union, has had low fertility for some time now. In the UAR, 33 percent of the population was under age 15 in 2000, compared with only 19 percent of the Latvian population. By contrast, in the UAR, only 2 percent of the population is 65 or older, compared to 14 percent in Latvia. What you now know, however, is that when fertility drops in the UAR to below replacement level, such as it currently is in Latvia, the age structure will automatically assume the barrel shape that currently characterizes Latvia, its average age will rise, and the young will decline as a fraction of the total, while the old increase as a fraction. This is the age transition at work in conjunction with the fertility transition component of the demographic transition.

Where Does Migration Fit In?

A population experiencing net in- or out-migration (and virtually all populations except the world as a whole experience one or the other) will almost certainly have its age and sex structure altered as a consequence. Since immigration has been especially important in the United States, it provides a good beginning for our analysis.

We can assess the potential impact of international migrants into the United States by looking at the age and sex distribution of legal immigrants into the United States in 2002 as compiled by the U.S. Office of Immigration Statistics. You can see in Figure 8.10 that men outnumber women only at the youngest ages, and then by about age 20, there are more female than male immigrants in the older ages. However, the most significant part of the picture is that both males and females exhibit the pattern of migration by age that I discussed in Chapter 7. While it is true that there are migrants at virtually every age, migrants are far more likely to be young adults than they are to be people of any other age. Although the United States has

Figure 8.10   Age Structure of Immigrants to United States, 2002

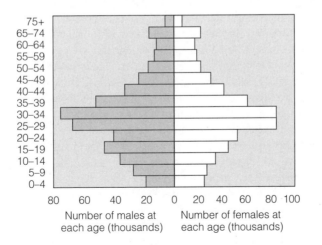

Source: Adapted from U.S. Office of Immigration Statistics, 2003. 2002 Yearbook of Immigration Statistics (Washington, DC: Department of Homeland Security), Table 6.

many more immigrants than emigrants, keep in mind that emigrants may be a bit older than immigrants, so that could have at least a small impact on the age structure (Bouvier, Poston, and Zhai 1997).

In the short run, immigration adds people, especially into the young adult ages, in the host area, and of course those people have been taken out of the age structure in the donor area. In the long run, the impact of migration is felt indirectly through its influence on reproduction, because immigrants typically are of prime reproductive ages. The number of immigrants to the United States was 1.2 million in 2002, but that still represented substantially less than 1 percent of the total U.S. population. Even in Canada, where immigrants represent a larger fraction of the population, the 250,000 immigrants per year still represent less than 1 percent of the total Canadian population. Yet, if you consider that all North Americans except the indigenous population historically are descendants of fairly recent migrants, it becomes apparent that the indirect effect of immigration has been enormous over the long run.

The long-run impact of immigration is illustrated in Figure 8.11, where I show the age and sex distributions for the United States in 2050 as projected by the U.S. Census Bureau (2000d) under two different scenarios. In one scenario, the level of net migration into the country between 1999 and 2050 is maintained at a level of nearly one million persons per year, and fertility- and mortality-level assumptions are consistent with the "middle" series, as I discussed earlier in the chapter. In the second scenario, fertility and mortality remain the same, but international migration is assumed to drop to zero. Despite the age structure of migrants you saw in Figure 8.10, you can observe in Figure 8.11 that the only real long-term difference in the zero-migration scenario compared to the continued-immigration assumption is that there are more people at every age if immigration continues than if it does not. The distribution of the population by age and sex is very similar, smoothed out over time by the fact that immigrants come into a population, some have babies, and then they themselves, along with their children, age with the rest of the population.

In the short run, however, migration affects local areas where migrants show up or depart more readily than it does national populations. This can be exemplified by looking at the differential impact of migration within a city. The age structure in a city is particularly vulnerable to migration when that area contains a social institution, such as a military base, college, or retirement community, that attracts a particular age group or sex. Examples of these situations are shown in Figure 8.12. Each of these age and sex distributions is drawn from areas in San Diego. The first example is the population in the Montezuma area near San Diego State University, where you can see the heavy bulges in the late teens and early twenties, with relatively few married couples with young children. This is not an area conducive to young families, probably because homes near the university have risen sharply in value since the demand for property has increased as the university's enrollment has grown to one of the largest in California. Furthermore, development in the area has emphasized the kind of amenities preferred by college students, rather than people with young children. Homeowners tend to be older residents who are aging in place (a phenomenon I will discuss in the next chapter), as can be seen by the bulges in the older ages. This age pattern is by no means unique, however. Most American college communities are characterized by this kind of bipolar distribution—young students

Figure 8.11   Age Pyramids for United States with and without International Migration between 2000 and 2050

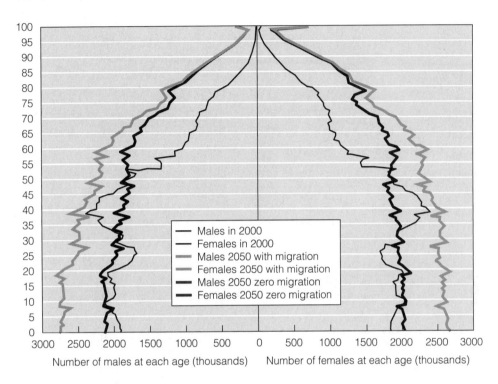

Males in 2000
Females in 2000
Males 2050 with migration
Females 2050 with migration
Males 2050 zero migration
Females 2050 zero migration

Number of males at each age (thousands)   Number of females at each age (thousands)

Source: U.S. Census Bureau, 2000d, Population Projections of the United States by Age, Sex, Race, Hispanic Origin, and Nativity, 1999 to 2100 (Washington, DC: U.S. Census Bureau Population Division).

living side by side with (although typically not in contact with) older residents (often retirees).

Families are located in adjacent parts of the city, which are slightly more suburban and of lower density than the area near the university. You can see in Figure 8.12 that in the Navajo area, across the freeway from the university, there are relatively few college-age people. The area is composed primarily of older adults and their children. The third example is the age and sex distribution for San Ysidro, which is a suburban community just north of the U.S.-Mexico border that is heavily Hispanic, including recent immigrants from Mexico. This area of the city is characterized by young families with their children and the age structure is consistent with an above-average level of fertility.

The fourth example is the most bizarre, yet it is typical of the downtown central business districts of many cities. Downtown San Diego is heavily adult male in all ages up to retirement. Virtually everyone residing in this area lives in a hotel or apartment, or one of the upscale gentrified condominiums. The excess of young males is a reflection partly of the presence of the Navy, with sailors living in nearby, but off-base, housing, and partly of a more transient population, including homeless people.

Figure 8.12   Age Pyramids in Different Areas of San Diego

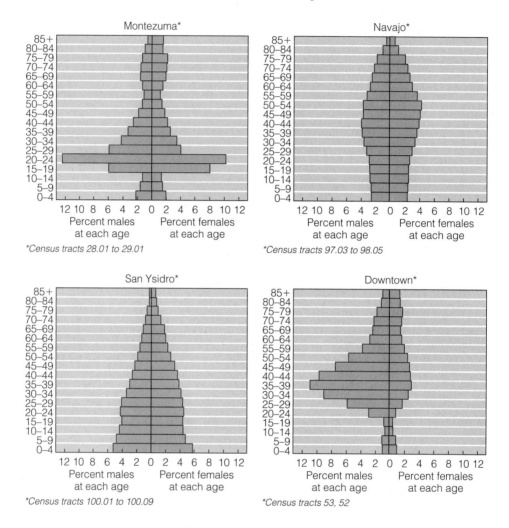

Source: San Diego Association of Governments, 2001, Regional Estimates and Forecasts (http://www. sandag.cog.ca.us; accesed 2001). Data refer to 2000.

The excess of middle-aged males probably also reflects a population of derelicts and other transients who move into the area for lack of anywhere else to go. At the same time, the newer residents in this area are higher-income people who have purchased expensive condominiums in high-rise buildings, and they may be predominantly male. At retirement ages, the slightly greater proportion of females is a consequence of the women's greater longevity and the fact that they are attracted to the downtown area by the construction of high-security, high-rise, low-rent apartments.

These age pyramids show how the age and sex distributions in an area reflect the history of that area and reveal important social differences. In each case, the major determinant of the age/sex structure was migration, either in or out, and this illustrates the dramatic effect that migration may have on an area. Migration in two instances acted to produce bulges in the young adult ages, and in another instance to produce a classic pyramid associated with the arrival of young immigrants having children. Migration, then, can affect either sex and can affect different ages to varying degrees. Of course, migration was not the only influence. Note that the abundance or lack of children in one area or another reflects high or low levels of fertility as well.

It is generally true that the more precisely you define a geographic area, the more likely it is that the age and sex structure will have been affected by migration, and the more likely it is that the area's "personality" will be affected by (and, of course, reciprocally influence) the age and sex structure. The state of California, for example, has an age distribution very similar to that for the United States as a whole. As you narrow in on San Diego, local variations show up as a result of the influence of the military and the three universities within the city limits. As you close in on specific sections of the city, as in Figure 8.12, the variations grow even wider. The character of each neighborhood begins to emerge, shaped by and helping to shape the age and sex distribution, which in turn affects and is affected by migration. The effects of the other two demographic processes, mortality and fertility, always go along with the effects of migration.

Having discussed the way in which the age and sex structure is built on the foundation of migration, mortality, and fertility rates, it is time now to see how the changing age structure influences life in human societies.

Age Transitions at Work

The mechanism of the age and sex structure translates all demographic changes into a force with which we must cope. A high birth rate does not simply mean more people: It means that a few years from now there will be more kids entering school than before; that 18 years from now there will be more new job hopefuls and college freshmen than before. An influx of young adult immigrants this year means a larger-than-average number of older people 30 to 40 years from now (and it may mean an immediate sudden rise in the number of births, with all the attendant consequences).

The Impact of Age Structure on Population Processes

As a population moves from a younger to an older age structure, the age grading of biological and social functions leaves its imprint on society. The actual pattern of population growth is of course determined by the combination of fertility, mortality, and migration, and the rate of each of these is influenced by the age structure—a fact alluded to in earlier chapters, but worth repeating. A bulge in the age structure at the young adult (reproductive) ages will tend to raise the crude birth rate by producing a large number of children relative to the total population. Even in otherwise low-fertility societies such as the United States and Canada, a large population of young women produces a substantial number of births. At the other extreme, a population with a relatively small proportion of young people, but with a high proportion of older people, will have a substantial number of deaths each year, even if life expectancy is high, just because there are so many people moving into those higher-risk years. This will generate a crude death rate higher than might otherwise be expected.

A young age structure produced by high fertility may also encourage a relatively large number of out-migrants. Each year such a population would produce a greater number of young adults, who are at greatest risk of migrating. Such migration, if it occurs, may well contribute to the sorts of massive changes almost invariably created by population growth. The question of social changes wrought by population processes leads us to examine the more general, and generally more interesting question: How do age and sex structures influence what a society is like? The first thing we have to recognize is that a young age structure has momentum built into it.

Population Momentum

Youthful age structures carry with them the potential for substantial **momentum of population growth.** This is a concept not unlike that idea that a heavy freight train takes longer to stop than does a light commuter train, or that a Boeing 747 requires a longer runway for landing than does a 737. The amount of momentum built into an age structure is determined by answering the following question: How much larger would the population eventually be, compared to now, if replacement fertility were instituted in the population right now? Put another way, how much larger would the stable population be based on replacement-level fertility compared to the current population? (Keyfitz 1971; Kim and Schoen 1997). You can probably see intuitively that if a population has a large fraction of women in their reproductive years, before the population stops growing, these women will contribute many additional babies to the future population even at replacement level, whereas an older population, with fewer women in their reproductive ages, will contribute fewer babies and thus will stop growing, with fewer people added. Preston and Guillot (1997) have shown, for example, that the populations of Africa and Asia would grow by a factor of 1.56 (they would be 56 percent larger in size when they finally stopped growing) if replacement-level fertility had been introduced in 1997. By contrast, the momentum factor for North America was calculated to be 1.10, indicating that the less-youthful age structure of North America would not grow by very much more even if replacement-level fertility had been achieved in 1997.

Most populations that now have above-replacement-level fertility will not immediately drop to replacement level. You know from Chapter 6 that the fertility transition does not normally work like that. So, researchers have devised formulas to calculate the momentum built into more gradual declines in fertility (Li and Tuljapurkar 1999; Schoen and Kim 1998). For example, a population that was growing at a rate of 3 percent per year (comparable to Guatemala in 2000) would be 1.7 times larger when it stopped growing if replacement-level fertility were achieved immediately. That would require the TFR to drop instantaneously from 4.4 to 2.1, which is very unlikely. More probable would be a decline over several decades. If the drop from 4.4 to 2.1 took 28 years, then the population would be 3.9 times larger when it stopped growing, and if it took as long as 56 years, the population would be 7.7 times larger when replacement level was finally reached (Schoen and Kim 1998).

Understanding population momentum allows you to appreciate that slowing down the rate of population growth requires a great deal of forethought. The payoff, which is a more favorable age structure to accompany the declining rate of demand for resources, is apt to be at least two decades down the road from the time that fertility begins to decline in earnest. Bongaarts and Bulatao (1999) have shown that in all regions of the world except sub-Saharan Africa, the momentum built into current age structures will be the principal component of future population growth. Only in sub-Saharan Africa is fertility still so high that it makes a significant independent contribution to future population growth. Thus, in that part of the world, both high fertility and the momentum of a young age structure are working to propel population growth, which is being held back only by the AIDS pandemic.

Reading an Age Structure

One way to illustrate the numerous and varied ways in which the age structure influences society is by "reading" an age structure. This involves looking at the tea leaves of history left behind in the age structure, and then using the logic inherent in population projection and modeling to suggest what the future holds in store. Let's do this for the United States.

We can trace the history of fertility (and recount the economic situation of the time) in the age pyramid for the United States in the year 2000 (Figure 8.13). In this age pyramid, I have labeled the vertical axis according to the birth cohort, rather than by the age in 2000. This allows us to talk more easily about how history has created the structure we see. I have also used single years of age to capture more detail. Keep in mind, of course, that above age 55 or 60, the effect of accelerating death rates obscures changes that might have occurred, so most of our discussion about the past will have to deal with people born since the end of World War II.

You can spot the baby boom generation as the birth cohorts of 1946 to 1964 in the United States, who were 35 to 54 years old in 2000. You can see that the infusion of the baby boomers into the age structure created a very pyramid-shaped structure for the cohorts born prior to the mid-1960s, when the baby boom came to a halt. But coming to a halt did not mean the end of the baby boom's influence on American society. The growth in the number of children meant that schools were hard-pressed to find facilities and teachers; later on there was a huge demand for housing and jobs; and we are now facing the question of how to finance retirement for this generation.

Let me illustrate the age structure impact of the baby boom with the following story. There were 48 percent more arrests in the United States in 1980 than in 1970, according to the FBI (data from U.S. Department of Justice 1971; 1981). Had society fallen apart? Had the nation lapsed into a police state? Neither one. What happened was that the baby boom generation was moving through the late teen and early adult years, when crime rates are highest. They swelled those ages and crime went up at least partly because there were more people "at risk" of committing a crime. At these younger ages, a person's fancy seems to turn not just to love, but

Figure 8.13 Age Structure of United States

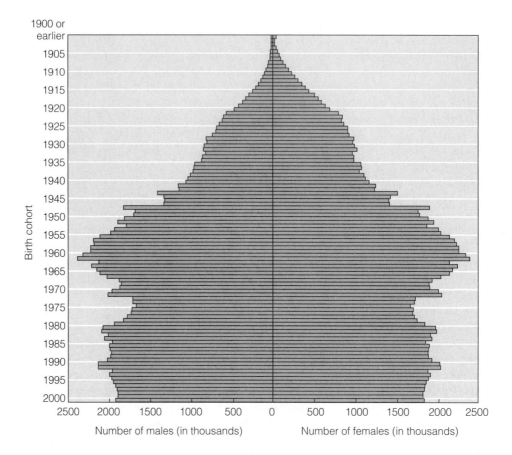

Note: The population pyramid for the United States in 2000 shows the impact of low birth rates during the Depression (approximated by the birth cohorts of 1930–1934), the higher birth rate of the baby boom generation (the birth cohorts of 1946–1964), the decline in fertility in the late 1960s and 1970s, the slight resurgence in fertility in the 1980s, and the resumption of a decline in the early 1990s, followed by an uptick in the late 1990s.

Source: Adapted from U.S. Census Bureau, 2000d, Population Projections of the United States by Age, Sex, Race, Hispanic Origin, and Nativity: 1999 to 2100 (Washington, DC: U.S. Census Bureau Population Division); and U.S. Census Bureau, 2001g, 2000 Census of Population, Table DP-1.

disproportionately to crime as well. In most societies, crime is the province of the young, and it stands to reason that a youthful age structure will produce higher overall crime rates (that is, crude crime rates) than will an older age structure (Cohen and Land 1987; Steffensmeier and Harer 1987).

Between 1970 and 1980, the arrest rate in the United States increased from 33 arrests per 1,000 people to 43 per 1,000. Nearly half of that increase was due to the fact that there were more young adults at the high-risk ages. This is very similar to the findings of the 1967 National Crime Commission, which concluded that 40–50 percent of the increase in the U.S. crime rate between 1960 and 1965 was due to the changing age structure—the early impact of the baby boom. A cohort-related explanation for rising crime rates even in the teen years themselves is that the number of children per adult rose during the baby boom, potentially lowering the amount of social control exercised over each child, thus leading to more deviant behavior that resulted in arrest. There is a fair amount of evidence to suggest that declining social control over teenagers (whether a function of the age structure or not) did play a role in rising crime rates in the United States when the baby boomers were teenagers (Osgood et al. 1996; Sampson and Laub 1990).

A less obvious, but potentially very important, aspect of the age structure's influence on crime is the fact that the age structure of potential victims may also affect the opportunities for crime and thus the crime rate itself. In the United States in the 1970s and 1980s, the proportion of the population that was fairly young, at work during the day, and affluent enough to own easily stolen and fenced goods rose dramatically, and thus the opportunity for crime rose. Cohen and Felson (1979) argue that this substantial increase in the opportunity to commit crime was, in fact, the most important explanation of the rise in the crime rate. If so, the age-structure effect was less direct, but no less influential. Gender equity has also played a role. In 1970, only 6 percent of those arrested were female, whereas by 1999 the figure had jumped to 22 percent (U.S. Department of Justice 2000). Overall, then, crime rates and trends cannot be completely explained by the age structure, but neither can they be understood without taking the age and sex structure into account (Blumstein and Wallman 2000).

Between 1961 and 1974, each birth cohort in the United States was typically smaller than the previous one, with the exception of 1970 and 1971. Those cohorts have also been added to by subsequent immigration of people born in those years. The birth cohort of 1974 represents the smallest cohort of people in the country since the cohort 30 years earlier, just before the baby boom. This was the height (so to speak) of the baby bust. That "birth dearth" or "baby bust" will be a feature of the age structure until the middle of this century and will be a long-lived reminder of the inflation, tight job market, energy shortage, and women's liberation from the late 1960s through the 1970s. The children of the baby bust (born from 1965–77) emerged into young adulthood in the 1990s and were labeled "Generation X"—the generation forgotten by America because it grew up in the shadow of the baby boom, coping with the detritus of the boomers' lifestyle choices.

The birth rate rebounded in the late 1970s and early 1980s (the "baby boomlet"). In some states, such as California, the immigrant population made important contributions to this increase and by the early years of the twenty-first century, these young adults represented a tidal wave of entrants into college, unprecedented since the days when the baby boomers swamped campuses from the late 1960s through the early 1980s.

In the 1980s, almost twice as many immigrants were admitted to the United States as in the previous decade. This contributed to the slight bulge in the size of the cohorts of people born in the early 1990s, but cohort size declined slightly in the mid-1990s, only to rise again slightly in the late 1990s. We can certainly expect that any population will experience fluctuations in cohort size, even when the rate of growth is very low. It is the more dramatic shifts that create the bigger crises for societies. By the year 2000, the baby boomers were approaching retirement. This has generated public angst about how best to deal with a much larger older population—a topic I will explore in more detail in the next chapter. The size of the older population will be complicated, of course, by the relative paucity of people in the generation younger than the boomers. The smaller birth cohorts of the 1970s will find that the retirement of the boomers will be both a blessing and a curse. They will almost certainly have more occupational flexibility and a higher demand for their services because they will reach their prime ages of economic productivity just as the baby boomers leave the labor market. But the younger generation will be forced to pay for the retirement of the baby boomers, and that will be a disproportionately heavy burden. The changing age structure guarantees us that old age pensions and health care for the elderly will be huge political issues for several decades to come. Since there will be so many older people, and since they are more likely to vote than younger people (another age-graded phenomenon), the baby boomers are likely to keep the issues of importance to them in front of the American public.

It is worth noting that the younger-old population in 2000, those people born between 1930 and 1940, are among the most advantaged cohorts in American history. Of course, no one could have guessed that when they were born during the Depression (and thus, you recall, there are not many of them), but they were too young to serve in World War II, although some were involved in the Korean War. They were advantaged primarily in an economic sense, since they entered the labor market during a time of relative economic expansion. Because there were fewer of them than there were of older cohorts, they were absorbed into the labor force more quickly and had less competition for career advancement. They were among the parents of the baby boom, and the boomers actually provided the economic impetus to carry their parents securely into old age. Overall, the rise in their standard of living from childhood to mid-adulthood was probably greater than for any other generation of Americans. Since they are retiring ahead of the baby boomers, they still have the baby boom generation paying the taxes to help support their Social Security payments.

At current levels of fertility, and with relatively unchanging levels of migration and slowly improving levels of mortality, the U.S. population is expected to continue growing up to and beyond the middle of this century. The United Nations middle series projection, mentioned earlier, suggests a total U.S. population of 409 million in the year 2050—124 million more than there were in 2000, which amounts to a 44-percent increase in population size in that 50-year period. This would be equivalent to adding the entire population of Japan to the United States over the period from 2000 to 2050. Under the assumptions used in these projections, you can see back in Figure 8.11 that the age distortions introduced especially by the baby boom and bust will be smoothed out. But of course no one predicted either the baby boom or the baby bust, so we will have to see what happens.

Some of what could happen includes, in particular, a return to tighter immigration laws and border enforcement. Some have argued that the country cannot, and what's more should not, tolerate additional people and that zero population growth or even negative population growth should be achieved immediately (Bouvier 2000). The principal objection to population growth is that Americans consume a radically disproportionate share of the world's resources and that adding to the U.S. population exacerbates the world's environmental problems and aggravates distorted income distribution in the world. I will discuss these ideas in Chapter 12, but here we should note that any such policy to immediately slow down the rate of growth would, of course, introduce new distortions into the age structure that American society would have to deal with down the road.

The Progression from a Young to an Old Age Structure

At the beginning of the demographic transition, every population had a young age structure with a characteristic pyramidal shape, as shown earlier in Panel A of Figure 8.9. At the end of the demographic transition, every population will have an old age structure, with a characteristic barrel shape, as shown in Panel B of Figure 8.9. In between, the age structure will undergo a period of time when it caves in at the younger ages as fertility declines, and there will be a bulge in the younger adult ages. The size and timing of that bulge will help tell the tale of how dramatically a society changes in the context of the age transition. A quick drop in fertility that occurs early in the mortality transition will have the most dramatic impact on a society, increasing the economic productivity of the adult population, with few children to deal with and also few older people to worry about.

A slower transition to an older age will be socially and economically less disruptive than a very sudden change. However, it may not be as conducive to the kind of revolutionary change a struggling economy may need in the face of the population growth that ensues as mortality declines and the demographic transition takes hold, as I discussed earlier in comparing age-structure changes in China and India. Transitions are not necessarily smooth, of course, and there is some evidence that the more variable the age transition, the more opportunities there are for inequities to creep into the social structure (O'Rand and Henretta 1999). I will return to this theme in the next chapter.

Crenshaw and his associates (1997) examined the pattern of age-specific growth rates and economic development for the period 1965–1990 and concluded that an increase in the child population (the impact of declining mortality in a high-fertility society) does indeed hinder economic progress. On the other hand, an increase in the adult population relative to other ages (the delayed effect of a decline in fertility) fosters economic development, producing what they call a "demographic windfall effect whereby the demographic transition allows a massive, one-time boost in economic development as rapid labor force growth occurs in the absence of burgeoning youth dependency" (p. 974).

Let me remind you of the situation in China, which profited from its rapid drop in fertility. In 1960, China had a population of 650 million, plagued by high mortality (a life expectancy around 45 years) and high fertility (a total fertility rate of more than five children per woman), although fertility was starting to decline. Its age structure exhibited the classic pyramidal shape, reflecting a high proportion of the population in the very young ages, as you can see in Figure 8.14. By 1990, China had transformed itself. Its population had nearly doubled to 1.1 billion, due especially to declining mortality, but fertility had declined dramatically. Most noteworthy about the age structure in 1990 was the big bulge in the young adult ages, combined with the small number of older people and the small number of children. This was an ideal age structure for an economy trying to take off, and the age structure helped

Figure 8.14   Changing Age Structures in China, United States, and Mexico Are Associated with Differing Patterns of Economic Change

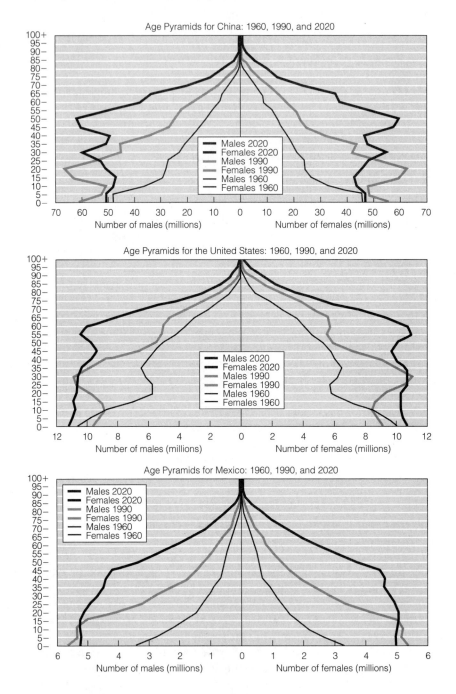

Source: Data for China and Mexico are from United Nations, 2000b; data for the United States are from U.S. Census Bureau 2000d.

accelerate the economic reforms that created the East Asian economic miracle. But the economy of China has cooled off somewhat since the 1990s, and the demographic momentum of the age transition in China will increasingly act as a brake on economic growth in that country.

The United Nations projects that by 2020 the population of China will have grown to about 1.4 billion. By this time, the bulge in the age structure will have moved into the middle adult ages, where productivity gains may be less than at the younger ages, but where consumption patterns tend to peak. Overall, by 2020, the population of China will be heavily weighted toward the older ages and the country will face important policy issues arising from a rapidly aging population (Adamchak 2001; Qiao 2001). China may also face internal pressures to increase the birth rate in order to create a pool of new workers whose infusion of resources into the population at younger ages will help pay the pensions of the older population.

The United States in 1960 had 181 million people, and the age structure for that year is noticeable for the effect of the baby boom generation at the younger ages, as seen in Figure 8.14. Of course, another way of viewing this graph is to see the baby boom as the continuation of trends that existed prior to the Depression, and to view the low fertility of the Depression as an aberration, rather than thinking of the baby boom as the aberration. Seen in this way, the U.S. in 1960 had a clear pyramidal age structure with a heavy burden coming from the young population. By 1990, the population had grown to 250 million, but fertility levels had declined dramatically—the total fertility rate dropping from 3.6 children per woman in 1960 down to 2.0 children per woman in 1990. Just as I noted above for China, the rapid drop in fertility produced a demographic windfall in the age structure that coincided nicely with the economic expansion of the 1990s. The U.S. age structure in 1990 bulged at the young adult ages, without the burden of huge numbers of either older or younger people.

The picture for 2020 in the U.S. is also remarkably similar to that in China. The population is expected to increase to 325 million, the big bulge in the age structure will be transitioning toward the older, higher-consuming years, and the number of younger people will be demographically overwhelmed by the number of older people (U.S. Census Bureau 2000d). This could be met by calls for a higher birth rate, but it will probably produce an increase in immigration, to bring in people at younger ages to keep the economy going.

Many of those expected immigrants to the U.S. will be from Mexico. In 1960, Mexico had 39 million people and a classic pyramidal age structure, although fertility in Mexico was higher that year than in either the U.S. or China (see Figure 8.14). Mortality has been steadily declining over time in Mexico and since the mid-1970s so has fertility. You can see in the graph for Mexico that by 1990, the younger ages had been truncated by the fertility decline. But you can also see that the slowness of the fertility decline and its more recent start than in either China or the U.S. meant that there was no bulge of population in the young adult ages—no demographic windfall to spur on an economy that was trying hard to grow, but was being weighed down by the demographic burden.

United Nations projections for the year 2020 in Mexico suggest that fertility will continue its slow drop, and Mexico will transition gradually to an older, barrel-shaped age structure, never having experienced the windfall that comes with a rapid drop in fertility. Thus, Mexico's economy will likely never receive the boost that the age structure provided in the 1990s for both China and the United States. Mexico's muddling toward low fertility has deprived it of a golden chance to use the age dividend to leap ahead economically.

Summary and Conclusion

The age and sex structure of a society is a subtle, commonly overlooked aspect of the social structure, yet it is one of the most influential drivers of social change in human society. The number of people at each age and of each sex is a very important factor in how a society is organized and how it operates, and for this reason the age transition that accompanies the mortality and fertility transitions is a critical force for change. Age composition is determined completely by the interaction of the three demographic processes. Migration can have a sizable impact, because migrants tend to be concentrated in particular age groups and, in addition, migration is sometimes selective for one sex or the other. Mortality has the smallest short-run impact on the age distribution, but when mortality declines suddenly (as in the less-developed nations), it makes the population more youthful. At the same time, a decline in mortality influences the sex structure at the older ages by producing increasingly greater numbers of females than males. Changes in fertility generally produce the biggest changes in a society's age structure, regardless of the level of mortality.

We study the effects of population processes on the age and sex structure by calculating stable and stationary population models, which are types of population projections—one of the most useful methodological tools created in the field of population studies. Looking backward through time, we can see how different birth cohorts have flowed through time, being influenced differentially at each age by social events, and in turn have shaped events by their passage through the age structure. Using population projections to look ahead, we can chart the potential course of change implied by different age structures. There are some fairly predictable changes that occur to a population in the context of the age transition from a younger population at the beginning of the demographic transition to an older population at the end of the demographic transition. The critical question is how quickly that transition occurs, because the distortions in the age structure that are part of the age transition can lead to changes in economic organization, political dominance, and social stability that must be dealt with by society—and they have the potential to be either positive or negative in their impact, depending on how society responds.

In all of the more developed societies in the world today, the fertility rate has been low for long enough that we are approaching the end of the age transition and population aging has become a major societal concern—not because we are afraid of old people, but rather because the demands on societal resources are very different for an older than for a younger population. "Life in an era of declining and ageing populations will be totally different from life in an era of rising and youthful populations. A new age calls for a new mindset" (Wallace 2001:220). In the next chapter, I help you scope out that mind-set.

Main Points

1. An age and sex structure represents the number of people of a given age and sex in society and is built from the input of births at age zero and from deaths and migration at every age.

2. The age transition is a predictable shift from a predominantly younger population when fertility is high (usually when mortality is high) to a predominantly older population when fertility is low (usually when mortality is also low).

3. The influences of fertility and mortality on the age and sex structure are usually understood by applying stable population models to real populations.

4. Population projections are developed from applying the age and sex distribution for a base year to sets of age-specific mortality, fertility, and migration rates for the interval between the base year and the target year.

5. The age structure is typically graphed as a population pyramid and measured by the median age or dependency ratio, and by age-specific rates of growth.

6. More male babies than females are generally born, but the age structure is further influenced by the fact that at almost every age, more males than females die.

7. A young population is one with a high proportion of young people, whereas an old population is one with a high proportion of elderly people.

8. The concepts of age stratification and cohort flow place age structures in their proper sociohistorical context.

9. Migration can have a very dramatic short-run impact on the age and sex structure of a society, especially in local areas.

10. In the short term, migration most drastically affects the number of young adults.

11. In the long term, migration mainly adds people in the host area and subtracts out-migrants and their future offspring from the donor area.

12. Mortality has very little long-run impact on the age structure, but in the short run a decline in mortality typically makes the population younger in medium- to high-mortality societies, and a little older in lower-mortality societies.

13. Declining mortality mainly operates to increase population size, rather than dramatically affecting the age and sex structure.

14. Fertility is the most important determinant of the shape of the age/sex structure.

15. High fertility produces a young age structure, whereas low fertility produces an older age structure.

16. The age composition of a society is a powerful stimulant to social change.

Suggested Readings

1. John I. Clarke, 2000, The Human Dichotomy: The Changing Numbers of Males and Females (Amsterdam: Pergamon).

A scholarly but very readable analysis of the way in which the sex ratio has been changing in tandem with the demographic transition and related global changes.

2. Leroy Stone, Editor, 1999, Cohort Flow and the Consequences of Population Ageing, An International Analysis and Review (Ottawa: Statistics Canada).

This is a CD-ROM-based volume, rather than a printed one, providing several chapters with supporting graphs depicting the impact of cohort flow on the way in which populations change over time.

3. John Bongaarts and Rodolfo A. Bulatao, Editors, 2000, Beyond Six Million: Forecasting the World's Population (Washington, DC: National Academy Press).

This is a report from the Panel on Population Projections of the Committee on Population of the National Research Council, and represents a good overview of the state of the art with respect to population projections.

4. Paul Wallace, 2001, Agequake: Riding the Demographic Rollercoaster Shaking Business, Finance, and Our World (London: Nicholas Brealey).

A popular and very readable (and also well-researched) examination of the impact of changing age structures on the modern world.

5. Ted Halsted, 1999, "A Politics for Generation X," The Atlantic Monthly 284(2), pp. 33–42.

The people born in the baby bust have been labeled Generation X and this article explores the political dilemmas facing this birth cohort that followed on the heels of the baby boom.

🕮 Web Sites of Interest

Remember that Web sites are not as permanent as books and journals, so I cannot guarantee that each of the following Web sites still exists at the moment you are reading this. It is usually possible, however, to retrieve the contents of old Web sites at http://www.archive.org:

1. http://www.census.gov/ipc/www/idbpyr.html
   At this Web site, the U.S. Bureau of the Census has a program that lets you draw an age pyramid "on the fly" for any country in the world. The site also has a dynamic pyramid that graphically changes the age and sex structure over time for any given country.

2. http://www.statcan.ca/english/kits/animat/pyca.htm
   Statistics Canada also has age pyramids that are available for each region of Canada, and the graphics change over time so that you can visualize the age transition in Canada.

3. http://esa.un.org/unpp/
   This is an online resource that lets you create your own summary tables of data from the United Nations Population Division's latest revisions of world population projections, including population data by age and sex.

4. http://www.rand.org/publications/MR/MR1274/
   This is a detailed report from RAND's Population Matters program, authored by David Bloom, David Canning, and Jaypee Sevilla, that details the "demographic dividend" inherent in the age transition.

5. http://www.statistics.gov.uk/populationestimates/svg_pyramid/default.htm
   Watch the age structure of England and Wales change over time in this interactive age pyramid.

CHAPTER 9
Population Aging

Figure 9.1 Percent of the Population that is Aged 65 and Older

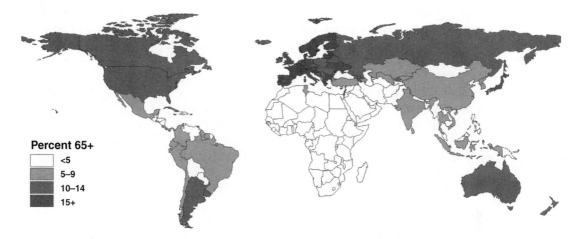

Percent 65+
- <5
- 5–9
- 10–14
- 15+

Source: Adapted by the author from medium variant projections in United Nations, World Population Prospects 2002.

Most old people are pretty much alike—they have no interest in sex, they are miserable most of the time, they cannot work as effectively as younger people, and they have incomes below the poverty level. Now, if you believe all that, I have some swampland in Florida you might want to buy! The truth is that older people are not all pretty much alike. On the contrary, the diversity of their life courses, exposure to several decades of history, and countless different patterns of social interaction tend to make older people even less alike than people at younger ages. Thus we find a wide range of interest in sex among older people, a wide range of life satisfaction (from miserable to extremely happy), a wide range of worker effectiveness, and a wide range of income and wealth. This means that you stereotype older people at your own risk, indeed, at an increasing risk, because the older population is growing at a rapid rate all over the world. As I discussed in the previous chapter, this is the seemingly inevitable end of the demographic transition.

The decline in mortality throughout the world has had even broader implications than more babies surviving to adulthood; it has meant more adults surviving to old age as well. As fertility rates have also declined, the *proportion* of the population that is older also increases, not just the *number* of older persons, as you will recall from the previous chapter. Migration also influences the proportion of the population that is older because younger people tend to migrate more than older people do. Because older people now represent the fastest-growing segment of the world's population, it is of considerable importance to understand the causes and consequences of the growth in their number and the increase in their percentage of the total population. I begin by defining what we mean by the "older" population—both in biological and in social scientific terms. Then I examine the growth and distribution of the older population, looking at the influence of mortality, fertility, and migration on these trends. Next, I review the basic sociodemographic characteristics of the older population, drawing on data especially for the United States, to see how the life course of older people has differed in important respects from the younger population. In the process of seeing what older people are like, I weave in a discussion of how the aging process will likely influence the future course of society, keeping in mind that as we move toward the end of the demographic transition, we are in largely uncharted territory.

What Is Old?

Age as we usually think of it is a social construct—something we talk about, define, and redefine on the basis of social categories, not purely biological ones. A good way to visualize this concept is to contemplate Satchel Paige's famous question: "How old would you be, if you didn't know how old you was?" which illustrates the point that age takes on its meaning from our interaction with other people in the social world. If we are defined by others as being old, we may be treated like an old person regardless of our own feelings about whether or not we are, in fact, old. We humans depend heavily on visual clues to assess the age of other people, and we learn quickly that there are certain kinds of outward physical changes typically associated with aging—the graying hair, the wrinkling skin, the muscle tone decline, and the changing shape caused by the redistribution of fat. These are taken as signs of physical decline and it is fair to say that most of us are, at best, ambivalent about the aging process—an attitude that is itself probably as old as human society (Warren 1998). In what might be thought of as a form of denial, research has shown that Americans are more prone than Germans, for example, to think of themselves as being younger than their biological age and that the discrepancy increases with age, especially among healthier older people (Westerhof, Barrett, and Steverink 2003).

Especially among women, the physical symptoms of aging may be associated with the end of reproduction. Indeed, humans are among the few species that have a substantial postreproductive existence (Albert and Cattell 1994). Still, these changes do not follow a rigid time schedule, and some can be successfully hidden or disguised (think of dyed hair, hair implants, Botox injections, face lifts, and liposuction). In general, then, there is no inherent chronological threshold to old age; however, in the United States and much of the world, old age has come to be defined as beginning at age 65. The number 65 assumes its almost mystical quality in the United States because it is the age at which important government-funded benefits such as Social Security and Medicare become fully available.

In 1935, when the present Social Security benefits were designed in the United States, the eligibility age was set at 65, "more because of custom than deliberate design. That age had become the normal retirement age under the few American pension plans then in existence and under the social insurance system in Germany" (Viscusi 1979:96). The Older Americans Act passed by the U.S. Congress in 1965 provided benefits in some programs for people aged 60 and over, and private pensions often permit benefits to be paid at age 55, but age 65 is a milestone firmly entrenched in the Western world, and so I, too, will adopt that definition in this chapter. However, remember that the number is arbitrary, and people obviously will not fit neatly into any senior citizen mold on reaching their 65th birthday. In fact, most people in low-mortality societies do not think of themselves as being old until well beyond that age, whereas most people in high-mortality societies would probably think of themselves as old well before age 65.

Our demographic interest in the older population comes from the fact that as the number and proportion of older people increase (as the *population* ages), changes are wrought in the organization of society. These changes are partly the result of the process of *individual* aging—people change biologically with age and societies react differently to older than to younger people, producing the social changes that we see accompanying population aging. These individual changes occur in the context of each birth cohort, however, because improvements in health are brought about largely by technological and environmental changes in life over which we have very little personal control.

I have outlined these concepts in Figure 9.2, where I point out that the exact nature of the social changes will depend partly on the social context, including aspects of society such as cohort-specific historical events that have affected different cohorts throughout the life course, but also societal levels of sexism, ageism, and racism. The impact of population aging will also differ according to the proportion of people who are young-old (the "third" age of life—in which the impact of aging on society is more social), compared with the proportion who are old-old (the "fourth" age of life—in which society must cope more with the individual biological aspects of aging). My goal in this chapter is to discuss the concepts shown in Figure 9.2,

Figure 9.2   Population Aging and Individual Aging Combine to Produce Change in Society

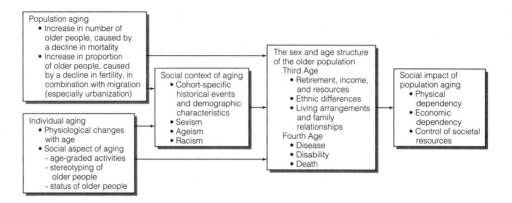

working from left to right in that figure. Let us begin by examining the dimensions of population aging.

Population Aging

The natural consequence of the demographic transition is for the population to age—for there to be more older people and a greater fraction of the population that is older. I discussed this general trend in the previous chapter, but let me here focus specifically on older people themselves, rather than on the entire age structure. Let us first examine how many older people there are in the world, and then ask where they are most apt to be found.

How Many Older People Are There?

There were 462 million people in the world aged 65 and older as of the year 2004, according to United Nations estimates (United Nations Population Division 2003c). If they all lived together under one flag, they would represent the third largest nation in the world. As a fraction of the total world population, the older population accounts for 7 percent, but this percentage varies considerably from one part of the world to another, as is illustrated in Figure 9.1. For example, in the year 2004, only 19 percent of the total population of the world was living in more developed nations, yet 41 percent of the world's population aged 65 or older was there, accounting for 14 percent of the total population of these wealthier countries. Still, that left the other 59 percent of people in the world aged 65 and older living in developing countries, even though the older population represented only 5 percent of those populations.

Table 9.1 provides detailed data on the growth of the older population compared to the total population for all major world regions between 1950 and 2050,

Table 9.1   The Number of Older People is Growing More Rapidly Than the Rest of the Population All Over the World

Region	1950	1975	2000	2025	2050
WORLD					
Total population	2,518,629	4,068,109	6,070,581	7,851,455	8,918,724
Population aged 65+	130,865	231,182	419,197	822,876	1,418,742
%65+	5.20%	5.68%	6.91%	10.48%	15.91%
r(total)		1.92%	1.60%	1.03%	0.51%
r(65+)		2.28%	2.38%	2.70%	2.18%
Europe					
Total population	547,403	675,542	727,986	696,036	631,938
Population aged 65+	44,957	77,265	107,178	148,256	176,553
%65+	8.21%	11.44%	14.72%	21.30%	27.94%
r(total)		0.84%	0.30%	-0.18%	-0.39%
r(65+)		2.17%	1.31%	1.30%	0.70%
North Africa					
Total population	53,302	97,543	173,615	254,150	306,046
Population aged 65+	1,844	3,657	7,316	17,704	42,350
%65+	3.46%	3.75%	4.21%	6.97%	13.84%
r(total)		2.42%	2.31%	1.52%	0.74%
r(65+)		2.74%	2.77%	3.53%	3.49%
Sub-Saharan Africa					
Total population	177,101	327,336	653,492	1,085,471	1,557,384
Population aged 65+	5,621	9,389	19,417	39,653	85,574
%65+	3.17%	2.87%	2.97%	3.65%	5.49%
r(total)		2.46%	2.77%	2.03%	1.44%
r(65+)		2.05%	2.91%	2.86%	3.08%
Western Asia					
Total population	50,922	99,975	192,222	301,699	400,822
Population aged 65+	2,206	4,190	8,549	20,044	48,604
%65+	4.33%	4.19%	4.45%	6.64%	12.13%
r(total)		2.70%	2.61%	1.80%	1.14%
r(65+)		2.57%	2.85%	3.41%	3.54%
South Central Asia					
Total population	498,508	879,443	1,486,049	2,097,128	2,463,916
Population aged 65+	18,582	33,450	68,325	153,576	323,606
%65+	3.73%	3.80%	4.60%	7.32%	13.13%
r(total)		2.27%	2.10%	1.38%	0.64%
r(65+)		2.35%	2.86%	3.24%	2.98%

(continued)

Table 9.1   (continued)

Region	1950	1975	2000	2025	2050
Southeastern Asia					
Total population	178,073	321,368	520,355	687,621	767,250
Population aged 65+	6,752	11,428	24,411	57,393	126,047
%65+	3.79%	3.56%	4.69%	8.35%	16.43%
r(total)		2.36%	1.93%	1.11%	0.44%
r(65+)		2.10%	3.04%	3.42%	3.15%
Eastern Asia					
Total population	670,985	1,096,726	1,481,110	1,655,784	1,590,070
Population aged 65+	29,976	51,767	114,662	243,506	381,372
%65+	4.47%	4.72%	7.74%	14.71%	23.98%
r(total)		1.97%	1.20%	0.45%	-0.16%
r(65+)		2.19%	3.18%	3.01%	1.79%
Oceania					
Total population	12,812	21,564	31,043	39,933	45,815
Population aged 65+	936	1,582	3,037	5,236	7,131
%65+	7.30%	7.34%	9.78%	13.11%	15.57%
r(total)		2.08%	1.46%	1.01%	0.55%
r(65+)		2.10%	2.61%	2.18%	1.24%
South America					
Total population	112,995	216,096	347,343	455,829	510,113
Population aged 65+	3,921	9,475	19,596	46,802	93,382
%65+	3.47%	4.38%	5.64%	10.27%	18.31%
r(total)		2.59%	1.90%	1.09%	0.45%
r(65+)		3.53%	2.91%	3.48%	2.76%
Central America					
Total population	37,065	78,657	135,213	186,441	211,758
Population aged 65+	1,513	2,969	6,225	16,049	37,492
%65+	4.08%	3.77%	4.60%	8.61%	17.71%
r(total)		3.01%	2.17%	1.29%	0.51%
r(65+)		2.70%	2.96%	3.79%	3.39%
North America					
Total population	171,616	243,425	315,915	394,312	447,931
Population aged 65+	14,102	25,005	38,901	71,289	91,746
%65+	8.22%	10.27%	12.31%	18.08%	20.48%
r(total)		1.40%	1.04%	0.89%	0.51%
r(65+)		2.29%	1.77%	2.42%	1.01%

Note: r(total) = average annual rate of growth of the total population; r(65+) = average annual rate of growth of the older population; population figures are in thousands.

Source: Adapted from United Nations, 2003, World Population Prospects: The 2002 Revision (New York: United Nations).

based on United Nations data for 1950–2000 and their projections for the period from 2000–2050. Since the end of World War II, but beginning before that, there has been an uninterrupted increase in the world population fraction that is aged 65 and older. However, in the less-developed nations, the percentage aged 65 and older actually went down between 1950 and 1970 before starting a sustained rise in 1980. In the period right after World War II, the death rate was declining, but the birth rate was not, and, as you remember from Chapter 5, the earlier declines in the death rate tend to favor the young, so that had the early effect of increasing the youthful population at the proportionate expense of the elderly. In 1950, only 5.2 percent of the world's population was aged 65 and older and by 2000 that had climbed a bit to 6.9 percent, as you can see in Table 9.1. However, that was the lull before the storm since the huge batch of babies born all over the world after World War II will begin moving into the older ages in the first half of this century: By 2025 the percentage 65+ is projected to rise to 10.5 percent and then by 2050 it will be 15.9 percent. By that time, the United Nations projects that world population growth overall will have slowed to less than 1 percent per year, but the older population will still be growing by more than 2 percent per year. By 2050, it is anticipated that there will be 1.4 billion older people in the world—more than the entire population of China as of 2004.

Where Are the Older Populations?

Given the data in Table 9.1, you can see that there are different answers to two seemingly similar questions. If the question is: "Where are the nations that have the highest percentage of older people?" then we point to the richer countries, especially those in Europe, for the answer. As of the year 2000, 14.7 percent of Europe's population was 65 and older, and it is projected that by 2050 that will have risen to 27.9 percent. North America was the only other region of the world that was in the double digits as of 2000, but by 2050 we can expect that 10 percent of the population in every region of the world except sub-Saharan Africa will be aged 65 and older.

On the other hand, if the question is: "Where are the people who are 65 and older?" we point to the less-developed countries. This is a result of where the world is with respect to the age transition. You can see that although Europe had more than 100 million older people in 2000, Eastern Asia (especially China, Japan, and South Korea) had a combined total that was higher than Europe's. By 2025, when Europe's older population will number nearly 150 million, there will likely be nearly a quarter of a billion in Eastern Asia, and more than 150 million in south central Asia (which includes India).

Let us focus on individual countries instead of regions. Table 9.2 shows the estimates for the year 2004 of the population (percentage and absolute number) aged 65 and older. Using the percentage of the population aged 65 and older as the index to an older population, nine of the top 10 countries are in Europe and the other one is Japan. You can see in Table 9.2 that the list is led by Italy (19.3 percent in 2004), followed by Japan (19.2 percent), Greece (18.7 percent), Germany (18.1 percent),

Table 9.2 The Top 10 Countries in Terms of Percentage Aged 65 and Older and in Terms of Number of People Aged 65 and Older as of 2004

Top 10 Countries in Terms of Percentage Aged 65 and Older

Rank	Country	Number 65+ (thousands)	Percent 65+
1	Italy	11,066	19.30
2	Japan	24,517	19.18
3	Greece	2,056	18.73
4	Germany	14,967	18.14
5	Belgium	1,817	17.58
6	Sweden	1,561	17.57
7	Spain	7,049	17.14
8	Switzerland	1,216	16.97
9	Croatia	743	16.83
10	Latvia	377	16.48

Top 10 Countries in Terms of Number of People Aged 65 and Older

Rank	Country	Number 65+ (thousands)	Percent 65+
1	China	97,150	7.40
2	India	56,509	5.23
3	United States of America	36,393	12.25
4	Japan	24,517	19.18
5	Russian Federation	19,893	13.97
6	Germany	14,967	18.14
7	Indonesia	11,885	5.34
8	Italy	11,066	19.30
9	Brazil	10,287	5.69
10	France	9,839	16.28

Source: Adapted from United Nations, 2003, World Population Prospects: The 2002 Revision (New York: United Nations).

and Belgium (17.6 percent) The United States, by the way, is 38th on this list, with 12.2 percent of its population aged 65 and older.

If we look at the total number of people aged 65 and older, the top 10 list looks more like the list you have seen earlier in terms of total population size: China has the largest number of older people as of 2004, followed by India and the United States. Only three countries show up on both of the top 10 lists: Japan, Germany, and Italy.

If we zoom into individual countries, we typically find spatial concentrations of the elderly in different parts of a country. Within the United States, for example, the state of Florida exceeds all others in its percentage of the population that is 65 or older (18 percent in 2000), followed by West Virginia and Pennsylvania (both at 15.6 percent), Iowa (15.2 percent) and North Dakota (15.0 percent). At the other extreme, the youngest states are, in order, Alaska, Utah, Georgia, California, and Texas. Canada's provinces are much less variable than the states in the U.S., but Manitoba has the highest percentage of the population aged 65 or older (13.0 percent in 2000), followed closely by British Columbia (13.0 percent) and Quebec (12.8 percent) (Statistics Canada 2001a). Canada's territories, which are populated especially by indigenous groups, have much lower fractions of older people, with Nunavut having the lowest at 2.6 percent.

The comparison of U.S. states with Canadian provinces is interesting because Florida has the highest percentage aged 65 and older largely as a result of the in-migration of retirees. By contrast, Manitoba has the highest percentage 65 and older largely as a result of the out-migration of young adults to other parts of Canada. Mexico's states all have much lower percentages of older people than in the United States or Canada. The highest is Zacatecas with 6.2 percent in 2000, followed by Yucatan, Nayarit, Oaxaca, and Distrito Federal at 5.9 percent (INEGI 2001b).

The Effect of Declining Mortality

As death rates drop and life expectancy increases, the older population increases partly because life expectancy goes up somewhat at the older ages, so people live longer when old. I mentioned in Chapter 5 that we do not know if there is a limit to the human life span, so we do not know whether the end of the age transition is a rectangularization of the age structure, or if people will just keep surviving to ever older ages. However, we do know that thus far in history the most important impact of mortality on aging (in terms of the number of older people) has been the fact that lower mortality increases the probability of surviving to old age. These relationships are shown for you in Table 9.3, using data for the United States.

In 1900, the life expectancy in the United States was lower than in almost every country in the world today. At birth, a male could expect to live an average of 46.3

Table 9.3 Declining Mortality Increases Likelihood of Survival to and within Old Age, United States 1900–98

Year	Life expectancy at birth		Proportion surviving from birth to age 65		Life expectancy at age 65		Proportion alive at age 65 who survive to age 80	
	Males	Females	Males	Females	Males	Females	Males	Females
1900	46.3	48.3	0.36	0.41	11.5	12.2	0.27	0.33
1950	65.6	71.1	0.66	0.82	12.8	15	0.38	0.47
2001	74.5	79.8	0.79	0.87	16.5	19.4	0.56	0.69
Change from 1990 to 2001	28.2	31.5	119%	112%	5	7.2	107%	109%

Sources: Data for 1900 and 1950 are from National Center for Health Statistics, 1997, Health, United States, 1996–97 (Hyattsville, MD: Public Health Service), Table 29, and author's own calculations of proportions surviving based on given life expectancies; data for 2001 are calculated from life tables in Chapter 5.

years, compared to 48.3 for females. These life expectancies were about the same as England at the time, but lower than most Scandinavian nations. At this level of mortality, a male baby had only a 36 percent chance of surviving from birth to age 65, and a female baby had a 41 percent chance of living that long. For those who reached 65, the pattern of mortality in 1900 produced a life expectancy at age 65 of 11.5 additional years for males, and 12.2 additional years for females. Scarcely one-fourth of males aged 65 and only about one-third of females aged 65 could expect to still be alive at age 80.

Significant improvements in mortality occurred in the United States and other developed countries between 1900 and 1950, as you can see in Table 9.3, and the improvements have continued, albeit at a slower pace, since then. By 2001, females in the United States had a life expectancy at birth of 79.8 years—31.5 years more than at the beginning of the twentieth century. This has more than doubled the proportion of female babies surviving from birth to age 65 and has been a major reason for the increase in the number of older persons in the United States. It is now commonplace for people to reach old age and we take it nearly for granted, as I discussed in Chapter 5. On reaching age 65, females can now expect to live an additional 19.4 years on average—an improvement of 7.2 years more beyond age 65 than was true in 1900. This translates into a doubling of the proportion of women age 65 who will still be alive at age 80. Indeed, in 2001 in the United States, more than two-thirds of the women alive at age 65 will still be alive at age 80. The situation is not quite so favorable for males, but it has been improving for them, too, as you can see in the table.

The Effect of Declining Fertility

Although declining mortality will always lead to an increase in the number of older people, the percentage age 65 and older will only increase noticeably if fertility declines, as it always has historically in the context of the demographic transition. We can use the stable population models I discussed in Chapter 8 to illustrate this. In Table 9.4, I have calculated the percentage of the population that would be aged 65 and older if a population maintained over time the various combinations of mortality and fertility that I have shown in the table. For example, a country whose life expectancy was only 30 years would have 3.9 percent of the population aged 65 and older if the total fertility rate (TFR) were five children per woman, and it would drop to 2.8 percent if the TFR went up to six (note that at a TFR of four or below, the country would be depopulating, so the percentage aged 65 and older would be temporarily high until everybody died off).

Let us stay focused for the moment on the total fertility rate of five children. You can see that as life expectancy increases from 30 to 60 years, the percentage aged 65 and older actually goes down slightly. I mentioned above that this had happened in Mexico. Mortality was declining in Mexico between 1950 and 1970, but fertility had not begun to decline, so the older population was declining as a percentage of the total population. However, you can also see in Table 9.4 that, as mortality continues to decline, eventually it gets low enough that the percentage representing the elderly begins to increase even if fertility does not change. Of course, a population with a life expectancy of 75 years and a total fertility rate of five children would be doubling in

Table 9.4 Percentage of Population 65 and Older Determined More by Fertility than by Mortality

Life Expectancy at Birth	Total Fertility Rate (TFR)				
	2	3	4	5	6
30	a	a	a	3.9	2.8
40	a	a	5.6	3.8	2.7
50	a	8.8	5.5	3.7	2.6
60	15.0	8.8	5.4	3.6	2.5
70	16.5	9.2	5.7	3.7	2.6
75	18.0	9.9	6.1	4.0	2.8

Source: Data are based on Coale–Demeny "West" Stable Population Models.

[a]No calculation made because this represents a situation of depopulation.

size every 20 years, and the number of older people eventually would be very large, even if they represented only 4 percent of the total.

If you look at any given level of life expectancy, you can see clearly that at each lower level of fertility, the percentage of the population that is aged 65 and older is higher. Now, finally, if you go down the diagonal of this table from the top right (low life expectancy and high fertility) to the bottom left (high life expectancy and low fertility), you can trace the typical path of the percentage aged 65 and older as a country passes through the demographic transition.

The Effect of Migration

The impact of migration on the growth of the older population is, as always, more complex than the effect of either mortality or fertility. Younger people are more likely to migrate than the elderly, but there is no biological limitation to migration and so we find that, although in general migration tends to leave the elderly behind, at other times the elderly move disproportionately to specific areas (such as Florida in the United States) in search of "amenity-rich communities with sunnier, warmer and recreationally more enjoyable environments" (Rogers 1992:3). Haas and Serow (1993) have pointed out that this is a type of pull migration in which people use their information sources to decide where to migrate, in a manner similar to patterns of international migration, especially from Mexico to the United States, as I discussed in Chapter 7. And, in a fashion that is also similar to Mexicans migrating to the United States, many older migrants are seasonal—perhaps we could call them "trans-communal"—having incorporated the movement between two communities (such as one in the Sun Belt and another in the Snow Belt) as a regular part of their life (Hogan and Steinnes 1998).

Migration affects population aging most by what older people do not do—they do not migrate very much. Throughout the world, older people tend to age in place. As a consequence, the urbanization part of the migration transition typically leaves a generation of older people abandoned in the countryside by their children who migrate to the cities to work. In richer nations, where most people now already live in cities, those children who migrated to the cities will age in place in city and may find themselves abandoned by their own children, who are more likely to move out to the suburbs. It is possible, I suppose, that at the end of the demographic transition, the older population will find that their children have remained close to them in the suburbs. These patterns affect the social relationships of older persons, as I discuss later. For now, it is enough to remember that out-migration tends to increase the percentage of the population that is older in a region, whereas in-migration has the opposite effect, except in those relatively rare cases where the in-migrants are themselves older persons.

Individual Aging as a Factor in Population Aging

"Are you hoping for a long life? Thought so. Are you looking forward to growing old? Thought not. People have wanted one without the other for thousands of years, and have invariably been disappointed" (*The Economist* 2000b:23). Population aging would be of considerably less consequence if there were no physical changes taking place in the aging body to which people and society in general can or should respond. So, to understand population aging, we must also understand what those physical changes are and how people react to them.

Individual aging has both a biological and a social component. The biological component refers to those physical changes that are the defining aspect of the aging process—the changes that tell us and others that we are, in fact, getting older. The social component refers to the way that people and social systems react to these physical symptoms of growing older.

Biological Aspects of Aging

Could you live forever if you were able to avoid fatal accidents and fatal communicable diseases, and if you had as healthy a lifestyle as possible? The answer seems to be no (Olshansky, Hayflick, and Carnes 2002). Biologists suggest to us that as we move past the reproductive years (past our biological "usefulness"), we undergo a set of concurrent processes know as **senescence**: a decline in physical viability accompanied by a rise in vulnerability to disease (Carnes and Olshansky 1993; Spence 1989). Some important generalizations about physical aging are: (1) the physiological changes that take place tend to be steady but gradual through adulthood into old age; (2) the more complex the bodily function, the more rapid its rate of decline; (3) individuals age at different rates, and different tissues and systems within one person may also age at different rates; (4) aging brings with it a lowered ability to respond to stress; and (5) aging brings a diminished ability to resist disease (Weg 1975).

Several theories are in vogue as to why people become susceptible to disease and death as age increases. These can be roughly divided into theories of "**wear and tear**" and "**planned obsolescence**." Wear and tear is one of the most popularly appealing theories of aging and likens humans to machines that eventually wear out due to the stresses and strains of constant use. But which biological mechanisms might actually account for the wearing out? One possibility is that errors occur in the synthesis of new proteins within the body. Protein synthesis involves a long and complex series of events, beginning with the DNA in the nucleus and ending with the production of new proteins. At several steps in this delicate process, it seems possible that molecular errors can occur that lead to irreversible damage (and thus aging) of a cell. Of special concern is the possibility that errors may occur in the body's immune system, so that the body begins to attack its own normal cells rather than just foreign invaders. This process is called autoimmunity. Alternatively, the immune system may lose its ability to attack the outside cells, leading to the functional equivalent of AIDS, in which the body no longer can fight off disease.

The planned obsolescence theories revolve around the idea that each of us has a built-in biological time clock that ticks for a predetermined length of time and then is still. It essentially proposes that you will die "when your number is up," because each cell in your body will regenerate only a certain number of times and no more. Which of these theories makes the most sense? Good question. If you had a solid answer, you could bottle it and retire. Short of that, let me remind you that current evidence points to two basic conclusions: (1) aging is much more complex than we have previously assumed, and different theories may fill in part of the puzzle; and (2) we have not yet discovered the basic, underlying mechanism of aging that (if it exists) would explain everything. The planned obsolescence theory could explain why animal species each have a different life span, whereas the various aspects of wear and tear seem better able to explain why members of the same species show so much variability in the actual aging process. Olshansky and Carnes (1997) have offered their opinion that, "Although there is probably not a genetic program for death, the basic biology of our species, shaped by the forces of evolution acting on us since our inception, places inherent limits on human longevity" (p. 76).

As we discuss human aging, it is important not to confuse senescence, the biological process of aging, with **senility.** Senility (or *senile dementia*, as it is known formally) has often been used as a derogatory term applied to forgetfulness or seemingly erratic behavior on the part of older people. We now know, however, that senility is not a normal part of aging. Instead it represents the consequences of several different organic brain disorders, especially **Alzheimer's disease,** which is a disease involving a change in the brain's neurons, producing behavioral shifts in its victim. This is a major cause of organic brain disorder among older people and a great deal of federally funded research has been aimed at identifying the exact mechanisms of Alzheimer's so that a treatment or cure can be found.

The physical changes that actually take place as we age are often exaggerated in our minds when we are younger. The gradualness with which changes occur usually provides ample time to learn to cope with them, and actual changes may well occur later or be different from what we suspected when we were younger. For example, one of the most blatant stereotypes of aging has to do with sexual functioning in old age. As a matter of fact, sexual activity tends to remain fairly strong well into old age, at least until death removes a person's partner (Atchley and Barusch 2004). The frequency and enjoyment of sexual relations in old age seem to be more closely related to individual preference than to age itself. Thus, people who enjoyed sex and were sexually active when younger will likely remain that way when older. although your physical capacity to engage in intercourse may diminish with age (perhaps enhanced by Viagra or Cialis if you are a man), your level of sexual activity will likely be influenced more by social factors than by biology.

Social Aspects of Aging

The social world of the elderly is different from that of younger people because virtually every society has some built-in system of age stratification—the assignment of social roles and status on the basis of age. As people age, their sets of social obligations and expectations change, and certain kinds of behavior are deemed appropriate for some ages but not for others. I discussed this phenomenon in general terms in the previous chapter, but here I will highlight the importance of age stratification for the older population.

One aspect of age stratification that crops up especially in Western societies is the stereotyping of older persons. Older people are often seen as being politically and socially more conservative than younger people, intolerant of others (especially younger people), and more prone to despondency and impairment. So pervasive are these stereotypes that older people themselves have them of other older people (Hummert et al. 1995). Stereotypes are most likely to arise when groups of people do not share a great deal of their life in common, and age stratification encourages this generational separation in modern societies.

Age stratification and generational separation have also been implicated as determinants of the social status of older people. Indicators of status such as access to economic resources, policy-making influence, and breadth of social relationships all suggest that the status of the elderly in Western nations underwent a decline for several decades until recently, when it seems to have risen once again. How and why have these changes occurred? The most widely accepted explanation of these events is based on the joint processes of the demographic transition and modernization.

Modernization and the Status of the Elderly   Most historians accept the idea that the status of older persons was relatively high in premodern, agrarian societies, even though there may be considerable variety in exactly how high that status was (see, for example, Kertzer 1995). However, things began to change as societies modernized and people began to live longer. The flip side of the increasing survivability of people to old age is the increasing likelihood that a young or middle-aged adult will have surviving parents. For example, under a constant mortality regime of 40 years of life expectancy, a person aged 30 has less than one chance in four of having both parents still alive, and there is about one chance in two that one parent will be still alive. A high level of mortality increases the odds that a younger person will be able to inherit the family farm or business, or will be able to move into some other position in society being vacated by the relatively early deaths of other people. However, at a life expectancy of 70 years, nearly two out of every three adults aged 30 can expect to have both parents still alive, and nearly eight in 10 can expect to have at least one parent still alive. This pattern essentially clogs up familial and societal mobility, and makes the younger population somewhat more redundant. The problem is alleviated as the birth rate drops, but in the middle of the demographic transition, when mortality is declining more rapidly than fertility, the pressure on population has historically produced a migration out of the less-developed, more agricultural areas and into urban areas, or into foreign countries, where younger people may have economic opportunities that would not otherwise have been available to them.

These ideas are summarized in Kingsley Davis's theory of demographic change and response (discussed in Chapter 3), although his focus is not on the older population. The conceptual tie to the elderly has been provided by Donald Cowgill (1979), and his framework is displayed in Figure 9.3. In Cowgill's model, four factors involved in modernization—health technology, economic technology, urbanization, and education—combine to lower the status of the elderly. These are all intertwined aspects of the demographic transition, so I have focused your attention on them.

The health technology accompanying modernization increases longevity, as you well know. That greater longevity in turn creates intergenerational competition for jobs, because people are no longer as likely to die and create room for the young to enter and advance within the labor market. This competition has led to the phenomenon we call retirement. Work itself is a valued activity in society, since work produces income. Because income is a major source of status, retirement leads inexorably to decreased status unless you retire with sufficient income—which most people have trouble doing.

The economic technology that accompanies modernization is closely associated with the creation of new, especially urban, occupations. Because the young are more likely to migrate to cities than are the old, and because urban occupations tend to pay more than those in rural areas (as I will discuss in Chapter 11), the result is "an inversion of status, with children achieving higher status than their parents instead

Figure 9.3   How Is Status of the Elderly Influenced by Demographic Transition?

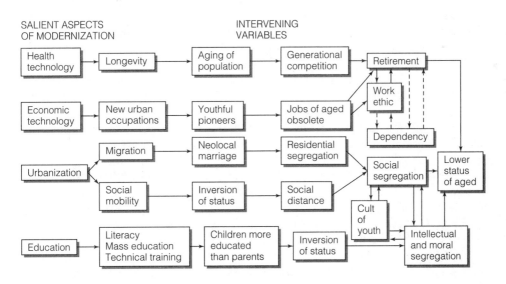

Source: Donald Cowgill, 1979, "Aging and Modernization: A Revision of the Theory," in J. Gubrium (ed.), Late Life: Communities and Environmental Policy (Springfield, Ill.: Charles C Thomas), Figure 5. Reprinted with permission.

of merely moving up to the status of their parents as in most pre-modern societies" (Cowgill 1979:59).

The third factor that Cowgill mentions as leading to a decline in the status of the elderly is urbanization itself. In addition to its relationship to mobility and the inversion of status, as I just mentioned, urbanization segregates the residences of the young and the old, which in turn produces a greater social distance between generations, relegating aging parents to a more peripheral role in the lives of their children. The greater physical distance between generations thus accelerates the already inverted social distance and further depresses the status of the elderly.

The process of modernization is closely bound up with an increase in formal education. In preliterate societies, education is based especially on the recitation of life's experiences by older people. The school of hard knocks prevails, and experience is the best teacher—a feature of life that enhances the status of the elderly. Modernization, though, tends to bring with it institutionalized, formal, mass education, and the content is more technical than experiential. The main targets of such educational programs are always the young, and this leads to the young being better educated than their parents. This further exacerbates the status of the older person because it leads to a greater intellectual and moral (or value) distance between the generations. Education brings with it (indeed is part of) changes in the value system of society, and the constant upward spiral of education has led to shifting values and intellectual levels from one generation to the next.

Status of the Elderly in the Context of the Demographic Transition   Cowgill's theory of loss of status for the elderly leaves them largely abandoned in the countryside, which reached a peak in the United States before World War II. However, eventually the countryside empties out—at least it did in the now-developed countries—and the vast majority of the population becomes urban. This has happened in the United States and other industrialized nations in the past few decades and seems to be an inevitable part of the demographic transition. The people who abandoned their parents by moving to the city have now grown old in the city, which has reduced the likelihood that they too will be abandoned by their children. More importantly, growing old in the city has given them broader access to social, economic, and political resources, and makes it less likely that they will be economically deprived and abandoned, even if their children do move somewhere else. Indeed, the end of the urban transition and a slowdown in the rate of growth of development affords the elderly an opportunity to regain control of resources, which implies a rise in their status.

The upshot of this is that the older households in the United States are those in which net worth tends to be the greatest. More specifically, we can find out who controls the United States by looking at the *Forbes Magazine* list of the 400 richest Americans to see what their ages are. Figure 9.4 shows the data for the year 2003; it is very clear that the elderly are far more likely to be very wealthy than are younger people. Of the wealthiest 400, you can see that 217, or 54 percent, were age 65 or older, despite the fact the elderly comprise less than 13 percent of the population. Not only are there more wealthy people at age 65 and older than at younger ages but they also command the greatest amount of wealth as a group. However, per-person wealth is highest for the 45–54 age group (who average $3.1 billion each),

Figure 9.4   U.S.'s Wealthiest 400 People in 2003 Disproportionately Aged 65 or Older

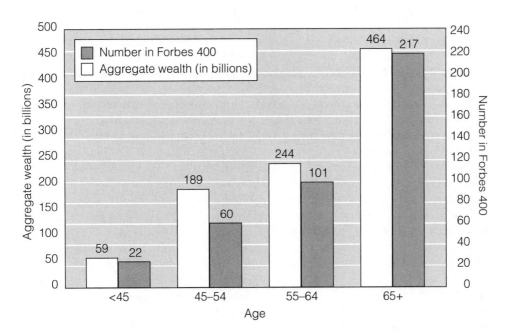

Source: Adapted from data at http://www.forbes.com, accessed 2004.

largely because the world's richest man, Bill Gates, was still only 47 in 2003. The average net worth among the 65-and-older group is "only" $2.1 billion.

Four of the wealthiest Americans in 2003 made their money in computers, and five of the other six are members of the Walton family, who founded Wal-Mart, but a large fraction of the very wealthiest inherited their wealth. As I discussed in the previous chapter, the cohort of people now aged 65 and older in the United States are better off than any that came before, and as they die, there will be a huge intergenerational transfer of wealth, probably the largest that the world has ever seen (Herman 1999). However, because people are living longer, the children who inherit that wealth will probably be fairly old themselves when they come into their money.

As the United States and other industrialized countries move into what may be the final stages of the demographic transition, the data suggest that the elderly have regained much, if not all, of the status they may have lost during the initial phases of modernization. At the same time, less-developed nations are now moving into those middle stages of the demographic transition, wherein a reduction in status of the elderly seems almost inevitable. Consistent with Cowgill's modernization theory, as less-developed societies go through the process of modernization, the status of their older population should drop, much as it did in the now-developed regions of the world. So it is that in country after country, the complaint is being heard that the elderly, increasing numerically with each new cohort, are being left behind by the younger generation and forgotten by the rest of society. At the beginning of the twenty-first century, nearly two out of every three older people in the world are in the less-developed nations, and their plight could become a significant factor in regional politics.

An anecdotal, but perhaps significant, bit of evidence about the status of the elderly in developing countries emerged in the People's Republic of China in 1997 when the government passed a law protecting the rights and interests of the elderly. In a nation that made famous the concept of *filial piety*—respect of children for their parents—a law now exists that forbids discrimination against the aged by "insulting, mistreating, or forsaking them" and the law calls for appropriate measures to be taken against anyone committing such abuse (*Global Aging Report* 1997). The result has been an explosion of lawsuits by parents against their children (Chang 2000). In 1997, Bolivia also put in place a law that guarantees older people legal and police protection, among other things. The necessity for such guarantees in Latin America is illustrated by a study in Costa Rica indicating that older family members were more likely to be abused if they were members of a larger, rather than a smaller, household—there was no safety in numbers. A report from Ethiopia indicated that widespread poverty in the capital city of Addis Ababa was particularly brutal for older people, some of whom were living in cardboard boxes with a plastic cover to keep out the rain (*Global Aging Report* 1997).

As the demographic transition progresses, the urban transition, combined with education and a redefinition of the labor force, tends to push the elderly aside. At the same time, older people themselves began some time ago to redefine their own expectations, anticipating less from their children in the way of old-age support (Coombs and Sun 1981). This crucial moment of change in the obligations and expectations that the old and the young have toward each other may be the tipping point in the generational flow of wealth, the perceived value of children, and, ultimately, the birth rate. Over the long run, of course, we expect the status of the elderly to rise again in the less-developed nations, as their achievement of higher standards of living and lower birth rates elevates the demographic and social profiles of the older population, as has happened in the more-developed nations of the world.

The Social Context of Aging and the Life Course

The fall and rise in the status of the elderly reminds us that the impact of aging on society is defined in large part by the social context in which it takes place. As historical forces shape and influence the status of the older population, those same forces are shaping the character of the older people themselves. This is the basis of the concepts of life course and cohort flow, which I discussed in Chapter 8. Thus, in an era of rapid social change, such as we have experienced throughout the last century, we should expect that each new generation of people moving into old age will be substantially different from the previous one. This has tremendous implications for society. Younger people may be more likely to engage in innovative behavior than older people are, but those innovations are taken with them into old age, transforming every age group as they proceed through the life course. Some of the innovations are responses to cohort-specific historical events.

Researchers at the University of Michigan found that people remember as most important in their lives those events that took place in late adolescence and early adulthood, and they carry those "generational memories" (and their responses to them) with them as they age (Schuman and Scott 1990). Is there life after high school? Yes, of course, but those experiences as teenagers are incredibly important to each cohort. This is not to be confused, however, with the idea that somehow older people *dwell* on the past. Older people remember the past and are influenced by it, but in the Berkeley Older Generation Study, the period of life mentioned most often by older people as bringing them the most satisfaction was "right now" (Field 1997).

It is these cohort-specific reactions to innovation and change that represent a major force of social change over time. Some events, such as current economic and political crises, may be reacted to in very similar ways by people of all ages (the "period" effect—remember the Lexis diagram of Chapter 8), but deeper social issues, such as sexism, racism, and even ageism, are more likely to be challenged and changed by younger people than by older people (the "age" effect). However, those attitudes do not melt away with age, so we can expect that higher levels of tolerance among younger people today will evolve into greater tolerance at older ages in the future (the "cohort" effect).

Similarly, the high level of political advocacy of youth in the 1960s is being translated into greater political advocacy among the older population as those people reach old age. Already, people aged 65 and older in the United States are the most likely to be registered to vote and most likely to actually vote in national elections, as you can see in Figure 9.5, where I have displayed voting data by age for national presidential elections between 1964 and 2000. That is the age effect. However, between the mid-1960s and the mid-1990s, the population aged 65 and older became more likely to vote in elections, while the population aged 18–24 became less likely to vote. This was a combination of period and cohort effects that, in effect, has put proportionately more political power in the hands of the older population, to go along with their increasing economic clout. Thus, not only are the elderly gaining in economic strength, they are increasingly determining the political direction of the country.

Another important change that is occurring within the older population all over the world is that it is becoming increasingly better educated. Of course, if we look at a cross section of society, we see that older people have lower levels of education than do younger adults. For example, in 2000, only 65 percent of people aged 75 and older in the United States were high school graduates, compared with 87 percent of those aged 25 to 29 (Newburger and Curry 2000). Similarly, only 19 percent of those 75 and older were college graduates, compared with 28 percent of those aged 25 to 29. Despite this cross-sectional, or period, difference, the cohort of people now in the older ages is the best educated ever to have occupied those ages. In fact, in the 30-year span between 1970 and 2000, the proportion of people aged 65 and older in the United States who were high school graduates rose from 28 percent to 65 percent. This rise was almost entirely a result of the higher educational levels of the people moving into the older years, not the recent, but still sparse, attempts to bring older people back into the educational system.

Figure 9.5   Older Americans More Likely to Vote than Younger People: 1964–2000

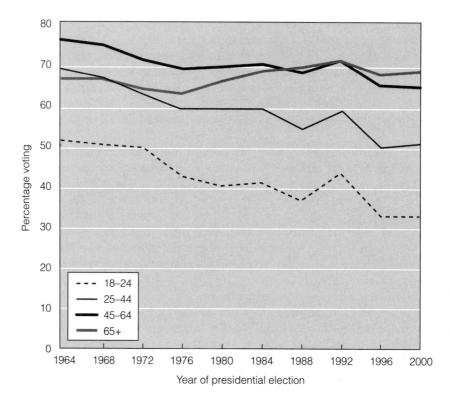

Source: Adapted from U.S. Census Bureau, Reported Voting and Registration by Race, Hispanic origin, Sex and Age Groups: November 1974 to 2000; http://www.census.gov/population/socdemo/voting/tabA-1.xls, accessed 2004.

On average, older people are not less well educated because they are less competent, less able to achieve higher levels of education, or even less ambitious. Rather, it is because the entire world has been experiencing an upward spiral of literacy and education for the last few hundred years, with the result that each generation tends to be better educated than the previous one. Because educational programs will almost certainly continue to be oriented largely to younger people, we can assess at least the minimum educational attainment of older people in the future by looking at today's younger cohorts. For example, the fact that 87 percent of people aged 25 to 29 in the United States in 2000 had a high school diploma foreshadows the year 2040, in which we can expect that nearly 90 percent of the population aged 65 to 69 and older in the United States will be high school graduates—an obvious increase from the 65 percent in 2000.

People carry a lot of personal and social baggage with them into the older years, but improved mortality at older ages means that people who are lucky enough to reach that point in life have more years than ever before in human history in which to enjoy, or at least cope with, the process of aging. This means that old age is more complicated than it used to be. Symptomatic of those complications is increasing diversity by sex and age.

The Sex and Age Structure of the Older Population

In the past, and still today in many less-developed societies, old age was implicitly that age at which a person could no longer make a full economic contribution to the household economy due to one or more disabilities that would eventually (and sooner rather than later, in most cases) lead to death. The same wealth in the modern world that has lowered death rates has also made it possible to separate the decline in economic productivity from a decline in physical functioning. This has allowed the institution of retirement to arise, creating what Laslett (1991) has called the "Third Age," distinguishing it from the later phase of aging—disability and death—which he has termed the "Fourth Age." Although I organize the discussion of sociodemographic characteristics of the older population around these two broad age categories, we first must recognize that the Fourth Age is more likely to be reached by women than by men, creating a phenomenon known as the feminization of old age.

The Feminization of Old Age

Women live longer than men in almost every human society and so they disproportionately populate the older ages. As recently as 1930 in the United States, there were equal numbers of males and females at age 65 and older (partly a result of the earlier influx of immigrant males) (Siegel 1993; Taeuber 1993), but by 2000 there were only 70 males for every 100 females at the older ages. In the United States, the ratio of males to females aged 65–74 declined steeply between 1950 and 1970, leveled off, and has risen a bit since 1980 as men have begun to close the gender gap in life expectancy. Both United Nations and U.S. Census Bureau projections assume that life expectancy gains will be greater for males than females (a reversal of historical trends), and that will help push up the ratio of men to women as time goes by. At age 75 and older, the higher mortality of males really has taken its toll: in 1950, there were 83 males per 100 females at that age in the United States, but this declined to only 54 males per 100 females by the 1990 census, with a slight rise since then. That means that at age 75 and older, two-thirds of the people alive in the United States are women. Again, the projections imply that this trend will be reversed over the next few decades, but women will continue to outnumber men by a large margin at the oldest ages for the foreseeable future.

The general pattern in the sex ratio at older ages has been similar in Canada and the United States, but the actual level of the sex ratio is consistently higher (albeit not by much) in Canada. This is probably due to the joint effects of immigration (immigrants account for a higher fraction of the Canadian than of the U.S. population), and the fact that the gender difference in life expectancy has been slightly less in Canada than in the United States.

In Mexico, the feminization of old age is clearly under way, especially in urban areas where death rates are lowest. Projections for Mexico suggest that by the year

2020, the ratio of males to females from ages 65–74 in Mexico will actually be lower than in the United States, as Mexico moves through that part of the demographic transition where these differences seem to widen for a while.

Looking outside of the Americas, we can note that in Japan both males and females live longer than in the United States and the difference between them in life expectancy is also somewhat less. As a result, the ratio of males to females in Japan has been slightly higher than in the United States since 1970, and it is estimated to be nearly identical to the sex ratio for Canada from about the year 2000 on. Between 1950 and 1970, the ratio was affected by the high mortality experienced by Japanese males in World War II.

The data for Pakistan tell a different story. Pakistan is one of several countries in the world where there are still more men than women at older ages. This is a result of the lower status of women, which increases their mortality rates compared to men over the entire life course. In 1950 in Pakistan, there were 142 men aged 65–74 for every 100 women that age, and the ratio was nearly that high at ages 75 and older. Nonetheless, progress is being made to bring about gender equality, and United Nations projections suggest that by the year 2010 there will be slightly fewer males than females at the older ages.

Even though Pakistan and several other countries may not yet have a female majority at the older ages, they are nonetheless moving in the direction of an increasing feminization of the elderly. This has a wide range of social ramifications for the economic well-being of older persons. Let us examine these in more detail.

The Third Age (The Young-Old)

You have probably grown up expecting, and probably looking forward to, a time interval between the end of your work career and death—a time of leisurely retirement. This is a very recent invention. In the earlier part of the twentieth century, the older population in the United States, for example, was concentrated in the ages 65–74, as you can see in Figure 9.6. High mortality meant that people at that age were pretty close to death, and most worked for as long as they were physically able. As mortality has declined, a greater fraction of people have survived into older ages, thus creating a new period between the traditional entrance into old age (approximately age 65) and the time when death begins to stalk us. We have taken advantage of this in society and have been able to act on the old adage that "youth is wasted on the young" by giving ourselves a youthlike carefree period toward the end of life. This is the so-called Third Age, a time when we are still healthy enough to engage in all of the normal activities of daily life, but are able to be free of regular economic activity.

Retirement, Income, and Resources   In industrialized societies, old age is now stereotypically a time of retirement from labor force activity. Indeed, when the Social Security Act was passed in the United States in the middle of the Great Depression in the 1930s, it was designed quite literally to encourage people to leave the labor force. At the time, the idea was to remove older workers from the workforce to replace them with younger workers and thus lower the rate of unemployment

Figure 9.6   Percentage of U.S. Old-Old Population Increasing over Time, 1900–2050

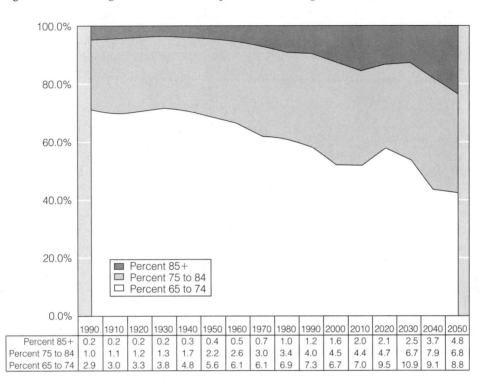

	1990	1910	1920	1930	1940	1950	1960	1970	1980	1990	2000	2010	2020	2030	2040	2050
Percent 85+	0.2	0.2	0.2	0.2	0.3	0.4	0.5	0.7	1.0	1.2	1.6	2.0	2.1	2.5	3.7	4.8
Percent 75 to 84	1.0	1.1	1.2	1.3	1.7	2.2	2.6	3.0	3.4	4.0	4.5	4.4	4.7	6.7	7.9	6.8
Percent 65 to 74	2.9	3.0	3.3	3.8	4.8	5.6	6.1	6.1	6.9	7.3	6.7	7.0	9.5	10.9	9.1	8.8

Note: The proportion of the older population that is aged 65 to 74 has declined steadily over time, as the older population has itself aged; this proces will be temporarily reversed early in the twenty-first century as the Baby Boom generation enters old age.

Source: Adapted from Frank B. Hobbs with Bonnie L. Damon, 1996, "65+ in the United States," U.S. Bureau of the Census, Current Population Reports, Special Studies, P23-190, Table 2-1.

among younger people. Most companies and government entities alike then turned age 65 into a mandatory age of retirement, forcing people out of the labor force whether or not they wanted to retire. People became increasingly interested in retiring before reaching age 65 if they could afford to do so, and the U.S. Congress facilitated this in the 1950s and 1960s by allowing reduced Social Security benefits to be available at age 62, which has been a popular option ever since.

At the same time, however, mandatory retirement ages have now been outlawed in the United States for employees working in companies with 20 or more employees, except for certain groups of people such as highly paid executives and protective services workers (police and firefighters, as well as military personnel). Although the general trend has been toward earlier retirement, that may have to be reversed in the future because the aging of the population means that there are proportionately fewer younger people to pay the taxes necessary to support older people in a lengthy retirement. I discuss this issue in the essay that accompanies this chapter.

WHO WILL PAY FOR BABY BOOMERS TO RETIRE IN THE RICHER COUNTRIES?

Back in 1935, when President Roosevelt's Committee on Economic Security was putting the finishing touches on the Social Security legislation in the United States, two committee members met to discuss the projections that had been made for Social Security expenditures for 1935 through 1980. Treasury Secretary Henry Morgenthau, Jr., and Harry Hopkins, head of the Federal Emergency Relief Administration, were aware of possible problems ahead, as is evidenced by their comments at the meeting (quoted in Graebner 1980:256):

Hopkins: Well, there are going to be twice as many old people thirty years from now, Henry, than there are now.

Morgenthau: Well, I've gotten a very good analysis of this thing . . . and I want to show them [other members of the committee] the bad curves.

Hopkins: That old age thing is a bad curve.

That "bad curve" referred to the ratio of workers to retirees, which, although quite favorable in the early years of Social Security, could be foreseen to worsen over the years as the small birth cohorts of the early 1930s tried to support the numerically larger older cohorts. Despite the fact that reference is often made to the term *trust fund,* you are probably aware that old age Social Security systems in most countries, including the United States, were never designed to have the government actually deposit money in an account with your name on it and have the money accrue principal and interest until you retire and start withdrawing your pension. Rather, almost every system is "pay as you go"—current benefits are paid from current revenue. The age curve looked bad in 1935, but in truth it has turned out to be even worse than expected in all of the wealthier countries of the world: Life expectancy has increased, and then the unexpected baby boom in Europe and North America has injected large cohorts that will soon have to be dealt with in retirement (but that, in the meantime, have helped delay the funding crisis because of their members' payroll tax contributions). On top of all that, the U.S. Congress has considerably expanded Social Security coverage and raised benefits since 1935.

The demographic impact on the U.S. Social Security system was felt keenly through the 1980s as the older population grew more rapidly than the number of younger workers. In 1990, for example, there were 10 percent more people aged 60 to 69 (people moving into retirement) in the United States than in 1980, yet there were 3 percent fewer people aged 20 to 29 (people moving into the labor force). These changes, of course, had been projected for some time, and in the mid-1970s (and again in the 1980s), Congress made adjustments to increase payroll taxes and cut back on the annual allowable increase in Social Security payments. These measures (along with a little borrowing from the disability and Medicare trust funds) allowed the system to survive the 1980s and early 1990s. The late 1990s saw a hiatus in the Social Security crisis because of the slowdown in the increase of new retirees as the Depression-era cohorts reached old age. For example, between 1990 and 2000, the number of people aged 60 to 69 in the United States declined by 6 percent. This eased the pressure of expenditures while revenues rose—the calm before a storm that promises to bring rain for the foreseeable future. Beginning in about 2010, the baby boomers will start retiring, and as you know, this will create a far larger older population than ever before in American history. Although they will be followed into retirement shortly thereafter by the baby bust generation, that was a pretty small dent in the age structure, and after that comes a long line of large cohorts that will keep pressing the resources available for retirement.

As we look ahead to the huge influx of baby boom retirees, we need to keep in mind that when the baby boomers were younger, Congress felt generous about retirement benefits because the baby boom cohort was supplying an influx of new workers to pay taxes, and inflation was showering Social Security with unexpected revenue. In 1972, Congress boosted retirement benefits by 20 percent and built in an automatic adjustment to keep benefits increasing each year along with inflation.

Back in the early 1980s, Robert Myers, former chief actuary of the Social Security Administration, worried that Congress would use the growing surplus of the 1990s to increase benefits, lower taxes, or pay off the national debt. Such a course of action could be disastrous, he felt, because around 2010, the baby boom generation will really crunch the pension system. In fact, in 2001, the surplus *was* used to lower taxes. Will that result in disaster? Quite possibly. Between the years 2000 and 2010, there will be nearly a 50 percent increase in the population aged 60 to 69—an unprecedented rise in the number of people who might be retiring—and this increase will continue until 2030. The Trustees of the Social Security Administration reported in 2003 that:

> For the next few years, annual surpluses for the OASDI and HI trust funds roughly offset the required 75-percent Federal general revenue contribution to SMI. But then, as the baby-boom generation begins to retire, these surpluses quickly dwindle and are replaced by growing annual cash deficits—such that by 2020 net Treasury payments to all four trust funds will total $495 billion, or nearly 2 percent of GDP. Since neither the interest paid on the Treasury bonds held in the HI and OASDI trust funds, nor their redemption, provides any net new income to the Treasury, the full amount of the required Treasury payments to these trust funds must be financed by increased taxation, increased Federal borrowing and debt, and/or a reduction in other government expenditures. Thus, these payments—along with the 75-percent general revenue contribution to SMI—will add greatly to pressures on the Federal budget much sooner than is generally appreciated (U.S. Social Security Administration 2003).

In the 1960s, there were nearly four workers for every Social Security retiree, but by 2030 that will have dropped to only two. The burden on the younger generation will obviously be intense. As a result, there may be considerable pressure on the elderly to be more self-sufficient—not only to work longer (retire later) but also to become involved in mutual self-help organizations that could relieve some of the burden on public agencies. It is ironic indeed that the Social Security system, which was designed in large part to encourage older people

to leave the work force, may in the future be bailed out because people can stay in the labor force longer. Congress has already thought of this, of course, and, as I mention in this chapter, people born after 1959 in the United States will have to wait until age 67 before receiving full Social Security benefits. They will still be able to retire as early as age 62, but only at a lower benefit level than currently prevails. Another hedge—less predictable, of course—is for the economy to grow fairly rapidly in the decade that the baby boom generation reaches retirement age. The kind of structural mobility (when everyone's economic situation is improving) that has typically accompanied rapid economic growth in the United States would make the transfer of money from the younger to the older generation a little less painful.

The experience of countries such as the United States has pushed less-developed countries to think of different ways of attempting to finance retirement for their own aging populations. The model popularized in the 1990s was the "Chile" model, crafted for that country by economists trained at the University of Chicago. The concept is that people must save for their own retirement, but they must be forced to do so by the government (or else the temptation is to spend the money on other things), while at the same time having a reasonably low level of risk of losing their money. In Chile (and now in many other countries as well), workers are required to put a certain fraction of their earnings into a governmentally regulated (but privately managed) set of mutual funds. The savings provide a pool of investment funds that is supposed to help the national economy develop, thus ensuring that the workers will have a nice pension benefit when they retire. The programs are so far too new for us to evaluate how successful they will be. For this reason (and many others), they have not yet been adopted in the United States as a way of dealing with Social Security.

Part of the problem in the United States and in most European nations, of course, is that people are already paying Social Security taxes, so if they are allowed to lower their tax payment to invest in the stock market, the Social Security system will be even worse off in the future. So, the only real way to

(continued)

implement a system of this kind is to force one generation of workers to pay double, to keep paying taxes to finance Social Security payments to the currently retired population while at the same time saving more money than ever before for their own retirement. Very few politicians are willing to propose this to voters, yet something will have to give. Bear in mind that, as critical as the situation is in the United States, it is even worse in most European nations (Lee 2000). Throughout Europe, the benefit levels for retirees are higher than in the United States, while the retirement age is younger, and the more rapid drop in fertility has created an even greater imbalance between workers and retirees. It is indeed a bad curve.

Europeans are now facing up to the fact that the bad curve can be dealt with only by making some fundamental changes in society. As I have discussed earlier in the book, the proposals for "replacement migration," which would bring in younger people from other countries to fill in the gaps created by the European baby bust, would not really solve the problem in the long run, and in the short term the immigrants have created

In general, the availability of an adequate retirement income is a strong inducement for people to leave the labor force. In 1890, 75 percent of all men remained in the labor force past age 64, whereas by 2003, only 19 percent were still employed at that age, although 32 percent of men from 65–69 were still in the labor force. The employment rate of older women declined from the 1960s until the mid-1980s, but has been slowly rising since, probably as a result of the entrance into the older years of cohorts of women who experienced higher levels of labor force participation at younger ages. In 2003, 11 percent of women aged 65 or older were working, although at ages 65–69 the rate was 23 percent.

The drop in employment at older ages is obviously a signal that the average age at retirement in the United States has dropped, especially for men, for whom it dropped from 66 years in the mid-1950s to just below 63 by the mid-1970s, but then leveled off by the mid-1990s (Gendell and Siegel 1996), and has since experienced minor ups and downs (Wiatrowski 2001). Workers in Germany and Sweden experienced even more rapid declines in their retirement age during the latter part of the twentieth century (Geudell 1998). Averages hide a good deal of variability, of course, and increasingly people are choosing more complex routes to retirement than in the past, some out of necessity and some by choice (O'Rand and Henretta 1999; Wiatrowski 2001). The necessity came from the popularity of corporate "downsizing" in the 1980s and 1990s, which forced many workers into retirement earlier than anticipated, and so they officially retired from one job, but maintained a presence in the labor force nevertheless. The choice comes from people who prefer a "blurred transition" to retirement rather than a "crisp exit" (Mutchler et al. 1997). A blurred transition involves, for example, taking retirement benefits from a previous full-time job, but continuing to work part-time, or even full-time but at a different job for less money. Indeed, it seems that money is often the reason for a person not wanting to retire fully.

This type of fuzzy transition has been encouraged in the United States by several changes over the years in Social Security benefits. If you were born before 1938, full Social Security benefits became available to you at age 65. But younger people will have to wait a little longer to receive those full benefits. If you were born more re-political issues for societies trying to cope with the inevitable cultural differences between older Europeans and younger immigrants who, by and large, are not from Europe. The fundamental changes now being discussed in Europe include especially the ideas that (1) workers are going to have to stay in the labor force longer, although this wouldn't necessarily imply a shorter retirement if, for example, retirement age increased in tandem with increases in life expectancy; and (2) pension plans will have to shift from defined benefit plans that guarantee a specific pension based on retire-ment age to a defined contribution plan, in which the pension ultimately depends on the amount saved and the investment results of the money saved. "In sum, the issues of demography and older workers go well beyond the reform of pension systems. They are a matter of social equity, not only between workers and pensioners, but also between generations. Without reform, and without a change in attitude, it will be our children and grandchildren who will pay the price" (Organization for Economic Co-operation and Development 2003:1).

cently than 1959, you won't be eligible for full Social Security benefits until you reach 67, although you can still retire between ages 62 and 67 and receive reduced benefits. Congress has also eliminated the penalty against people who worked after age 65 while receiving Social Security. Benefits used to be denied if you earned more than a limited amount of money beyond your Social Security, but that is no longer true for anybody who is at least 65 (U.S. Social Security Administration 2001b). This makes it easier for people to accept part-time employment or even to stay in a job that pays less than they might wish it did, because their employment is, in a sense, subsidized by their Social Security check (Schultz 2000).

Needless to say, when people drop out of the labor force, their income is likely to decline. This is certainly true for older people. The minimum Social Security benefit is below the poverty level, although in 1974, establishment of Supplemental Security Income (SSI) for older people meant that most old people in the United States are now guaranteed an income no lower than the poverty level. Indeed, between 1959 and 1998, the percentage of people aged 65 and older with poverty-level incomes in the United States dropped from 35 percent to 10 percent (Federal Interagency Forum on Aging-Related Issues 2000).

How much money do older Americans have to live on? In 2002, the median income of households in which the head of the household was 65 years or older was $28,173 per year, scarcely more than half the $49,510 average for households whose heads are less than 65, and well below the median of $59,845 in households headed by someone aged 45 to 49 (U.S. Census Bureau 2003b). The difference, of course, stems in large part from the fact that older householders are not likely to be in the labor force. Note that in 2002, the highest median income earned by men who were still year-round full-time workers was among those aged 70–74 ($54,071), and there was a steady increase in income at each successive age from youngest to oldest for those who were full-time members of the labor force. By contrast, the 2002 average income from Social Security benefits among older men was $12,084.

In the United States, more than 90 percent of all households headed by someone aged 65 or older receives a Social Security check, and almost all families are able to combine Social Security with other sources of income such as a pension, dividends or interest income, property income, earnings from a job (including self-employment), or public assistance (principally SSI). In-kind benefits such as food stamps and Medicare are available to shore up impoverished elders, and for the better off, security in old age may be bolstered by home ownership and other assets. In fact, in 2000, the age at which median net worth peaked was 70 to 74, and even people 75 and older had higher net worth than those aged 45 to 54 (Orzechowski and Sepiella 2003).

How can a person most likely arrive at an advantageous position in old age? The answer is to be in an advantageous position when you are young and to plan well for your future. Educational attainment, for example, is closely related to the economic assets you have to draw upon in old age, and is positively associated with your health in old age (Lauderdale 2001). If we look at net worth among households where the head was 65 or older in 1999, we find that if the head of household had at least gone to college, the net worth was $301,000, whereas it was $157,000 in households where the head had only a high school diploma, and only $63,000 in those households where the head had less than a high school education (Federal Interagency Forum on Aging-Related Issues 2000). Another important lesson from researchers is that poverty in old age is positively associated with retiring early. Since early retirement almost always is associated with a lower benefit level from Social Security and other pensions, the lower income can become a disadvantage for the rest of a retiree's life (Haveman et al. 2003).

For at least one segment of the American population, however, a decent living in old age has been made more difficult because opportunities to do well in younger years were not always available. These are older members of minority groups.

Ethnic Differences   People who are both old and members of an ethnic minority group are said to be in double jeopardy. As minority group members, they spent their lives dealing with prejudice and discrimination; as older people, they also face the prejudice and discrimination that may befall a person just because of being old. In fact, higher mortality rates at younger ages among African-Americans, in particular, means that the probability of reaching old age is less for these groups than for white non-Hispanics or Asians, for whom life expectancy is higher. Indeed, the higher mortality and slightly higher fertility among African-Americans than non-Hispanic whites means that in 2000 only 8.3 percent of the African-American population in the United States was aged 65 and older, compared with 14.8 percent of the non-Hispanic white population. Once they do reach old age, minority elders are less likely than their counterparts in the majority to be doing well economically.

The economic plight of older minority group members can perhaps be best appreciated by noting the percentage below the poverty level. In 1999, when 8.2 percent of non-Hispanic whites aged 65 and over lived below the poverty line, 26.4 percent of African-American elders lived at that level, as did 21.0 percent of older Hispanics (Federal Interagency Forum on Aging-Related Issues 2000).

In terms of income, the patterns for older people by race and ethnicity are similar to those for younger people, but the levels are lower. From the Current Population Survey in the United States, we have data for the two most populous ethnic minority groups—African-Americans and Hispanics. In 1999, the average (median) family income for Hispanics with a householder aged 65 and over was $16,858, compared with $16,312 for older African-American households. As a comparison, white householders aged 65 or older had a median household income of $23,344. We might say then that in the United States, there is a $6,000 per year penalty among older householders for not being a non-Hispanic white.

Lower incomes in old age are largely a consequence of lower incomes at younger ages, and the effects cumulate over a lifetime, producing a considerable inequality in health (Schoenbaum and Waidmann 1997) and wealth (monetary assets) in old age (Smith 1997). Housing is an important asset and minority group members are less likely to be well housed at younger ages, so it is no surprise that they have the least adequate housing among older persons. Housing inadequacy is especially noticeable among blacks living in the south (Markham and Gilderbloom 1998).

Living Arrangements and Family Relationships   In the past, and still today in many less-developed countries, the higher status of the elderly was tied partly to the fact that as old age approached, they were situated in their own housing unit. Even if they lived with their children, it was likely that the children (typically a son with his wife and children) were actually living in the parental home, rather than the other way around (Kertzer 1995). The concept of filial piety, of respect for one's parents, has been a traditional value in most cultures, encouraging children to take care of their parents when the need arises. Of course, in high-mortality societies, the probability that parents would survive to old age (and the probability that their children would survive to help them) was low enough so that relatively few people ever had to make good on that concept, as I mentioned previously.

In the modern world, society after society has bemoaned the fact that the multigenerational family has been a victim of "the movement toward smaller families, the expansion of the female labor market, the geographic mobility of villagers, and the tendency of the young toward more individualistic life styles" (Sung 1995). That description was applied to Korea, but it is echoed in many other places. Older people are no longer assured that they will live out their days nestled in the bosom of their family. To be sure, not all older people necessarily want to live with their children, especially if they are forced to be dependent on the children. A worldwide phenomenon has been emerging of older people wanting "intimacy at a distance." In Malaysia, for example, survey results have shown that the more economically advantaged older people are, the less likely they are to live with adult children. Those who coreside with children do so out of necessity, not necessarily because they prefer that arrangement (DaVanzo and Chan 1994).

As I have already mentioned, although some older people move around the time of retirement, most people age in place. This has led to the creation of what are sometimes called "naturally occurring retirement communities" (NORCs). The term was originally applied to places that were attractive to older migrants, and then became retirement communities (Hunt 1984), but over time it has been applied to any apartment building or neighborhood where a high percentage (50 percent or more) of the residents are more than 60 years of age. This is the spatial component of the life course. People are attracted to neighborhoods and particular apartment buildings because of the similarity of other people's demographic characteristics, including the stage of life. Then, perhaps even without people realizing it, the population ages together and becomes a retirement community.

It is probably most accurate to say that diversity in living arrangements is as much a part of the lives of older people as it is among the young (which I will discuss in more detail in the next chapter). In more developed societies, the diversity of living arrangements among the elderly is compounded by patterns of marriage, divorce, and remarriage in combination with differences in mortality between males and females. The unbalanced sex ratio at older ages in most societies signals a change in marital status, which in turn means a change in living arrangements for many people as they grow older. In the United States in 1998, 75.1 percent of men aged 65 and older were currently married, compared to only 42.9 percent of women. By ages 85 and older, 49.9 percent of men were married, but a scant 13.4 percent of women that age were still married, whereas 77.3 percent were widowed (Federal Interagency Forum on Aging-Related Issues 2000). The United States tends to mirror much of the world in these terms. Males, of course, are less likely to experience a change in marital status as they grow older, because they are more likely to be outlived by their wives and more likely than older women to remarry.

Does a change in marital status affect living arrangements? The answer is yes, and it means especially that more women wind up living alone. In fact, the proportion of older people living alone in the United States has been constantly increasing for the past several decades (Choi 1991; Federal Interagency Forum on Aging-Related Issues 2000). Men are not nearly as likely as women to live alone, of course, since they are more likely to be married. At ages 65–74 in 1998, for example, 14 percent of men lived alone, compared to 30 percent of women. By ages 75 and older, the percentage had climbed to only 22 percent for males, but reached 53 percent for females. This is in contrast to the still common notion that old people are packed off to the "home." Only a very small fraction (less than 5 percent) of the older population in the United States lives in an institution designed specifically for the elderly, and most who do are there for health reasons, as I will discuss below (Hobbs and Damon 1996). This is also in contrast to the pattern earlier in this century, when widowhood in old age was more apt to mean that a woman would be living with one or more of her children (Harevan and Uhlenberg 1995).

In American and European societies, older people are much more separate from their children than in Asian societies such as Japan (Kinsella 1995). In Japan, it has long been the norm for older parents to live with a child (usually the eldest son). Will the Japanese become Westernized and alter the pattern of elders living with their children? Maybe, but it is likely to be a slow process, since the respect and dignity afforded older people in Japan are not likely to be eroded in a single generation. The proportion of older parents living with children in Japan is decreasing, especially in urban areas and among the highly educated (as we might expect from Cowgill's theory), but the 1995 Census of Japan still found that only 6 percent of males and 16 percent of females aged 65 and older were living alone (Statistics Bureau of Japan 1998)—far lower percentages than the ones I noted above for the United States. And while these same values prevail in China, they will be severely put to the test as the post-World War II baby boom moves into old age in the next two decades. There will be so many older people who had very small families that the living arrangements of the elderly, and all other aspects of their life, will become increasingly complicated (Zeng and George 2000).

Similar trends have been experienced by ethnic minority groups in the United States. For example, there is a strong tradition in the Mexican-origin community in the United States for family members to care for older parents, but the reality is that in the United States, as in Mexico, the demographic and social trends have been changing the patterns of intergenerational relations (Dietz 1995). Older Mexican-Americans are increasingly living independently of their children, although they, like African-Americans, are still less likely than non-Hispanic whites to be homeowners. Data from the 1995 American Housing Survey show that among Hispanics, 60 percent of households headed by a person aged 65 or older were owner-occupied, slightly lower than the 65 percent for African-Americans. These numbers are in contrast to the 81 percent ownership rate among older non-Hispanic white households (U.S. Department of Housing and Urban Development 1999).

The Fourth Age (The Old-Old)

Definitions of the old-old population vary (Laslett 1995), although Suzman and associates (Suzman, Willis, and Manton 1992) have suggested that being 85 or older is a reasonable criterion for the *oldest*-old. The distinguishing characteristic of this stage of life is an increasing susceptibility to senescence—increases in the incidence of chronic disease and associated disabilities, and, of course, death. It is the age that Shakespeare described as the second infancy of "mewling and puking" (Mabry and Bengston 2001). If life expectancy refers to when we die, the concept of **disability-free life expectancy** refers essentially to when we arrive at the "Fourth Age"—when the rest of our life will be increasingly consumed by coping with the health effects of old age. Fortunately, this age keeps being pushed back to later in life (Cambois, Robine, and Hayward 2001; Egidi 2003).

To be sure, one of the greatest fears that many younger people have about growing old is that they will become frail, decrepit, and dependent and then be shuffled off to a nursing home, where they will be neglected and die. However, data from the U.S. National Nursing Home Survey indicate that in 1999 only 4.3 percent of the population 65 and older was living in a nursing home, and that percentage has been steadily *declining* over time, not going up (National Center for Health Statistics 2003). Among the younger old (ages 65 to 74) in the United States, only 1 percent of people were residing in a nursing home, whereas the percentage rose to 19 percent at ages 85 and older. If you survive to age 90 or older, the chances increase dramatically that you will spend at least some time in a nursing home—46 percent for women and 31 percent for men (Feinleib, Cunningham, and Short 1994). These numbers are fairly similar to other wealthier countries such as France (Désesquelles and Brouard 2003), Canada, the United Kingdom, Switzerland, and Finland (Légaré and Martel 2003), so the pattern is not peculiar to the United States.

Many of the functions of nursing homes are gradually being replaced by home health care and hospice care, but the move to a long-term care facility is usually occasioned by disabilities that limit a person in his or her ability to carry out the activities of daily living (ADLs, as they are usually called). I mentioned earlier that this is often associated with the progression of Alzheimer's disease, although there are, of course, other causes of disability severe enough to limit one's activities. At ages 65 to 69, nearly three-fourths of males and females in the United States report that their health is good or excellent and that they have no difficulties with the activities of daily living. You tend to be the best judge of your own health, by the way, so self-reported health is a very good index of your physical well-being (Ferraro and Farmer 1999; Ferraro and Su 2000). By ages 85 and older, about two-thirds of men and women report their health to be good or excellent, and so it is not surprising that about one-third of men and one-half of women at the oldest ages report at least one difficulty with ADLs (Mermelstein et al. 1993).

Without question, the risk of chronic disease and disability increases as we get older. It is easy to exaggerate that in your mind when you are younger, however. If we mentally associate age with disability, then we are more likely to notice older people who are disabled and less apt to see that most older people are functioning in a normal fashion. Furthermore, and perhaps even more importantly, we will overestimate our own risk of morbidity in old age by looking at the current cohort of the oldest-old. Each group of people moving into old age is healthier than the previous one (Manton and Land 2000), so if you have a few decades to go before you reach your 80s, the health of the oldest may by then be markedly better. We might even have redefined what it means to be old.

Centenarians and Rectangularization—Is This the End of the Age Transition?

At the beginning of the twenty-first century, it has to be recognized that most people do not survive to be 100 years old. Nonetheless, this is one of the fastest-growing ages in the population, reminding us that the life course in the future is going to be very different from what it was just a few decades ago. At the end of World War II, there were only a few hundred centenarians in England and Wales; now there are several thousand, most of whom are women (Thatcher 2001). In the United States, there were probably about 2,300 centenarians in 1950, after adjusting for likely age exaggeration (Krach and Velkoff 1999). In 1990, that number had increased to more than 30,000, and there were more than 50,000 counted in Census 2000, most of whom were women. Mexico's census in 2000 enumerated nearly 20,000 centenarians, the majority of whom were women. By 2025, United Nations projections suggest that there will be as many centenarians in China as there are people of all ages in Mexico right now (nearly 100 million); by 2045, Chinese centenarians will equal the current population of the United States (nearly 300 million).

The numbers for the present are still pretty small, but you can see the trend: It looks like the end of the age transition is associated with increasing numbers of people at the highest ages. Thus far, then, we have not seen a clear pattern of rectangularization which, as I discussed in Chapter 5, would mean that people essentially live to an advanced age and then quickly die off. The emerging pattern is that we are continuing to push the envelope, so to speak, of old age. This is, in reality, one way in which population growth can be maintained even in the face of below-replacement fertility. Or, put more accurately, if people live to ever older ages, we will have to redefine what below-replacement-level fertility really is, because additional years lived at old age essentially compensate for some of the babies who are not being born—at least in terms of the total population size.

Summary and Conclusion

The older population is the most rapidly growing segment of the population almost everywhere in the world. This is a result of declining mortality. In addition, as fertility drops, the percentage of the population that is elderly also begins to rise. Changes occur in many aspects of social organization as a population ages, because the aging process brings with it numerous changes in individuals themselves, both biological and social. The biological changes are related to the gradual decline in physical functioning and the concomitant rise in susceptibility to disease. Social changes are related especially to the system of age stratification, and in the context of the demographic transition this seems to generate a decline in the status of older people as a society moves into the middle of the transition. The downgrading of status seems to be a combined result of greater longevity, which has led to retirement (which has lower status than work); economic technology, which renders the skills of the elderly obsolete; urbanization, which segregates the generations and reinforces the inversion of status between generations engendered by the upward spiral of technology; and improvements in education, which means that children tend to be better educated than their parents. Related to these historical changes is the fact that each cohort is unique in its social and historical experiences as it moves through the life course. This introduces a dynamic element into the process of aging, signaling future changes in the demographic characteristics and lifestyles of the elderly. As the demographic transition moves toward the end stages, the older population is in a position to regain some or all of its former economic and political clout, and thus increase its social status.

Demographically, the older population in the United States is characterized by an unbalanced sex ratio, because male mortality is higher than female mortality. This means that as women age they are increasingly susceptible to widowhood and the prospect of living alone. Old age also brings with it a fairly dramatic drop in income as people are encouraged or forced to leave the labor force. Minority elders are in double jeopardy because they, as younger people, were less likely to have had access to the higher educational and occupational levels associated with high incomes among the elderly.

The future prospects for the elderly differ according to a society's development level. In industrialized societies, the options available to older people are increasing. These include the ability to work longer, the growth in lifelong learning opportunities—college courses designed for people of all ages, and travel discounts for seniors that may open up previously closed opportunities. These trends are likely to continue almost unabated well into this century as a result of the previously mentioned cohort flow. For example, greater participation of women in the labor force and the resulting increase in female financial independence will likely mean a greater sense of personal freedom for women as they grow older. Higher life expectancy in old age is associated with improved levels of overall health and well-being to increasingly older ages. In developing societies, however, the situation seems to be moving in the opposite direction for older people, because of the different timing of the demographic transition. The demographic transition is widening the gap between the older and younger generations in some of the basic sociodemographic characteristics, including education, occupation, and income, and, in the process, pushing the older population to the margins of society. Although I have discussed these sociodemographic characteristics in this chapter as they relate to the older population, it is time to expand that analysis to all ages, and I do so in the next chapter.

Main Points

1. The population aged 65 and older in the world numbers more than 400 million and represents the fastest-growing segment of the world's population.

2. The growth of the older population is actually more rapid in less-developed than in more-developed countries.

3. The increased number of older people is caused by the long-term drop in death rates throughout the world.

4. The percentage of the population that is 65 and older is greatest in more-developed countries, because its increase largely depends on a decline in the birth rate.

5. Most people age in place, often being left behind by younger people who migrate. However, the migration of older people can be an important source of demographic change in selected areas, especially those attractive to retirees.

6. Biological aging refers to a decline in physical abilities accompanied by an increased risk of illness.

7. Socially, aging carries with it a changing set of obligations and role expectations.

8. Increased longevity, higher levels of education, urbanization, and economic technology may all have conspired to alter the status of older people in industrialized nations.

9. The status of older people in developed nations currently is on the rise, moving in the opposite direction of the situation in less-developed nations.

10. The upward spiral of literacy and education over the last few hundred years has meant that each generation of older people tends to be better educated than the previous generation.

11. In most societies, females outnumber males at older ages. The exceptions are nations in which the social status of women tends to be clearly inferior to that of men.

12. The age structure of the elderly can usefully be divided between the "Third Age" (when social forces are most important) and the "Fourth Age" (characterized by disease, disability, and death).

13. The older ages are typically characterized by an increase in widowhood and an increase in the proportion of people (especially women) who live alone.

14. In industrialized nations, old age is characterized by retirement from full-time employment and by a subsequent drop in income.

15. Older members of minority groups are said to be in double jeopardy, discriminated against both for their minority status and for their age.

16. Despite the fears of many younger people about their own aging, most older people in more-developed nations live independently and only require nursing home care at very advanced ages.

Suggested Readings

1. Angela M. O'Rand and John C. Henretta, 1999, Age and Inequality: Diverse Pathways Through Later Life (Boulder, CO: Westview Press).

   The authors apply the cohort flow and life course perspectives to the structure of aging to show how different life courses can lead to considerable diversity and inequality at older ages.

2. David Kertzer and Peter Laslett, Editors, 1995, Aging in the Past: Demography, Society, and Old Age (Berkeley and Los Angeles: University of California Press).

   Although this is nominally a demographic history of aging, most of the chapters are highly relevant to the current understanding of aging in the world, and the book as a whole provides a great deal of insight into the demographic aspects of aging in society.

3. Jay Sokolovsky, Editor, 1997, The Cultural Context of Aging: Worldwide Perspectives, Second Edition (Westport, CT: Bergin & Garvey).

   This book, which pulls together a wide range of authors bringing a global (and largely anthropological) perspective to the study of aging, will round out your perspective on the aging process.

4. Peter G. Peterson, 1999, Gray Dawn: How the Coming Age Wave Will Transform America—and the World (New York: Times Books).

   Although the author of this book is a former chairman of Lehmans, not an academic, the book contains a great deal of interesting information about the impact of aging on society. The book received considerable publicity when it was published.

5. Henry Aaron and Robert Reischauer, 2001, Countdown to Reform: The Great Social Security Debate, Revised and Updated Edition (New York: Century Foundation Press).

   Two economists argue in this book that the "bad curve" (see the essay accompanying this chapter) may not be as bad as we think, and that there are relatively straightforward solutions to the Social Security problem in the United States.

✺ Web Sites of Interest

Remember that Web sites are not as permanent as books and journals, so I cannot guarantee that each of the following Web sites still exists at the moment you are reading this. It is usually possible, however, to retrieve the contents of old Web sites at http://www.archive.org:

1. http://www.census.gov/population/www/socdemo/age.html#elderly
   The U.S. Bureau of the Census regularly updates its analyses of the demographics of aging in the United States and its current offerings should be available at this Web site.

2. http://www.ssa.gov
   The Social Security Administration (SSA) is the U.S. agency charged with administering the Social Security system, including pension benefits. The demographic changes taking place in American society have been of concern to them for a long time and are fully addressed in the annual report of the Trustees of the SSA, which can be found at: http://www.ssa.gov/OACT/TRSUM/trsummary.html

3. http://www.nih.gov/nia
   The National Institute on Aging (NIA) supports much of the research taking place on the older population of the United States. The agency's Web site includes links to publications, data, and other aging-related Web sites that may interest you.

4. http://www.worldhealth.net
   This is the Web site of the American Academy of Anti-Aging Medicine, which is a medical society devoted to treating aging as a preventable disease. It sees itself as somewhat anti-establishment in this regard.

5. http://www.retireonyourterms.com
   This is the site of a commercial firm, so be aware of that as you go there. The site includes a very detailed and educational "Retirement Quiz" that you might find valuable to take, along with educational material in case you should get an answer wrong.

CHAPTER 10
The Family and Household Transition

Figure 10.1 Percent of Women 15–19 Who Are Married

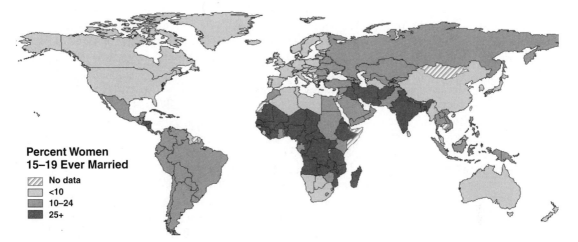

**Percent Women
15–19 Ever Married**

▨ No data
☐ <10
▨ 10–24
▨ 25+

Source: Adapted by the author from data in United Nations Population Division, World Marriage Patterns 2000.

DEFINING FAMILY DEMOGRAPHY AND LIFE
  CHANCES

THE FAMILY AND HOUSEHOLD TRANSITION
  Household Composition and Family
    Structure
  Gender Equity and the Empowerment of
    Women

PROXIMATE DETERMINANTS OF FAMILY
  AND HOUSEHOLD CHANGES
  Delayed Marriage Accompanied by Leaving
    the Parental Nest
  Cohabitation
  Out-of-Wedlock Births
  Childlessness
  Divorce
  Widowhood
  The Combination of These Determinants

CHANGING LIFE CHANCES
  Education
  Labor Force Participation

Female Labor Force Participation and
    Fertility
Female Labor Force Participation and
    Empowerment
Occupation
Income
    Income Disparities by Race/Ethnicity
    Income Disparities by Gender
    Poverty
Wealth
Race and Ethnicity
    Race and Ethnicity in the United States
    Ethnicity in Canada and Mexico
    Ethnicity in Other Countries
Religion

THE INTERSECTION OF CHANGING LIFE
  CHANCES AND THE FAMILY AND
  HOUSEHOLD TRANSITION

ESSAY: How Does the Age Transition Promote
  a Delay in Marriage?

Households used to be created by marriage and dissolved by death—in between there were children. Throughout the world this pattern has been transformed by what some have called the "second demographic transition," which I will discuss as the "family and household transition" in the context of the broader demographic transition. "The demographic transition is in essence a transition in family strategies: the reactive, largely biological family-building decision rules appropriate to highly uncertain environments come eventually to be supplanted by more deliberate and forward-looking strategies that require longer time horizons" (Cohen and Montgomery 1998:6); that is, the transition from "family building by fate" to "family building by design" (Lloyd and Ivanov 1988:141) that I mentioned in Chapter 6 as being key to the fertility transition.

In general terms, we can describe this as the shift in family and household structure occasioned by people living longer, with fewer children born, increasingly in urban settings, and subject to higher standards of living, all as part of the demographic transition. Households no longer depend on marriage for their creation, nor do they depend on death to dissolve them, and children are encountered in a wide array of household and living arrangements. "The family is in crisis, as witnessed by increasing instability of unions, fluidity of the 'marital' home and the economic stress experienced by women and children of disrupted marriages" (Makinwa-Adebusoye 1994:48). Although that certainly sounds like the subject matter of books and talk shows in the United States, the author of that quote is referring to sub-Saharan Africa. "Marriage is becoming rarer and the age at first marriage is increasing; the number of couples who cohabit before and without marrying is rapidly increasing; and so as a consequence, are births out of wedlock" (Blossfeld and de Rose 1992:73). That, too, could be a description of the United States, but in this instance the authors are talking about Italy.

All over the world changes in household formation and living arrangements are being discussed and discoursed, and in the United States it has been suggested that "one of the most widely debated issues in contemporary sociology has been how to interpret patterns of family change . . ." (Brooks 2002:191). Curiously, however, these debates rarely include a review of the underlying demographic changes that helped spawn this massive social shift, and so my purpose in this chapter is to rectify that deficiency for you. The changes we see occurring all over the world are the inevitable result of powerful social forces unleashed by the demographic transition. I am not going to go so far as to suggest that we can predict exactly which changes will take place at a given time in a given society, but I will suggest to you that no social system could remain unchanged in the face of massive declines in mortality, followed by massive declines in fertility, and accompanied by massive migration and a dramatic change in the age structure—all of the changes we have discussed in the previous chapters.

I begin the chapter with a discussion of exactly how the structure of households and living arrangements have, in fact, changed over time—how big is this transition in the United States and elsewhere? Particularly noteworthy is the change in the status of women. I have referred to that repeatedly in previous chapters, but we need to keep reminding ourselves that gender equity is central to the well-being of any society. Then I turn to the specific demographic changes that have contributed to the increasing diversity in household structure. A critical element is the changing set of **life**

chances that people are experiencing in the United States and all over the world—changes in the population (or demographic) characteristics that influence how your life will turn out. These include especially education, labor force participation, occupation, and income, which in turn affect **gender roles** (the social roles considered appropriate for males or females) and marital status. All of these things have influenced the changing family and household structure, although differently for some cultural groups than others. Indeed, race and ethnicity, along with religion, mediate the impact of life chances in every human society. The intersection of your population characteristics and family and household structure is a crucial determinant of what life will be like for you. Similarly, at the societal level, the distribution of the population by different characteristics and by family and household structure will be influenced by where a society is in the demographic transition, so we are in a position to say something about the future by fitting all of these pieces together.

Defining Family Demography and Life Chances

In virtually every human society ever studied, people have organized their lives around a family unit. In a general sense, a **family** is any group of people who are related to one another by marriage, birth, or adoption. The nature of the family, then, is that it is a *kinship* unit (Ryder 1987). But it is also like a minisociety, a micropopulation that experiences births, deaths, and migration, as well as changing age structures as it goes through its own life course. The changes that occur in the broader population—the subject of this book in general—mainly occur within the context of the family unit, and so the study of population necessitates that we study the family (see, for example, Goldscheider 1995). Implicit in the definition of a family is that its members share a sense of social bonding: the mutual acceptance of reciprocal rights and obligations, and of responsibility for each other's well-being. We usually make a distinction between the **nuclear family** (at least one parent and their/his/her children) and the **extended family,** which can extend to other generations (add in grandparents and maybe even great-grandparents) and can also extend laterally to other people within each generation (aunts and uncles, cousins, and so forth).

The next question of interest to us is: Where do these people live? People live in a **housing unit,** which is the physical space used as separate living quarters for people. It may be a house, an apartment, a mobile home or trailer, or even a single room or group of rooms. People who share a housing unit are said to have formed a **household.** The household is thus a *residential unit* and so you can see that a **family household** is a housing or residential unit occupied by people who are all related to one another. More specifically, we can say that a family household is a household in which the **householder** (defined by the U.S. Census Bureau as the person in whose name the house is owned or rented) is living with one or more persons related to her or him by birth, marriage, or adoption. On the other hand, a **nonfamily household** is a housing unit that includes only a person who lives alone, or consists of people living with nonfamily coresidents, such as friends living together, a single householder who rents out rooms, or cohabiting couples. Especially important to the concept of the family household is that the family part of it makes it a kinship unit, as noted above, while the household part of it makes it a *consumption* unit (Wall,

Robin, and Laslett 1983). Part of the responsibility that family members have when they live in the same household is to produce goods and services that are shared by, and for the mutual benefit of, the family members who live together. So, family members do not necessarily share a household, and household members are not necessarily family members. But when family members share a housing unit, we have the most powerful kinship and consumer unit that we are likely to find in any society.

Family demography is concerned largely with the study and analysis of family households: their formation, their change over time, and their dissolution. Families represent the *fusion* of people from other families and long before a family household dissolves, it is likely to have *fissured* into other families, as children born into the family grow up and leave the family household of their parents to create (fuse) their own households. All humans grow up in, and typically live for all of their lives, in social groups that represent some sort of family, so our lives are shaped and bounded by our membership in the group. We cannot understand the changes taking place elsewhere in society unless we connect those changes to what is happening to the family.

We can describe a family in terms of its *geographic location* because where you are in the world will influence the kinds of social, cultural, economic, and physical resources that will be available to the family. We can also describe a family in terms of its *social location* (where it is positioned in the local social system) because that will influence the family's access to whatever local resources exist on which the family can draw. And we can describe the family in terms of its own *social structure,* which refers to the number of people within the family, their age and gender, and their relationship to each other. Each of these characteristics of the family will influence the life chances of family members. Your own life chances refer, for example, to your probability of having a particular set of demographic characteristics, such as having a high-prestige job, lots of money, a stable marriage or not marrying at all, and a small family or no family at all. These differences in life chances, of course, are not necessarily a reflection of your worth as an individual, but they are reflections of the social and economic makeup of society—indicators of the demographic characteristics that help to define what a society and its members are like.

We are born with certain **ascribed characteristics,** such as sex or gender and race and ethnicity, over which we have essentially no control (except in extreme cases). These characteristics affect life chances in very important ways because virtually every society uses such identifiable human attributes to the advantage of some people and the disadvantage of others. Religion is not exactly an ascribed characteristic, but worldwide it is typically a function of race or ethnicity and, as with ascribed characteristics, it is often a focal point for prejudice and discrimination, which influence life chances.

Life chances are also directly related to **achieved characteristics** or your personal human capital, those sociodemographic characteristics, such as education, occupation, labor force participation, income, and marital status, over which you do exercise some degree of control. For example, the better educated you are, the higher your occupational status is apt to be, and thus the higher your level of income will likely be. Indeed, income is a crass, but widely accepted, index of how your life is turning out. Ascribed characteristics affect your life chances primarily by affecting your access to achieved characteristics, which then become major ingredients of social status—education, occupation, and income. Population characteristics affect your own demographic behavior, especially family formation and fertility, although they also influence mortality and migration, as I have already discussed in Chapters 5 and 7, respectively.

The demographics of your family, in turn, affect life chances through the possession or acquisition of social capital—the ability to facilitate or retard your access to opportunities for higher education, a higher-status occupation, or a better-paying job. All of these aspects of population characteristics and their influence on life chances converge to affect the kind of family we choose to create and the type of household we form.

The Family and Household Transition

The "traditional" family household of a married couple and their children is no longer the statistical norm in North America and in many other parts of the world, even if it remains the ideal type of household in the minds of many people. Families headed by females, especially with no husband present, are increasingly common, as are "nontraditional" households inhabited by unmarried people (including never-married, divorced, widowed, and cohabiting couples), by older adults raising their grandchildren, or by married couples with no children. You are almost certainly aware of these societal shifts through personal experience or the mass media. What may be less obvious is that these changes are closely linked to demographic trends. The variability from place to place may be culturally and societally specific, but the basic reasons for change are distinctly demographic in nature. Let us look first at the changes that are taking place, focusing especially on the United States.

Household Composition and Family Structure

The total number of households in the United States increased from 63 million in 1970 to 106 million in 2000, but within that increase was a dramatic change in the composition of the American household, as Figure 10.2 illustrates. In 1970, the classic "married with children" households accounted for 40 percent of all households in the United States. Married couples without children (either before building a family or after the kids were grown) accounted for another 30 percent. The "other" families include male- and, disproportionately, female-headed households in which other family members (usually children) are living with the householder. In Figure 10.2, you can see that the light shading represents all family households (a household in which the householder is living with one or more persons related to her or him by birth or marriage). In 1970, they comprised 81 percent of all households—a drop from 90 percent in 1940, when the Census Bureau first began to compile these data (Fields 2001). By 2000, it had dropped even further to 68 percent, and by then less than one in four households included a married couple with children.

As married couples with children have become less common, they have been replaced most often by female-headed, mother-only families, and by nonfamily households, including people who live alone, and nonfamily coresidents (friends living

Figure 10.2   U.S. Households Have Become Increasingly Diverse

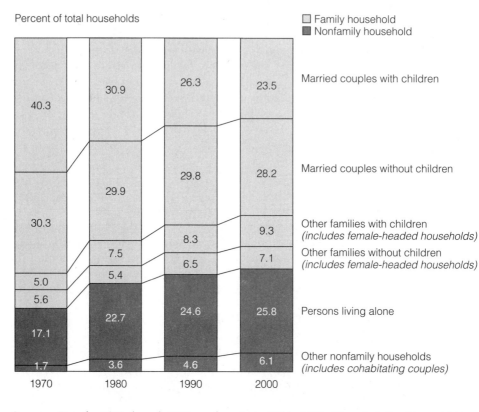

Percent of total households

☐ Family household
■ Nonfamily household

Sources: Data for 1970 through 1990 are from Jason Fields, 2001, "America's Families and Living Arrangements: 2000." Current Population Reports P20-537, Figure 1; data for 2000 are from U.S. Census Bureau, 2001h, Census 2000, Table DP-1, Profile of General Demographic Characteristics, 2000.

together, a single householder who rents out rooms, cohabiting couples, etc.). The phrase that best describes the changes in household composition as shown in Figure 10.2 is increased diversity or "pluralization" (meaning that no single category captures a majority of households). Although I focus here on the United States, Canada and most European nations have also experienced a decline in the relative importance of households composed of a married couple with children, along with an increase in female-headed households with children present (McLanahan and Casper 1995), although the rise in the latter is more pronounced in the U.S. than in Europe (Kuijsten 1999). Indeed, in Italy almost all children are born to a married mother and there is very little divorce among parents with children in the home (Anderson 2002). At the same time, Italy has one of the world's highest rates of delayed marriage and lowest rates of fertility, so there is undoubtedly a connection here.

Increasing diversity is a result of the several elements of the demographic transition. The demographic transition promotes a diversity of family types and household types because (1) people are living longer and this means that they are likely to be widowed, more likely to tire of the current spouse and divorce, and less likely to feel pressure to marry early and begin childbearing; (2) the latter pressure is relieved by the decline in fertility, which means that women, in particular, do not need to begin childbearing at such a young age, and both men and women have many years of life after the children are grown; and (3) an increasingly urban population is presented with many acceptable lifestyle options besides marriage and family-building. I will discuss these in more detail later in the chapter.

The rise in female-headed households is not a necessary component of the changes that occur during the demographic transition, but, as I will discuss below, it is tied at least transitionally to the rise in the status of women. Between 1970 and 2000, the number of female-headed families with children (mother-only families) went up from 2.8 million to 7.6 million in the United States, whereas the number of married-couple-with-children families went down from 25.5 million to 24.8 million. In 1970, female-headed households accounted for 10 percent of families with children, but by 2000 that had jumped to 22 percent (see Figure 10.3). Even those numbers hide the true scope of the transformation, because a married-couple family in 2000 was more likely than in 1970 to be a recombined family, involving previously

Figure 10.3   U.S. Racial/Ethnic Differences in Percentage of Households with Children That Are Mother-Only Families

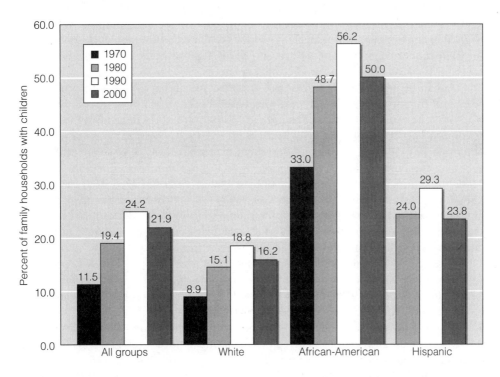

Sources: Data for 1970 through 1990 are from S. Rawlings, 1994, "Household and Family Characteristics: March 1993," Current Population Reports P20-477, Table F; data for 2000 are from Jason Fields, 2001, "America's Families and Living Arrangements: 2000." Current Population Reports, P20-537, Table F.

married spouses and children from other unions. At the end of this chapter, I discuss the social impact of these transformations, but my goal at present is to describe the changes. One obvious point to make is that female-headed households represented a smaller fraction of all family households with children in 2000 than had been true in 1990.

Widely discussed in public debate is the fact that substantial racial/ethnic differences exist with respect to female-headed households, especially among families with children, as shown in Figure 10.3. Although the rise in mother-only families has been experienced by all groups in American society, in 1970, African-American families were already more apt to be headed by a female than were white or Hispanic households at any time between 1970 and 2000. Ruggles (1994a) has found that, since at least 1880, African-American children have been far more likely to reside with only their mothers (or with neither parent) than white children. Nonetheless, the data in Figure 10.3 show that the percentage of African-American mother-only families has increased considerably just in the past few decades, although it too declined a bit between 1990 and 2000.

In 2000, nearly three out of 10 households in the United States were non-family households, as you can see looking back at Figure 10.2. This is part of the trend away from what is often thought of as the traditional family, enshrined by old TV sitcoms—a family in which a married couple live together with their children and the husband works full-time while the wife cares for the children and attends to domestic chores (Hernandez 1993). In fact, this type of "Leave It to Beaver" family is a relatively new phenomenon historically—a product of the demographic transition. High mortality alone (but especially when combined with high fertility and an agrarian economic environment) prevented this type of household from being the norm for most of human history. Let me explain.

Human beings are by nature social animals. We prefer to live with others and nearly all human economic activities are based on cooperation and collaboration with other humans. Our identity as individuals paradoxically depends on our interaction with other people. We only know who we are by measuring the reaction of other people to us, and we depend on others to teach us how to behave and how to negotiate the physical and social worlds. Furthermore, humans are completely dependent creatures at birth, and every known society has organized itself into social units (households/families) to ensure the survival of children and the reproduction of society. In high-mortality societies, the rules about who can and should be part of that social unit must be a bit flexible because death can take a mother, father, or other important household member at almost any time.

I referred in Chapter 5 to the Nuatl-speaking Mexican families in Morelos in the sixteenth century, who lived in a "demographic hell," where high mortality produced high rates of orphan- and widowhood. In response to these "vagaries of severe mortality," they developed a complex household structure that was "extremely fluid and in constant flux. Headship and household composition shifted rapidly because marriages and death occurred at what must have been a dizzying pace" (McCaa 1994:10). Zhao (1994) has used genealogies from China from the thirteenth through the nineteenth centuries to show that high mortality kept most Chinese from actually living in a multigenerational family at any particular time, although, as in Mexico, the high death rate produced a complex form of the family because of shifting membership.

The bottom line here is that what we think of as the "traditional" family is, in fact, not very traditional at all. It depends on low mortality, which is a historically recent phenomenon, in combination with a fairly young structure, characterized by young adults with their children. This combination of demographic processes is found largely in the middle phase of the demographic transition, when mortality has dropped, but fertility is still above replacement level. Over time, as mortality remains low and fertility drops to low levels, the population ages, as I discussed in the previous two chapters, older married couples are left without children any longer in the household, and then women are left without husbands in the household. But, at younger ages, people are still marrying and having children, marrying and not having children, not marrying but having children, and neither marrying nor having children. All of those things are possible in a low-mortality, low-fertility society with a barrel-shaped age structure.

Given the fact that societies have historically changed in response to demographic conditions, it should be no surprise to you that since the end of World War II, with demographic conditions undergoing tremendous change all over the globe, the status of women is one of the facets of social organization undergoing a significant transition.

Gender Equity and the Empowerment of Women

The demographic transition does not inherently produce gender equity and the empowerment of women, but it creates the conditions under which they are much more likely to happen. The combination of longer life and lower fertility, even if achieved in an environment in which women are still oppressed, opens the eyes of society—including women themselves—to the fact that women are in a position to contribute to society in the same way that men do when not burdened by full-time parenting responsibility. And, as I have already mentioned, the fact that this combination of low mortality and low fertility typically occurs in an urban setting means that women have many more opportunities than would exist in rural areas to achieve the same kind of economic and social independence that has been largely the province of men for much of human history.

Let me suggest to you that at no time in human history has there been a good justification for the domination of women by men, but the demographic conditions that prevailed for most of human history did at least facilitate that domination. Demographic conditions no longer provide that prop and in most of the world, the impediment to full social, economic, and political empowerment of women is simply the attitude of men, often aided by women who have grown up as "codependents" in the system of male domination. At the beginning of the twentieth century, it would have been unthinkable for a woman in America to go out to dinner without being accompanied by a man, and equally unthinkable for her to drive one of the cars that were making their first appearance at that time. At the beginning of the twenty-first century, those things are still unthinkable to most women in Saudi

Arabia, but just as demographic changes were occurring in the United States 100 years ago, so they are occurring in Saudi Arabia today, as I mentioned in the essay in Chapter 8. The changes under way mean that the life chances for women are beginning to equal those of men, and thus women are in a position to voluntarily head their own household if they want to, or to do so successfully even if forced into that position. It is certainly no coincidence that those countries still in the earlier stages of the demographic transition are also those where the rights of women are most trampled. In Kenya and Tanzania, as in much of sub-Saharan Africa and South Asia, women's access to economic resources is restricted by severe limitations on their ability to inherit property and own land (World Bank 1999). In India, women's access to the formal labor market is restricted, and so women wind up in the informal sector, where they are far more likely to be economically exploited (Dunlop and Velkoff 1999).

Countries at the later stages of the demographic transition have generally discovered the benefits of unleashing the resources of the half of the population that had previously been excluded from full societal participation. Since 1979, the United Nations has been encouraging all countries to sign on to the Convention on the Elimination of All Forms of Discrimination Against Women (CEDAW) and as of 2004, 175 countries had ratified it, although the United States was not one of them at the time of this writing (United Nations Division for the Advancement of Women 2004). The empowerment of women contributes to further change in society by expanding women's life chances, which can expand economic opportunity and enrich society and households. All of these transformations contribute to the diversity of family and household structure. Let us now examine some of the direct proximate causes of the household structure transformation societies have been experiencing.

Proximate Determinants of Family and Household Changes

The increasing diversity in household structure is a result of several interdependent trends taking place in society (see, for example, United Nations Population Division 2003b). This is especially due to a delay in marriage, accompanied by young people leaving their parents' home (which in most more-developed nations has increased the incidence of cohabitation), and an increase in divorce (which also contributes to cohabitation); whereas at the older ages, the greater survivability of women over men has increased the incidence of widowhood, which has an obvious impact on family and household structure. In between the younger and older ages, the smaller number of children in each family means that a much shorter period of time in each parent's life is devoted to activities directly related to childbearing. The life course of the family has thus been revolutionized.

Delayed Marriage Accompanied by Leaving the Parental Nest

One of the most important mechanisms preventing women from achieving equality with men is early marriage. When a girl is encouraged or even forced to marry at a young age, she is likely to be immediately drawn into a life of childbearing and family-building that makes it very difficult, if not impossible, for her to contemplate other options in life. This is one of the principal reasons why high fertility is so closely associated with low status for women. The map in Figure 10.1 at the beginning of this chapter shows the world pattern in the percentage of women who are already married by the time they reach their 20th birthday. You will not be surprised to notice that the highest percentage married at young ages occurs in those countries where fertility is highest and where the status of women is known to be low. By contrast, it is no coincidence that in the low-fertility regions of North America, Europe, Oceania, and Eastern Asia, the percentage of women who marry at young ages is very low and the status of women is higher than in other places in the world.

Even though women typically marry at young ages in high-fertility societies, men tend to be under less constraint on that score. This means that in those places where women are young when they marry, the chance is good that their husband is several years older. This further contributes to the ability of a man to dominate his spouse. "A girl with minimal education, raised to be submissive and subservient, married to an older man, has little ability to negotiate sexual activity, the number of children she will bear or how she spends her time" (Gupta 1998:22). The flip side of this is that a rising age at marriage in sub-Saharan Africa since the 1970s has had the beneficial side-effect of reducing the rate of polygynous marriages, which normally involve a young woman marrying an older man as his second or third wife—a situation that typically will diminish a woman's life chances (Chojnacka 2000).

Table 10.1 shows that as the percentage of women who are married at ages 15–19 goes up, the difference in age between husband and wife also increases. Thus, in those countries where less than 10 percent of women are married at ages 15–19, a man is on average 2.7 years older than the woman he marries. At the other end of the continuum, in those countries where 40 percent or more of women are married at ages 15–19, the average difference between bride and groom is 6.5 years. Although husbands tend to be older than their wives in the United States, data for the year 2000 show that 32 percent of married couples are within one year of each other, and in 12 percent of couples, the wife is older than the husband by at least

Table 10.1 Average Difference in Age of Brides and Grooms Declines as Percent of Women Married at Ages 15–19 Goes Down

% of women married at ages 15–19	Average difference in age between bride and groom	Number of countries
40 or higher	6.5	15
30–39	5.8	11
20–29	4.2	21
10–19	3.3	53
Less than 10	2.7	93
Total	*4.5*	*193*

Source: Calculated from data in United Nations Population Division, 2000, World Marriage Patterns, 2000 (New York: United Nations).

HOW DOES THE AGE TRANSITION PROMOTE A DELAY IN MARRIAGE?

In this chapter, I point out that the rise in the status of women in the United States and other industrialized nations has been aided by a series of long-term demographic events, but that a catalytic force such as the women's movement was necessary to unleash the power built into demographic forces. Ironically, many analysts (for example, Davis and van den Oever 1982; Heer and Grossbard-Schechtman 1981) have argued that the women's movement itself was triggered by one of the major demographic phenomena of the baby boom—the marriage squeeze. The marriage squeeze is "an evocative though imprecise term used to describe the effects of an imbalance between the number of males and females in the prime marriage ages" (Schoen 1983:61). It is a by-product of the combination of different-sized cohorts and the fact that women do not usually marry men their own age. The former is, of course, built into the age transition; whereas the latter is a remnant of tradition in most societies that doesn't serve any practical purpose.

You may recall from Chapter 8 that, although more boy babies are born than girl babies, higher male mortality evens the sex ratio by early adulthood. Thus a young woman expecting to marry someone her own age would have virtually a 100 percent chance of finding someone. However, in nearly every human society, women typically marry someone older than themselves. Now, if there are fewer older males, due to the facts that the number of births was lower prior to the woman's year of birth or the death rate was declining and each year there are more survivors per birth cohort than in the previous year, then her probability drops of finding someone of the usual marriageable age. Since men her own age may not be expecting to marry for another few years, she too may be forced to postpone marriage.

American and Canadian women, on average, marry a man who is two years older, and over time the vast majority of American women have preferred marriage (at least at some point in life) to remaining single. Assuming that people will continue to want to marry, the imbalance in the number of males or females can lead to a delay in marriage for the sex with excess numbers. This is exactly what has happened to women of the baby boom. In 1950, there were 98 men aged 22–26 for every 100 women two years younger (aged 20–24). Thus, there was near equality in the number of males and females of typical marriageable ages at that time. If we assume that women preferred men three years older than themselves (the age gap in 1950 was three years, which has since dropped to two), the ratio was even closer to equity. There were 99 men aged 23–27 in 1950 per 100 women aged 20–24. But that ratio dropped precipitously through the years as the pre-baby boom and baby boom cohorts reached marriage age, because in nearly every year from 1937–1957, there were more people born in the United States than in the previous year. Women looking at the age group two to three years older kept seeing fewer men than women of their own age. In 1960, there were only 84–86 men per women of the usual marriageable age, and in 1970 the ratio was in the 85–87 range. The result was that many women were forced to postpone marriage, a phenomenon I discussed in this chapter that has also been well documented for Canada (Foot 1996).

By 1980, relative equity had been achieved again in the sex ratio at the usual marriage ages, but by that time some remarkable changes had already taken place in American gender roles. "The Women's Liberation Movement may be interpreted, on the one hand, as a collective means by which women helped themselves to reorient each other to the new lower compensation for the traditional female role and, on the other, as the means by which the increasing number of women outside of the traditional wife-mother role sought to combat the discrimination meted out to women in the job world" (Heer and Grossbard-Schechtman 1981:49). Either interpretation can be viewed as women's responses to demographic forces at work in the marriage market (Grossbard-Schechtman 1985; 1993). In the late 1980s and early 1990s, North American males were experiencing a marriage squeeze as the declining number of births through the 1960s and early 1970s produced a dearth of women of the usual marriage age. This led to a better bargaining position for women in the marriage market, and almost certainly helped in some measure to improve gender equality.

Guttentag and Secord (1983) used the concept of sex ratio to develop a theory of gender relations, based on Becker's new household economics theory. In general, their hypothesis was that a shortage of women relative to men encourages domesticity, marital stability, and a maintenance of male dominance over women. Conversely, a low sex ratio (more women than men) encourages women to find alternatives to traditional role models, since there are too few men of marriageable age. Guttentag and Secord used nineteenth-century America as an example. On the frontier, the sex ratio was high and female roles were typically very traditional, with an emphasis on family building and family life. At the same time, the New England states had a much lower sex ratio and it was in that part of the country, perhaps not coincidentally, that the suffrage movement began.

African-Americans have for several decades had a sex ratio lower than that of whites because of higher male mortality from birth on. The sex ratio has been further lowered because of the higher proportion of black than white males in prison. Guttentag and Secord argued that this low sex ratio led to greater independence among black women and to lower marital stability as well, because a relative surplus of women allows men to be choosier and thus more assertive.

It is probably dangerous to be too deterministic with respect to demography, but there can be little question that as long as age differences in marriage exist between men and women, then the changing size of cohorts in the context of the age transition will upset the previous ratios and force changes in behavior. Bhrolcháin (2001) has noted quite correctly that most marriage squeezes could be readily alleviated by greater flexibility in the ages at which people marry. In the main, this would mean that the age of wives and husbands would converge, which has been the pattern in North America and the United States. To do that, however, requires that men be willing to give up some of the marital power that seems to accrue to the older partner in a relationship, and that in itself would be clear evidence of a move toward greater gender equality.

In the 1980s, Schoen (1983) noted that a number of developing countries were increasingly beginning to experience marriage squeezes resulting from the declines in infant mortality since the end of World War II that have produced a demographic effect very similar to the American baby boom. The relative scarcity of husbands for women, produced by the age transition, is almost certainly one of the reasons for delays in marriage in almost every less-developed nation. In societies that allow polygamous marriages, such a situation could have been dealt with in the past by forcing a woman to become the second or higher-order wife of an already married man. However, other changes taking place in the world, especially including international efforts to improve the status of women, are helping change the way women are likely to perceive the marriage situation. Part of the international message is that a delay in marriage is crucial for women to avoid the oppression of early motherhood, as well as male domination.

In India, where the status of women remains low by international standards, a marriage squeeze as described by Schoen has been implicated as a factor that forces families to use ever more of their limited resources for a dowry in order to marry their daughters in an increasingly competitive marriage market (Rao 1993). Goodkind (1997) has even discovered what he calls a "double marriage squeeze" for the Vietnamese. In Vietnam, the combination of population change, war, and excess male migration produced a shortage of men and thus a marriage squeeze for females, leading to a rise in the age at marriage and a decrease in the marriage rate. At the same time, the male migrants who left the country faced an overseas marriage squeeze because there are not enough overseas Vietnamese women to go around, so the men, too, have had to delay or forego marriage.

Overall, then, the lesson is one that I have repeated throughout the book, which is that almost everything of consequence in the social world has an important demographic component, and that to understand how the world is really working we must comprehend the demographic forces that quietly, but impressively, shape our lives.

two years (Fields 2001). An increasing body of evidence from Demographic and Health Surveys indicates that the age at marriage is on the rise in many developing countries, signaling a potential change in fertility, female empowerment, and family change in those parts of the world (Westoff 2003).

In the United States, as in many northern and western European countries, the early decline in fertility more than 100 years ago was accomplished especially by a delay in marriage. It is thus not too surprising that it was in these countries that some of the early feminist movements were able to take root. At a time when very few effective contraceptives were available, and when it was extremely difficult to get a divorce, postponement of marriage was the principal route by which women were able to increase their options in life. In 1890, more than one-third of all women aged 14 and older (34 percent) and close to one-half of all men (44 percent) were single in the United States. Between 1890 and 1960, being single became progressively less common as women, and especially men, married at earlier ages. Only since the 1960s has there been a resurgence of delayed marriage. By 2000, the average (median) age at marriage for females had risen to 25.1, the highest level in U.S. history, while for males it had increased to 26.8, also higher than the level of 1890 (Fields 2001). In fact, in 2000, 39 percent of all American women aged 25–29 were still single, compared with only 11 percent in 1970, and 52 percent of all men of that age had not yet married, a huge increase from only 19 percent in 1970. Changes in the popularity of early marriage have been roughly similar for both African-Americans and whites, although African-Americans have been more likely than whites to delay marriage or remain single (McLanahan and Casper 1995).

Since the 1960s, the slowdown of economic growth, accompanied by stiff competition for jobs brought about by baby boom children growing up, has made it more advantageous for couples to postpone marriage to take maximum advantage of educational and career opportunities. And, very importantly, the contraceptive revolution, especially the birth control pill, has allowed people to disconnect marriage from sexual intercourse. People have known about and used birth control methods for a long time (refer to Chapter 6 if you need a review). However, the failure rate of all of those prepill methods was significantly higher than that of the pill, and a couple engaging in intercourse ran a clear risk of an unintended pregnancy. Prior to 1973 and the legalization of abortion in the United States, an American woman could end an unintended pregnancy only by flying to a country such as Sweden, where abortion was legal, or by seeking an illegal (and often dangerous) abortion.

In more traditional societies (including the United States and Canada until the 1960s), an unintended pregnancy was most apt to lead to marriage, although a woman might also bear the child quietly and give it up for adoption. Illegitimacy was widely stigmatized and having a child out of wedlock was the course of last resort. Marriage was the only genuinely acceptable route to regular sexual activity, and only married couples were routinely granted access to available methods of birth control. That situation still prevails in most of the world's predominantly Muslim nations.

In the late nineteenth century, the older age at marriage already alluded to in North America and Europe had been accomplished by a delay in the onset of regular sexual activity—the Malthusian approach to life. Intercourse was delayed until marriage, and in this way nuptiality was the main determinant of the birth rate: Early marriage meant a higher birth rate, and delayed marriage meant a lower one (Wrigley and Schofield 1981). A variety of social and economic conditions might discourage an early marriage. The societal expectation that a man should be able to provide economic support for his wife and children tended to delay marriage for men until those expectations could be met. Under conditions of rising material expectations, as was the case in the late nineteenth century, marriage had to be delayed even a bit longer than in previous generations because the bar had been raised higher than before. Delayed marriage typically meant that young people stayed with their parents in order to save enough money to get ahead financially and thus be able to afford marriage. Staying with parents also minimized the opportunities for younger people to be able to engage in premarital sexual intercourse, which might lead to an unintended pregnancy and destroy plans for the future. Thus, prior to the latter half of the twentieth century, delayed marriage did not typically lead younger people to leave home and set up their own independent household prior to marriage.

In the early post-World War II period, economic robustness allowed a young person to leave the parental home at an earlier age without an economic penalty and, since the risk of pregnancy meant that intercourse was still tied closely to marriage, the age at marriage reached historic lows in the United States and Europe. In discussing the situation in Germany, Blossfeld and de Rose (1992) have argued that "until the late 1960s, the opportunity for children to leave their parental home had increased remarkably because of the improvement of economic conditions. But the social norm that they had to be married if they wanted to live together with a partner of the opposite sex was still valid, so that age at entry into marriage was decreasing until the end of the 1960s" (p. 75). Similar arguments could be made for the United States and for other European countries, with the exceptions of Italy and Spain, where out-of-wedlock births still tend to be socially unacceptable, so a delay in marriage has meant lower fertility than in other European countries because women are not compensating for delaying marriage by having children outside of marriage. In southern Europe, women may also be reluctant to marry and have children because employment opportunities are more fragile than in other parts of Europe and once they leave the labor force to have a child they find it difficult to return (Adsera 2004).

Modern contraception has allowed sex to be disconnected from marriage and this has been accompanied by a dramatic rise in the fraction of young people who leave their parental home before marriage, even while delaying marriage. In 1940, 82 percent of unmarried males and females aged 18–24 lived with their parents (Goldscheider and Goldscheider 1993); by 2000, that had declined to only 42 percent for women and 56 percent for men (Fields 2001). "By the 1980s, leaving home before marriage had evidently become institutionalized in the United States" (Goldscheider and Goldscheider 1993:34). Goldscheider and Goldscheider (1994) analyzed data from the National Survey of Families and Households to show that until the early 1960s, the majority of women who left home after age 18 were doing so to get married. That pattern changed in the 1960s, and by the late 1980s, only one-third of women were leaving home to marry. Men, however, have been leaving home for reasons other than marriage for a long time. So, the recent delay in marriage and nest-leaving behavior has been more significant as a force of change for women than for men.

You may have asked yourself how the data on increasing independence of young people can square with the wide array of stories about these same people returning home to live with their parents. The answer is that as the fraction of young people who leave home to live independently increases, the absolute volume of people available to move back in with parents increases. Children graduating from college and those returning from the military are especially likely to return home for a short while, whereas those who left home to marry are the least likely to return home (Goldscheider and Goldscheider 1999). Nonetheless, because children do not necessarily return for very long, a snapshot photo of society (such as the census or a survey) does not capture many young people who have returned home, even though a fairly high percentage may do so at some time or another. Data from Canada show that from 1981 to 1996 the percentage of young people remaining in their parents' home increased (Boyd and Norris 2000), but even this is consistent with the fact that delayed marriage increases the options available to young people and allows both an increase in leaving home for reasons other than marriage and an increase in staying in or returning to their parental home.

Cohabitation

The delay in marriage has not necessarily meant that young people have been avoiding a familylike situation, nor that they have necessarily avoided having children out of wedlock. When leaving the parental home, young people may set up an independent household either by living alone or sharing a household with nonfamily members, or they may move into nonhousehold group quarters such as a college dormitory. From these vantage points, many then proceed to **cohabitation,** which can be defined as the sharing of a household by unmarried persons who have a sexual relationship (Cherlin 1999). As this trend was unfolding in the 1970s, the Census Bureau created estimates of its extent from data on household composition. The resulting measure was "partners of the opposite sex sharing living quarters," which became widely known as POSSLQ (pronounced "PA-sul-cue"). As the measure has been refined based on more direct questions about unmarried partners, researchers have found that this was, in fact, a pretty good measure, although it probably underestimated cohabiting couples with children (Casper and Cohen 2000).

In 1970, when the average age at marriage for women was 20.3, there were about 500,000 cohabiting couples in the United States, and the ratio of cohabiting to married couples was 1 to 100. By 2000, when the average age at marriage for women had climbed to 25.1, the number of cohabiting couples had increased to 3.8 million and the ratio of cohabiting to married couples also had jumped to 6 per 100 (Fields 2001). That number almost certainly underestimated the importance of cohabitation, however, because it has become a widely accepted part of the life course in many low-fertility societies. Rather than being an alternative to marriage (a "poor person's marriage"), it has become a stepping-stone to marriage for many, as well as a step back from marriage after a divorce for others. Although only a small fraction of couples are cohabiting at any one time, the data for the United States suggest that nearly half of all people under age 35 have cohabited at least once, usually for a fairly short period of time (Graefe and Lichter 1999). Survey data from France suggest that in 1965 only 8 percent of couples cohabited before marriage, but by 1995 that figure had jumped to 90 percent (Toulemon 1997).

The desire to have children may determine the timing of formal marriage for cohabiting couples, although in many instances the birth of the child may precede the marriage. Bumpass and Lu (2000) have estimated that half of the out-of-wedlock births in the United States are born to cohabiting couples. Nonetheless, cohabiting couples seem to have lower levels of fertility than married couples and, once married, couples who had cohabited prior to marriage appear to follow essentially the same pattern of family building as couples who had not cohabited (Manning 1995).

Out-of-Wedlock Births

Given the delay in marriage, which has been accompanied by high rates of premarital sexual activity (aided by the fact that young people have been getting out of the parental home before marriage), it should not be surprising that the United States and some of the other low-fertility nations have been experiencing an increase in the proportion of out-of-wedlock births. This is an event, of course, that immediately transforms a woman living alone, or an unmarried couple living together, from a nonfamily to a family household. In France, the percentage of births outside marriage increased from 6 percent in 1965 to 40 percent in 1998 (Muñoz-Peréz and Prioux 2000). Between 1970 and 2002, the proportion of babies in the United States who were born out-of-wedlock increased from 11 percent to 34 percent, according to data from live birth records (Martin et al. 2003; National Center for Health Statistics 1994).

Data from the Current Population Survey provide a broader portrait of the women who are bearing a child out-of-wedlock, as shown in Figure 10.4. You can see that 89 percent of births to women aged 15–19 in 2002 were born out-of-wedlock, and the fraction drops rapidly with increasing age. To be sure, when you get above the average age at marriage, almost all births are to married women. Among younger women, the nonuse of contraception and lack of local access to abortion may push up the likelihood of young women getting pregnant and bearing a child. The United States is more restrictive than most low-fertility societies are in providing teenagers with easy and inexpensive access to methods of fertility limitation. Prior to the 1970s, most young women conceiving out-of-wedlock would have married prior to the baby's birth (Bachu 1999), and so the illegitimacy ratio (the ratio of out-of-wedlock births to all births) would have been much lower, even with the same level of premarital conceptions. Of course, the odds were also very high that the marriage would have ended in divorce after only a few years, so neither scenario is particularly rosy for the mother and her baby.

A disproportionate share of younger women bearing children out-of-wedlock are African American. Since birth rates overall are nearly as low for blacks as they are for whites in the United States, the proportion of out-of-wedlock births represents a different pattern of parenting, not a different overall level of childbearing. The pattern is to have children at a younger age than the rest of the population, and then to stop having them at a younger age as well. Thus, in 2002, the age-specific birth rates for black women were quite a bit higher than for whites at ages younger

Figure 10.4   Profile of Women Bearing Children Out-of-Wedlock in U.S.

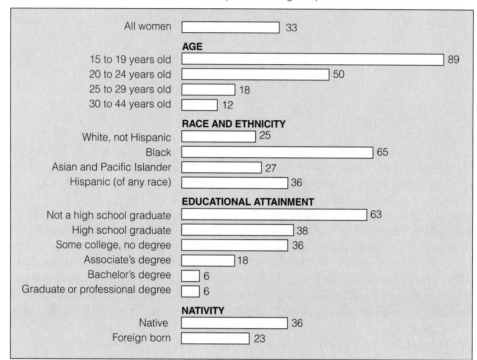

Source: Barbara Downs, "Fertility of American Women: June 2002," Current Population Reports, P20–548: Figure 1.

than 25, but then lower than for whites at ages 25 and higher (Martin et al. 2003). It is not clear why this pattern exists, but even in the 1930s, when only 6 percent of white women in the United States were having a baby out-of-wedlock, the percentage among blacks was 31 (Bachu 1999).

Regardless of the reasons, the data suggest that only 15 percent of black mothers raise their children in an intact two-parent family, compared with 60 percent of non-Hispanic white mothers (Rendell 2000). Furthermore, it is not just that the children are born outside of marriage; the fact that 22 percent of black children are born to teenage mothers (compared to 11 percent of whites) negatively affects their life chances. Even in Sweden, childbearing in adolescence disadvantages both the mother and child in later life (Olausson et al. 2001).

These data have fueled enormous public debate about the social cost of out-of-wedlock births (measured monetarily by welfare benefits and socioculturally by the deprivations suffered by fatherless children) and they have been interpreted as signs of imminent cultural decay. But, as McLanahan and Casper (1995) point out, the situation is not quite what it seems:

The increase in the illegitimacy ratio between 1960 and 1975 was due to two factors: the decline in the fertility of married women and the delaying of marriage. But not to an increasing birth rate among unmarried women. Beginning in the mid-1970s, marital fertility rates stopped declining, nonmarital fertility rates began to rise, and the age at first marriage continued to rise. After 1975, the rise in the illegitimacy ratio was due to increases in nonmarital fertility as well as to increases in the number of women at risk of having a nonmarital birth.

Although in the past few years the probability has increased that an unmarried woman will bear a child, part of the rise in the illegitimacy ratio was actually because married women were having fewer or no children, thus tilting the ratio of births in the direction of those born out-of-wedlock.

Childlessness

About 10 percent of couples may remain childless because of impaired fecundity, as was discussed in Chapter 6. In the 1970s, data from the Current Population Survey suggested that, in fact, this percentage of women (10 percent) was reaching ages 40–44 without having had a child. Since then, however, the percentage has slowly but steadily increased to 18 percent by 2002 (Downs 2003). Many of these women have decided that they did not want to have any children. Some fraction of them probably "drifted" toward childlessness by continually postponing the first child, since we know that most reproductive decisions, in both high- and low-fertility societies, are made sequentially from one birth to the next (Morgan 1996). This also means that some women who start out in life thinking that they do not want to have children wind up later changing their mind.

An important consequence of gender equality is that there is less pressure on a woman to have children, even if she is married, particularly if she has a rewarding career that she doesn't want to interrupt or, as certainly happens in some cases, she and her husband simply prefer a life without children. Childlessness increases household diversity by increasing the percentage of households represented either by someone living alone, or a married couple with no children of their own, or a cohabiting couple with no children, or a multiple-person nonfamily household.

Divorce

Not only has marriage been increasingly pushed to a later age, but once accomplished, marriages are also more likely to end in divorce than at any previous time in history (Schoen and Weinick 1993). This trend reflects many things. An obvious reason is that changes in divorce laws since the 1970s have made it easier for either partner to end the marriage at any time for any reason (Waite 2000). Friedberg (1998) concluded that divorce laws alone can explain about 17 percent of divorces in the United States. For the other 83 percent of the explanation, we can start by asking ourselves why legislators were motivated to make those changes; for answers, we can look to the loosening hold of men over women and the longer lives we are leading, both of which may produce greater conflict within marriage. In 1857 in the United States, there was only a 27 percent chance that a husband aged 25 and a wife aged 22 would both still be alive when the wife reached 65, but for couples marrying in the early twenty-first century, the chances have rocketed to 60 percent. Conversely, only about 5 percent of marriages contracted in 1867 ended in divorce (Ruggles 1997), whereas it has been estimated that about half of the marriages contracted since the 1970s will end in divorce (Kreider and Fields 2002; Raley and Bumpass 2003).

The United States is certainly not unique in experiencing an increase in divorce probabilities. Goode (1993) compiled data showing that throughout Europe the percentage of marriages ending in divorce virtually doubled between 1970 and the mid-1980s. For example, in Germany in 1970, it was estimated that 16 percent of marriages would end in divorce, increasing to 30 percent in 1985. In France, the increase went from 12 percent to 31 percent during that same time period. Australia has experienced similar trends (Bracher et al. 1993).

Cherlin (1999) summarizes the major risk factors for divorce as including low income for the couple (which causes stresses and tension), early age at marriage (which often means a poorer job of choosing a spouse), spouse's similarity (known as *homogamy,* referring to the fact that people who are more similar to one another are more likely to get along with each other or, conversely, those who are less similar will be more likely to divorce), parental divorce (the "copy-cat" phenomenon, in which people whose parents divorced are more likely themselves to divorce), and cohabitation. Given the previous discussion of cohabitation, it is of some interest to note that, contrary to popular belief in the value of "trial" marriage, cohabitation before marriage appears to be one of the factors that increases the odds of a marriage ending in divorce, at least in North America. Several studies have suggested that cohabitation is selective of people who are mistrustful of marriage, probably because of their own "social inheritance of divorce" (Diekmann and Englehardt 1999). If cohabitation leads to marriage among these people, they may thus be more prone to end that marriage by divorce (Axinn and Thornton 1992; Hall and Zhao 1995).

Many marriages that in earlier days would have been dissolved by death are now dissolved by divorce. This seems apparent from the fact that the annual combined rate of marital dissolution from both the death of one spouse and divorce remained remarkably constant for more than a century. The overall rate of marital dissolution was essentially unchanged between 1860 and 1970 (Davis 1972a). As widowhood declined, divorce rose proportionately. Only with the rapid increase in divorce during the 1970s did that pattern begin to diverge. So dramatic was the rise in divorce in the 1970s that in the mid-1960s the elimination of divorce would have added an additional 6.7 years to the average marriage, whereas by the mid-1970s its elimination would have added 17.2 years (Goldman 1984).

Widowhood

As death has receded to older ages, the incidence of widowhood has steadily been pushed to older years as well. Divorce is a more important cause of not being married than is widowhood up to age 65, beyond which widowhood increases geometrically because of the higher death rate of men, undoubtedly compounded by the tendency of divorced women to change their status to widow upon the death of a former husband. As is true with so many social facts in the United States, African-American women are disadvantaged compared to whites in terms of marital status. At every age, blacks are more likely to be either divorced or widowed.

The Combination of These Determinants

As the demographic transition unfolds, then, we are finding that people are waiting longer to marry and when they do marry, their marriage is more likely to end in divorce than in widowhood. Schoen and Weinick (1993) used life table methodology to try to quantify the relative importance of these changes in family demography. Their results showed, for example, that between 1970 and 1988, the proportion of women surviving to age 15 who could expect ever to marry declined from 96 percent to 88 percent. At the same time, the average age at marriage was increasing, the percentage of marriages ending in divorce was increasing, and the percentage of marriages ending in widowhood was declining. Furthermore, as life expectancy increases while the average duration of a marriage shortens, and the percentage of divorced people remarrying goes down, the percentage of a person's life spent being married declines, thus adding to the individual diversity of household types in which a person might live during an entire lifetime.

Having described the principal features of the transformation of families and households in industrialized societies, let us see how they have been influenced by changing life chances and how, in turn, life chances interact with family formation and household structure.

Changing Life Chances

The leading explanations for the shift in household structure in Western nations combine elements of the demographic transition perspective with the life course perspective. The demographic transition perspective relates changing demographic conditions (especially declining mortality, declining fertility, and urbanization) to the rise in women's status. This was aided especially by increasing age at marriage, which encouraged higher levels of educational attainment. In turn, this has increased a woman's ability to enter the labor force and earn sufficient income to have the economic and social freedom to choose her own pattern of living. As these changes have unfolded, women's differing life chances have contributed to the transformation of families and households. Again, I emphasize that these demographic conditions are probably necessary, but not sufficient, to initiate the current rise in the status of women in industrialized societies. What is also required is some change in circumstance to act as a catalyst for the underlying demographic factors. This is where the life course perspective is drawn on, because women who have grown up in a different demographic and social milieu, and thus see the world differently than did earlier generations of women, have the potential to generate change in society. It is easy to know where to begin the discussion of changes in life chances, because nothing is more important than education.

Education

Becoming educated is probably the most dramatic and significant change you can introduce into your life. It is the locomotive that drives much of the world's economic development, and it is a vehicle for personal success used by generation after generation of people in the highly developed nations of the world. Still, the relative recency with which advanced education has taken root in American society can be seen in Table 10.2. In 1940, less than one in four Americans aged 25 or older had graduated from high school, although women were more likely than men to have done so. Slightly more than 5 percent of men and less than 4 percent of women were college graduates. You can see that the median number of years of education was 8.6 for men and 8.7 for women. A historically short six decades later, in 2000, 84 percent of both men and women were high school graduates and about one in four Americans had graduated from college—with men still being more likely than women to be in that category.

If we look at the youngest ages, we can get a better sense of the most recent trends. Among people aged 25–34 in 2000, 89 percent of women were high school graduates compared to 87 percent of men, and 30 percent of women were college graduates, compared to 29 percent of men. In Canada in 1996, among people aged 20–44, 76 percent of men and 79 percent of women were high school graduates, while 19 percent of men and 20 percent of women were college graduates. Thus, gender equality exists in education in Canada as in the U.S., although educational levels are not quite as high in Canada. From the 2000 census in Mexico, we have data on the median number of years of school completed and the data show an average of 7.8 for men and 7.3 for women—levels below those that prevailed in the U.S. in 1940.

Table 10.2   U.S. Educational Attainment Has Increased Significantly Over Time

Year	Male			Female		
	% High School Graduate or More	% College Graduate	Median Years of Education	% High School Graduate or More	% College Graduate	Median Years of Education
2000	84.2	27.8	13.2	84.0	23.6	13.0
1990	77.7	24.4	12.8	77.5	18.4	12.7
1980	69.2	20.9	12.6	68.1	13.6	12.4
1970	55.0	14.1	12.2	55.4	8.2	12.1
1960	39.4	9.6	10.3	42.5	5.8	10.9
1950	31.5	7.1	9.0	35.1	5.0	9.6
1940	22.3	5.4	8.6	25.9	3.7	8.7

Sources: Data for 1940 to 1990 are from U.S. Census Bureau, 2000, Educational Attainment Historical Table A-1 (http://www.census.gov, accessed 2001); data for 2000 are from U.S. Census Bureau, 2000, Educational Attainment of the Population 15 Years and Over, by Marital Status, Age, Sex, Race, and Hispanic Origin; March 2000, Table 3. Median year calculations for 2000 are by the author.

The world as a whole has been experiencing an increasing equalization of education among males and females—an important component in raising the global status of women. The ratio of females per males attending secondary school has been steadily increasing worldwide from since at least the 1970s, according to World Bank estimates. In some areas of the world, such as North America and Europe, women have already achieved parity with men, or are even more likely than men to be enrolled in secondary school (the pattern in central and eastern Europe). In sub-Saharan Africa and south Asia, as well as in northern Africa and western Asia, where the status of women has been notably low and where education for girls continues to lag behind that of boys, there have nonetheless been notable improvements in the ratio of girls to boys attending school. These trends will affect the world early in the twenty-first century as the gender gap is closed in education. Fargues (1995) has noted that in the middle of the twentieth century, most of the countries of northern Africa and western Asia had gender equity with respect to education in the sense that most people—males and females alike—were illiterate. Education was extended first to young men, creating for a while both a gender and a generation gap in education. Now, by the beginning of this century, as more generations of children have been educated, and as education has been offered to girls as well as boys, both of those gaps are closing and this could help accelerate the process of social and economic development.

The comments about closing the gap should not be taken to mean that we have arrived at gender equity in education everywhere on the planet. Table 10.3 lists the 30 countries in the world (representing 29 percent of the world's population) in which illiteracy among young women (aged 15–24) was at least 10 percentage points higher than for males that age as of 2001. Thus, in Yemen, a western Asian nation (just south of Saudi Arabia) of 20 million people, 16 percent of men aged 15–24 are unable to read or write, compared to 51 percent of women that age, for a difference of 35 percentage points. Yemen is tied with Benin—a western African nation of nearly seven million people, for this "honor" of lowest level of gender equity with respect to literacy. Twenty of the 30 countries are on the African continent, nine are in Asia, and one (Guatemala) is in Central America.

Our interest in education lies especially in the fact that by altering your world view, education tends to influence nearly every aspect of your demographic behavior and outcomes, as I have discussed to varying degrees in the previous chapters. Data from censuses and from sources such as the Demographic and Health Surveys show that nearly anywhere you go in the world, the more educated a woman is, the fewer children she will have. Not that education is inherently antinatalist; rather, it opens up new vistas—new opportunities and alternative approaches to life, other than simply building a family—and in so doing, it delays the onset of childbearing, which is a crucial factor in setting the tone for subsequent fertility (Marini 1984; Rindfuss, Morgan, and Offutt 1996). So, education tends to lower fertility, or to keep it lower than it might otherwise be, and this contributes to the variety of household and family structures we will see.

The fact that education alters the way you view the world also has implications for the marriage market in the United States. For much of American history, a major concern in choosing a marriage partner was to pick someone who shared your religious background (social scientists call this "religious homogamy"). Over time,

Table 10.3   World Educational "Hot Spots": Countries Where Illiteracy Rate Among Young (Aged 15–24) Women is at Least 10 Percentage Points Higher Than for Young Men, 2001

Country	Region	Population	Youth (15–24 illiteracy) 2001		
			Boys	Girls	Excess Illiteracy of Girls
Yemen	Western Asia	20,010	16	51	35
Benin	Western Africa	6,736	28	63	35
Nepal	South Central Asia	25,164	23	56	33
Liberia	Western Africa	3,367	14	46	32
Iraq	Western Asia	25,175	40	70	30
Pakistan	South Central Asia	153,578	28	57	29
Guinea-Bissau	Western Africa	1,493	26	54	28
Mozambique	Eastern Africa	18,863	24	52	28
Togo	Western Africa	4,909	12	35	23
Burkina Faso	Western Africa	13,002	53	75	22
Mali	Western Africa	13,007	52	74	22
Eritrea	Eastern Africa	4,141	19	39	20
Niger	Western Africa	11,972	67	86	19
Malawi	Eastern Africa	12,105	19	38	19
Côte d'Ivoire	Western Africa	16,631	29	46	17
Senegal	Western Africa	10,095	40	57	17
Bangladesh	South Central Asia	146,736	43	60	17
Morocco	Northern Africa	30,566	23	40	17
Georgia	Western Asia	5,126	33	49	16
Syrian Arab Republic	Western Asia	17,800	4	20	16
Mauritania	Western Africa	2,893	43	59	16
Central African Republic	Middle Africa	3,865	23	39	16
India	South Central Asia	1,065,462	20	34	14
Lao People's Democratic Republic	Southeastern Asia	5,657	15	28	13
Egypt	Northern Africa	71,931	23	36	13
Chad	Middle Africa	8,598	25	38	13
Uganda	Eastern Africa	25,827	14	27	13
Guatemala	Central America	12,347	14	27	13
Ethiopia	Eastern Africa	70,678	38	50	12
Sudan	Northern Africa	33,610	17	27	10

Source: World Bank, World Development Indicators 2003 (Washington, DC: World Bank Group).

however, the salience of religion has given way to "educational homogamy"—people want to marry someone with similar amounts of education, and so education has been replacing religion as an important factor in spouse selection (Kalmijn 1991; Mare 1991; Smits, Ultee, and Lammers 1998).

Table 10.4   Better-Educated Workers Have Higher Incomes

Year-Round Full-Time Workers Aged 25+		Males	Females	Ratio of Female Income to Male income
Total		$41,616	$31,356	0.75
High School	Not high school graduate	$24,364	$18,096	0.74
	Graduate	$34,723	$25,302	0.73
College	Some college, no degree	$41,045	$30,418	0.74
	Associate's degree	$42,776	$32,152	0.75
	Bachelor's degree	$55,929	$40,994	0.73
	Master's degree	$70,898	$50,668	0.71
	Professional degree	$100,000	$61,747	0.62
	Doctorate degree	$86,965	$62,122	0.71

Source: Adapted from U.S. Census Bureau, 2003, "Educational Attainment in the United States, March 2002, Detailed Tables," Table 8.

The greater proportion of people going to college has altered the lifestyles of many young Americans. It has been accompanied by delayed marriage, delayed and a diminished amount of childbearing, and, consequently, higher per-person income among young adult householders. Young people today have more personal disposable income than any previous generation. In 2002, women consistently earned less than men, but for both men and women, the data in Table 10.4 show that those with a postgraduate degree were earning more than twice as much per year as those who were only high school graduates. This has become known as the "college premium," and its existence has almost certainly contributed to increased education (investment in human capital) and increased economic productivity. However, for the investment in education to pay off for people, households, and society more generally, it must lead to higher levels of labor force participation and more productive occupations. Let us examine these aspects of your life chances.

Labor Force Participation

As education increases, so does the chance of being in the labor force. Among both males and females in the United States in 2000, the higher the level of education attainment among people aged 25 and older, the higher the percentage of people who were currently in the labor force. Women are less likely than men to be in the labor force at any given level of education, but it is nonetheless true that the pattern over time has been for women to be working more, and men to be working less. Prior to the 1970s, for example, women who worked typically did so only before they married or became pregnant. Thus, the labor force participation rates by age peaked in the early 20s and declined after that. That is still a common pattern in many less-developed countries, but it no longer characterizes women in the more-developed nations; labor force participation rates by age are now very similar for males and females.

Keep in mind as we talk about labor force participation rates that most countries include unemployed persons as being in the labor force. Thus, if you are looking for work, even though you are not actually working or even if you have never before held a job, you are considered to be in the labor force. Unemployment rates are strongly related to age—the older the age, the lower the rate. At younger ages, considerable numbers of people are looking for work even if they haven't found it yet, whereas at older ages, people are more likely to give up on employment and seek a retirement pension as soon as it is available if they experience difficulty finding a job. Women also tend to have lower unemployment rates than men.

By far the biggest gain in employment over the past few decades has been the movement of baby boom women, especially married women, into the labor market. They literally burst their way into the workforce in the 1970s, and by 2000, 78 percent of women aged 35–54 (the baby boomers) had a job. In 1960, just before the baby boomers came of age, labor force participation rates among married women in the United States were well below those for single women at every age from 20 through 64. By 1970, as the first wave of baby boomers had reached adulthood, the rates for married women were clearly on the rise, and by 2000 they had nearly reached those of single women.

Female Labor Force Participation and Fertility   Working, as I have mentioned before, cuts down on fertility under normal circumstances. It is certainly no coincidence that the birth rate in the United States began to drop at about the same time that labor force participation rates for married women began to rise. Even in 2002, when fertility was generally very low, it was still true that those who were in the labor force had slightly fewer children than those who were not (Downs 2003). Some of that difference is due to the fact that working women, even those currently married, are more likely to be childless than women who are not in the labor force. Overall, the highest levels of fertility in the United States are found among poor women who do not work, whereas the lowest levels of fertility are among those who do work and are well paid.

The ability of married women to work helps bring fertility down and maintain it at low levels, but, as I mentioned in Chapter 6, it can also help keep fertility from dropping to below-replacement levels. When women are able to combine having a family, even if small, and a career, they are more likely to choose both. But, in European countries such as Italy and Spain, where family values discourage that combination, women seem to choose in favor of career over family, contributing to very low levels of fertility. The pattern that has emerged in more-developed countries is that a high proportion of women work before marriage, remain working after marriage, and adjust their fertility downward to accommodate their working. How far down they adjust may depend on how much familial and societal support they receive in making that accommodation. If child care is readily available and if husbands share domestic chores, we may well expect that fertility will be slightly higher than if a woman is expected to shoulder all of those burdens.

In Mexico, an increase in labor force participation has accompanied a decline in fertility, as we would expect. In 1979, when the total fertility rate in Mexico was 5.0, only 21.5 percent of women were in the labor force; by the mid-1990s, total fertility had dropped to 3.4 and female labor force participation had climbed to 31.4

percent (Fleck and Sorrentino 1994). It is unlikely that one directly caused the other; rather, each change helped stimulate the other. An initial decline in fertility, for example, may lead to a small increase in the percentage of women in the labor force, which may encourage other women to limit fertility, which may promote other opportunities in the labor force. Each event cumulatively causes the other, and when two things occur together, it is difficult to sort out cause and effect.

In a similar vein, a study of two seemingly different villages in India revealed that in the village where women were employed rolling cigarettes for a local contractor, the employment gave women, very serendipitously, substantial autonomy and increased their use of contraception. In another village without employment opportunities, the researchers expected to find that contraceptive use had remained low. Yet, in that village, the time spent not working had been spent instead on additional education for women and that, it seems, had increased contraceptive use (Dharmalingam and Morgan 1996). There are multiple and cumulative paths to low fertility and to improved social situations for women.

We have already discussed the new household economics as an approach to explaining why fertility is kept low in developed societies and why households might encourage family members to migrate. Now, we can call on it again to explain why the rise in the status of women and increased female labor force participation might generate the household transformations we have been reviewing in this chapter. The basic idea is that "the rises in women's employment opportunities and earning power have reduced the benefits of marriage and made divorce and single life more attractive. While marriage still offers women the benefits associated with sharing income and household costs with spouses, for some women these benefits do not outweigh other costs, whatever these may be" (McLanahan and Casper 1995:33).

Female Labor Force Participation and Empowerment In most social systems, people who can take care of themselves and have enough money to be self-reliant have higher status and greater freedom than those who depend economically on others. Further, a pecking order tends to exist among those who are economically independent, with higher incomes being associated with higher status than are low incomes. Being independent, though, is definitely the starting point, and an increasing number of women in the world are arriving at that point. However much you might take for granted the fact that in a rich nation women are as readily employable in the paid labor force as men, it is actually a rather recent phenonomen.

Although mortality and fertility have been declining since the nineteenth century in the United States and urbanization has been occurring throughout that time, it was during World War II that the particular combination of demographic and economic circumstances arose to provide the leading edge of a shift toward labor force equality of males and females. The demand for armaments and other goods of war in the early 1940s came at the same time that men were moving out of civilian jobs into the military, and there was an increasing demand for civilian labor of almost every type. Earlier in American history, the demand for labor would have been met by foreign workers migrating into the country, but the Immigration Act passed in the 1920s (see Chapter 13) had set up national quotas that severely limited immigration. The only quotas large enough to have made a difference were those for immigrants from countries also involved in the war and thus not a potential source of labor.

With neither males nor immigrants to meet the labor demand, women were called into the labor force. Indeed, not just women per se, but more significantly, married women, and even more specifically, married women with children. Single women had been consistently employable and employed since at least the beginning of the century, as each year 45–50 percent of them had been economically active, as I pointed out earlier. But in the early 1940s, there were not enough young single women to meet labor needs, partly because the improved economy was also making it easier for young couples to get married and start a family. It was older women, past their childbearing years, who were particularly responsive to making up the deficit in the labor force (Oppenheimer 1967; Oppenheimer 1994).

These were the women who broke new ground in female employment in America, with the biggest increase in labor force participation between 1940 and 1950 coming from women aged 45 to 54, and because more than 92 percent of those women were married, this obviously represented a break with the past. Who were these women? They were the mothers of the Depression, mothers who had sacrificed the larger families they wanted (as I noted in Chapter 6) to scrape by during one of America's worst economic crises. They were women who had smaller families than their mothers and thus were more easily able to participate in the labor force. However, the ideal family size remained more than three children, and the improved economy permitted the low fertility of the 1930s to give way to higher levels in the 1940s and 1950s. Women with small families from the Depression actually opened the door to employment for married women, but younger women were not ready to respond to those opportunities in the 1940s and 1950s. Indeed, between 1940 and 1950 the labor force activity rates of women aged 25 to 34 actually declined.

Things changed, as you know, and so between 1950 and 2000, there was a substantial increase in the number and proportion of American women who were in the labor force and earning independent incomes. In 1950, for example, there were 29 female, year-round, full-time workers for every 100 males in that category; by 2000, there were 70 females working full-time, year-round per 100 male workers. This increase in labor force activity was accomplished initially by younger women; the younger the cohort, the more rapidly did labor force involvement increase, especially beyond ages 25 to 34, when children tend to be in school.

Getting a job is one thing, of course, but the kind of job you get—your occupation—depends heavily on education, and is also influenced by factors such as gender and race/ethnicity.

Occupation

Occupation is without question one of the most defining aspects of a person's social identity in an industrialized society. It is a clue to education, income, and residence—in general, a clue to lifestyle and an indicator of social status, pointing to a person's position in the social hierarchy. From a social point of view, occupation is so important that it is often the first (and occasionally the only) question a stranger may ask about you. It provides information about what kind of behavior can be expected from you, as well as how others will be expected to behave toward you. Although such a comment may offend you if you believe that "people are people," it is nonetheless true that there is no society in which all people are actually treated equally.

Since there are literally thousands of different occupations in every country, we need a way of fitting occupations into a few slots. Organizations like the U.S. Census Bureau and the International Labour Organization have devised classification schemes to divide occupations into several mutually exclusive categories. In Table 10.5, I have listed the occupational distribution of employed males and females in the United States as measured in the Current Population Survey in 1970 and 2002. The occupational categories changed a bit over time, but you can still see that the distribution of occupations for women looked more like that for men in 2002 than it had back in 1970. Indeed, the index of dissimilarity (sometimes also known as the Gini coefficient) was .43 in 1970 and had been reduced to .32 in 2002. We would interpret that to mean that in 2002, you would had to have moved 32 percent of women into other occupational categories to have exactly the same distribution as for men. That does not yet signal equality, but it was a clear improvement over the situation in 1970. In both years, there was a clear tendency for women to be disproportionately in clerical occupations, while men were disproportionately in laborer occupations.

There is a global tendency for women to be stuck in less desirable jobs than men, and there are three important issues that the International Labour Organization sees as still needing considerable improvement in order to achieve gender equality in the workplace: (1) the "glass ceiling" (women being less likely than men to make it to top management), (2) the gender pay gap (worldwide, women earn an average of about two-thirds what men earn), and (3) the "sticky floor" (women tend to get stuck in the lowest-paid jobs) (International Labour Organization 2003).

People holding the higher-status occupations are more likely to think of themselves as having a career as opposed to just a job, and they are apt to derive more intrinsic satisfaction from their work. They are also less likely to be working in what Kalleberg and his associates (Kalleberg, Reskin, and Hudson 2000) have labeled as "bad jobs"—those with low pay and no health or pension benefits. Nearly one in seven jobs in the United States fits this categorization, and they are the types of jobs more likely to be held by women than by men, especially if the woman is not a college graduate (McCall 2000). The kind of job you have, and the income you earn from it, will be your key to economic and social independence.

Income

The average CEO (chief executive officer) at the 200 biggest companies in the United States made $11.4 million in 2001, including salary, bonuses, and stock gains (Lowenstein 2001). This works out to more than $40,000 a day, so if we assume a normal work day, this average CEO has already made $10,000 by the time the doughnuts arrive for his (yes, "his") morning coffee break. The CEO's average daily salary, by the way, was nearly equivalent to the entire annual income for the average American household ($43,052 in 2002). There is a good deal of controversy about whether the average CEO is worth that kind of money, but for the rest of us, income is at least partially a consequence of the way in which we have parlayed a good education into a good job.

Table 10.5   U.S. Occupational Distributions Still Different for Males and Females

Employed civilians 18–64	Percentage Distribution by Occupational Category—2002	
	Females	Males
TOTAL	100.0	100.0
Executive, administrative, and managerial	15.2	15.8
Professional specialty	19.2	13.9
Technical and related support	4.0	2.8
Sales	11.8	10.9
Administrative support, including clerical	23.0	5.4
Private household service	0.9	0.0
Other service	16.5	10.4
Precision production, crafts, and repair	2.0	18.9
Machine operators, assemblers, and inspectors	3.7	6.0
Transport and material moving	0.9	7.0
Handlers, equipment cleaners, and laborers	1.6	5.7
Farming, forestry, and fishing	1.0	3.1

Employed persons 25–64	Percentage Distribution by Occupational Category—1970	
	Females	Males
TOTAL	100.0	100.0
Professional, technical, and kindred workers	16.1	15.2
Managers and administrators, except farm	5.2	16.6
Sales workers	6.4	5.3
Clerical and kindred workers	31.3	6.7
Craft and kindred workers	1.3	22.0
Operatives, including transport	16.4	19.2
Laborers, except farm	0.4	5.2
Service workers, including private household	21.3	5.4
Farmers and farm managers	0.3	3.3
Farm laborers and supervisors	1.3	1.1

Sources: Data for 2002 are from U.S. Census Bureau, 2003, "Educational Attainment in the United States, March 2002, Detailed Tables"; for 1970, the data are from the 1970 Current Population Survey, Table 6, http://www.census.gov, accessed 2004.

Occupation may be the primary clue that people have about our social standing, yet our well-being is thought by most people to be a product of our income level. People are too polite to ask you how much you make, but by knowing your occupation, they will have important clues about your income level. Little has changed since the late 1970s, when Coleman and Rainwater concluded that "money, far more than anything else, is what Americans associate with the idea of social class" (Coleman and Rainwater 1978:29). It is not just having the money; rather it is how you spend it that signals to others where you stand in society. The principal indicators of having money include the kind of house you live in, the way your home is furnished, the car or cars you drive, the clothes you wear, the vacations and recreations (the "toys") you can afford, and even the charities you support.

It is no mystery that there is an uneven distribution of income in virtually every human society, including the United States. In 2002 in the United States, the richest 20 percent of families earned 50 percent of the nation's total income, while the poorest 20 percent earned 3 percent (DeNavas-Walt, Cleveland, and Webster 2003). This was a continued deterioration of the distribution that had prevailed in the late 1960s, when the top 20 percent commanded 43 percent, while the bottom 20 percent shared more than 4 percent of the nation's income. These changes have produced the "hourglass" economy, which bulges with households at the high and low ends of incomes and is squeezed in the middle. Three broad explanations have been offered for the increasing inequality: (1) public policy changes, such as tax "reforms" that benefit some groups more than others; (2) labor market changes (occurring throughout the world, not just in the United States), such as an increasing mismatch between jobs and the skills of the labor force, or a polarization of jobs into those that require high skills and those that require few skills, with little in between; and (3) changes in demographic structure, such as the increasing fraction of households headed by females. All three of these changes probably have played a role in increasing income inequality in the United States (Danziger and Gottschalk 1993; Morris, Bernhardt, and Handcock 1994), but let us focus for the moment on the third explanation, relating to demographic change.

Income Disparities by Race/Ethnicity   Family income rose steadily for American families during the 1950s and 1960s, leveled off in the 1970s and early 1980s, and then rose, albeit unevenly, between the mid-1980s and the early 2000s. These patterns can be seen in Figure 10.5. In 1947, a white family in the United States had an income equivalent to $20,870 at 2001 prices; that had nearly tripled to $54,067 by 2001. African American families also experienced not just a tripling, but nearly a quadrupling, of income since the end of World War II, from $9,644 (in 2001 dollars) in 1947 to $33,598 in 2001. In relative terms, income for black families has grown slightly faster than that for white families. Thus "the ratio" in Figure 10.5 shows that in 1947 African-American family income was less than half that of whites (a ratio of 0.46). By 2001, that ratio had gone up to 0.62. Yet, despite the rise in family income and a narrowing of the income gap with whites in relative terms, blacks were actually losing ground to white families in absolute terms. In 1947, the "dollar gap" between African American and white families (the difference between the two shaded areas in Figure 10.5) had been $11,226, but by 2001 it had grown to $20,469.

Thus African Americans have been in the peculiar position of having their incomes rise faster in percentage terms than whites, but in dollar terms (the actual money available to spend on family members) falling further behind. This is one of the paradoxes that results from **structural mobility**—that situation in which an entire society is improving its situation economically, even though some groups may be

Figure 10.5   Gap in White and African American Family Income Widened in Second Half of Twentieth Century

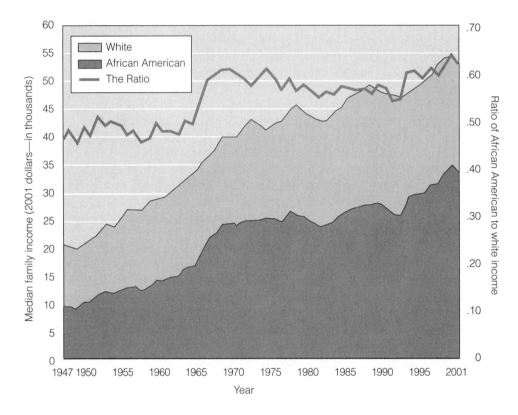

Sources: Adapted from data in U.S. Census Bureau, 1992, Current Population Reports, P60–180: Table B–7; and 2003, P60–221: Table 6. Data are for family households and are expressed in 2001 dollars.

gaining at a faster rate than others. That is the only time in which one group can improve itself socially or economically without forcing an absolute sacrifice from another group. There was considerable economic mobility from the end of World War II until 1970, but when structural mobility slowed down in the 1970s, the gains of African Americans also hit the brakes, because at that point any rise in income would have been the result of a deliberate policy of income redistribution among ethnic groups.

I should add a word of caution about the comparison of family incomes of whites and African Americans in the United States. African Americans have slightly fewer earners per family than do whites, which will reduce family income, all other things being equal. Furthermore, a much larger proportion of African American families are headed by a female than is true for whites, and since females tend to earn on average only about three-fourths of what men make (see Table 10.4), that too lowers African-American family income compared to whites. We can control for these factors by looking separately at the incomes of only those people who are year-round, full-time earners.

If we were to assume that every family in the United States had both a male and a female year-round, full-time worker, each earning the median income for his or her gender and race, then we would find that in 1955 (the first year for which such data are available), white families would have had an average family income of $47,163 (in 2001 dollars), compared with $27,377 for African American families. Thus, under these controlled circumstances, African American families would have had $19,786 less money to spend each year—less than two-thirds of the income of white families. By 2001, however, the situation had definitely improved for everybody. Non-Hispanic white two-earner families would have been making an average of $75,067, compared with $59,590 for non-Hispanic black families. That amounts to a difference of $15,477, with African American family income 79 percent that of whites. Between 1955 and 2001, the average income of these hypothetical non-Hispanic white couples would have increased 60 percent, while the increase for the hypothetical African American couples during this time amounted to 118 percent! Furthermore, if all other things had been equal, the actual dollar income gap between African Americans and whites would have been cut by a few thousand dollars. Since we controlled for full-time employment, number of workers, and gender, one of the major things left to explain the gap, of course, would be racial discrimination.

Income Disparities by Gender   The top CEOs in the United States make more money than rock stars, but there is a huge gender gap in this regard. A survey in 1997 showed that only two of the CEOs of Fortune 500 companies (the largest companies in the United States) were women (Lublin 1998). Although in that year the highest-ranking female executive was averaging $518,000 in total compensation, that was still 32 percent less than the highest ranking male executive in the same company. Let us consider in more detail the issue of income disparities by gender. In 2001, the average female wage earner, working year-round, full-time, earned only 75 percent of the income garnered by males (as you will recall from Table 10.4). This was an improvement from 1977, however, when pay for women was 58 percent of that for men. Some of this difference is due to the fact that women are likely to have been in the labor force less time than men and in their current job for a shorter period of time, and that women are more likely than men to delay the completion of their education (which delays their reaching better-paid occupational levels). Yet, these "compositional" differences between males and females are unable to fully account for the gender gap.

Why is there a differential in income by gender? The obvious answer is that there has been a history of discriminating against women in the labor market in terms of what kinds of jobs they are hired for and what pay they receive (Marini and Fan 1997). This is true almost everywhere you go in the world. It has been quipped that "Japan is the land of the rising sun, but only the son rises." In the German Democratic Republic right up to the period before the Berlin Wall collapsed, communism was supposed to guarantee gender equality. Yet, women with the same education as men, working at the same jobs as men, were receiving less pay than men (Sörenson and Trappe 1995). One of the few countries where true gender equality appears to have been reached, at least outside the home, is in Sweden (Casper, McLanahan, and Garfinkel 1994).

The good news is that in the United States the data clearly seem to suggest that the gap in status between men and women is closing for younger people (Bianchi 1995), and part of the explanation for this is the fact that women's wages have been rising more quickly than those for younger men (Bianchi 1995; Oppenheimer 1994). The U.S. economy, for example, had to swallow a "Big Gulp" to find employment for the baby boomers, since not only were there more of them but a larger proportion of females than ever before were looking for jobs. Although the economy did eventually absorb these people, it did so without offering much improvement in earnings over previous generations. This is partly because male wages were bound to suffer somewhat from the competition with women for spots in the labor force (Waite 2000). Economic improvement for households has required that two-earner households become the norm, and this has certainly contributed to the delay in marriage and the rise in divorce.

Poverty   If you have several children, the odds increase that your income will be below average, and if, on top of that, you are a divorced mother, the chance skyrockets that you will be living below the poverty level. In 2002, 28 percent of people in an American family headed by a woman with no husband present were living below the poverty level, compared to 12 percent of the total population that was living below the poverty line (Proctor and Dalaker 2003). However, if we look at female-headed households with children under the age of five, we find that one out of two such families (49 percent) were below the poverty level in 2002 and two out of three (65 percent) of those families were within 150 percent of the poverty line.

To imagine the struggle it is to manage successfully in the United States on so little money, it is necessary only to review the definition of the poverty level. The **poverty index** was devised initially in 1964 by Mollie Orshansky of the U.S. Social Security Administration. It was a measure of need based on the finding of a 1955 Department of Agriculture study showing that approximately one-third of a poor family's income was spent on food, and on a 1961 Department of Agriculture estimate of the cost of an "economy food plan"—a plan defined as a minimally nutritious diet for emergency or temporary use (Orshansky 1969). By calculating the cost of an economy food plan and multiplying it by three, the poverty level was born. It has been revised along the way, but the idea has remained the same, and since 1964 it has been raised at the same rate as the consumer price index.

In 2002, the poverty threshold for a single person under the age of 65 was $9,359. This was the equivalent of earning $4.23 an hour if you were a year-round, full-time worker, but keep in mind that the federal minimum wage in that year was $5.15. A single parent with two children under the age of 18 could be earning $14,494 (the equivalent of $6.56 per hour for a year-round, full-time worker) and still be right at the poverty-level threshold. Between 1960 and 1973, the percentage of Americans living below the poverty level was cut in half, from 22 percent to 11 percent. Since that time, the poverty level has never again been as low as 11 percent, nor has it been higher than 15 percent; it was 12 percent in 2002.

Canada has adopted a strategy similar to that of the U.S. for measuring the lower end of the income scale, but the Canadian government has tried to avoid controversy by not officially defining a poverty threshold. Rather, Statistics Canada produces what are labeled "low income cut-offs." As of 2001, 14 percent of Canadians lived below that cut-off, but 45 percent of families with a female lone parent and children under the age of 18 were living with low income (Statistics Canada 2003b).

On the basis of global comparisons, it might be argued that very few people in North America are poor in absolute terms—it is the relative deprivation that is socially and morally degrading. Organizations such as the International Labour Organization (ILO, a specialized agency within the United Nations) and the World Bank have adopted an international standard that defines poverty as an income of less than $1.08 per day—an astonishingly small amount of money on which to try to survive. If $1.08 seems like an odd number, it is only because the index started a few years ago at $1 per day and has since been adjusted upward for inflation. Using this definition, the World Bank has estimated that in the year 2000, there were 1.1 billion people in the world (17 percent of the global population, but 22 percent in developing countries) living below that poverty level (World Bank 2003). If we stretch the poverty line to $2.15 a day, we find that 2.7 billion live at that level, and account for 54 percent of the people in developing countries. These data are summarized in Table 10.6.

"We live not as we wish, but as we can," is how a southern Indian peasant described life (quoted by Hockings 1999:213). One index of poverty used in India is the number of meals per day that a person can afford. If you can afford at least two meals a day, you are above the poverty line, and by that definition only about 20 percent of Indians were below the poverty level in 1996 (Jordan 1996), but in fact 44 percent of Indians in 2000 lived on less than $1.08 a day and 86 percent lived on less than $2.15 per day. South Asia is matched in its overall poverty only by sub-Saharan Africa. As you can see in Table 10.6, nearly everybody (91 percent of the population) in Nigeria is estimated to live on less than $2.15 per day. And, of course, since HIV/AIDS has been ravaging sub-Saharan Africa, this crushing level of poverty has actually gotten worse, not better, since 1990. Compared to that, just living at the poverty threshold in the United States (which for one person would equal $26 a day) would mean a life of considerable luxury.

Wealth

Poverty implies not only the lack of adequate income from any and all sources, but also the lack of any other assets on which a person might draw sustenance. As people obtain and build assets, they create wealth. An asset is something that retains value or has the potential to increase in value over time. Every generation produces its share of self-made people like Bill Gates of Microsoft, but his three children will inherit that wealth, rather than having to produce it for themselves. Wealth and its attendant high income are essentially ascribed characteristics for those born into families that own huge homes, large amounts of real estate, and tremendous interests in stocks and bonds or other business assets. For most people, a home is the most important asset they will acquire in a lifetime, but assets can also include

Table 10.6   Huge Percentage of People in Developing Countries Live on Less than $2.15 Per Day

	Poverty Rate (% below)				Number of Poor (1,000,000)			
	$1.08/day		$2.15/day		$1.08/day		$2.15/day	
	1990	2000	1990	2000	1990	2000	1990	2000
East Asia & the Pacific	29.4	14.5	68.5	48.3	470.1	261.4	1094.4	872.6
China	31.5	16.1	69.9	47.3	360.6	204.4	799.6	599.5
Indonesia		7.2		55.4				
East Europe & Central Asia	1.4	4.2	6.8	21.3	6.3	19.9	31.1	101.3
Russia		6.1		23.8				
Tajikistan		10.3		50.8				
Latin America	11.0	10.8	27.6	26.3	48.4	55.6	121.1	135.7
Mexico		8.0		24.3				
Colombia		14.4		26.5				
Middle East & North Africa	2.1	2.8	21.0	24.4	5.1	8.2	49.8	71.9
Egypt		3.1		43.9				
Yemen		15.7		45.2				
South Asia	41.5	31.9	86.3	77.7	466.5	432.1	970.9	1051.9
India		44.2		86.2				
Pakistan		13.4		65.6				
Sub-Saharan Africa	47.4	49.0	76.0	76.5	241.0	322.9	386.1	504.0
Ghana		44.8		78.5				
Nigeria		70.2		90.8				
Total	28.3	21.6	60.8	53.6	1237.3	1100.2	2653.3	2737.3

Source: Adapted from data in World Bank, 2003, Global Poverty Monitoring, http://www.worldbank.org/research/povmonitor, accessed 2004.

personal property such as jewelry or other collectibles, stock in companies or mutual funds, savings accounts in banks, rental property, or ownership in a business venture. Typically, wealth is measured as net worth—the difference between the value of assets and the money owed on those assets. If your only asset is the house you just bought for $225,000 (the average sales price in the United States in 2004), you are in the process of building wealth, but your net worth may be close to zero because you still may owe as much on the mortgage as the house is worth. There are three basic ways to generate wealth: (1) inherit assets from your parents or other relatives (the easiest way), (2) save part of your income to purchase assets (the hardest way), and (3) borrow money to purchase assets (the riskiest way).

Since most of us do not have fabulously wealthy parents, our ability to inherit enough from our parents so we can build on to create our own wealth will, in fact, be determined importantly by a basic demographic characteristic—how many siblings we have (Keister 2003). The fewer people with whom we have to share our parents' inheritance, the more there is for us to use ourselves, so those groups with the lowest fertility are likely to have a higher proportion of people who are able to accumulate wealth. The principal sources of data on the wealth of Americans are household surveys, which always present problems of accurate reporting. The most representative source of information is almost certainly a set of questions asked periodically in the Survey of Income and Program Participation (SIPP) conducted by the U.S. Bureau of the Census. These data show that net worth among Americans in 2000 averaged $50,000 per household (Orzechowski and Sepiella 2003). Marriage is an important ingredient in accumulating wealth (as long as the couple stays married) and thus we find that the highest level of net worth occurs among older (65 and older) married-couple households, whose net worth in 2000 of $351,000 was more than twice the net worth of unmarried male householders ($110,000) or unmarried female householders ($85,000) of that age. Note that the median net worth of unmarried female householders is almost exactly three-fourths that of unmarried male householders, matching the gender gap with respect to income.

Figure 10.6 also reminds you of the value of a college education. Householders with at least a bachelor's degree had assets worth $313,000 in 2000, which was

Figure 10.6   2000 U.S. Median Net Worth Highest for Older Married Couples

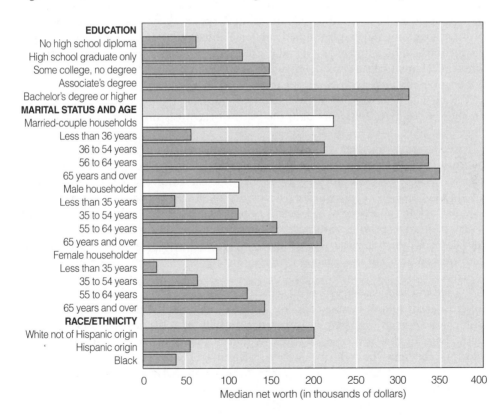

Source: Adapted from Shawna Orzechowski and Peter Sepiella, 2003, "Net Worth and Asset Ownership of Households: 1998 and 2000," Current Population Reports, Table 5: pp. 70–88.

almost three times the level of householders who had not gone beyond high school. However, there can be little question that one of the most striking comparisons is by race and ethnicity. In 2000, the average non-Hispanic white household had a net worth of $198,000 compared to $52,000 for Hispanics and only $35,000 for blacks. Some of this difference is due to differences in the age structure and some is due to differences in marital status, but much of it is due to the greater difficulty that racial and ethnic minority household members have in generating enough income for long enough to be able to acquire higher levels of wealth.

Race and Ethnicity

The assimilationalist model of immigration, which I discussed in Chapter 7, assumes that a nation is a melting pot where everyone eventually shares cultural values and norms and ultimately every person becomes pretty much like everyone else. This is sometimes referred to as the "North American Model" of race and ethnic relations, which is aimed at combating racial discrimination and ethnic inequality (Haug 2000). The multicultural model, on the other hand, assumes a salad bowl where everybody stays different but gets along just fine. European nations, with the notable exception of France, have tended to prefer this approach, and it has actually been adopted as the main American model in the past two decades, represented by a positive governmental emphasis on diversity. The assimilationist model assumes that distinctions of race and ethnicity will eventually be wiped out by intermarriage, whereas the multicultural model assumes not only that things won't work like that, but that people prefer to remain separate. The United States has historically fluctuated somewhere between those two extremes, but from the moment of the country's creation more than 200 years ago, the issue of race was on the table. It has been there ever since.

Recognizing that race and ethnicity are important issues in a society does not necessarily mean that they will be easy to measure. They are not easy to measure, of course, because they are not easy to define and so measurement becomes more of an art than a science. Indeed, the "science" part is scary because it can too easily lead us back to the Eugenics movement at the beginning of the twentieth century in which people were trying to measure genetic differences among people who, in fact, only looked different without having any other distinguishing characteristics. Our cultural heritage, not our genetic heritage, distinguishes us, and that is not easy to measure.

Race and ethnicity represent human differences with some type of physical manifestation that allows people to identify and be identified with a particular group. The characteristics may be physical in nature such as skin pigmentation, hair texture, shape of the eyes or nose (these would normally fall under the category of **race**), or they may be more behavioral, such as language, or identification with a particular ancestry and geographic place (these would normally fall under the category of **ethnicity**).

The history of racism in the world suggests that anything that distinguishes you can, and probably will, be used against you. Therefore, to be a member of a subordinate (not in political control) racial, ethnic, or religious group in any society is to be at jeopardy of impaired life chances. African Americans, Hispanics, Asians, and American Indians in the United States are well aware of this, as are indigenous people in Mexico, Tamils in Sri Lanka, Muslims in Israel or India, Indians in Malaysia, and virtually any foreigner in Japan or China.

Race and Ethnicity in the United States   In Census 2000, 69.1 percent of the population indicated that they belonged to one racial category and that category was white, and furthermore they were not Hispanic. I mentioned in Chapter 4 that Census 2000 permitted people for the first time ever to choose more than one racial category. The underlying purpose was to be able to capture more accurate information about multiracial individuals. If you accept the assimilation theory, then you would expect an increasing fraction of the population to identify with more than one racial/ethnic group, whereas the multicultural perspective would not expect this to be happening. In the United States, the major racial categories are defined as "white," "Black or African American," "Native American or Alaska Native," "Asian," and "Native Hawaiian and other Pacific Islander" (Grieco and Cassidy 2001). You will also see tables from the census that include a number of people in the "some other race" category, but almost all of them are Hispanic, which represents the principal ethnic category asked about in the census. A person of "Hispanic or Latino" origin is defined as one who is "of Cuban, Mexican, Puerto Rican, South or Central American, or other Spanish culture or origin regardless of race."

It turned out that 2.4 percent of respondents to the census (6.8 million people) indicated that they identified with more than one race ("multiracial"). Of some interest was that 4 percent of children under age 18 were listed as multiracial, compared to only 2 percent of adults. This could reflect an increase over time in multiracial marriages, but since parents typically answer the census questionnaire on behalf of their children, it could also reflect a different sense that the parent has of the child's identity than the child might have.

In the United States, blacks have been the largest minority group for literally all of the nation's history, but Census 2000 data show that the Hispanic or Latino population has essentially caught up with the black population in size. With an enumerated population of 35.3 million (12.5 percent of the total population), the Hispanic population is slightly larger than the number of people who indicated only one race, with that race being black or African American (34.7 million people or 12.3 percent of the population). On the other hand, if you include people who indicated black in combination with some other race, then the black population numbers 36.4 million, or 12.9 percent of the population. But, if you remove those people who said they were both black and Hispanic and put them in the Hispanic category (as the Census Bureau normally does when it derives its "Hispanic-exclusive" categories), then the number of Hispanics slightly exceeds the number of blacks. In all events, given the fact that the Hispanic population is growing quickly and the black population is not, there can be little question that by Census 2010, Hispanics will be the largest ethnic minority group in the United States (and remember that they are already the largest ethnic group in California, exceeding the white non-Hispanic group).

No longer being the numerically largest minority group in the country does not necessarily make life easier. Being of African origin in the United States is associated with higher probabilities of death, lower levels of education, lower levels of occupational status, lower incomes, and higher levels of marital disruption than for the non-Hispanic white population. McDaniel (1996) has argued that the different life chances of whites and blacks in American society are due to **racial stratification,** which he defines as a socially constructed system that characterizes one or more groups as being distinctly different. Your membership in a group defined as different from the others then creates, in essence, a different social world for you than for those who are in other groups, and this affects your behavior and your life chances in society, because there is no genuine societal expectation that you will be assimilated into the rest of the society. He offers an illustration:

> The ability of a group to be assimilated depends on whether it is considered an ethnic or racial group . . . For example, immigrants from Nigeria and Ghana assimilate into the African American race, and immigrants from Sweden and Ireland assimilate into the European American race . . . Ethnic groups such as those from Europe and segments of the Hispanic population show definite signs of assimilability; however, the African-descendent population continues to be blocked from assimilation within the United States (McDaniel 1996:139).

Earlier in this chapter, I noted that in recent decades income levels of African-Americans have actually risen at a faster rate than those of whites in the United States. This trend, combined with a wide array of legal changes that have protected the rights of minority group members, induced Wilson (1978) to talk of the "declining significance of race" in the United States: "Race relations in America have undergone fundamental changes in recent years, so much so that now the life chances of individual blacks have more to do with their economic class position than with their day-to-day encounters with whites" (Wilson 1978:1). Hout (1984) used data from the Occupational Changes in a Generation Surveys of 1962 and 1973 to support Wilson's claims, showing, in particular, that the public sector provided substantial new opportunities for black males to move into the middle class. Bianchi (1981) and Farley (1984) cautioned us to be wary of overestimating the gains that have been made by African Americans, but Sakamoto and associates (Sakamoto, Wu, and Tzeng 2000) have shown that the wage disadvantages for African American, Asian, and American Indian men diminished between 1950 and 1990.

It is good news that progress is being made, but that does not mean the end of racial stratification, especially for blacks (see especially Anderson and Massey 2001). If you are African American, there is still a good chance that you will be earning less money than if you were white, even if everything else is the same. For example, in 2001, income for an African American male aged 35–44 who was a year-round full-time worker and had a college degree was $20,000 less than for a comparable non-Hispanic white male. Why the difference? The obvious reason is discrimination, which looms as a significant (although difficult-to-measure) factor, accounting for the fact that members of a minority group may not earn as much money as whites even when they are the same age and equipped with comparable educational skills (Grodsky and Pager 2001).

A classic attempt to measure the effect of discrimination (albeit indirectly and inferentially) was made by Hauser and Featherman (1974). They noted that in 1962, white males aged 35–44 earned $3,755 more per year on average than

African American males of the same age. By 1972, the income differential had dropped to $3,195. They then controlled for factors such as family background (father's education, father's occupation, and farm background); the number of siblings (too large a family might have cut down opportunity); education of the respondent; and occupational status, to see whether differences in those social structural factors could explain all the differences in income. They could not. In 1962 and 1972, nearly 40 percent of the income differential was left unexplained, and the researchers speculated that discrimination might be the missing factor.

Official statistics in the United States make extensive use of the racial and ethnic categories I have just discussed, but on an everyday basis, people are also conscious of ethnicity in a broader context, as measured by the question about ancestry that has appeared in the last several U.S. censuses. You can see back in Figure 4.2 that this was question 10 on the Census long form and people were asked their ancestry or ethnic origin. People were given two lines, so they could put in two categories, if they so chose. Overall, the most-often-recorded ancestry was German (listed by 42 million people—15 percent of the U.S. population). Next most often listed was Irish—30 million in America, which is pretty remarkable when you consider that there are only four million people living in Ireland, and not even all of them are Irish! English, American, and Italian rounded out the top five in 2000.

Ethnicity in Canada and Mexico   The United States does not have a corner on the racial and ethnic minority market, nor are demographic differences by race and ethnicity peculiar to the United States. As befits a multiracial, multiethnic, officially bilingual society, Canada has several ways to measure diversity. What would be called "race" in the United States is labeled "visible minority" in Canada. They accounted for 13.4 percent of Canada's population in the 2001 census, up from 11.2 percent in 1996. The most populous group was Chinese, followed by South Asians (largely from India) and blacks (largely from the Caribbean). Ethnicity is essentially a geographic concept, based on a place with which you identify, similar to the concept of "ancestry" as measured by the U.S. Census. The single biggest ethnic group in Canada is, guess what—Canadian. However, the largest non-Canadian origin identified was the British Isles. Another identifier in Canada is whether or not you are a member of the aboriginal population, which accounts for 3 percent of the country's total population, and coincided with the ethnic identification of North American Indian. Language is the other major identifier in Canadian society, with English being listed most often (58 percent of the population), followed by French (23 percent, of whom 86 percent live in the province of Quebec), and then Chinese (3 percent).

Language is a divisive-enough issue that it led the Francophone (French-speaking) population in the eastern edge of the country (Quebec) to attempt to secede from the Anglophone (English-speaking) remainder of the country in the 1990s. Demographics played a role in defeating the referendum on separation held in 1995, however, because the traditionally Catholic Francophone population now has very low levels of fertility—probably the lowest anywhere in North America. French Canadians are not replacing themselves and non-Francophones (especially recent immigrants) generally do not support separation, leading Canadian demographers correctly to predict that separation would not be approved by the voters (Kaplan 1994; Samuel

1994). Still, the controversy over language in Canada underscores the power of society to turn any population characteristic into a sign of difference, from which prejudice and discrimination often follow.

Language is also an issue in Mexico, where the lowest stratum of society tends to be occupied by those who speak an indigenous language (linguistically related to Aztec and Mayan languages), rather than Spanish. According to the 2000 census, 7 percent of Mexicans speak an Indian language, but in the Yucatan, the percentage is 38 percent; in Oaxaca it's 37 percent; and in Chiapas, where Indian-speaking people have been challenging the authority of the central government since the 1990s, it's 27 percent. These are also the Mexican states where people are poorest and fertility is highest. Language minorities thus represent both the geographic and demographic extremes in North America, from Francophone Quebec in the northeast, with very low fertility, to Mayan-language Chiapas in the south, with very high fertility.

Ethnicity in Other Countries Goldscheider (1996) has examined the demographics of ethnic pluralism in Israel and found that the Muslim Arab population in that country has been growing at twice the rate of the Jewish population. In 1950, Muslims represented 8 percent of the total population, but that percentage has grown consistently, and by 1995, 14 percent of Israel's population was Muslim and nearly one in five residents of Israel was non-Jewish (Israel Central Bureau of Statistics 2001). Like minority group members in the United States, the Muslim minority in Israel is less educated, holds lower-status jobs on average, and has a lower life expectancy and higher fertility levels: "In most cases, Jews in Israel believe that the State should prefer Jews to Arabs in all arenas of public policy. Only a small proportion of Jews subscribe to the notion of 'equal opportunity' for all groups in areas of higher education, jobs, and financial support from the government" (Goldscheider 1989:23).

Fertility has declined among Arabs in Israel since 1960, but the drop has been modest, and since Jewish fertility has also declined, the gap in fertility levels remains fairly wide. In 1998, the total fertility rate for the Jewish population was 2.7 children per woman, compared with 4.8 for the Muslim population (Israel Central Bureau of Statistics 2001). However, high fertility in Israel is not simply a function of being Arab, because the Christian Arab minority (representing 14 percent of the total Arab population) has the lowest total fertility rate (just under the Jewish average at 2.6 children per woman) of any group in Israel. The contrasts among Muslims, Jews, and Christians take us next into the realm of religion.

Religion

Virtually everyone is born into some kind of religious context, which is why I have likened religion to an ascribed characteristic. Yet, people can willingly change their religious preference during their lifetime, so it is akin to an achieved status. Despite the appearance of choice, however, most people do not alter religious affiliation, so it is a nearly permanent feature of their social world. People may become more or less religious within their particular group, but they are unlikely to change the major affiliation. Like race and ethnicity, religion sets people apart from one another and has historically been a common source of intergroup conflict throughout the world (Choucri 1984). Because it is an often-discussed sociodemographic characteristic, religion has regularly come under the demographer's microscope, with particular attention being paid to its potential influence on fertility.

America's history of **religious pluralism,** in which a wide variety of religious preferences have existed side by side, perhaps sensitized American demographers to the role of religion in influencing people's lives. A good deal of attention has been focused over the years on the comparison between Protestants and Catholics. During the past century at least, Catholics have tended to want and have more children than Protestants have in the United States, and internationally it has been true that predominantly Protestant areas (such as the United States and northern Europe) experienced low fertility sooner than did predominantly Catholic areas (such as southern and eastern Europe). However, data from the 1988 National Fertility Study in the United States were among the first to suggest that the long-time differences between Protestant and Catholic fertility levels in the United States have essentially disappeared (Goldscheider and Mosher 1991). This appears to be closely related to the substantial drop in church attendance over time, especially among Catholic women (Williams and Zimmer 1990).

Is religion less important now than it used to be? Are there other factors besides religion that explain the seeming differences sometimes noted among religious groups? Obviously, the relationship between religion and fertility is not a simple one, and, in fact, there are three major themes that run through the literature: (1) religion plays its most important role in the middle stage of the demographic transition; (2) religiosity may be more important than actual religious belief; and (3) religion and ethnicity are inextricably bound up with each other. Let us examine each of these ideas.

In a study of religious differentials in Lebanese fertility, Chamie (1981) concluded that a major effect of religion may be to retard the adoption of more modern, lower-fertility attitudes during the transitional phase of the demographic revolution. Adherents to religious beliefs that have been traditionally associated with high fertility will be slower to give ground than will people whose religious beliefs are more flexible with respect to fertility. In the United States, Jews have generally had lower fertility levels than the rest of the population. Trends in Jewish fertility have followed the American pattern (a decline in the Depression, a rise with the baby boom, a drop with the baby bust), but at a consistently lower level. "Widespread secularization processes, upward social mobility, a value system emphasizing individual achievement, and awareness of minority status have all been indicated as factors that are both typical of American Jews and conducive to low fertility" (DellaPergola 1980:261). Indeed, it is not just American Jews whose fertility is low. DellaPergola (1980) points out that Jewish communities in central and western Europe have also been characterized by low fertility since as early as the second half of the nineteenth century, largely because contraception is readily accepted in the Jewish normative system (at least among non-Orthodox Jews). Leasure (1982) also found that, in the United States, fertility declined earliest in those areas dominated by more secularized religious groups. People who are more traditional in their religious beliefs tend to be less educated, have less income, and are thus more prone to higher fertility.

The former Yugoslavia is the site of centuries-old ethnic battles, with religion as one of the important ethnic identifiers. Courbage (1992) has shown that until only a couple of decades ago, the Muslims in that region of Europe (who are predominantly Slavic in ethnic origin, not Arab) had considerably higher fertility levels than Christians did. However, by the time of the 1981 census in Yugoslovia, an interesting pattern had emerged, in which people who identified themselves as "Muslim nationals" (the population that now comprises the Muslim part of Bosnia) had lowered their fertility to "European" levels, similar to the Slavic Christian population, whereas those who identified themselves as "Albanians" were Muslims whose fertility levels remained closer to Middle Eastern levels, rather than European levels. And, consistent with that idea, Albania, which is just to the south of Yugoslavia, has the highest fertility rate in Europe (the only country in Europe with fertility above replacement level), and is also Europe's only Muslim-majority nation. These differences in fertility are important, of course, because they in turn affect the life chances of family members.

In looking at religious differentials in fertility, it may be at least as important to examine the religiosity of people as to know their specific religious beliefs. It is likely, in fact, that fertility differences by religion will always exist in any society to some extent just because some groups inspire greater religious fervor than others and, in almost all instances, religious zealotry is associated with a desire for larger-than-average families because there is usually a desire to maintain or return to more traditional value systems in which large families are the norm. This may also lead religious fundamentalism to have a negative influence on educational attainment among its adherents (Darnell and Sherkat 1997), which of course will likely keep fertility higher than it might otherwise be. In many respects, this is the flip side of saying that secularism causes a decline in fertility. To eschew education and secularism generally means to maintain high fertility.

For the United States, I have already discussed the high fertility of the Hutterites (see Chapter 6). Another, larger religious group whose higher-than-average level of religiosity seems to contribute to higher-than-average fertility is the Mormon Church. Mormons dominate the social, political, and demographic fabric of the state of Utah, explaining the fact that Utah's birth rate is the highest of any state—twice the national average. Interestingly, Provo, Utah, which is the site of Brigham Young University, the principal Mormon institution of higher education, has the highest birth rate of any city in the United States. Census 2000 data for Provo show that 43 percent of that city's households are married couples with children under age 18, compared with the national average of 24 percent. The average family size in Provo in 2000 was 3.86 persons, compared to the national average of 3.14. Provo is also not very diverse with respect to race and ethnicity. Of the 369,000 people counted in Census 2000, 89 percent were non-Hispanic white.

The Intersection of Changing Life Chances and the Family and Household Transition

At the beginning of this chapter, I noted that the family is usually thought of as both a kinship and an economic unit. The kinship part provides its members with social capital (the connection to networks of people who may be in a position to help you out in life), and the economic part provides its members with human capital (access to resources such as education). These familial resources play a role in each person's life and throughout each person's life, but they are especially crucial when you are a child. Thus, societies tend to pay particular attention to the type of household in which children are growing up, no matter how diverse the household structure may be in society as a whole. Are children being raised in a household environment that maximizes the children's opportunities to acquire human and social capital, and thus to increase the odds of success in the next generation? This is how family and household structures intersect with life chances.

Westoff (1978) has suggested that the institutions of marriage and the family were showing signs of change back in the 1970s because "the economic transformation of society has been accomplished by a decline in traditional and religious authority, the diffusion of an ethos of rationality and individualism, the universal education of both sexes, the increasing equality of women, the increasing survival of children and the emergence of a consumer-oriented culture that is increasingly aimed at maximizing personal gratification" (p. 53). It has been argued that many of these cultural changes have followed, rather than preceded, the changes in household structure. They may not have initiated the trends, but they have reinforced the transformations and ensured their spread within each country and from one country to another. Global changes are, however, occurring unevenly.

In predominantly Muslim nations, as well as in South Asia, the gender gap between men and women is especially wide (Westley 2002), and may at least temporarily widen as development brings about an increase in the breadwinner system (remember the earlier comments by Davis [1984b]), which tends to increase the economic and political dependence of women. Charlton (1984) contends that in India, the early stages of modernization were accompanied by women's declining status, and this is a story corroborated by the Report of the National Committee on the Status of Women in India, which concluded that "to our dismay, we found that such disabilities [centers of resistance to change in the status of women] have sometimes been aggravated by the process of development itself" (National Committee on the Status of Women 1975:2).

Worldwide economic changes have emphasized the value of nondomestic labor and this has marginalized the economic contribution that women make to the family economy. It has also aggravated the problems associated with cultural practices such as the dowry that Indian families are expected to pay to their new in-laws on the marriage of each daughter. Because in India girls traditionally do not receive an inheritance on the death of their parents, the dowry is in essence an "upfront" payment of a potential inheritance. However, from a practical point of view, it means that any investment in a girl will accrue to her husband's family, so parents are not necessarily motivated to do much for daughters. Indeed, it has been suggested that child labor is used in India as a way for girls to defray the costs of the dowry (*The Economist* 1994). A solution to such a problem of "cultural entrapments embroidered by female submissiveness" (Mhloyi 1994) was signed into law in the Indian state of Maharashtra in 1994 when the legislators of that state passed laws that granted women the same rights of inheritance as men, and also reserved up to 30 percent of government jobs for women.

The cultural model prevalent for the past two to three decades in industrialized nations has been that self-fulfillment and individual autonomy are the most important values in life and serve to justify scrapping a marriage. If women are approaching the level of economic independence previously reserved for men, perhaps the value of marriage has been permanently eroded, and marriage will (or has) become only one option among many from which people may reasonably choose. One of the complaints about marriage often registered by women is that the move toward gender equality in the division of labor in the formal marketplace has not necessarily been translated into equity in the division of labor within the household. Women in the Western world are able to operate in society independently of a husband or other male patriarch or protector, but they may not have the same ability to have an equal relationship at home with a husband. Data from the National Survey of Families and Households show that married couples are the least egalitarian of all households with respect to the division of domestic chores. Marriage is the household setting in which women do the greatest amount of domestic work (regardless of their own labor force status), while men do the least (Shelton 2000; South and Spitze 1994).

So, does marriage matter? An increasing body of evidence suggests that marriage matters very much even in a modern, industrialized society—it enhances household income and wealth and promotes the well-being of spouses and children, while adding to sexual gratification in the bargain. Waite (1995) and Waite and Gallagher (2000) have reviewed the literature and analyzed numerous data sets in order to draw the following conclusions about the benefits of marriage: (1) married couples have higher household income than the unmarried; (2) married couples save more of their income than the unmarried; and therefore (3) married couples have more wealth than the unmarried; (4) married men and women live longer than the unmarried, and engage in fewer high-risk behaviors; (5) children in a marriage are better off financially than those in a one-parent family; (6) children in a marriage are less likely to drop out of school, less likely to have a teenage pregnancy, and less likely to be "idle" (out of both school and work) as a young adult than children in a one-parent family; and (7) married couples have sex more often and derive greater satisfaction from it than the unmarried do.

The social impact of marriage derives from these personal benefits. Perhaps most compelling is the fact that the family remains the primary social unit in which society is reproduced—in which children are taught the rights and reciprocal obligations of membership in human society. The evidence seems to suggest that this is accomplished most efficiently in a household/family unit that includes both parents of the children in question. The evidence is persuasive that children derive few, if any, positive benefits from growing up without a father and, indeed, tend to suffer both short- and long-term ill effects if fatherless; the same is probably true for motherless families, although we have fewer studies of such family settings (Furstenberg and Cherlin 1991; McLanahan and Sandefur 1994). "The data clearly indicate that a healthy two-parent family optimizes both the economic well-being and the physical and mental health of children" (Angel and Angel 1993:199). If this is correct, then the implications are that diversity of household composition is problematic only if children are involved. The negative effects of "nontraditional" families seem to fall disproportionately on children, and as these consequences become more widely understood in society, it is possible that new attitudes may lead to new behavior (and, of course, vice versa). Part of the complication is that the traditional family was not necessarily a bed of roses, either, and it is probably harder on children to be in an intact family where the parents are constantly battling each other than to be in some other type of household situation where there is less conflict (Amato and Sobolewski 2001; Furstenberg 1998).

It appears that the diversification of households in industrialized societies has leveled off. There was much less change from the mid-1980s through the 1990s than there had been in the 1970s and early 1980s. Waite and her associates found that the increasing independence of women with respect to living arrangements helps shape a change in attitudes among women toward the division of labor within the family, as well as in society generally (Waite, Goldscheider, and Witsberger 1986). As men also adopt those attitudes, their satisfaction with marriage appears to increase (Amato and Booth 1995). Thus we may venture to guess, with Goldscheider and Waite (1991), that "couples with modern attitudes toward both women's work outside and inside the family will create more egalitarian families in which children can grow up and later emulate" (p. 141). These are, of course, the challenges that will be confronting families all over the world as other countries work their way through the demographic transition, in a process of family and household change that is already well under way in developing nations (Bongaarts 2001).

Smaller families and longer lives seem inevitably to lead to diversification in the household and family structure of a society. There is little controversy in that proposition. The controversy arises when children are involved, because some family situations and household structures seem more likely to offer children the social and human capital resources most societies seem concerned that children have as they grow up. These were some of the uncharted waters that more-developed countries wandered into in the course of the demographic transition, but family demographers have been working hard to map the territory for you.

Summary and Conclusion

The past few decades have witnessed a fundamental shift in household structure in the United States and other industrialized nations, and these changes are beginning to evolve all over the world as the family and household transition accompanies the mortality, fertility, migration, and age transitions. Married-couple households with children have become less common, being replaced by a combination of married couples without children, cohabiting couples with and without children, lone parents with children, people living alone, people living with nonfamily members, and just about anything else that you can think of. This greater diversity in household structure is a direct result of the trends in marriage and divorce, which are themselves influenced by trends in mortality and fertility. Marriage has been increasingly delayed (although most people do eventually marry), but people are leaving the parental nest to live independently by themselves, with friends, or in a cohabiting relationship prior to marriage, a pattern greatly facilitated by access to effective methods of contraception. Once married, there is an increased tendency to dissolve the marriage. Some of this is due to the fact that spouses are far less likely to die than in earlier times, and some is due to the fact that divorce laws have accommodated the changing relationships between men and women. Accompanying these trends has been an increase in the proportion of children born out of wedlock, contributing to the increased percentage of children who are living with only one parent.

Less directly, but no less importantly, the transformation of family and household structure has been a result of changing population characteristics, especially the improvement in women's life chances. Women have become less dependent on men as they have begun to live longer and spend more of their lives without children in an urban environment, where there are alternatives to childbearing and family life. Throughout the world, women are closing the education gap between themselves and men, entering the paid labor force, and moving up the occupational ladder. These new opportunities to be more fully engaged in all aspects of social, economic, and political life have been simultaneously the cause and consequence of declining fertility and improved life expectancy. They have enabled women to delay marriage while becoming educated and establishing a career, choose marriage or not (most do), choose children or not (most choose two), and, if married, to choose to stay married (only about half do). Therein lie the principal explanations for the increased diversity of families and households.

Not all people have equal access to societal resources such as advanced education, a well-paying job, and other assets with which to build wealth. In the United States, blacks, Hispanics, and American Indians are less likely than others to be highly educated, and this may contribute to their relative social and economic disadvantage in American society, although racial prejudice and discrimination continue to play a role. On the other hand, Asians tend to have higher levels of education than other groups, which may help account for their higher levels of income (and higher life expectancies, as well). In most countries of the world we find one or more groups who, for reasons of discrimination beyond their control, are disadvantaged compared to the dominant group.

The impact of the trends toward greater family and household diversity falls disproportionately on children. Growing up in a household that is other than a two-parent family lowers household income, increases the odds of health and social problems in childhood and young adulthood, and generally increases the risk that your life chances may be limited. However, another group also bears the brunt of dissolved marriages. Today's divorced women could become the biggest group of elderly poor in the future. This is a trend we will have to watch over time, but it will be easier to track than in previous generations because most households with older people in industrialized nations are now in urban areas where they may be more visible than the elderly poor in rural areas of less-developed nations. This transition from a predominantly rural to a predominantly urban society is an important feature of the trends inherent in the demographic transition, and I turn to that topic in the next chapter.

Main Points

1. Married-couple households with children are declining as a fraction of all households, being replaced by a variety of other family and nonfamily household types.

2. The direct causes of these changes in household composition are a delay in marriage, a rise in the propensity to divorce, and, to a lesser extent, widowhood in the older population.

3. The underlying indirect causes of these changes are declining mortality, declining fertility, and the underlying age structure changes brought about over time by declining mortality and fertility.

4. Delayed marriage has been accompanied by an increase in leaving the parental home to live independently before marriage; this increasingly leads to cohabitation prior to marriage.

5. Declining mortality has helped contribute to a freeing of women from childbearing responsibility and has provided more years of prereproductive and postreproductive time.

6. Urbanization has helped improve the status of women by providing opportunities for nonfamilial activities.

7. The transformation of families and households has accompanied improved life chances for women, including higher levels of education, labor force participation, occupation, and income.

8. Average educational attainment has increased substantially over time in most countries, and, especially in industrialized nations, women have been rapidly closing the gender gap in education.

9. In the United States during World War II, a combination of demand for labor and too few traditional labor force entrants created an opening for married women to move into jobs previously denied them.

10. Since 1940, the rates of labor force participation have risen for women, especially married women, while declining for men.

11. Women represent nearly 50 percent of all workers, but they are still concentrated disproportionately in administrative support, sales, and service occupations.

12. The better educated you are, the more money you can expect to earn in your lifetime.

13. Americans of almost all statuses are wealthier in real absolute terms now than they were in the 1940s, but there have been only minor changes in the relative status of most groups.

14. Even at comparable occupational levels, with similar educational backgrounds, blacks on average earn less money in the United States than whites.

15. Even at comparable occupational levels, with similar educational backgrounds, women on average earn less money than men.

16. Race may be just "a pigment of your imagination," but blacks, in particular, tend to be disadvantaged compared to whites in American society.

17. The diversity of households seems to have plateaued, and the future of married-couple families may depend on the willingness of men to adopt more egalitarian attitudes toward domestic roles within the household.

18. Family demographers can prove that the average person in Miami, Florida, is born Cuban and dies Jewish.

Suggested Readings

1. Martha Shirk, Neil G. Bennett, and J. Lawrence Aber, 1999, Lives on the Line: American Families and the Struggle to Make Ends Meet (Boulder, CO: Westview Press).

   The diversification of American families has created new dilemmas in how to deal with the emotional and financial costs of raising children. This book offers data and case studies to illustrate the complexities of the problem.

2. Daniel T. Lichter and Martha L. Crowley, 2002, "Poverty in America: Beyond Welfare Reform," Population Bulletin 57(2).

   Although poverty levels have dropped in the United States over the past few decades, a disproportionate number of children are growing up in poverty. This bulletin examines this and other family-related issues in the context of welfare reform legislation.

3. Harriet B. Presser and Gita Sen, 2000, Women's Empowerment and Demographic Processes: Moving Beyond Cairo (Oxford: Oxford University Press).

   The papers in this volume are aimed at helping put the issue of the empowerment of women squarely in the middle of demographic theory and research.

4. Linda J. Waite and Maggie Gallagher, 2000, The Case for Marriage (New York: Doubleday).

   Experimentation with diverse models of family-building since the 1970s has led social scientists to reexamine the place that marriage has in a modern developed society. This book does that by building on academic research, although it is intended for a general audience.

5. Jane Lewis, 2001, The End of Marriage? Individualism and Intimate Relations (Northampton, MA: Edward Elgar).

   The United Kingdom is wrestling with the same issues of the "decline of marriage" as is the United States and this book puts the changes into historical perspective.

🌐 Web Sites of Interest

Remember that Web Sites are not as permanent as books and journals, so I cannot guarantee that each of the following Web Sites still exists at the moment you are reading this. It is usually possible, however, to retrieve the contents of old Web Sites at http://www.archive.org.

1. http://www.umich.edu/~psid
   The Panel Study of Income Dynamics (PSID), which began in 1968 at the University of Michigan, is a longitudinal study emphasizing the dynamic (changing) aspects of demographic and economic behavior in American society.

2. http://www.contemporaryfamilies.org/public/about.php
   The Council on Contemporary Families is a nonprofit organization established by university-based social scientists to provide a forum for accurate information about American families. For example, this site includes a summary of research results from the Fragile Families and Child Wellbeing Study at Princeton University.

3. http://www.irc.essex.ac.uk
   The Economic and Social Research Council (ESRC) Research Centre on Micro-Social Change is the home of the British Household Panel Study, a longitudinal study begun in 1989.

4. http://www.unfpa.org/support/friends/34million.htm
   The United Nations Population Fund (UNFPA) has been active for a long time in promoting the well-being and empowerment of women all over the world. When the United States

government in 2002 withdrew its pledge of $34 million in support of the work of the UNFPA, a group of volunteers set about collecting the money from private sources instead.

5. http://www.unicef.org/statistics/index.html
The United Nations Childrens Fund (UNICEF) maintains a database for all countries in which you can obtain a profile of the basic indicators of the life chances of the average child. Combine that with a visit to the World Bank's Web site (http://www.worldbank.org), where you can click on the Poverty Net link and obtain poverty level data for each country.

CHAPTER 11
The Urban Transition

Figure 11.1 Percent of the Population Living in Urban Areas

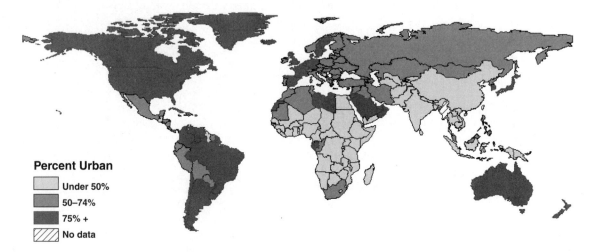

Percent Urban
- Under 50%
- 50–74%
- 75% +
- No data

Source: Adapted from United Nations 2003, World Population Prospects 2002. Based on medium variant projections for 2000.

452

The world is rapidly becoming urban. Consider that as recently as 1850 only 2 percent of the entire population of the world lived in cities of 100,000 or more people. By 1900, that figure had edged up to 6 percent, and it had risen to 16 percent by 1950 (Davis 1972b). So for most of human history almost no one had lived in a city—cities were small islands in a sea of rurality. But within just one century, from 1850 to 1950, cities had jumped out to grab one in six human beings, and a short 50 years after that, in the year 2000, virtually one in every two people was living in a place labeled as urban. By the middle of this century, that fraction will probably increase to two out of every three people living in urban areas (United Nations Population Division 2002c).

> The present historical epoch, then, is marked by population redistribution as well as by population increase. The consequences of this redistribution—this "urban transition" from a predominantly rural, agricultural world to a predominantly urban, nonagricultural world—are likely to be of the same order of magnitude as those of the more widely-heralded increase in world population (Firebaugh 1979: 199)

The majority of Americans live in—indeed were born in—cities, and most people in the richer countries of the world share that urban experience. Some of us take the city for granted, some curse it, some find its attractions irresistible, but no one denies that urban life is the center of modern civilization. Cities, of course, are nothing new, and their influence on society is not a uniquely modern feature of life; however, the widespread emergence of urban life—the explosive growth of the urban population—is very much a recent feature of human existence.

This urban transition is one of the most significant demographic movements in world history partly because it is intimately tied to population growth and the demographic transition. It is not too much of an exaggeration to say that world population growth is occurring in the countryside, but showing up in the cities. What are the demographic components of urbanization, and what are the demographic consequences for society of an ever-increasing concentration of people in urban areas? These are the questions I will focus on in this chapter. Clearly, cities are implicated in a wide range of problems, issues, and triumphs in all societies, but my intention here is not to review what life is like in a city (which could, and has, filled volumes). Rather, I want to provide you with a demographic perspective on **urbanization** (a word I will use interchangeably with the term urban transition). To accomplish that, I begin with an overview of urbanization, putting it into broad historical perspective. Then I turn to an analysis of the demographic components of the urban transition, looking especially at the way in which migration, mortality, and fertility interact to produce the process of urbanization. In this context, we see how the urban transition fits into the overall pattern of the demographic transition and how the rise of cities as the center of social and economic life has created new worldwide interlinkages among urban areas. Finally, I examine some of the spatial processes taking place in urban areas that are both causes and consequences of many of the demographic trends discussed in the previous chapters. These include trends in metropolitanization, suburbanization, patterns of residential segregation within urban areas, and the response of humans to being crowded into urban places.

What Is Urban?

An urban place can be defined as a spatial concentration of people whose lives are organized around nonagricultural activities; the essential characteristic here is that **urban** means nonagricultural, whereas **rural** means any place that is not urban. A farming village of 5,000 people should not be called urban, whereas a tourist spa or an artist colony of 2,500 people may well be correctly designated an urban place. You can appreciate, then, that "urban" is a fairly complex concept. It is a function of (1) sheer population size, (2) space (land area), (3) the ratio of population to space (density or concentration), and (4) economic and social organization.

As the number and fraction of people living in urban places has increased, the importance of urban life obviously becomes more important. The study of human society is increasingly the study of urban society, and the variability across space and time in the urban environment is a crucial part of the changes occurring in every society. The urban environment, however, is a combination of social and built environments. The concept of *urban* is, at root, a place-based idea (Weeks 2004). Places are typically defined as "urban," and on the basis of that definition the people living there are thought of as being part of the urban population. But, we don't usually apply the term "urban" to a person. The personal adjective "urbane," still occasionally used to describe a person, is defined by the Oxford English Dictionary as "having the qualities or characteristics associated with town or city life; esp. elegant and refined in manners, courteous, suave, sophisticated" (Brown 1993:3527). My own anecdotal experience suggests that this would describe only a small minority of people living in urban places in the twenty-first century.

The definitions of *urban* used in most demographic research unfortunately rarely encompass all the above ingredients. Due to limitations in available data and sometimes simply for expediency, researchers (and government bureaucrats as well) typically define urban places on the basis of population size alone, implying that density is the major criterion. Thus, all places with a population of 2,000, 5,000, 10,000, or more (the lower limit varies) might be considered urban for research purposes. A variation on that theme is to designate cities and towns as being, by definition, urban and everything else as rural: "Of the 228 countries for which the United Nations (UN) compiles data, roughly half use administrative considerations—such as residing in the capital of the country or of a province—to designate people as urban dwellers. Among the other countries, 51 distinguish urban and rural populations based on the size or density of locales, 39 rely on functional characteristics such as the main economic activity of an area, 22 have no definition of "urban," and eight countries define all (Singapore, for example) or none (several countries in Polynesia) of their populations as living in urban areas" (Brockerhoff 2000: Box 1).

Of course, you should recognize that any arbitrary cutoff disguises a lot of variation in human behavior. Although the difference between rural and urban areas may at first appear to be a dichotomy, it is really a continuum in which we might find an aboriginal hunter-gatherer near one end and an apartment dweller in Manhattan near the other. In between will be varying shades of difference. The next time you drive from the city to the country (or the other way around), you might ask yourself where you would arbitrarily make a dividing line between the two. In the United States in the nineteenth and early twentieth centuries, rural turned into urban when you reached streets laid out in a grid. Today, such clearly defined transitions are rare and, besides, even living in a rural area in most industrialized societies does not preclude your participation in urban life. The flexibility of the automobile combined with the power of telecommunications puts most people in touch with as much of urban life (and what is left of rural life) as they might want. In the most remote areas of developing countries, radio and satellite-relayed television broadcasts can make rural villagers knowledgeable about urban life, even if they have never seen it in person (Critchfield 1994).

For Census 2000, the U.S. Census Bureau used the power of geographic information systems (GIS) to create a more flexible definition than ever before of what is urban (U.S. Census Bureau 2001a). To be called urban in the United States, a place must be part of an **urban area (UA)** or an **urban cluster (UC),** defined as follows:

> For Census 2000, an urbanized area (UA) will consist of a densely settled core of census block groups (BGs) and census blocks that meet minimum population density requirements, along with adjacent densely settled surrounding census blocks that together encompass a population of at least 50,000 people, at least 35,000 of whom live in an area that is not part of a military installation. For Census 2000, an urban cluster (UC) will consist of a densely settled core of census BGs and census blocks that meet minimum population density requirements, along with adjacent densely settled surrounding census blocks that together encompass a population of at least 2,500 people, but fewer than 50,000 people, or greater than 50,000 people if fewer than 35,000 of them live in an area that is not part of a military installation (U.S. Census Bureau 2001b:17018).

It begins with a core census block or block group (an area that is part of a census tract) that has a population density of at least 1,000 persons per square mile, and then any contiguous areas that have at least 500 persons per square mile are added to the core to comprise the urban area. As long as the total population of this combined area is at least 2,500 people, the area is defined as urban, and any smaller place within that urban area is called urban. Any place that is not within an urban area or urban cluster is defined as rural.

Canada defines an urban place in a similar, but not identical fashion. An *urban place* or *urban area,* as defined by Statistics Canada, has a population of at least 1,000 concentrated within a continuously built-up area, at a density of at least 400 per square kilometre" (equivalent to about 1,000 persons per square mile) (Statistics Canada 2001b). Mexico has traditionally defined *urban* as being any locality that has at least 2,500 inhabitants (INEGI 2001a), although Mexico, like Canada, typically defines a city as a place that has at least 100,000 inhabitants.

You can see that the United States and Canada use a combination of population size and density to define urban, while Mexico relies largely on population size. No consideration is given to the economic and social characteristics of a place. Yet, an essential ingredient of being urban is economic and social life organized around nonagricultural activities. There is an explicit recognition that urban people order their lives differently than rural people do; they perceive the world differently and behave differently, and this is why being in an urban place is important. You should keep in mind that there is a discontinuity between the *concept* of *urban* and the *definition* of *urban* as I turn now to a discussion of the demographic aspects of the process whereby a society is transformed from rural to urban—the urban transition, or the process of urbanization.

An Overview of Urbanization

Urbanization refers to the change in the proportion of a population living in urban places; it is a relative measure ranging from 0 percent, if a population is entirely rural or agricultural, to 100 percent, if a population is entirely urban. It is a phenomenon as recent as, and part of, the demographic transition. The earliest cities were not very large, because most of them were not demographically self-sustaining. The ancient city of Babylon might have had 50,000 people, Athens possibly 80,000, and Rome as many as 500,000; but they represented a tiny fragment of the total population. They were symbols of civilization, visible centers that were written about, discussed by travelers, and densely enough settled to be dug up later by archaeologists. Our view of ancient history is colored by the fact that our knowledge of societal detail is limited primarily to the cities, although we can be sure that most people actually lived in the countryside.

Precursors to Modern Urbanization

Early cities had to be constantly replenished by migrants from the hinterlands, because they had higher death rates and lower birth rates than did the countryside, usually resulting in an annual excess of deaths over births. The self-sustaining character of modern urban areas began with the transformation of economies based on agriculture (produced in the countryside) to those based on manufactured goods (produced in the city), to those based on servicing the rest of the economy (and often located in the suburbs). Control of the economy made it far easier for cities to dominate rural areas politically and thus ensure their own continued existence in economic terms. A crucial transition in this process came between about 1500 and 1800 with the European discovery of "new" lands, the rise of mercantilistic states (that is, based on goods rather than landholdings), and the inception of the Industrial Revolution. These events were inextricably intertwined, and they added up to a diversity of trade that gave a powerful stimulus to the European economy. This was a period of building a base for subsequent industrialization, but it was still a preindustrial and largely preurban era. During this time, for example, cities in England were growing at only a slightly higher rate than the total population, and thus the urban population was rising only very slowly as a proportion of the total. Between 1600 and 1800, London grew from about 200,000 people to slightly less than one million (Wrigley 1987)—an average rate of growth considerably less than 1 percent per year; also during this span of 200 years, London's population increased from 2 percent of the total population of England to 10 percent—significant, but not necessarily remarkable, especially considering that in 1800 London was the largest city in Europe. In 1801, only 18 percent of the population in England lived in cities of 30,000 people or more, and nearly two-thirds of those urban residents were concentrated in London. Thus, on the eve of the Industrial Revolution, Europe (like the rest of the world) was predominantly agrarian.

Neither England nor any other country was urbanizing with any speed at that time, because industry had not yet grown sufficiently to demand a sizable urban population and because cities could not yet sustain their populations through natural increase. Early competitive, laissez-faire capitalism characterized the economies of Europe and North America from the late eighteenth through the mid-nineteenth centuries and cities were still largely commercial in nature. In the nineteenth century, urbanization began in earnest, the timing of which was closely tied to industrialization and the decline in mortality that triggered population growth.

Urban factory jobs were the classic magnets sucking young people out of the countryside in the nineteenth century. This happened earliest in England, and in Figure 11.2 you can see the rapid rise in urbanization in the United Kingdom in response to early industrialization. Japan and Russia entered the industrial world later than the United Kingdom or the United States, and you can see in Figure 11.2 that their pattern of urbanization was therefore delayed. Once started, however, urbanization proceeded quickly in these countries.

Figure 11.2   Industrialized Nations Are Highly Urbanized

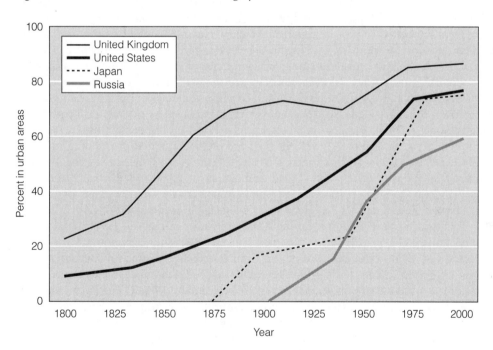

Note: Until the post–World War II period, the process of urbanization was closely associated with industrialization. England and Wales (comprising most of the United Kingdom) began to industrialize before the United States did, and so their urban fraction increased sooner. Japan began to industrialize and urbanize in the late nineteenth century, whereas those two processes were delayed in Russia until after the revolution.

Source: Kinglsey Davis, 1965, "The Urbanization of the Human Population," Scientific American 213(3):47, copyright 1965 by Scientific American, inc., all rights reserved; updated by the author from United Nations, 1991, World Urbanization Prospect 1990 (New York: United Nations), Tables A.2 and A.4.

Current Patterns of Urbanization

In the contemporary post-industrial world characterized by what is often called "advanced capitalism," the function of cities is changing again. In the developed, already urbanized part of the world, cities are losing their industrial base and are increasingly service centers to economic activities occurring in the hinterlands of the same country, or in another country altogether. In less-developed countries, commercial and industrial activities combine with historically unprecedented rates of city growth to generate patterns of urbanization somewhat different from that which occurred in the now-developed nations.

Figure 11.1 maps the countries of the world according to the percentage of the population living in urban places as of the beginning of the twenty-first century. These data are based on each individual country's own definition of "urban," but typically it refers to places with 2,000 people or more. A little less than one-fourth of the countries of the world (24 percent) have less than 50 percent of the population living in urban places. As you can see from the map, these include the two most populous nations, China and India, as well as the rest of the Indian subcontinent, most of Southeast Asia, and nearly half of the sub-Saharan nations. More than four in 10 of the world's nations (43 percent) have between 50 and 74 percent of the population living in urban places. In the Americas, these include countries in Central and South America with large indigenous populations. Sub-Saharan Africa is sprinkled with countries in this range, and most of the former Soviet republics in Central Asia are in this category. Some European countries are also in this range, as you can see from the map, including Spain and Portugal, as well as the region from Greece northward through the Balkan states up into Finland. The remaining one-third of the world's nations have 75 percent or more of their population residing in urban places. This includes most countries in North and South America, as well as most European nations, along with Japan and Korea in Asia, and Australia and New Zealand in Oceania. In essence, the highest percentage of the urban population tends to be found among European and "overseas" European nations.

The pace of urbanization can be seen graphically in Figure 11.3. In 1950, more than half of the population (55 percent) in more-developed regions of the world already lived in urban places, but less than one in five people (18 percent) in the less-developed nations were urban. By 2000, three-fourths (75 percent) of the population in more-developed regions was urban, compared to 40 percent in the less-developed regions. By 2030, the United Nations projects that 83 percent of the population in more-developed regions will be urban, and that 56 percent in less-developed regions will be urban, producing a world total of 60 percent urban (United Nations Population Division 2002c).

Impressive as they are, the data in Figure 11.3 nonetheless understate the real growth in the urban population. Since most of the world's population growth since 1950 has occurred in less-developed nations, the increase in the percentage that is urban has meant a staggering rise in the urban population of those countries. Table 11.1 summarizes the United Nations' estimates of the size of the urban and rural populations by major geographic region of the world. Especially noteworthy is Asia, which in 1950 had 244 million people in urban areas, but by 2000 had 1.4 billion people in urban places, with an expected doubling of that number by 2030. In

Chapter 11 The Urban Transition

Figure 11.3   World Urbanizing at Rapid Pace since 1950

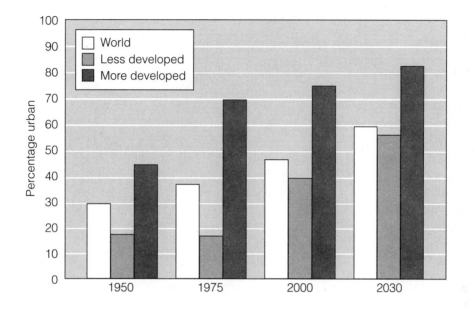

Source: Adapted from United Nations Population Division, 2002, World Urbanization Prospects: The 2001 Revision (New York: United Nations), Table 1.

Africa, the urban population in 2000 was 10 times its size in 1950. In Latin America and the Caribbean, the urban population in 2000 was nearly six times what it had been in 1950.

The United Nations expects that the world's rural population in 2030 will be about the same number as in 2000 (3.2 billion people) and that will represent about twice as many people as there had been in 1950 (1.8 billion) (United Nations Population Division 2002c). However, the urban population of the world in 2030 (4.9 billion) will be 70 percent larger than it was in 2000 (2.9 billion), and that will be a sevenfold increase over the world's urban population of 1950 (which was 750 million). As you watch the number and percentage of the population that is urban climb over time, you may assume that the explanation is very easy—people move out of rural areas into urban areas. That is part of the story, of course, but not the whole story.

The Proximate Determinants of the Urban Transition

Urbanization can occur not only as a result of internal rural-to-urban migration, but also through natural increase, international urban migration, reclassification of places from rural to urban, and combinations of these processes. Another important thing to keep in mind is that the urban transition may end when nearly everybody is living in an urban area, but the urban *evolution*—changes taking place within urban areas—may continue forever, as I discuss later in the chapter.

Table 11.1 Asian, African, and Latin American Urban Population Increasing at Staggering Rate since 1950

	Population (in millions)			
	1950	1975	2000	2030
Urban population				
North America	110	180	243	335
Latin America and Caribbean	69	197	391	608
Europe	287	455	535	540
Oceania	8	15	23	32
Africa	32	102	295	787
Asia	244	594	1,376	2,679
WORLD	750	1,542	2,861	4,981
Rural population				
North America	62	64	71	61
Latin America and Caribbean	98	125	127	116
Europe	260	221	193	131
Oceania	5	6	8	10
Africa	188	304	498	702
Asia	1,158	1,812	2,297	2,271
WORLD	1,769	2,523	3,195	3,289

	Percentage Urban			
North America	64%	74%	77%	85%
Latin America and Caribbean	41%	61%	75%	84%
Europe	52%	67%	73%	81%
Oceania	62%	71%	74%	77%
Africa	14%	25%	37%	53%
Asia	17%	25%	38%	54%
WORLD	30%	38%	47%	60%

Source: Adapted from United Nations Population Division, 2002, World Urbanization Prospects: The 2001 Revisions (New York: United Nations), Tables A2, A3, and A4.

Internal Rural-to-Urban Migration

The migration of people within a country from rural to urban places represents the classic definition of the urban transition because it is intuitively the most obvious way by which a population can be shifted from countryside to curbside. There is no question that in the developed countries, rural-to-urban migration was a major force in the process of urbanization. Over time, the agricultural population of these countries has tended to decline in absolute numbers, as well as in relative terms, even in the face of overall population growth. In less-developed countries, though, rural-to-urban migration has been occurring in large absolute terms, but without a consequent depopulation of rural areas—although such a depopulation is projected for later in this century, as you can see in Table 11.1. The reason, of course, is the difference in the rates of natural increase in less-developed countries compared with rates in developed nations.

Had it not been for migration, cities of the nineteenth century and before could not have grown in population size. In fact, in the absence of migration, the excess of deaths over births would actually have produced deurbanization. Of course, migration did occur, because economic development created a demand for an urban population that was largely met by migrants from rural areas. Industrial cities drew the largest crowds, but commercial cities, even in nonindustrial countries, also generated a demand for jobs and created opportunities for people to move from agrarian to urban areas. The cities of most previously colonized countries bear witness to this fact. For example, migration accounted for 75–100 percent of the total growth of nineteenth-century cities in Latin America (Weller, Macisco, and Martine 1971). The growth of cities in southern Asia and western Africa was also stimulated by the commercial contacts of an expanding European economy. Naturally, in the richer, highly urbanized countries, the agricultural population is so small that cities (also nations) now depend on the natural increase of urban areas rather than migration for population growth. This helps to explain why Europe—the world's most urban region—is facing depopulation.

Natural Increase

The underlying source of urbanization throughout the world is the rate of natural increase of the rural population. The decline in death rates in rural places, without a commensurate drop in the birth rate, has led to genuine overpopulation in rural areas (too many people for the available number of jobs) and causes people to seek employment elsewhere (the now-familiar tale of demographic change and response). If there were no opportunities for rural-to-urban migration, then the result might simply be that the death rate would eventually rise again in rural areas to achieve a balance between population and resources (the "Malthusian" solution). However, opportunities typically have existed elsewhere in urban places precisely because the innovations that led to a drop in the rural death rate have originated in the cities—the site of technological and material progress and the source of economic development.

The speed of urbanization—the number of years it takes to go from low-percentage urban to high-percentage urban—depends partly on the difference in the rates of natural increase between urban and rural areas. In turn, the rate of natural increase depends on trends in both mortality and fertility, and these patterns have changed dramatically over time.

Urbanization and Mortality   Kingsley Davis (1973) has estimated that in the city of Stockholm, Sweden, in 1861–70, the average life expectancy at birth was only 28 years; for the country as a whole at that time, the life expectancy was 45 years. I discussed in Chapter 5 that the ability to resist death has been passed to the rest of the world by the industrialized nations, and the diffusion of death control has usually started in the cities and spread from there to the countryside. Thus, there has been a crucial reversal in the urban-rural difference in mortality. When the now-industrialized nations were urbanizing, death rates were higher in the city than in the countryside (see, for example, Williams and Galley 1995) and this helped keep the rate of natural increase in the city lower than in rural areas. In turn, that meant that rural-to-urban migration was a more important factor influencing the urban percentage in a country.

For the past several decades, however, death rates have been lower in the city than in the countryside in virtually every part of the world. As a consequence, the process of urbanization in less-developed countries is taking place in the context of historically high rates of urban natural increase. Furthermore, when mortality declines as a response to economic development, structural changes also take place that tend to reduce fertility; but when death control is introduced independently of economic development, mortality and fertility declines lose their common source, and mortality decreases whereas fertility may take quite a while to respond. This results in fertility levels being higher today in less-developed countries (urban and rural places alike) than they were at a comparable stage of mortality decline in the currently advanced countries.

Urbanization and Fertility   It is almost an axiom in population studies that urban fertility levels are lower than rural levels; it is also true, of course, that fertility is higher in less-developed than in developed nations. Putting these two generalizations together, you can conclude that urban fertility in less-developed nations will be lower than rural fertility but still higher than the urban fertility of cities in the industrialized nations. Cities in developing countries, especially in Africa and western Asia, have fertility levels that are probably higher than those ever experienced in European cities. Data from Demographic and Health Surveys in developing countries in the 1990s show total fertility rates (TFRs) in urban areas in 16 surveys that are above 5.0 and an additional 14 surveys in which urban areas have TFRs between 4.0 and 4.9. Without exception, the countries with urban TFRs of 4.0 or above are in sub-Saharan Africa or western Asia (Weeks et al. 2004).

Table 11.2 summarizes data available from surveys completed in the latter half of the 1990s. You can see in that table that sub-Saharan Africa is especially noteworthy for high urban fertility rates—which are still lower than the very high rural rates. In Mali in 1996, the total fertility rate (TFR) among urban women was 5.4 children each, although that was lower than the 7.3 children per woman in rural areas. At the lower end of fertility was Ghana in 1998, with an urban TFR of 3.0 children, compared to the rural rate of 5.3 children. Countries in Asia tended to have the lowest levels of fertility in both urban and rural areas, whereas countries such as Jordan and Egypt tended to have relatively small differences between urban and rural fertility. Among Latin American countries surveyed, you can see that Guatemala had the highest urban fertility at 4.1 children per woman, while the greatest spread between urban and rural fertility was found in Bolivia, which had levels of urban and rural fertility almost identical to those in Togo in sub-Saharan Africa. In both of these countries, rural women were having twice as many children as women in urban places.

Table 11.2   Urban-Rural Fertility Differences Still Pronounced in Most
Developing Countries

	Urban Fertility Rates	Fertility Rate 15–49	Rural Fertility Rates	Fertility Rate 15–49	Difference (Rural-Urban)
Sub-Saharan Africa		4.4		6.4	2.0
Eritrea 1995	4.2		7.0		2.8
Kenya 1998	3.1		5.2		2.1
Comoros 1996	3.8		5.0		1.2
Madagascar 1997	4.2		6.7		2.5
Tanzania 1996	4.1		6.3		2.2
Uganda 1995	5.0		7.2		2.2
Mozambique 1997	4.6		5.3		0.7
Zambia 1996	5.1		6.9		1.8
Burkina Faso 1999	3.9		6.9		3.0
Benin 1996	4.9		6.7		1.8
Cameroon 1998	3.8		5.4		1.6
Ghana 1998	3.0		5.3		2.3
Guinea 1999	4.4		6.1		1.7
Mali 1996	5.4		7.3		1.9
Niger 1998	5.6		7.6		2.0
Senegal 1997	4.3		6.7		2.4
Chad 1997	5.9		6.5		0.6
Togo 1998	3.2		6.3		3.1
Near East/North Africa		4.1		5.4	1.3
Jordan 1997	4.2		5.0		0.8
Yemen 1997	5.0		7.0		2.0
Egypt 1995	3.0		4.2		1.2
Europe/Eurasia		2.4		3.5	1.1
Kazakhstan 1995	2.0		3.1		1.1
Kyrgyz Republic 1997	2.3		3.9		1.6
Uzbekistan 1996	2.7		3.7		1.0
Turkey 1998	2.4		3.1		0.7
Asia		2.4		3.7	1.3
Bangladesh 1997	2.1		3.4		1.3
Nepal 1996	2.9		4.8		1.9
Indonesia 1997	2.4		3.0		0.6
Philippines 1998	3.0		4.7		1.7
Vietnam 1997	1.6		2.5		0.9
Latin America and Caribbean		3.0		4.9	2.0
Dominican Republic 1996	2.8		4.0		1.2
Guatemala 1999	4.1		5.8		1.7
Nicaragua 1997	2.9		5.0		2.1
Bolivia 1998	3.3		6.4		3.1
Brazil 1996	2.3		3.5		1.2
Colombia 1995	2.5		4.3		1.8
Peru 1996	2.8		5.6		2.8

Source: Macro International, 2001, Demographic and Health Surveys, STATcompiler, http://www.measuredhs.com/data/, accessed 2001.

You would have to go back to 1940 in the United States to find a time when, at every age over 19, rural farm women had at least twice as many children as urban women. By the early 1990s, fewer than 2 percent of women in the United States of reproductive age were living on farms, and their fertility was only slightly higher than the other 98 percent of the population (Bachu 1993). By the late 1990s, farm residence was not even reported in the data for the United States, although women who worked in occupations listed as "farming, forestry, and fishing" did have higher fertility levels than any other group of employed women (Bachu and O'Connell 2000). Still, the differences were so small that Long and Nucci (1996) were willing to conclude that "the traditional urban-rural fertility differential in the United States has ended, at least temporarily" (p. 19).

We can usually anticipate that people residing in urban areas will have fairly distinctive ways of behaving compared with rural dwellers. So important and obvious are these differences demographically that urban and rural differentials in fertility are among the most well documented in the literature of demographic research. John Graunt, the seventeenth-century English demographer whose name I first mentioned in Chapter 3, concluded that London marriages were less fruitful than those in the country because of "the intemperance in feeding, and especially the Adulteries and Fornications, supposed more frequent in London than elsewhere . . . and . . . the minds of men in London are more thoughtful and full of business than in the Country" (quoted by Eversley 1959:38).

In rural areas, large families may be useful (for the labor power), but even if they are not, a family can "take care of" too many members by encouraging migration to the city. Once in the city, people have to cope more immediately with the problems that large families might create and, besides, the city offers many more alternatives to family life than do rural areas. In recent decades, the once wide divergence in urban and rural fertility levels in more-developed nations has narrowed as rural fertility has decreased compared to urban levels, reflecting the growing dependence of rural places on urban production and an urban lifestyle.

Urban fertility levels are also related to migration, since migrants tend to be young adults of reproductive ages. Furthermore, migrants from rural areas typically wind up having levels of fertility lower than people in the rural areas they left, but they still have higher fertility levels than those in the urban areas to which they have moved (Goldstein and Goldstein 1981; Ritchey and Stokes 1972; Zarate and de Zarate 1975). To some extent, of course, the fertility impact of migrants will depend on whether males or females (or neither) predominate in the migration stream.

Sex Ratios in Cities   Males are generally more likely to migrate than females, as I mentioned in Chapter 7, and from this you can deduce that migrants to cities are more likely to be men than women. Although this pattern is generally found around the world, there is considerable regional variability (Brockerhoff 2000; Clarke 2000). Where the status of women is low, the movement of women tends to be more restricted, and so males are more dominant among rural-urban migrants. This pattern generally holds in Africa, the Middle East, and Asia. Where women have more social and economic freedom, such as in Europe and the Americas, men are less dominant among migrants. A higher proportion of men than women at the young adult ages in cities could be expected to reduce fertility somewhat below what it might otherwise be; and it could also raise death rates a bit, since mortality rates are higher for males than for females. However, the complexity of migration, especially within a country, where both men and women may migrate to the city only temporarily, or may go back and forth between the countryside and the city, make it very difficult to analyze the impact of sex ratios on the demographics of the urban transition.

International Urbanward Migration

International migration also operates to increase the level of urbanization, because most international migrants move to cities in the host area regardless of where they lived in the donor area. From the standpoint of the host area, then, the impact of international migration is naturally to add to the urban population without adding significantly to the rural population, thereby shifting a greater proportion of the total to urban places. Currently, more than 90 percent of immigrants to the United States wind up as urban residents in big cities or their suburbs (Gober 2000).

In spite of the massive increase in international migration in the past few decades and the fact that most international migrants live in urban areas, the actual impact on the process of urbanization throughout the globe has been less dramatic than it might seem. The reason for this is that many international migrants are headed toward cities in countries that are already highly urbanized (Berry 1993).

Reclassification

It is also possible for "in-place" urbanization to occur. This happens when the absolute size of a place grows so large, whether by migration, natural increase, or both, that it reaches or exceeds the minimum size criterion used to distinguish urban from rural places. Note that reclassification is more of an administrative phenomenon than anything else and is based on a unidimensional (size-only) definition of urban places, rather than also incorporating any concept of economic and social activity. Of course, it is quite probable that as a place grows in absolute size it will at the same time diversify economically and socially, probably away from agricultural activities into more urban enterprises. This tends to be part of the social change that occurs everywhere in response to an increase in population size; an agricultural population can quickly become redundant and the lure of urban activities (such as industry, commerce, and services) may be strong under those conditions.

Another administrative trick that can lead to rapid city growth is annexation. City boundaries are almost constantly changing, and the effect is often to bring into the city limits people who might otherwise be classified as rural. Urban growth rates can thus be misleading. For example, "The city of Houston grew 29 percent during the 1970s—one of the most rapidly growing large cities in the country. But the city also annexed a quarter of a million people. Without the annexation, the city would have grown only modestly" (Miller 2004:31). In the United States, this kind of phenomenon is associated with urban sprawl, which I discuss in the essay that accompanies this chapter. In general, this kind of sprawl is associated with ever larger urban areas.

NIMBY AND BNANA—THE POLITICS OF URBAN SPRAWL IN AMERICA

Where will suburbanization end? How far away from a city's center are people willing to live? Does the concept of a city center even mean much any more? These are the kinds of questions that are inspired by **urban sprawl**—"the straggling expansion of an urban area into the adjoining countryside" (Brown 1993:3002). Urban lives are increasingly complicated by two-earner households and by greater movement of people from job to job. There is evidence from England that these changes have caused an increasing number of people to accept a longer commute in lieu of moving (Green, Hogarth, and Shackleton 1999). So, we can attribute part of the sprawl to the willingness of some people to endure a long commute. But a greater fraction is almost certainly due to the conflict between people wanting to live in a low-density area yet be part of the urban scene. The combination of NIMBY and BNANA attitudes leads almost inevitably to urban sprawl.

NIMBY, which stands for Not In My Back Yard, refers to the idea that whatever is proposed to be built should be built somewhere else besides your neighborhood. You do not mind that it (whatever it is) is built, you just don't want it near you. BNANA, which stands for Build Nothing Anywhere Near Anyone, is more extreme. This represents a generalized antigrowth attitude expressed by people who essentially want to close the urban door behind them and let nothing and no one else in. The problem with both attitudes is that there is worldwide pressure for an increase in urban areas—the urban transition is an inevitable consequence of population growth everywhere in the world. So, if new homes and businesses are not built near you, they will nonetheless be built somewhere else near the urban area, in a rural area that will soon become part of the urban area (no matter how hard the BNANAs may protest) and that will contribute to urban sprawl, which contributes to the demise of the countryside, widespread traffic gridlock, and a lowering of the perceived quality of life.

Since sprawl occurs in the absence of regional planning, even in the presence of protests, planning movements have arisen in metropolitan areas around the world to create "smart growth":

> In general, smart growth invests time, attention, and resources in restoring community and vitality to center cities and older suburbs. New smart growth is more town-centered, is transit and pedestrian oriented, and has a greater mix of housing, commercial and retail uses. It also preserves open space and many

Metropolitanization and Agglomeration

In countries such as the United States, cities have grown so large and their influence has extended so far that a distinction is often made between metropolitan and nonmetropolitan areas, rather than the more vague distinction between urban and rural. In 1949, the U.S. Census Bureau developed the standard metropolitan area (SMA), consisting of a county with a core city of at least 50,000 people and a population density of at least 1,000 people per square mile. The concept proved useful in conjunction with the 1950 census and was subsequently revised to be called the **standard metropolitan statistical area (SMSA)**.

The basic idea of a metropolitan statistical area "is that of an area containing a recognized population nucleus and adjacent communities that have a high degree of integration with that nucleus" (U.S. Office of Management and Budget 2000:82228). Over time, there have been modifications to the calculation of the SMSA, ordered always by the U.S. Office of Management and Budget (OMB), which uses these classifications for a variety of government purposes. In the 1990s, the U.S. was divided into **MSAs (metropolitan statistical areas), CMSAs (consoli-**

other environmental amenities. But there is no "one-size-fits-all" solution. Successful communities do tend to have one thing in common—a vision of where they want to go and of what things they value in their community—and their plans for development reflect these values (Smart Growth Network 2001)

Smart growth is thus about containing growth spatially. This implies higher population densities, but within a context in which communities are rethought and well-thought-out. Higher, but smarter, densities might mean improved public transportation, more small but well-planned urban open spaces, and the creation of urban villages that attempt to recreate the (largely mythical) atmosphere of small towns in the past.

Two important aspects of American public policy have contributed to urban sprawl and are addressed by smart growth policies: (1) massive public spending on highways, and (2) local government authority over land use and taxation. If the government is willing to help subsidize the building of highways, then it is relatively easy to keep pushing the urban zone into the countryside, all other things being equal. What makes this possible, however, is the ability and willingness of local gov-

ernments to zone land for urban uses, often as a way of increasing local tax revenue—which may then be used to improve local infrastructure (water, sewage, electricity, communications, etc.), which of course stimulates even more urban development. However, if a state or other regional authority is able to draw an urban boundary line, beyond which urban uses are not permitted no matter what local governments might otherwise be willing to tolerate, then smart growth might have a chance—although the issues are extremely complex (see, for example, Ackerman 1999).

One state that has passed a smart growth initiative is Maryland, which did so in 2000 under a Democratic governor. The essence of the program is that certain areas are designated for growth and only within those areas will the state of Maryland provide any funding for infrastructure for water, sewers, highways, schools, and economic development projects (Bouvier and Stein 2000). However, by 2004, the Maryland Office of Smart Growth was slated to be closed by the new governor (a Republican) and its functions transferred to the Department of Planning. In late 2000, voters in the states of Arizona and Colorado rejected similar kinds of smart growth plans, so the message seems to be: "Have some sprawl, y'all."

---

dated metropolitan statistical areas), and **PMSAs (primary metropolitan statistical areas).** An MSA is defined as a county that has a core city with at least 50,000 people. The MSA then includes that core population and the population of contiguous areas that meet certain size and density criteria.

At the time Census 2000 was taken, there were 261 "stand-alone" MSAs, 19 CMSAs (each with several MSAs of their own), and 76 PMSAs, including three MSAs, one CMSA, and three PMSAs in Puerto Rico. These metropolitan areas encompassed within their boundaries about 80 percent of the total U.S. population. An example of a stand-alone MSA would be the Raleigh-Durham-Chapel Hill MSA in North Carolina, which was tagged by *Money* magazine in the 1990s as the best community in America in which to live. Although the MSA comprises the three places in its title, along with several smaller **suburban** cities and areas, it is not part of any larger unit. By contrast, the nation's most populous MSA, New York City (which, incidentally, was 123rd on the list of best places to live), is only one of 15 MSAs that make up the "New York–Northern New Jersey–Long Island, NY-NJ-CT-PA" CMSA. Furthermore, each one of those constituent MSAs is known as a PMSA (I'm not making this up).

Many people were unhappy with this way of classifying metropolitan areas because it seemed inadequate to the task of defining the newly emerging patterns of urban growth in the U.S. (Adams, VanDrasek, and Phillips 1999). Among the problems were that it used counties as the basic building block and thus wound up defining as "metropolitan" a lot of areas that would not normally be thought of as urban, and it also failed to take into account the way in which contiguous counties were or were not integrated with one another (Morrill, Cromartie, and Hart 1999). So, in the 1990s, the OMB ordered that the MSA definitions be revised, yet again, for use with the Census 2000 data. The Metropolitan Area Standards Review Project completed its work in December 2000 and decided that counties should continue to be the building blocks of a redefined **Core-Based Statistical Area (CBSA)** classification scheme (U.S. Office of Management and Budget 2000). The CBSA is similar in concept to the SMSA. Each CBSA must begin with a Census Bureau-defined urbanized area of at least 50,000 people or a Census Bureau-defined urban cluster of at least 10,000 people. Contiguous counties are then added to the CBSA if they meet specific criteria of connectivity: Either 25 percent or more of the employed residents of the county work in the central county of the CBSA or at least 25 percent of employment in the county is accounted for by workers who reside in the central county of the CBSA. There are myriad details to these definitions and they will be phased in over time (for more on this, see Fitzsimmons and Ratcliffe 2004). In the meantime, researchers continue to use data from smaller geographic building blocks, especially census tracts, to try to develop a more meaningful urban continuum.

In Canada, the definition of metropolitan is similar, although not identical, to that in the United States. The Canadian **census metropolitan area (CMA)** has a core urban area of at least 100,000 and includes those adjacent urban and rural areas that have a high degree of economic and social integration with the urban core. In Mexico, the government agencies have not defined metropolitan areas quite as precisely as in Canada and the United States (Garza 2004).

Another level of aggregation is the **urban agglomeration,** a term used largely by the United Nations, but consistent with the more popular term **megalopolis.** The UN defines an urban agglomeration as "the population contained within the contours of contiguous territory inhabited at urban levels of residential density without regard to administrative boundaries" (United Nations 1996b). The concept accepts a country's own definition of what is urban, and then puts together (agglomerates) all of the contiguous urban areas. It has the advantage of providing international comparisons. In Table 11.3, I have listed the largest metropolitan areas in the United States (based on U.S. Census Bureau definitions of metropolitan areas), and Canada and Mexico (based on the United Nations' definition of urban agglomeration). In general, the United Nations' definitions are slightly more limited geographically than those of the U.S. census, but are nearly identical to those of the Canadian definitions. By UN definitions, Mexico City is just slightly more populous than the New York metropolitan area, although the U.S. census definition of the CMSA of New York includes 3.5 million more people than the UN definition. Toronto is Canada's most populous metropolitan area and its size would make it the fifth largest metro area in the United States and the second largest in Mexico.

Table 11.3 Largest Metropolitan Areas in United States, Canada, and Mexico

Country	Metropolitan Areas with Population of at Least 1 Million	Population (millions) in 2005
United States	New York-Northern New Jersey-Long Island, NY-NJ-PA	18.3
	Los Angeles-Long Beach-Santa Ana, CA	12.4
	Chicago-Naperville-Joliet, IL-IN-WI	9.1
	Philadelphia-Camden-Wilmington, PA-NJ-DE	5.7
	Dallas-Fort Worth-Arlington, TX	5.2
	Miami-Fort Lauderdale-Miami Beach, FL	5.0
	Washington-Arlington-Alexandria, DC-VA-MD	4.8
	Houston-Baytown-Sugar Land, TX	4.7
	Detroit-Warren-Livonia, MI	4.5
	Boston-Cambridge-Quincy, MA-NH	4.4
	Atlanta-Sandy Springs-Marietta, GA	4.2
	San Francisco-Oakland-Fremont, CA	4.1
	Riverside-San Bernardino-Ontario, CA	3.3
	Phoenix-Mesa-Scottsdale, AZ	3.3
	Seattle-Tacoma-Bellevue, WA	3.0
	Minneapolis-St. Paul-Bloomington, MN-WI	3.0
	San Diego-Carlsbad-San Marcos, CA	2.8
	St. Louis, MO-IL	2.7
	Baltimore-Towson, MD	2.6
	Pittsburgh, PA	2.4
	Tampa-St. Petersburg-Clearwater, FL	2.4
	Denver-Aurora, CO	2.2
	Cleveland-Elyria-Mentor, OH	2.1
	Cincinnati-Middletown, OH-KY-IN	2.0
	Portland-Vancouver-Beaverton, OR-WA	1.9
	Kansas City, MO-KS	1.8
	Sacramento-Arden-Arcade-Roseville, CA	1.8
	San Jose-Sunnyvale-Santa Clara, CA	1.7
	San Antonio, TX	1.7
	Orlando, FL	1.6
	Columbus, OH	1.6
	Providence-New Bedford-Fall River, RI-MA	1.6
	Virginia Beach-Norfolk-Newport News, VA-NC	1.6
	Indianapolis, IN	1.5
	Milwaukee-Waukesha-West Allis, WI	1.5
	Las Vegas-Paradise, NV	1.4
	Charlotte-Gastonia-Concord, NC-SC	1.3

(continued)

Table 11.3   (continued)

Country	Metropolitan Areas with Population of at Least 1 Million	Population (millions) in 2005
	New Orleans-Metairie-Kenner, LA	1.3
	Nashville-Davidson-Murfreesboro, TN	1.3
	Austin-Round Rock, TX	1.2
	Memphis, TN-MS-AR	1.2
	Buffalo-Niagara Falls, NY	1.2
	Louisville, KY-IN	1.2
	Hartford-West Hartford-East Hartford, CT	1.1
	Jacksonville, FL	1.1
	Richmond, VA	1.1
	Oklahoma City, OK	1.1
	Birmingham-Hoover, AL	1.1
	Rochester, NY	1.0
Canada	Toronto	5.2
	Montreal	3.6
	Vancouver	2.2
	Ottawa	1.1
	Calgary	1.0
	Edmonton	1.0
Mexico	Mexico City	18.9
	Guadalajara	3.9
	Monterrey	3.5
	Toluca	1.9
	Puebla	1.9
	Tijuana	1.5
	Ciudad Juarez	1.4
	Leon	1.4
	Torreon	1.1

Sources: U.S. Census Bureau, Census 2000 PHC-T-29, Ranking Tables for Population of Metropolitan Statistical Areas, Micropolitan Statistical Areas, Combined Statistical Areas, New England City and Town Areas, and Combined New England City and Town Areas: 1990 and 2000 (Areas defined by the Office of Management and Budget as of June 6, 2003), Table 3b; United Nations Population Division, 2002, World Urbanization Prospects: The 2001 Revision (New York: United Nations), Table A12.

As a further "refinement," the United Nations refers to any urban agglomeration with more than 10 million people as a **mega-city.** By this definition, there are projected to be 19 mega-cities in the world in 2005. Figure 11.4 lists these mega-cities, or large urban agglomerations, as of 2000. Tokyo is the largest urban agglomeration, with 26.4 million people, followed by São Paulo, Mexico City, Mumbai (formerly Bombay), Paulo, and New York City. The bottom panel of Figure 11.4

Figure 11.4   World's Largest Urban Agglomerations Changed Dramatically from 1950–2005

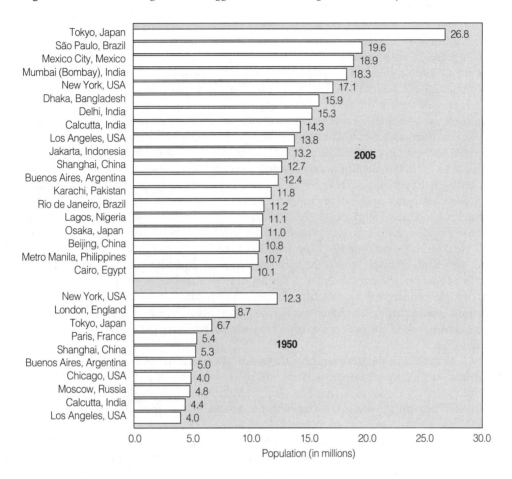

Source: Adapted from United Nations Population Division, 2002, World Urbanization Prospects: The 2001 Revision (New York: United Nations), Table A11. Population figures are in millions.

shows what the 10 largest urban agglomerations were back in 1950. In that year, only three of the top cities were in less-developed countries—Shanghai, China; Buenos Aires, Argentina; and Calcutta, India. By 2000, only three of the 10 largest cities were in more-developed nations—Tokyo, New York, and Los Angeles. This is because the cities of the less-developed nations have simply skyrocketed in their population size.

The rapid growth of cities in the context of rapid population growth in developing countries deserves more scrutiny on our part because it represents a potent source of social change with which societies must cope. To understand that change, however, we must first examine some major theoretical perspectives about the place of cities in the global economic system.

The Urban Transition in the Context of the Demographic Transition

As economic development occurred, cities grew because they were economically efficient places. Commercial centers bring together in one place the buyers and sellers of goods and services. Likewise, industrial centers bring together raw materials, laborers, and the financial capital necessary for the profitable production of goods. They are efficient politically because they centralize power and thus make the administrative activities of the power base that supports them more efficient. In sum, cities perform most functions of society more efficiently than is possible when people are spatially spread out. Mumford says it well: "There is indeed no single urban activity that has not been performed successfully in isolated units in the open country. But there is one function that the city alone can perform, namely the synthesis and synergy of the many separate parts by continually bringing them together in a common meeting place where direct fact-to-face intercourse is possible. The office of the city, then, is to increase the variety, the velocity, the extent, and the continuity of human intercourse" (Mumford 1968:447).

Cities are efficient partly because they reduce costs by congregating together both producers and consumers of a variety of goods and services. By reducing costs, urban places increase the benefits accruing to industry—meaning, naturally, higher profits. Those profits translate into higher standards of living, and that is why cities have flourished in the modern world. They are part and parcel of the modern rise of capitalism.

The theory of the demographic transition was derived originally from the modernization theory, which focused on the role played by cities. The basic thesis of modernization theory was that economic development is built on the efficiencies of cities. Cities are the engines of growth because of the concentration of capital (to build industry), labor (to perform the industrial tasks), and the financial, governmental, and administrative services necessary to manufacture, distribute, sell, and regulate the goods that comprise the essence of economic development. Once economic growth begins, the principle of cumulative causation (another familiar concept by now) kicks in to promote further local development, often at the expense of other regions.

Modernization theory focused attention on individual nations, examining the role that cities played in the development process in that particular country. The fact that cities are crucial to modernization led many developing nations such as Chile, Kenya, Malaysia, and Mexico to establish industrial cities precisely to mimic the process of modernization that occurred in the more developed nations (Potter 1992). The results have been mixed at best.

Modernization theory offers a generally good explanation of what happened in the now-developed countries, especially in the period up to World War II, as individual nation building was taking place, but transportation and communication still limited the amount of direct interaction between countries. Most of the still-quoted theories of modernization were developed in the 1950s (such as Hirschman 1958; Myrdal 1957), at a time when the "Third World" had not yet been defined as such, and when urbanization in developing countries was still in its early stages. Patterns of urbanization in different parts of the world may reflect where a country entered the modern economic system, which adds globalization to the process of urbanization (Douglass 2000), creating new types of city systems and urban hierarchies.

The Urban Hierarchy

In virtually every nation of the world there is one city that stands out as the leading urban center and is noticeably more populous than other cities in the region. Such a place is called a **primate city**—a disproportionately large leading city holding a central place in the economy of the country. Other cities are less important, but may have their own pecking order, so to speak. In the 1930s, Christaller (1966) developed the concept of central place theory to describe why and how some cities were territorially central and thus in a position to control markets and the regional economy.

Early empirical studies of city systems within countries suggested that a common pattern of cities by size could be expressed by the **rank-size rule.** As set out by Zipf (1949), this says that the population size of a given city (P_i) within a country will be approximately equal to the population of the largest city (P_1) divided by the rank of city i by population size (R_i), or:

$$\text{rank-size rule: } P_i = P_1/R_i$$

So, if the largest city in a country has a population of five million, then the second largest city should have a population size of approximately 5/2 or 2.5 million, whereas the third largest city should have a population of 5/3 or 1.7 million people.

Many countries generally follow a pattern that is basically similar to the rank-size rule, but there are important exceptions. For example, in Mexico, the largest city (Mexico City) has a population of 18.9 million people, so the rank-size rule would predict a population of 18.9/2, or 9.5 million, for the second largest city (Guadalajara) and a population of 18.9/3, or 6.3 million, for the third largest city (Monterrey). However, the estimated actual sizes of Guadalajara and Monterrey in 2005, as you can see in Table 11.3, are only 3.9 million and 3.5 million, respectively. Clearly, the urban hierarchy in Mexico follows a different pattern than that expressed by the rank-size rule.

Even though an empirical generalization such as the rank-size rule does not fit all countries, the importance of such ideas is that they led to the realization among scholars that most countries had a somewhat predictable system of cities that might be amenable to a consistent theoretical interpretation. An important set of such theoretical perspectives was the **core-periphery model** put forth by Friedmann in 1966 (Potter 1992). Prior to economic development, a series of independent cities may exist in a region, but development tends to begin in, and will be concentrated in, one major site (the primate city). This is especially apt to happen in less-developed countries that have a history of colonial domination, in which colonial functions were centralized in one city. Over time, the development process diffuses to other cities, but this happens unequally because the primate city (the core) controls the resources, and the smaller cities (the periphery) are dependent on the larger city. It was only a small step, then, to apply these ideas of the core and periphery to a world system of countries and cities.

The **world systems theory** is based on the notion that inequality is part of the global economic structure, and has been since at least the rise of capitalism 500 years ago (Wallerstein 1974). "Core" countries are defined as the highly developed nations that dominate the world economy and were the former colonizers of much of the rest of the world. The "periphery" is composed of those countries that, in order to become a part of the global economic system (the alternative to which is to remain isolated and undeveloped), have been forced to be dependent on the core nations, which control the basic resources required for development and a higher standard of living. Because countries tend to be dominated by cities, the world systems theory predicts that cities of core countries will control global resources, operating especially through multinational corporations headquartered in core cities (Chase-Dunn 1989; Chase-Dunn and Hall 1997), and cities in peripheral countries will be dependent on the core cities for their own growth and development (and, of course, that will filter down from the primate city to the other cities in a nation's own city system). These models suggest that the core cities are networked globally and emphasize services over manufacturing, whereas peripheral cities tend to be linked to more local markets and emphasize manufacturing over services (Maume 1983).

As is so often the case, no single model seems capable of explaining all of the complexity in the process of the urban transition. Several tests of modernization and world systems theories have suggested that variations of both contribute to our understanding of the real world experience of urbanization (see, for example, Bradshaw 1987; Firebaugh and Beck 1994; London 1987). Part of the complexity is created by the fact that the sheer size of a city does not guarantee its importance in the system of world cities. To be a global city means to have economic power, which is often characterized by the presence of headquarters or major subsidiary locations of multinational corporations (Godfrey and Zhou 1999; Sassen 2001). However, sheer size is important at the national or regional level. The larger the city, the greater the economic opportunity is likely to be for its residents and the greater the attraction will be to potential migrants from rural areas. Empirical data show that cities in the United States are indeed rank-ordered with respect to the income of workers. In general, the more global a city is, the higher the income, and the lower down in the ranking a city is, the higher the likelihood that its residents will be involved in low-income work (Elliott 1999).

Keep these complexities in mind as I illustrate the urban transition with examples from Mexico and China.

An Illustration from Mexico

The impact of population processes on the urban transition is illustrated from the perspective of rural areas by what has happened to people in one village in Mexico—Tzintzuntzan (which in the indigenous language means "the place of hummingbirds"), situated midway between Guadalajara and Mexico City in the state of Michoacán. Historically, the site was the capital of the Tarascan empire (Brandes 1990), but today it is a village of artisans, farmers, merchants, and teachers. For nearly 400 years, the population of Tzintzuntzan stayed right at about 1,000 people (Foster 1967). In the mid-1940s, as Foster began studying the village, the population size was starting to climb slowly, because death rates had started to decline in the late 1930s at about the time a government project gave the village electricity, running water, and a hard-surfaced highway connecting it to the outside world (Kemper and Foster 1975). In 1940, the population was 1,077 and the death rate was about 30 per 1,000, while the birth rate of 47 per 1,000 was leading to a rate of natural increase of 17 per 1,000. For some time, there had been small-scale, local out-migration from the village to keep its population in balance with the limited local resources, but by 1950 the death rate was down to 17 per 1,000 and the birth rate had risen. Better medical care had reduced the incidence of miscarriage and stillbirth, and in 1950 the village had 1,336 people (Foster 1967). By 1970 the population had reached about 2,200 (twice the 1940 size); however, were it not for out-migration draining away virtually all of the natural increase of Tzintzuntzan, the population would again have doubled in about 20 years (Kemper and Foster 1975).

What, you ask, does growth in a small Mexican village have to do with the urban transition? The answer, of course, is that population growth meant that the local population was larger than the local economy could handle, and so some of these villagers were forced to migrate out in search of work. The migrants headed for the cities, and therein lies the tale of the urban transition. Poverty is extensive in rural Mexico, and almost all of those people who leave Tzintzuntzan go to urban places where better opportunities exist—with Mexico City (230 miles away) having been the most popular destination (Kemper and Foster 1975). One of the easiest demographic responses that people can make to population pressure is migration, and in Mexico, as in most countries of the world, the city has been the receiving ground. Furthermore, the demographic characteristics of those who go to the city are what you would expect; they tend to be younger, slightly better educated, of higher occupational status, and more innovative than nonmigrants (Kemper 1977).

For Tzintzuntzeños, migration to Mexico City has raised the standard of living of migrant families, altered the world view of both adults and their children toward greater independence and achievement, and, indirectly, "urbanized" the village they left behind. This last effect is due to the fact that having friends and relatives in Mexico City is one factor that leads the villagers to be aware of their participation in a wider world. This makes it easier for each successive generation to make the move to Mexico City, because they know what to expect when they arrive and they know people who can help them.

Tzintzuntzeños have also been attracted to the United States, initially recruited through the Bracero program (which I mentioned in Chapter 7), and the same patterns of mutual assistance have encouraged the flow of money and ideas from cities in the United States to this small village in the interior of Mexico (Kemper 1991; 1996). International migration can be an extremely important source of income in rural Mexico (de Janvry, Gordillo, and Sadoulet 1997), and migrants have brought back to the village many of the accoutrements of urban life, from new stoves and sewing machines to stereos and television sets, in effect urbanizing what was once a remote village. This in-place urbanization ultimately has had an effect on fertility. With rising education and incomes, birth rates appear finally to be dropping here as they are throughout Mexico. In the 2000 census, the village was enumerated at just over 3,000 people, which was about the same as in 1990 and only three times the 1940 population, compared with a more than tenfold increase in the size of Mexico City during that interval.

From a theoretical perspective, we can see that modernization is the key to the transformation of the lives of Tzintzuntzeños. However, the process of modernization was not endogenous—it sought the villagers out, rather than the other way around. In a very literal sense, the modernization of the village and its inhabitants depended on what was happening elsewhere. Government leaders in Mexico City (the core) made the decision to provide rural areas (the periphery) with health care, electricity, and paved highways. The rest, as they say, is history, because few villagers, when given the choice, turn down the opportunity for a higher standard of living (Critchfield 1994).

An Illustration from China

China is a very interesting case of urbanization because it is one of the few countries in the world where the government fairly successfully "kept them down on the farm." The Chinese Communist Party officially adopted an anti-urban policy when it came to power in 1949, believing that cities were a negative "Western" influence, and Chinese government policies in the 1960s and 1970s were designed to counteract the urban transition occurring in most of the rest of the world. These policies attempted to "promote wider income distribution, reduce regional inequalities, and create a more balanced urban hierarchy, which would lead to a greater decentralization of economic activities. In doing so, the intention was to slow population growth in the largest cities, while allowing continued increases in medium-sized and smaller urban centers" (Goldstein 1988, as quoted in Bradshaw and Fraser 1989:989). As wonderful as that may sound, the basic policy was actually reinforced (or enforced) by a rigid system of household registration called *hukou*, which created a type of occupational apartheid in China. "Anyone in a rural county is automatically registered as a farmer, anyone in a city as a non-farmer; and the distinction is near rigid. A city-dwelling woman (though not a man) who marries a farmer loses the right to urban life" (*The Economist* 1998a:42). Various reforms since the 1980s have diminished the importance of the *hukou* system, although it is still officially in place (Hutzler and Lawrence 2003; Zhu 2004).

Government policies in China thus prevented the high rate of natural increase in the rural areas from spilling over disproportionately into migration to urban areas. This was mitigated partly by the governmental emphasis on heavy industry, often located in rural areas, rather than on the light and service industries that typify newer cities in the world today (Hsu 1994). Of course, that does not mean that the urban population was not growing. Quite the contrary. Between 1953 and 1990, 326 new cities were created in China (Hsu 1994), and urban growth occurred primarily in small to medium-sized places (Han and Wong 1994), just as the government had planned. Also according to plan, urbanization is low by world standards—a bit more than one in three Chinese (36 percent) lived in urban places in 2000, but the rate climbed quickly during the 1990s (Chan and Hu 2003).

In the 1980s, the Chinese government relaxed some of its restrictions, permitting rural residents to become temporary "guest" workers in urban areas, although without the right to become permanent residents. In 1993, it was suggested that "freed from the strictures of a command economy, residents in places like Humen discovered they could make more money in light industry, property speculation or running sing-along bars than they could on their farms" (Brauchli 1993:A12). It has been estimated that a "floating population" of as many as 80 to 110 million Chinese are working in cities even though they technically belong in the countryside (Chang 1996), and the government has been trying to deal with this by offering legal urban residency and a job to rural migrants who agree to stay in the smaller "buffer" cities rather than seek work in one of China's mega-cities (Johnson 1997).

China has thus been able administratively to create a much flatter urban hierarchy than exists in most modern nations (Fan 2000). The obvious examples are that Beijing is the politically most important city, Shanghai is the most populous city, and Hong Kong is the richest city (although it was already that way when China annexed it in 1997). But, more important has been the creation of new cities and the growth of existing small to mid-sized cities, rather than huge increases in the larger urban agglomerations. Of particular importance to China's recent urbanization has been the government's emphasis on creating nonagricultural jobs in rural areas, which leads to a reclassification of areas from rural to urban (Zhu 2004).

The Urban Evolution That Accompanies the Urban Transition

It is not an exaggeration to suggest that the urban transition represents a revolutionary shift in society. In 1964, Reissman described it this way:

> Urbanization is social change on a vast scale. It means deep and irrevocable changes that alter all sectors of a society. In our own history [the United States] the shift from an agricultural to an industrial society has altered every aspect of social life . . . the whole institutional structure was affected as a consequence of our urban development. Apparently, the process is irreversible once begun. The impetus of urbanization upon society is such that society gives way to urban institutions, urban values, and urban demands (154).

As the richer countries approach a situation where almost everybody lives in urban places, it is important to remember that the end of the urban transition does not necessarily signal the end of the process of urban *evolution* (Pumain 2004). Evolutionary processes include patterns of suburbanization, exurbanization, periurbanization and even counterurbanization—the shift of urban populations from more dense to less dense regions (Champion 1989; Frey 2004). The mere fact that people are increasingly likely to live in places defined as urban does not mean that the urban environment itself stops changing and evolving across time and space. Indeed, there is probably more variability among urban places, and within the populations in urban places, than ever before in human history.

The benefits of cities, of course, are what make them attractive, and they at least partially explain the massive transformation of countries like the United States and Canada from predominantly rural to primarily urban nations within a few generations. But there are also costs involved in living in cities, and the evolution of cities is partly a result of people trying to mitigate the downside of city life. The negative impact of urbanization on the human condition represents the set of unintended consequences that may prevent the city from being as attractive as it might otherwise be.

The efficiencies of the city have generally been translated into higher incomes for city dwellers than for farmers and, as I just mentioned, the larger the city, the higher the wages tend to be. Wage differentials undoubtedly have been and continue to be prime motivations for individuals to move to cities and stay there. That is not to say that people necessarily prefer to live in cities; the reverse may actually be true. Throughout American history, the sins and foibles of urban life have been decried, and the city is often compared unfavorably with a pastoral existence (Fischer 1984). Of course, people recognize the advantages of the city, and Americans prefer to be near a city although not right in it. Fuguitt and Zuiches have pointed out that public opinion polls since 1948 have shown that a vast majority of Americans indicate a preference for living in rural areas or small cities and towns. However, in the 1970s when they asked for the first time a survey question about the desire to be near a large city, those rural preferences became more specific. Their data showed that of all the people who said they preferred living in rural areas or in small cities, 61 percent also wanted to be within 30 miles of a central city (Fuguitt and Zuiches 1975). A replication of the survey in 1988 revealed a remarkable consistency over time in those residential preferences (Fuguitt and Brown 1990). In general, Americans like it both ways. They aspire to the freedom of space in the country but also prefer the economic and social advantages of the city. The compromise, of course, is the suburb, and it turns out that the average American is already living in this preferred location, leading Frey (1995) to declare that "America is in the suburbs. The suburbs are America" (p. 314). The rest of the world is generally headed in the same direction. Tourists to Paris, London, or Tokyo may spend most of their time near the older center of the city, but most of the people in these global cities now live and work out in the suburbs.

Suburbanization

More than a century has passed since Adna Weber (1899) noted that American cities were beginning to **suburbanize**—to grow in the outlying rings of the city. It was not until the 1920s, however, that suburbanization really took off. Hawley (1972) calculated that between 1900 and 1920 people were still concentrating in the centers of cities, but after 1920 the suburbs began regularly to grow in population at a faster pace than the central cities. Two factors related to suburbanization are people's desire to live in the less-crowded environment of the outlying areas and their ability to do so—a result of increasing wealth and the availability of transportation, especially automobiles. Such transportation has added an element of geographic flexibility not possible when the early suburbanites depended on fixed-rail trolleys to transport them between home in the suburbs and work in the central city.

From the 1920s through the 1960s, the process of suburbanization continued almost unabated in the United States (as indeed in most cities of the world). Admittedly, the process was hurried along by automobile manufacturers and tire companies that bought local trolley lines in order to dismantle and replace them with gasoline-powered buses (Kunstler 1993). Nonetheless, the advantages of the automobile are numerous and it was inevitable that cars would influence the shape of urban areas. By 2000 in the United States, 76 percent of all workers got to work by driving alone in their automobile (car, truck, or van)—a higher fraction than in 1990—and an additional 12 percent carpooled to work in an automobile, meaning that altogether 88 percent of American workers get to work by driving. In 2000, there were nearly as many people who walked to work (3 percent) in the United States as used public transportation (5 percent).

The fact that North America has become predominantly suburban adds new complexity to the metropolitan structure. Several trends are worth commenting on: (1) There has been a decided westward tilt to urbanization in the United States, which has facilitated suburbanization through the creation of new places; (2) Many of those new places are **edge cities** within the suburbs, replacing the functions of the old central city; and (3) Older parts of cities have been gentrified. Let us examine each of these in a bit more detail, remembering that, although my comments are directed primarily at the United States, many of these same trends are being seen in other developed nations as well. Suburbanization has, in fact, become an integral part of the urban transition and is now fully under way in China, among other large countries (Zhou and Ma 2000).

The western United States, the land of open spaces, became the most highly metropolitanized area of the country in the 1980s (Abbott 1993). A higher fraction of residents lives in metropolitan areas in the West than in any other part of the country. This has happened because the flow of migration in the United States has been consistently westward, especially since the end of World War II, as you will recall from Chapter 7, and migration in the modern world is almost always toward or between urban places. People and jobs have been moving west (as well as south). Kasarda (1995) has noted that, as recently as 1960, 25 percent of all Fortune 500 firms were headquartered in New York City, but by 1990 that had declined to 8 percent. Companies have shifted their operations to Sunbelt cities, but not necessarily the central parts of those cities. The suburbs are increasingly the sites of company headquarters, congregating near major highways and regional airports. It is perhaps a sign of the times that the richest person in the United States (Bill Gates) runs a company (Microsoft) located in the suburbs (Redmond, Washington) of a western city (Seattle).

Increasing suburbanization has meant greater metropolitan complexity, as new areas spring up on the edges of cities, competing with each other for jobs and amenities (Frey 1995; Frey 2004; Hughes 1993). Garreau (1991) coined the term "edge city" to describe the suburban entities that have emerged in the rings and beltways of metropolitan areas and are replicating, if not replacing, the functions of older central cities. Some of the edge cities are actually within the same city limits as the central city, but are distinct from it. Furthermore, larger metropolitan areas may have several edge cities, each with its own pattern of dominance over specific economic functions (such as high technology or financial services) in conjunction with a full range of retail shops and dining and entertainment establishments. Muller (1997) argues, in fact, that it is precisely in these edge cities that the globalization of American cities is taking place.

The growth of edge cities and the increasing economic and social complexity of the suburbs help to explain the shift in commuting patterns in the United States. In essence, the flexibility of the automobile allows people to live and work almost anywhere else within the same general area. The number of commuters going from one suburban area to another far exceeds the number of commuters going from the suburbs to the central city (Kasarda 1995). Those cars on the freeway in the morning are not all headed downtown—they are headed every which way.

These trends in metropolitan complexity and diversity have tended to leave the central cities with a daytime population of "suits" who commute downtown to work at various service companies (especially government administration and financial services industries) that have remained in the central city. At the same time, shopping centers, corporate headquarters, many new high-technology industries, and traffic gridlock have all relocated to the suburbs.

The baby boom generation in the United States grew up in the suburbs to a greater extent than any previous cohort, but as they reached an age to buy homes, baby boomers found themselves caught in the midst of spiraling housing costs amid increasing density in the suburbs. One alternative was to head even farther out of town, to what sometimes are called the **exurbs**—the suburbs of the suburbs. Another term for this is the **peri-urban region**—the periphery of the urban zone that looks rural to the naked eye but houses people who are essentially urban. "The peri-urban region is a distinctive zone that spans the landscape between contiguous urban development and the rural countryside" (Ford 1999:298).

As the urban population sprawls deeper into the countryside, some people have reversed that trend and moved back into the central city, where gentrification of buildings and of whole neighborhoods has been taking place in some, although not all, older cities of developed nations. Typically, this process removes the lower-income population that had settled into the areas previously abandoned by higher-income households, and dramatically alters the social structure of the surrounding area. Because these innovative renovators tend to be white and upwardly mobile (and often without children), they have been likened to the gentry moving back into the city, and thus the term **gentrification** is applied, referring to the restoration and habitation of older homes in central city areas by urban or suburban elites. The popularity of downtown living has expanded the scope of renovation in an increasing number of cities to include tearing down entire blocks of older buildings and erecting new high-rise, high-priced condominiums (Ford 2003). Nonetheless, despite the radical transformation of neighborhoods in places like New York City, Chicago, Baltimore, and Washington, DC, as well as Paris and London (Carpenter and Lees 1995), gentrification has not yet involved enough people to counter the continuing movement of people out of the central cities and into the suburbs and the peri-urban region. Those who are unable to leave the older sections of urban places find themselves residentially segregated and left out of the mainstream of the economy.

Residential Segregation

Although suburbia has become a legendary part of American society and seems to be a major force of urban evolution, suburbanization can be viewed as an innovation that started with some people and then diffused to others. In particular, it is a residential transformation that disproportionately involved whites in the United States until the 1970s. For example, in 1970 in 15 large areas studied by Farley (1976), 58 percent of whites lived in the suburbs compared with 17 percent of nonwhites. Beginning in the 1930s, the proportion of whites living in central cities declined steadily and the proportion of African Americans rose steeply (Schnore, Andre, and Sharp 1976); the African American population was undergoing a very rapid urbanization at the same time that whites were suburbanizing.

During the period 1910–30, there was a substantial movement of African Americans out of the South destined for the cities of the North and the West. The urban population of African Americans grew by more than 3 percent per year during that 20-year period, whereas the rural population declined not only relatively but in absolute terms as well. The reasons for migration out of rural areas were primarily economic, with the decline in the world demand for southern agricultural products providing the push out of the South. But there were concurrent pull factors as well in the form of demands in northern and western cities for labor, which could be met cheaply by blacks moving from the South (Farley 1970). During the Depression, there was a slowdown in the urbanization of African Americans, but by the beginning of World War II, half of the nation's African Americans lived in cities, reaching that level of urbanization 30 years later than whites had. After World War II, the urbanization and rural depopulation of blacks resumed at an even higher level than after World War I, and by 1960, the African American population was 58 percent urban in the South and 95 percent urban in the North and West. Urbanization was associated not only with the economic recovery after the war but also with severely restricted international migration which, until the law was changed in 1965, meant that foreigners no longer were entering the labor force to take newly created jobs, thus providing a market for African American labor. The consequence of the urbanization of African Americans, their urban rates of natural increase, and the relatively higher rates of suburbanization of whites has been the segregation of black and white populations within metropolitan areas.

The segregation of people into different neighborhoods on the basis of different social characteristics (such as ethnicity, occupation, or income) is a fairly common feature of human society (Bobo et al. 2000; Zlaff 1973). However, in the United States residential segregation by race is much more intense than segregation by any other measurable category. Residential segregation of blacks in the United States has been called an "American apartheid system" (Massey and Denton 1993), and the maintenance of this pattern until fairly recently has been explained by Farley and Frey (1994) as being due especially to the following factors: (1) Mortgage lending policies were discriminatory; (2) African Americans who sought housing in white areas faced intimidation and violence similar to that occurring during World War I; (3) After World War II, suburbs developed strategies for keeping African Americans out; and (4) Federally sponsored public housing encouraged segregation in many cities.

It is not clear whether these trends have eased over time, or whether patterns of residential segregation have simply changed. The 1965 changes in the Immigration Act (see Chapter 7) have diversified the ethnic structure of the country, and data suggest that Asians and Hispanics have a greater propensity or ability to suburbanize than do African Americans (Denton and Massey 1991; Farley, Danziger, and Holzer 2000; Logan and Alba 1993; Logan et al. 1996). The slower rate of suburbanization of African Americans continues to be the result of discrimination on the part of whites (Emerson, Yancy, and Chai 2001). "Thus, decades after the passage of the Civil Rights Act of 1964, the Voting Rights Act of 1965, the Immigration Act Amendments

(Hart-Celler) of 1965, and the Fair Housing Act of 1968, our society continues to be riven by a historical legacy of deep racism, spasms of nativisim, and the current condition of persistent racial and ehtnic inequality in life chances" (Bobo et al. 2000:32).

From the standpoint of demographic characteristics, the suburbs are composed especially of higher-income married-couple families (Alba and Logan 1991)—a pattern that disproportionately works against blacks who have both lower incomes and lower marriage rates than whites. But demographic components of suburbanization do not explain residential segregation; they merely point to its existence. The explanations are essentially social in nature, and one of the prevailing ones is based on the idea that "status rankings are operationalized in society through the imposition of social distance" (Berry et al. 1976:249). In race relations, the social status of blacks has been historically lower than that of whites. That status ranking used to be maintained symbolically by such devices as uniforms, separate facilities, and so forth, which were obvious enough to allow social distance even though blacks and whites lived in close proximity to each other. However, as African Americans left the South and moved into industrial urban settings, many of those negative status symbols were also left behind. As a result, spatial segregation has served as a means of maintaining social distance "where 'etiquette'—the recognition of social distance symbols—breaks down" (Berry et al. 1976:249). Thus as blacks have improved in education, income, and occupational status, whites have maintained social distance by means of residential segregation facilitated by suburbanization.

Differential suburbanization by race has done more than just keep whites separate from blacks. It has also separated blacks from job opportunities. Jobs have followed the population to the suburbs, making it increasingly difficult for people living in the central city—who may have inadequate access to transportation—to find employment. This "spatial mismatch" has been shown to be associated with higher unemployment rates for blacks living in Detroit and Chicago (Mouw 2000). Massey (1996a) has suggested that the spatial isolation of poverty and especially of low-income blacks portends future community instability and violence in America. However, there is some evidence that a trend toward desegregation does exist and it appears to be related to the westerly drift of the population and the increasing suburbanization taking place especially in the South and West. Comparing data from the 1980 and 1990 censuses, Frey and Farley (1996) have concluded that the largest decreases in segregation scores by residential area have occurred in the more recently built suburbs of southern and western metropolitan areas.

European cities are also characterized by a certain amount of residential segregation, largely with respect to the ethnic minority groups that have comprised the guest-worker populations. However, in many parts of Europe a large segment of the housing market for working-class families is subsidized and controlled by the government, and this has limited the scope of residential segregation (Huttman, Blauw, and Saltman 1991).

The Impact of Urban Crowding

For centuries, the **crowding** of people into cities was doubtless harmful to existence. Packing people together into unsanitary houses in dirty cities raised death rates. Furthermore, as is so often the case, as cities grew to unprecedented sizes in nineteenth-century Europe, death struck unevenly within the population. Mortality went down faster for the better off, leaving the slums as the places where lower-income people were crowded into areas "with their sickening odor of disease, vice and crime" (Weber 1899:414).

When early students of the effects of urbanization such as Weber and Bertillon discussed crowding and overcrowding, they had in mind a relatively simple concept of density—the number of people per room, per block, or per square mile. Thus Weber quotes the 1891 census of England, "regarding as overcrowded all the 'ordinary tenements that had more than two occupants to a room, bedrooms and sitting rooms included'" (1899:416). The prescription for the ill effects (literally) of overcrowding was fairly straightforward as far as Weber was concerned. "The requirement of a definite amount of air space to each occupant of a room will prevent some of the worst evils of overcrowding; plenty of water, good paving, drainage, etc. will render the sanitary conditions good."

Crime and vice are also often believed to be linked to urban life and, as a matter of fact, crime rates are higher in cities than in the countryside, at least in the United States. But what is it about crowding that might lead to differences in social behavior between urban and rural people? To examine that question, you have to ask more specifically what crowding is. The simplest definition of crowding is essentially demographic and refers to **density**—the ratio of people to physical space. As more and more people occupy a given area, the density increases and it therefore becomes relatively more crowded. Under these conditions, what changes in behavior can you expect? In a 1905 essay, Georg Simmel suggested that the result of crowding was an "intensification of nervous stimulation" (1905:48), which produced stress and, in turn, was adapted to by people reacting with their heads rather than their hearts. "This means that urban dwellers tend to become intellectual, rational, calculating, and emotionally distant from one another" (Fischer 1976:30). Here were the early murmurs of the **urbanism** concept—that the crowding of people into cities changes behavior—a concept often expressed with negative overtones.

Perhaps the most famous expression of the negative consequences of the city is Louis Wirth's paper "Urbanism as a Way of Life" (1938), in which he argued that urbanism will result in isolation and the disorganization of social life. Density, Wirth argued, encourages impersonality and leads to people exploiting each other. For two decades, there was little questioning of Wirth's thesis and, as Hawley put it, "in one short paper, Wirth determined the interpretation of density for an entire generation of social scientists" (1972:524). The idea that increased population density had harmful side effects lay idle for a while, but it was revived with considerable enthusiasm in the 1960s following a report by Calhoun on the behavior of rats under crowded conditions.

Although he initiated his studies of crowding among rats in 1947, it was not until 1958 that Calhoun (1962) began his most famous experiments (which later helped inspire the popular stories about the Rats of NIMH). In a barn in Rockville, Maryland, he designed a series of experiments in which rat populations could build up freely under conditions that would permit detailed observations without humans influencing the behavior of the rats relating to each other. Calhoun built four pens, each with all the accoutrements for normal rat life and divided by electrified partitions. Initially, eight infant rats were placed in each pen, and when they reached maturity, Calhoun installed ramps between each pen. At that point, the experiment took its own course in terms of the effects of population growth in a limited area. Normally, rats have a fairly simple form of social organization, characterized by groups of 10 to 12 hierarchically ranked rats defending their common territory. There is usually one male dominating the group, and status is indicated by the amount of territory open to an individual.

As Calhoun's rat population grew from the original 32 to 60, one dominant male took over each of the two end pens and established harems of eight to 10 females. The remaining rats were congregated in the two middle pens, where problems developed over congestion at the feeding hoppers. As the population grew from 60 to 80, behavior patterns developed into what Calhoun called a "behavioral sink"—gross distortions of behavior resulting from animal crowding. Behavior remained fairly normal in the two end pens, where each dominant male defended his territory by sleeping at the end of the ramp, but in the two middle pens, there were severe changes in sexual, nesting, and territorial behavior. Some of the males became sexually passive; others became sexually hyperactive, chasing females unmercifully; and still another group of males was observed mounting other males, as well as females. Females became disorganized in their nesting habits, building very poor nests, getting litters mixed up, and losing track of their young. Infant mortality rose significantly. Finally, males appeared to alter their concept of territoriality. With no space to defend, the males in the two middle pens substituted time for territory, and three times a day, the males fought at the eating bin.

Calhoun's study can be summarized by noting that, among his rats, crowding (an increase in the number of rats within a fixed amount of space) led to the disruption of important social functions and to social disorganization. Related to these changes in social behavior were signs of physiological stress, such as changes in their hormonal systems that made it difficult for females to bring pregnancies to term and care for their young. Other studies have shown that not only rats but also monkeys, hares, fish, shrews, elephants, and house mice tend to respond to higher density by reducing their fertility (Galle, Gove, and McPherson 1972).

Although the severe distortions of behavior that Calhoun witnessed among rats have never been replicated among humans, research has suggested that at the macro (group) level, there may be some fairly predictable consequences of increasing population density (mainly as a result of increases in population size). For example, Mayhew and Levinger have argued that violent interaction can be expected to increase as population size increases; "the opportunity structure for murder, robbery, and aggravated assault increases at an increasing rate with aggregate size" (Mayhew and Levinger 1976:98). There are more people with whom to have conflict, and an increasingly small proportion of people over whom we exercise direct social control (which would lessen the likelihood of conflict leading to violence). Increasing size leads to greater superficiality and to more transitory human interaction, that is, greater anonymity. Mayhew and Levinger point out that "since humans are by nature finite organisms with a finite amount of time to devote to the total stream of incoming signals, it is necessarily the case that the average amount of time they can devote to the increasing volume of contacts . . . is a decreasing function of aggregate size. This will occur by chance alone" (1976:100).

Because no person has the time to develop deeply personal relationships (primary relations) with more than a few people, the more people there are entering a person's life, the smaller the proportion one can deal with in depth. This leads to the appearance that people in cities are more estranged from each other than in rural settings, but Fischer (1981) offers evidence that in all settings, people are distrustful of strangers—we just encounter more of them in the city. This may lead to personal stress as people try to sort out the vast array of human contacts, since the more people there are, the greater the variety of both expectations others have of you and obligations you have toward others. The problems of not enough time to go around and of contradictory expectations lead to "role strain"—a perceived difficulty in fulfilling role obligations. Most of these problems of size seem to arise naturally in large metropolitan areas and, indeed, the problems are most intense there.

A study of rural villages in India has suggested that even in a rural area, high density may dampen fertility levels (Firebaugh 1982)—a finding consistent with studies among other animals, but a study in Chicago suggested that the effects of density on fertility are trivial (Loftin and Ward 1983). An extensive survey of the effects of crowding in Bangkok, which is far more crowded than any North American city, suggested that crowding, although not necessarily a good thing, had little effect on sexual behavior or on health in general (Edwards et al. 1994). On a more positive note, data from the General Social Survey of the National Opinion Research Center have been used to suggest that moving to a big city increases your tolerance for other human beings, rather than the other way around (Wilson 1991).

It might be fair to say that the most reasonable reaction to the negative aspects of city life is to figure out how best to cope with them and minimize them. The most common response has been to get as far away from high-density city life as possible, while also staying close enough to it so that you can participate in the benefits of city life. This is probably the underlying reason for the evolution in urban places I have been discussing in this chapter. Most people do not really prefer to be packed together with others, and the increasing sophistication of transportation and information technologies has made it increasingly easy to spread out.

The problem is that not everybody can spread out in a world where human existence is increasingly tied to the city. Thus, there has been a renewed emphasis in the literature on policies and plans to restructure and "humanize" the city environment in order to improve the quality of everyday life for city residents (Ford 2000). Many of these ideas are designed to counteract the negative effects of urban sprawl, which I discuss in the essay that accompanies this chapter. Another related and very important part of the city environment is the environmental impact of cities—are cities sustainable as places for humans to live?

Cities as Sustainable Environments

"Poor countries' cities are bursting at the seams, yet rural migrants are coming in faster than ever; a social and environmental meltdown is waiting to happen" (*The Economist* 1996:44). This is how one writer described the rationale for the United Nations Conference on Human Settlements (Habitat II) held in Istanbul in 1996 (Habitat I had been held in 1976 in Vancouver). The opportunities that cities offer to rural peasants in less-developed nations may seem meager to those of us raised in a wealthier society. Most third-world cities have long since outgrown their infrastructure and, as a result, drinkable water may be scarce, sewage is probably not properly disposed of, housing is hard to find, transportation is inadequate, and electricity may be only sporadically available (Montgomery et al. 2003). This is not a pretty picture, but it is still an improvement on the average rural village! Thus, "despite these problems, the flood of migrants to the cities continues apace. Why? The answer lies in the natural population increase in rural areas, limited rural economic development, and the decision-making calculus of urban migrants. . . . What this all means, of course, is that the primary cause of what some have termed 'overurbanization' (more urban residents than the economies of cities can sustain) is increasingly severe 'overruralization' (more rural residents than the economies of rural areas can sustain)" (Dogan and Kasarda 1988:19).

The most obvious symbols of crowding and overurbanization in cities are the squatter settlements that line the outer ring of almost every large city in developing countries. They may be called *colonias* in Mexico or *favelas* in Brazil, but the meaning is the same—slums in which people live in small dirt-floored huts with no running water, electricity, or sewage disposal systems. This is typically a result of cities growing faster than governments can afford to build an urban infrastructure. In other places, such as Mumbai (formerly Bombay), the government may also be to blame for the inner-city slums, but for a different reason. It has been reported that "a decent-sized three-bedroom flat in southern Bombay costs $2 million. Meanwhile, two thirds of the city's 10 million residents live either in one-room shacks or on the pavement" (*The Economist* 1995b). Government regulations controlling rent and land use have contributed to this situation, and, of course, this has encouraged population growth in the suburbs without relieving the inner city crowding.

Two decades ago, Mexico City was the second most populous city in the world and many people projected it to continue growing almost forever. However, the serious environmental problems in the Valley of Mexico created by population growth convinced the Mexican government to undertake a concerted effort to move industry out of the area and divert migrants to other metropolitan areas. The effort clearly paid off, because the 1990 census counted fewer people than anticipated and, on that basis, the United Nations demographers revised their projections of population growth in Mexico City downward. Still, the 18.1 million people living in Mexico City in 2000 was a huge increase from the 3.1 million in 1950, and the United Nations projects that the population of Mexico City will top 20 million by the year 2015. Terrible smog, a dwindling water supply, and an increase in crime rates are all features of modern Mexico City—yet there is more opportunity for the average rural peasant of Mexico than there is in the countryside, so people continue to move there.

Evaluations of the quality of life in the largest cities of the world reveal, not surprisingly, that the metropolitan areas of the richer countries are better places to live. A study done in 2002 by a London-based human resources firm that places people in jobs all over the world concluded that Zurich, Switzerland, was the best city in the world in which to live. Overall, their assessment was that cities in Europe, Australia, and New Zealand offered the highest quality of life, followed by cities in North America. Not a single one of their top 50 cities was in a developing country.

An older, but very comprehensive, survey by Population Action International (1990) assembled data on crime, health, income, education, infrastructure, traffic, and related indexes of the quality of urban life to produce an "urban living standards score." Of 21 cities ranked as "very good," the list was led by Melbourne, Montreal, and Seattle-Tacoma. Only three of the 23 cities categorized as "good" were in developing countries. On the other hand, 25 of 26 cities rated as "fair," and all 28 cities listed as "poor" were in the third world. The two lowest-rated of the 100 largest cities were both in sub-Saharan Africa—Kinshasa, the capital of Congo (formerly Zaire), and Lagos, the capital of Nigeria. Indeed, the Second African Population Conference in 1984 concluded that African cities are "plagued by the adverse effects of rapid growth—urban sprawl, unemployment, delinquency, inadequate social services, traffic congestion, and poor housing" (Goliber 1985:17). A major reason for this situation is that urban areas in Africa have consistently been growing at more rapid rates than urban areas anywhere else in the world, as you can see by referring back to Table 11.1.

I keep coming back to the same theme, however, that, as dismal as urban life is in developing countries, it is a step up from rural areas. The United Nations estimates that in the 1990s, 96 percent of urban residents in developing countries had access to health care, compared with only 76 percent of rural residents; 87 percent of urban residents had access to water, compared to only 60 percent in rural areas; and 72 percent of urban residents had access to sanitation services compared with only 20 percent living in the countryside (United Nations Development Programme 1997). Cities are where economic development is occurring in the world, and it is where infrastructure and housing will continue to be built. The demand for housing itself is a function of the relationship between the population of young adults wanting to form their own family household and the number of people who are dying and thus presumably freeing up existing housing. The high rates of growth and attendant youthful age structures of the developing countries tell us that the worldwide demand for housing will be predominantly in the cities of less-developed nations. Along with that housing will be needed infrastructure improvements—water, sewers, electricity, roads, and public transportation, all of which make urban life possible. Between the beginning and the middle of the twenty-first century, the world will likely need to shelter an additional two billion families—the majority of them in third-world cities (Weeks and Fuller 1996). This is a daunting task, considering that the volume of housing needed is unprecedented in world history and, of course, we have not come even close to adequately sheltering the current generation of people. Although it is contrary to the usual perspective, let me suggest that the benefit of these people moving into the city is that they bring their poverty with them to a place where it is exposed to public view and, as a consequence, is more likely to be acted upon by governments and NGOs (non-governmental organizations). On the other hand, improving life for billions of people living in cities will use a tremendous amount of resources. A related and very important cost of the urban transition is that, as we gather ourselves into cities, we lose perspective of where our resources come from, and where our waste goes. It seems like magic on both ends, but of course it is not. We must learn how to deal successfully with the earth's limited resources, because there are no viable alternatives to the urban transition.

Summary and Conclusion

The urban transition describes the process whereby a society shifts from being largely bound to the country to being bound by the city. It is a process that historically has been the close companion of economic development, which, of itself, suggests the close theoretical connection of urbanization with demographic processes. Nearly half of the human population now resides in urban places, and that percentage is climbing steadily. Although rural-to-urban migration is a major contributor to the urban transition, mortality and fertility are importantly associated as well, as both cause and consequence. Population pressures created by declining mortality in rural areas, combined with the economic opportunities offered by cities, have been historically linked to urbanization. On the other hand, mortality is now almost always lower in cities than in rural areas, which permits higher rates of urban natural increase than in the past.

The development of industrialized countries is replete with examples of how urban life helps generate or ignite the first two of Ansley Coale's three preconditions for a fertility decline—the acceptance of calculated choice as an element in personal family size decisions and the perception of advantages to small families (see Chapter 6). Nonetheless, in the cities of less-developed nations, urban fertility, which is lower than rural fertility, is still typically much higher than in cities of the developed world. As a result, cities in less-developed nations are the most rapidly growing places on earth.

The urban transition is associated with increasing differentiation among cities, leading to identifiable urban hierarchies throughout the world. Every country has its own rank-ordering of cities, but there is also a world ranking, leading to the designation of some cities as being world or global cities. These are generally thought of as being core cities; the other peripheral cities depend on the core for their place in the global economic scheme of things.

As richer countries reach the point where the vast majority of people are already residing in urban places (which we might define as the end of the urban transition), a new set of urban evolution processes takes place. This is exemplified most visibly by suburbanization, which is now a worldwide phenomenon, but other trends such as exurbanization, development in the peri-urban region, counterurbanization, and gentrification are all part of the process, as well.

The most readily identifiable beneficial consequences of living in urban places are higher standards of living compared with rural areas. The detrimental consequences that lead to lower quality of life in some of the world's largest cities seem primarily to be the result of the economic inability of cities to adequately absorb the tremendous numbers of newcomers. Because urban residents have higher incomes, they also use more resources without being regularly reminded of where those resources came from or where their waste products go. It is dangerous to forget our link to nature, because there is a very real possibility that we have already overshot our capacity to sustain life for everybody in the world at a level even approaching that of the average urban resident of the United States or Canada. In the next chapter, I discuss the relationship between population growth and the environment.

Main Points

1. The world is at the point at which one of every two people lives in an urban area; by the middle of this century, nearly two out three will be there.

2. Urbanization refers to the increase in the proportion of people living in urban places.

3. The urban transition reflects the process whereby population growth in rural areas (fueled by death control invented in cities) forces people to find jobs elsewhere—in the cities, contributing to an increase in the number and percentage of people living in urban places.

4. One of the most striking features of urbanization is its recency in world history. Highly urban nations like England and the United States were almost entirely agricultural at the beginning of the nineteenth century.

5. Cities are related to the development process because they are centers of economic efficiency.

6. Urbanization can occur as a result of internal rural-to-urban migration, higher rates of natural increase in urban than in rural areas, international urban migration, and reclassification of places from rural to urban.

7. Until the twentieth century, death rates in cities were so high and fertility was low enough that cities could not have grown had it not been for migration of people from the countryside.

8. In virtually every society, fertility levels are lower in cities than in rural areas.

9. Despite lower fertility in cities, cities in less-developed nations almost always have higher fertility levels than cities in developed nations.

10. Eighty percent of the U.S. population resides within the boundaries of the nation's metropolitan areas.

11. Tokyo is the largest urban agglomeration in the world, followed in size by São Paulo, Mexico City, and Bombay (Mumbai).

12. Cities throughout the world are arranged in urban hierarchies that are often described in terms of the core and periphery model.

13. The urban transition is morphing into the urban evolution, as urban places become increasingly spread out and complex.

14. Urbanization in the United States has now turned into suburbanization, with most Americans living in suburbs—a trend that is spreading to the rest of the world.

15. Suburbanization is increasingly associated with urban sprawl.

16. The suburbanization process has heightened residential segregation in the United States, just as it has, to a lesser extent, in other developed nations.

17. Population growth in cities has given rise to fears about the potential harmful effects of crowding, and especially to concerns about the environmental sustainability of cities.

Suggested Readings

1. Martin Brockerhoff, 2000, "An Urbanizing World," Population Bulletin 55(3).

   An excellent overview of the urban transition, focusing on the world as a whole, with discussion of regional variability in urban processes.

2. Mark R. Montgomery, Richard Stren, Barney Cohen, and Holly Reed, Editors, 2003, Cities Transformed: Demographic Change and Its Implication in the Developing World (Washington, DC: National Academy Press).

   Developed by the Committee on Urban Population Dynamics of the National Research Council, this book provides an excellent overview of virtually all topics covered in this chapter, with an emphasis on cities in developing countries.

3. Tony Champion and Graeme Hugo, Editors, 2004, New Forms of Urbanization: Beyond the Urban-Rural Dichotomy (London: Ashgate Publishing).

   The chapters in this book are based on papers prepared for a conference in Bellagio, Italy, organized by the International Union for the Scientific Study of Population (IUSSP) and represent state-of-the-art developments in the study of the urban transition.

4. Saskia Sassen, 2001, The Global City: New York, London, Tokyo, 2nd Edition (Princeton, NJ: Princeton University Press).

   Sassen is not studying the urban transition per se, but her books have been very influential in shaping the way people think about city systems, and about the role cities play in the broader structure of society.

5. Reynolds Farley, Sheldon Danziger, and Harry J. Holzer, 2000, Detroit Divided (a volume in the Multi-City Study of Urban Inequality) (New York: Russell Sage Foundation).

   Farley has been a pioneer in research on residential segregation and this volume takes a new close look at one of America's most segregated cities.

⊛ Web Sites of Interest

Remember that Web sites are not as permanent as books and journals, so I cannot guarantee that each of the following Web sites still exists at the moment you are reading this. It is usually possible, however, to retrieve the contents of old Web sites at http://www.archive.org.

1. http://www.unchs.org/
   UN-Habitat, more formally the United Nations Human Settlements Programme, monitors the cities of the world, with a focus on those in developing countries.

2. http://www.worldbank.org/html/schools/issues/urban.htm
   The World Bank has created this resource for studying urbanization, including links to its latest data on cities and urban growth.

3. http://www.smartgrowth.org
   Do you want to help slow down urban sprawl? Learn more about smart growth at this Web site.

4. http://www.un.org/esa/population/unpop.htm
   This Web site has information on the United Nations Population Division's latest estimates and projections of the world's urban and rural populations.

5. http://www.iied.org/human
   The International Institute for Environment and Development, based in London, has a Human Settlements program that has been very influential in the area of environment and health in cities.

PART FOUR
Using the Demographic Perspective

CHAPTER 12
Population and the Environment

Figure 12.1 Per Capita Purchasing Power Parity (PPP) by Country

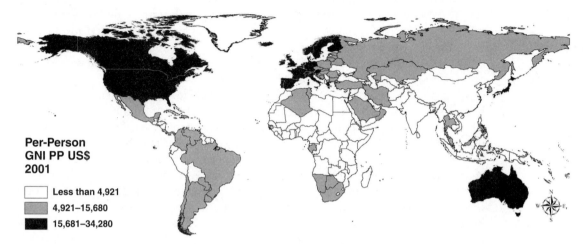

Per-Person GNI PP US$ 2001

- Less than 4,921
- 4,921–15,680
- 15,681–34,280

Source: Adapted by the author from data courtesy of the Populatioin Reference Bureau, 2003 World Population Data Sheet.

It is elementary, my dear Watsons: Humans cannot survive without food and water. Those favored few of us in the world who can rely on water from the tap and groceries from the supermarket deal with this principle pretty much on a theoretical level. We know intellectually that some areas of the world have regularly been faced with the prospect of famine and drought. We also know that more than 200 years ago Malthus was already stewing about population growth outstripping the food supply. Although it is certainly a shame that all people cannot find a seat on the gravy train, the fact is that Malthus was wrong. Right? After all, food production has actually outpaced population growth over the past 200 years. It is a fact, and there are **boomsters** who believe that population growth stimulates economic development and that the food record speaks for itself—we can grow it as we need it (in this context the boomsters are known as **cornucopians**). The logical extension of this perspective is the idea that somehow we will be able to find the magical formula whereby everybody is better off in the future and we can all live happily ever after—that is the promise of the obviously popular concept of **sustainable development.**

However, look a little closer—the picture is less rosy, even for people fortunate enough to live in wealthier nations. The clues increasingly point to the grim reality that we will all be paying a very heavy price for coaxing ever higher yields from our increasingly overburdened planet. In trying not only to feed but also to improve the lives of an ever larger population, we are polluting the land, affecting our global climate, and using up our supply of fresh water. The plot of our mystery has taken a turn. Maybe Malthus and the **doomsters** are right. Although the formula for ultimate disaster was more complicated than he knew, critical resources such as land and water are finite. At some point, we may exhaust the earth's capacity to produce—then everybody loses.

In the United States, these issues are often seen as juxtaposing the views of the doomsters with those of the boomsters. Doomsters are the neo-Malthusians, exemplified most famously by Paul Ehrlich, who has argued for decades that continued population growth will lead to certain economic and environmental collapse in a worldwide tragedy of the commons (see, for example, Ehrlich 1968; Ehrlich and Ehrlich 1972; Ehrlich and Ehrlich 1990; Ehrlich, Ehrlich, and Daily 1997). For Ehrlich, the policy choices have always been clear—population control must be a part of any development strategy, or that strategy will fail.

The boomsters have been most famously influenced by the late Julian Simon, who argued for decades that population growth stimulates development, rather than slowing it down (see Simon 1981; Simon 1992). For Simon, the policy choices were also clear—development strategies should not deliberately slow down population growth because such growth is both a cause and a symptom of economic development. The boomster view recognizes that population cannot grow indefinitely, but argues that people will lower their fertility when they see an advantage from doing so, which means lifting them out of poverty through free trade and globalization (World Bank 2000). These have been the central ingredients in the idea of sustainable development, which I will discuss in greater detail later in the chapter.

Outside of the United States a related perspective is often put forth—a neo-Marxian view that population growth has nothing to do with economic development at all. Economic development, where it lags, is held back by the injustice of the world system that creates dependency by the periphery on the dominant core countries. Like the boomster view, this perspective recognizes that population cannot grow indefinitely, and similarly argues that people will lower their fertility when they see an advantage from doing so, which means lifting them out of poverty. The difference (and it is a huge one) is that the neo-Marxian policy perspective is that this should be accomplished by dismantling multinational corporations and putting the money into the hands of local populations, where it can be distributed equitably to relieve poverty and improve the human condition.

Each of these perspectives—boomster, doomster, neo-Marxian—carries with it a very different set of policy prescriptions and a different range of environmental consequences, so it matters which one is right. I will not leave you in suspense on this—each of these perspectives has some merit. To understand this, though, we need some definitions.

Economic Development—The Use and Abuse of Resources

Economic development represents a growth in average income, usually defined as **per capita (per person) income.** A closely related idea is that economic development occurs when the output per worker is increasing; since more output should lead to higher incomes, you can appreciate that they are really two sides of the same coin. Of course, if you are holding down two jobs this year just to keep afloat financially, you know that producing more will not necessarily improve your economic situation. Rather, it may only keep it from getting worse. Thus, a more meaningful definition of economic development refers to a rise in real income—an increase in the amount of goods and services you can actually buy.

An important aspect of development is that it is concerned with improving the welfare of human beings. It includes more than just increased productivity; it includes the resulting rise in the ability of people to consume (either buy or have available to them) the things they need to improve their level of living. Included in the list of improvements might be higher income, stable employment, more education, better health, consumption of more and healthier food, better housing, and increased public services such as water, sewerage, power, transportation, entertainment, and police and fire protection. Naturally, these improvements in human welfare help increase economic productivity because the relationship is synergistic.

The starting point of economic development is the investment of financial capital. Capital represents a stock of goods used for the production of other things rather than for immediate enjoyment. Although capital may be money spent on heavy machinery or on an assembly line or on infrastructure such as highways and telecommunication, it can better be thought of as anything we invest today to yield income tomorrow (Spengler 1974). This means not only equipment and construction, but also investments in human capital—education, health, and, in general, the accumulation and application of knowledge. For an economy to grow, the level of capital investment must grow. Clearly, the higher the rate of population growth, the higher the rate of investment must be; this is what Leibenstein (1957) called the "population hurdle." If a population is growing so fast that it overreaches the rate of investment, then it will be stuck in a vicious Malthusian cycle of poverty; the economic growth will have been enough to feed more mouths, but not enough to escape from poverty.

Crucial to our understanding of economic development is that this increase in well-being typically requires that we use more of the earth's resources—especially more energy and more water, not to mention more minerals and timber. How efficiently we are able to use these resources influences how widely they can be spread out among the entire (and growing) population. The use of every resource leads to waste products, and our efficiency in reducing waste and dealing with it effectively influences the extent to which we can minimize damage to the environment and thus sustain a larger population, or sustain a smaller population at a higher level of living.

Economic Development Compared to Economic Growth

Economic growth refers to an increase in the total amount of productivity or income in a nation (or whatever your geographic unit of analysis might be) without regard to the total number of people, whereas economic development relates that amount of income to the number of people. But how do we measure that income? The most commonly used index of a nation's income is the gross national income, or GNI (the term now preferred by the World Bank), which is also more famously known as gross national product, or GNP. The World Bank somewhat obscurely defines this as "the sum of the gross value added by all resident producers in the economy plus any product taxes and minus any subsidies not included in the value of the products" (World Bank 2003a). Basically, if you add up the value of all of the paid work that goes on in a country, and then add in the money received from other countries, you have the measure of gross national product. If you exclude the money from abroad and just include the income generated within a country's own geographic boundaries, you have **gross domestic product** (GDP).

Measuring GNI and Purchasing Power Parity

Gross national income is the most widely used measure of economic well-being in the world, but it is important to keep in mind the things that GNI does not measure: (1) It does not take into account the depletion and degradation of natural resources; (2) It does not make any deduction for depreciation of manufactured assets such as infrastructure; (3) It does not measure the value of unpaid domestic labor such as that generated especially by women in developing nations; and (4) It does not account adequately for regional or national differences in purchasing power.

This latter limitation is one that the World Bank has been particularly interested in. Although GNI figures are usually expressed in terms of U.S. dollars, a dollar may go farther in Ghana than it will in England, even when exchange rates have been taken into account. The United Nations and the World Bank have sponsored a number of household expenditure surveys in developing countries to try to estimate actual differences in the standard of living, in order to produce more meaningful income comparisons. The wealthier nations have also been encouraged to conduct such surveys, along the lines of the U.S. Bureau of Labor Statistics' Consumer Expenditure Survey. The product of these efforts is a measure called **purchasing power parity** (PPP), defined as "the number of units of a country's currency required to buy the same amount of goods and services in the domestic market as one dollar would buy in the United States" (World Bank 1994b:246). One way of expressing this concept is through the use of what *The Economist* calls its "Big Mac Index." McDonald's sells its hamburgers in nearly 120 countries, and in each country, the sandwich must conform to essentially the same standards of ingredients and preparation. If the Big Mac costs $2.71 in the United States, then it should cost the same in real terms anywhere else in the world. So, if you go to Santiago, Chile, and discover that you're paying 1,400 pesos for a Big Mac, that should tell you that there are 1,400 / 2.71 = 516.6 Chilean pesos per U.S. dollar, in terms of the "real" cost of living (Ong 2003). By expressing gross national income in terms of purchasing power parity (rather than official exchange rates), the result is the **gross national income in PPP (GNI PPP).**

In 2001, the gross national income in the United States was $9.8 trillion, as you can see in Figure 12.2 which shows the GNI PPP for the world's 20 largest economies. The U.S. economy is so big that it accounts for 22 percent of the world's entire income, despite the fact that the U.S. population is less than 5 percent of the total. The other two most populous countries of the world, China and India, are also in the top five of economies, and those top five (which also include Japan and

Figure 12.2   World's 20 Largest Economies, Based on Gross National Income Measured in Purchasing Power Parity, 2001

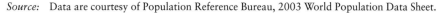

Source: Data are courtesy of Population Reference Bureau, 2003 World Population Data Sheet.

Germany) represent more than 50 percent of the entire economic product of the world. The 20 countries shown in Figure 12.2 account for 82 percent of the world's economic output, but only 64 percent of the world's total population. However, if we take the U.S. off the list, the remaining 19 countries account for 60 percent of the world's economy and 59 percent of the world's population.

If we divide total national income by the number of people, we obtain a measure of per capita income, which gives a sense of the relative well-being of people in one country compared to another. In Table 12.1, I have listed the per-person income for selected countries according to gross national income based on purchasing power parity. China may have one of the largest economies in the world, but its huge population means that its per-person income is only a tiny fraction of that in the United States. On a per capita basis, the Chinese have only 12 percent of the average American's income—or, put another way, the average American is eight times better off than the average person in China, and 12 times better off than the average person in India. All but one of the countries at the bottom of the income ladder are in sub-Saharan Africa and you can see that the average person in those countries has only about 2 percent of the of the average American's income.

Where does the income come from that goes into these measures? An important part of it comes from the transformation of natural resources into things that are more useful to us—converting a tree into a house and furniture; converting the "fruit" of cotton plants into shirts and dresses; converting minerals found in rocks into the steel body of an automobile; transforming a hidden pool of underground oil into fuel used by machines. Then, these products must be packaged, delivered, and sold; people have to coordinate all of that and make sure that the infrastructure exists to do everything that needs to be done. We can divide the resources that go into producing income into two broad categories—**natural resources** (what is given to us on the planet), and **human resources** (how clever and successful we are in making something of those natural resources). Together, this combination of resources can be thought of as the **wealth of a nation** (Dixon and Hamilton 1996). Measuring these things is not easy, as you can imagine, although researchers at the World Bank have worked on it (applying their human resources to this issue, as it were) and their results suggest, for example, that Canada has a higher per-person level of wealth than the United States, based on the natural resources in Canada, and that the value of human resources in the United States exceeds the natural resources wealth of the country (World Bank 1995). More importantly, their analysis suggests that natural capital is distributed fairly evenly around the inhabitable portions of the globe, so the variable factors in global economic well-being tend to be the level of human resources in one area compared with another, and the number of humans in one area compared with another among whom these resources are shared. Either way, demography seems to be playing a role.

How Is Population Related to Economic Development?

There is a nearly indisputable, albeit somewhat complex, statistical association between economic development and population growth; that is, when one changes, the other also tends to change. As you no doubt already know, though, two things may

Table 12.1   Top 10 and Bottom 10 Countries and Other Selected Countries in Terms of GNI PPP ($US) 2001

Country Top 10 Countries	Per Person GNI PP ($US) 2001	Ratio to United States
United States	34280	1.00
Switzerland	30970	0.90
Norway	29340	0.86
Netherlands	27390	0.80
Ireland	27170	0.79
Canada	26530	0.77
Austria	26380	0.77
Belgium	26150	0.76
Japan	25550	0.75
Germany	25240	0.74
Other Selected Countries		
United Kingdom	24340	0.71
France	24080	0.70
Chile	8840	0.26
Mexico	8240	0.24
Russia	6880	0.20
China	3950	0.12
Indonesia	2830	0.08
India	2820	0.08
Bottom 10 Countries		
Ethiopia	800	0.02
Nigeria	790	0.02
Mali	770	0.02
Zambia	750	0.02
Yemen	730	0.02
Burundi	680	0.02
Congo	680	0.02
Congo, DRC	630	0.02
Tanzania	520	0.02
Sierra Leone	460	0.01

Source: Data are adapted from John L. Allen and Elizabeth J. Leppman, 2004, Student Atlas of World Politics, 6th Edition (New York: McGraw-Hill/Dushkin), Table E.

be related to each other without one causing the other. Furthermore, the patterns of cause and effect can conceivably change over time. Does population growth promote economic development? Are population growth and economic development only coincidentally associated with each other? Or is population growth a hindrance to economic development?

The problem is that the data presently available lend themselves to a variety of interpretations. If we look at the global pattern of per-person income (measured with per capita GNI based on PPP) you can see in Figure 12.1 at the beginning of this chapter that the poorest countries are in sub-Saharan Africa and South Asia, whereas the richest countries are the European and "overseas" European countries (the United States, Canada, Australia, and New Zealand), as well as Japan. The rest of the world generally falls in between, and includes most of Latin American, northern Africa, western Asia, and eastern Europe.

There is a high correlation between being a high-income country by World Bank standards and being a "core" country as defined by **world systems** analysts. "Core nations are linked to other core countries and to semiperipheral and peripheral nations; peripheral nations are linked to core nations but not to each other, and semiperipheral nations are linked to core countries and other semiperipheral nations, but are only weakly linked to peripheral nations" (Bollen and Appold 1993:286–87). The upper-middle-income countries tend to be the "semiperipheral" nations and the lower-middle- and low-income nations tend to be the "peripheral" countries.

The geographic pattern shown in Figure 12.1 is similar to maps showing the components of the demographic transition. Those places in the world where incomes are highest also tend to have the lowest levels of fertility, the lowest mortality levels, the lowest overall rates of population growth, the oldest age structures, the highest rates of immigration, and the highest levels of urbanization. The picture is fairly compelling, buttressed by the data in Figure 12.3. In that graph, I have compiled data for the 140 countries for which the World Bank provides estimates of the per-person income measured in terms of GNI PPP—the same data that were used for Table 12.1. The income data refer to 2001, while the population growth data represent the average annual rate of growth during the previous 20-year period. You can see that in general, countries in which average income levels are low tend to be clustered around the high end of population growth rates. I have partitioned the graph into four parts. The dashed horizontal line represents the average per-person income for all countries ($7,986) and the dashed vertical line represents the average rate of population growth between 1981 and 2001 (1.73 percent). Thus Jordan is a nation that is below the world average in terms of income but well above the world average with respect to population growth. Joining Jordan in that quadrant are countries like India and most of the non-oil-producing nations of Africa, Asia, and Latin America.

At the other extreme are the wealthier countries with low rates of population growth, exemplified by the United States, Japan, Switzerland, and the other countries of western and northern Europe. The countries that do not fit the expected pattern include the wealthy oil-producing Middle Eastern nations such as Saudi Arabia and the United Arab Emirates, where fertility rates have only recently begun to decline. Also off the track are those countries that have not been able to translate low rates of population growth into higher standards of living. Most of these countries, like Bulgaria, are just emerging from decades of state socialism.

Clearly, a low rate of population growth is no assurance of a high income, but that point is obvious, since for most of human history the rate of population growth was low and so was the overall standard of living. The data in Figure 12.3 also show

Figure 12.3   Countries with High Per Capita Income Tend to Have Low Rates of Population Growth

Per capita GNI at purchasing power parity, 2001 (in $US)

Note: There is a clear tendency for countries with high per capita gross national product to have had low rates of population growth during the previous 20 years. The horizontal axis represents the average income for all countries, and the vertical axis indicates the average rate of population growth for all countries.

Source: Income data are courtesy of the Population Reference Bureau, 2003 World Population Data Sheet, and rates of population growth are from the United Nations World Population Prospects: The 2002 Revision.

that, at least in the short run, countries with sufficient resources (especially oil) can achieve high levels of income even in the face of rapid population growth. Nonetheless, the data also exhibit a reasonably strong negative relationship as evidenced by the correlation coefficient *(r)* of -.53. Mexico, for example, has been sliding down the slope toward a combination of lower rates of population growth and higher incomes in spite of the occasional setbacks experienced by the Mexican economy. Furthermore, China has been taking economic advantage of the "age dividend" brought about by its slowdown in population growth over the past three decades, as I discussed in Chapter 8.

Despite the intuitive appeal of the negative association shown in Figure 12.3 between population growth and levels of national income, it is not a perfect association, and it does not allow us to establish a direct cause-and-effect relationship.

First, let us consider the possibility that population growth might itself be a stimulus to economic development.

Is Population Growth a Stimulus to Economic Development?

An early proponent of the idea that population growth could be the trigger of economic development was the Danish economist Ester Boserup. In a set of extremely influential writings (see, especially, 1965; 1981), she advanced the idea that, in the long run, a growing population is more likely than either a nongrowing or a declining population to lead to economic development. The history of Europe shows that the Industrial Revolution and the increase in agricultural production were accompanied almost universally by population growth. Boserup's argument (also advanced by Clark 1967) is based on the thesis that population growth is the motivating force that brings about the clearing of uncultivated land, the draining of swamps, and the development of new crops, fertilizers, and irrigation techniques, all of which are linked to revolutions in agriculture. The kernel of the argument has been well stated by British agricultural economist Colin Clark:

> [Population growth] is the only force powerful enough to make such communities change their methods, and in the long run transforms them into much more advanced and productive societies. The world has immense physical resources for agriculture and mineral production still unused. In industrial communities, the beneficial economic effects of large and expanding markets are abundantly clear. The principal problems created by population growth are not those of poverty, but of exceptionally rapid increase of wealth in certain favoured regions of growing population, their attraction of further population by migration, and the unmanageable spread of their cities (Clark 1967:Preface).

This same line of reasoning is part of a strategy of development forwarded by Hirschman (1958), who argued as follows: (1) An increase in population size will lower a population's standard of living unless people reorganize their lives to increase production; (2) It is "a fundamental psychological postulate" that people will resist a lowering in their standard of living; and (3) ". . . the activity undertaken by the community in resisting a decline in its standard of living causes an increase in its ability to control its environment and to organize itself for development. As a result, the community will now be able to exploit the opportunities for economic growth that existed previously but were left unutilized" (1958:177).

The thesis that population growth is beneficial to economic development does have some foundation in fact. In Europe and the United States, development may well have been stimulated by population increase. Indeed, some historians regard preindustrial declines in death rates in Europe, associated partly with the disappearance of the plague (perhaps also with the introduction of the potato), as the spark that set off the Industrial Revolution. The reasoning goes that the lowered death rates created a rise in the rate of population growth, which then created a demand for more resources (see Clark 1967 for a review). An analogous example of population growth influencing development is the case of the American railroad, which opened up the frontier and hastened resource development in the United States. Fishlow (1965) demonstrated that the railroad (which helped accelerate the economic development of the western states) was actually following people westward rather than the other way around.

Although history may show that population growth was good for development in the now highly industrialized nations, statistics also reveal very important differences between the European-American experience and that of modern, less-developed nations. The less-developed countries today are not, in general, retracing the steps of the currently developed nations. For example, less-developed nations are building from a base of much lower levels of living than those that prevailed in either Europe or the United States in the early phases of economic development (Kuznets 1965; Boserup 1981). Furthermore, although the rate of economic growth in many underdeveloped countries has recently been higher than at comparable periods in the history of developed nations, population growth is also significantly higher. Underdeveloped countries have experienced higher rates of population growth than European or American countries ever did, with the possible exception of America's colonial period. In fact, over the past half century the rates of population growth in the underdeveloped world have been virtually unparalleled in human history.

It appears that population growth may have helped stimulate economic growth in developed countries "by forcing men out of their natural torpor and inducing innovation and technical change, or by speeding up the replacement of the labour force with better educated labour" (Ohlin 1976:9). The less-developed nations of today, however, do not seem to require any kind of internal stimulation to be innovative. They can see in the world around them the fruits of economic development, and quite naturally they want to share in as many of those goodies as possible—a situation often referred to as "the revolution of rising expectations." People in less-developed nations today know what economic development is, and by studying the history of the highly industrialized nations, they can see at least how it used to be achievable.

If it was ever true that more people meant a greater chance of producing the genius who would solve the world's problems, it is a difficult argument to sustain today. As Nathan Keyfitz has pointed out, "the England that produced Shakespeare and shortly after that Newton held in all 5 million people, and probably not more than one million of these could read or write. . . . The thought that with more people there will be more talent for politics, for administration, for enterprise, for technological advance, is best dismissed. . . . For the most part, innovation comes from those who are comfortably located and have plenty of resources at their disposal" (quoted in United Nations Population Fund 1987:16).

It seems unlikely that a spark such as population growth is necessary any longer to stimulate economic development, but nonetheless in the 1980s, Julian Simon popularized his thesis that a growing human population is the "ultimate resource" in the search for economic improvement. Eschewing the Malthusian idea that resources are finite, Simon suggested that resources are limited only by our ability to invent them and that, in essence, such inventiveness increases in proportion to the number of brains trying to solve problems. Coal replaced wood as a source of energy only to be replaced by oil, which may ultimately be replaced by solar energy—if we can figure out how to do it properly. From Simon's vantage point, innovation goes hand in hand with population growth, although he was quick to point out that moderate, rather than fast (or very slow) population growth is most conducive to an improvement in human welfare (Simon 1981).

Simon, however, made a crucial assumption: To be beneficial, population growth must occur in an environment in which people are free to be expressive and creative. To him, that meant a free market or capitalist system. Ironically, the idea that the important element in economic development is the type of economic system that prevails draws him into the same perspective on population growth and economic development as that shared by Marxists and **neo-Marxists**.

Is Population Growth Unrelated to Economic Development?

In his later writings, Simon moved toward the position that population growth is far less important an issue in economic development than is the marketplace itself. He suggested that "the key factor in a country's economic development is its economic and political system. . . . Misplaced attention to population growth has resulted in disastrously unsound economic advice being given to developing nations" (Simon 1992:xiii). With respect to the relationship between population growth and economic development, this is not far from the usual Marxist view that population problems will disappear when other problems are solved, and that economic development can occur readily in a socialist society. Marx (and Engels) believed that each country at each historical period has its own law of population, and that economic development is related to the political-economic structure of society, not at all to population growth. Indeed, Marx seemed to be arguing that whether population grew as a nation advanced economically was due to the nature of social organization. In an exploitive capitalist society, the government would encourage population increase to keep wages low, whereas in a socialist state there would be no such encouragement. Socialists argue that every member of society is born with the means to provide his or her own subsistence; thus economic development should proportionately benefit every person. The only reason why it might not is if society is organized to exploit workers by letting capitalists take large profits, thereby depriving laborers of a full share of their earnings.

It is increasingly difficult to find economists who believe that socialism is a more successful route to economic development than capitalism is, but neo-Marxists share many ideas in common with the world systems approach to understanding the global economic situation. The world system, it is argued, works much as Marx described capitalism—except that the scope is global, not country-specific. The developed nations of the West "are charged with buying raw materials cheap from developing countries and selling manufactured goods dear, thus putting developing countries permanently in the role of debtors and dependents" (Walsh 1974:1144). If the economic power of developed nations could be reduced and that of developing nations enhanced, the boost to developing nations would dissipate problems such as hunger and poverty that are believed to be a result of too many people. At such time, the population problem will disappear because, it is argued, it is not really a problem after all. When all other social problems (primarily economic in origin) are taken care of, people will deal easily with any population problem if, indeed, one occurs. This was obviously the attitude of Friedrich Engels, who wrote in a letter in

1881, "If at some stage communist society finds itself obligated to regulate the production of human beings . . . it will be precisely this and this society alone, which can carry this out without difficulty" (quoted in Hansen 1970:47).

The evidence from countries such as Russia, Cuba, and, indeed, China suggests that a revolution may alter the demographic picture of a nation, but the relationship to economic development is somewhat cloudy. In the previously Marxist, centrally planned economies of eastern Europe and Russia, low rates of population growth did not translate into commensurately high levels of living. On the contrary, there is widespread speculation that low birth rates in those countries were a response to the economic limits placed on the family—especially scarce housing and limited consumer goods—although female education and labor force participation had been emphasized.

The idea that population growth and economic development may be only tenuously related to each other is also reflected in empirical work that seems to support a non-Marxist, but still "neutral," view of the relationship between population growth and per capita income. Using data for developing societies for the period 1965–84, Bloom and Freeman (1986) concluded that despite rapid population growth, the labor markets in most developing countries were able to absorb large population increases at the same time that per-worker incomes were rising and productivity was increasing. In other words, just as Davis had pointed out in the theory of demographic change and response, a society's initial response to rapid population growth is to work harder to support its new members. But can that be sustained? Preston (1986) argued that it could be in those areas that have sufficient natural resources and, more importantly, are making increasingly efficient use of their major societal resource—human capital. This means not simply more people (as Simon seemed to infer) but a better-educated and better-managed labor force, combined with improved methods of communication and transportation (the economic infrastructure).

Is Population Growth Detrimental to Economic Development?

The neo-Malthusian position that economic development is hindered by rapid population growth is a simple proposition in its basic form. Regardless of the reason for an economy starting to grow, that growth will not be translated into development unless the population is growing slower than the economy. An analogy can be made to business. A storekeeper will make a profit only if expenses (overhead) add up to less than gross sales. For an economy, the addition of people involves expenses (**demographic overhead**) in terms of feeding, clothing, sheltering, and providing education and other goods and services for those people, and if demographic overhead equals or exceeds national income, there will be no improvement (profit) in the overall standard of living. If overhead exceeds income, then, just like a family, disaster can be averted for a while by borrowing money, but eventually that money has to be repaid and if the loan is used simply to pay the overhead, rather than being invested in human capital, it is unlikely that there will be money available to pay the loan when it comes due. So, the loan is extended or refinanced, and disaster is averted for just a little while longer . . .

Let me illustrate the point further with a few numbers. Between 1980 and 2000, the population of Nigeria increased from 71 million to 123 million people, representing an average annual rate of population growth of 2.7 percent per year. The economy was also growing, thanks to substantial oil resources. In fact, at a rate of 2.3 percent per year, Nigeria's economy was growing at almost exactly the same clip as that of most European nations during 1980–92 (World Bank 1994a). However, that was not fast enough to keep up with population growth in Nigeria and, as a consequence, the per capita gross national product (in constant dollars) in Nigeria in 2000 was less than it had been in 1980. Despite a growing economy, the average person was worse off. It is easy to ignore the impact of population growth when you are dealing with a country such as Nigeria, where an oppressive and corrupt government has skimmed off much of the country's wealth for the benefit of a small elite group. This is a blatant example of what Leibenstein (1957) politely called "organizational inefficiencies"—the existence of corruption and bribery in an environment without a stable banking or insurance system. In such a context, it is easy to see why people would pay little attention to population growth, but we can be sure that population growth has been compounding an already bad situation.

The more positive side of this equation is that lower mortality and fertility can lead to higher incomes, in turn encouraging still lower mortality and fertility, which in turn foster higher levels of income (Birdsall and Sinding 2001). This reciprocal relationship is at least partially a result of the transitions put in place by the demographic transition. As I noted in Chapter 8, the age transition can provide a demographic windfall or bonus for countries by creating a period of time when an increasingly larger fraction of the population is in the economically productive ages. Since this is a function largely of declining fertility, it is also usually accompanied by a delay in marriage and childbearing among women (part of the family and household transition) that, as I discussed in Chapter 10, can lead to increased levels of education and labor force participation for women. The movement of women into the paid labor force may then have the effect of noticeably increasing the overall level of productivity, which contributes to economic development.

The Bottom Line for the Future: Can Billions More People Be Fed?

It seems intuitively obvious to neo-Malthusians (and a lot of other people who do not think of themselves as neo-Malthusians) that population growth can make a difference in how many resources are consumed in the world. This discussion typically starts with the bottom-line question: Can we feed the population of the world? Despite our concern with the overall standard of living, the issue of having enough food to sustain a large and growing population is still a huge problem for the world.

The Food and Agricultural Organization (FAO) of the United Nations estimates that more than 800 million people in the world have inadequate access to food (FAO 2002)—almost the same number of people alive when Malthus first wrote his *Essay on Population* in 1798. During the next minute, as you read this page, 11 children under the age of five will die of diseases related to malnutrition, although their places will be more than taken by the 264 babies who will be born during that same minute. The good news is that the number of undernourished humans in the early 2000s is less than in the 1970s, although not by much. The question is whether we can sufficiently increase the food supply to feed the more than 80 million additional people on the planet every year while also improving everyone's diet in the process. We do not really know. Do we have enough fresh water to support all those people? We do not know that either. How did we get to this point and what can we learn from the past that might help us in the future?

The Relationship between Economic Development and Food

Roughly 10,000 years ago, humans began to seriously domesticate plants and animals, thereby making it possible to grow food and settle down in permanent villages. The domestication of plants, of course, hinged on the use of tools to work the ground near the settlement site, and the invention of those tools and their application to farming can be traced to many different areas of the world. Some of the earliest known sites are in the Dead Sea region of the Middle East (Cipolla 1965; Diamond 1997), where the Agricultural Revolution apparently took place around 8000 B.C. From the eastern end of the Mediterranean, agricultural innovations spread slowly westward through Europe (being picked up in the British Isles around 3000 B.C.) and east through Asia.

Plants and animals were also domesticated in the Western Hemisphere several thousand years ago (Harlan 1976), resulting in an increase in the amount of food that could be produced per person. As you know by now, increased food production was associated with population growth (possibly as a cause, possibly as a consequence). Overall, the Agricultural Revolution "created an economy which, by . . . giving men a more reliable supply of food, permitted them to multiply to a hitherto unknown degree" (Sanchez-Albornoz 1974:24). The classic Malthusian view, of course, is that cultivating land caused population increase by lowering mortality and possibly raising fertility. However, the Boserupian view is that independent increases in population size among hunter-gatherers, perhaps through a long-run excess of births over deaths, led to a need for more innovative ways of obtaining food, and so, of necessity, the revolution in agriculture gradually occurred. Seen from this perspective, the Agricultural Revolution was the result of a "resource crisis," in which population growth, slow though it may have been 10,000 years ago, generated more people than could be fed just by hunting and gathering. The crisis led to a revolution in human control over the environment—humans began to deliberately produce food, rather than just take what nature provided. In turn, this had the cumulative effect of sustaining slow but steady population growth in most areas of the world for several thousand years preceding the Industrial Revolution.

Industrialization and economic development in general require a massive increase in energy use. If everyone is consuming more, it is because production per person has increased, and that comes about by applying nonhuman energy to tasks previously done less efficiently by people, or not done at all because people could not do them. Wood has served as the major energy source for most of human history, but in Europe a few hundred years ago, population growth and the beginnings of industrialization led inexorably to deforestation, producing an "energy crisis" (Harrison

1993). That crisis forced a new way of thinking about energy sources—a new way of controlling the environment—and the result was the Industrial Revolution.

Keep in mind that the Agricultural and Industrial Revolutions are linked together. The Industrial Revolution of the nineteenth century was preceded, and indeed made possible, by important changes in agriculture that significantly improved output (Clough 1968). Through most of human history, including in Malthus's day, increases in the food supply depended largely on **extensification of agriculture**—putting more land under production. However, since we have essentially run out of new land that can be farmed, the modern rise in agricultural output has come about through **intensification of agriculture**—getting more out of the land than we used to. In Europe and North America, the factors helping to increase agricultural productivity in a relatively short time included (1) mechanization of cultivating and harvesting processes, (2) increased use of fertilizers and irrigation, and (3) reorganization of land holdings (Poleman 1975; Walsh 1975).

The Industrial Revolution generated a host of mechanical devices, especially mechanical reapers, to greatly speed up harvesting. Drawn first by horses or oxen, reapers were pulled later by an even more efficient energy converter—the tractor. Like most early engines of industrialization, tractors were driven by steam. Their thirst for fuel was quenched by wood as long as it lasted, but the use of coal became necessary as a result of deforestation. The idea behind the steam engine, by the way, has been around for a long time, just waiting for the right moment to be adapted to something dramatically useful. In the ancient world, "Greek mechanics invented amusing steam-operated automata but never developed the steam engine; the crankshaft and connecting rod were not invented until the middle ages, and without a crankshaft it is impossible to transform longitudinal into circular motion" (Veyne 1987:137). Overall, the mechanization of agriculture vastly increased the number of acres that one or a few people could farm, and also increased the amount of land that could be devoted to more than one crop per growing season, since land could be cultivated and harvested so much more easily.

Although mechanization was certainly a prime mover of increased productivity in agriculture, especially in North America where land was plentiful in relation to people, it is not an absolute requirement. In North America, where population density was low and labor was scarce, the increase in energy needed to intensify agriculture came from mechanical devices. In Japan, on the other hand, where even at the beginning of industrialization labor was at a surplus, the initial increase in energy came from people—people working harder and more efficiently on the land (Gordon 1975).

Many agricultural innovations have also been made possible by reorganizing agricultural land and developing better policies for land use (Dyson 1996). Collecting farms into large units and using meadows and pastures for cultivation rather than extensive grazing have increased production, particularly in the United States and Europe, since large farms introduce economies of scale that permit investment in expensive tractors, harvesters, fertilizers, irrigation systems, and the like. In the United States, this process has a long history and is still continuing; for example, between 1950 and 1997, the number of small farms (less than 180 acres) in the United States decreased from 4.1 million to 1.2 million, a 71 percent decline (U.S. Census Bureau 2001b). In the 1990s, the total number of farms in the U.S. dropped below two million for the first time since 1850, after reaching a peak of 6.8 million in 1935. This does not mean that the number of acres under cultivation has declined much (it has changed very little), but rather that there is a trend toward large commercial farms and away from small family farms.

Although it may be intuitively obvious, it bears repeating that industrial expansion cannot occur unless agricultural production increases proportionately. Industrialization is typically associated with people migrating out of rural and into urban areas, naturally resulting in a shift of workers out of agriculture and into industry. Therefore, those workers left behind must be able to produce more—enough for themselves and also for the nonagricultural sector of the population. The flip side of that is that as the population and demand for food grows, the need to mechanize agriculture to increase production leads to a declining demand for agricultural workers. Thus you can see that the Industrial Revolution would have been impossible if agricultural production had not increased—and vice versa. Adam Smith, the classical economist, once remarked that "when by the improvement and cultivation of land . . . the labor of half the society becomes sufficient to provide food for the whole, the other half . . . can be employed . . . in satisfying the other wants and fancies of mankind" (quoted in Nicholls 1970:296).

So, let us ask ourselves how can food production be increased? There are two aspects to feeding the world's growing population. The first is the technical problem of growing enough food, and the second is the organizational problem of getting it to the people who need it. As I have already mentioned, food production can be increased though either extensification or intensification of agriculture. Let us examine each approach.

Extensification—Increasing Farmland

Water covers about 71 percent of the earth's surface, leaving the remaining 29 percent for us to scratch out our respective livings. Only 11 percent of the world's land surface is readily suitable for crop production, and an additional 26 percent is devoted to permanent pasture. Forests and woodlands cover about 32 percent of the land surface, and the remaining 31 percent is too hot or too cold for any of those things, or is used for other purposes (such as cities and highways). Most of the land that could be fairly readily cultivated is already cultivated; the rest is covered by ice, or is too dry, too wet, too hot, too cold, or too steep, or has soil unsuitable for growing crops (Bruinsma 2003). Of course, as Lester Brown, Founder and Board Chairman of Worldwatch Institute, once commented, "If you are willing to pay the price, you can farm the slope of Mt. Everest" (*Newsweek* 1974:62).

In 1860, there were an estimated 572 million hectares of land in the world cleared for agricultural use (Revelle 1984). As the populations of Europe and North America expanded in the late nineteenth century, the amount of farmland in these regions virtually doubled. More recently, the population pressure in developing countries has been accompanied by an expansion of farmland in those parts of the world, often generated by slash-and-burn techniques in relatively fragile ecosystems. All of this adds up to a total of 1.5 billion hectares of farmland in the world today—nearly triple that of 1860. This seems to be the real limit of decent-quality farmland. Table 12.2 traces

Table 12.2   Inputs to Agriculture Not Keeping Pace with Population Growth

Year	Total Population (1,000)	Arable and Permanent Cropland (1,000ha)	Irrigated Agric Area (1,000ha)	Fertilizer Consumption (MT)	Tractors in Use	Fish Catch (MT)
1961	3,078,867	1,356,722	139,134	31,182,244	11,318,243	39,182,424
1971	3,765,507	1,405,053	171,807	73,310,242	16,524,147	65,876,902
1981	4,505,510	1,442,386	213,532	115,147,224	22,311,077	75,093,428
1991	5,338,524	1,504,117	248,724	134,606,391	26,248,089	98,286,362
2001	6,134,138	1,532,090	273,052	137,729,730	26,854,002	126,178,607
			Per Person			
1961		0.44	0.045	0.010	0.0037	0.013
1971		*0.37*	0.046	0.019	0.0044	0.017
1981		*0.32*	0.047	0.026	0.0050	0.017
1991		*0.28*	0.047	*0.025*	*0.0049*	0.018
2001		*0.25*	*0.045*	*0.022*	*0.0044*	0.021

Source: Food and Agricultural Organization of the United Nations, FAOSTATS database, http://www.fao.org, accessed 2003. Figures in italics indicate per person rates that were lower than in the previous time period.

the data on population and agriculture for the world between 1961 and 2001. You can see that the amount of agricultural land in use in the world increased slightly from 1.4 billion hectares in 1961 to 1.5 billion in 2001. However, the population was increasing from 3.1 to 6.1 billion during that 40-year period, so the amount of farmland per person was steadily declining.

In reaching the limits of readily cultivable land, we have been encroaching on land that supports plant and animal habitats that we really cannot do without. We are only now beginning to discover how much we depend on biological diversity (Miller 2004), and we threaten our environment as we search for more land on which to grow food. If that list of problems is not dismal enough, consider that the amount of good farmland is actually shrinking. In some parts of the world, this is a result of soil erosion or desertification, whereas in many other places it is a consequence of urban sprawl. Most major cities are in abundant agricultural regions that can provide fresh food daily to the city populations; only recently have transportation and refrigeration lessened (but not eliminated) that need. As cities have grown in size, nearby agricultural land increasingly has been graded and paved for higher-profit residential or business uses. About 1 percent of the earth's land surface is devoted to urban use (World Resources Institute 1998), which may sound like a small amount, but it looms large next to the 11 percent that is devoted to crops, because of the fact that so much cropland is in the vicinity of cities.

The allure of farmland to the developer is clear: It is generally flat and well-drained, thus good for building. It is probably outside town limits and therefore taxed at cheaper county rates. And the land may be family owned, with the potential that nonfarming family members will pressure their other family members to sell it as a means of converting the property to other assets. Such conversion is extremely difficult to reverse: "When farmland goes, food goes. Asphalt is the land's last crop" (The Environmental Fund 1981:1).

Soil erosion is a major problem throughout the world, as more topsoil is washed or blown away each year, a result of overplowing, overgrazing, and deforestation, all of which leave the ground unprotected from rain and wind (Brown 1998). In the world as a whole, it appears that we are losing soil to erosion faster than nature can rebuild the supply. Thus, we are losing ground (quite literally) at the same time that few places in the world still await the plow. It has also been suggested that a viable source of "land" is the sea—**mariculture,** or the "blue revolution," as some have called it. Farming the sea includes both fishing and harvesting kelp and algae for human consumption, although the expense of growing kelp and other plants is so great that it does not appear to be an economically viable alternative to cultivating land. Farming fish (including shellfish), or **aquaculture,** has been steadily increasing as a source of fish for food. It appears that we have reached the level of the ocean's sustainable fish catch, so any increase in fish production will of necessity be through aquaculture. Just in the five years between 1994 and 1999, aquaculture increased as a percentage of fish production from less than 19 percent to more than 26 percent (Food and Agricultural Organization of the United Nations 2001).

It seems very doubtful that either extending agricultural land or farming the ocean will produce the amount of food needed by the world in the next century, given current world population projections. The output per acre of land under production must continue to increase if we are to feed billions more.

Intensification—Increasing Per-Acre Yield

There are several different ways to increase output from the land, and often methods must be combined if substantial success is to be realized. Those methods include plant breeding, increased irrigation, and increased use of pesticides and fertilizers. In combination, they add up to the **Green Revolution.**

The Green Revolution is a term coined by the U.S. Agency for International Development back in the 1960s, but it began quietly in the 1940s in Mexico at the Rockefeller Foundation's International Maize and Wheat Improvement Center. The goal was to provide a means to increase grain production, and under the direction of Norman Borlaug (who received a Nobel Peace Prize for his work), new **high-yield varieties (HYV)** of wheat were developed. Known as dwarf types, they have shorter stems that produce more stalks than do most traditional varieties. In the mid-1960s, these varieties of wheat were introduced into a number of countries, notably India and Pakistan, with spectacular early success—a result that had been anticipated after what the researchers had seen in Mexico (Chandler 1971). In 1954, the best wheat yields in Mexico had been about three metric tons per hectare, but the introduction of the HYV wheat (now used in almost all of Mexico's wheatland) raised yields to six or even eight tons per hectare if crops were carefully managed. A major difference was that more traditional varieties were too tall and tended to lodge (fall over) prior to harvest, thus raising the loss per acre, whereas the dwarf varieties (being shorter) prevented lodging. This is critical, because lodging can be devastating; it destroys some ears of grain and damages others. Furthermore, resistance to lodging makes possible the heavy fertilization and irrigation that are necessary for high yields.

The Green Revolution was not restricted to high-yield wheat and maize. In 1962, the Ford Foundation began to research rice breeding at the International Rice Research Institute in the Philippines. In a few short years, a high-yield variety of dwarf rice had been developed that, like HYV wheat, dramatically raised per-acre yields. Rice production increased in India and Pakistan, as well as in the Philippines, Indonesia, South Vietnam, and several other less-developed countries. China and India have both embraced the Green Revolution as a means for ensuring the food security of their people. **Food security,** by the way, is a United Nations term meaning that people have physical and economic access to the basic food they need in order to work and function normally; that is, the food is there, and they can afford to buy it.

There is, at it turns out, a huge set of costs attached to the Green Revolution—success requires more than simply planting a new type of seed. These plants require fertilizers, pesticides, and irrigation in rather large amounts, a problem compounded by the fact that fertilizer and pesticides are normally petroleum-based and the irrigation system requires fuel for pumping. These are expensive items and usually demand that large amounts of adjacent land be devoted to the same crops and the same methods of farming, which in turn often means using tractors and other farm machinery in place of less-efficient human labor. This is the true meaning of the revolution.

The plants involved in the Green Revolution have principally been wheat, maize, and rice, but considerable research has also gone into the development of HYV soybeans, peanuts, and other high-protein plants. Other research includes breeding synthetic species such as triticale, a very hardy and high-yield cross of wheat and rye that has mainly been used as forage for feed animals in developed countries (especially Europe) rather than for people in less-developed nations. Another candidate for continued development is the winged bean (or goa bean), sometimes known as "a supermarket on a stalk" because the plant combines the desirable nutritional characteristics of the green bean, garden pea, spinach, mushroom, soybean, bean sprout, and potato—all in one plant that is almost entirely edible, save the stalk. It is becoming a staple in poorer regions of Africa and South Asia because it grows quickly, is disease resistant, and is high in protein (Herbst 1995).

At least as important as the nutritional aspect of plant breeding is the development of disease- and pest-resistance. The rapid change in pest populations requires constant surveillance and alteration of seed strains. Insects are very much our competitors for the world food supply, and are a problem both before and after crops are harvested. Efforts to control pest damage have focused on designing seed varieties that are resistant to pests, or that are resistant to pesticides and herbicides, so that these chemicals can be used on plants without destroying the plant itself. However, in the United States and Europe there has been a consumer backlash against these kinds of genetically modified seeds, so food manufacturers have become cautious about using them (Kilman 2001).

High-yield seeds generally require substantial amounts of water to be successful, and irrigation is the only way to ensure that they get it (nature is a bit too fickle). This is because they, like all crops, grow best with a controlled supply of water and also because irrigation can increase the opportunities for multiple cropping. In 2001, only 18 percent of the world's cropland was irrigated, as you can see in Table 12.2 by dividing the irrigated acreage by the arable and permanent cropland, but irrigated land has been estimated to account for 40 percent of all the world's food (Postel 1996), so this is a very important input to agriculture. Irrigation, of course, requires a water source (typically a reservoir created by damming a river), an initial capital investment to dig canals and install pipes, and energy to drive the pumps. Each of these elements represents an expensive resource. "There is widespread agreement that the future supply of water for agriculture represents a much more significant constraint to raising food production than do any likely foreseeable difficulties relating to soil or land" (Dyson 1996:149).

The problem is that almost all of the choice dam sites have already been exploited, most within the historically short time span since the end of World War II. To give you some sense of the magnitude of the water issue, it takes about half a million gallons of water to grow an acre of rice, and irrigated agriculture accounts for about 70 percent of the water consumed worldwide (Falkenmark and Widstrand 1992). The tremendous expense of providing irrigation imposes serious limits to any sizable future increase in the amount of land being irrigated in developing nations. Although only 18 percent of the world's cropland is irrigated (the remainder is fed by rain), that was a major increase from 10 percent in 1961 and there is also considerable variability in the world. As a country, Egypt leads the world in having virtually all of its cropland irrigated, but the highest overall levels of irrigation (34 percent of cropland in 2001) are in Asia. Japan, for example, uses irrigation on 63 percent of its cropland, and China has 52 percent of its cropland under irrigation. Europe can get by with very little irrigation because the growing season is associated with considerable rainfall. That is not usually the case in sub-Saharan Africa, however, where nonetheless only a small fraction of cropland is irrigated.

In order to maximize yields, plants must be fed (fertilized) and farm machinery used. These are key ingredients in the success of the Green Revolution. Table 12.2 shows that both of these inputs increased substantially between 1961 and 2001. The increases were especially rapid in Asia—representing major investments in agriculture in both China and India. At the beginning of the twenty-first century, Asia led the major regions of the world in terms of fertilizer use per 1,000 hectares and it had surpassed Latin America in mechanization per hectare. Europe continues to be the most mechanized agricultural region of the world.

Pesticides represent another important input to the Green Revolutions, but their use is decidedly a two-edged sword. Although heavy use of pesticides initially killed insects and increased per-acre yield, pesticides can also kill beneficial predators of insects and diseases that feed off the crops. Pesticide production has increased steadily worldwide, but its use has become more judicious—too much pesticide in the short term actually lowers crop productivity in the long term.

The Green Revolution, to be effective in all parts of the world, would require major changes in the way social life is organized in rural areas, not just a change in the plants grown or the fertilizers used. This is because the Green Revolution is based on Western (especially American and Canadian) methods of farming, in which the emphasis is on using expensive supplies and equipment and on the high-risk, high-profit principle of economies of scale—plant one crop in high volume and do it well.

The data in Table 12.2 show that all of the inputs to agriculture increased in absolute terms between 1961 and 2001, so that was good news. But the bad news was that none of the major inputs kept pace with population growth. I have already pointed out that the amount of cropland per person declined between 1961 and 2001, which is why intensification had to replace extensification in agricultural output. However, despite the increase in the amount of land irrigated, the per-person acreage in irrigated land declined between 1991 and 2001. Furthermore, even though fertilizer consumption has been steadily rising in the world, it has steadily declined on a per-person basis since 1981. Tractor use has followed the same pattern.

Little global progress has been made in agricultural output since the mid 1990s and this has forced the Food and Agricultural Organization to recognize that its goal of halving the number of hungry people in the world by 2015 is unlikely to be realized (Food and Agricultural Organization of the United Nations 2003). Has the Green Revolution simply run its course? Not necessarily, and this gets us back to the issue of genetically modified foods. In its 2001 Human Development Report, the United Nations argued strongly that the world's richest nations must get over the fear of genetically modified foods if poverty and food insecurity are to be eradicated in poorer nations (United Nations 2001). This sentiment was echoed by the Green Revolution's "founder," Norman Borlaug, who pointed clearly to the link between population change and sustainability of the food supply:

> I've spent the past 20 years trying to bring the Green Revolution to Africa—where the farmers use traditional seeds and the organic farming systems that some call "sustainable." But low-yield farming is only sustainable for people with high death rates, and thanks to better medical care, more babies are surviving . . . Africa desperately needs the simple, effective high-yield farming systems that have made the First World's food supply safe and secure (Borlaug 2002:A16).

Africa almost certainly needs a major investment in agriculture to overcome its food insecurity problem, but elsewhere in the world a subtle, yet effective, way of getting more out of each acre of food production is to waste less. Governments that help farmers use water and fertilizer more efficiently will clearly have more of those resources to spread around. But a great deal of waste occurs once food is grown, both in its storage (where it may become spoiled or eaten by other creatures) and in the hands of consumers. One way of wasting less food is to use some form of preservative. Especially in developed nations, chemical substances are added to food to protect its nutritional value, to lengthen its shelf life (preservatives), and to change or enhance flavors and colors. Additives can greatly aid the process of feeding people by keeping food from spoiling and helping preserve its value. This has aided in the mass distribution of food and has made it possible for people to live considerable distances from food sources. The use of preservatives is one means of deterring food spoilage by microorganisms, and more widespread use of preservatives could at least partially alleviate worldwide food shortages. For example, the World Health Organization estimates that about 20 percent of the world's food supply is lost to microorganism spoilage. Nonetheless, preservatives have increasingly been attacked as potential cancer agents. Sodium nitrate, used for centuries in curing meat to prevent botulism, is now suspected of being a carcinogen, at least in some dietary combinations. In 1984, the U.S. Food and Drug Administration approved the use of low doses of irradiation as a form of food preservation and data suggest that it is about as safe as food prepared in a microwave oven. Currently, spices, vegetables, fruit, poultry, eggs, and meat can be irradiated. The World Health Organization and the American Medical Association endorse the practice. Nonetheless, consumers in the U.S. have been far more suspicious of so-called nuked food than have people in most of the rest of the world, so its use is somewhat limited in the U.S.

With respect to consumption, it is certainly true that Americans, for example, could eat less meat and still be well nourished, and there is increasing pressure in that direction. It takes several pounds of grain to produce one pound of red meat, and there are other, more efficient ways to get protein (such as soybeans, peanuts, peas, and beans). Cutting back on animal protein could then free up the production of grain for human rather than animal consumption. Of course, most Americans do not welcome a suggestion that they should eat less meat, since eating beef, especially, is as much a part of American culture as not eating meat is in India. Thus in the United States there is resistance to the idea of meatless meals, just as in India many people resist killing cows, monkeys, and even rats, in the belief that all living things are sacred.

How Many People Can Be Fed?

We are approaching the limits of exploitable land and water, but per-acre yield can still be increased, and we can reduce waste. Could this combination produce enough food to meet the needs of the nine billion or more people whom we anticipate will inhabit the planet by the middle of this century? We know how dangerous it is to try to predict the future, but the value of trying to do so is that we may be able to invent a future that is more to our liking. In 1968, Paul Ehrlich wrote in *The Population Bomb* that the world population situation "boils down to a few elementary facts. There is not enough food today. How much there will be tomorrow is open to debate. If the optimists are correct, today's level of misery will be perpetuated for perhaps two decades into the future. If the pessimists are correct, massive famines will occur soon, possibly in the early 1970's, certainly by the early 1980's" (Ehrlich 1968:44). Over the years, many people have derided Ehrlich for being so wrong, but of course he noted in the final chapter of his book that this is a situation in which the "penalty" for being wrong is that fewer people will be starving than expected, and that perhaps the dire warnings about the problems of population growth and food will have helped to spur action to avoid that consequence.

Bearing these things in mind, others step forward periodically to assess the world's potential for feeding itself. Vaclav Smil, a Canadian geographer, did so in 1994. He began by reviewing estimates made during the past 100 years that range from Ehrlich's low estimate of two billion people being sustainable from the world's food supply (Ehrlich 1968) to Simon's conclusion that the food supply has no upper limit (Simon 1981). Most other estimates are between 6 billion and 40 billion (see also Cohen 1995). Smil then reminded us that no reasonable calculation of the earth's capacity for growing food generates an estimate even close to the idea that the six billion plus alive today would have a diet similar to that of the average American (Smil 1994). We have, for all intents and purposes, exceeded the carrying capacity with respect to the average American diet. Smil then points out that this is not as big a problem as it might seem, however, because Americans are overfed and very wasteful. To be sure, throughout the world there is tremendous slack ("recoverable inefficiencies") in the way food is produced and eaten. Smil calculated that by improving agricultural practices, reducing waste, and promoting a healthier diet (limiting fat intake to 30 percent of total energy, especially by reducing meat intake), the world in 1990 could have had a 60 percent gain in the efficiency of food production without putting a single additional acre under production. This would have fed an additional 3.1 billion in 1990, raising the supportable total in 1990 to 8.4 billion rather than the actual 5.3 billion.

Supportability in 1990 does not necessarily imply sustainability down the road. Smil introduced a series of conservative assumptions to suggest how a population of 10–11 billion could be supported during the next century, even without assuming some kind of magical technological fix. Applying the same logic of reducing inefficiencies, he calculated that the biggest gains (47 percent) could come from increasing the per-acre yield, followed by a continued extension of cultivated land, cultivating idle land, using high-efficiency irrigation, reducing beef production, irrigating some crops with salt water, and farming the sea. Underlying these estimates are assumptions that all populations will have a healthier (not necessarily a higher-calorie) diet, and that the food will be grown where it can be and distributed to where the demand is. These are huge assumptions and tell us only that in the best of circumstances it might be possible to feed a much larger population than we currently have.

The assumptions of changing diets and food distribution take us from the realm of technology to social organization and culture. People are remarkably adaptable if they want to be, but "culture" often intervenes to prevent the "rational" response to situations. For example, the idea that nations need to be self-sufficient with respect to food (which underlies the idea of food security) may wind up wasting a nation's resources that could be used to produce something else that can be sold in order to buy food. Most of us living in developed nations are not personally self-sufficient with respect to either food or water. We rely on the good faith of strangers to provide us with what we need because we are willing to pay for it. The billions of people who will be sharing the planet with us later in this century will be properly fed only if the entire world adopts that same trading principle. Trade and a surplus for aid are necessary antidotes to the maldistribution of the physical and social resources required for growing food. At the same time, we must remember that trade implies trust and the world is not always a trustworthy place, so countries that depend on others for food resources are naturally going to be anxious about the producers of food "blackmailing" them by withholding food in exchange for some other kind of concession.

Keep in mind that, since much of the world depends on rain-fed agriculture, the weather is still an important issue in our food supply. In a review of evidence linking weather to population changes in the preindustrial world, Galloway concluded that "an important driving force behind long term fluctuations in populations may very well be long term variations in climate and its effects on agricultural yields and vital rates" (Galloway 1984:27). Historical data for China and Europe suggest that, much as you would expect, bad weather contributes to poor harvests, higher mortality, and slower population growth, whereas good weather has the opposite effect. In the African sahel south of the Sahara, **drought** (a prolonged period of less-than-average rainfall) occurs with devastating predictability, and for the thousands of years that people have lived there, they have learned to combat drought and high mortality by having large families.

Drought is only one of several adverse weather conditions that can lead to a **famine**—defined as "food shortage accompanied by a significant increase in deaths" (Dyson 1991:5). Famine periodically hits South Asia either as a result of a drought caused by the lack of monsoon rains or by flooding caused by over-heavy rains. Either extreme can devastate crops and lead to an increase in the death rate. As in Africa, the high death rates are typically compensated for by high birth rates, and the famine-struck regions of the world continue to increase in population size. That maintains the pressure on the agricultural sector to improve productivity all it can. Still, we always have to come back to the fact that in the long run the only solution is to halt population growth; at some point, the finite limit to resources will close the gate on population growth.

One factor already at work to limit future food production, which will also endanger human health and thus produce a deteriorating human condition, is degradation of the environment. This is related to the attempt to be ever more productive agriculturally, as well as industrially, and to produce ever more energy to mobilize that productivity. It is ironic that some of the techniques that have seemed to offer the greatest hope for increasing the food supply and improving standards of living may be changing the very ecosystem upon which food production depends.

A growing body of evidence suggests that if we stay on our current course of environmental degradation, we will permanently lower the sustainable level of living on the planet. We are in the midst of what Paul Harrison (1993) called "the pollution crisis," and we will either work our way through this on a global scale or be faced with a major eco-catastrophe that could greatly diminish the quality of life for all of us. The World Bank, reporting on the situation in Africa, put it this way:

> With their economies largely linked to agricultural production, most West African countries must battle simultaneously to alleviate widespread poverty, ensure food security and achieve environmentally sustainable development. This has to be accomplished against a background of high illiteracy rates, rapidly growing populations, low and erratic rainfall, inherently infertile soils, and development strategies which have had a strong urban bias. Under such conditions, traditional production systems are unable to sustain the population. Without significant change, land degradation will accelerate and the natural resource base on which agriculture production depends will continue to decline (World Bank 1998).

By-Products of Development— Degradation of the Environment

The environmental issues that confront the world today deal largely with the side effects of trying to feed and otherwise raise the standard of living of an immense number of human beings. In coping with an ever-increasing number of humans, we are damaging the lithosphere, the hydrosphere, and the atmosphere. This is because the reality of the environment is that everything is connected:

> The web of life is seamless, and the consequences of disruption to one part of the ecosystem ripple throughout the whole; soil erosion in the Himalayas contributes to massive flooding in Bangladesh; the deforestation of the Amazon may alter the atmospheric balance over the whole globe; and chemicals and gases produced in the richer industrialized countries are destroying the ozone layer that protects everyone, rich and poor alike (Seager 1990:Preface)

In order to examine these interconnections as they relate to the environmental by-products of development, we need to remind ourselves of a few basic concepts and definitions.

Environmental Concepts and Definitions

The world inhabited by humans is known to scientists as the biosphere—the zone of Earth in which life is found (Miller 2004). As mentioned in Chapter 1, the **biosphere** consists of three major parts: (1) the lower part of the **atmosphere** (known as the **troposphere**—the first 11 miles or so of the atmosphere above the surface of the earth); (2) the **hydrosphere** (most surface water and groundwater); and (3) the **lithosphere** (the upper part of the earth's crust containing the soils, minerals, and fuels that plants and animals require for life). Within the biosphere are **ecosystems** representing communities of species interacting with each other and with the inanimate world. All of the world's ecosystems then represent the **ecosphere,** which is the living portion of the biosphere.

All living organisms in the biosphere require three basic things: (1) resources (food, water, and energy); (2) space to live; and (3) space to "dump waste." The carrying capacity of the biosphere, or of any ecosystem within the ecosphere, is the number of organisms that can be sustained indefinitely—the number for whom there are renewable resources, sufficient space to live, and sufficient space to get rid of waste products (all forms of life generate waste products). If the population exceeds an ecosystem's carrying capacity in any one of these categories, we have a situation of **overshoot** or **overpopulation.**

Damage to the Lithosphere—Polluting the Ground

We survive on the thin crust of the earth's surface. Actually, we live on only 29 percent of the surface. The rest is covered with water, especially the oceans, which we tend to treat as open sewers, but we also exploit resources that are in the ground under that water. The land surface of the earth is where most things we humans are interested in grow, and the damage we do to this part of the environment has the potential to lower the ability of plants and animals to survive. We have been busy doing damage such as: (1) soil erosion; (2) soil degradation from excess salts and water; (3) desertification; (4) deforestation; (5) loss of biodiversity; (6) strip mining for energy resources; and (7) dumping hazardous waste.

Almost every step of improving agricultural productivity has its environmental costs—from irrigation to the use of fertilizers and pesticides to the creation of energy sources and the production of machinery. "In spite of the increasing pace of world industrialization and urbanization, it is ploughing and pastoralism which are responsible for many of our most serious environmental problems and which are still causing some of our most widespread changes in the landscape" (Goudie 1990:323). If not carefully managed, farming can lead to an actual destruction of the land. For example, improper irrigation is one of several causes of soil erosion, to which valuable farm land is lost every year. In the United States alone it is estimated that during the past 200 years, at least one-third of the topsoil on croplands has been lost, ruining as much as 100 million acres of cultivated land (Brown and Mitchell 1998; Pimentel et al. 1976). Even if cropland is not ruined, its productivity is lowered by erosion, because few good chemical additives exist that can adequately replace the nutrition of natural topsoil. Unfortunately, the push for greater yield per acre may lead a farmer to achieve short-term rises in productivity without concern for the longer-term ability of the land to remain productive:

> In many human cultures, agriculture is practiced as an extractive industry and soils continue to be degraded throughout the world. Continuation of the observed rate of soil degradation from 1945 to 1990 suggests an effective half-life of the vegetated soils of the earth of about 182 years. Such conversion of land to agricultural purposes alters the entire ecosystem, and the resulting impact on soil structure and fertility, quality and quantity of both surface and groundwater and the biodiversity of both terrestrial and aquatic communities diminishes both present and future productivity (Vanderpool 1995:2).

Crop rotation and the application of livestock manure help to reduce soil erosion, but in some parts of the world the land is robbed of even cow dung by the need of growing populations for something to burn as fuel for cooking and staying warm. The eroded soil has to go somewhere, of course, and its usual destinations are river beds and lake bottoms, where it often causes secondary problems by choking reservoirs (Seitz 1995). Desertification and deforestation are ecological disasters associated with the pressure of population growth on the environment. The southern portion of the Sahara desert has been growing in size as overgrazing (complicated by drought) has denuded wide swaths of land (Goliber 1985).

At the dawn of human civilization, forests covered about half of the earth's land surface (excepting Greenland and Antarctica). Only about half of that forest is left and "each year another 16 million hectares of forest disappear as land is cleared by timber operations or converted to other uses, such as cattle ranches, plantations, or small farms" (Abramovitz 1998). Nearly 80 percent of that deforestation has been attributed to the impact of population growth, even if indirectly (United Nations Population Fund 1991). In Brazil, population pressure has led the government to encourage people to head for the Amazon Basin in search of land. The land they find is covered with a rain forest, which they have been cutting down at a prodigious rate. Deforestation then contributes to soil erosion by removing physical barriers that prevent wind and water from carrying the topsoil away.

Forests are also susceptible to the effects of air pollution, which can damage the vegetation and lessen the plant's resistance to disease. In their turn, fewer trees and less healthy trees may alter the climate because the forests play a key role in the **hydrologic cycle** as well as in the **carbon cycle.** In the hydrologic cycle, water is being continuously converted from one status to another as it rotates from the ocean, the air, the land, through living organisms, and then back to the ocean. Solar energy causes evaporation of water from the oceans and from land, and it condenses into liquid as clouds, from whence comes rain, sleet, and snow to return water to the ground. Trees are important in this cycle both directly, because water transpires through the plants and is evaporated into the air, and indirectly, because the trees slow down the runoff and heighten the local land's absorption of the water. More than half of the moisture in the air above a forest comes from the forest itself (Miller 2004), so when the forest is gone, the local climate will become drier. These changes can mean that an area once covered by lush and biologically diverse tropical forest can be converted into a sparse grassland or even a desert.

The carbon cycle is that process through which carbons, central to life on the planet, are exchanged between living organisms and inanimate matter. Plants play an important role in this cycle through photosynthesis and forests are sometimes called the earth's "lungs." Deforestation thus has the effect of reducing the planet's lung capacity, so to speak, and that contributes to global warming because it increases the amount of greenhouse gases that, in the right number, otherwise keep us at just the right temperature for normal existence.

Damage to the Atmosphere—Polluting the Air

The atmosphere is the mixture of gases surrounding the planet, and it is a layered affair (each layer being a "sphere"). We spend our life in the troposphere, that part of the atmosphere near the surface, where all the weather takes place. But other layers are of importance as well, such as the ozone in the stratosphere that protects us from the ultraviolet radiation from the sun. Most famous of the gases are the **greenhouse gases** (mainly carbon dioxide and water, but also ozone, methane, nitrous oxide, and chlorofluorocarbons) which allow light and infrared radiation from the sun to pass through the troposphere and warm the earth's surface, from which it then rises back into the troposphere. Some of it just escapes back into space, but some of this heat is trapped by the greenhouse gases and this has the effect of warming the air, which radiates the heat back to the earth (Drake 2000). In general, the greenhouse effect is a good thing, because without it the average temperature on the planet would be zero degrees Fahrenheit (−18°C) and life would not exist in its present form, but too many greenhouse gases have the effect of **global warming**—an increase in the global temperature.

Global warming has the potential to change climatic zones, warm up and expand the oceans, and melt ice caps. The result would be a rise in average sea level, inundating coastal areas (where a disproportionate share of humans live), and a shift in the zones of the world where agriculture is most productive. The evidence is virtually overwhelming that we have been adding to greenhouse gases and that human activity is contributing to a rise in global temperature (Drake 2000; Gelbspan 1997; Middleton 1999). This has happened as a polluting side effect of trying to support more humans, and to do so at a higher standard of living.

Population growth, the intensification of agriculture, and the overall increase in people's standard of living have been made possible by substantial increases in the amount of energy we use. Holdren (1990) has estimated that in 1890, when the world's population was 1.5 billion, the annual world energy use was 1.0 terawatts. (A terawatt is equal to five billion barrels of oil.) One hundred years later, in 1990, when the world's population was at 5.3 billion, total world energy use had rocketed to 13.7 terawatts. This is an important number because *energy supply accounts for a major share of human impact on the global environment*" (Holdren 1990:159, emphasis added).

The by-products of our energy use (especially carbon dioxide and methane) wind up disproportionately in the atmosphere and contribute to global warming. As you can see from Table 12.3, we know who is pumping carbon dioxide (CO_2) into the atmosphere and the enemy is us. The United States leads the list of CO_2 producers in absolute terms and is third (after the United Arab Emirates and Kuwait) in terms of per-person emissions. Canada is eighth on the list in absolute emissions but fifth in terms of per-person emissions. Mexico is eleventh in terms of total output of carbon dioxide emissions, but its volume is only a tiny fraction of the United States' volume and its per capita use does not make the world's top 20 list, as shown in Table 12.3.

The list of countries by CO_2 emissions is very similar to the list of the largest economies of the world, which I showed you earlier in Figure 12.1 and Table 12.1. This is no coincidence, of course, since emissions are directly related to energy use, which is directly related to levels of income. Indeed, the United States has 22 percent of the total income in the world and produces 25 percent of the world's CO_2 emissions. The top four countries on the list in Table 12.3 of total emissions (United States, China, Russia, and Japan) actually account for exactly half of all the world's carbon dioxide emissions. It is fair to say that those with the biggest incomes make the biggest mess.

Other gases that we send into the environment—especially chlorofluorocarbons—have the potential to thin the **ozone layer**, which protects us from deadly ultraviolet light. These "holes" in the ozone layer, which have been documented especially in the southern hemisphere, can damage crops and livestock and, of course, humans as well. Although the switch from wood to coal for creating steam may have helped save forests, the by-product of burning coal is "acid rain"—sulfur particles trapped in the air, which then cause damage by killing plants, undermining animal habitat, and eroding human-built structures, especially those made of marble and limestone. Photochemical smog produced by automobile and industrial emissions creates a wide band of air pollution known to be harmful to humans, other animals, and plants as well. In a variation on the theme "what goes up, must come down," the gases and particles that we pump into the atmosphere come back to haunt us in myriad ways, none of them beneficial.

Damage to the Hydrosphere—Water Supply and Water Quality

Water is an amazing liquid. It covers 71 percent of the earth's surface, including almost all of the southern hemisphere and nearly half of the northern hemisphere. You are full of it—about 65 percent of your weight is water. Despite all of that water, only a small fraction—3 percent—of it is the fresh water that humans, other animals, and plants need. Furthermore, most of that 3 percent is water that is locked up

Table 12.3 Emissions of Carbon Dioxide by Country

Country	Total CO$_2$ Emissions in Millions of Metric Tons (1999)	Country	Per Capita CO$_2$ Emissions (metric tons per 1,000 population)
United States	5495.4	United Arab Emirates	29.38
China	2825.0	Kuwait	19.04
Russia	1437.3	United States	18.69
Japan	1155.2	Australia	17.46
India	1077.0	Canada	13.92
Germany	792.2	Singapore	12.77
United Kingdom	539.3	Estonia	12.24
Canada	438.6	Finland	11.22
Italy	422.7	Czech Republic	10.64
South Korea	393.5	Ireland	10.21
Mexico	378.5	Belgium	10.12
Ukraine	374.3	Russia	10.03
France	359.7	Saudi Arabia	9.72
Australia	344.4	Germany	9.61
South Africa	334.6	Israel	9.50
Poland	314.4	Denmark	9.27
Iran	301.4	North Korea	9.21
Brazil	300.7	United Kingdom	9.10
Spain	273.7	Japan	9.05
Indonesia	235.6	Norway	8.54

Source: World Bank, World Development Indicators 2003 (Washington, DC: World Bank): Table 3.8.

as ice in the poles and glaciers or in extremely deep groundwater. Only about 0.003 percent of the total volume of water on the planet is fresh water readily available to us in lakes, soil moisture, exploitable groundwater, atmospheric water vapor, and streams (Falkenmark and Widstrand 1992). Although fixed in amount, the water supply is constantly renewed in the hydrologic cycle of evaporation, condensation, and precipitation. The principal issues with respect to water have to do with its management (distributing it where it is needed), purity from disease (in order to be drinkable), and pollution.

It has been estimated that in 1850 the freshwater resources in the world were equivalent to 33,000 cubic meters per person per year (United Nations Population Fund 1991), but by 2000 that had shrunk to scarcely more than 7,000 (World Resources Institute 2000). Within the ecosphere, salt water is converted to fresh water through the hydrologic cycle, but it is very expensive to mimic nature. In fact, it has been joked that the two most difficult things to get out of water are politics and salt. Most desalination plants are based on a process of distillation that imitates the water cycle by heating water to produce vapor that is then condensed to produce fresh, potable (drinkable) water. The problem is that it is very costly to heat the water and, as a result, desalinated water is typically several times more expensive than drinkable local water. Reverse osmosis as a desalination process may hold some promise, but it seems unlikely that anything but naturally generated fresh water will be able to supply human needs for the foreseeable future, and we will have to survive by using that resource more efficiently than in the past.

All over the globe more people are competing for water even as water consumption per person has been on the rise, and, all the while, we have been sending pollutants into the water, degrading the already limited supply. Some of the pollution goes directly into the water, and some goes into the ground where it seeps into the water supply or into the air where it then falls on us as acid rain. We know as well that polluted water can alter marine life, killing fish and other sources of marine food. Ironically, one of the sources of water pollution is the chemicals we add to the soil to improve agricultural productivity, and this is aggravated by using irrigation, which increases the amount of water exposed to the chemicals.

Irrigation requires dams, of course, and there has been a worldwide movement to stop the construction of dams as we learn more about the ecological damage caused upstream, downstream, and on the cropland itself by dams and the irrigation water, not to mention the millions of people who have been displaced around the world because their home was going to be underwater in the reservoir behind the dam. I mentioned earlier in the chapter that most of the choice dam sites have already been taken, but not all. Asia has a higher percentage of its land under irrigation than any other area of the world, and China, in particular, is adding to the total. As of the early 2000s, China was in the midst of building three major dams, the largest and most famous of which is the Three Gorges Dam on the Yangtze River. When finished in 2009, it will supply irrigation water, generate hydroelectric power (estimated at more than 10 percent of China's total), and will have forced the migration of more than one million people.

The World Bank has very succinctly summarized the world's water resources situation as follows:

> During the past century, while the world's population tripled, the aggregate use of water increased sixfold, with irrigation consuming over 70 percent of available water. These increases have come at high environmental costs: half of the world's wetlands disappeared over the last century, with some rivers now no longer reaching the sea, and 20 percent of freshwater fish now endangered or extinct. If current trends continue, 4 billion people will live under conditions of severe water stress by 2025, particularly in Africa, the Middle East, and South Asia (Shafik and Johnson 2003:2).

Assessing the Damage Attributable to Population Growth

The role of population in environmental degradation differs from place to place, from time to time, and depends on what type of degradation we are discussing. In general terms, however, environmental degradation can be seen as the combined result of population growth, the growth in production (transformation of products of the natural environment for human use) that we call economic development, and the technology applied to that transformation process. Ehrlich and Ehrlich (1990) have summarized this relationship in their **impact (IPAT) equation:**

$$\text{Impact } (I) = \text{Population } (P) \times \text{Affluence } (A) \times \text{Technology } (T)$$

Impact refers to the amount of a particular kind of environmental degradation; population refers to the absolute size of the population; affluence refers to per-person income; and technology refers to the environmentally damaging properties of the particular techniques by which goods are produced (measured per unit of a good produced). "Technology is double-edged. An increase in the technical armoury sometimes increases environmental impact, sometimes decreases it. When throwaway cans replaced reusable bottles, technology change increased environmental impact. When fuel efficiency in cars was increased, impact was reduced" (Harrison 1993:237).

Barry Commoner (1972; 1994) proposed a slight variation on the Ehrlich and Ehrlich formula that better allows the researcher to assess the relative contributions that population growth, affluence, and technology might have on environmental degradation by examining specific types of pollutants:

$$\text{Pollution} = \text{Population} \times (\text{good/population}) \times (\text{pollution/good})$$

If we want to measure the relative impact of population on the pollution generated by motor vehicle traffic (the "good"), we measure the (good/population) as being equal to the number of vehicle miles per person; whereas the (pollution/good) is measured as carbon monoxide emissions per vehicle mile driven. For example, between 1970 and 1987 carbon monoxide emissions declined in the United States by 42 percent, due to a combination of increased regulation and higher fuel costs that spurred automobile manufacturers to lower emissions levels and improve the fuel efficiency of cars. Commoner (1994) has shown that technology change (a 66 percent reduction in pollution per automobile mile driven) lowered the overall pollution from carbon monoxide, despite the fact that population in the United States increased 19 percent during that time and the number of miles driven per person increased by 45 percent. Population and affluence were pushing pollution upward, but that was counteracted by technology change. That was a hopeful trend.

Although those of us living in developed nations still consume a vastly disproportionate share of the earth's resources and thus contribute disproportionately to the pollution crisis, the rates of population growth and economic development in developing countries mean that the global impact is shifting increasingly in that direction. For example, the data in Table 12.3 show that China and India are among the five biggest CO_2 polluters in the world, despite very low per-person rates in those two relatively poor countries. Secondly, notice that technological improvements are already operating to dampen the environmentally degrading impact of consumption, but population growth has been exerting continual upward pressure on degradation.

You can see the dilemma here: Just to maintain the current impact on the environment, technology must completely counteract the impact of population growth and increasing affluence. Much of the affluence in developed nations has come at the expense of the rest of the world—we have used resources without paying for them because the cost of goods we purchased did not typically include the environmental costs associated with their production and consumption (Brown and Mitchell 1998; Daly 1996; Turner, Pearce, and Bateman 1993). We cannot continue indefinitely to draw down the "capital" of nature to supplement our income. The price of goods increasingly will have to include some measure of the cost of dealing with the environmental impact of making that product (the pollution from the manufacturing process) and the cost of getting rid of the product when it is used up (the pollution from waste). Measuring the cost of goods in this way may slow down the rate of economic development, measured in a purely economic way, but it should increase overall human well-being by balancing economic growth with its environmental impact. Nowhere in this set of equations can it be concluded, however, that increased population is beneficial. Population growth is something that must be coped with at the same time that we continue to try to slow it down, because "rational people do not pursue collective doom; they organize to avoid it" (Stephen Sandford, quoted by Harrison, 1993:264). In the world as a whole we expect there to be close to nine billion people by the middle of this century—all of whom will likely be hoping for a good diet and a reasonably high standard of living. Is it possible not only to provide that kind of development, but to sustain it?

Sustainable Development—Possibility or Oxymoron?

In 1987, the United Nations World Commission on Environment and Development issued its extremely influential report "Our Common Future" (World Commission on Environment and Development 1987). This is usually called the report of the Brundtland Commission, named for its chair, Gro Harlem Brundtland, who was then Prime Minister of Norway and later became the Director of the World Health Organization. The Commission defined the now-popular term **sustainable development** as "... development that meets the needs of the present without compromising the ability of future generations to meet their own needs" (p. 43). At the same time, and for the first time in the world arena, the issue of environment and equity was laid on the table. It was made clear that part of the environmental problem is that some (rich) nations are consuming too much, while at the other end of the continuum, environmental problems are caused by people living in poverty who use the environment unsustainably because their own survival is otherwise at stake. Within the concept of sustainable development, the Commission recommended that "overriding priority" should be given to the essential needs of the world's poor.

The Brundtland Commission defined sustainable development in a deliberately vague way. That has had the advantage of building a worldwide constituency for the concept, with the attendant disadvantage that everybody defines it the way they want to. One of the most popular ways to define sustainable development is to translate it to mean "let's sustain development," implying that economic growth is the best solution to all of the world's problems (Chambers, Simmons, and Wackernagel 2002; Daly 1996; Wackernagel and Rees 1996). In particular, economic growth is viewed as the way to salvation for the world's poor. The World Bank and the United Nations have taken up the theme of eliminating poverty as one of the world's important missions. The World Bank's Development Report for 2000/2001 was subtitled "Attacking Poverty" (World Bank 2000), and the World Bank and the United Nations collaborated to design a set of Millennium Development Goals, which are listed for you in Table 12.4. The goal of reducing poverty in developing nations is without question an important one, but it sidesteps the issue that continues to drive poverty in developing nations—population growth and its aftermath (which is the impact of the age transition).

The Brundtland Commission report led directly to the first Earth Summit—the United Nations Conference on Environment and Development in Rio de Janeiro in 1992, with a series of follow-up meetings, especially the one in Kyoto in 1997 at which a framework was established for a worldwide treaty to limit long-term carbon emissions (a treaty the Bush Administration refused to sign when it came into power in the U.S. in 2001). Another follow-up meeting of the United Nations Framework Convention on Climate Change was held in 2000 in the Hague, but ended with no firm agreements being reached beyond those agreed to in Kyoto. The UN meetings have focused largely on climate change, continuing the concern over the polluting aspect of resource consumption. This is a good thing, but it is only part of the picture because it, too, sidesteps the aspirations of a growing population to have access to the very resources that are causing these environmental problems.

The World Bank's Development Report for 2003 was about sustainability (World Bank 2003b) and was written in response to the United Nations World Summit on Sustainability (the so-called Earth Summit 2002) held in Johannesburg, South Africa, in 2002. The report did address the need to monitor the age and urban transitions, but population was not the central focus. The concern among demographers that population issues might be buried in the Earth Summit discussions was expressed very well at the time by the Global Science Panel on Population & Environment: "If we do not put the human population at the core of the sustainable development agenda, our efforts to improve human well-being and preserve the quality of the environment will fail" (Lutz and Shah 2002:1).

The Malthusian specter of sheer numbers of people potentially exhausting available resources is rather overwhelming, and, as a result, it disguises other, more subtle consequences that population growth has for economic development and degradation. The consequences to which I refer are those associated with each of the

Table 12.4 Millennium Development Goals for World Community

1. Eradicate extreme poverty and hunger
2. Achieve universal primary education
3. Promote gender equity and empower women
4. Reduce childhood mortality
5. Improve maternal health
6. Combat HIV/AIDS, malaria, and other diseases
7. Ensure environmental sustainability
8. Develop global partnership of development

Source: http://www.developmentgoals.org, accessed 2003.

transitions that nations pass through during the course of the overall demographic transition.

The Impact of Each of the Transitions

The first transition I discussed was the *mortality transition* (which incorporated the epidemiological transition and the nutrition transition). Declining mortality means more than people surviving longer; it is associated with healthier people who consume more food. Improved nutrition brings with it an increase in the size of the average human, and in the number of daily calories used per person (Fogel and Helmchen 2002). Table 12.5 shows the change in daily caloric intake in the world in the 40 years between 1961 and 2001. You can see that there was a 24 percent increase in the world as a whole, with a higher increase in calories from animal products (mainly meat) than from vegetable products. So, the food capacity of the world was obviously outpacing population growth during this time period, because the average person was consuming more in 2001 than in 1961. The increase was most noticeable among residents of developing countries, where there was a 39 percent increase overall in caloric intake, composed of a 30 percent increase in vegetable products and a 157 percent increase in the agriculturally less-efficient animal products.

The increase in body mass is a good thing for economic productivity—bigger, healthier people work better than smaller, sicklier people—but we have already seen that there was a slowdown in the last part of the twentieth century in inputs to agriculture. That raises the obvious question of whether or not it will be possible to grow enough food for a population that is increasing both in number and in body mass.

The second transition we discussed was the *fertility transition*. Declining fertility is beneficial for both individuals and societies. At the individual level, it

Table 12.5   Differences in Per-Person Daily Caloric Intake Over Time

	Calories per day per person		
	1961	2001	Percent increase
World—all food products	2255	2807	24%
From vegetable products	1916	2347	22%
From animals products	338	460	36%
Developed countries—all food products	2947	3285	11%
From vegetable products	2179	2428	11%
From animals products	768	856	11%
Developing countries—all food products	1929	2675	39%
From vegetable products	1793	2325	30%
From animals products	136	350	157%

Source: Food and Agricultural Organization of the United Nations, FAOSTATS, http://www.fao.org, accessed 2003.

encourages women to delay marriage, improve their education, and become socially more empowered and economically more productive. Greater productivity translates into higher incomes and more savings, which encourages economic development. At the societal level, of course, the fertility decline produces the now-famous demographic windfall, which I will mention again below. The only issue with respect to greater productivity is whether or not it can be sustained and whether or not we will be able to control the degrading by-products of a vastly greater transformation of products of nature into products for human consumption.

The third transition was the *migration transition*, which increasingly is bound up with the *urban transition*. Migration tends to move people from economically less well-off places to better-off places, and of course urbanization does that very explicitly. This can accelerate the process of development, which in general is a good thing for the people involved, but may not be so good in terms of its impact on the environment (York, Rosa, and Dietz 2003). On the other hand, it may be that moving people into urban places will actually lead to improved environmental management of natural resources, as a reaction to the increased concentration of problems that the urban transition brings with it. When people are spread out in rural areas, it is less obvious that they are degrading the environment, whereas the impact is far easier to see and thus act on in urban areas.

The *age transition* is the most obvious of the transitions in terms of its effect on economic development. A rapidly growing population has a young age structure, as you will remember from Chapter 8. This youthfulness leads to a high level of dependency, which puts severe strains on the economy's ability to generate savings for the investment needed for industry to create the jobs sought by an ever-increasing number of new entrants into the labor force. A major theme of the very influential Coale and Hoover (1958) study of economic development was that a high rate of population growth leads to a situation in which the ratio of workers (people of working age) to dependents (people either too young or too old to work) is much lower than if a population is growing slowly. This means that in a rapidly growing society, each worker will have to produce more goods (that is, work harder) just to maintain the same level of living for each person as in a more slowly growing society. I have made this point before: The parents of six children will have to earn more money than the parents of three just to keep the family living at the same level as the smaller family. But it goes deeper than that. A nation depends at least partially on savings from within its population to generate investment capital with which to expand the economy, regardless of the kind of political system that exists. With a very young age structure, money gets siphoned off into taking care of more people (buying more food, and so on) rather than into savings per se. This forces countries to borrow from wealthier nations in order to build the infrastructure needed for economic development.

At the same time, as we have seen, there are some distinct economic payoffs from fertility decline. One demographic part of the "Asian Economic Miracle" (the rapid economic development in east Asia during the 1990s) was the fact that declining fertility in that region has been associated with an increase in savings (which generates capital for investment in an economy) (Higgins and Williamson 1997; Mason 1997). An analysis of data in Thailand has shown that even in rural populations in a poor country, couples with fewer children are better able to accumulate wealth than are couples with large families (Havanon, Knodel, and Sittitrai 1992). Importantly, the evidence suggests that the faster the decline in fertility, and thus the more rapid the age transition, the sooner savings rates rise, and the longer they stay at a high rate (Lee, Mason, and Miller 2001).

As countries move through the epidemiological and fertility transitions, the age structure inevitably becomes older and, as Mason (1988) has pointed out, an older age structure may also be conducive to lower levels of saving, since in retirement people may be taking money out rather than putting it in. In between the young and the old age structure—in between the start and the end of the age transition—is the period in which the age structure helps promote economic development. All of this suggests that there is a curvilinear, rather than a straight-line, relationship between the demographic transition and its relationship to economic development (and thus to environmental degradation). This implies that the end of the age transition (if there is such a thing—after all, fertility could rise again) may be associated with lower levels of economic productivity than the middle stages of the transition.

Finally, the *family and household transition* plays a role that can be thought of as generally detrimental to the environment. Diversity of living arrangements may encourage the kind of independence that promotes innovation and higher productivity, but it may also lead to a less-than-efficient use of resources. As the number of households increases because the number of people per household is declining, people are less likely to share the use of basic resources such as water and energy. Heating a house, for example, uses about the same amount of energy whether the house is occupied by one or five people. A car uses about the same amount of energy whether it is occupied by one or five people. Washing dishes or clothes for one person may use just about the same amount of water as for two people. You get the idea.

In general, the data suggest that the overall rate of population growth is less closely related to economic development than is the rate of change in each of the several transitions that comprise the demographic transition. No one has yet put together a computer simulation model that incorporates all of these elements into it in order to assess the independent effects of each transition, but underlying the complexity of such a model is the fact that demographic change is an important contributor to the sustainability of our level of living. Of course, demographic change is not the only important variable in economic development and environmental degradation. Demographic opportunities won't necessarily be converted to economic improvement unless other important nondemographic factors are in place (Cincotta and Engelman 1997). Individual freedom, property rights, free markets, and democratic governments may well be the keys to whether a country will continue to develop and achieve high levels of economic development, or will falter at this point and stagnate economically (see, for example, De Soto 2000; World Bank 2003b). However, to reach a high level of economic development it seems reasonable to suggest that a country must bring its rate of population growth down to a very low level, and that must happen much more quickly for peripheral nations than it did in the history of the core nations, because the less-developed nations have experienced economic development in the context of much more rapid rates of population growth than the now-developed countries did. Furthermore, the now-developed countries have achieved high levels of living (which means high levels of consumption of resources), to which most other nations seem to aspire. Will the combination of population growth for another several decades and our continuing quest for higher standards of living push us beyond the point of sustainability?

Are We Overshooting Our Carrying Capacity?

In animal populations, overshoot occurs with a certain regularity in some ecosystems, and the consequence is a die-back of animals to a level consistent with resources, or at the extreme, a complete die-off in that area. A good rain one winter may produce an abundance of food for one species, creating an abundance of food for its predators, and so on. In classic Malthusian fashion, each well-fed species breeds beyond the region's carrying capacity, and when normal rainfall returns the following season there is not enough food to go around, and the death rate goes up from one end of the food chain to the other. Biologists have been documenting such stories for a long time.

Premodern humans were susceptible to the same phenomena. The apocryphal story is told of the goat that destroyed a civilization. A civilization existed that depended heavily on goats for meat and milk. "The goat population thrived, vegetation disappeared, erosion destroyed the arable land, sedimentation clogged what once had been a highly efficient irrigation system. The final result was no water to drink or food to eat. It did not happen overnight, but gradually the people had to leave to survive and the civilization perished" (Freeman 1992:3). In general form, this is apparently what happened to the great Mesopotamian civilizations of Sumeria and Babylon that flourished in Western Asia nearly 9,000 years ago. The region at the time was covered with productive forests and grasslands, but each generation over time made greater and greater modifications to the environment—deforesting the area and building great irrigation canals. Around 1900 B.C., it appears that the population peaked at a level that greatly exceeded the ecosystem's carrying capacity (Simmons 1993). A combination of environmental degradation, climate change, drought, and a series of invading armies led to a long-term decline in population in the region (Miller 2004), and the area became the barren desert that today makes up parts of Iran and Iraq. In more recent history, the Mayan civilization in Central America reached a peak of population size about the year A.D. 800, but the civilization collapsed as the population overshot the region's agricultural capacities, perhaps aggravated by a severe drought (Hodell, Curtis, and Brenner 1995).

These are only a few among many stories of premodern humans exceeding the carrying capacity of a region. Human life has survived these catastrophes, but human civilizations have not. Carrying capacity is, to be sure, a moving target. We know that the carrying capacity of the earth is greater than Malthus thought because we have discovered that certain kinds of technological and organizational improvements can improve the productivity of the land. Cornucopians (the boomsters in farmer attire) assume that human ingenuity will permit a continued expansion of carrying capacity up to the point at which the world's population stops growing of its own accord which, according to the medium variant of the United Nations projections, could be in the year 2050, with a population of just less than nine billion (United Nations Population Division 2003c). However, the stories of Mesopotamia and of the Mayans remind us that factors generally beyond our control, such as the weather, can also reduce the carrying capacity of a region—and perhaps of the entire planet. Furthermore, as you know, we as humans have been in the process of trashing our environment to the point that we might ourselves be lowering the carrying capacity of the earth, rather than increasing it. This phenomenon is less visible at the local level than it used to be because, especially in the highly industrialized nations, we use up resources that are very distant from where we live, and we try to dump our pollution as far away as we can. I discuss this more in the essay that accompanies this chapter, dealing with the concept of the **ecological footprint.**

Let us assume that for all of human history up to the beginning of the nineteenth century, the carrying capacity of the globe was essentially fixed (even if it fluctuated over time from region to region) and was greater than the existing global population, but that the gap between population and sustainable resources had been narrowing—the perception of which spurred Malthus to write his *Essay on Population.* The Industrial Revolution was associated with increasing population growth, of course, but also with innovations in agriculture that allowed food production to stay ahead of that population growth, as I discussed earlier, and at least some of these innovations in growing food are certainly sustainable, so it is reasonable to assume (even if cannot be proven) that the carrying capacity is greater today than it has ever been in history. Thus, the assumption is that, since the dawn of the Industrial Revolution, both population size and the global carrying capacity have increased. But the problem is that we don't know whether or not we have now already exceeded that carrying capacity. If we have exceeded carrying capacity, then we are in a period of global overshoot and will face a catastrophe down the road.

There is strong evidence that we have, indeed, exceeded the global carrying capacity for sustaining more than about two billion people at the current North American standard of living (Chambers et al. 2002; Cohen 1995; Pimentel et al. 1997). If you and everyone else in the world were content to live at the level of the typical South Asian peasant, then the number of humans the world could carry would be considerably larger than if everyone were trying to live like the board of directors of General Motors. Indeed, it is improbable that the world has enough resources for six billion people (much less nine billion) to ever approach the standard of living of a successful business executive. That implies that we are doomed to global inequality with respect to the consumption of resources, unless we in the highly developed world are willing (or are forced) to dramatically lower our standard of living. Chase-Dunn (1998) put it this way: "[I]f the Chinese try to eat as much meat and eggs and drive as many cars (per capita) as the Americans the biosphere will fry" (p. xxi). A headline in *The Wall Street Journal* in late 2003 noted ominously that "China's Growing Thirst for Oil Remakes the Global Market" (Wonacott, Whalen, and Bahree 2003:A1), and that followed a story from earlier in 2003 about "The Great Leap Forward: Demand for Cars in China Is Accelerating at a Remarkable Rate" (*The Economist* 2003b).

But, even if we assume that we have not yet exceeded the global carrying capacity, how much additional room do we have before we do? Are we headed in the same direction as Mesopotamia and the Mayan Civilization? One of the most elaborate and well-known empirical investigations to examine this question was the "Club of Rome study," *Limits to Growth* (Meadows 1974; Meadows et al. 1972). This study addressed the population size that would enable the earth to maximize

HOW BIG IS YOUR ECOLOGICAL FOOTPRINT?

Demographers are sometimes at a loss to explain why the relationship between population growth and change does not show up more clearly in the statistics of world development. It is intuitively obvious that more people consuming resources and leaving behind the detritus of the industrial world are detrimental to the long-term health of the planet. But it is maddeningly difficult to show that population growth in Mexico, for example, is more or less damaging to the earth than population growth in Indonesia. One of the problems is that those of us in the highly industrialized countries don't always make the biggest mess where we live—we are able to get someone else somewhere else to make our mess for us, which means that they (usually in less-developed nations) in turn have an environmental impact that really should not be directly attributed to them. The best way to visualize this is with the concept of the *ecological footprint.*

An ecological footprint has been defined as the land and water area required to support indefinitely the material standard of living of a given human population, using prevailing technology (Chambers et al. 2002; Rees 1996; Rees and Wackernagel 1994; Wackernagel and Rees 1996). For most of human history, this was easy to figure out. If you farmed 10 acres and all of your needs were met from the resources within those 10 acres and all of your waste products were deposited within that acreage, then you did not influence life outside that zone in any demonstrable way. But urbanization changed all that because, as I noted in Chapter 11, an urban population requires that someone else in the countryside grow their food, cut their wood, gather their sources of energy, and stow their trash. Thus, it is not easy to determine how big an impact an urban resident has on the resources of the earth, since the resources are drawn from multiple sources and the waste is spread out in multiple directions. Urban

Ecological Footprint by Country

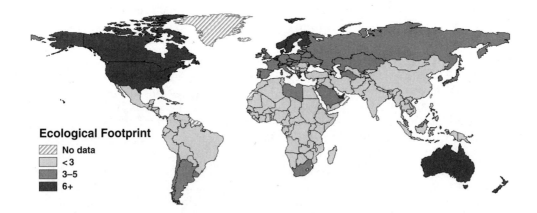

Source: Adapted from data in WWF International, 2002, *Living Planet Report 2002,* Table 2 (data are for 1999); http://www.panda.org/downloads/general/lpr2002.pdf, accessed 2004.

areas are not sustainable on their own; they must borrow their carrying capacity from elsewhere, and this is why it has been so important historically for cities to establish political and economic dominance over the countryside. Who are city residents borrowing from, and how much are they borrowing? Those are questions that William Rees of the University of British Columbia sought to answer, in collaboration with Mathis Wackernagel, then one of his graduate students. They devised a set of calculations to estimate the total land and water required to generate the materials used by humans (including materials for food, shelter, consumer goods, energy, and so forth) and the land and water needed by each human to deposit the waste products generated by consumption of those materials. The benchmark for the late 1990s was that 1.9 hectares of biocapacity were available to each human being for their use (see the accompanying table).

The wealth of cities virtually ensures that they will exceed the average footprint. Rees calculated that the 472,000 residents of his home city of Vancouver, British Columbia, generated an average ecological footprint of more than four hectares per person, which meant that the city had a footprint of more than two million hectares. Vancouver itself comprises only 11,400 hectares, so Rees points out that "the ecological locations of cities no longer coincide with their locations on the map" (Rees 1996:2). A city like Tokyo, for example, has an ecological footprint that would cover a large section of southeast Asia if it were aggregated all in one place. The same analysis can be applied to countries. Many wealthy countries exceed their own carrying capacity by borrowing ecological resources from other parts of the globe. In the long run, sustainability means that those countries that are running an ecological deficit (we might call them the

Ecological Footprints of Nations

Country	Ecological footprint in hectares per person	Available biocapacity per person within the nation	Ecological deficit (if negative) in hectares per capita
United States	9.7	5.3	−4.4
Canada	8.8	14.2	5.4
New Zealand	8.7	23.0	14.0
Australia	7.6	14.6	7.0
United Kingdom	5.3	1.6	−3.7
France	5.3	2.9	−2.4
Russia	4.5	1.6	−2.9
Mexico	2.5	1.7	−0.8
China	1.5	1.0	−0.5
Indonesia	1.1	1.8	0.7
India	0.8	0.7	-0.1
WORLD	2.3	1.9	-0.4

Source: Redefining Progress, Oxford, UK, "Sustainability Indicators Program/Ecological Footprint Accounts/1999 National Account Results." http://www.redefiningprogress.org/programs/sustainabilityindicators/ef/projects/1999_results.html (accessed 2003).

Note: A hectare is 10,000m², or approximately 2.5 acres as measured in the U.S.; one acre is approximately the size of an American football field.

(continued)

HOW BIG IS YOUR ECOLOGICAL FOOTPRINT? (CONTINUED)

"exploiters") must be offset by those who have an ecological surplus. The numbers in the accompanying table summarize data for several countries of the world, and the accompanying map gives you a global perspective. It probably will not surprise you to learn that the average resident of the United States has the largest ecological footprint in the world, at 9.7 hectares per person, which is equivalent to about 24 football fields per person used up to sustain the lifestyle of the average American. The vastness of the country's size and resources offers the average American an ecological capac-

ity of 5.3 hectares per person—well above the world average of 1.9. Yet, Americans still exceed that capacity by 4.4 hectares per person. Canada, on the other hand, has an estimated available ecological capacity of 14.6 hectares, and Canadians use 8.8 hectares each, leaving surplus ecological capacity for the moment, but of course that surplus is essentially being "used" by others, especially by residents of the United States.

The wealth of most European countries causes them to exceed their capacity, but even Mexico was estimated to be exceeding its ecological ca-

the socioeconomic well-being of its citizens. After building a computer model simulating various paths of population growth and capital investment in resource development, this team of social scientists came to the conclusion that the world's population was already so large in the 1970s and was consuming resources at such a prodigious rate that by the year 2100 resources will be exhausted, the world economy will collapse, and the world's population size will plummet. After introducing their most optimistic assumptions into the model, the Meadows' team described the potential result in the following way:

> Resources are fully exploited, and 75 percent of those used are recycled. Pollution generation is reduced to one-fourth of its 1970 value. Land yields are doubled, and effective methods of birth control are made available to the world population. The result is a temporary achievement of a constant population with a world average income per capita that reaches nearly the present U.S. level. Finally, though, industrial growth is halted, and the death rate rises as resources are depleted, pollution accumulates, and food production declines (1972:147).

The authors freely acknowledged that their models did not replicate the complexities of the real world, nor were they attempting to predict the future. Nonetheless, the study demonstrated the possibility that, for the world as a whole, the optimum population was perhaps no larger than the level in the 1970s (much less the level in the early 2000s).

Meadows and associates discussed the need for "dynamic equilibrium" in which population and capital remain constant, while other "desirable and satisfying activities of man—education, art, music, religion, basic scientific research, athletics, and social institutions . . ." also flourish (1972:180). More recently, these same ideas (which in fact hearken back to the nineteenth-century writings of John Stuart Mill— see Chapter 3) have been put forth by Herman Daly (1996), who argues that when we think of sustainable development, we must think of development in qualitative, not quantitative terms. We need to think about improving the quality of our life in pacity as of 1999, and so were China and India. In fact, the bottom line of the table shows that, by these calculations, the average person in the world in the late 1990s required the constant production of 2.3 hectares in order to maintain his or her standard of living, whereas only 1.9 hectares were calculated to be available on a sustainable basis. If these data represent the real impact of humans on the environment, then we have already overshot our carrying capacity, like people who continue to charge things on their credit card without knowing if or how they will pay back that indebtedness.

York, Rosa, and Dietz (2003) examined cross-national comparisons of ecological footprints (using data such as that provided in the accompanying table) and concluded that "[B]asic material conditions, such as population, economic production, urbanization, and geographical factors all affect the environment and explain the vast majority of cross-national variation in environmental impact . . . Taken together, these findings suggest societies cannot be sanguine about achieving sustainability via a continuation of current trends in economic growth and institutional change" (p. 279).

ways that put less burden on the earth's resources, rather than in ways that just use and abuse the earth.

In 1977, in response to the *Limits to Growth* study, President Carter directed that a study be conducted "of the probable changes in the world's population, natural resources, and environment through the end of the century" (Council on Environmental Quality 1980). The result was the "Global 2000" report, released in 1980, and containing the following dreary conclusions:

> If present trends continue, the world in 2000 will be more crowded, more polluted, less stable ecologically, and more vulnerable to disruption than the world we live in now. . . . Barring revolutionary advances in technology, life for most people on earth will be more precarious in 2000 than it is now . . . (Council on Environmental Quality 1980:1–3).

You tell me. Was the world in 2000 "more crowded, more polluted, less stable ecologically, and more vulnerable to disruption" than the world in 1980? The answer has to be yes, although warnings such as these have helped shape policies designed to mitigate the worst effects of population growth and environmental degradation. Still, there is little comfort here for countries not yet fully developed, since for them, the implication is that they need to follow the dual policies of stopping population growth as soon as possible and then hoping for a redistribution of income from wealthier nations.

Summary and Conclusion

If people did not think that population growth was related in some way to economic development, there would be far less interest in population issues than exists today in the world. Although most people seem to share the view that too much population growth is not good, it has been frustratingly difficult to prove a cause-and-effect relationship. Into this void of conclusions have leapt several competing perspectives. The doomsters are neo-Malthusians who argue that population growth must slow down or the economic well-being of the planet will deteriorate over time. Boomsters are those who believe that population growth is a good thing—it stimulates development and is a sign of rather than a threat to, well-being. A third perspective is shared by Marxists, neo-Marxists, and adherents (whether Marxist or not) to the world systems perspective. This position argues that population growth is irrelevant to economic development. It is the relationships among national economies, and a nation's place in the global economic system, that determine the pace and pattern of economic development.

There can be little question that population growth creates long-term pressures on societal resources that must be dealt with. In the final analysis, each of the several perspectives of the relationship between population growth and economic development probably has merit—it is just that each is describing a different part of a complicated process, one that is unfolding differently for today's less-developed nations than it did historically for the now-developed nations. A major difference is the experience with much more rapid rates of population growth among developing countries over the past few decades than were ever encountered in the history of wealthier nations. This creates more opportunities for commerce than might otherwise exist (more feet to be shod, more food to be processed, more people to buy TVs and mobile phones), but it also means that we have been consuming resources at an historically unprecedented rate. Early capitalists were accustomed to saving to build up capital (hence the name), which was crucial as the building block for future success. However, rapid population growth has meant that we have been dipping into our capital—not just the financial capital, but the resource base of the planet. We have been using up our environmental resources and changing the very nature of human life in the process, and that raises some very legitimate concerns about how long we can sustain both development and increasing population size.

The world's rapidly growing population naturally requires an equally rapid increase in food production. Since the world is almost out of land that can be readily cultivated, increases in yield per acre seem to offer the principal hope for the future. Indeed, that is what the Green Revolution has been all about—combining plant genetics with pesticides, fertilizer, irrigation, crop rotation, land reorganization, and multiple cropping to get more food out of each acre. At current levels of technology, it may be reasonable to suppose that the world's population could be fed for many years to come if food can be properly distributed, if farmers in less-developed nations are able to reach their potential for production, and if environmental degradation does not intervene to limit productivity. Whether that happens is more a political, social, and economic question than anything else. What does seem clear is that it is almost inconceivable that all of the world's people will ever be able to eat as Americans currently do. We have almost certainly exceeded the carrying capacity for that level of living.

A major constraint in providing food and a better life for the growing billions of newcomers to the planet is that, in the process of providing for our sustenance, we are degrading the environment—perhaps irreparably. In the early 1990s, Paul Harrison (1993:270) offered a summary that is still accurate:

We are in the embrace of an environmental crisis that is coiling around more and more regions and ecosystems. Accelerating deforestation in the South, forest death in the North, red tides, the ozone hole, the threat of global warming: all have arrived over the space of a mere fifteen years. Underlying them is the long attrition of biological diversity, and the progressive degradation of land. These problems arose when perhaps no more than 1.4 billion people were consuming at levels of at least moderate material affluence. Yet ahead lie four decades of the fastest growth in human numbers in history.

Demographers, of course, cannot provide solutions to the problems of feeding the population or of halting environmental degradation, but they have wrestled mightily with the task of coping with population change, as I discuss in the next, and final, chapter.

Main Points

1. Economic development represents a growth in average income—a rise in the material well-being of people in a society.

2. Economic growth often occurs in conjunction with population growth, but economic development may be hampered by a high rate of population growth.

3. Data for the world indicate that higher levels of average income typically are associated with lower rates of population growth, whereas higher rates of population growth generally are accompanied by lower levels of average income.

4. Ester Boserup and Julian Simon led the "boomster" argument that population growth serves as a stimulus for technological development and economic advancement.

5. Those concerned about international economic relationships (such as proponents of the world systems theory) argue that population growth is irrelevant to the process of economic development, and that no meaningful relationship exists.

6. Neo-Malthusians express the "doomster" view that population growth is detrimental to economic development.

7. Food production can be increased by increasing farmland or per-acre yield. The latter can be accomplished by continued plant breeding and increased use of irrigation, fertilizers, and pesticides.

8. The Green Revolution involved an increase in grain yields through the development of new strains of plants, along with the application of irrigation, fertilizer, pesticides, and often new patterns of land tenure.

9. It is often said that if you give a man a fish, he will eat for a day, but if you teach him how to fish, he will sit in a boat and drink beer all day.

10. It is estimated that the world could not grow enough food for the entire population to eat an average American diet.

11. Remember to practice safe eating—always use condiments.

12. It is perhaps within the realm of possibility that 11 to 12 billion people could be sustained at current levels of agricultural technology if food production and distribution worked far more efficiently and if environmental degradation did not intervene to lower agricultural productivity.

13. In trying to feed ever more people and attempting to elevate world standards of living, we have managed to do a great deal of damage to the environment.

14. We are degrading the lithosphere especially through soil erosion, desertification, deforestation, and dumping waste on the land.

15. We are degrading the hydrosphere especially through depletion of aquifers and pollution of groundwater.

16. We are degrading the atmosphere especially by sending gases into the air that generate photochemical smog, acid rain, ozone holes, and global warming.

17. The immediate or proximate causes of environmental degradation are population growth, economic development, and technology. The relative contribution of each varies from place to place, from time to time, and according to the specific type of degradation.

18. Economic development and the environmental degradation that follows in its wake are influenced in myriad ways by each of the transitions that comprise the overall demographic transition.

19. There is a very real possibility that we have overshot the carrying capacity for life at the average American level; achieving sustainable development will require a different definition of a desirable standard of living and/or acceptance of permanent inequalities in levels of living around the world.

Suggested Readings

1. Ansley Coale and Edgar Hoover, 1958, Population Growth and Economic Development in Low-Income Countries (Princeton, NJ: Princeton University Press).

   This book is such a classic in the area of the demography of economic development that it is the basis of comparison for evaluating virtually all theories and research findings.

2. Nancy Birdsall, Allen C. Kelley, and Steven W. Sinding, Editors, 2001, Population Matters: Demographic Change, Economic Growth, and Poverty in the Developed World (New York: Oxford University Press).

   At the behest of the United Nations Population Fund, a workshop was convened in Bellagio, Italy, in 1998 to reassess the relationship between population growth and economic development. The results of that very important workshop are the chapters in this book, accompanied by an excellent overview chapter by the editors.

3. Glenn Firebaugh, 2003, The New Geography of Global Income Inequality (Cambridge, MA: Harvard University Press).

   Firebaugh reviews data on national income levels over time and concludes that the world has been experiencing a transition in global income inequality in which inequalities between countries are decreasing while inequality within countries is increasing. He discusses this in the context of population changes.

4. Herman E. Daly, 1996, Beyond Growth: The Economics of Sustainable Development (Boston: Beacon Press).

   Daly was one of the first economists to suggest the possible inconsistencies in the global push for sustainable development—especially the general lack of consideration given to population growth in most discussions of sustainable development.

5. AMBIO: A Journal of the Human Environment, Special Issue on Population, Consumption, Environment, 2002 (Volume XXXI, Number 4, June).

   This special issue focuses on the interaction of human populations and coastal ecosystems, but the numerous case studies (accompanied by an excellent overview article) provide the kind of nuanced insight necessary to appreciate how important our interaction with the environment is.

🌐 Web Sites of Interest

Remember that Web sites are not as permanent as books and journals, so I cannot guarantee that each of the following Web sites still exists at the moment you are reading this. It is usually possible, however, to retrieve the contents of old Web sites at http://www.archive.org.

1. http://www.earthday.net/footprint/index.asp
   Calculate your own ecological footprint at this Web site, and estimate your impact on the earth's resources.

2. http://www.povertymap.net/
   The poverty mapping network is a joint effort of the CGIAR (Consultative Group on International Agricultural Research), the UN's Food and Agricultural Organization, and the United Nations Environment Programme. It offers a valuable set of resources for examining the relationships among poverty, food insecurity, and environmental issues.

3. http://www.worldbank.org
   Over the years, the World Bank has replaced the United Nations as the best single source for comparative world data that relate to topics such as population growth and economic development. This Web site summarizes the most recent World Development Indicators (WDI), and contains several tables of data that can be viewed online or downloaded and analyzed on your own computer.

4. http://www.fao.org
   The Food and Agricultural Organization (FAO) of the United Nations is the principal source of data on land use and agricultural production. The data from its annual Production Yearbook are available online at this Web site in an interactive mode that lets you choose the information you want and the format you want it in. Search the site carefully because there are a lot of other things there.

5. http://www.ucsusa.org/index.cfm
   The Union of Concerned Scientists (UCS) is a nationwide nonprofit organization devoted to presenting what it believes is the most objective and trustworthy information about global environmental issues, and promoting policy initiatives to improve the relationship between people and the planet.

CHAPTER 13
Coping With Demographic Change

Figure 13.1 Government View on Overall Rate of Population Growth

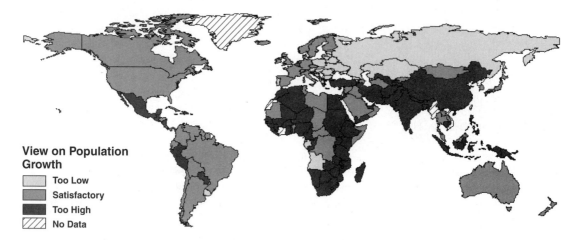

View on Population Growth
- Too Low
- Satisfactory
- Too High
- No Data

Source: Adapted from data in United Nations, 2004, World Population Policies 2003 at http://www.un.org/esa/population/publications/wpp2003/wpp2003.htm, accessed in 2004.

The world's population has been increasing since the days when, more than 200 years ago, Malthus brought it to the western world's attention more famously than had anyone before him. World population growth throughout the nineteenth century was largely taking place in the now-rich countries and you might say that the "boomster" view prevailed that population increase was a good thing and was a sign of prosperity and well-being. Early in the twentieth century, however—after the end of World War I—it seemed increasingly obvious to many observers that population growth was slowing in the rich countries because of declining fertility and was starting to pick up in poorer countries because of declining mortality. This was not seen by everybody as a good thing. Instead, it was a potential problem to be coped with, and it is probably no coincidence that, during this time, the world's two largest professional organizations of demographers were founded—the International Union for the Scientific Study of Population (IUSSP—based in Paris), and the Population Association of America (PAA—based in Washington, DC). Nor is it a coincidence that Margaret Sanger, the founder of the global family planning movement (as I noted in Chapter 6), was involved in the creation of both organizations (Weeks 1996).

The time after World War I was, however, heavily consumed by the worldwide Depression and the rising military threats from Germany and Japan that culminated, of course, in World War II. I mentioned earlier in the book that World War II was a turning point in demographic history because it led to the spread of death control technology all over the world, unleashing a tidal wave of human numbers that still threatens to overwhelm us. Ever since the end of the World War II, we have faced a genuinely unprecedented increase in the number of human beings, and we have spent the time since then coping with that growth. Earlier concerns about low fertility in richer countries were replaced by concerns about the rise in the birth rate—the baby booms—in most of those countries. At the same time, the more rapidly declining mortality in less-developed countries created concerns about when fertility might begin to drop in order to slow the soaring rate of population growth.

Coping tends to be a three-stage process: (1) understanding the fact that unfolding events represent a potential (or real) problem; (2) establishing the motivation to do something about the problem; and (3) generating the means to do something about it. This is, not coincidentally, the same sequence of stages that you saw earlier in the book when we were discussing the preconditions for a fertility transition—understanding that fertility is within individual control, being motivated to limit the number of children, and then having the means available to limit fertility.

In the 1960s and 1970s, it was relatively easy to see that the population of the world was growing too rapidly and that something needed to be done about it. Things were done about it, through direct and indirect policies, and so now, in the twenty-first century, we find that the *rate* of population growth is slowing down, and Europe—where the demographic transition started in the first place—is on the verge of a declining population. But the tremendous momentum built into the world's age structure means that a huge number of people are still being added to the world's total each day—and this will probably continue for the rest of your life. In the process, the implications of population growth and change have grown increasingly complex, requiring new policies and new approaches to policy implementation. In this chapter, your demographic perspective will be put to work looking at how people and nations have tried, and continue to try, to influence demographic events. This is an important use to which a demographic perspective can be put—employing your understanding of the causes and consequences of population growth to improve the human condition, including your own. I conclude this chapter, and thus the book, with an assessment of what may lie ahead for the world, given the trajectory of all of the transitions (mortality, fertility, migration, age, family and household, and urban) working their way through time country by country.

Policies Designed to Influence the Transitions

A policy is a formalized set of procedures designed to guide behavior. Its purpose is to either maintain consistency in behavior or alter behavior to achieve a specified goal. It is, in essence, a management strategy, similar to what might be implemented for a business to achieve a desired profit or an NGO to achieve its social mission (Magretta 2002). **Population policy** represents a strategy for achieving a particular pattern of population change. The strategy may consist of only one specific component—a single-purpose goal—such as to reduce fertility to replacement level by a specific date. Or it may be multifaceted, such as an attempt to improve the reproductive health of women. Naturally, in both cases, the objective requires a policy only if there is some indication that the goal may not be achieved unless a policy is implemented. Note that in the foregoing situations I am referring to a **direct population policy,** one aimed specifically at altering demographic behavior. There are also **indirect population policies,** which are not necessarily designed just to influence population changes but do so by changing other aspects of life. For example, a policy to increase the educational level of women will improve the quality of life of affected women in many ways, and it will also increase the likelihood that they will take control of their reproductive behavior and limit family size.

I have outlined the basic elements of analyzing policy formulations in Figure 13.2. Your first step is to carefully assess the current demographic situation, a technique I have tried to illustrate in virtually every chapter of this book. This is obviously a crucial task, since you have to know where you are now if you expect to

Figure 13.2   Formulating a Population Policy

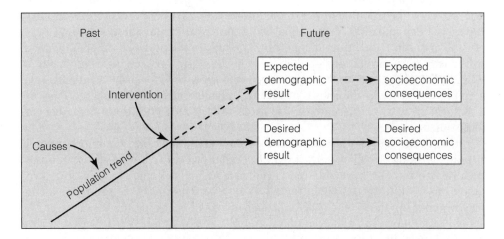

Note: The formulation of a population policy requires an assessment of current trends (requiring an understanding of the causes of population change) and an evaluation of the expected consequences of current trends. If the expected consequences differ from the desired result, then a policy may be implemented to alter the course of demographic events.

Source: Kingsley Davis, 1975, "Demographic Reality and Policy in Nepal's Future," Workshop Conference on Population, Family Planning, and Development in Nepal, University of California, Berkeley, p. 2. Used by permission.

chart a course for the future. Assuming that you can accurately measure (or even carefully estimate) the present situation, your next step is to analyze what the future would bring if society were left to its own devices.

Assessing the Future

We humans have been preoccupied with looking into the future for centuries, since knowledge of what is coming gives us the power to prepare for it or possibly to change it. Predicting or forecasting the future, though, is an almost impossible task except in very general terms, because of the amazing complexity of the social and physical world. As a result, demographers rarely stretch their necks out to predict the future, but rely instead on projections, which are statements about what would happen under a given, specified set of conditions, as I discussed in Chapter 8. For example, when I say that in the year 2004, the world's population was growing so fast that it would double in size in 53 years, that is a projection. In other words, if neither birth rates nor death rates changed from their levels in 2004, twice as many people would be living in the world in the middle of the twenty-first century as were alive at the end of the twentieth century. In fact, we do not *expect* the world's population ever to double again, because we think that fertility rates are going to continue to decline. But that, too, is only a projection—albeit a more reasonable projection than the idea that things will not change.

Even if the world's population never again doubles, it will continue to grow for the next several decades and we know that in some parts of the world the population will very likely double in size in the future, even as other parts of the world may experience depopulation. Projections also permit you to ask questions about other possible paths into the future if conditions change. In the previous chapter, I emphasized the precarious relationship that exists in the world between population and resources. Whether the population is expected to grow by a lot or by a little makes a very big difference in the future impact of humans on the environment. Thus, by looking at an array of alternatives, you can lay the groundwork for anticipating various courses of events.

It is at this stage, in deciding what the future might look like, that most attempts at policy making (whether in the area of population or anything else) break down. Because most of us have different insights and perceptions of the world (even though those differences may be small), it is sometimes impossible to reach a consensus about anything much beyond how many 19-year-olds there will be in the United States next year, and even that is complicated by differing estimates of migration. However, if you try to avoid the issue of assessing the future, you void the possibility of implementing a policy, because you won't have established a case for it. If a policy is to be implemented with the expectation of producing predictable results (and controllable side effects), you must get a bead on the most likely future. If you do, then not only is your policy well grounded but you are also in a position to decide later if your assessment was wrong, and then either abandon your policy as incorrect or alter it in accordance with your revised estimate of where your society is going.

Establishing a Goal

Once you have an idea of what the future may be, or at least a range of reasonable alternatives, you are in a position to compare that with what you aspire to in demographic and social terms. Establishing a goal is not an easy task, and it grows more difficult as the number of people involved in setting the goals grows. As a result, goals are usually general and idealistic in nature; those related to population issues might include improving the standard of living, reducing economic inequalities, promoting gender equality, eliminating hunger and racial/ethnic tension, reducing environmental degradation, preserving international peace, and increasing personal freedom. For example, in Cairo in 1994, delegates to the United Nations International Conference on Population and Development (ICPD) agreed to a set of 15 principles embracing the concept that population-related goals and policies are integral parts of cultural, economic, and social development, all of which are aimed at improving the quality of life for all people. These principles set the framework for specific policy making (as I will discuss later in the chapter), and I have reproduced them as Table 13.1.

The demographic future is assessed primarily with an eye toward determining whether projected demographic trends will enhance or detract from the ability to achieve other broad goals. In other words, population control is rarely an end in itself but rather an "implementing strategy" that helps achieve other goals. This is

Table 13.1   The 15 Principles Underlying Population Policymaking as Established at the Cairo Population Conference

The implementation of the recommendations contained in the Programme of Action is the sovereign right of each country, consistent with national laws and development priorities, with full respect for the various religious and ethical values and cultural backgrounds of its people, and in conformity with universally recognized international human rights.

International cooperation and universal solidarity, guided by the principles of the Charter of the United Nations, and in a spirit of partnership, are crucial in order to improve the quality of life of the peoples of the world.

In addressing the mandate of the International Conference on Population and Development and its overall theme, the interrelationships between population, sustained economic growth and sustainable development, and in their deliberations, the participants were and will continue to be guided by the following set of principles:

Principle 1

All human beings are born free and equal in dignity and rights. Everyone is entitled to all the rights and freedoms set forth in the Universal Declaration of Human Rights, without distinction of any kind, such as race, colour, sex, language, religion, political or other opinion, national or social origin, property, birth or other status. Everyone has the right to life, liberty and security of person.

Principle 2

Human beings are at the centre of concerns for sustainable development. They are entitled to a healthy and productive life in harmony with nature. People are the most important and valuable resource of any nation. Countries should ensure that all individuals are given the opportunity to make the most of their potential. They have the right to an adequate standard of living for themselves and their families, including adequate food, clothing, housing, water and sanitation.

Principle 3

The right to development is a universal and inalienable right and an integral part of fundamental human rights, and the human person is the central subject of development. While development facilitates the enjoyment of all human rights, the lack of development may not be invoked to justify the abridgement of internationally recognized human rights. The right to development must be fulfilled so as to equitably meet the population, development and environment needs of present and future generations.

Principle 4

Advancing gender equality and equity and the empowerment of women, and the elimination of all kinds of violence against women, and ensuring women's ability to control their own fertility, are cornerstones of population and development-related programmes. The human rights of women and the girl child are an inalienable, integral and indivisible part of universal human rights. The full and equal participation of women in civil, cultural, economic, political and social life, at the national, regional and international levels, and the eradication of all forms of discrimination on grounds of sex, are priority objectives of the international community.

Principle 5

Population-related goals and policies are integral parts of cultural, economic and social development, the principal aim of which is to improve the quality of life of all people.

Principle 6

Sustainable development as a means to ensure human well-being, equitably shared by all people today and in the future, requires that the interrelationships between population, resources, the environment and development should be fully recognized, properly managed and brought into harmonious, dynamic balance. To achieve sustainable development and a higher quality of life for all people, States should reduce and eliminate unsustainable patterns of production and consumption and promote appropriate policies, including population-related policies, in order to meet the needs of current generations without compromising the ability of future generations to meet their own needs.

(continued)

Table 13.1   (continued)

Principle 7

All States and all people shall cooperate in the essential task of eradicating poverty as an indispensable requirement for sustainable development, in order to decrease the disparities in standards of living and better meet the needs of the majority of the people of the world. The special situation and needs of developing countries, particularly the least developed, shall be given special priority. Countries with economies in transition, as well as all other countries, need to be fully integrated into the world economy.

Principle 8

Everyone has the right to the enjoyment of the highest attainable standard of physical and mental health. States should take all appropriate measures to ensure, on a basis of equality of men and women, universal access to health-care services, including those related to reproductive health care, which includes family planning and sexual health. Reproductive health-care programmes should provide the widest range of services without any form of coercion. All couples and individuals have the basic right to decide freely and responsibly the number and spacing of their children and to have the information, education and means to do so.

Principle 9

The family is the basic unit of society and as such should be strengthened. It is entitled to receive comprehensive protection and support. In different cultural, political and social systems, various forms of the family exist. Marriage must be entered into with the free consent of the intending spouses, and husband and wife should be equal partners.

Principle 10

Everyone has the right to education, which shall be directed to the full development of human resources, and human dignity and potential, with particular attention to women and the girl child. Education should be designed to strengthen respect for human rights and fundamental freedoms, including those relating to population and development. The best interests of the child shall be the guiding principle of those responsible for his or her education and guidance; that responsibility lies in the first place with the parents.

Principle 11

All States and families should give the highest possible priority to children. The child has the right to standards of living adequate for its well-being and the right to the highest attainable standards of health, and the right to education. The child has the right to be cared for, guided and supported by parents, families and society and to be protected by appropriate legislative, administrative, social and educational measures from all forms of physical or mental violence, injury or abuse, neglect or negligent treatment, maltreatment or exploitation, including sale, trafficking, sexual abuse, and trafficking in its organs.

Principle 12

Countries receiving documented migrants should provide proper treatment and adequate social welfare services for them and their families, and should ensure their physical safety and security, bearing in mind the special circumstances and needs of countries, in particular developing countries, attempting to meet these objectives or requirements with regard to undocumented migrants, in conformity with the provisions of relevant conventions and international instruments and documents. Countries should guarantee to all migrants all basic human rights as included in the Universal Declaration of Human Rights.

Principle 13

Everyone has the right to seek and to enjoy in other countries asylum from persecution. States have responsibilities with respect to refugees as set forth in the Geneva Convention on the Status of Refugees and its 1967 Protocol.

Principle 14

In considering the population and development needs of indigenous people, States should recognize and support their identity, culture and interests, and enable them to participate fully in the economic, political and social life of the country, particularly where their health, education and well-being are affected.

Table 13.1   (continued)

Principle 15

Sustained economic growth, in the context of sustainable development, and social progress require that
growth be broadly based, offering equal opportunities to all people. All countries should recognize their
common but differentiated responsibilities. The developed countries acknowledge the responsibility that
they bear in the international pursuit of sustainable development, and should continue to improve their
efforts to promote sustained economic growth and to narrow imbalances in a manner that can benefit all
countries, particularly the developing countries.

Source: United Nations Population Fund, ICPD Key Documents, http://www.unfpa.org/icpd/icpd_poa.htm#ch2, accessed 2004.

analogous to my comment in Chapter 6, that, for individuals, having children is typically a means to other ends, rather than being intrinsically a desired goal. Thus, when you look at the future course of demographic events, you might ask whether projected population growth will undermine the ability of an economy to develop. Will projected shifts in the age/sex distribution affect the ability of an economy to provide jobs, thereby leading to lower incomes or a greater welfare burden? Will projected growth and urbanization in one ethnic group or another lead to greater intergroup tension and hostility? Will the projected growth of the population lead to a catastrophic economic-demographic collapse that will drastically restructure world politics? Is continued population growth consistent with a goal of sustainable development? Will declining fertility lead to a huge unmanageable population of older people?

Because population policies are only a means to one or more of these other ends, it is easy for population policies to be "hijacked" by the proponents or opponents of these other goals (Basu 1997), but whatever your goal, if it is discrepant with the projected future, you can use demographic knowledge to propose specific policies to avert unhappy consequences. Your work does not end there, of course, because once your policies are implemented, you have to continually evaluate them to see whether they are accomplishing what you had hoped and to make sure they are not producing undesirable side effects.

The process of policy formulation outlined in Figure 13.2 assumes a country or a group of people oriented toward the future and anticipating change. If a population or its government leadership is "traditional" and expects tomorrow to be just like today, policies aimed at altering behavior are not likely to be adopted. If demographic policies exist at all in such a country, they may have as a goal the maintenance of the status quo and are probably inherently **pronatalist;** that is, they might forbid divorce or abortion, impede the progress of women, and so forth. In general, they would probably discourage the kind of innovative behavior that might lead to demographic change. To be sure, that type of attitude may even be associated with poor perceptions of reality which, in their turn, help to preserve the status quo. Randall (1996) provides an interesting example of a traditionally nomadic population in Mali in sub-Saharan Africa. Their fertility levels have been lower than those of sedentary farmers in the area in which they reside, and demographic analysis revealed that this is due to the influence of monogamous marriages in a society in which husbands are substantially older than wives, and where high adult mortality produces a marriage squeeze that reduces the exposure of women to reproduction. However, the local perception is that women have a very high rate of sub-fecundity and that this is a problem that needs to be solved. Any policy implemented on the basis of that perception would likely have very different consequences than a policy based on demographic reality.

The biggest obstacle to population policy in the world is probably gender inequality. In societies where the legal, social, and economic status of women is inferior to men, there is apt to be greater resistance to implementation of policies that lead indirectly to lower fertility by empowering women to have greater control over their own health and well-being. Thus, in the twenty-first century, the world has focused its sights less on achieving specific demographic targets and more on achieving gender equality.

It is also easy to misperceive the important role that demography plays in the modern world. It is, as Westoff (1997) has noted, a large problem with low visibility. As you can appreciate from having read this book, there are few corners of the earth and few aspects of human existence that are not influenced by demography, but the demographic roots of social and technological change are not always recognized for what they are, and so their influence may be denigrated or ignored (Smail 1997).

No country in the world today can completely ignore the issue of population policy. This is partly because population growth is a world renowned issue, even if its visibility may be lower than it should be, and partly because the United Nations Population Division regularly queries each country of the world about its policy position on population growth. The first survey was conducted in 1974, and it has been repeated every few years since. The number of governments saying that their policy is to slow down the rate of population growth has been steadily growing over time. Looking at government attitudes toward the overall rate of growth as of 2001, Figure 13.1, at the beginning of the chapter, shows you the worldwide distribution of countries according to whether their government thinks the overall rate of population growth is too high, too low, or satisfactory. Governments that think the rate of population growth is too high are likely to have a policy to try to lower it. These countries are located in Latin America, Africa, and Asia. The countries thinking that their rate of growth is too low are mainly in Europe and very few of these countries are really trying to do much to increase it. Outside Europe, the countries that think their rates of growth are too low include Israel (for the obvious reason that the Palestinian population is one of the fastest-growing populations in the world), Uruguay, Angola, and Gabon. The latter two countries, both on the west coast of Africa, share in common a very high birth rate, life expectancy that is well below the world average, and oil reserves that may give their respective governments some confidence that larger populations can be afforded.

The group of nations in which the government is satisfied with the growth rate, and more specifically with the birth rate, is divided into those that are satisfied with a low birth rate (North America, the southern nations of South America, Northern Europe, China, and Australia) and those that are satisfied with a high rate (including countries in the northern part of South America, Africa, and Asia). The government of Somalia, for example, says that it is satisfied with the current birth rate—which translates into an average of 7.2 children per woman, and Saudi Arabia's government is similarly satisfied with its birth rate, which produces an average of 5.7 children per woman (Population Reference Bureau 2003).

If a government is so inclined, several different basic policy orientations can be adopted to either close the gap or maintain the fit between goals and projections, based on combinations of managing the mortality, fertility, and migration transitions. Each policy orientation offers a wide variety of specific means that can be implemented to achieve the desired kind of demographic future. These different paths to population growth may seem fairly obvious to us now, but it took us a long time to arrive at this point.

World Population Conferences as Policy Tools

The basic idea of having a world conference to discuss population policy is essentially a variation on the diffusion-of-innovations model. If you gather together experts on the demographic situation and on how human behavior can be transformed to achieve differing demographic goals, then presumably the best practices will emerge and they can be agreed on and copied by everyone.

World Population Conferences of Population Experts

The first of a series of world conferences included mainly population experts and was held in Geneva in 1927. It was organized primarily by Margaret Sanger, with funding that included money from her husband along with a grant from the Rockefeller Foundation, and she hoped it would be a forum at which the issue of birth control could be discussed (Sanger 1938). Perhaps sensing that this nearly taboo subject would be brought to the table, the Conference chairman (Sir Bernard Mallet, president of Britain's Royal Statistical Society) asked her to remove her name from the program (which she reluctantly agreed to do), and neither she nor birth control were mentioned during the conference. Nonetheless, she later judged the conference to be a great success in that it brought together for the first time an international group of experts to discuss world population issues. Furthermore, she memorialized her role in the conference by editing the conference proceedings and arranging for their publication, with her name listed in the title as editor (World Population Conference 1927).

Despite the success of that conference (it launched the International Union for the Scientific Study of Population, as I noted earlier), it would be more than 25 years before it would be repeated. In 1954, the United Nations organized a World Population Conference in Rome, jointly sponsored by the IUSSP, which is generally credited with alerting the world's demographic community about the very high rates of growth in less-developed nations (Harkavy 1995). Following this conference, the Ford Foundation began its deep involvement in worldwide population issues. In 1965, the United Nations joined forces again with the IUSSP for a follow-up World

Population Conference in Belgrade. Up to this point, the United Nations had been reluctant to get involved in policymaking per se, because policy would certainly involve birth control and that was opposed by a seemingly unlikely combination of the Vatican, the French government, and the Soviet Union and other communist countries (Harkavy 1995). The Vatican and the French government (which was officially pronatalist) were opposed to birth control on moral grounds, whereas the communist countries were still swayed by the Marxist idea that overpopulation was impossible in a socialist society, as I discussed in Chapter 3. The papers from the 1965 conference made very plain, however, that population growth was potentially out of control and that concerted world action was advisable. The United Nations responded in 1969 with the creation of the United Nations Fund for Population Activities (UNFPA), which is now known more simply as the United Nations Population Fund.

The 1974 World Population Conference—Bucharest

Unlike earlier world population conferences, which had been organized for professional demographers (and those kinds of meetings still take place), the UNFPA decided to organize World Population Conferences that brought together policy makers from government and from nongovernmental organizations, with the explicit goal of developing world population policy. This had the advantage of bringing to the table representatives of agencies that could actually implement policies, but with the disadvantage that underlying population concerns could be lost in the midst of political battles that had nothing to do with demographic change (Finkle and McIntosh 2002). The first of these policy-oriented conferences was held in Bucharest in 1974. After nearly two weeks of "often acrimonious debate" (Demeny 1994:5), the delegates to the Conference (government representatives from all over the world) agreed to a World Population Plan of Action. The plan was important for several reasons: (1) This was the first time that an international body had established anything approximating a world population policy; (2) The 1984 and 1994 Population Conferences were designed in advance to evaluate and alter this plan as necessary; and (3) The Plan incorporated elements from both sides of the "acrimonious debate," which, in essence, represented the U.S. perspective on the one hand (with the general support of other industrialized nations) and the perspective of developing countries on the other (Finkle and Crane 1975).

Key U.S. government representatives went to the Bucharest meeting with a very clear perspective: Developing countries should establish population growth rate targets, and those targets should be met by the provision of family planning (Donaldson 1990). Highlights of the Conference included a statement that countries are "invited to set quantitative goals," and another statement that countries should "encourage appropriate education concerning responsible parenthood."

In contrast to the U.S. position, leaders of developing countries arrived in Bucharest with the growing conviction that they were the less-advantaged participants in a world economic system organized by and for the benefit of the industrialized nations. The UN itself had tacitly endorsed that view just before the 1974 World Population Conference with a General Assembly resolution calling for a

"New International Economic Order." Developing nations were thus interested in seeing a world plan that integrated population policy with economic development programs. Indeed, it was at this Conference that a delegate from India coined the phrase "development is the best contraceptive" (Visaria and Chari 1998).

The 1984 International Population Conference—Mexico City

In the 10 years between the meetings in Bucharest and Mexico City, most developing countries came to the conclusion that rapid population growth hindered their efforts to develop economically, and many of those countries implemented family planning programs to help slow down population growth. Thus, the stage was set for the 1984 conference to be a better meeting of the minds. However, the United States threw cold water on that prospect by proposing, instead, that "population growth is, by itself, a neutral phenomenon." This was a "choice piece of nonsense," as Demeny (1994:10) has remarked, and it was inspired by Julian Simon's influence within the Reagan Administration (and abetted by a coalition of groups opposed to abortion). Simon's basic premise, as you will recall from Chapter 12, is that population growth can help stimulate development, which, in fact, is what the developing countries have in mind as their ultimate goal. The implication is that if you are focusing your energies on population control rather than economic development (especially establishing free markets and democratic institutions), you may never achieve your goal of improving the well-being of your citizens.

It has been said that "in Bucharest, the Chinese had declared that overpopulation was no problem under socialism. In 1984, the Americans declared that overpopulation could be solved by capitalist development" (Donaldson 1990:130). Neither of those views is actually supportable by evidence from the real world, but the turnabout of attitudes among delegates meant that the 1984 Conference did not move the world population policy agenda along in any important way. Demeny (1994) has succinctly characterized the outcome of the Mexico City conference as follows:

> The "Recommendations for the Further Implementation of the World Population Plan of Action" adopted at Mexico City amplified the language of Bucharest but added no qualitatively novel elements. The 88 Recommendations placed added emphasis and urgency on concerns with the environment, the role and status of women, meeting "unmet needs" for family planning, and education (1994:12).

At the conclusion of the conference, the delegates agreed to a Declaration on Population and Development, summarizing the major accomplishments of the conference. The consensus was that the World Population Plan of Action adopted at Bucharest was still valid, even if countries had not yet done very much to implement their national population policies.

One of the most significant events at the Mexico City Conference was the announcement by the United States that any organization in the world that received money from the U.S. government was prohibited from providing abortion, abortion counseling, or abortion-related services. This was called the "Mexico City Policy" and later dubbed the "Global Gag Rule." The Reagan Administration, which had devised this policy, later decided that the UNFPA was in violation (despite repeated denials by the United Nations) and the Reagan and then the Bush administrations withheld all U.S. funding from the UNFPA from 1987–92. U.S. funding was reinstated by the Clinton administration in 1993, but withheld again when George W. Bush took office in 2001.

The Build-Up to the Cairo Conference

In the early 1990s, as people began to anticipate the United Nations' plans for a 1994 Population Conference, the discussion splintered into the three corners of the debate about the relationship between population growth and economic development outlined in Chapter 12. Representing the view that the world has a tremendous population problem that we must collectively continue to try to resolve were experts in the social and environmental sciences who study the issues, as well as people in the public and medical health service fields who deal regularly with the repercussions of rapid population growth (see, for example, Westoff 1994).

Representing the view that population growth was not the issue—that building free markets and capitalism were the important tasks—was, once again, Julian Simon and other "boomsters" (see, for example, Eberstadt 1994a). From this group's perspective, there was no particular reason to have a population conference, and that view was shared by the Vatican, which has been opposed to all of the world population conferences, given their emphasis on the provision of family planning.

The newest and loudest voices added to the debate over world population policy were those of feminists, whose arguments centered on the idea that abstract population targets represent additional ways in which men try to exploit women for their own purposes. In particular, the idea that "population control is a euphemism for the control of women" (Seager 1993:216) drew evidence from the coercive policies put into place in China. Although in agreement on this point, the feminist movement split into two differing views on the importance of population issues and thus on the relevance of a population conference. One view was an essentially neo-Marxist position that the problems in the world are caused not by population growth but by inequities in the distribution of power and resources (Hartmann 1995). The other feminist perspective did not deny the importance of population growth as a constraint to human well-being, but argued that the most important issue for the world is to improve the levels of women's rights and reproductive health on a global basis (see, for example, Chesler 1994; Dixon-Mueller 1993). In so doing, it was argued, the "solution" to the high birth rate would be effected.

The Role of Nongovernmental Organizations (NGOs)

The world's experience over the past several decades suggests that when governments are committed to a program of reducing population growth, that goal is likely to be accomplished. The reality is, however, that most governments of the world prefer to tread lightly on issues of human reproduction. Furthermore, the collapse of the Soviet Union and the rise of global market forces have turned attention away from the role that government plays in life and have focused attention on the influence of the private sector. The social service equivalent of the private sector is represented by **nongovernmental organizations,** or so-called NGOs. Jain (1998) argues that NGOs are "among the most influential actors in the policymaking process" (p. 11). NGOs include private donor agencies, such as the Rockefeller Foundation, the Ford Foundation, the Andrew J. Mellon Foundation, the Packard Foundation, the Hewlett Foundation, and the Wellcome Trust. Other NGOs work with their own money and/or that of donor agencies (both private and public) to provide direct reproductive health services in developing countries. These include organizations such as the International Planned Parenthood Federation (IPPF), Pathfinder International, John Snow, Inc., and Management Sciences for Health. Another group of NGOs is largely devoted to providing information and education materials and resources that can be used to promote population policy messages wherever they might be useful. Organizations in this category include the Futures Group, Population Action International, the Population Institute, The Population Reference Bureau, the Population Resource Center, the World Population Foundation, the World Resources Institute, and Zero Population Growth. These are not meant to be exhaustive listings, but they give you an idea of which organizations are NGOs in the field of population.

The value of NGOs is that they are not burdened by a government bureaucracy, and they have well-developed policy initiatives, so they can help set the agenda for a nation's population program. They provide technical assistance that government personnel might otherwise lack, and they can also fund demonstration projects to convince government leaders of the value of population programs (Jain 1998). For example, in the 1950s and 1960s, the Ford Foundation was especially instrumental in organizing population policy in India, and its activities helped draw the United States government into the population movement (Harkavy 1995). The work of NGOs in the population field tends to focus on reducing both infant and maternal mortality while at the same promoting gender equality and family planning. These programs are thus not always devoted exclusively and explicitly to retarding population growth, but that is typically an underlying motivation. NGOs emerged as important policy drivers at the 1994 Population Conference.

The 1994 International Conference on Population and Development (ICPD)—Cairo

The International Conference on Population and Development was preceded by several preparatory meetings (PrepComs) sponsored by the United Nations Population Division and the United Nations Population Fund, as well as meetings sponsored by NGOs, and by a plethora of books, professional journal and magazine articles, newspaper columns and editorials, talk radio and TV shows, and any other imaginable medium of human communication. In the process, the World Population Plan of Action from the 1974 Bucharest meeting was essentially scrapped, and in its place a new Programme of Action was drafted and circulated in printed form to the delegates and made available to the rest of us on the Internet.

Most of the content of the Programme of Action had been agreed to ahead of time at the PrepComs, but controversial material was bracketed (literally—the text had brackets around it) for discussion by the delegates in Cairo. The most controversial issue was abortion, and the Vatican created worldwide public attention with its opinion of the matter. After five days of debate and special committee meetings, the Vatican did succeed in altering some of the language regarding safe abortions.

As approved at the conclusion of the Cairo conference, the Programme of Action covers an extremely wide range of topics, in recognition of the fact that the causes and consequences of population growth reach into almost every aspect of human existence. Some critics have complained that many of the action items are too vague to be effective—drafted that way to avoid controversy. Indeed, the conference report stayed away from China's controversial population policies by simply never mentioning China at all! Still, the conference regalvanized world interest in and concern about population growth. Nafis Sadik, then the Executive Director of the United Nations Population Fund, served as secretary-general of the conference, and she summarized the work of the conference as follows:

> The delegates have crafted a Programme of Action for the next 20 years which starts from the reality of the world we live in, and shows us the path to a better reality. The Programme contains highly specific goals and recommendations in the mutually reinforcing areas of infant and maternal mortality, education, and reproductive health and family planning, but its effect will be more wide-ranging than that. This Programme of Action has the potential to change the world.
>
> Implementing the Programme of Action recognizes that healthy families are created by choice, not chance.
>
> The Programme of Action recognizes that poverty is the most formidable enemy of choice. One of its most important effects will be to draw women into the mainstream of development. Better health and education, and freedom to plan their family's future, will widen women's economic choices; but it will also liberate their minds and spirits. As the leader of the Zimbabwe delegation put it, it will empower women, not with the power to fight, but with the power to decide. That power of decision alone will ensure many changes in the post-Cairo world (Sadik 1994:4–5).

Delegates agreed to a series of action items that I will discuss in more detail later in the chapter. Some of the action items had numeric target goals associated with them, and the major target agreed to was to try to limit world population to no more than 9.8 billion as of 2050 (this was the United Nations' medium variant projection at the time of Conference). Perhaps most importantly, the concerns that feminists had about the role of women in society were addressed very specifically in a variety of places within the Programme of Action and many were specifically endorsed for funding by organizations such as the U.S. Agency for International Development (Daulaire 1995) and the World Bank (World Bank 1994a). It is in this context of reproductive decision making that the language of the 1994 International Conference on Population and Development differs most from its predecessors, and because of the range and scope of its recommendations, the United Nations agreed to a follow-up meeting in 1999, rather than waiting a full 10 years to assess progress on world population policy.

The 1999 Cairo + 5 Meeting—The Hague

In 1999, the United Nations convened an international forum in The Hague, Netherlands, to review country-by-country accomplishments in achieving the Programme of Action laid out at the 1994 ICPD. This session laid the groundwork for a Special Session of the United Nations General Assembly held in July 1999 to consider these accomplishments. That session was kicked off with an important address by Kofi Annan, UN Secretary-General, in which he noted that the ICPD "was part of a process, going back twenty-five years or more, during which we have all learned that every society's hopes of social and economic development are intimately linked to its demography." He added that "the world does understand . . . that we have to stabilize the population of this planet. Quite simply, there is a limit to the pressures our global environment can stand" (Annan 1999:632–33). You will recognize these as ideas I have spent this entire book trying to explore with you.

The General Assembly then went on to accept a proposal by Brazil that public health systems should train and equip health service providers and should take other measures to ensure that abortion is safe and accessible in countries where abortion is legal. The Vatican objected to this, of course, but Brazil—the proposer of this policy—is a predominantly Catholic country, so there was little other dissent. The General Assembly also approved a new set of benchmarks for the Programme of Action to be monitored by the UNFPA.

No More Conferences?

The 1994 Conference had established goals for the world that extended 20 years out from that date, but it was still expected that a 2004 conference would be held, as a way of reinforcing the importance of those goals. However, for a variety of reasons, including the fact that the Bush administration withdrew all U.S. funding for the UNFPA, the decision was made to not have a 2004 World Population Conference, but rather to maintain a focus on the Programme of Action, especially since many of those action items had been incorporated into the Millennium Goals, which I discussed in Chapter 12. The UNFPA received money from the Swiss government to help its regional offices coordinate regional and country-by-country reviews of progress toward the ICPD goals, and The Population Institute, a U.S.-based NGO, organized a global conference in 2004 designed to be a replacement of sorts for the conference the United Nations decided not to hold.

Population Conferences in the Context of Demographic Theory

All three World Population Conferences (excluding ICPD+5, which was just an extension of the 1994 meeting) were built on the basic premise that population growth is a potential hindrance to a country's ability to develop economically, and that a plan for the improved well-being of a population needs to include a strategy for limiting population growth (even if some delegates did not necessarily agree with that premise). All three conferences further accepted the idea that mortality reduction should always be a high priority, despite the fact that reductions in mortality put added pressure on population growth. All three conferences accepted the idea that migration should be controlled, although none of the conferences were very specific about how to do that.

The biggest problem that each conference dealt with was how to cope with high fertility. If we follow the U.S. perspective over time, we see that each of Coale's three preconditions for a fertility decline has been presented as the predominant U.S. policy prescription, except that they have been presented in reverse order. In 1974, the United States thought that family planning (Coale's third precondition) was the most important aspect of fertility on which to concentrate policy initiatives. In 1984, the emphasis on development implicitly recognized that the motivation to reduce fertility (Coale's second precondition) must accompany (indeed, precede) the demand for family planning services. Only by 1994 did the U.S. position get to the first precondition—that couples (especially women) must be empowered to take control of their own reproductive decisions. Only when that happens will the other preconditions become relevant for policy. To be sure, each of these elements was listed in the Action Plans of all three conferences, but with respect to fertility, in particular, the emphasis is important because that influences what will likely be funded by international donors and put into place by countries facing rapid population growth.

Finally, then, by 1994 it seemed as if the world community may have gotten it right in terms of a global perspective on population policy. Of course, setting a course is a lot easier than taking the trip, and policy making and policy implementation are both extremely difficult tasks. As Rumbaut (1995a) reminds us, policy makers are always faced with a difficult task: "Condemned to try to control a future they cannot predict by reacting to a past that will not be repeated, policy makers are nonetheless faced with an imperative need to act that cannot be ignored as a practical or political matter" (p. 311).

Population Policy in the Twenty-First Century—Managing the Transitions

The Programme of Action approved at the 1994 International Conference on Population and Development (ICPD) in Cairo set out policy prescriptions aimed at each of the several transitions I have discussed throughout the book. Let us look at each of them separately, beginning with the mortality transition, which is both the root cause of population growth and, as I pointed out in Chapter 5, the single most revolutionary transformation in human existence.

Managing the Mortality Transition—A Perpetual Engine of Growth

In general, we would all like to live as long as possible, and the only real issue there is that the longer we live, the lower the birth rate must be in order to keep a lid on population growth. In the 1960s and 1970s, when rates of population growth were very high and fertility had not yet begun to decline in most developing nations, people such as Paul Ehrlich (1968) issued the Malthusian warning that death rates might have to rise if birth rates did not fall. The "lifeboat ethic" and "triage" are two perspectives on how this might be implemented. The **lifeboat ethic** is based on the premise that, since a lifeboat holds only so many people and any more than that will cause the whole boat to sink, only those with a reasonable chance of survival (those with low fertility) should be allowed into the lifeboat. Withholding food and medical supplies could drastically raise death rates in less-developed nations and thus provide a longer voyage for those wealthier nations already riding in the lifeboat. A closely related doctrine is that of *triage,* which is the French word for sorting or picking, and refers to an army hospital practice of sorting the wounded into three groups—those who are in sufficiently good shape that they can survive without immediate treatment, those who will survive if they are treated without delay, and those "basket cases" who will die regardless of what treatment might be applied. As with the lifeboat ethic, it translates into selectivity in providing food and economic aid should the day come when supplies of each are far less than the demand. It means sending aid only to those countries that show promise of being able to bring their rates of population growth under control and abandoning those nations that are not likely to improve.

Lowering Infant, Child, and Maternal Mortality   Most people probably share the opinion that raising mortality is better grist for science fiction than for population policy. Obviously, the goal of the entire world community is to improve the health of all humans—which has the effect of lowering mortality, and actions to lower mortality are paramount among the goals of the ICPD, as shown in Table 13.2, where I have shown a selection of the action items related to managing mortality. Despite a reduction in mortality in most countries of the world, death rates are still commonly higher in rapidly growing, less-developed nations than in more highly advanced countries. This is particularly true in sub-Saharan Africa, where HIV-AIDS is ravaging the adult population.

I discussed in Chapter 6 the intuitively appealing idea that fertility will not drop until parents are sure their children will survive, so the beginning point of any policy to lower fertility should be to lower infant and childhood mortality. If women actually do bear more children than they want just to overcome high mortality, then it is reasonable to propose that lowering infant mortality will lower the need for children. This would hopefully be translated into a demand for birth control. Although that argument certainly makes sense in general, there is very little consistent evidence to support it at the individual family level (Ahn and Shariff 1994; Preston 1978).

The difficulty with lowering infant mortality as a means of ultimately lowering fertility is that it assumes that people are having more children than they actually want simply to overcome the effect of high mortality. The data suggest a more complex relationship . Life expectancy always improves in the context of broader social and economic changes that affect much more than just the survivability of children. All of these changes, including the drop in infant mortality rates, influence the attitudes of couples regarding family size (Ross and Frankenberg 1993).

The consequence is that a further decline in infant mortality could have the unintended (but certainly predictable) effect of raising, not lowering, the rate of

Table 13.2   Mortality Management in ICPD Programme of Action

Programme of Action Chapter and Paragraph	Action Item
6.5	Aim to reduce high levels of infant, child and maternal mortality so as to lessen the need for high fertility and reduce the occurrence of high-risk births.
8.5	Countries should aim to achieve by 2005 a life expectancy at birth greater than 70 years and by 2015 a life expectancy at birth greater than 75 years. Countries with the highest levels of mortality should aim to achieve by 2005 a life expectancy at birth greater than 65 years and by 2015 a life expectancy at birth greater than 70 years. Efforts to ensure a longer and healthier life for all should emphasize the reduction of morbidity and mortality differentials between males and females as well as among geographical regions, social classes and indigenous and ethnic groups.
8.16	By 2005, countries with intermediate mortality levels should aim to achieve an infant mortality rate below 50 deaths per 1,000 and an under-5 mortality rate below 60 deaths per 1,000 births. By 2015, all countries should aim to achieve an infant mortality rate below 35 per 1,000 live births and an under-5 mortality rate below 45 per 1,000. Countries that achieve these levels earlier should strive to lower them further.
8.21	Countries should strive to effect significant reductions in maternal mortality by the year 2015: a reduction in maternal mortality by one half of the 1990 levels by the year 2000 and a further one half by 2015.
8.32 to 8.35	Governments should mobilize all segments of society to control the AIDS pandemic; The international community should mobilize the human and financial resources required to reduce the rate of transmission of HIV infection; Governments should develop policies and guidelines to protect the individual rights of and eliminate discrimination against persons infected with HIV and their families; Responsible sexual behaviour, including voluntary sexual abstinence, for the prevention of HIV infection should be promoted and included in education and information programmes.

Source: United Nations Population Fund, ICPD Key Documents, http://www.unfpa.org/icpd/icpd_poa.htm#ch2, accessed 2004.

population growth. Indeed, this is what happened in one of India's largest states between 1951 and 1975, when infant and childhood mortality declined noticeably. Even with an official family planning program, the fertility levels showed little sign of change (Srinivasan, Reddy, and Raju 1978). Data for Guatemala suggest that it may take at least two generations for the reduction in infant and child mortality to influence the level of fertility (Pebley, Delgado, and Brinemann 1979). In fact, it may be that the relationship works the other way around—that the best way to lower infant mortality is to develop a comprehensive family planning program. Couples throughout the world have known for a long time that the health of both mothers and infants is improved by greater spacing between births (Gadalla 1978), and research in Gambia has shown that women there are not shy about using modern contraceptive techniques to regulate the spacing of children in order to achieve higher rather than lower fertility (Bledsoe and Hill 1998; Bledsoe et al. 1994).

The World With and Without AIDS   Of all the action items listed in Table 13.2, none receives greater attention than the control of AIDS. It has altered the social fabric of sub-Saharan Africa, still threatens much of Southeast Asia, and is an emerging issue in China, India, and Russia, where governments have not yet been willing to acknowledge the potential for a major spread of HIV. So serious is AIDS that the United Nations Population Division made two population projections— one that took AIDS into account, and one that looked at the world without AIDS. Table 13.3 shows you the numbers, which suggest that between 2000 and 2050 AIDS will be responsible for the deaths of 278 million more people than would have otherwise died. When you take into account the babies that will not be born because the mother died of AIDS or had impaired fecundity from HIV, the overall impact of AIDS could be as high as 479 million between 2000 and 2050. Put another way, were it not for AIDS, the projected population in the year 2050 would actually be half a billion more people.

Managing the Fertility Transition—Can We Take the Chance That It Will Happen Without Help?

For the past century, the biggest single issue surrounding population growth was whether or not fertility rates would drop to levels that would bring the rate of population growth back closer to zero. We now know not only that they can drop that

Table 13.3   Difference in Projected Population With and Without AIDS and Percentage Difference for Groups of Affected Countries in Major Areas, 2000, 2015, and 2050

	2000		2015		2050	
Country group	Population difference (thousands)	Percentage difference	Population difference (thousands)	Percentage difference	Population difference (thousands)	Percentage difference
All 53 affected countries	23,001	1	128,524	3	479,374	8
38 countries in Africa	16,087	3	91,310	10	319,919	19
5 countries in Asia	4,836	0	28,533	1	136,921	4
8 countries in Latin America and the Caribbean	908	0	3,646	2	9,118	3
7 countries with prevalence of 20 percent or more	3,329	3	26,233	19	77,502	36

Source: Population Division of the Department of Economic and Social Affairs of the United Nations Secretariat, 2003, World Population Prospects: The 2002 Revision. Highlights, New York: United Nations.

low but also that in many countries they have dipped well below replacement level. We assume (perhaps at our peril) that developing countries will follow the European pattern—that once a fertility decline is under way it will indeed continue down toward replacement. The issue for the world is: Can we take a chance that this will happen automatically? The delegates at the ICPD thought not and set out a number of action items related to the management of fertility decline, as you can see in Table 13.4. Some of these policy initiatives are aimed directly at influencing demographic behavior, while others are oriented toward trying to change social behavior, which will then indirectly have an impact on fertility. You can see that the action items under Chapter 4 all relate to gender equity and empowering women to control their reproduction.

A Broader Mandate to Empower Women   Lifting pronatalist pressures will especially involve a change in gender roles taught to boys and girls, giving equal treatment to the sexes in the educational and occupational spheres. If a woman's adulthood and femininity are expressed in other ways besides childbearing, then the pressure lessens to bear children as a means of forcing social recognition. Likewise, if a man's role is viewed as less domineering, then the establishment of a family may be less essential to him as a means of forcing social recognition. Required, of course, are alternatives to children and families in general as bonds holding social relationships together, indeed as centers of everyday life. This does not require the abolition of the family, but it does involve playing down the importance of a large extended kin network, which may (perhaps I should say "will") require massive social change—a virtual revolution in the way social life is organized in much of the world.

Any policy aimed at affecting motivation will, by definition, have to alter the way people perceive the social world and how they deal with their environment on an everyday basis. It will have to involve a restructuring of power relationships within the family, a reordering of priorities with respect to gender roles, a reorganization of the economic structure to enhance the participation of women, a concerted effort to raise the level of education for all people in society, and economic and political stability, which allows people to plan for a future rather than just cope with today's survival problems. History suggests that most of these changes have evolved somewhat naturally in the course of economic development, at least in Western nations. However, they do not inherently depend on development and, as a result, could be part of a deliberate policy even before large-scale development occurs (see, for example, Das Gupta 1995). In fact, it is probable that those are the very kinds of social changes that would help accelerate economic development, leading, as they would, to a substantial improvement in the human condition, at least by Western standards.

Lifting the penalties for antinatalist behavior also starts with redefining of gender roles—a more positive evaluation of single or childless persons. The centrally planned economies of China, Cuba, the former Soviet Union, and eastern Europe (prior to 1990) have shared the same demographic fate—the policies introduced to socialize the economy led to generally unintended declines in fertility. In China, the government ultimately decided it wanted yet a lower level of fertility and enacted policies to implement that. However, Russia and the former Soviet-bloc nations of eastern Europe had a long history of dissatisfaction with low rates of population

Table 13.4   Fertility Management in the ICPD Programme of Action

Programme of Action Chapter and Paragraph	Action Item
4.4	Countries should act to empower women and should take steps to eliminate inequalities between men and women as soon as possible by: (a) Establishing mechanisms for women's equal participation and equitable representation at all levels of the political process and public life in each community and society and enabling women to articulate their concerns and needs; (b) Promoting the fulfilment of women's potential through education, skill development and employment, giving paramount importance to the elimination of poverty, illiteracy and ill health among women; (c) Eliminating all practices that discriminate against women; assisting women to establish and realize their rights, including those that relate to reproductive and sexual health; (d) Adopting appropriate measures to improve women's ability to earn income beyond traditional occupations, achieve economic self-reliance, and ensure women's equal access to the labour market and social security systems; (e) Eliminating violence against women; (f) Eliminating discriminatory practices by employers against women, such as those based on proof of contraceptive use or pregnancy status; (g) Making it possible, through laws, regulations and other appropriate measures, for women to combine the roles of child-bearing, breast-feeding and child-rearing with participation in the workforce.
4.17	Overall, the value of girl children to both their family and society must be expanded beyond their definition as potential child-bearers and caretakers and reinforced through the adoption and implementation of educational and social policies that encourage their full participation in the development of the societies in which they live.
4.26	The equal participation of women and men in all areas of family and household responsibilities, including family planning, child-rearing and housework, should be promoted and encouraged by Governments.
7.15	Governments and the international community should use the full means at their disposal to support the principle of voluntary choice in family planning.
7.16	All countries should take steps to meet the family-planning needs of their populations as soon as possible and should, in all cases by the year 2015, seek to provide universal access to a full range of safe and reliable family-planning methods and to related reproductive health services which are not against the law.

Source: United Nations Population Fund, ICPD Key Documents, http://www.unfpa.org/icpd/icpd_poa.htm#ch2, accessed 2004.

growth. There are, of course, numerous population policy lessons in the experiences of these countries, although they are by now familiar themes. The socialist centrally planned societies taught us that fertility can be indirectly influenced to decline through a combination of the following factors: (1) educating women; (2) providing women with access to the paid labor force; (3) legalizing the equality in status of males and females; (4) legalizing abortion and/or making contraceptives freely available; (5) slowly raising the standard of living, while at the same time making housing and major consumer items difficult to obtain, forcing a couple to ask themselves: "Kicsi or kosci?" (as they say in Hungarian—"A child or a car?").

Making Family Planning Available   The fertility management action items in Chapter 7 of the ICPD Programme of Action relate to family planning, which involves the provision of birth prevention information, services, and appliances. It also involves teaching women (and increasingly men as well) about their bodies and teaching them how to prevent births, usually with contraceptives but sometimes also with abortion or sterilization. An important component added to many of these programs in recent years has been teaching about the transmission of HIV-AIDS and other sexually-transmitted infections (STIs) and how to protect yourself from such infections (Upadhyay and Robey 1999).

An early assumption of family planning programs was that women were having large families because they were uninformed about birth prevention or lacked access to the means for preventing births. It was assumed that a demand existed for family planning, and that meeting that demand would lower the birth rate. Not surprisingly, perhaps, the world turned out to be more complex than that, but even so family planning programs have remained the most popular means of implementing a policy to slow down population growth. A major reason for the widespread prevalence of family planning is the fact that it is usually associated with health programs, which are almost universally acceptable. This is an important point, because many issues surrounding reproduction are very sensitive politically, if not socially.

The spread of family planning services and technology around the world has taken place fairly rapidly since about 1965:

> In the mid-1960s, developing countries began to adopt policies to support family planning as a means of slowing population growth. By the late 1960s, family planning had become a worldwide social movement that involved international organizations such as the United Nations Fund for Population Activities, government agencies such as the United States Agency for International Development, private American philanthropic groups such as the Ford and Rockefeller foundations, nonprofit organizations such as the affiliates of the International Planned Parenthood Federation, and a host of individuals, many with backgrounds in medicine and public health (Donaldson and Tsui 1990:4).

In 1965, only 21 countries in the world admitted to actively supporting family planning programs (Isaacs and Cook 1984). By 1974, the United Nations had counted 86 governments that provided direct support for family planning, and by 1989 the number was up to 123, covering 91 percent of the world's population (United Nations 1989).

Family planning programs are implicitly designed to close the gap of **unmet needs** in a population. Unmet needs refer to sexually active women who would prefer not to get pregnant but are nevertheless not using any method of contraception (Robey, Ross, and Bhushan 1996), the strongest indicator of which is the percentage of women who say they would prefer not to have any more children, but are not using any method of contraception (Westoff and Bankole 1996). Although the basic goal of family planning programs is to make sure that people have access to methods of birth control, family planning programs generally do far more than this.

Researchers at the Carolina Population Center at the University of North Carolina, Chapel Hill, spent several years trying to understand how family planning programs work. Several ingredients emerged that might be thought of as the seven secrets of successful family programs, including:

1. Easy accessibility of services is very important in the early stages of a fertility decline.

2. A community-based approach (as opposed to only having clinics to which people must go) increases continuation rates.

3. High-quality services increase the prevalence rate.

4. A greater variety of methods increases the prevalence rate.

5. Mass-media messages increase the prevalence rate.

6. Informal discussions by users with nonusers increases the prevalence rate.

7. Women tend to prefer contraception to abortion (Samara, Bucker, and Tsui 1996).

Case Studies in Fertility Management   Table 13.4 highlighted some of the major ways by which the ICPD suggested that fertility could be managed. In Table 13.5, I have organized a fuller range of options open to any country, based on the three preconditions for a fertility decline discussed in Chapter 6. One set of policies aims to promote the empowerment of people to exercise their control over reproduction, a second set of policies aims to motivate people to limit fertility, and the third set aims to provide people with the means to have the number of children they want. Let me illustrate some of these options with examples from different countries.

Although family planning programs may be run by government agencies, the role of NGOs is also increasingly important, as I mentioned earlier. Governmentally supported family planning programs have themselves often evolved from the efforts of private citizens seeing a need for such services. In Egypt, for example, private voluntary organizations opened family planning clinics in urban areas as early as the 1950s. These clinics distributed contraceptive foams and jellies to women as long as they met three preconditions: (1) They already had three children; (2) They had their husband's permission; and (3) They could show a health or economic reason that established the need for birth control (Gadalla 1978). As quaint as that sounds, it helped set the stage for the far more massive government effort currently under way in Egypt, in which government health clinics dispense pills, IUDs, and other contraceptives, and outreach workers distribute contraceptives in rural areas.

Table 13.5   A Range of Policies to Limit Fertility

Precondition for Which Intervention Is Desired	Examples of Policies	
	Direct	Indirect
Rational choice	Provide full legal rights to women	Promote secular education
	Increase legal age at marriage for women	Promote communication between spouses
Motivation for smaller families	*Incentives*	*Incentives*
	Payments for not having children	Economic development
	Priorities in jobs, housing, education for small families	Increased educational opportunities for women
	Community improvements for achievement of low birth rate	Increased labor force opportunities for women
		Peer pressure campaigns
	Disincentives	Lower infant and child mortality rates
	Higher taxes for each additional child	
	Higher maternity and educational costs for each additional child ("user fees")	*Disincentives*
		Child labor laws
		Compulsory education for children
		Peer pressure campaigns
Availability of means for limiting family size	Legalize abortion	Community birth quotas
	Legalize sterilization	
	Legalize all other forms of fertility control	Public campaigns to promote knowledge and use of birth control
	Train family planning program workers	Politicians speaking out in favor of birth control
	Manufacture or buy contraceptive supplies	
	Distribute birth control methods at all health clinics	
	Make birth control methods available through local vendors	
	Establish systems of community-based distribution	

Source: John R. Weeks, 1992, "How to Influence Fertility: The Experience So Far." In Lindsey Grant (ed.), The Elephants in the Volkswagen: How Many People Can Fit Into America? (New York: W. H. Freeman Company), Table 15.1.

Since the mid-1980s, the Egyptian government has combined the provision of family planning with a concerted drive to increase the motivation of couples to use birth control. This has included speeches by President Hosni Mubarak, published interviews with Islamic scholars, mass-media messages, and activities by local community volunteers (Gilbar 1994; Ibrahim and Ibrahim 1998). To be sure, the total fertility rate dropped from a high of 7.1 in 1960 to 3.3 in the early 2000s, but it is not clear how much of that decline can be attributed to family efforts, per se. Fargues (1997) notes that changes in the birth rate have been very closely tied to other government policies that affect fertility less directly—especially economic policies and the effort to improve educational levels of women, and in conjunction with rising education it appears that an increase in marriage age has played an important role in the decline (Weeks et al. 2000; Weeks et al. 2004). Still, the lessons of Chapter 6 suggest that without a family planning program in place, the motivation to limit fertility will have less chance of being converted to lower birth rates.

In 1952, India began experimenting with family planning programs to keep the population "at a level consistent with requirements of the economy" (Samuel 1966:54). Although initial progress was pretty slow—the Indian government was moving cautiously—by 1961, there were about 1,500 clinics in operation, providing condoms, diaphragms, jellies, foam tablets, and other services, largely free (Demerath 1976). In 1963, the birth rate still was not responding and the government, now with the advice of the Ford Foundation, reorganized its family planning effort, mainly in an outreach effort to spread the birth control message to more people. Still, the birth rate did not go down, so in 1966 there was another reorganization, and in 1967 Dr. S. Sripati Chandrasekhar, a demographer, was named Minister of State for Health and Family Planning. Chandrasekhar boosted the use of the mass media, offered transistor radios to men undergoing vasectomy, pushed through a legalization of abortion, and generally brought about more family planning action than had occurred in all the previous years of family planning. Male sterilization, being both permanent and the cheapest method of birth control, has been especially pushed by the Indian government, and between 1967 and 1973 (when Chandrasekhar left his position) 13 million men had been sterilized in India. The birth rate in India declined ever so slightly between 1961 and 1971, but the death rate sank even more, so by the mid-1970s, the Indian population was growing faster than ever before. Why had the family planning efforts failed? Because family planners had not taken into account the broader social context within which reproduction occurs. It was naive to think that family planners could, of their own accord, generate the kind of social and cultural revolution required to make small-family norms an everyday practice throughout India (Demerath 1976).

By 1976, Indira Gandhi's government in India was beginning to show signs of desperation at being unable to slow the rate of population growth, and it became the first government in the world to lean toward compulsory measures. Dr. D. N. Pai, director of family planning in Bombay (now Mumbai), has reportedly said, "Ninety percent of the people have no stake in life. How do you motivate them? They have nothing to lose. The only way out of the situation is compulsion" (Rosenhause 1976). The government implemented a now-famous, if not infamous, program of enforced sterilization. "Officially, there was no coercion, but the elaborate system of 'disincentives' amounted to the same thing. Government employees had to produce two or more candidates for sterilization. For such civil servants, or for anybody who was being pressured into submitting to sterilization himself, it was usually possible to hire a stand-in for about 200 rupees ($22). For those not in government service all sorts of privileges—such as licenses for guns, shops, ration cards—were denied unless the applicant could produce a sterilization certificate" (Guihati 1977).

As a result of the policy of enforced sterilization, the number of people (largely male) who were sterilized in India jumped from 13 million in 1976 to 22 million in 1977. By contrast, I should note, there were fewer than four million users of the IUD or the pill—only 3 percent of the total population of married women (U.S. Census Bureau 1978a). As it turned out, however, the average male who was sterilized was in his 30s and already had several children (Guihati 1977), so the program was less effective in reducing fertility than those numbers might imply.

Although her campaign effectively raised the level of sterilization, Indira Gandhi's policies generated violent hostility, and in 1977 she was defeated in her bid for reelection. Her successor, 82-year-old Morarji Desai, professed support for family planning, but only half-facetiously suggested that self-control (that is, abstinence), rather than birth control, was the solution to India's high rate of population growth (Bhatia 1978). Gandhi was elected once again in 1979, but she kept a relatively low profile with respect to her support for family planning, right up to the time of her assassination in 1984. Interestingly enough, that was a period of time in which the birth rate did appear to be dropping, although survey results have shown that no change in desired family size took place between 1970 and 1980 (Khan and Prasad 1985).

In 1984, Indira Gandhi's son, Rajiv, was elected prime minister of India. He had been actively involved in the family planning movement during his mother's administration and in 1985 he declared that the nation was on "war footing" to reduce its rate of population growth to a two-child family norm by 2000 (Population Action Council 1985:1). A five-year, $3.6-billion program was unveiled that included monetary rewards to women who limited their family size (see the discussion later in this chapter of these motivational schemes) and a broadening of the scope and quality of the nation's family planning services. He was voted out of office in 1989 and later assassinated, although neither event seemed to be related specifically to his views on population limitation.

One of the distinguishing characteristics of population policy in India until recently was its emphasis on family planning targets:

> Annual fixed targets for all methods, imposed from the top down, are set for family planning workers at all levels, and their performance is judged in terms of the fulfillment of the target. The pressure to achieve the targets starts at the top and works its way down to the field worker, who ultimately bears the brunt of the challenge. No excuses are accepted when targets are not achieved, and punitive action is taken in the form of salary freezes. This approach was most severe during the emergency period [in the 1970s]. (Visaria and Chari 1998:90)

Targets such as these came under sharp attack at the 1994 United Nations International Conference on Population and Development, and it was probably no coincidence that within two years after that conference the government of India moved its family planning approach away from method-specific targets and more toward an emphasis on reproductive and child health (Visaria and Chari 1998). Family planning in India still depends on sterilization, although the emphasis is now on tubal ligation for women. Other contraceptive methods represent only a small fraction of family planning effort in India (Pathak, Feeney, and Luther 1998).

The small (four million, but growing) republic of Singapore began emphasizing disincentives to large families in 1969. Singapore is an island city-state on the Malaysian peninsula whose recent history was influenced by the British, who developed the harbor, built the city, and controlled the area (as they did all of Malaysia) for more than a century until World War II. Japan briefly took over during the war, but after the war, as Great Britain was divesting itself of its colonies, the Malaysian Federation, including both Malaysia and Singapore, was formed. In 1963, Singapore seceded from the Federation, preferring to go its own way, and in 1965 it gained its full independence.

Singapore has an ethnically diverse population, of whom more than three-fourths are ethnic Chinese, with the remaining population being Malay (who are mainly Muslim), Indian (who are mainly Hindu), and European (who are mainly Christian), and upon independence in 1965, the government of Singapore realized that its rapid rate of population growth at the time would make it impossible to continue developing economically, since the country has few natural resources and practically no arable land of its own. Its principle resource is its population, and so population policy was seen as a form of human resource management (Drakakis-Smith and Graham 1996).

At first, the policy incorporated the family planning program approach, emphasizing the distribution of contraception through clinics and hospitals (Salaff and Wong 1978). After four years of family planning, however, the government grew restless at the slow pace of the decline in the birth rate. The result, in 1969, was a liberalization of the abortion law and concomitant establishment of some economic disincentives to encourage adoption of the idea that "two is enough." These measures included steeply rising maternity costs for each additional child, low school enrollment priorities for third and higher-order children, withdrawal of paid two-month maternity leave for civil service and union women after the second child, low public housing priority for large families, and no income tax allowance for more than three children.

Did these basically coercive measures work? Apparently so. Between 1966 and 1985 the total fertility rate in Singapore dropped from 4.5 children per woman to 1.4, with a bigger drop occurring among wealthier, educated women than among poorer women, as you might expect. This classic case of differential fertility worried Singapore's prime minister Lee Kuan Yew. In 1983, he became concerned that not enough bright Singaporans were being born and he argued, "In some way or other, we must ensure that the next generation will not be too depleted of the talented. Get our educated women fully into the life cycle. Get them to replace themselves" (quoted by Leung 1983:20). There was also concern that the rapid drop in fertility would lead to an aging population in which there would be only two workers for every retired person. Thus, the government moved from an antinatalist policy driven by disincentives to a pronatalist policy driven by incentives.

Singapore flatly rejected the idea that immigration might solve that problem and instead, in 1987, the government instituted the New Population Policy, in which the philosophy was "have three or more children, if you can afford them" (Yap 1995):

> Today, Singapore's population policy is pro-family, providing strong support to working mothers. Singapore recognizes the difficulties working women face in balancing their multiple roles as workers, mothers and wives. To assist them, we have introduced generous tax incentives for working mothers, employers are encouraged to grant special leave for working mothers with sick children, and subsidized child care is widely available. With the increasing female labour participation rate now exceeding 51 per cent, these measures have been effective in helping working mothers to realize their desired family size and to balance their multiple roles. (Tarmugi 1999)

The goal was to raise the birth rate back up to replacement level by the year 1995, but that did not happen. By the late 1990s, the total fertility rate had risen to 1.8, but the increase from 1.4 in the mid-1980s seemed largely due to the fact that women who had postponed children at an earlier age had gone ahead and had them, and by 2003, the fertility rate was back down to 1.4 children per woman. The government has responded by lowering some of its immigration restrictions, and by organizing social events where the best and brightest young Singaporeans can meet and, hopefully, match up for marriage and childbearing (Saywell 2003).

Romania has the dubious distinction of having done the most in recent decades to try to raise the birth rate. In 1966, the Romanian government decided to halt the downward trend in the birth rate (which was being accomplished primarily by abortion) by establishing a policy that made abortions illegal except under extreme circumstances. The government also discouraged the use of other contraceptives by stopping their importation and making them available only for medical reasons. The result was that the birth rate skyrocketed from 14 per 1,000 population in 1966 to 27 per 1,000 in 1967. Although it did begin to fall again after 1967, as women resorted to illegal abortions and found other means of birth control (primarily rhythm and withdrawal), by 1989, the total fertility rate of 2.3 was still higher than the 1.9 level of 1966. But 1989 brought a significant event—the change in government. In December 1989, Nicolae Ceausescu, longtime dictator of Romania, was overthrown and executed. One of the first legislative acts of the new government was to legalize abortion in order to reduce the high maternal mortality rate that had resulted from botched illegal abortions during the prior two decades (Serbanescu et al. 1995). By early 2003, the total fertility rate had dropped to 1.2, one of the world's lowest.

France, which has experienced a low birth rate for longer than almost any other existing society, has also maintained its long-standing policy of providing monthly allowances to couples who have a second or higher-order child, despite the lack of evidence that such allowances have had any measurable impact on the birth rate. Single mothers are also assured a monthly allowance, and all mothers in France have access to nursery school placement for their children by age three, so they can return to work (Bergman 1996). Although these measures have not been enough to raise the birth rate in France, they may contribute to the fact that, at 1.9 children per woman, France's fertility has not dropped too far below replacement and is now one of the higher fertility rates in all of Europe.

Managing the Migration/Urban Transition—The Changing Face of the World

Migration should be the most easily controlled of the three population processes, at least in theory. You cannot legislate against death (except for laws prohibiting homicide or suicide), and few countries outside China have dared try to legislate directly against babies. But you can set up legal and even physical barriers to migration—keeping people in (as in the former Soviet Union) or out (as practiced by most countries in the world). In reality, of course, controlling migration can be very difficult if people are highly motivated to move. Undocumented migration from Mexico to the United States is a leading example. Perhaps for these reasons, the action items from the ICPD that relate to migration and urbanization are somewhat vague, as you can see in Table 13.6.

Labor migration has become commonplace in the world today, as I discussed in Chapter 7, and countries have accepted guest workers on the naive assumption that they can be sent home when the need arises. This is never easy, regardless of whether the host country is in the "north" or the "south." Germany is dealing with immigrants from eastern Europe and Turkey (and other less developed nations); the United Kingdom accepts about 50,000 immigrants a year, almost exclusively from former colonies, especially India and Pakistan (Coleman 1995). Even Italy—long a source of migrants itself—was forced in 1986 to enact its first law controlling immigration into the country (Martin 1993). About 15,000 immigrants enter Denmark each year for family unification, but after Denmark found that 90 percent of Danish Turks (mainly men) find their spouses in Turkey, legislation was enacted in 2000 to deter any immigrant younger than 25 from bringing a foreign spouse to Denmark (*Migration News* 2001a)

To remind you that this is not just an issue with more-developed nations, the Malaysian and Indonesian governments agreed in 2000 that illegal migrants to Malaysia who are unemployed would be returned to Indonesia, with both countries sharing the cost of repatriation. An estimated 65,000 Indonesians had arrived in Malaysia illegally in the first nine months of 2000 (*Migration News* 2000). In another instance, in fall 2000, Pakistan began restricting the entry of Afghans through the Khyber Pass: Afghans were fleeing continued fighting as well as drought, but Pakistan was not in a position to help them (*Migration News* 2001b).

To date, the success of attempts to limit immigration has been highly variable at best, and Massey (1996b) has explained this with what he has called his "perverse laws of international migration":

1. Immigration is a lot easier to start than it is to stop.

2. Actions taken to restrict immigration often have the opposite effect.

3. The fundamental causes of immigration may be outside the control of policy makers.

4. Immigrants understand immigration better than politicians and academicians.

5. Because they understand immigration better than policy makers, immigrants are often able to circumvent policies aimed at stopping them.

In the final analysis, most attempts to limit immigration are motivated less by a desire to limit population growth in general, and more by a desire to limit the

Table 13.6   Migration/Urbanization Management in the ICPD Programme of Action

Programme of Action Chapter and Paragraph	Action Item
9.4	In order to achieve a balanced spatial distribution of production employment and population, countries should adopt sustainable regional development strategies and strategies for the encouragement of urban consolidation, the growth of small or medium-sized urban centres and the sustainable development of rural areas, including the adoption of labour-intensive projects, training for non-farming jobs for youth and effective transport and communication systems.
9.14	Governments should increase the capacity and competence of city and municipal authorities to manage urban development, to safeguard the environment, to respond to the need of all citizens, including urban squatters, for personal safety, basic infrastructure and services, to eliminate health and social problems, including problems of drugs and criminality, and problems resulting from overcrowding and disasters, and to provide people with alternatives to living in areas prone to natural and man-made disasters.
9.21	Countries should address the causes of internal displacement, including environmental degradation, natural disasters, armed conflict and forced resettlement, and establish the necessary mechanisms to protect and assist displaced persons, including, where possible, compensation for damages, especially those who are not able to return to their normal place of residence in the short term.
10.3	Governments of countries of origin and of countries of destination should seek to make the option of remaining in one's country viable for all people.
10.11	Governments of receiving countries are urged to consider extending to documented migrants who meet appropriate length-of-stay requirements, and to members of their families whose stay in the receiving country is regular, treatment equal to that accorded their own nationals with regard to the enjoyment of basic human rights, including equality of opportunity and treatment in respect of religious practices, working conditions, social security, participation in trade unions, access to health, education, cultural and other social services, as well as equal access to the judicial system and equal treatment before the law.
10.17	Governments of countries of origin and countries of destination are urged to cooperate in reducing the causes of undocumented migration, safeguarding the basic human rights of undocumented migrants including the right to seek and to enjoy in other countries asylum from persecution, and preventing their exploitation. Governments should identify the causes of undocumented migration and its economic, social and demographic impact as well as its implications for the formulation of social, economic and international migration policies.
10.23	Governments are urged to address the root causes of movements of refugees and displaced persons by taking appropriate measures, particularly with respect to conflict resolution; the promotion of peace and reconciliation; respect for human rights, including those of persons belonging to minorities; respect for independence, territorial integrity and sovereignty of States.

Source: United Nations Population Fund, ICPD Key Documents, http://www.unfpa.org/icpd/icpd_poa. htm#ch2, accessed 2004.

entry of certain kinds of people into the country (no matter what country we talking about). The greater the social and cultural differences between receiving and sending societies, the more likely it is that attempts will be made to slow down the pace of immigration (Castles and Miller 1998).

Managing the Age Transition—The Perpetual Engine of Change

The next phase in world population policy planning is to prepare for a world in which the haves and have-nots—the developed and developing countries—are divided not only by income and rates of population growth, but by distinctly different population structures. The legacy of several decades of low fertility and low mortality in the developed countries is an increasingly older population, whereas a much later start to the fertility decline, combined with ever lower mortality, keeps the populations of developing areas (countries "in transition," as they are sometimes called) with significantly younger population age structures.

In Figure 13.3, you can see that by the year 2025 there will be important differences in the age structure of different regions of the world. These data draw on the United Nations medium variant projections—their most likely scenario. The upper left pyramid shows the age structure for Europe, which is projected to be the oldest population in the world in 2025. Actually, it has been the oldest for a long time, but the projected pattern of continued low fertility produces genuine vacancies in the younger ages, which won't be compensated for by immigration, at least not at current levels (which are built into the projection). By contrast, some of Europe's major immigrant sources are shown in Figure 13.3, and you can see that northern Africa is expected to have a bulge in the teenage populations in 2025, although the projected decline in fertility is noticeable at the very youngest ages. This large group of teenagers is not necessarily a good sign for the stability of the region, of course, unless the economies of these countries are sufficiently strong to absorb them into the labor force. The situation looks even less sanguine in sub-Saharan Africa and western Asia, where current projections suggest a gradual decline in fertility that will leave the age structures still very young and increasing numerically at a very high rate. These regions of the world could dramatically improve their social and economic prospects if the fertility transition were speeded up over what the United Nations has projected.

Looking at North America (the United States and Canada) in the year 2025, the projections show a much more rectangular shape than in Europe. The United States is not facing depopulation in the way that Europe is, largely as a result of several decades of strong immigration. The expectation is that the children of immigrants will have levels of fertility that are near, but not much below, replacement level. It is still true that North America is aging, but not as precipitously as Europe. Central America, which is dominated by Mexico, is a major supplier of immigrants to the United States. You can see in Figure 13.3 that Central America's projected number of children in the youngest ages is expected to have declined by 2025, and so by that time the pressure to migrate to the United States might be somewhat less intense than it is now. Nonetheless, the younger age structure of Central America "fits" the older age structure of North America for the next few decades.

Figure 13.3   Age Structure of Different World Regions in 2025, According to United Nations Medium Variant Population Projection

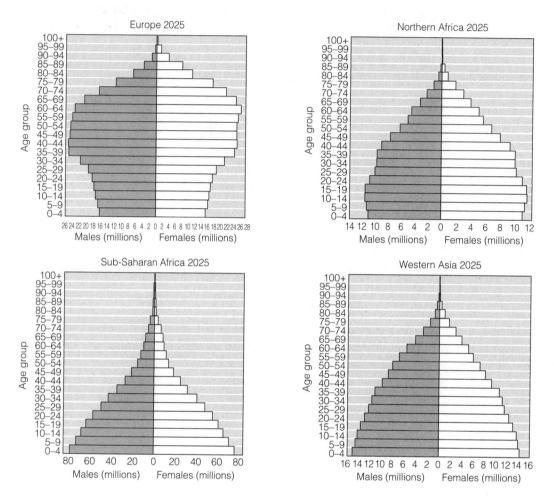

You can also see that, despite the worries in North America about an aging population, the situation is likely to be more intense in eastern Asia by 2025, because its age structure is projected to look more like Europe than like North America. The continued drop in fertility in China and Korea, along with an already very low fertility in Japan, signals a decline in the number of younger people at the same time that the older population will be steadily increasing. In southeastern Asia (which includes Indonesia and the Philippines) and south central Asia (which includes India, Pakistan, and Bangladesh), the slower pace of the fertility transition means that the age structure will remain younger for much longer than in eastern Asia.

Overall, then, policy makers in North America, Europe, eastern Asia, and parts of South America are busily at work trying to figure out how to cope with an aging

Figure 13.3   (continued)

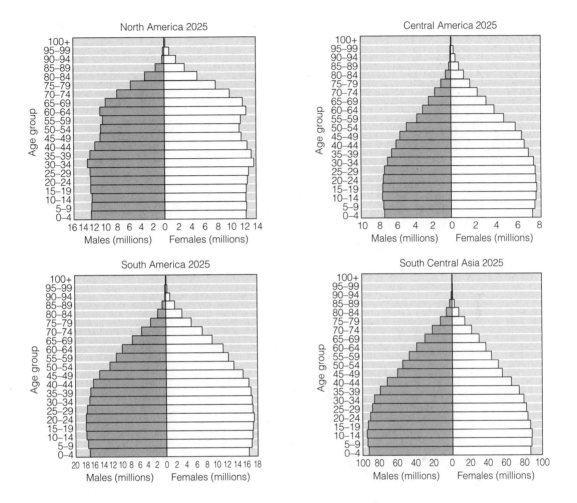

North America 2025

South America 2025

Central America 2025

South Central Asia 2025

(continued)

population. The kinds of changes in family size and household structure that produced such a high fraction of older people also has led to the near disappearance of the traditional family structure that in earlier generations might have taken care of the elderly (Ogawa and Retherford 1997). The elderly in more-developed nations will have fewer children to call upon for support, but they are also growing older with unprecedentedly high levels of living, which they will, of course, hope to maintain. Have they saved enough money themselves to do that, will they work to an older age (reversing the current trend of earlier retirement), or will they count on communal support—from taxes on younger people? (Wise 1997). How will the economies of more-developed nations continue to function with a potentially shrinking labor force? It seems likely that North America, Europe, and possibly even

Figure 13.3   (continued)

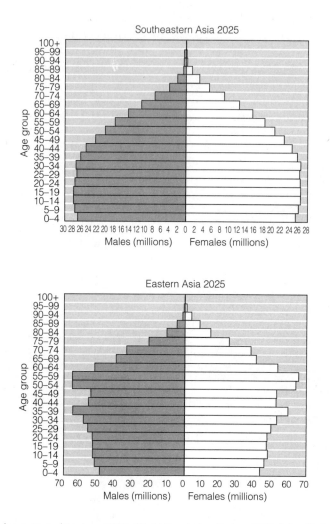

Source: Adapted from data in United Nations, 2003, World Population Project, 2002 (New York: United Nations). From United Nations Variant Population Projection.

Japan will draw increasingly upon immigrants to support the older population (Espenshade and Gurcak 1996; Lesthaeghe and Willems 1999; United Nations 2000), beyond the numbers shown in the United Nations pictured in Figure 13.3. Or, maybe the birth rate will go back up. That is one predicted consequence of improving the "private" status of women in Europe, in particular. If women in Europe can gain the same equity at home that they have in public, they may be more interested and willing to combine a career and children.

In the meantime, less-developed nations will continue to struggle to meet the payroll, so to speak—needing to keep finding jobs for the numerous young people in their midst, which will continue to put pressure on people in these countries to seek employment opportunities elsewhere. Absent such opportunities, young age structures can present societies with real problems. Hagan and Foster (2001:874) remind us that "[A]dolescence is a time of expanding vulnerabilities and exposures to violence that can be self-destructive as well as destructive of others." You might have that in mind as you read that Central American countries, for example, have faced a wave of gang activity exacerbated by a very young age structure in an environment where there are not enough jobs to go around (Millman 2003). Society is forced to deal with the problem no matter what its cause, but knowing that the age structure can produce such effects is the first step toward preventing problems from getting out of hand.

From a global perspective, you can appreciate that the diversity of age structures prevents a worldwide consensus on what to do. Indeed, the age transition is largely managed in an indirect fashion by the speed of the fertility decline, and so the main policy items that are separate from fertility relate to things the world has never really confronted before—a large number of societies with rapidly aging populations. Thus, the action items shown in Table 13.7 relate solely to aspects of aging.

Managing the Family and Household Transition—The Engine of Freedom

The ICPD Programme of Action gives voice to the idea that family transformations are part and parcel of the other demographic changes taking place in society, and as you can see in Table 13.7, the management issues relate especially to issues of gender equity and protecting the rights of women in families and households. Of particular importance is the idea that women should have access to education and the paid labor force, with governments assisting women with this by providing access to child care. As you might guess, this action item is not simply for developing countries. It can apply as well to the countries of southern and eastern Europe, where women are educated and are in the labor force, but where traditional family values still suggest that when a woman marries she should leave the labor force and take care of the children (and the husband, as well!). This is an oft-cited reason not only for the delay in marriage but also for the low level of fertility within marriage in eastern and southern Europe. Without greater gender equity in all aspects of family life, we can see that fertility can be "too low" as well as "too high."

What Is the Population Policy of the United States?

In response to the United Nations surveys on population policies, the United States has consistently indicated that the government is satisfied with current growth and fertility rates and has no plans for intervention. Explicitly, the closest that the country has come to a population policy was under President Nixon, who in 1969 proposed a commission (that Congress thereafter legislated into existence) to examine population growth in the United States, assess that growth's impact on the U.S. future, and "make recommendations on how the nation can best cope with that impact" (Commission on Population Growth and the American Future 1972:Preface).

Table 13.7   Age and Family/Household Management in the ICPD Programme of Action

Programme of Action Chapter and Paragraph	Action Item
6.18	All levels of government in medium- and long-term socio-economic planning should take into account the increasing numbers and proportions of elderly people in the population.
6.19	Governments should seek to enhance the self-reliance of elderly people to facilitate their continued participation in society.
6.20	Governments, in collaboration with non-governmental organizations and the private sector, should strengthen formal and informal support systems and safety nets for elderly people and eliminate all forms of violence and discrimination against elderly people in all countries, paying special attention to the needs of elderly women.
5.3	Governments, in cooperation with employers, should provide and promote means to facilitate compatibility between labour force participation and parental responsibilities, especially for single-parent households with young children. Such means could include health insurance and social security, day- care centres and facilities for breast-feeding mothers within the work premises, kindergartens, part-time jobs, paid parental leave, paid maternity leave, flexible work schedules, and reproductive and child health services.
5.4	When formulating socio-economic development policies, special consideration should be given to increasing the earning power of all adult members of economically deprived families, including the elderly and women who work in the home, and to enabling children to be educated rather than compelled to work.
5.5	Governments should take effective action to eliminate all forms of coercion and discrimination in policies and practices. Measures should be adopted and enforced to eliminate child marriages and female genital mutilation.
5.9	Governments should formulate family-sensitive policies in the field of housing, work, health, social security and education in order to create an environment supportive of the family, taking into account its various forms and functions, and should support educational programmes concerning parental roles, parental skills and child development.

Source: United Nations Population Fund, ICPD Key Documents, http://www.unfpa.org/icpd/icpd_poa.htm#ch2, accessed 2004.

Before producing its findings in 1972, the commission spent three years gathering information, listening to the public and experts, and deliberating. Its essential conclusion was that "no substantial benefits would result from continued growth of the nation's population" (Commission, 1972:1), and its principal recommendation was that the United States pursue a policy of population stabilization. From these facts might flow the facile conclusion that there is no population policy in the United States. On the contrary, the United States has an implicit domestic population policy and an explicit international population policy.

Domestic Policy Regarding Fertility

Implicitly, the United States supports the ability of Americans to maintain low fertility through the availability of contraceptives, voluntary sterilization, and abortion, although it hasn't always been that way. Fertility declined in the United States up to the 1940s without a widespread family planning program. Nonetheless, the family planning or planned parenthood movement in the United States in fact has a fairly long history, dating back to the nineteenth century, when in 1832 Charles Knowlton, a physician in Massachusetts, published his pamphlet on the "Fruits of Philosophy," which named the ways then known by which couples could attempt to limit their fertility. These methods included withdrawal (coitus interruptus), a penile sheath (condom), a vaginal sponge (not unlike a diaphragm), and various concoctions to use as a vaginal wash (douche) (Knowlton 1832 [1981]). As I mentioned in Chapter 3, this pamphlet landed Knowlton in jail for publishing obscene materials and later produced the same result for Annie Besant and Charles Bradlaugh in England when they published a subsequent edition. But all the publicity encouraged the spread of contraceptive information, even though the methods were limited. The two most effective methods were the condom and the diaphragm—then known as a French pessary. In 1873, after production of the condom had begun and in the midst of a general decline in U.S. fertility, Congress passed the Comstock Law, which prohibited the distribution of contraceptive information and devices through the mails. According to Westoff and Westoff, "Anthony Comstock, for whom the law was named, formed the New York Society for the Suppression of Vice . . . (and) was responsible for seven hundred arrests, the suicide of a woman abortionist, and the seizure of thousands of books and contraceptives" (Westoff and Westoff 1971:47). The Comstock Law was not repealed until 1970, although it had been toned down by court order in 1936 (Jaffe 1971), in response to the pioneering efforts of Margaret Sanger, as I discussed in Chapter 6. In 1958, the ban on prescribing contraceptives in public hospitals was lifted in New York City and that opened the way for publicly financed health institutions to provide family planning services. In 1965, the U.S. Supreme Court struck down a Connecticut statute barring the use of contraceptives and a number of states subsequently repealed their restrictive laws (Jaffe 1971:119).

Congress has provided federal funding for family planning programs in the United States since 1967, but the Family Planning Services and Population Research Act of 1970 specifically forbade funds to any organization that provided abortion services, even though several states already had liberalized abortion laws at that time. In 1973, the Supreme Court struck down restrictive abortion laws in the United States and one of the last obstacles to free choice in birth prevention was hurdled. Of course, there are ongoing attempts in the United States to reverse that decision and once again restrict the access of American women to legal abortion, but in the United States, as in other parts of the world, a decline in fertility did not wait for the legal restrictions to be removed from contraception and abortion; nor did it await government programs to develop new contraceptive technologies such as the IUD and the pill (both were privately financed), or for the government to provide free or subsidized birth control measures. Fertility declined dramatically from the nineteenth century to World War II, went up briefly, and then declined

IS THERE A DEMOGRAPHIC DIMENSION TO SUSTAINING SOCIALIST POLITICAL/ECONOMIC SYSTEMS?

Classical economic theories, such as those of Adam Smith, that espouse the value of capitalism and free markets, emerged at a time late in the eighteenth century when we still had little control over mortality and fertility. This was just before the demographic transition began, but at a time when the population was growing in Europe because of external factors (especially changes in the weather that improved crop production and the introduction of the potato) that had lowered mortality without any real human involvement in that process. The policies arising from these theories, which emphasize merit and personal accomplishment, are bound to amplify inequalities in society, but this is less noticeable when there is structural mobility—when almost everyone is better off than before, even if the gap between top and bottom is widening. The rich exploit the less well off, but everyone gains, even if not equally (see, for example, Firebaugh 2003). This is the essential ingredient of the "boomster" approach that characterized writers such as Julian Simon (1981) in the second half of the twentieth century, as I discussed in the previous chapter.

Those very inequalities, emerging in full force in the nineteenth century and documented most notoriously by Karl Marx (but also by novelists such as Charles Dickens), account heavily for the rise among many modern nations of some sort of socialist or welfare approach to economic systems. Such systems represent a type of intergenerational social contract. Those of working age agree to give part of their wages to others who have less income because (a) They received such benefits from others when they were younger and could not fend for themselves, and/or (b) They expect to receive such benefits from others when they are older and can no longer fend for themselves. The contract is implemented by an agreed-on tax rate that is, presumably, not unduly burdensome to the worker, but at the same time provides for the needs of the other members of society. Contracts tend to work best when they stay in force without a need for change, and the socialist or welfare state thus works best under a steady state in both the number of people and the tax rate and benefits received from the state over time. If the number of dependents increases, but the number of workers does not, then the per-person benefits of dependents (typically the poorer segments of society) will drop and they may complain. If taxes are raised to meet this demand, then workers (typically the richer segments of society) will complain. Could it be, then, that the age transition in societies is associated with different probabilities of sustaining socialist/welfare economies? That is one of the intriguing questions that arises as we ask how societies are transformed by the various aspects of the demographic transition and how they cope with those transitions.

Let us define a socialist/welfare economy as one in which all or major segments of the population are guaranteed certain levels of income (directly and/or in the form of subsidized goods and services), paid for by a tax on the economically active population. A well-balanced system will be one in which people are receiving benefits they feel are appropriate, while those paying taxes are not perceiving themselves to be overtaxed. However, if the economically active perceive the tax burden as oppressive, they may well demand tax relief. On the other hand, if people consider themselves to be in hardship because guaranteed government benefits drop, they may well demand reform to increase the benefits. Either or both of these scenarios can pave a road of trouble for the government.

A major cause of imbalance in the system is one that is by now familiar to you—changes in the age structure that may have relatively little to do with the underlying political or economic structure of a society. Any economy must be organized around the principle that the productive members of society (whether that be a solitary family, a tribe, or a nation) must be able to provide for the unproductive members of the group, as well as for themselves. But, as the ratio of producers to dependents changes, both the burden on producers and the needs of dependents may change and whatever system is in place to allocate resources may become unsustainable.

Let us focus our attention on nations. Although there are multiple paths by which taxes might be raised to oppressive levels (causing the "workers" to rebel) and/or benefits may by cut by the government (causing the "beneficiaries" to rebel), the nearly silent role of the age transition in this process has only recently been recognized as a cause of such change because of the trauma facing Europeans as that region's population ages in the wake of a protracted decline in fertility without, as yet, a compensating influx of immigrants. Yet, silent though it may be, the age transition is an inevitable consequences of the demographic transition and may help us, after the fact at least, to explain failed efforts to establish socialist governments in developing countries with young populations.

A socialist or welfare-based economy, in the context of political freedom, may represent one option among many for a country whose age structure is stable. But it may not be sustainable over the entire course of the age transition unless a government recognizes the interaction of the age transition and the economy and makes structural changes to accommodate that relationship. This is a result of the fact that, as mortality drops and population begins to grow, people are being added largely at the youngest ages, where the population is least economically productive. You will recall from Davis's theory of demographic change and response that the first thing that families (and by inference nations as well) are likely to do in the face of increasing numbers of surviving children is to work harder to support those people. But, socialist-welfare states typically represent an economic system that discourages private innovation and harder work because of the high tax rate on profits and income. In such a setting, the increasing numbers of young people can present the economy with an unsustainable burden because the number of young dependents will be increasing, but economic productivity may not be growing sufficiently to support an increasing number of people at some governmentally agreed-on level. Keep in mind that this mismatch between people and resources is unlikely to be felt immediately, since children are not very expensive for their first few years of life. The impact of this rise in the dependency burden will likely begin to be felt after a lag of about 5 to 20 years, as the young population matures to progressively "more expensive" ages where society has to deal first with feeding and clothing and educating a greater number of children, and then—most expensively—has to find them jobs and homes.

This dilemma could be (or, when talking about the past, could have been) resolved by an aggressive birth control campaign that lowers the rate of population growth and thus reduces the dependency burden; or it could be (could have been) dealt with by liberalizing the economy to increase output, thus lowering the per-person burden on the economically productive segment of the population. Neoliberal economic policies can be seen as one type of response of the wealthy to the growth in the number of the poor. Enough! We can't take it any more! Will no one rid me of this burden? But neoliberalism falls back on the idea that growing populations can be exploited to generate low wages for the production of goods, so that the rich get richer and the poor also become a little better off—structural economic mobility. Of course, the situation could also be handled, albeit not solved, by a repressive government that forces people to accept the inevitable drop in per-person income that occurs when the population increases but economic output does not.

It is possible that these dynamics were at work in Chile when Salvador Allende's government was overthrown by the Chilean military in 1973. This is not meant to suggest that either Allende's election as president or his subsequent overthrow by Augusto Pinochet and the military was determined by demographic events, but it is possible that demography was working for Allende in his rise in power, but against him as he tried to achieve the kind of socialism that he envisioned for Chile even if the military had not intervened as it did (for a review of these events, see Weeks 2003). When Allende was elected president in 1970, the United States and the Chilean military alike were worried about Chile becoming a communist country. But, in fact, the population of Chile had been getting progressively younger since the end of World War II, due to a combination of declining infant mortality and an increase in fertility (Weeks 1970), and it is well known that a younger population tends to exaggerate inequality in a society (Gustafsson and Johannson 1999).

The changing age structure certainly did not convince Allende that social welfare reform was necessary in Chile, but the growing inequality that was exacerbated by the age transition would certainly increase the audience for his ideas. The

(continued)

IS THERE A DEMOGRAPHIC DIMENSION TO SUSTAINING SOCIALIST POLITICAL/ECONOMIC SYSTEMS? (CONTINUED)

accompanying graph shows the ratio of successive generations during the second half of the twentieth century in Chile. As I mentioned in Chapter 6, we typically think of the average age of childbearing (about 25 in most societies) as the length between generations. So, the figure shows the relative size of successive "generations" over time in Chile. From 1955 to 1980, the ratio of young people aged 15–24 to adults of the next generation (ages 40–49) was steadily increasing, putting ever more pressure on the economy and thus on the older generation to educate and find employment for this growing younger generation.

This younger group, in particular, would have been a growing audience for the social welfare reforms being proposed by Allende, but it is easy to imagine that the taxpayers in the older generation were not going to be overly pleased by a set of plans that almost certainly was going to noticeably increase the tax rate. Another way to deal with the growing number of young people would be to open up the economy to greater foreign investment that might produce jobs for younger people, thus exploiting the growing labor force in that way. These are, in fact, the kinds of neoliberal reforms that Pinochet instituted after he and his military allies completed their coup d'etat in 1973 by bombing La Moneda (the presidential palace), leading to the death of President Allende in the process of Pinochet's establishment of a military dictatorship that lasted until he relinquished control back to civilian rule in 1990.

The Pinochet government revealed that it did not understand the underlying dynamics of the age structure, since during the late 1970s and early 1980s, it was calling for a significant increase in population to protect the country from population growth in neighboring Latin American nations (Chile has had a war with Peru and the threat of a war with Argentina). This would have increased the population of young people, increasing the power base that had supported Allende. However, events worked in the opposite direction. After overthrowing Allende, the Pinochet regime imposed a level of austerity in government spending that reduced incomes and raised levels of unemployment. This encouraged out-migration, postponement of marriage, and the widespread use of contraception within marriage, speeding up a decline in fertility that had already gotten under way shortly before Allende was elected president. The TFR dropped from its peak of 5.3 in 1965 to below 3.0 by 1980, and has since dropped even further to 2.4 by the beginning of the twenty-first century—a level with which the current government is satisfied. As you now know, economies are able to reap an age dividend from a rapid drop in fertility such as that occurring in Chile. The neoliberal economic reforms played their role, to be sure, but the shifts in the demographic structure following the Pinochet takeover were working away silently in the background to allow those reforms to be successful.

Would the Chilean economy also have turned around had Allende remained as president? We will obviously never know, but you can see in the accompanying figure that the generational age ratios were ready to turn around during the Pinochet regime because the birth rate had started to fall even before Allende was elected. Furthermore, you can probably appreciate that a social welfare system is easiest to maintain in the middle stages of the age transition, especially if fertility has dropped quickly to produce the demographic dividend. In that middle stage of the age transition, the population in the productive ages is very large compared to both the younger and older populations and so the dependent population could be maintained with relatively low tax rates on the average worker. In fact, under this scenario, it may be possible to simultaneously lower the tax rate on the economically productive and improve benefits offered to societal members, as happened in the United States in the last two decades of the twentieth century. The big

dramatically again. The first decline in the U.S., as in England, was probably due as much to the "culture of abstinence"—reflecting a strong motivation to limit fertility—as to specific birth control methods (Szreter 1996). Only during the second decline was a wide range of birth prevention methods readily available to women; in essence, the methods were part of the rise in the standard of living.

question in Chile is whether or not it was the climate of change and repression in the Pinochet years that accelerated the fertility decline, or whether fertility would have declined as it did even if Allende had been allowed to remain in office. Perhaps it is only a coincidence that fertility essentially stopped declining once Pinochet left office in 1990 and the country returned to civilian rule.

No matter how rosy the situation is for social welfare programs during the middle of the age transition, that situation will be very difficult to sustain as the population ages and people begin to retire, because the swelling older population will now be withdrawing savings from the capital market and also seeking benefits to be paid from a shrinking number of workers—the problem currently facing Europe, Japan, and, to a lesser extent, the U.S. and Canada. Finally, at the end of the age transition, when the age structure is again stable (unless destabilized by a rise in the birth rate, the death rate, or both), the attractiveness of social welfare policies may again increase in step with that stability.

Changing Sizes of Generational Cohorts in Chile in Comparison to Political Events, 1950–2000

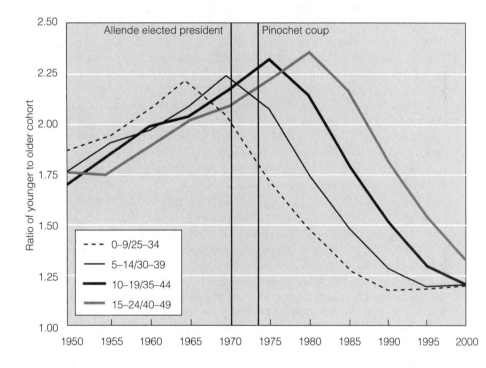

Source: Adapted by the author from data in United Nations Population Division, 2003, World Population Prospects: The 2002 Revision.

Domestic Policy Regarding Gender Equity

At the beginning of the twentieth century, it was unusual for women to go out in public alone and they were not eligible to vote. They gained the right to vote only in 1920 when the Nineteenth Amendment was passed. More recently, attempts have been made to add a constitutional amendment that would further establish gender equity in the United States. The proposed Equal Rights Amendment (ERA) to the Constitution would have guaranteed that women have the same legal rights as males, thereby potentially expanding their ability to be financially and socially independent. The ERA was initially passed by Congress in 1972 and required approval by 38 states before it could become part of the U.S. Constitution. After 10 years without ratification, the ERA expired in 1982, only to be reintroduced in the House and then defeated there in 1983. President Reagan argued that a constitutional amendment was not needed to fight sexual bias, and a majority of congressional members apparently agreed. The issue has not been raised since.

Domestic Policy Regarding Immigration

Immigration may be the sincerest form of flattery, but few countries encourage it. Migration between countries is fraught with the potential for conflict among people of different cultural backgrounds and, as technology has made transportation and thus migration easier, policies to deal with international migration have come into sharper focus. Nowhere is this more true than in the United States, where international immigration has been a way of life for more than two centuries and toward which a large fraction of the annual volume of the world's international migrants still heads. Prior to World War I, there were few restrictions on migration into the United States, so the number of migrants was determined more by the desire of people to come than anything else. Particularly important as a stimulus to migration, of course, was the drop in the death rate in Europe during the nineteenth century, which launched a long period of population growth. Free migration from Europe to the temperate zones of the world—especially North and South America and Oceania—represents one of the most significant movements of people across international boundaries in history. The social, cultural, economic, and demographic impacts of this migration were enormous (Davis 1974).

The opening up of new land in the United States coincided with a variety of political and economic problems in Europe, and that helped to generate a great deal of labor migration to the United States in the nineteenth century, particularly after the end of the Civil War. The United States responded with its first immigration restrictions in the form of the Immigration Act of 1882. This law levied a head tax of 50 cents on each immigrant and blocked the entry of idiots, lunatics, convicts, and persons likely to become public charges (U.S. Immigration and Naturalization Service 1991).

Meanwhile, the discovery of gold in California had prompted a demand for labor—for railroad building and farming—that had been met in part by the migration of indentured Chinese laborers. However, in 1869, after the completion of the transcontinental railroad, American workers moved west more readily and the Chinese showed up in the East on several occasions as strikebreakers. Resentment against the Chinese built to the point that in 1882 Congress was willing to break a recently signed treaty with China and suspend Chinese immigration for 10 years (Stephenson 1964). The law was challenged unsuccessfully in the courts and, over time, restrictions on the Chinese, even those residing in the United States, were tightened (indeed, the Chinese Exclusion Acts passed in 1882 were not repealed until

1943). The exclusion of the Chinese led to an increase in Japanese immigration in the 1880s and 1890s, but by the turn of the century hostility was building against them, too (the Japanese Exclusion Act was passed in 1924), and against several other immigrant groups.

In Chapter 7, I discussed the fact that by the late nineteenth century, the ethnic distribution of immigrants to North America had shifted away from the predominance of northern and western Europeans toward an increase in migration from southern Europe, especially Italy. In 1890, 86 percent of all foreign-born persons in the United States were of European origin, but only 2 percent were from southern Europe—almost exclusively Italy. Only 30 years later, in 1920, it was still true that 86 percent of the foreign-born were Europeans, but 14 percent were southern Europeans—a sevenfold increase (U.S. Census Bureau 1975).

The immigrant processing center in New York City was moved to Ellis Island in 1892 to help screen people entering the United States from foreign countries (Stephenson 1964), since the changing mix of ethnicity had led to public demands for greater control over who could enter the country. In 1891, Congress legislated that aliens were not to be allowed into the country if they suffered from "a loathsome or dangerous contagious disease" (Auerbach 1961:5) or if they were criminals. Insanity was added to the unacceptable list in 1903, and tuberculosis joined it in 1907. In 1917, a highly controversial provision was passed that established a literacy requirement, thus excluding aliens over age 16 who were unable to read.

Recognition of the problems created by still relatively free migration led to a new era of restrictions right after World War I. Europe was unsettled and in the midst of economic chaos, and there was a widespread belief that "millions of war-torn Europeans were about to descend on the United States—a veritable flood which would completely subvert the traditional American way of life" (Divine 1957:6). The United States and Canada both passed restrictive legislation in step with the eugenics movement that had gained popularity throughout Europe and North America in the 1920s (Boyd 1976). As it was applied to migration, the ideology was ethnic purity, and the sentiment of the time about migrants is perhaps best expressed as "not too tired, not too poor, and not too many."

In 1921, Congress passed the first act in American history to put a numeric limit on immigrants. The Quota Law of 1921 "limited the number of aliens of any nationality to three percent of foreign-born persons of that nationality who lived in the U.S. in 1910" (Auerbach 1961:9). For example, in 1910, 11,498 people in the United States had been born in Bulgaria (U.S. Census Bureau 1975), so 3 percent of that number, or 345, would be permitted to enter each year from Bulgaria. Under the law, about 350,000 people could enter the United States each year as quota immigrants, although close relatives of American citizens and people in certain professions (for example, artists, nurses, professors, and domestic servants) were not affected by the quotas. The law of 1921 remained in effect only until 1924, when it was replaced by the Immigration Quota Act. The 1924 law was even more restrictive than that of 1921, because public debate over immigration had unfortunately led to a popularization of racist theories claiming that "Nordics [people from northwestern Europe] were genetically superior to others" (Divine 1957:14).

To avoid the charge that the immigration law was deliberately discriminatory, though, a new quota system—the National Origins Quota—was adopted in 1929

(Auerbach 1961). This was a complex scheme in which a special Quota Board took the percentage of each nationality group in the United States in 1790 (the first U.S. Census) and then traced "the additions to that number of subsequent immigration" (Divine 1957:28). The task was not an easy one, since by and large the necessary data did not exist, so a lot of arbitrary assumptions and questionable estimates were made in the process. Once the national origins restriction had been established, the actual number of immigrants allowed from each country each year was calculated as a proportion of 150,000, which was established as the maximum number of all immigrants. Thus if 60 percent of the population was of English origin, then 60 percent of the 150,000 immigrants, or 90,000, could be from England. The number turned out to be slightly more than 150,000, because every country was allowed a minimum of 100 visas. Furthermore, close relatives of American citizens continued to be exempt from the quotas. In Canada, a similar immigration act was passed in 1927 (Boyd 1976). Congress, of course, retained the ability to override those quotas if the need arose, as it did during and after World War II, when refugees from Europe were accommodated.

In 1952, in the middle of the anti-Communist McCarthy era, another attempt was made in the United States to control immigration by increasing the "compatibility" of migrants and established U.S. society. The McCarran-Walter Act (the Immigration and Naturalization Act of 1952) retained the system of national origin quotas and added to it a system of preferences based largely on occupation (Keely 1971). The McCarran-Walter Act permitted up to 50 percent of the visas from each country to be taken by highly skilled persons whose services were urgently needed. Relatives of American citizens were ranked next, followed by people with no salable skills and no relatives who were citizens of the United States. Thus, the freedom of migration into the United States was severely restricted, even from those countries with an advantage according to the national origins quota system. The same was true in Canada, which passed similar legislation in the same year.

Canada, I should note, had at least two reasons for echoing the immigration policies of the United States. In the first place, it shares much of the sociocultural heritage of the United States, and, second, of course, it shares a border with the United States. This cultural similarity and close proximity would have left Canada inundated with migrants excluded from the United States had Canada not passed its own restrictive laws.

In the 1960s, the ethnically discriminatory aspects of North America's immigration policy ended, but its restrictive aspects were maintained. The Immigration Act of 1965 ended the nearly half-century of national origins as the principal determinant of who could enter this country from non-Western Hemisphere nations. Again, related changes had occurred in Canada in 1962. Although the criterion of national origins is gone, restrictions on the numbers of immigrants remain, including a limit on immigrants from Western Hemisphere as well as non-Western Hemisphere nations. A system of preference was retained, but modified to give first crack at immigration to relatives of American citizens. Parents of U.S. citizens could migrate regardless of the quota. In addition, a certification by the U.S. Labor Department is now required for occupational preference applicants to establish that their skills are required in the United States. In 1976, the law was amended so that parents of U.S. citizens had highest priority only if their child was at least 21 years old. The intent of that change was to eliminate the fairly frequent ploy of a pregnant woman entering the country illegally, bearing her child in the United States (the child then being a U.S. citizen), and then applying for citizenship on the basis of being a parent of a U.S. citizen.

It is almost certain that the restriction of immigration in the Western Hemisphere contributed to the number of people who have become undocumented workers or illegal immigrants to the United States, especially from Mexico. The requirement of labor certification, as well as a numeric quota on Western Hemisphere nations (dominated by Mexico), came at a time when the population of Mexico was growing much more rapidly than the Mexican economy could handle, especially in terms of jobs. Thus, there are distinct push factors in Mexico. Migration of unskilled labor to the United States from Mexico had been an alternative from the 1950s through the 1960s as part of the *bracero* program, which California growers had pushed as a means of obtaining cheap labor. In the 1960s, however, Mexican Americans lobbied successfully for a halt to the bracero program, and that followed on the heels of the more restrictive labor requirements. Thus, it is no surprise that illegal migration from Mexico has been high ever since.

As undocumented migration from Mexico continued, the cry was often heard that the southern border of the United States was "out of control," and it was widely believed that undocumented workers were taking jobs from U.S. citizens and were draining the welfare system. Typically, reality was less dramatic. The number of undocumented workers in the United States is less than many people imagined (as I discussed in Chapter 7), yet the widespread public perception of negative consequences of undocumented migration helped push Congress to pass the 1986 Immigration Reform and Control Act (IRCA). From the standpoint of the undocumented immigrant, this was a "good news/bad news" piece of legislation. The good news was that the law offered "amnesty"—relief from the threat of deportation and the prospect of legal resident status for undocumented workers who had been living continuously in the United States since before January 1, 1982. The bad news was that in order to curtail new workers from entering the country without documentation, it is now unlawful for an employer to knowingly hire such a person. The teeth of the law include fines for employers beginning at $250, with the prospect of much larger fines and even jail sentences for employers who are repeatedly found to hire undocumented workers. Notice that this law was aimed entirely at curtailing illegal immigration and had no impact on the pattern of legal migration. In fact, in the 1990s, the U.S. opened the door to legal immigration a little wider, as I mentioned in Chapter 7.

Since the 1990s, the United States has also been spending millions of dollars on new border fences and walls, designed to make it harder for unauthorized people to cross from Mexico into the United States, and to make it easier to apprehend people who try to do so. The impact has been that undocumented immigrants try to enter the U.S. in places where fences do not exist, but there is little evidence that the actual number of migrants is much different than it would otherwise have been (Loaeza Tovar and Martin 1997).

Over time the immigration policy of the United States, and to a lesser extent Canada, has been that of gatekeeping—keeping out those people considered undesirable. Those people generally considered most desirable are professionals: physicians, scientists, engineers, teachers, lawyers, and others. Indeed, one interesting aspect of the abolition of the national origins criterion is that the proportion of all legal immigrants who are professionals has actually increased. A major reason for this seems to be a type of "brain drain" from countries whose people were previously excluded, or nearly so, from the United States.

International Policy

The United States has played a key role in explicitly encouraging governments of developing nations to slow down their rate of population growth, and it has provided a great deal of money to establish and maintain family planning programs all over the world. Donaldson (1990) has chronicled the growing American consciousness from the 1940s through the 1960s that rapid population growth in less-developed nations might not be good either for those nations specifically or for the United States more generally. When President Kennedy established the Agency for International Development (USAID) within the Department of State in 1961, there was an early recognition that population was an important factor in development. By 1967, that idea bore fruit when Congress earmarked USAID funds for the purpose of providing population assistance along with economic assistance. For the next 18 years, the United States played a crucial part in helping to slow down the rate of population growth in the world—an effort that may have reduced the world population by nearly a half-billion people compared with the growth that would have occurred in the absence of organized family planning efforts in developing countries (Bongaarts, Mauldin, and Phillips 1990).

Things changed in 1985, however, as a result of Reagan administration policies (the "Mexico City Policy" I mentioned earlier) that were carried over into the subsequent Bush administrations, with an interstitial eight-year reprieve during the Clinton administration. The opposition to abortion created controversy in both the implicit domestic policy, because of the push to reverse *Roe v. Wade,* and in the international policy, because in 1985, U.S. opposition to abortions in China was used as a pretext for cutting off all United States government funding for the United Nations Population Fund, as I discussed above.

The halt in U.S. funding for the UNFPA has not meant the end of population assistance, however. USAID has remained the single largest source of international support for family planning programs. Such assistance is typically provided through other organizations working in various parts of the world. USAID also funds important programs in training and research activities. Furthermore, other countries, especially Germany, Japan, and Norway, have picked up some of the slack left by the United States.

A majority of Americans are in favor of providing aid to help slow the growth of population in developing nations, according to results from a 1998 survey conducted by the Rand Corporation (Adamson et al. 2000). These results are consistent with an earlier Gallup Poll of American attitudes (Newport and Saad 1992). Americans tend to be fairly ignorant of the actual size of the world's population—only 14 percent of the Rand survey respondents knew that the world population was about six billion, and 40 percent simply did not know. Nonetheless, a majority believe that the world is overpopulated, and that this contributes to environmental problems, civil strife, and economic problems in developing countries; the vast majority support the fundamental right of individuals to choose the number of children they want, and a majority are in favor of offering family planning assistance to developing countries. In this way, then, the majority of Americans seem to support the basic thrust of the implicit U.S. international population policy.

What Lies Ahead?

No matter what the best-laid population projections may suggest, it is good advice to "never underestimate the stupidity factor as a determinant of history" (Schreyer 1980:1). Humans have never shied away from doing stupid things, and some of history's turning points—those unpredictable events that set the course of history down a different path—are attributable to something less than rational choice. Nonetheless, it is reasonable to ask what the current transitions in the world portend for the future, because coping successfully with these changes may reduce the risk of something stupid happening.

"Demographic transition theory might seem to imply that, after a period of exceptional population growth resulting from the time lag between mortality and fertility declines, every population, and then the whole world population will stabilize and, consequently, no more acute population problems will appear. Does the claim, recently gaining credibility, that the end of the transition is at hand actually imply a stage without major population problems? Nothing is less sure" (Vallin 2002:105). On the contrary, we can be pretty sure that we are in for some rough times between now and whenever we might reach the end of the demographic transition. The major reason for that, as I have tried to emphasize, is that different countries are in different stages of their various transitions and so it is almost impossible to generalize about the global population situation. The fact that the population is not growing where you are does not mean that there are no population problems in the world. In the United States, for example, the population of the middle states is scarcely growing at all and there are relatively few immigrants. On the two coasts, by contrast, there is much more rapid population growth, much of it fueled by immigrants.

Between now and the middle of this century, we expect that at least three billion more people will be added to the six billion alive at the beginning of the century—even assuming that birth rates continue to fall around the world. Only a dramatic drop in fertility (or a dramatic rise in the death rate) will prevent the population from being much larger in a few decades than it is right now. Those new inhabitants of the earth are going to need to be fed, housed, clothed, educated, and employed. This is going to take resources, and the use of those resources will create waste that we are going to have to deal with. At this point, we really do not know where those resources are going to come from, nor do we know how we will deal with the by-products of so many more people using so many resources. How are we going to deal with the demands on the global commons, especially the use of the atmosphere as a sink for greenhouses gases that are emitted as a by-product of our greater economic productivity (McNicoll 1999)? The challenges of population and environmental policy will be to continue to keep people alive longer, making sure that they have as few children as is reasonable, and that they are living in places where they are able to contribute the most to the sustainability of human existence.

Sustainability means that we will have figured out ways to generate a renewable supply of water and food for nine billion people, but to have nine billion living healthier lives than the current six billion means that the amount of water and food per person will have to be higher than it currently is. To generate the wealth that will make that possible will almost certainly demand the use of vastly greater quantities of energy than we currently consume, and more energy consumption will bring with it a greater amount of waste—including contributions to global warming and ozone depletion. To minimize the impact of many more people using many more resources will almost certainly require that we alter our thinking about where people live and how they manage their local environments. We will have to reconcile the huge inconsistencies in what we want to accomplish. Our desire to lower mortality leads to policies that promote population growth. These have to be constantly countered by policies that promote fertility limitation. Our desire that more people should be able to live not only longer but also at a higher standard of living is probably doomed to failure if the birth rate does not drop more quickly than is currently projected. In the meantime, the search for a better life will lead ever more people to move to those economic environments where that better life seems like more than just a distant dream. Just dealing with the globalization of migration will present an extraordinarily difficult set of policy dilemmas for virtually all of the wealthier countries of the world, but in the midst of it all, we must not lose sight of the importance of maintaining every effort to keep bringing fertility levels down to manageable, sustainable levels.

Summary and Conclusion

One of the major uses of demographic science is as a tool for shaping the future, for trying to improve the conditions, both social and material, of human existence. To do that requires a demographic perspective—an understanding of how the causes of population change are related to the consequences. Throughout the previous 12 chapters, I have explored the ins and outs of those causes and consequences: how and why mortality, fertility, and migration change; how they affect the age and sex structure of society; how population characteristics are influenced by and in turn affect demographic changes; how women's status can influence fertility levels; how demographic changes affect the process of growing old in developed societies; and how population growth affects the urban environment, economic development, food resources, and environmental degradation. Implicit in all those discussions is the idea that an understanding of what has happened in the past and what is happening now will provide you with insights about the future, which is, of course, your first step toward advocating or promoting a population policy, and toward managing your own personal demographic transitions.

On the basis of what you now know about population change, you should be able to make some reckoning of what the future may be. If what you see concerns you, then I have tried to provide you with ideas about what can be done to cope with the population changes put into play by demographic transitions all over the world. If what you see does not worry you, then perhaps you should reread the book.

Main Points

1. A population policy is a strategy for achieving a particular pattern of demographic change.

2. Population policies are rarely ends in themselves but are designed to bring about desired social changes.

3. The 1974 World Population Conference produced a World Population Plan of Action that was largely reaffirmed at the 1984 International Conference on Population in Mexico City.

4. The 1984 World Population Conference is especially famous because of the imposition by the United States of a "global gag rule" with respect to the use of funds for anything having to do with abortion.

5. The 1994 International Conference on Population and Development produced a new Programme of Action emphasizing the importance of women's social and reproductive rights.

6. The Programme of Action proposed that the mortality transition be managed by lowering infant, child, and maternal mortality and by bringing the AIDS pandemic under control.

7. Without AIDS, it is projected that the world would have half a billion more people by midcentury.

8. The management of the fertility transition is largely devoted to empowering women at home, in society, and with respect to the availability of acceptable means of family planning.

9. Managing the migration and urban transitions involves trying to minimize the need for people to move, and trying to make them as safe and healthy as possible when they do move.

10. The age transition is largely managed through the indirect influence of the fertility transition.

11. Coping with the future demographic change especially means dealing with changing age structures and with very different age structures from region to region in the world.

12. Although the United States does not have an explicit population policy, it has strong implicit domestic policies regarding fertility and immigration, and a strong implicit international policy that is largely focused on the fertility transition, although assistance is also offered to other countries with respect to health issues in general.

13. If you read this book continuously at the pace of one chapter per hour, nearly 108,000 people were added to the world's population between the time you

started reading Chapter 1 and now. That is the equivalent of about 330 full flights on a Boeing 777, or more than 25 flight loads of people per chapter.

Suggested Readings

1. Louise Lassonde, 1997, Coping with Population Challenges (translated by Graham Grayston) (London: Earthscan Publications Ltd).

   This book was written after the 1994 population conference in Cairo as a way of addressing the many issues left unresolved by the conference. It is also interesting because it is written by a French demographer with a European perspective on how to cope with population change.

2. Nicholas Polunin, Editor, 1998, Population and Global Security (Cambridge: Cambridge University Press).

   This book is another "post-Cairo" assessment of where the world is in terms of population growth and the relationship between population and the environment.

3. East-West Center, 2002, The Future of Population in Asia (Honolulu: East-West Center).

   Asia has been the world's most populous region for all of recorded history and its future is obviously of considerable importance to the global population picture. This volume provides a thorough review of where Asia is headed demographically.

4. Oscar Harkavy, 1995, Curbing Population Growth: An Insider's Perspective on the Population Movement (New York: Plenum Press).

   Oscar Harkavy was an official with the Ford Foundation who helped to map the path of world population policy in the 1950s and 1960s—helping to get us to where we are today.

5. National Research Council, 2001, Preparing for an Aging World: The Case for Cross-National Research (Washington, DC: National Academy Press).

   The aging of populations is really less of a "crisis" than it is simply a reality—the inevitable consequence of the demographic transition. This volume is devoted to examining the ways in which societies need to think about the future, which will be occupied by an older population than we have today.

Web Sites of Interest

Remember that Web sites are not as permanent as books and journals, so I cannot guarantee that each of the following Web sites still exists at the moment you are reading this. It is usually possible, however, to retrieve the contents of old Web sites at http://www.archive.org.

1. http://www.unfpa.org/icpd/index.htm
   The United Nations Population Fund (UNFPA) was the organizer of the 1994 ICPD and the ICPD+5. This site includes the full text of the 20-year Programme of Action, as well as the documents from the ICPD+5, and has links as well to the Millennium Goals.

2. http://www.tfgi.com
   The Futures Group is a well-established NGO that works with countries all over the world on projects related to population growth and sustainability. It is perhaps most famous for its RAPID software, which models the socioeconomic impact of rapid population growth on a country. You can try out a demonstration copy of the software at this site.

3. http://www.populationinstitute.org
   The Population Institute in Washington, DC, is an NGO whose goal is to disseminate information about population issues and mobilize activities in support of programs that seek to lower rates of population growth and reduce human impact on the environment.

4. http://www.npg.org/
   Negative Population Growth is an organization devoted to slowing down the rate of population growth (leading eventually to fewer people), with an emphasis on the United States, but with a global focus as well, especially in terms of the impact of population on the environment.

5. http://www.africa2000.com
   Not all the voices on the Internet are in favor of implementing population policies and this Web site offers some dissent, although in the process it also includes some good material in favor of population policies that the authors of the site then try to debunk. In particular, visit the resource called X (for xenophobia) files.

APPENDIX
Population Data for the World as of 2004, Keyed to Figure 2.1

Map ID in Figure 2.1	Country	Population in millions	Crude Death Rate per 1000	Crude Birth Rate per 1000	Nert Migration Rate per 1000	Rate of Natural Increase per 1000	Rate of Population Growth per 1000	Infant Mortality Rate	Life Expectancy at Birth for Males	Life Expectancy at Birth for Females	Percent Under Age 15	Percent Age 65+
1	China	1,299	13.0	6.9	-0.4	6.1	5.7	25.3	70	74	22	8
2	India	1,065	22.8	8.4	-0.1	14.4	14.4	57.9	63	65	32	5
3	United States	293	14.1	8.3	3.4	5.8	9.2	6.6	75	80	21	12
4	Indonesia	238	21.1	6.3	0.0	14.9	14.9	36.8	67	72	29	5
5	Brazil	184	17.3	6.1	0.0	11.1	11.1	30.7	67	76	27	6
6	Pakistan	159	31.2	8.7	-2.8	22.6	19.8	74.4	62	64	40	4
7	Russia	144	9.6	15.2	1.0	-5.5	-4.5	17.0	60	73	15	14
8	Bangladesh	141	30.0	8.5	-0.7	21.5	20.8	64.3	62	62	34	3
9	Nigeria	137	38.2	14.0	0.3	24.3	24.5	70.5	50	51	43	3
10	Japan	127	9.6	8.8	0.0	0.8	0.8	3.3	78	85	14	19
11	Mexico	105	21.4	4.7	-4.9	16.7	11.8	21.7	72	78	32	6
12	Philippines	86	25.8	5.5	-1.5	20.3	18.8	24.2	67	73	36	4
13	Germany	82	8.5	10.4	2.2	-2.0	0.2	4.2	76	82	15	18
14	Vietnam	82	19.6	6.1	-0.5	13.4	13.0	29.9	68	73	29	6
15	Egypt	76	23.8	5.3	-0.2	18.5	18.3	33.9	68	73	33	4
16	Ethiopia	68	39.2	20.4	0.0	18.9	18.9	102.1	40	42	45	3
17	Turkey	68	17.2	6.0	0.0	11.3	11.3	42.6	70	75	27	7
18	Iran	68	17.1	5.6	-5.0	11.5	6.5	42.9	68	71	28	5
19	Thailand	65	16.0	6.9	0.0	9.1	9.1	21.1	69	74	24	7
20	France	60	12.3	9.1	0.7	3.3	3.9	4.3	76	83	19	16
21	United Kingdom	60	10.9	10.2	2.2	0.7	2.9	5.2	76	81	18	16
22	Italy	58	9.1	10.2	2.1	-1.2	0.9	6.1	77	83	14	19
23	Congo, Democratic Republic of	58	44.7	14.6	-0.2	30.1	29.9	94.7	47	51	48	3
24	Ukraine	48	10.2	16.4	-0.4	-6.2	-6.6	20.6	61	72	16	15
25	Korea, South	48	12.3	6.1	0.0	6.2	6.2	7.2	72	80	20	8
26	South Africa	43	18.4	20.5	-0.3	-2.2	-2.5	62.2	44	44	30	5
27	Myanmar	43	18.6	12.2	-1.8	6.5	4.7	68.8	54	58	28	5
28	Colombia	42	21.2	5.6	-0.3	15.6	15.3	21.7	68	75	31	5
29	Spain	40	10.1	9.6	1.0	0.6	1.6	4.5	76	83	14	18
30	Poland	39	10.6	10.0	-0.5	0.7	0.2	8.7	70	79	17	13

Map ID in Figure 2.1	Country	Population in millions	Crude Death Rate per 1000	Crude Birth Rate per 1000	Nert Migration Rate per 1000	Rate of Natural Increase per 1000	Rate of Population Growth per 1000	Infant Mortality Rate	Life Expectancy at Birth for Males	Life Expectancy at Birth for Females	Percent Under Age 15	Percent Age 65+
31	Argentina	39	17.2	7.6	0.6	9.6	10.2	15.7	72	80	26	11
32	Tanzania	37	39.0	17.5	-2.1	21.6	19.5	102.1	43	46	44	3
33	Sudan	37	35.8	9.4	0.0	26.4	26.4	64.1	57	59	44	2
34	Algeria	32	17.8	4.6	-0.4	13.2	12.8	32.2	71	74	30	5
35	Canada	32	10.9	7.7	6.0	3.2	9.2	4.8	77	84	18	13
36	Morocco	32	22.8	5.7	-1.0	17.1	16.1	43.3	68	73	33	5
37	Kenya	32	27.8	16.3	-0.1	11.5	11.4	62.6	45	45	41	3
38	Peru	28	21.3	6.3	-1.1	15.0	13.9	33.0	67	71	32	5
39	Afghanistan	28	47.3	21.1	23.1	26.2	49.2	166.0	42	43	45	2
40	Nepal	27	32.0	9.7	0.0	22.3	22.3	68.8	60	59	39	4
41	Uzbekistan	26	26.1	8.0	-1.7	18.2	16.5	71.3	61	68	34	5
42	Uganda	26	46.3	16.6	0.0	29.7	29.7	86.2	44	47	51	2
43	Venezuela	25	19.3	4.9	0.0	14.4	14.4	23.0	71	77	31	5
44	Iraq	25	33.1	5.7	0.0	27.4	27.4	52.7	67	69	40	3
45	Saudi Arabia	25	29.7	2.7	-2.7	27.1	24.4	13.7	73	77	38	2
46	Malaysia	24	23.4	5.1	0.0	18.3	18.3	18.4	69	75	33	5
47	Taiwan	23	12.7	6.3	0.0	6.4	6.4	6.5	74	80	20	9
48	Romania	22	10.7	11.7	-0.1	-1.0	-1.1	27.2	68	75	16	14
49	Korea, North	22	16.8	7.0	0.0	9.8	9.8	24.8	68	74	25	8
50	Ghana	21	24.9	10.7	-0.6	14.2	13.6	52.2	55	57	38	4
51	Mozambique	20	36.1	23.9	0.0	12.2	12.2	137.1	38	36	44	3
52	Sri Lanka	20	15.9	6.5	-1.3	9.4	8.1	14.8	70	76	25	7
53	Australia	20	12.4	7.4	4.0	5.0	9.0	4.8	77	83	20	13
54	Yemen	20	43.2	8.8	0.0	34.4	34.4	63.3	60	63	47	3
55	Syria	18	28.9	5.0	0.0	24.0	24.0	30.6	68	71	38	3
56	Côte d'Ivoire	17	39.6	18.5	-0.1	21.2	21.1	97.1	40	45	45	2
57	Kazakhstan	17	15.5	9.6	-3.4	5.9	2.6	30.5	61	72	24	8
58	Madagascar	17	41.9	11.6	0.0	30.3	30.3	78.5	54	59	45	3
59	Cameroon	16	35.1	15.3	0.0	19.7	19.7	69.2	47	49	42	3
60	Netherlands	16	11.4	8.7	2.9	2.7	5.7	5.1	76	81	18	14
61	Chile	16	15.8	5.7	0.0	10.1	10.1	9.1	73	80	26	8

Map ID in Figure 2.1	Country	Population in millions	Crude Death Rate per 1000	Crude Birth Rate per 1000	Nert Migration Rate per 1000	Rate of Natural Increase per 1000	Rate of Population Growth per 1000	Infant Mortality Rate	Life Expectancy at Birth for Males	Life Expectancy at Birth for Females	Percent Under Age 15	Percent Age 65+
62	Ecuador	13	23.2	4.3	-8.6	18.9	10.3	24.5	73	79	34	5
63	Guatemala	13	34.6	6.8	-1.7	27.8	26.1	36.9	64	66	43	3
64	Cambodia	13	27.1	9.1	0.0	18.0	18.0	73.7	56	61	38	3
65	Burkina Faso	13	44.5	18.8	0.0	25.7	25.7	98.7	43	46	46	3
66	Zimbabwe	13	30.1	23.3	0.0	6.8	6.8	67.1	39	37	39	4
67	Mali	12	47.3	19.1	-0.3	28.2	27.8	118.0	45	46	47	3
68	Cuba	11	12.2	7.2	-1.6	5.0	3.4	6.5	75	79	20	10
69	Malawi	11	44.4	23.0	0.0	21.3	21.4	104.2	37	38	47	3
70	Greece	11	9.7	10.1	2.4	-0.4	2.0	5.6	76	82	15	19
71	Niger	11	48.9	21.5	-0.7	27.4	26.7	122.7	42	42	48	2
72	Angola	11	45.1	25.9	0.0	19.3	19.3	192.5	36	38	44	3
73	Senegal	11	35.7	10.7	0.2	25.0	25.2	56.5	55	58	43	3
74	Belarus	10	10.5	14.1	2.5	-3.6	-1.1	13.6	63	75	16	15
75	Belgium	10	10.6	10.2	1.2	0.4	1.6	4.8	75	82	17	17
76	Czech Republic	10	9.1	10.5	1.0	-1.4	-0.5	4.0	73	79	15	14
77	Portugal	10	10.9	10.4	3.6	0.5	4.1	5.1	74	81	17	17
78	Hungary	10	9.8	13.2	0.9	-3.4	-2.5	8.7	68	77	16	15
79	Serbia and Montenegro	10	12.1	10.5	-1.3	1.6	0.3	13.4	72	77	18	15
80	Zambia	10	39.0	24.4	0.0	14.6	14.7	98.4	35	35	46	3
81	Tunisia	10	15.7	5.1	-0.6	10.7	10.1	25.8	73	76	26	7
82	Chad	10	46.5	16.4	-0.1	30.1	30.0	94.8	47	50	48	3
83	Sweden	9	10.5	10.4	1.7	0.1	1.8	2.8	78	83	18	17
84	Dominican Republic	9	23.6	7.1	-3.2	16.5	13.3	33.3	66	69	33	5
85	Bolivia	9	24.7	7.8	-1.3	16.9	15.6	54.6	63	68	36	5
86	Austria	8	8.9	9.6	2.0	-0.7	1.4	4.7	76	82	16	16
87	Azerbaijan	8	19.8	9.8	-4.9	10.1	5.2	82.1	59	68	27	8
88	Guinea	8	42.3	15.5	-3.1	26.7	23.7	91.8	48	51	44	3
89	Somalia	8	46.0	17.3	5.4	28.7	34.1	118.5	46	49	45	3
90	Bulgaria	8	9.7	14.3	-4.6	-4.6	-9.2	21.3	68	76	14	17
91	Rwanda	8	40.0	21.9	0.0	18.2	18.2	101.7	38	40	42	3
92	Hong Kong S.A.R.	7	7.2	6.0	5.2	1.3	6.5	3.0	79	84	14	13

Map ID in Figure 2.1	Country	Population in millions	Crude Death Rate per 1000	Crude Birth Rate per 1000	Nert Migration Rate per 1000	Rate of Natural Increase per 1000	Rate of Population Growth per 1000	Infant Mortality Rate	Life Expectancy at Birth for Males	Life Expectancy at Birth for Females	Percent Under Age 15	Percent Age 65+
93	Switzerland	7	9.8	8.4	4.1	1.4	5.4	4.4	78	83	17	15
94	Haiti	7	33.8	13.2	-3.4	20.6	17.1	74.4	51	53	42	4
95	Benin	7	42.6	13.7	0.0	28.9	28.9	85.9	50	51	47	2
96	Tajikistan	7	32.6	8.4	-2.9	24.2	21.4	112.1	62	68	39	5
97	Honduras	7	31.0	6.6	-2.0	24.4	22.4	29.6	65	67	41	4
98	Burundi	6	39.7	17.6	-0.1	22.1	22.0	70.4	43	44	46	3
99	El Salvador	6	27.5	5.9	-3.7	21.6	17.8	25.9	67	75	37	5
100	Israel	6	18.5	6.2	0.7	12.3	12.9	7.2	77	81	27	10
101	Paraguay	6	29.8	4.6	-0.1	25.2	25.1	26.7	72	77	38	5
102	Laos	6	36.5	12.1	0.0	24.4	24.4	87.1	53	57	42	3
103	Sierra Leone	6	43.3	20.6	0.0	22.7	22.7	145.2	40	45	45	3
104	Slovakia	5	10.6	9.5	0.3	1.1	1.4	7.6	70	78	18	12
105	Denmark	5	11.6	10.5	2.5	1.1	3.5	4.6	75	80	19	15
106	Libya	5	27.2	3.5	0.0	23.7	23.7	25.7	74	79	34	4
107	Jordan	5	22.7	2.6	6.6	20.1	26.7	18.1	76	81	35	4
108	Togo	5	34.4	11.6	0.0	22.7	22.7	67.7	51	55	44	3
109	Finland	5	10.6	9.7	1.0	0.9	1.8	3.6	75	82	18	16
110	Papua New Guinea	5	30.5	7.5	0.0	23.0	23.0	53.2	62	67	38	4
111	Nicaragua	5	25.5	4.5	-1.2	21.0	19.7	30.2	68	72	38	3
112	Georgia	5	10.1	9.0	-4.7	1.1	-3.6	19.3	72	79	19	16
113	Kyrgyzstan	5	22.1	7.2	-2.5	14.9	12.5	36.8	64	72	32	6
114	Turkmenistan	5	27.8	8.8	-0.9	19.0	18.1	73.1	58	65	36	4
115	Norway	5	11.9	9.5	1.7	2.4	4.1	3.7	77	82	20	15
116	Eritrea	4	39.0	13.4	0.0	25.7	25.7	75.6	51	54	45	3
117	Singapore	4	9.6	4.1	11.5	5.6	17.1	2.3	79	84	17	8
118	Moldova	4	14.8	12.8	-0.3	2.1	1.8	41.0	61	69	21	10
119	Croatia	4	9.5	11.3	1.6	-1.8	-0.2	7.0	70	78	17	16
120	Bosnia and Herzegovina	4	12.6	8.3	0.3	4.2	4.5	21.9	70	76	19	11
121	Puerto Rico	4	14.1	7.8	-1.5	6.3	4.9	8.4	73	82	22	12
122	New Zealand	4	14.0	7.5	4.1	6.5	10.5	6.0	76	82	22	12
123	Ireland	4	14.5	7.9	5.0	6.6	11.6	5.5	75	80	21	12

Map ID in Figure 2.1	Country	Population in millions	Crude Death Rate per 1000	Crude Birth Rate per 1000	Nert Migration Rate per 1000	Rate of Natural Increase per 1000	Rate of Population Growth per 1000	Infant Mortality Rate	Life Expectancy at Birth for Males	Life Expectancy at Birth for Females	Percent Under Age 15	Percent Age 65+
124	Costa Rica	4	19.0	4.3	0.5	14.7	15.2	10.3	74	79	30	6
125	Lebanon	4	19.3	6.3	0.0	13.0	13.0	25.5	70	75	27	7
126	Central African Republic	4	35.6	20.0	0.0	15.6	15.6	92.2	40	43	43	3
127	Lithuania	4	8.5	11.0	-0.7	-2.5	-3.3	7.1	68	79	17	15
128	Albania	4	15.1	5.0	-4.9	10.1	5.1	22.3	74	80	26	8
129	Uruguay	3	14.4	9.1	-0.3	5.4	5.1	12.3	73	79	24	13
130	Armenia	3	11.4	8.1	-6.5	3.3	-3.2	24.2	68	75	23	11
131	Liberia	3	44.8	17.9	0.0	27.0	27.0	130.5	47	49	43	4
132	Congo (Brazzaville)	3	28.7	14.5	0.0	14.2	14.2	93.9	49	51	38	4
133	Panama	3	20.4	6.4	-0.9	14.0	13.1	21.0	70	75	30	6
134	Mauritania	3	41.8	12.7	0.0	29.1	29.1	72.4	50	55	46	2
135	Oman	3	37.1	3.9	0.3	33.2	33.5	20.3	71	75	42	3
136	Mongolia	3	21.4	7.1	0.0	14.3	14.3	55.5	62	66	30	4
137	Jamaica	3	16.9	5.4	-4.9	11.5	6.6	12.8	74	78	28	7
138	United Arab Emirates	2	18.7	4.1	1.0	14.5	15.5	15.1	73	78	26	3
139	Latvia	2	8.9	13.7	-2.2	-4.9	-7.1	9.7	66	76	15	16
140	Lesotho	2	26.9	24.8	-0.7	2.1	1.4	85.2	37	37	37	6
141	West Bank	2	33.2	4.1	3.0	29.1	32.1	20.2	71	75	44	4
142	Kuwait	2	21.9	2.4	14.2	19.4	33.6	10.3	76	78	28	3
143	Bhutan	2	34.4	13.2	0.0	21.2	21.2	102.6	54	54	39	4
144	Macedonia	2	13.1	7.8	-1.5	5.3	3.9	11.7	72	77	22	11
145	Slovenia	2	8.9	10.2	1.1	-1.3	-0.1	4.5	72	80	14	15
146	Namibia	2	33.5	21.0	0.0	12.5	12.5	69.6	42	39	42	4
147	Botswana	2	24.7	33.6	0.0	-8.9	-8.9	70.0	31	31	39	5
148	Gambia, The	2	40.3	12.1	1.6	28.2	29.8	73.5	53	57	45	3
149	Estonia	1	9.8	13.3	-3.2	-3.5	-6.6	8.1	66	77	16	17
150	Guinea-Bissau	1	38.0	16.6	-1.6	21.5	19.9	108.7	45	49	42	3

Source: U.S. Census Bureau, International Data Base: http://www.census.gov/ipc/www/, accessed 2004.

GLOSSARY

This glossary contains words or terms that appeared in boldface type in the text. I have tried to include terms that are central to an understanding of the study of population. The chapter notation in parentheses refers to the chapter(s) in which the term is discussed in detail.

abortion   the induced or spontaneous premature expulsion of a fetus (Chapter 6).

abridged life table   a life table (see definition) in which ages are grouped into categories (usually five-year age groupings) (Chapter 5).

accidental death   loss of life unrelated to disease but attributable to the physical, social, or economic environment (Chapter 5).

acculturation   a process undergone by immigrants in which they adopt the host language, bring their diet more into line with the host culture's diet, and participate in cultural activities of the host society (Chapter 7).

achieved characteristics   those sociodemographic characteristics such as education, occupation, income, marital status, and labor force participation over which we have some degree of control (Chapter 10).

adaptation   a process undergone by immigrants in which they adjust to the new physical and social environment of the host society (Chapter 7).

administrative data   demographic information derived from administrative records, including tax returns, utility records, school enrollment, and participation in government programs (Chapter 4).

administrative records   forms filled out for each person entering the U.S. from abroad that are then collected and tabulated by the U.S. Citizenship and Immigration Service (chapter 4).

age and sex structure   the number (or percentage) of people in a population distributed by age and sex (Chapter 8).

age pyramid   graph of the number (or percentage) of people in a population distributed by age and sex (Chapter 8).

age-sex-specific death rate (ASDR)   the number of people of a given age and sex who died in a given year divided by the total (average midyear) number of people of that age and sex (Chapter 5).

age-specific fertility rate (ASFR)   the number of children born to women of a given age divided by the total number of women that age (Chapter 6).

age stratification   the assignment of social roles and social status on the basis of age (Chapter 8).

age structure   the distribution of people in a population by age; a "young" population is one with more than 35 percent of the population under age 15; an "old" population is one with more than 12 percent aged 65 and older (Chapter 8).

age transition   the shift from a predominantly younger to a predominantly older population as a society moves through the demographic transition (Chapters 3 and 8).

Agricultural Revolution   change that took place roughly 10,000 years ago when humans first began to domesticate plants and animals, thereby making it easier to settle in permanent establishments (Chapter 2).

Alzheimer's disease   a disease involving a change in the brain's neurons, producing behavioral shifts; a major cause of organic brain disorder among older persons (Chapter 9).

ambivalent   an attitude of being caught between competing pressures and thus being uncertain about how to behave properly (Chapter 6).

amenorrhea   temporary absence or suppression of menstruation (Chapter 6).

American Community Survey   an ongoing "continuous measurement" survey conducted by the U.S. Census Bureau to track the detailed population characteristics of every American community; designed to allow the long form to be dropped from the decennial census (Chapter 4).

anovulatory   pertaining to a menstrual cycle in which no ovum (egg) is released (Chapter 6).

antinatalist   based on an ideological position that discourages childbearing (Chapters 3 and 13).

applied demography   see *demographics* (Chapter 1).

apportionment   the use of census data to determine the number of seats in the U.S. Congress that will be allocated to each state (Chapter 1).

aquaculture   farming fish (including shellfish); a steadily increasing source of fish for food (Chapter 12).

arable   land that is suitable for agricultural purposes (Chapter 12).

ascribed characteristics   sociodemographic characteristics such as sex or gender, race, and ethnicity that we are born with and over which we have essentially no control (Chapter 10).

assimilate   a process undergone by immigrants in which they not only accept the outer trappings of the host culture, but also assume the behaviors and attitudes of the host culture (Chapter 7).

asylee   a person who has been forced out of his or her country of nationality and who is seeking legal refuge (permanent residency) in the country to which he or she has moved (Chapter 7).

atmosphere   the whole mass of air surrounding the earth (Chapters 1 and 12).

average age of a population   one measure of the age distribution of a population—may be calculated as either the mean or the median (Chapter 8).

Baby Boom   the dramatic rise in the birth rate following World War II. In the United States it refers to people born between 1946 and 1964; in Canada it refers to people born between 1947 and 1966 (Chapters 2 and 6).

base year   the beginning year of a population projection (Chapter 8).

biosphere   the zone of Earth where life is found, including the atmosphere, hydrosphere, and lithosphere (Chapter 12).

birth control   regulation of the number of children you have through the deliberate prevention of conception; note that birth control does not necessarily include abortion, although it is often used in the same sense as the terms contraception, family planning, and fertility control (Chapter 6).

birth rate   see *crude birth rate* (Chapter 6).

boomsters   a nickname given to people who believe that population growth stimulates economic development (Chapter 12).

bracero   a Mexican laborer admitted legally into the United States for a short time to perform seasonal, usually agricultural, labor (Chapter 7).

breast-feeding   temporarily impairs fecundity by prolonging postpartum amenorrhea and suppressing ovulation (Chapter 6).

capital   a stock of goods used for the production of other goods rather than for immediate enjoyment; anything invested to yield income in the future (Chapter 12).

capitalism   an economic system in which the means of production, distribution, and exchange of wealth are maintained chiefly by private individuals or corporations, as contrasted to government ownership (Chapter 3).

carbon cycle   that process through which carbons, central to life on the planet, are exchanged between living organisms and inanimate matter (Chapter 12).

cardiovascular disease   a disease of the heart or blood vessels (Chapter 5).

carrying capacity   the size of population that could theoretically be maintained indefinitely at a given level of living with a given type of economic system (Chapter 2).

celibacy   permanent nonmarriage, implying as well that a celibate is a person who never enters a sexual union (Chapter 6).

census metropolitan area (CMA)   metropolian areas as defined in Canada, including a core urban area of at least 100,000 and adjacent urban and rural areas that have a high degree of economic and social integration with that core urban area (Chapter 11).

census of population an official enumeration of an entire population, usually with details as to age, sex, occupation, and other population characteristics; defined by the United Nations as "the total process of collecting, compiling and publishing demographic, economic and social data pertaining, at a specified time or times, to all persons in a country or delimited territory" (Chapter 4).

chain migration the process whereby migrants are part of an established flow from a common origin to a prepared destination where others have previously migrated (Chapter 7).

checks to growth factors that, according to Malthus, keep population from growing in size, including positive checks and preventive checks (Chapter 3).

child control the practice of controlling family size after the birth of children (postnatally), through the mechanisms of infanticide, fosterage, and orphanage (Chapter 6).

child–woman ratio a census-based measure of fertility, calculated as the ratio of children aged 0–4 to the number of women aged 15–49 (Chapter 6).

children ever born births to date for a particular cohort of women at a particular point in time (Chapter 6).

chronic disease disease that continues for a long time or recurs frequently (as opposed to acute)—often associated with degeneration (Chapter 5).

cluster marketing identifying neighborhoods on the basis of a whole set of shared sociodemographic characteristics and using that information to market goods and services to people in the identified geographic areas (Chapter 1).

cohabitation the sharing of a household by unmarried people who have a sexual relationship (Chapter 10).

cohort people who share something in common; in demography, this is most often the year (or grouped years) of birth (Chapter 6).

cohort component method of population projection a population projection made by applying age-specific survival rates, age-specific fertility rates, and age-specific measures of migration to the base year population in order to project the population to the target year (Chapter 8).

cohort flow the movement through time of a group of people born in the same year (Chapter 8).

cohort measures of fertility following the fertility of groups of women as they proceed through their childbearing years (Chapter 6).

Columbian Exchange the exchange of food, products, people, and diseases between Europe and the Americas as a result of explorations by Columbus and others (Chapters 3 and 5).

communicable disease (also called infectious disease) a disease capable of being communicated or transmitted from person to person (Chapter 5).

completed fertility rate (CFR) the cohort measure of fertility, calculated as the number of children ever born to women who have reached the end of their reproductive career (Chapter 6).

components of change or residual method of migration estimation a method of measuring net migration between two dates by comparing the estimate of total population with that which would have resulted solely from the components of birth and death, with the residual attributable to migration (Chapter 7).

components of growth a method of estimating and/or projecting population size by adding births, subtracting deaths, and adding net migration occurring in an interval of time, then adding the result to the population at the beginning of the interval (Chapter 8).

condom a thin sheath, usually of rubber, worn over the penis during sexual intercourse to prevent conception or venereal disease (Chapter 6).

consolidated metropolitan statistical areas (CMSAs) groupings of the very largest Metropolitan Statistical Areas in the United States (Chapter 11).

content error an inaccuracy in the data obtained in a census; possibly an error in reporting, editing, or tabulating (Chapter 4).

contraception the prevention of conception or impregnation by any of various techniques or devices (Chapter 6).

contraceptive device mechanical or chemical means of preventing conception (Chapter 6).

contraceptive prevalence the percentage of "at risk" women of reproductive age (15 to 44 or 15 to

49) who are using a method of contraception (Chapter 6).

Core-Based Statistical Area (CBSA) A method for identifying metropolitan areas adopted by the U.S. Office of Management and Budget in 2000 and implemented by the U.S. Census Bureau in connection with Census 2000 data (Chapter 11).

core-periphery model a model of city systems in which the primate city (the core) controls the resources, and the smaller cities (the periphery) depend on the larger city (Chapter 11).

cornucopians "boomsters" who believe that we can always grow enough food to feed whatever size population we have (Chapter 12).

coverage error the combination of undercount (the percentage of a particular group or total population that is inadvertently not counted in a census) and overcount (people who are counted more than once in the census) (Chapter 4).

crowding the gathering of a large number of people closely together; the number of people per space per unit of time (Chapter 11).

crude birth rate the number of births in a given year divided by the total midyear population in that year (Chapter 6).

crude death rate (CDR) the number of deaths in a given year divided by the total midyear population in that year (Chapter 5).

crude net migration rate (CNMR) a measure of migration calculated as the number of in-migrants minus the number of out-migrants divided by the total midyear population (Chapter 7).

cumulated cohort fertility rate see *children ever born* (Chapter 6).

de facto population the people actually in a given territory on the census day (Chapter 4).

degeneration the biological deterioration of a body (Chapter 5).

de jure population the people who legally "belong" in a given area whether or not they are there on census day (Chapter 4).

demographic analysis (DA) a method of evaluating the accuracy of a census by estimating the demographic components of change since the previous census and comparing it with the new census count (Chapter 4).

demographic balancing equation the formula that shows that the population at time 2 is equal to the population at time 1 plus the births between time 1 and 2 minus the deaths between time 1 and 2 plus the in-migrants between time 1 and 2 minus the out-migrants between time 1 and 2 (Chapter 4).

demographic change and response the theory that the response made by individuals to population pressures is determined by the means available to them (Chapter 3).

demographic characteristics see *population characteristics* (Chapter 10).

demographic overhead the general cost of adding people to a population caused by the necessity of providing goods and services (Chapter 12).

demographic perspective a way of relating basic information to theories about how the world operates demographically (Chapter 3).

demographic processes see *population processes* (Chapter 1).

demographic transition the process whereby a country moves from high birth and high death rates to low birth and low death rates with an interstitial spurt in population growth (Chapter 3).

demographics the application of demographic science to practical problems; any applied use of population statistics (Chapter 1).

demographics of human resource management using demographic information to help manage an organization's labor force (Chapter 1).

demography the scientific study of human populations (Chapter 1).

density the ratio of people to physical space (Chapter 11).

dependency ratio the ratio of people of dependent age (usually considered to be 0–14 and 65+) to people of economically active ages (15–64) (Chapter 8).

diaphragm a barrier method of fertility control— a thin, dome-shaped device, usually of rubber, inserted in the vagina and worn over the uterine cervix to prevent conception during sexual intercourse (Chapter 6).

differential undercount the situation that occurs in a census when some groups of people are more likely to be underenumerated than other groups (Chapter 4).

direct population policy a population policy that is aimed specifically at altering demographic behavior (Chapter 13).

disability-free life expectancy the period of time up until a person's arrival at the "Fourth Age," when the rest of his or her life is increasingly consumed by coping with the health effects of old age (Chapter 9).

doctrine a principle laid down as true and beyond dispute (Chapter 3).

donor area the area from which migrants come (Chapter 7).

doomster a nickname given to people who believe that population growth retards economic development (Chapter 12).

douche (as a contraceptive) washing of the vaginal area after intercourse to prevent conception (Chapter 6).

drought a prolonged period of less-than-average rainfall (Chapter 12).

dual-system estimation (DSE) a method of evaluating a census by comparing respondents in the census with respondents in a carefully selected postenumeration survey or through a matching with other records (Chapter 4).

Easterlin relative cohort size hypothesis the perspective that fertility is influenced less by absolute levels of income than by relative levels of well-being produced by generational changes in cohort size (Chapter 3).

ecological footprint the total area of productive land and water required to produce the resources for and assimilate the waste from a given population (Chapter 12).

economic development a rise in the average standard of living associated with economic growth; a rise in per capita income (Chapter 12).

economic growth an increase in the total amount of income produced by a nation or region without regard to the total number of people (Chapter 12).

ecosphere all of the earth's ecosystems; the living portion of the biosphere (Chapter 12).

ecosystem communities of species interacting with one another and with the inanimate world (Chapter 12).

edge cities cities that have been created in suburban, often unincorporated, areas and that replicate most of the functions of the older central city (Chapter 11).

emergency contraceptive pills methods designed to avert pregnancy within 72 hours after intercourse, usually by taking a large dosage of the same hormones contained in the contraceptive pill (Chapter 6).

emigrant a person who migrates from one country to settle permanently in another (Chapter 7).

enclave a place within a larger community within which members of a particular subgroup tend to concentrate (Chapter 7).

endogenous factors those things that are within the scope of (internal to) one's own control (Chapter 6).

epidemiologic transition the pattern of long-term shifts in health and disease patterns as mortality moves from high levels (dominated by death at young ages from communicable diseases) to low levels (dominated by death at older ages from degenerative diseases) (Chapter 5).

ethnic group a group of people of the same race or nationality who share a common and distinctive culture while living within a larger society (Chapter 10).

ethnicity the ancestral origins of a particular group, typically manifested in certain kinds of attitudes and behaviors (Chapter 10).

ethnocentric characterized by a belief in the inherent superiority of one's own group and culture accompanied by a feeling of contempt for other groups and cultures (Chapter 3).

exclusion dealing with immigrants to an area by keeping them separate from most members of the host society, forcing them into separate enclaves or ghettos (Chapter 7).

exogenous factors those things that are beyond the control of (external to) the average person (Chapter 6).

expectation of life at birth see *life expectancy* (Chapter 5).

extended family family members beyond the nuclear family (Chapter 10).

extensification of agriculture increasing agricultural output by putting more land into production (Chapter 12).

exurbs nonrural population beyond the suburbs (Chapter 11).

family a group of people who are related to each other by birth, marriage, or adoption (Chapter 10).

family control ways of limiting family size after the birth of children (Chapter 6).

family demography the study and analysis of family households: their formation, their change over time, and their dissolution (Chapter 10).

family household a household in which the householder is living with one or more persons related to her or him by birth, marriage, or adoption (Chapter 10).

family and household transition the shift in family and household structure occasioned in societies by people living longer, with fewer children born, increasingly in urban settings, and subject to higher standards of living, all as part of the demographic transition (Chapters 3 and 10).

famine food shortage accompanied by a significant increase in deaths (Chapter 12).

fecundity the physical capacity to reproduce (Chapter 6).

female genital mutilation (FGM) sometimes known as female circumcision, which typically involves removing a woman's clitoris, thus lessening her enjoyment of sexual intercourse (Chapter 6).

fertility reproductive performance rather than the mere capacity to reproduce; one of the three basic demographic processes (Chapter 6).

fertility awareness methods of birth control methods of natural family planning that are combined with a barrier method to protect against conception during the fertile period (Chapter 6).

fertility differential a variable in which people show clear differences in fertility according to their categorization by that variable (Chapter 6).

fertility rate see general fertility rate (Chapter 6).

fertility transition the shift from "natural" fertility (high levels of fertility) to fertility limitation (low levels of fertility) (Chapters 3 and 6).

food security a term meaning that people have physical and economic access to the basic food they need in order to work and function normally (Chapter 12).

force of mortality the factors that prevent people from living to their biological maximum (Chapter 5).

forced migrant someone who has been forced to leave his or her home because of a real or perceived threat to life and well-being (Chapter 7).

forward survival (or residual) method of migration estimation a method of estimating migration between two censuses by combining census data with life table probabilities of survival between the two censuses (Chapter 7).

fosterage the practice of placing an "excess" child in someone else's home (Chapter 6).

gender role a social role considered appropriate for males or females (Chapter 10).

general fertility rate the total number of births in a year divided by the total midyear number of women of childbearing age (Chapter 6).

generational replacement a net reproduction rate of one, which indicates that each generation of females has the potential to just replace itself (Chapter 6).

gentrification restoration and habitation of older homes in central city areas by urban or suburban elites (Chapter 11).

geodemographics analysis of demographic data that have been georeferenced to specific geographic places (Chapter 1).

geographic information system (GIS) computer-based system that allows the user to combine maps with data that refer to particular places on those maps and then to analyze those data and display the results as thematic maps or some other graphic format (Chapter 4).

geo-referenced a piece of information that includes some form of geographic identification such as precise latitude-longitude coordinates, a street address, ZIP code, census tract, county, state, or country (Chapter 4).

global warming an increase in the global temperature caused by a build-up of greenhouse gases (Chapter 12).

Graunt, John Seventeenth century London haberdasher who has become known as the "father of demography" for his pioneering studies of the regular patterns of death in London (Chapters 3 and 5).

Green Revolution an improvement in agricultural production begun in the 1940s based on high-yield-variety strains of grain and increased use of fertilizers, pesticides, and irrigation (Chapter 12).

greenhouse gases those atmospheric gases (especially carbon dioxide and water) that trap radiated heat from the sun and warm the surface of the earth (Chapter 12).

gross domestic product the total value of goods and services produced within the geographic boundaries of a nation in a given year, without reference to international trade (Chapter 12).

gross national income in PPP (GNI PPP) gross national product expressed in terms of Purchasing Power Parity (rather than the official exchange rates) (Chapter 12).

gross national product (GNP) the total output of goods and services produced by a country, including income earned from abroad (Chapter 12).

gross (or crude) rate of in-migration the total number of in-migrants divided by the total midyear population in the area of destination (Chapter 7).

gross (or crude) rate of out-migration the total number of out-migrants divided by the total mid-year population in the area of origin (Chapter 7).

gross reproduction rate the total fertility rate multiplied by the proportion of all births that are girls. It is generally interpreted as the number of female children that a female just born may expect to have in her lifetime, assuming that birth rates stay the same and ignoring her chances of survival through her reproductive years (Chapter 6).

growth rates by age a way of measuring population growth by calculating age-specific rates of growth (Chapter 8).

high growth potential the first stage in the demographic transition, in which a population has a pattern of high birth and death rates (Chapter 3).

high-yield varieties (HYV) dwarf types of grains that have shorter stems and produce more stalks than most traditional varieties (Chapter 12).

historical sources data derived from sources such as early censuses, genealogies, family reconstitution, grave sites, and archeological findings (Chapter 4).

host area the destination area of migrants; the area into which they migrate (Chapter 7).

household all of the people who occupy a housing unit (Chapter 10).

householder the person in whose name the house is owned or rented (sometimes called the head of household) (Chapter 10).

housing unit a house, apartment, mobile home or trailer, group of rooms (or even a single room if occupied as a separate living quarters) intended for occupancy as a separate living quarters (Chapter 10).

human capital investments in individuals that can improve their economic productivity and thus their overall standard of living; including things such as education and job-training, and often enhanced by migration (Chapters 7 and 10).

human resource management see *demographics of human resource management* (Chapter 1).

human resources the application of human ingenuity to convert natural resources to uses not originally intended by nature (Chapter 12).

hydrologic cycle the water cycle by which ocean and land water evaporates into the air, is condensed, and then returns to the ground as precipitation (Chapter 12).

hydrosphere the earth's water resources, including water in oceans, lakes, rivers, groundwater, and water in glaciers and ice caps (Chapters 1 and 12).

illegal (undocumented) migrants immigrants who lack government permission to reside in the country to which they have moved (Chapter 7).

immigrant a person who moves into a country of which he or she is not a native for the purpose of taking up permanent residence (Chapter 7).

impact (IPAT) equation a formula developed by Paul Ehrlich and associates to express the relationship between population growth, affluence and technology and their impact on the environment: Impact (I) = Population (P) x Affluence (A) x Technology (T) (Chapter 12).

impaired fecundity a reduced ability to reproduce, defined as a woman who believes that it is impossible for her to have a baby; or a physician has told her not to become pregnant because the pregnancy would pose a health risk for her or her baby; or she has been continuously married for at least 36 months, has not used contraception, and yet has not gotten pregnant (Chapter 6).

implementing strategy a possible means (such as migration) whereby a goal (such as an improvement in income) might be attained (Chapter 7).

incipient decline the third (final) stage in the demographic transition when a country has moved from having a very high rate of natural increase to having a very low (possibly negative) rate of increase (Chapter 3).

index of fertility product of the proportion of the female population that is married times the index of marital fertility (Chapter 6).

indirect population policy a population policy that is not necessarily designed just to influence population changes but instead does so by changing other aspects of life (Chapter 13).

Industrial Revolution the totality of the changes in economic and social organization that began about 1750 in England and later in other countries; characterized chiefly by the replacement of hand tools with power-driven machines and by the concentration of industry in large establishments (Chapter 2).

infant mortality death during the first year of life (Chapter 5).

infant mortality rate the number of deaths to infants under one year of age divided by the number of live births in that year (and usually multiplied by 1,000) (Chapter 5).

infanticide the deliberate killing or abandonment of an infant; a method of "family control" in many premodern and some modern societies (Chapter 2).

infecundity the inability to produce offspring, synonymous with sterility (Chapter 6).

in-migrant a person who migrates permanently into an area from somewhere else. This term usually refers to an internal migrant; an international in-migrant is an *immigrant* (Chapter 7).

integration incorporating immigrants into the receiving society through the mechanism of mutual accommodation (Chapter 7).

intensification of agriculture the process of increasing crop yield by any means—mechanical, chemical, or otherwise (Chapter 12).

intercensal the period between the taking of censuses (Chapter 4).

intermediate variables means for regulating fertility; the variables through which any social factors influencing the level of fertility must operate (Chapter 6).

internal migration permanent change in residence within national boundaries (Chapter 7).

international migration permanent change of residence involving movement from one country to another (Chapter 7).

intervening obstacles factors that may inhibit migration even if a person is motivated to migrate (Chapter 7).

intrauterine device (IUD) any small, mechanical device for semipermanent insertion in the uterus as a contraceptive (Chapter 6).

investment demographics basing investment decisions at least partly on the analysis of projected population changes (Chapter 1).

legal immigrants immigrants who have legal permission to be permanent residents of the country to which they have moved (Chapter 7).

life chances the probability of having a particular set of demographic characteristics, such as having a high-prestige job, lots of money, a stable marriage or not marrying at all, and a small family or no family at all (Chapter 10).

life expectancy the average duration of life beyond any age of people who have attained that age, calculated from a life table (Chapter 5).

life span the longest period over which an organism or species may live (Chapter 5).

life table an actuarial table showing the number of people who die at any given age from which life expectancy is calculated (Chapter 5).

lifeboat ethic the population policy perspective that suggests that health assistance should only be offered to countries that have a reasonable chance of success in limiting population growth (Chapter 13).

lithosphere the upper part of the earth's crust, which contains the soils, minerals, and fuels that plants and animals require for life (Chapter 1 and 12).

logarithm the exponent indicating the power to which a fixed number, the base, must be raised to produce a given number; related to the concept that human populations have the capacity to grow in a logarithmic fashion, increasing geometrically in size from one generation to the next (Chapter 2).

long-term immigrant an international migrant whose stay in the place of destination is more than one year (Chapter 7).

longevity the ability to resist death, measured as the average age at death (Chapter 5).

Malthusian pertaining to the theories of Malthus, which state that population tends to increase at a geometric rate, while the means of subsistence increase at an arithmetic rate, resulting in an inadequate supply of the goods supporting life, unless a catastrophe occurs to reduce (check) the population or the increase of population is checked by sexual restraint (Chapter 3).

mariculture Farming the sea, including both fishing and harvesting kelp and algae for human consumption (Chapter 12).

marital status the state of being single, married, separated, divorced, widowed, or living in a consensual union (cohabiting) (Chapter 10).

marketing demographics the use of demographic information to improve the marketing of a product or service (Chapter 1).

marriage squeeze an imbalance between the numbers of males and females in the prime marriage ages (Chapter 10).

Marxian the theory of Karl Marx, which rejects Malthusian theory and argues instead that each society at each point in history has its own law of population that determines the consequences of population growth (Chapter 3).

maternal mortality the death of a woman as a result of pregnancy or childbearing (Chapter 5).

maternal mortality ratio the number of deaths to women due to pregnancy and childbirth divided by the number of live births in a given year (and usually multiplied by 100,000) (Chapter 5).

mean length of generation the average age at childbearing (Chapter 6).

means of subsistence the amount of resources (especially food) available to a population (Chapter 3).

mega-city a term used by the United Nations to denote any urban agglomeration with more than 10 million people (Chapter 11).

megalopolis see *urban agglomeration* (Chapter 11).

menarche the onset of menstruation, usually occurring when a woman is in her teens (Chapter 6).

menopause the time when menstruation ceases permanently, usually between the ages of 45 and 50 (Chapter 6).

mercantilism the view that a nation's wealth depended on its store of precious metals and that generating this kind of wealth was facilitated by population growth (Chapter 3).

metropolitan area an urban place extending beyond a core city (Chapter 11).

metropolitan statistical area (MSA) a "stand-alone" metropolitan area composed of a county with an urbanized area of at least 50,000 people (Chapter 11).

migrant a person who makes a permanent change of residence substantial enough in distance to involve a shift in that individual's round of social activities (Chapter 7).

migrant stock the number of people in a region who have migrated there from somewhere else (Chapter 7).

migration the process of permanently changing residence from one geographic location to another; one of the three basic demographic processes (Chapter 7).

migration effectiveness the crude net migration rate divided by the total migration rate (Chapter 7).

migration evolution the current state of migration, with the population largely urban-based and people moving between and within urban places (Chapter 7).

migration flow the movement of people between regions (Chapter 7).

migration ratio the ratio of the net number of migrants (in-migrants minus out-migrants) to the difference between the number of births and deaths—measuring the contribution that migration makes to overall population growth (Chapter 7).

migration transition the shift of people from rural to urban places (see urban transition), and the shift to higher levels of international migration (Chapters 3 and 7).

migration turnover rate the total migration rate divided by the crude net migration rate (Chapter 7).

mobility geographic movement that is either not permanent, or is of sufficiently short distance that it is not considered to be migration (Chapter 7).

model stable population a population whose age and sex structure is implied by a given set of mortality rates and a particular rate of population growth (Chapter 8).

modernization the process of societal development involving urbanization, industrialization, rising standards of living, better education, and improved health that is typically associated with a "Western" lifestyle and world view and was the basis for early explanations of the demographic transition (Chapter 3).

momentum of population growth   the potential for a future increase in population size that is inherent in any present age and sex structure even if fertility levels were to drop to replacement level (Chapter 8).

moral restraint   according to Malthus, the avoidance of sexual intercourse prior to marriage and the delay of marriage until a man can afford all the children his wife might bear; a desirable preventive check on population growth (Chapter 3).

mortality   deaths in a population; one of the three basic demographic processes (Chapter 5).

mortality transition   the shift from deaths at younger ages due to communicable diseases to deaths at older ages due to degenerative diseases (Chapter 3).

mover   a person who moves within the same county and thus, according to the U.S. Census Bureau definitions, has not moved far enough to become a migrant (Chapter 7).

multiculturalism   incorporating immigrants into a host society in a manner that allows the immigrants to retain their ethnic communities but share the same legal rights as other members of the host society (Chapter 7).

natural fertility   fertility levels that exist in the absence of deliberate, or at least modern, fertility control (Chapter 6).

natural increase   the excess of births over deaths; the difference between the crude birth rate and the crude death rate is the rate of natural increase (Chapter 2).

natural resources   those resources available to us on the planet (Chapter 12).

neo-Malthusian   a person who accepts the basic Malthusian premise that population growth tends to outstrip resources, but (unlike Malthus) believes that birth control measures are appropriate checks to population growth (Chapter 3).

neo-Marxist   a person who accepts the basic principle of Marx that societal problems are created by an unjust and inequitable distribution of resources of any and all kinds, without necessarily believing that communism is the answer to those problems (Chapter 12).

net census undercount   the difference between the undercount and the overcount (Chapter 4).

net migration   the difference between those who move in and those who move out of a particular region in a given period of time (Chapter 7).

net reproduction rate (NRR)   a measure of generational replacement; specifically, the average number of female children that will be born to the female babies who were themselves born in a given year, assuming no change in the age-specific fertility and mortality rates and ignoring the effect of migration (Chapter 6).

nonfamily household   one that includes people who live alone, or with nonfamily coresidents (friends living together, a single householder who rents out rooms, cohabiting couples, etc.) (Chapter 10).

nongovernmental organizations (NGOs)   private social service agencies that support the implementation of population policies (Chapter 13).

nonsampling error   an error that occurs in the enumeration process as a result of missing people who should be counted, counting people more than once, respondents providing inaccurate information, or recording or processing information inaccurately (Chapter 4).

nuclear family   at least one parent and his/her/their children (Chapter 10).

nutritional transition   a predictable shift in diet that may accompany the stages of the epidemiological transition (Chapters 5 and 12).

opportunity costs   with respect to fertility, the things foregone in order to have children (Chapter 6).

optimum population size   the number of people that would provide the best balance of people and resources for a desired standard of living (Chapter 12).

oral rehydration therapy   an inexpensive glucose and electrolyte solution that is very effective in controlling diarrhea (Chapter 5).

orphanage   the practice of abandoning children in such a way that they are likely to be cared for by strangers (Chapter 6).

out-migrant   a person who permanently leaves an area and migrates someplace else. This term usually refers to internal migration, whereas *emigrant* refers to an international migrant (Chapter 7).

overpopulation a situation in which the population has overshot a region's carrying capacity (Chapter 12).

overshoot exceed a region's carrying capacity (Chapter 12).

ozone layer the region of the earth's upper atmosphere that protects the earth from the sun's ultraviolet rays (Chapter 12).

parity progression ratio the proportion of woman with a given number of children (parity refers to how many children have already been born) who "progress" to having another child (Chapter 6).

per-capita (per-person) income a common measure of average income, calculated by dividing the total value of goods and services produced (GNP or GDP or PPP) by the total population size (Chapter 12).

period data population data that refer to a particular year and represent a cross section of population at one specific time (Chapter 6).

period rates rates referring to a specific, limited period of time, usually one year (Chapter 6).

peri-urban region the periphery of the urban zone that looks rural to the naked eye but houses people who are essentially urban (Chapter 11).

physiocratic the philosophy that the real wealth of a nation is in the land, not in the number of people (Chapter 3).

planned obsolesence theories of aging based on the idea that each person has a built-in programmed biological time clock; barring death from accidents or disease, cells still die because each is programmed to reproduce itself only a fixed number of times (Chapter 9).

population (or demographic) characteristics those demographic traits or qualities that differentiate one individual or group from another, including age, sex, race, ethnicity, marital status, occupation, education, income, wealth, and urban-rural residence (Chapters 1 and 8 through 11).

population distribution where people are located and why (Chapter 1).

population explosion a popular term referring to a rapid increase in the size of the world's population, especially the increase since World War II (Chapter 2 and 3).

population forecast a statement about what you expect the future population to be; distinguished from a projection, which is a statement about what the future population could be under a given set of assumptions (Chapter 8).

population growth or decline how the number of people in a particular place is changing over time (Chapter 1).

population implosion a popular term referring (somewhat misleadingly) to the end of the population explosion (Chapter 2).

population momentum see *momentum of population growth* (Chapter 8).

population policy a formalized set of procedures designed to achieve a particular pattern of population change (Chapter 13).

population processes fertility, mortality, and migration; the dynamic elements of demographic analysis (Chapter 1).

population projection the calculation of the number of people we can expect to be alive at a future date, given the number now alive and given reasonable assumptions about age-specific mortality and fertility rates and migration (Chapter 8).

population pyramid see *age pyramid* (Chapter 8).

population register a list of all people in a country on which are recorded all vital events for each individual, typically birth, death, marriage, divorce, and change of residence (Chapter 4).

population size how many people are in a given place (Chapter 1).

population structure how many males and females there are of each age (Chahpter 1).

positive checks a term used by Malthus to refer to factors (essentially mortality) that limit the size of human populations by "weakening" or "destroying the human frame" (Chapter 3).

postpartum following childbirth (Chapter 6).

poverty index a measure of need that in the United States is based on the premise that one third of a family's income is spent on food; the cost of an economy food plan multiplied by three. Since 1964, it has increased at the same rate as the consumer price index (Chapter 10).

preconditions for a substantial fertility decline Ansley Coale's theory as to how an individual would have to perceive the world on a daily basis if fertility were to be consciously limited (Chapter 6).

preventive checks in Malthus's writings, any limits to birth, among which Malthus himself preferred moral restraint (Chapter 3).

primary metropolitan statistical area (PMSA) MSA within a CMSA (Chapter 11).

primate city a disproportionately large leading city (Chapter 11).

principle of population the Malthusian theory that human population increases geometrically whereas the available food supply increases only arithmetically, leading constantly to "misery" (Chapter 3).

pronatalist an attitude, doctrine, or policy that favors a high birth rate; also known as "populationist" (Chapter 13).

proximate determinants of fertility a renaming of the intermediate variables (defined previously) with an emphasis on age at entry into marriage and proportions married, use of contraception, use of abortion, and prevalence of breast-feeding (Chapter 6).

prudential restraint a Malthusian concept referring to delaying marriage without necessarily avoiding premarital intercourse (Chapter 3).

purchasing power parity (PPP) a refinement of GDP that is defined as the number of units of a country's currency required to buy the same amounts of goods and services in the domestic market as one dollar would buy in the United States (Chapter 12).

push–pull theory a theory of migration that says some people move because they are pushed out of their former location, whereas others move because they have been pulled, or attracted, to another location (Chapter 7).

race a group of people characterized by a more or less distinctive combination of inheritable physical traits (Chapter 10).

racial stratification a socially constructed system that characterizes one or more groups as being distinctly different (Chapter 10).

radix the initial hypothetical group of 100,000 babies that is used as a starting point for life table calculations (Chapter 5).

rank–size rule a hypothesis derived from studies of city–systems that says that the population size of a given city in a country will be approximately equal to the population of the largest city divided by the city's rank in the city–system (Chapter 11).

rational choice theory any theory based on the idea that human behavior is the result of individuals making calculated cost–benefit analyses about how to act (Chapter 3).

rectangularization refers to the process whereby the continuing decline in death rates at older ages means that the proportion of people surviving to any given age begins to square off at the oldest ages, rather than dropping off smoothly over all ages (Chapter 5).

redistricting reconfigurating U.S. Congressional districts (geographic areas) that each seat in Congress represents (Chapter 1).

refugee a person who has been forced out of his or her country of nationality (Chapter 7).

religious pluralism the existence of two or more religious groups side by side in society without one group dominating the other (Chapter 10).

remittances money sent by migrants back to family members in their country of origin (Chapter 7).

residential mobility the process of changing residence over a short or a long distance (Chapter 7).

rural of, or pertaining to, the countryside. Rural populations are generally defined as those that are nonurban in character (Chapter 11).

sample surveys a method of collecting data by obtaining information from a sample of the total population, rather than by a complete census (Chapter 4).

sampling error error that occurs in sampling due to the fact that a sample is rarely identical in every way to the population from which it was drawn (Chapter 4).

secularization a spirit of autonomy from other worldly powers; a sense of responsibility for one's own well-being (Chapter 3).

segmentation manufacturing and packaging products or providing services that appeal to specific sociodemographically identifiable groups within the population (Chapter 1).

segmented assimilation a situation in which the children of immigrants either adopt the host language and behavior, but are prevented from fully participating in society by their identification with a racial/ethnic minority group, or assimilate economically in the new society, but retain strong attachments to their own ethnic/racial group (Chapter 7).

senescence a decline in physical viability accompanied by a rise in vulnerability to disease (Chapter 9).

senility an outdated term implying a loss of mental abilities (Chapter 9).

sex ratio the number of males per the number of females in a population (usually multiplied by 100 to get rid of the decimal point) (Chapter 8).

sex structure see *age and sex structure* (Chapter 8).

site selection demographics using demographic information to help decide where to locate a business (Chapter 1).

social capillarity Arsène Dumont's term for the desire of a person to rise on the social scale to increase one's individuality as well as one's personal wealth (Chapter 3).

social capital the network of family, friends, and acquaintances that increases a person's chances of success in life (Chapters 7 and 10).

social institutions sets of procedures (norms, laws, etc.) that organize behavior in society in fairly predictable and ongoing ways (Chapter 3).

social roles the set of obligations and expectations that characterize a particular position within society (Chapter 8).

social status relative position or standing in society (Chapter 8).

socialism an economic system whereby the community as a whole (i.e., the government) owns the means of production; a social system that minimizes social stratification (Chapter 3).

socialization the process of learning the behavior appropriate to particular social roles (Chapter 8).

sojourner an international mover seeking paid employment in another country but never really setting up a permanent residence in the new location (Chapter 7).

spermicide a chemical agent that kills sperm (Chapter 6).

stable population a population in which the percentage of people at each age and sex eventually stabilizes (no longer changes) because age-specific

rates of fertility, mortality, and migration remain constant over a long period of time (Chapter 8).

standard metropolitan statistical area (SMSA) a term used by the U.S. Census Bureau to define a county or group of contiguous counties that contain at least one city of 50,000 inhabitants or more and any contiguous counties that are socially and economically integrated with the central city or cities (Chapter 11).

stationary population a type of stable population in which the birth rate equals the death rate, so that the number of people remains the same, as does the age/sex distribution (Chapter 8).

step migration the process whereby a migrant moves in stages progressively farther away from his or her place of origin (Chapter 7).

sterilization the process (either voluntarily—surgically—or involuntarily) of rendering a person incapable of reproducing (Chapter 6).

structural mobility the situation in which most, if not all, people in an entire society experience an improvement in living levels, even though some people may be improving faster than others (Chapter 10).

subfecundity see *impaired fecundity* (Chapter 6).

suburban pertaining to populations in low-density areas close to and integrated with central cities (Chapter 11).

suburbanize to become suburban—a city suburbanizes by growing in its outer rings (Chapter 11).

supply-demand framework a version of neoclassical economics in which it is assumed that couples attempt to maintain a balance between the potential supply of and demand for children, taking into account the costs of fertility regulation (Chapter 6).

surgical contraception permanent methods of contraception, including tubal ligation for women and vasectomy for men (Chapter 6).

sustainable development defined by the Brundtland Commission as "development that meets the needs of the present without compromising the ability of future generations to meet their own needs" (Chapter 12).

synthetic cohort a measurement obtained by treating period data as though they represented a cohort (Chapter 6).

target year   the future year to a population projection is made (Chapter 8).

targeting   a marketing technique of picking out particular sociodemographic characteristics and appealing to differences in consumer tastes and behavior reflected in those particular characteristics (Chapter 1).

theoretical effectiveness   with respect to birth control methods, the probability of preventing a pregnancy if a method is used exactly as it should be (Chapter 6).

theory   a system of assumptions, accepted principles, and rules of procedure devised to analyze, predict, or otherwise explain a set of phenomena (Chapter 3).

total fertility rate (TFR)   a synthetic cohort estimate of the average number of children who would be born to each woman if the current age-specific birth rates remained constant (Chapter 6).

total or gross migration rate   the sum of in-migrants plus out-migrants, divided by the total midyear population (Chapter 7).

transitional growth   the second (middle) stage of the demographic transition when death rates have dropped but birth rates are still high. During this time, population size increases steadily—this is the essence of the "population explosion" (Chapter 3).

transnational migrant   an international migrant who maintains close ties in both his or her country of origin and his or her country of destination (Chapter 7).

troposphere   that portion of the atmosphere which is closest to (within about 11 miles of) the earth's surface (Chapter 12).

tubal ligation   method of female sterilization (surgical contraception) in which the fallopian tubes are "tied" off with rings or by some other method (Chapter 6).

unmet need   as applied to family planning, the number of sexually active women who would prefer not to get pregnant but are nevertheless not using any method of contraception (Chapter 13).

urban   describes a spatial concentration of people whose lives are organized around nonagricultural activities (Chapter 11).

urban agglomeration   according to the United Nations, the population contained within the contours of contiguous territory inhabited at urban levels of residential density without regard to administrative boundaries (Chapter 11).

urban cluster   part of the U.S. Census Bureau's definition of a place considered to be "urban"; an urban cluster has 2,500–50,000 people (Chapter 11).

urban sprawl   the straggling expansion of an urban area into the adjoining countryside (Chapter 11).

urban transition   the shift over time from a largely rural population to a largely urban population (see *urbanization*) (Chapters 3 and 11).

urbanism   the changes that occur in lifestyle and social interaction as a result of living in urban places (Chapter 11).

urbanization   the process whereby the proportion of people in a population who live in urban places increases (Chapter 11).

urbanized area   part of the U.S. Census Bureau's definition of a place considered to be "urban"; an urbanized area has at least 50,000 people (Chapter 11).

use–effectiveness   the actual pregnancy prevention performance associated with using a particular fertility control measure (Chapter 6).

usual residence   the concept of including people in the census on the basis of where they usually reside (Chapter 4).

vasectomy   a technique of male sterilization (surgical contraception) in which each vas deferens is cut and tied, thus preventing sperm from being ejaculated during intercourse (Chapter 6).

vital statistics   data referring to the so-called vital events of life, especially birth and death, but usually also including marriage, divorce, and sometimes abortion (Chapter 4).

wealth flow   a term coined by Caldwell to refer to the intergenerational transfer of income (Chapter 3).

wealth of a nation   the sum of known natural resources and our human capacity to transform those resources into something useful (Chapter 12).

wear-and-tear theory   the theory of aging that argues that humans are like machines that eventually wear out due to the stresses and strains of constant use (Chapter 9).

withdrawal a form of fertility control that requires the male to withdraw his penis from his partner's vagina prior to ejaculation; also called *coitus interruptus* (Chapter 6).

world–systems theory the theory that since the sixteenth century the world market has developed into a set of core nations and a set of peripheral countries dependent on the core (Chapter 11 and 12).

xenophobia fear of strangers, which often leads to discrimination against migrants (Chapter 7).

zero population growth (ZPG) a situation in which a population is not changing in size from year to year, as a result of the combination of births, deaths, and migration (Chapter 8).

BIBLIOGRAPHY

Abbott, C. 1993. *The Metropolitan Frontier: Cities in the Modern American West.* Tucson: University of Arizona Press.

Abernethy, Virginia. 1979. *Population Pressure and Cultural Adjustment.* New York: Human Sciences Press.

Abma, J., A. Chandra, W. Mosher, L. Peterson, and L. Piccinino. 1997. "Fertility, Family Planning, and Women's Health: New Data From the 1995 National Survey of Family Growth." *National Center for Health Statistics Vital Health Statistics* 23 (19).

Abma, Joyce and Freya L. Sonenstein. 2001. "Sexual Activity and Contraceptive Practices Among Teenagers in the United States, 1988 and 1995." *Vital and Health Statistics* Series 23 (21).

Abramovitz, Janet N. 1998. "Sustaining the World's Forests." In *State of the World 1998,* edited by L. Brown, C. Flavin, and H. French. New York: W. W. Norton.

Ackerman, William V. 1999. "Growth Control Versus the Growth Machine in Redlands, California: Conflict in Urban Land Use." *Urban Geography* 20 (2):146–167.

Adamchak, Donald J. 2001. "The Effects of Age Structure on the Labor Force and Retirement in China." *The Social Science Journal* 38 (1):1–11.

Adams, John S., Barbara J. VanDrasek, and Eric G. Phillips. 1999. "Metropolitan Area Definition in the United States." *Urban Geography* 20 (8):695-726.

Adamson, David M., Nancy Belden, Julie DaVanzo, and Sally Patterson. 2000. *How Americans View World Population Issues: A Survey of Public Opinion.* Santa Monica: Rand.

Adelman, C. 1982. "Saving Babies With a Signature." *The Wall Street Journal* 28 July.

Adlakha, Arjun and Judith Banister. 1995. "Demographic Perspectives on China and India." *Journal of Biosocial Science* 27:163–178.

Adsera, Alicia. 2004. "Changing Fertility Rates in Developed Countries: The Impact of Labor Market Insitutions." *Journal of Population Economics* 17 (1):17–43.

Aguirre, B. E., Rogelio Saenz, John Edmiston, Nan Yang, Elsa Agramonte, and Dietra Stuart. 1993. "The Human Ecology of Tornadoes." *Demography* 30 (4):623–33.

Ahlburg, D. and M. Schapiro. 1984. "Socioeconomic Ramifications of Changing Cohort Size: An Analysis of U.S. Post-War Suicide Rates by Age and Sex." *Demography* 21 (1):97–105.

Ahn, N. and A. Shariff. 1994. "A Comparative Study of Socioeconomic and Demographic Determinants of Fertility in Togo and Uganda." *International Family Planning Perspectives* 20:14–17.

Akin, J., R. Bilsbarrow, D. Guilkey, B. Popkin, D. Benoit, P. Cantrelle, M. Garenne, and P. Levi. 1981. "The Determinants of Breast-Feeding in Sri Lanka." *Demography* 18 (3):287–308.

Alba, Richard and John R. Logan. 1991. "Variations on Two Themes: Racial and Ethnic Patterns in the Attainment of Suburban Residence." *Demography* 28:431–53.

Alba, Richard and Victor Nee. 1997. "Rethinking Assimilation Theory for a New Era of Immigration." *International Migration Review* 31 (4):826–74.

Albert, Steven M. and Maria G. Cattell. 1994. *Old Age in Global Perspective.* New York: G.K. Hall.

Albert, Susan. 1998. "Site Selection's Essential Ingredients." *Business Geographics* (May):44.

Alchon, Suzanne Austin. 1997. "The Great Killers in Precolumbian America: A Hemispheric Perspective." *Latin American Population History Bulletin* (27):2–11.

Alonso, William and Paul Starr. 1982. "The Political Economy of National Statistics." *Social Science Research Council Items* 36 (3):29–35.

Amato, Paul R. and A. Booth. 1995. "Changes in Gender Role Attitudes and Perceived Marital Quality." *American Sociological Review* 60 (1):58–66.

Amato, Paul R. and Juliana M. Sobolewski. 2001. "The Effects of Divorce and Marital Discord on Adult Children's Psychological Well-Being." *American Sociological Review* 66:900–921.

American Health Consultants. 2001. "Make the Distinction: EC Prevents Pregnancy." *Contraceptive Technology Update* 22 (1):4.

Anderson, Elijah and Douglas S. Massey, Ed. 2001. *Problem of the Century: Racial Stratification in the United States*. New York: Russell Sage Foundation.

Anderson, Gunnar. 2002. "Children's Experience of Family Disruption and Family Formation: Evidence From 16 FFS Countries." *http://www.demographic-research.org/*, accessed 2004.

Anderson, Margo J. and Stephen E. Fienberg. 1999. *Who Counts? The Politics of Census-Taking in Contemporary America*. New York: Russell Sage Foundation.

Anderson, Robert N. 1999. "Method for Constructing Complete Annual U.S. Life Tables." *Vital and Health Statistics* (129).

Anderson, Robert N., Kenneth D. Kochanek, and Sherry L. Murphy. 1997. "Report of Final Mortality Statistics 1995." *Monthly Vital Statistics Report* 45 (11, Supplement 2).

Anderson, Robert N. and Betty L. Smith. 2003. "Deaths: Leading Causes for 2001." *National Vital Statistics Reports* 52 (9).

Angel, Ronald J. and Jacqueline L. Angel. 1993. *Painful Inheritance: Health and the New Generation of Fatherless Families*. Madison, WI: University of Wisconsin Press.

Annan, Kofi. 1999. "Address of Kofi Annan, UN Secretary-General, to the Special Session of the UN General Assembly on the Follow-Up to the International Conference on Population and Development." *Population and Development Review* 25 (3):632–634.

Appleman, Philip. 1965. *The Silent Explosion: With a Foreward by Julian Huxley*. Boston: Beacon Press.

Appleyard, Reginald and Patrick Taran. 2000. "Introduction to Special Issue on Human Rights." *International Migration* 38 (6):3–6.

Arabicnews.com. 2003. "Saudi Arabia's Unemployment Reaches 30 Percent." *http://www.arabicnews.com/ansub/Daily/Day/030305/2003030513.html*, accessed 2004.

Arias, Elizabeth. 2002. "United States Life Tables, 2000." *National Vital Statistics Reports* 51 (3).

Aries, P. 1962. *Centuries of Childhood*. New York: Vintage Books.

Arnold, F. 1988. "The Effect of Sex Preferences on Fertility and Family Planning: Empirical Evidence." *Population Bulletin of the United Nations* 23/24:44–56.

Arnold, Fred, Minja Kim Choe, and T. K. Roy. 1998. "Son Preference, the Family-Building Process and Child Mortality in India." *Population Studies* 52:301–315.

Arriaga, Eduardo. 1970. *Mortality Decline and Its Demographic Effects in Latin America*. Berkeley: University of California Institute of International Studies.

——. 1982. "Changing Trends in Mortality in Developing Countries." In *Paper Presented at the Annual Meeting of the Population Association of America*. San Diego.

Associated Press. 1980. "Study on Old Age Released in China." *San Diego Union*. San Diego, CA. 10 November. P. A3.

Atchley, Robert C. and Amanda Barusch. 2004. *Social Forces and Aging, 10th Edition*. Belmont, CA: Wadsworth Publishing.

Attané, Isabell. 2001. "Chinese Fertility on the Eve of the 21st Century: Fact and Uncertainty." *Population: An English Selection* 13 (2):71–100.

——. 2002. "China's Family Planning Policy: An Overview of Its Past and Future." *Studies in Family Planning* 33 (1):103–113.

Auerbach, F. 1961. *Immigration Laws of the United States, 2nd Edition*. New York: Bobbs-Merrill.

Avdeev, Alexandre and Alain Monnier. 1995. "A Survey of Modern Russian Fertility." *Population: An English Selection* 7:1–38.

Axelrod, P. 1990. "Cultural and Historical Factors in the Population Decline of the Parsis of India." *Population Studies* 44:401–20.

Axinn, W. G. and Arland Thornton. 1992. "The Relationship Between Cohabitation and Divorce: Selectivity or Causal Influence?" *Demography* 29 (3):357–74.

Axinn, William G. and Jennifer S. Barber. 2001. "Mass Education and Fertility Transition." *American Sociological Review* 66:481–505.

Bachu, Amara. 1993. "Fertility of American Women: June 1992." *Current Population Reports* Series P20 (No. 470).

——. 1999. "Trends in Premarital Childbearing: 1930 to 1994." *Current Population Reports* (23–197).

Bachu, Amara and Martin O'Connell. 2000. "Fertility of American Women: June 1998." *Current Population Reports* 20–526.

Baird, Donna D., Clarice R. Weinberg, Lynda F. Voigt, and Janet R. Daling. 1996. "Vaginal Douching and Reduced Fertility." *American Journal of Public Health* 86 (6):844–850.

Baldauf, Scott. 2001. "In India, The Decline of a Middle-Class Icon." *Christian Science Monitor*: 30 August. *http://www.csmonitor.com*, accessed 2003.

Banister, Judith. 1987. *China's Changing Population*. Stanford, CA: Stanford University Press.

Banks, J. A. 1954. *Prosperity and Parenthood: A Study of Family Planning Among the Victorian Middle Classes*. London: Routledge & Kegan Paul.

Banks, J. A. and O. Banks. 1964. *Feminism and Family Planning in Victorian England*. Liverpool: Liverpool University Press.

Barrett, Paul M. 1996. "High-Court Ruling on Census Sets Back Some Cities That Disputed '90 Numbers." *The Wall Street Journal*, 21 March, p. B6.

Basu, A. M. 1993. "How Pervasive Are Sex Differentials in Childhood Nutritional Levels in South Asia?" *Social Biology* 40 (1–2):25–37.

Basu, Alaka Malwade. 1997. "The 'Politicization' of Fertility to Achieve Non-Demographic Objectives." *Population Studies* 51 (1):5–18.

BBC News. 2001. "China's Population Growth 'Slowing'." *http://news.bbc.co.uk/hi/english/world/asia-pacific/newsid_1246000/1246731.stm*, accessed 2001.

Bean, Frank and Gray Swicegood. 1985. *Mexican American Fertility Patterns.* Austin: University of Texas Press.

Bean, Frank D., C. Gray Swicegood, and Ruth Berg. 2000. "Mexican-Origin Fertility: New Patterns and Interpretations." *Social Science Quarterly* 81 (1): 404–420.

Becker, Gary. 1960. "An Economic Analysis of Fertility." In *Demographic Change and Economic Change in Developed Countries*, edited by National Bureau of Economic Research. Princeton: Princeton University Press.

Becker, Jasper. 1997. *Hungry Ghosts: Mao's Secret Famine.* New York: Free Press.

Behrman, J. and P. Taubman. 1989. "A Test of the Easterlin Fertility Model Using Income For Two Generations and a Comparison with the Becker Model." *Demography* 26:117–23.

Benedict, B. 1972. "Social Regulation of Fertility." In *The Structure of Human Population*, edited by G. A. Harrison and A. J. Boyce. Oxford: Clarendon Press.

Bergman, Barbara R. 1996. *Saving Our Children From Poverty: What the United States Can Learn From France.* New York: Russell Sage Foundation.

Berry, Brian, C. Goodwin, R. Lake, and K. Smith. 1976. "Attitudes Toward Integration: The Role of Status." In *The Changing Face of the Suburbs*, edited by B. Schwartz. Chicago: University of Chicago Press.

Berry, Brian J. L. 1993. "Transnational Urbanward Migration, 1830–1980." *Annals of the Association of American Geographers* 83 (3):389–405.

Bhatia, P. 1978. "Abstinence—Desai Solution for Baby Boom." *San Francisco Chronicle* 11 June, p. A4.

Bhrolcháin, Máire Ní. 2001. "Flexibility in the Marriage Market." *Population: An English Selection* 13 (2): 9–48.

Bianchi, S. 1981. *Household Composition and Racial Inequality.* New Brunswick: Rutgers University Press.

———. 1995. "Changing Economic Roles of Women and Men." In *State of the Union: America in the 1990s, Volume One, Economic Trends*, edited by R. Farley. New York: Russell Sage Foundation.

Biraben, Jean-Noël. 1979. "Essai Sur L'Évolution Du Nombre Des Hommes." *Population* 34 (1):13–24.

Birdsall, Nancy and Steven W. Sinding. 2001. "How and Why Population Matters: New Findings, New Issues." In *Population Matters: Demographic Change, Economic Growth, and Poverty in the Developing World*, edited by N. Birdsall, A. C. Kelley, and S. W. Sinding. New York: Oxford University Press.

Blackburn, R. D., J. A. Cunkelman, and V.M. Zlidar. 2000. "Oral Contraceptives—An Update." *Population Reports* Series A (9).

Blake, Judith. 1967. "Family Size in the 1960s—A Baffling Fad." *Eugenics Quarterly* 14:60–74.

———. 1974. "Can We Believe Recent Data in the United States." *Demography* 11 (1):25–44.

———. 1979. "Personal Communication."

Blayo, Y. 1992. "Political Events and Fertility in China Since 1950." *Population: An English Selection* 4: 209–32.

Bledsoe, Caroline and Allan G. Hill. 1998. "Social Norms, Natural Fertility, and the Resumption of Post-Partum 'Contact' in The Gambia." Ch 12 in *The Methods and Uses of Anthropological Demography*, edited by A. M. Basu and P. Aaby. Oxford: Clarendon Press.

Bledsoe, Caroline, Allan G. Hill, Umberto D'Alessandro, and Patricia Langerock. 1994. "Constructing Natural Fertility: The Use of Western Contraceptive Technologies in Rural Gambia." *Population and Development Review* 20 (1):81–113.

Bloom, D. and R. Freeman. 1986. "The Effects of Rapid Population Growth on Labor Supply and Employment in Developing Countries." *Population and Development Review* 12 (3):381–414.

Blossfeld, H. and A. de Rose. 1992. "Educational Expansion and Changes in Entry Into Marriage and Motherhood, the Experience of Italian Women." *Genus* XLVIII (3–4):73–91.

Blumstein, Alfred and Joel Wallman, Ed. 2000. *The Crime Drop in America.* New York: Cambridge University Press.

Bobo, Lawrence D., Melvin L. Oliver, James H. Johnson, Jr., and Able Valenzuela, Jr. 2000. "Analyzing Inequality in Los Angeles.", edited by L. D. Bobo, M. L. Oliver, J. H. Johnson, Jr., and A. Valenzuela, Jr. New York: Russell Sage Foundation.

Bollen, K. A. and S. J. Appold. 1993. "National Industrial Structure and the Global System." *American Sociological Review* 58:283–301.

Bolton, C. and J. William Leasure. 1979. "Evolution Politique et Baisse de la Fecondite en Occident." *Population* 34:825–44.

Bongaarts, John. 1978. "A Framework for Analyzing the Proximate Determinants of Fertility." *Population and Development Review* 4 (1):105–32.

———. 1982. "The Fertility-Inhibiting Effects of the Intermediate Fertility Variables." *Studies in Family Planning* 13:179–89.

———. 1993. "The Supply-Demand Framework for the Determinants of Fertility: An Alternative Implementation." *Population Studies* 47:437–456.

——. 1996. "Global Trends in AIDS Mortality." *Population and Development Review* 22 (1):21–46.

——. 2001. "Household Size and Composition in the Developing World in the 1990s." *Population Studies* 55:263–279.

——. 2003. "Completing the Fertility Transition in the Developing World: The Role of Educational Differences and Fertility Preferences." *Population Studies* 57 (3):321–336.

Bongaarts, John and Rodolfo A. Bulatao, Ed. 2000. *Beyond Six Billion: Forecasting the World's Population.* Washington, DC: National Academy Press.

Bongaarts, John and Griffith Feeney. 1998. "On the Quantum and Tempo of Fertility." *Population and Development Review* 24 (2):271–292.

——. 2003. "Estimating Mean Lifetime." Population Council Working Paper No. 179. New York.

Bongaarts, John and Charles F. Westoff. 2000. "The Potential Role of Contraception in Reducing Abortion." *Studies in Family Planning* 31 (3):193–202.

Borlaug, Norman. 2002. "We Can Feed the World. Here's How." *The Wall Street Journal.* New York. 13 May. P. A16.

Boserup, Ester. 1965. *The Conditions of Agricultural Growth.* Chicago: Aldine.

——. 1970. *Woman's Role in Economic Development.* New York: St. Martin's Press.

——. 1981. *Population and Technological Change: A Study of Long-Term Trends.* Chicago: University of Chicago Press.

Bounds, Wendy. 2000. "Rushing to Cash in on the New Baby Boom." *The Wall Street Journal.* P. B1.

Bouvier, Leon F. 2000. *Toward a Stationary U.S. Population.* Washington, DC: Negative Population Growth.

Bouvier, Leon F. and Sharon McCloe Stein. 2000. *Is Smart Growth Enough? Maryland's Population in 2050.* Washington, DC: Negative Population Growth.

Bouvier, Leon, Dudley L. Poston, and Nanbin Benjamin Zhai. 1997. "Population Growth Impacts of Zero Net International Migration." *International Migration Review* 31 (2):294–311.

Boyd, Monica. 1976. "Immigration Policies and Trends: A Comparison of Canada and the United States." *Demography* 13:83–104.

Boyd, Monica, Gustave Goldmann, and Pamela White. 2000. "Race in the Canadian Census." In *Race and Racism: Canada's Challenge,* edited by L. Driedger and S. S. Halli. Montreal: McGill/Queen's University Press.

Boyd, Monica and Elizabeth M. Grieco. 1998. "Triumphant Transitions: Socioeconomic Achievements of the Second Generation in Canada." *International Migration Review* 32 (4):853–876.

Boyd, Monica and Doug Norris. 2000. "Demographic Change and Young Adults Living with Parents, 1981–1996." *Canadian Studies in Population* 27 (2):267–281.

Boyd, Monica and Michael Vickers. 2000. "100 Years of Immigration in Canada." *Canadian Social Trends* Autumn:2–11.

Boyle, Paul, Thomas J. Cooke, Keith Halfacree, and Darren Smith. 2001. "A Cross-National Comparison of the Impact of Family Migration on Women's Employment Status." *Demography* 38 (2):201–213.

Bracher, M., G. Santow, S. P. Morgan, and J. Trussell. 1993. "Marriage Dissolution in Australia: Models and Explanations." *Population Studies* 47 (3):403–26.

Brackett, James. 1967. "The Evolution of Marxist Theories of Population: Marxism Recognizes the Population Problem." In *Annual Meeting of the Population Association of America.* Cincinnati, Ohio.

Bradshaw, Benjamin and W. Parker Frisbie. 1992. "Mortality of Mexican Americans and Mexican Immigrants: Comparisons With Mexico." In *Demographic Dynamics Along the US-Mexico Border,* edited by J. R. Weeks and R. Ham-Chande. El Paso: Texas Western Press of the University of Texas at El Paso.

Bradshaw, York. 1987. "Urbanization and Underdevelopment: A Global Study of Modernization, Urban Bias, and Economic Dependency." *American Sociological Review* 52 (2):224–39.

Bradshaw, York and E. Fraser. 1989. "City Size, Economic Development, and Quality of Life in China: New Empirical Evidence." *American Sociological Review* 54:986–1003.

Brandes, Stanley H. 1975. *Migration, Kinship, and Community.* New York: Academic Press.

Brandes, Stanley H. 1990. "Ritual Eating and Drinking in Tzintzuntzan: A Contribution to the Study of Mexican Foodways." *Western Folklore* 49:163–75.

Brauchli, M.W. 1993. "Migrants Fuel China's Economic Growth." *The Wall Street Journal* 17 August, p. A12.

Brentano, Ludwig. 1910. "The Doctrine of Malthus and the Increase of Population During the Last Decade." *Economic Journal* 20:371–393.

Brockerhoff, Martin. 2000. "An Urbanizing World." *Population Bulletin* 55 (3).

Brodie, Janet Farrell. 1994. *Contraception and Abortion in Nineteenth-Century America.* Ithaca: Cornell University Press.

Brooks, Clem. 2002. "Religious Influence and the Politics of Family Decline Concern: Trends, Sources, and U.S. Political Behavior." *American Sociological Review* 67:191–211.

Brown, Lawrence A. 1981. *Innovation Diffusion: A New Perspective.* London and New York: Methuen.

Brown, Lesley, Ed. 1993. *The New Shorter Oxford English Dictionary on Historical Principles.* Oxford: Clarendon Press.

Brown, Lester. 1998. "The Future of Growth." In *State of the World 1998,* edited by L. Brown, C. Flavin, and H. French. New York: W. W. Norton.

Brown, Lester and Jennifer Mitchell. 1998. "Building a New Economy." In *State of the World 1998,* edited by L. Brown, C. Flavin, and H. French. New York: W.W. Norton.

Brownson, R. C., J. S. Reif, M. C. Alavanja, and D. G. Bal. 1993. "Cancer." In *Chronic Disease Epidemiology and Control,* edited by R. C. Brownson, P. L. Remington, and J. R. Davis. Washington: American Public Health Association.

Bruinsma, Jelle. 2003. *World Agriculture: Towards 2015/2030, An FAO Perspective.* London: Earthscan.

Bulatao, R. and R. Lee. 1983. *Determinants of Fertility in Developing Countries, Volume 1, Supply and Demand for Children.* New York: Academic Press.

Bumpass, Larry and Hsien-Hen Lu. 2000. "Trends in Cohabitation and Implications for Children's Family Contexts in the United States." *Population Studies* 54:29–41.

Burch, Thomas. 1975. "Theories of Fertility as Guides to Population Policy." *Social Forces* 54 (1):126–39.

Butz, W. and M. Ward. 1979. "Will U.S. Fertility Remain Low? A New Economic Interpretation." *Population and Development Review* 5:663–88.

Cain, M. 1981. "Risk and Fertility in India and Bangladesh." *Population and Development Review* 7:435–74.

Caldas de Castro, Marcia and Burton H. Singer. 2000. "Agricultural Colonization, Environmental Changes, and Patterns of Malaria Transmission in the Tropical Rain Forest: The Case of Machadinho d'Oeste, Rondonia, Brazil." Annual Meeting of the Population Association of America, Los Angeles.

Caldwell, John. 1976. "Toward a Restatement of Demographic Transition Theory." *Population and Development Review* 2 (3–4):321–66.

——. 1980. "Mass Education as a Determinant of the Timing of Fertility Decline." *Population and Development Review* 6 (2):225–56.

——. 1982. *Theory of Fertility Decline.* New York: Academic Press.

——. 1986. "Routes to Low Mortality in Poor Countries." *Population and Development Review* 12 (2):171–220.

Caldwell, John C. 2000. "Rethinking the African AIDS Epidemic." *Population and Development Review* 26 (1):117–135.

Calhoun, John. 1962. "Population Density and Social Pathology." *Scientific American* (May):118–25.

California Department of Finance. 1994. "Recession Slows California's Growth." *News Release of the Demographic Research Unit.* 7 February.

——. 1998. *Race/Ethnic Population with Age and Sex Detail, 1970–2040.* Sacramento: Demographic Research Unit.

California Department of Finance. 2002. *California Public Postsecondary Enrollment Projections, 2002 Series.* Sacramento: Demographic Research Unit.

Cambois, Emmanuelle, Jean-Marie Robine, and Mark D. Hayward. 2001. "Social Inequalities in Disability-Free Life Expectancy in the French Male Populaiton, 1980–1991." *Demography* 38 (4):513–524.

Campani, Giovanna. 1995. "Women Migrants: From Marginal Subjects to Social Actors." In *The Cambridge Survey of World Migration,* edited by R. Cohen. Cambridge: Cambridge University Press.

Cann, Rebecca L. and Allan C. Wilson. 2003. "The Recent African Genesis of Humans." *Scientific American Special Edition* 13 (2):54–61.

Cantillon, Richard. 1755 (1964). *Essai sur la Nature du Commerce en General, Edited with an English Translation by Henry Higgs.* New York: A. M. Kelley.

Cantor, Norman F. 2001. *In the Wake of the Plague: The Black Death & the World it Made.* New York: Free Press.

Carey, James R. and Debra S. Judge. 2001. "Principles of Biodemography with Special Reference to Longevity." *Population: An English Selection* 13 (1):9–40.

Carlson, Elwood. 1980. "Divorce Rate Fluctuations as a Cohort Phenomenon." *Population Studies* 33:523–36.

Carlson, Elwood and S. Tsvetarsky. 1992. "Concentration of Rising Bulgarian Mortaltiy Among Manual Workers." *Sociology and Social Research* 76 (2): 81–84.

Carlson, Elwood and M. Watson. 1990. "The Family and the State: Rising Hungarian Death Rates." In *Aiding and Aging: The Coming Crisis in Support for the Elderly by Kin and State,* edited by J. Mogey. Westport, CT: Greenwood Press.

Carnes, B. A. and S. J. Olshansky. 1993. "Evolutionary Perspectives on Human Senescence." *Population and Development Review* 19 (4):793–806.

Carpenter, J. and L. Lees. 1995. "Gentrification in New York, London and Paris: An International Comparison." *International Journal of Urban and Regional Research* 19 (2):286–304.

Carr-Saunders, A. M. 1936. *World Population: Past Growth and Present Trends.* Oxford: Clarendon Press.

Caselli, G. and V. Egidi. 1991. "A New Insight Into Morbidity and Mortality in Italy." *Genus* XLVII (3–4): 1–30.

Caselli, Graziella, France Mesle, and Jacques Vallin. 2002. "Epidemiologic Transition Theory Exceptions." *Genus* 58 (1):9–51.

Caselli, Graziella and Jacques Vallin. 2001. "Demographic Trends: Beyond the Limits?" *Population: An English Selection* 13 (1):41–72.

Casper, L.M., S. McLanahan, and I. Garfinkel. 1994. "The Gender-Poverty Gap: What Can We Learn From Other Countries?" *American Sociological Review* 59: 594–605.

Casper, Lynne M. and Philip N. Cohen. 2000. "How Does POSSLQ Measure Up? Historical Estimates of Cohabitation." *Demography* 37 (2):237–245.

Cassedy, J. 1969. *Demography in Early America: Beginnings of the Statistical Mind, 1600–1800*. Cambridge: Harvard University Press.

Castles, Stephen and Mark J. Miller. 1998. *The Age of Migration, Second Edition*. New York: Guilford Press.

Catasús Cervera, Sonia and Juan C. A. Fraga. 1996. "The Fertility Transition in Cuba." In *The Fertility Transition in Latin America*, edited by J. M. Guzmán, S. Singh, G. Rodríguez, and E. Pantiledes. Oxford: Clarendon Press.

Census 2000 Initiative. 2001. "Commerce Secretary Chooses Unadjusted Census Data After Census Bureau Recommends Against Statistical Correction." *http://www.census2000.org*, accessed 2001.

Centers for Disease Control and Prevention. 1992. "The Management of Acute Diarrhea in Children: Oral Rehydration, Maintenance, and Nutritional Therapy." *MMWR* 41 (RR–16).

Centers for Disease Control and Prevention. 2001. "The Global HIV and AIDS Epidemic, 2001." *Mortality and Morbidity Weekly Report* 50 (21):434–439.

Cerrutti, Marcela and Douglas S. Massey. 2001. "On the Auspices of Female Migration From Mexico to the United States." *Demography* 38 (2):187–200.

Chambers, Nicky, Craig Simmons, and Mathis Wackernagel. 2002. *Sharing Nature's Interest*. London: Earthscan.

Chamie, Joseph. 1981. "Religion and Fertility." Cambridge: Cambridge University Press.

Champion, Anthony G. 1989. "Counterurbanization: The Conceptual and Methodological Challenge." In *Counterurbanization: The Changing Pace and Nature of Population Deconcentration*, edited by A. G. Champion. London: Edward Arnold.

Champion, Marc. 2002. "Immigration Moves to Front Burner in Europe." *The Wall Street Journal*. New York. 24 April. P. A20.

Chan, Kam Wing and Ying Hu. 2003. "Urbanization in China in the 1990s: New Definition, Different Series, and Revised Trends." *The China Review* 3 (2):49–71.

Chandler, R. F. 1971. "The Scientific Basis for the Increased Yield Potential of Rice and Wheat." In *Food, Population and Employments: The Impact of the Green Revolution*, edited by T. Poleman and D. Freebain. New York: Praeger.

Chandler, T. and G. Fox. 1974. *3000 Years of Urban Growth*. New York: Academic Press.

Chandra, Anjani and Elizabeth Hervey Stephen. 1998. "Impaired Fecundity in the United States: 1982–95." *Family Planning Perspectives* 30 (1):34–42.

Chandrasekhar, S. 1979. *"A Dirty, Filthy Book"—The Writings of Charles Knowlton and Annie Besant on Reproductive Physiology and Birth Control and an Account of the Bradlaugh-Besant Trial*. Berkeley: University of California Press.

Chang, Leslie. 2000a. "Confucius Said: Sons, Care for Your Elders; Elders Say: We Sue." *The Wall Street Journal*. New York. 3 April. P. A1.

——. 2000b. "In China, A Headache of a Head Count." *The Wall Street Journal*. New York. 2 November. P. A21.

Chang, Sen-dou. 1996. "The Floating Population: An Informal Process of Urbanisation in China." *International Journal of Population Geography* 2:197–214.

Chant, Sylvia and Sarah Radcliffe. 1992. "Migration and Development: The Importance of Gender." In *Gender and Migration in Developing Countries*, edited by S. Chant. London: Bellhaven Press.

Charbit, Yves. 2002. "The Platonic City: History and Utopia." *Population-E* 57 (2):207–236.

Charles, Enid. 1936. *The Twilight of Parenthood*. London: Watts.

Charlton, S. 1984. *Women in Third World Development*. Boulder, CO: Westview Press.

Chase-Dunn, Christopher K. 1989. *Global Formation: Structure of the World-Economy*. Cambridge, MA: Blackwell.

Chase-Dunn, Christopher K. 1998. *Global Formation: Structures of the World Economy, Updated Edition*. Lanham, MD: Rowman and Littlefield.

Chase-Dunn, Christopher K. and Thomas D. Hall. 1997. *Rise and Demise: Comparing World Systems*. Boulder, CO: Westview Press.

Chelala, Cesar. 1998. "A Critical Move Against Female Genital Mutilation." *Populi* 25 1:13–15.

Chen, Lincoln C., Friederike Wittgenstein, and Elizabeth McKeon. 1996. "The Upsurge Mortality in Russia: Causes and Policy Implications." *Population and Development Review* 22 (3):457–482.

Chen, Muhua. 1979. "Birth Planning in China." *International Family Planning Perspectives* 5 (3):92–101.

Cherfas, J. 1980. "The World Fertility Survey Conference: Population Bomb Revisited." *Science80* 1 (November):11–14.

Cherlin, Andrew J. 1999. *Public and Private Families: An Introduction, Second Edition*. New York: McGraw-Hill College.

Chesler, E. 1994. "No, the First Priority Is: Stop Coercing Women." *The New York Times Magazine* 6 February.

Chesnais, Jean-Claude. 1996. "Fertility, Family, and Social Policy in Contemporary Western Europe." *Population and Development Review* 22 (4):729–740.

——. 2000. "Homicide and Suicide in the Industrial World." *Population: An English Selection* 12 331–334.

Chiang, C. 1984. *The Life Table and Its Applications.* Melbourne, FL: Krieger.

Chin, James E. 1999. *Control of Communicable Diseases Manual, 17th Edition.* Washington: American Public Health Association.

Choi, N. 1991. "Racial Differences in the Determinants of Living Arrangements of Widowed and Divorced Elderly Women." *The Gerontologist* 34:496–504.

Choiniére, Robert. 1993. "Les Inégalités Socio-Économiques et Culturelles de la Mortalité à Montréal à la Fin des Années 1980." *Cahiers Québécois de Démographie* 22 (2):339–362.

Chojnacka, Helena. 2000. "Early Marriage and Polygyny: Feature Characteristics of Nuptiality in Africa." *Genus* LVI (3–4):179–208.

Choldin, Harvey. 1994. *Looking For the Last Percent: The Controversy Over the Census Undercounts.* New Brunswick, NJ: Rutgers University Press.

Choucri, Nazli. 1984. "Perspectives on Population and Conflict." In *Multidisciplinary Perspectives on Population and Conflict,* edited by N. Choucri. Syracuse: Syracuse University Press.

Christaller, Walter. 1966. *Central Places in Southern Germany: Translated from from Die zentralen Orte in Süddeutschland by Carlisle W. Baskin.* Englewood Cliffs, NJ: Prentice-Hall.

Christenfeld, Nicholas, David P. Phillips, and Laura M. Glynn. 1999. "What's in a Name: Mortality and the Power of Symbols." *Journal of Psychosomatic Research* 47 (3):241–54.

Cicourel, Aaron. 1974. *Theory and Method in a Study of Argentina.* New York: Wiley.

Cincotta, Richard P. and Robert Engelman. 1997. *Economics and Rapid Change: The Influence of Population Growth.* Washington: Population Action International.

Cipolla, Carlo. 1965. *The Economic History of World Population.* London: Penguin.

——. 1981. *Fighting the Plague in Seventeenth-Century Italy.* Madison: University of Wisconsin Press.

Clark, Colin. 1967. *Population Growth and Land Use.* New York: St. Martin's Press.

Clark, Jon R. and Darlene Moul. 2003. "Coverage Improvement in Census 2000 Enumeration." U.S. Census Bureau Census 2000 Testing, Experimentation, and Evaluation Program, Topic Report Series, No. 10, Washington, DC.

Clarke, John I. 2000. *The Human Dichotomy: The Changing Numbers of Males and Females.* Amsterdam: Pergamon.

Clarke, Sally C. 1995. "Advance Report of Final Marriage Statistics, 1989 and 1990." *Monthly Vital Statistics Report* 43 (12, Supplement).

Cleland, John and Christopher Wilson. 1987. "Demand Theories of the Fertility Transition: An Iconoclastic View." *Population Studies* 41:5–30.

Cliff, Andrew D. and Peter Haggett. 1996. "The Impact of GIS on Epidemiological Mapping and Modelling." In *Spatial Analysis: Modelling in a GIS Environment,* edited by P. Longley and M. Batty. New York: John Wiley.

Clough, S. 1968. *European Economic History.* New York: Walker.

Coale, Ansley. 1973. "The Demographic Transition." In *International Population Conference,* Vol. I, edited by IUSSP. Liege, Belgium: IUSSP.

——. 1974. "The History of the Human Population." *Scientific American* 231 (3):41–51.

——. 1986. "Preface." in *The Decline of Fertility in Europe,* edited by A. Coale and S. C. Watkins. Princeton: Princeton University Press.

Coale, Ansley and Judith Banister. 1994. "Five Decades of Missing Females in China." *Demography* 31 (3):459–80.

Coale, Ansley and Paul Demeny. 1983. *Regional Model Life Tables and Stable Populations.* New York: Academic Press.

Coale, Ansley and James Trussell. 1974. "Model Fertility Schedules: Variations in the Age Structure of Child-Bearing in Human Populations." *Population Index* 40 (2):185–256.

Coale, Ansley and Melvin Zelnick. 1963. *Estimates of Fertility and Population in the United States.* Princeton: Princeton University Press.

Cohen, Barney and Mark R. Montgomery. 1998. "Introduction." In *From Birth to Death: Mortality Decline and Reproductive Change,* edited by M. R. Montgomery and B. Cohen. Washington: National Academy Press.

Cohen, Joel. 1995. *How Many People Can the Earth Support.* New York: Norton.

Cohen, L. and M. Felson. 1979. "Social Change and Crime Rate Trends." *American Sociological Review* 44 (4):588–607.

Cohen, L. and K. Land. 1987. "Age Structure and Crime." *American Sociological Review* 52 (2):170–83.

Cohen, Mark Nathan. 1977. *The Food Crisis in Prehistory: Overpopulation and the Origins of Agriculture.* New Haven: Yale University Press.

Cole, W. A. and P. Deane. 1965. "The Growth of National Incomes." In *The Cambridge Economic History of Europe, Vol 6, Part 1,* edited by H. H. Habakkuk and M. Postan. Cambridge: Cambridge University Press.

Coleman, D. 1995. "International Migration: Demographic and Socioeconomic Consequences in the United Kingdom and Europe." *International Migration Review* 29 (1):155–206.

Coleman, R. and Lee Rainwater. 1978. *Social Standing in America.* New York: Basic Books.

Comfort, Alex. 1979. *The Biology of Senescence*. New York: Elsevier.

Commission on Population Growth and the American Future. 1972. *Population and the American Future.* Washington: Government Printing Office.

Commoner, Barry. 1972. "The Environmental Cost of Economic Growth." *Chemistry in Britain* 8 (2):52–65.

——. 1994. "Population, Development and the Environment: Trends and Key Issues in the Developed Countries." In *Population, Environment and Development: Proceedings of the United Nations Expert Group Meeting on Populations, Environment and Development,* edited by United Nations Population Division. New York: United Nations.

Condorcet, Marie Jean Antoine Nicolas de Caritat. 1795 (1955). *Sketch for an Historical Picture of the Progress of the Human Mind, translated from the French by June Baraclough*. London: Weidenfield and Nicholson.

Coombs, L. and T. Sun. 1981. "Familial Values in a Developing Society: A Decade of Change in Taiwan." *Social Forces* 59 (4):1229–55.

Cornelius, Wayne A. 2002. "Impacts of NAFTA on Mexico-to-U.S. Migration." In *NAFTA in the New Millenium,* edited by P. H. Smith and T. Chambers. La Jolla, CA: UCSD Center for U.S.-Mexican Studies.

Corona Vásquez, Rodolfo. 1991. "Confiabilidad de los Resultados Preliminares del XI Censo General de Población y Vivienda de 1990." *Estudios Demográficos y Urbanos* 6 (1):33–68.

Cosio-Zavala, Maria Eugenia, Ed. 1997. *Women and Families: Evolution of the Status of Women as a Factor and Consequence of Changes in Family Dynamics*. Paris: CICRED.

Council on Environmental Quality. 1980. *The Global 2000 Report to the President, Volume 1*. Washington: Government Printing Office.

Courbage, Yousef. 1992. "Demographic Transition Among Muslims in Eastern Europe." *Population: An English Selection* 4:161–86.

——. 2000. "The Future Population of Israel and Palestine." *Population & Sociétés* 362 (November):1–4.

Cowgill, Donald. 1979. "Aging and Modernization: A Revision of the Theory." In *Late Life: Communities and Environmental Policy,* edited by J. Gubrium. Springfield, IL: Charles C. Thomas.

Crenshaw, Edward M., Ansari Z. Ameen, and M. Christenson. 1997. "Population Dynamics and Economic Development: Age-Specific Population Growth Rates and Economic Growth in Developing Countries, 1965 to 1990." *American Sociological Review* 62:974–84.

Crispin, Shawn W. 2004. "Experts Confer on Bird-Flu Outbreak." *The Wall Street Journal,* 28 January, P. D4.

Critchfield, Richard. 1994. *The Villagers: Changed Values, Altered Lives: The Closing of the Urban-Rural Gap*. New York: Anchor Books.

Crosby, Alfred W. 1972. *The Columbian Exchange: Biological and Cultural Consequences of 1492*. Westport, CT: Greenwood Press.

Crosby, Alfred W. 1989. *America's Forgotten Pandemic: The Influenza of 1918*. New York: Cambridge University Press.

Crosby, Richard A. and William L. Yarber. 2001. "Perceived Versus Actual Knowledge About Correct Condom Use Among U.S. Adolescents: Results From a National Study." *Journal of Adolescent Health* 28 (5):415–420.

Cross, A., W. Obungu, and P. Kizito. 1991. "Evidence of a Transition to Lower Fertility in Kenya." *International Family Planning Perspectives* 17:4–7.

Cutright, Phillips and Robert M. Fernquist. 2000. "Effects of Societal Integration, Period, Region, and Culture of Suicide on Male Age-Specific Suicide Rates: 20 Developed Countries, 1955–1989." *Social Science Research* 29:148–172.

Dalla Zuanna, Gianpiero. 2001. "The Banquet of Aeolus: A Familistic Interpretation of Italy's Lowest Low Fertility." *Demographic Research* 4 (5):133–162.

Daly, H. 1971. "A Marxian-Malthusian View of Poverty and Development." *Population Studies* 25 (1):25–37.

Daly, Herman. 1996. *Beyond Growth: The Economics of Sustainable Development*. Boston: Beacon Press.

Danziger, S. and P. Gottschalk, Ed. 1993. *Uneven Tides: Rising Inequality in America*. New York: Russell Sage Foundation.

Darnell, Alfred and Darren Sherkat. 1997. "The Impact of Protestant Fundamentalism on Educational Attainment." *American Sociological Review* 62:306–315.

Das Gupta, Monica. 1995. "Fertility Decline in Punjab, India: Parallels with Historical Europe." *Population Studies* 49 (3):481–500.

Daulaire, N. 1995. "The Road From Cairo: Perspectives From Participants." In *Panel Session at the Annual Meeting of the Population Association of America*. San Francisco.

DaVanzo, Julie and A. Chan. 1994. "Living Arrangements of Older Malaysians: Who Coresides With Their Adult Children?" *Demography* 31 (1):95–113.

Davis, Karin. 1998. "Ancient Byproduct of War Returns: Foreigners Pay to Free Sudanese Slaves." *The San Diego Union-Tribune,* 8 February, p. A2.

Davis, Kingsley. 1945. "The World Demographic Transition." *The Annals of the American Academy of Political and Social Science* 237 (January):1–11.

——. 1949. *Human Society*. New York: Macmillan.

——. 1955a. "Institutional Patterns Favoring High Fertility in Underdeveloped Areas." *Eugenics Quarterly* II (Section 2):33–39.

——. 1955b. "Malthus and the Theory of Population." In *The Language of Social Research,* edited by P. Lazarsfeld and M. Rosenberg. New York: Free Press.

——. 1963. "The Theory of Change and Response in Modern Demographic History." *Population Index* 29 (4):345–66.

——. 1967. "Population Policy: Will Current Programs Succeed?" *Science* 158:730–39.

——. 1972a. "The American Family in Relation to Demographic Change." In *U.S. Commission on Population Growth and the American Future, Volume 1, Demographic and Social Aspects of Population Growth,* edited by C. Westoff and R. Parke. Washington: Government Printing Office.

——. 1972b. *World Urbanization 1950–1970, Volume 2: Analysis of Trends, Relationships, and Development.* Berkeley: Institute of International Studies, University of California.

——. 1973. *Cities and Mortality.* Liège: IUSSP.

——. 1974. "The Migration of Human Populations." *Scientific American* 231:92–105.

——. 1984a. "Demographic Imbalances and International Conflict." In *Annual Meeting of the American Sociological Association.* San Antonio, Texas.

——. 1984b. "Wives and Work: The Sex Role Revolution and its Consequences." *Population and Development Review* 10 (3):397–418.

——. 1986. "Low Fertility in Evolutionary Perspective." In *Below-Replacement Fertility in Industrial Societies: Causes, Consequences, Policies,* Supplement to Population and Development Review, Volume 12, edited by K. Davis, M. S. Bernstam, and R. Ricardo-Campbell. New York: Population Council.

——. 1988. "Social Science Approaches to International Migration." In *Population and Resources in Western Intellectual Tradition,* edited by M. Teitelbaum and J. Winter. New York: Population Council.

Davis, Kingsley and Judith Blake. 1955. "Social Structure and Fertility: An Analytic Framework." *Economic Development and Cultural Change* 4:211–35.

Davis, Kingsley and P. van den Oever. 1982. "Demographic Foundations of New Sex Roles." *Population and Development Review* 8 (3):495–511.

de Beauvoir, Simone. 1953. *The Second Sex; Translated and Edited by H.M. Parshley.* New York: Knopf.

de Groot, Natasja G., Nel Otting, Gaby G. M. Doxiadis, Sunita S. Balla-Jhagjhoorsingh, Jonathan L. Heeney, Jon J. van Rood, Pascal Gagneux, and Ronald E. Bontrop. 2002. "Evidence for an Ancient Selective Sweep in the MHC Class I Gene Repertoire of Chimpanzees." *Proceedings of the National Academy of Sciences* 99 (18):11742–11747.

de Janvry, Alain, Gustavo Gordillo, and Elisabeth Sadoulet. 1997. *Mexico's Second Agrarian Reform: Household and Community Responses.* La Jolla: Center for U.S.-Mexican Studies, University of California, San Diego.

De Jong, G. and J. Fawcett. 1981. "Motivations for Migration: An Assessment on a Value-Expectancy Model." In *Migration Decision Making,* edited by G. De Jong and R. Gardner. New York: Pergamon Press.

De Jong, Gordon. 2000. "Expectations, Gender, and Norms in Migration Decision-Making." *Population Studies* 54:307–319.

de Sandre, Paolo. 2000. "Patterns of Fertility in Italy and Factors of its Decline." *Genus* LVI (1–2):19–54.

De Santis, Solange 1996. "Canada Modifies 1996 Census Form After Angry Homemaker's Campaign." *The Wall Street Journal,* 9 August 9. P. A5C.

de Savigny, Don. 2003. "Integrating Census and Survey Data." In *Conference on Migration, Urbanization and Health: The Role of Remote Sensing (RS) and Geographic Information Systems (GIS) for Monitoring, Surveillance, and Program Evaluation.* Princeton: Princeton University Program in Urbanization and Migration.

De Soto, Hernando. 2000. *The Mystery of Capital: Why Capitalism Triumphs in the West and Fails Everywhere Else.* New York: Basic Books.

DeAre, D. 1990. "Longitudinal Migration Data From the Survey of Income and Program Participation." *Current Population Reports, Series P20–166.*

Deevey, E. 1960. "The Human Population." *Scientific American* 203 (3):194–205.

DellaPergola, S. 1980. "Patterns of American Jewish Fertility." *Demography* 17 (3):261-73.

Demeny, Paul. 1968. "Early Fertility Decline in Austria-Hungary: A Lesson in Demographic Transition." *Daedalus* 97 (2):502–22.

——. 1988. "Social Science and Population Policy." *Population and Development Review* 14:451–79.

——. 1994. *Population and Development.* Liège: International Union for the Scientific Study of Population.

Demerath, N. 1976. *Birth Control and Foreign Policy: The Alternatives to Family Planning.* New York: Harper & Row.

Demographic Research Unit. 2003. "County Population Estimates and Detailed Components of Change, July 1, 2000–2002, with Historical 1999 Estimates and Census 2000 Counts." Demographic Research Unit of the California Department of Finance: *http://www.dof.ca.gov/HTML/DEMOGRAP/e-6cnty.htm,* accessed 2004.

Demos, J. 1965. "Notes on Life in Plymouth Colony." *William and Mary Quarterly* 22:264–86.

DeNavas-Walt, Carmen, Robert W. Cleveland, and Bruce H. Webster, Jr. 2003. "Income in the United States: 2002." *Current Population Reports:* 60–221.

Denton, Nancy and Douglas Massey. 1991. "Patterns of Neighborhood Transition in a Multiethnic World: U.S. Metropolitan Areas 1970–80." *Demography* 28:41–63.

Désesquelles, Aline and Nicolas Brouard. 2003. "The Family Networks of Peopole Aged 60 and Over Living at Home or in an Institution." *Population-E* 58 (2):181–206.

Desmond, A., F. Wellemeyer, and F. Lorimer. 1962. "How Many People Have Ever Lived on Earth?" *Population Bulletin* 18 (1).

Dharmalingam, A. 1994. "Old Age Support: Expectations and Experiences in a South Indian Village." *Population Studies* 48:5–19.

Dharmalingam, A. and S. Philip Morgan. 1996. "Women's Work, Autonomy, and Birth Control: Evidence From Two South Indian Villages." *Population Studies* 50 (2):187–202.

Diamond, Jared. 1997. *Guns, Germs, and Steel*. New York: Norton.

Diaz-Briquets, S. and L. Perez. 1981. "Cuba: The Demography of Revolution." *Population Bulletin* 36 (1).

Diekmann, Andreas and Henriette Englehardt. 1999. "The Social Inheritance of Divorce: Effects of Parent's Family Type in Postwar Germany." *American Sociological Review* 64:783–793.

Dietz, T.L. 1995. "Patterns of Intergenerational Assistance Within the Mexican American Family." *Journal of Family Issues* 16 (3):344–56.

Diminescu, Dana. 1996. "A New Pattern of Migration: Oas Peasants' Travels Abroad." *Review of Social Research (Romania)* 3 (2).

Divine, R. 1957. *American Immigration Policy, 1924–1952*. New Haven: Yale Univeristy Press.

Dixon, John A. and Kirk Hamilton. 1996. *Expanding the Measure of Wealth*, edited by the World Bank, Washington, DC.

Dixon-Mueller, Ruth. 1993. *Population Policy and Women's Rights: Transforming Reproductive Choice*. Westport, CT: Praeger.

——. 2001. "Female Empowerment and Demographic Processes: Moving Beyond Cairo." In *IUSSP Contributions to Gender Research*, edited by International Union for the Scientific Study of Population. Paris: IUSSP.

Djerassi, Carl. 1981. *The Politics of Contraception*. San Francisco: Freeman.

Dogan, M. and John Kasarda. 1988. "Introduction: How Giant Cities Will Multiply and Grow." In *The Metropolis Era, Volume 1, A World of Giant Cities*, edited by M. Dogan and J. Kasarda. Newbury Park, CA: Sage.

Domschke, E. and D. Goyer. 1986. *The Handbook of National Population Censuses: Africa and Asia*. Westport, CT: Greenwood Press.

Donaldson, Peter J. 1990. *Nature Against Us: The United States and the World Population Crisis, 1965–1980*. Chapel Hill: University of North Carolina Press.

Donaldson, Peter and Amy O. Tsui. 1990. "The International Family Planning Movement." *Population Bulletin* 45 (3).

Doornbus, G. and D. Kromhout. 1991. "Educational Level and Mortality in a 32-Year Follow-Up Study of 18-Year-Old Men in the Netherlands." *International Journal of Epidemiology* 19:374–9.

Douglas, Emily Taft. 1970. *Margaret Sanger: Pioneer of the Future*. New York: Holt, Rinehart and Winston.

Douglas, M. 1966. "Population Control in Primitive Groups." *British Journal of Sociology* 17:263–73.

Douglass, Mike. 2000. "Mega-Urban Regions and World City Formation: Globalisation, the Economic Crisic and Urban Policy Issues in Pacific Asia." *Urban Studies* 37 (12):2315–2335.

Downs, Barbara. 2003. "Fertility of American Women, June 2002." *Current Population Reports* 20–548.

Drakakis-Smith, David and Elspeth Graham. 1996. "Shaping the Nation State: Class and the New Population Policy in Singapore." *International Journal of Population Geography* 2 (1):69–90.

Drake, Francis. 2000. *Global Warming: The Science of Climate Change*. London: Arnold.

Dublin, Louis, Alfred Lotka, and Mortimer Spiegleman. 1949. *Length of Life: A Study of the Life Table*. New York: Ronald.

Duff, Christina. 1997. "Plans for Census 'Sampling' Anger GOP in Congress." *The Wall Street Journal* 12 June, P. A20.

Duleep, H. 1989. "Measuring Socioeconomic Mortality Differentials Over Time." *Demography* 26:345–51.

Dumond, D. 1975. "The Limitation of Human Population: A Natural History." *Science* 232:713–20.

Dumont, Arsène. 1890. *Depopulation et Civilisation: Etude Demographique*. Paris: Lecrosnier & Babe.

Dunlop, John E. and Victoria A. Velkoff. 1999. "Women and the Economy in India, Report WID/98-2." Washington: U.S. Census Bureau.

Durand, John D. 1967. "The Modern Expansion of World Population." *Proceedings of the American Philosophical Society* 3 (June):137–140.

Durkheim, Emile. 1893 (1933). *The Division of Labor in Society, translated by George Simpson*. Glencoe: Free Press.

Dyson, Tim. 1991. "On the Demography of South Asian Famines." *Population Studies* 45:5–25.

——. 1996. *Population and Food: Global Trends and Future Prospects*. London: Routledge.

Easterlin, Richard A. 1968. *Population, Labor Force, and Long Swings in Economic Growth*. New York: National Bureau of Economic Research.

Easterlin, Richard A. 1978. "The Economics and Sociology of Fertility: A Synthesis." In *Historical Studies of Changing Fertility*, edited by C. Tilly. Princeton: Princeton University Press.

Easterlin, Richard A. and Eileen Crimmins. 1985. *The Fertility Revolution: A Supply-Demand Analysis*. Chicago: University of Chicago Press.

Easterlin, Richard, C. M. Schaeffer, and D. J. Macunovich. 1993. "Will the Baby Boomers Be Less Well Off Than Their Parents? Income, Wealth, and Family Circumstances Over the Life Cycle in the United

States." *Population and Development Review* 19 (3):497–522.

Eaton, J. and A. Mayer. 1954. *Man's Capacity to Reproduce.* Glencoe: Free Press.

Eberstadt, N. 1994a. "What You Won't Hear at Cairo." *The Wall Street Journal,* 6 September, P. B1.

Eberstadt, Nicholas. 1994b. "Demographic Shocks After Communism: Eastern Germany, 1989–93." *Population and Development Review* 20 (1):137–52.

——. 2001. "The Population Implosion." *Foreign Policy* (March-April).

Edmonston, Barry and Charles Schultze, Ed. 1995. *Modernizing the U.S. Census.* Washington: National Academy Press.

Edwards, John N., Theodore D. Fuller, Sairudee Vorakitphokatorn, and Santhar Sermsri. 1994. *Household Crowding and Its Consequences.* Boulder, CO: Westview Press.

Egidi, Viviana. 2003. "Health Status of Older People." *Genus* LIX (1):169–200.

Ehrlich, Paul. 1968. *The Population Bomb.* New York: Ballantine Books.

——. 1971. *The Population Bomb, Second Edition.* New York: Sierra Club/Ballantine Books.

Ehrlich, Paul and Anne Ehrlich. 1972. *Population, Resources, and Environment, 2nd Edition.* San Francisco: Freeman.

——. 1990. *The Population Explosion.* New York: Simon & Schuster.

Ehrlich, Paul R., Anne H. Ehrlich, and Gretchen C. Daily. 1997. *The Stork and the Plow: The Equity Answer to the Human Dilemma.* New Haven: Yale University Press.

Ellertson, Charlotte. 1996. "History and Efficacy of Emergency Contraception: Beyond Coca-Cola." *International Family Planning Perspectives* 22 (2):52–56.

Elliott, James R. 1999. "Putting 'Global Cities' in Their Place: Urban Hierarchy and Low-Income Employment During the Post-War Era." *Urban Geography* 20 (2):95–115.

Emerson, Michael O., George Yancy, and Karen J. Chai. 2001. "Does Race Matter in Residential Segregation? Exploring the Preferences of White Americans." *American Sociological Review* 66:922–935.

Engels, Frederich. 1844 (1953). *Outlines of a Critique of Political Economy, Reprinted in R. L. Meek, 1953, Marx and Engels on Malthus.* London: Lawrence and Wishart.

Englehart, R. 1994. "Alcohol Plays a Part in Many Bike Accidents." *The Wall Street Journal* 15 April, p. B2.

Epstein, Paul R. 2000. "Is Global Warming Harmful to Health?" *Scientific American* 282 (August):50–57.

Espenshade, Thomas J. and Jessica C. Gurcak. 1996. "Are More Immigrants the Answer to U.S. Population Aging?" *Population Today* (December):4–5.

Eversley, D. 1959. *Social Theories of Malthus and the Malthusian Debate.* Oxford: Clarendon Press.

Ewald, P. W. 1994. *Evolution of Infectious Disease.* New York: Oxford University Press.

Exter, Thomas G. 1996. "Buying Demographics in the Late 1990s." *Business Geographics* (March):18.

Ezzati, Majid and Alan D. Lopez. 2003. "Estimates of Global Mortality Attributable to Smoking in 2000." *The Lancet* 362 (9387):847–849.

Fagan, Brian. 2000. *The Little Ice Age: How Climate Made History 1300–1850.* New York: Basic Books.

Falkenmark, M. and C. Widstrand. 1992. "Population and Water Resources: A Delicate Balance." *Population Bulletin* 47 (3).

Fan, C. Cindy. 2000. "The Vertical and Horizontal Expansions of China's City System." *Urban Geography* 20 (6):493–515.

Faour, M. 1991. "The Demography of Lebanon: A Reappraisal." *Middle Eastern Studies* 27 (4):631–41.

Fargues, Philippe. 1994. "Demographic Explosion or Social Upheaval." In *Democracy Without Democrats? The Renewal of Politics in the Muslim World,* edited by G. Salamé. London: I.B. Taurus.

——. 1995. "Changing Hierarchies of Gender and Generation in the Arab World." In *Family, Gender, and Population in the Middle East,* edited by C. M. Obermeyer. Cairo: American University in Cairo.

——. 1997. "State Policies and the Birth Rate in Egypt: From Socialism to Liberalism." *Population and Development Review* 23 (1):115–138.

——. 2000a. *Générations Arabes.* Paris: Fayard.

——. 2000b. "Protracted National Conflict and Fertility Change: Palestinians and Israelis in the Twentieth Century." *Population and Development Review* 26 (3):441–482.

Farley, Reynolds. 1970. *Growth of the Black Population.* Chicago: Markham.

——. 1976. "Components of Suburban Population Growth." In *The Changing Face of the Suburbs,* edited by B. Schwartz. Chicago: University of Chicago Press.

——. 1984. *Blacks and Whites: Narrowing the Gap?* Cambridge: Harvard University Press.

Farley, Reynolds, Sheldon Danziger, and Harry J. Holzer. 2000. *Detroit Divided.* New York: Russell Sage Foundation.

Farley, Reynolds and William H. Frey. 1994. "Changes in the Segregation of Whites From Blacks During the 1980s: Small Steps Toward a More Integrated Society." *American Sociological Review* 59:23–45.

Fearnside, Philip M. 1997. "Transmigration in Indonesia: Lessons from its Environmental and Social Impacts." *Environmental Management* 21 (4):553–570.

Federal Interagency Forum on Aging-Related Issues. 2000. *Older Americans 2000: Key Indicators of Well-Being.* Washington: Government Printing Office.

Feinleib, S., P. Cunningham, and P. Short. 1994. *Use of Nursing and Personal Care Homes by the Civilian Population, 1987*. Rockville, MD: Public Health Services.

Fenker, Richard M. 1996. "The Art and Science of Building Site Evaluation Models." *Business Geographics* (July/August):28–31.

Fenwick, R. and C. M. Barresi. 1981. "Health Consequences of Marital-Status Change Among the Elderly: A Comparison of Cross-Sectional and Longitudinal Analyses." *Journal of Health and Social Behavior* 22:106–16.

Ferraro, Kenneth F. and Melissa M. Farmer. 1999. "Utility of Health Data From Social Surveys: Is There a Gold Standard for Measuring Morbidity?" *American Sociological Review* 64:303–315.

Ferraro, Kenneth F. and Ya-ping Su. 2000. "Physician-Evaluated and Self-Reported Morbidity for Predicting Disability." *American Journal of Public Health* 90 (1):103–108.

Feshbach, Murray and Alfred Friendly. 1992. *Ecocide in the USSR: Health and Nature Under Siege*. New York: Basic Books.

Fetto, John. 2000. "It Ain't Easy Eating Green." *American Demographics* 22 (5):13.

Field, Dorothy. 1997. "'Looking Back, What Period of Your Life Brought You the Most Satisfaction'." *International Journal of Aging and Human Development* 45 (3):169–94.

Fields, Jason. 2001. "America's Families and Living Arrangements: 2000." *Current Population Reports* 20–537.

Findlay, Allan M. 1995. "Skilled Transients: The Invisible Phenomenon." In *The Cambridge Survey of World Migration*, edited by R. Cohen. Cambridge: Cambridge University Press.

Findley, Sally E. 1999. *Women on the Move: Perspectives on Gender Changes in Latin America*. Liège: International Union for the Scientific Study of Population.

Findley, Sally and J. Ford. 1993. "Reversals in the Epidemiological Transition in Harlem." In *Paper Presented at the Annual Meeting of the Population Association of America*. Cincinnati, Ohio.

Finkle, Jason and B. Crane. 1975. "The Politics of Bucharest: Population Development and the New International Economic Order." *Population and Development Review* 1 (1):87–114.

Finkle, Jason L. and C. Alison McIntosh. 2002. "United Nations Population Conferences: Shaping the Policy Agenda for the Twenty-First Century." *Studies in Family Planning* 33 (1):11–23.

Firebaugh, Glenn. 1979. "Structural Determinants of Urbanization in Asia and Latin America, 1950–1970." *American Sociological Review* 44:199–215.

——. 1982. "Population Density and Fertility in 22 Indian Villages." *Demography* 19 (4):481–94.

——. 2003. *The New Geography of Global Income Inequality*. Cambridge: Harvard University Press.

Firebaugh, Glenn and F. D. Beck. 1994. "Does Economic Growth Benefit the Masses? Growth, Dependence, and Welfare in the Third World." *American Sociological Review* 59:631–53.

Fischer, Claude. 1976. *The Urban Experience*. New York: Harcourt Brace Jovanovich.

——. 1981. "The Public and Private Worlds of City Life." *American Sociological Review* 46:306–316.

——. 1984. *The Urban Experience, 2nd Edition*. San Diego: Harcourt Brace Jovanovich.

Fishlow, A. 1965. *American Railroads and the Transformation of the Antebellum Economy*. Cambridge: Harvard University Press.

Fishlow, Harriet. 1997. "Enrollment Projection in a Multicampus University System." In *Demographics: A Casebook for Business and Government*, edited by H. Kintner, J. T. W. Merrick, P. A. Morrison, and P. R. Voss. Santa Monica, CA: RAND.

Fitzsimmons, James D. and Michael R. Ratcliffe. 2004. "Reflections on the Review of Metropolitan Area Standards in the United States, 1990–2000." In *New Forms of Urbanization: Beyond the Urban-Rural Dichotomy*, edited by A. G. Champion and G. Hugo. London: Ashgate.

Fleck, S. and C. Sorrentino. 1994. "Employment and Unemployment in Mexico's Labor Force." *Monthly Labor Review* 117 (11):3–31.

Fogel, Robert. 1994. "The Relevance of Malthus for the Study of Mortality Today: Long-Run Influences on Health, Mortality, Labour Force Participation, and Population Growth." In *Population, Economic Development, and the Environment*, edited by K. Lindahl-Kiessling and H. Landberg. Oxford: Oxford University Press.

Fogel, Robert W. and Lorens A. Helmchen. 2002. "Economic and Technological Development and Their Relationships to Body Size and Productivity." In *The Nutrition Transition: Diet and Disease in the Developing World*, edited by B. Caballero and B. M. Popkin. San Diego: Academic Press.

Foner, Anne. 1975. "Age in Society: Structure and Change." *American Behavioral Scientist* 19 (2):144–65.

Food and Agricultural Organization of the United Nations. 2001. *State of World Fisheries and Aquaculture 2000*. Rome: FAO.

——. 2002. "The State of World Food Insecurity in the World 2002." FAO: *http://www.fao.org/docrep/005/y7352e/y7352e00.htm*, accessed 2003.

——. 2003. *The State of Food Insecurity in the World 2003*. Rome: Food and Agricultural Organization of the United Nations.

Foot, David K. 1996. *Boom, Bust & Echo: How to Profit From the Coming Demographic Shift*. Toronto: Macfarlane Walter & Ross.

Ford, Larry R. 2000. *The Spaces Between Buildings.* Baltimore and London: Johns Hopkins University Press.

———. 2003. *America's New Downtowns: Reinvention or Revitalization.* Baltimore: Johns Hopkins University Press.

Ford, Tania. 1999. "Understanding Population Growth in the Peri-Urban Region." *International Journal of Population Geography* 5:297–311.

Foster, George. 1967. *Tzintzuntzan: Mexican Peasants in a Changing World.* Boston: Little, Brown.

Francese, Peter. 1979. "The 1980 Census: The Counting of America." *Population Bulletin* 34 (4).

Freedman, A. 1991. "Nestlé to Restrict Baby-Food Supply to Third World." *The Wall Street Journal,* 30 January, P. B3.

Freedman, Alix. 1998. "Why Teenage Girls Love "the Shot"; Why Others Aren't So Sure." *The Wall Street Journal,* New York, 14 October, P. 1.

Freeman, O.L. 1992. "Perspectives and Prospects." *Agricultural History* 66 (2):3–11.

Frejka, Tomas and Lucille Atkin. 1996. "The Role of Induced Abortion in the Fertility Transition of Latin America." In *The Fertility Transition in Latin America,* edited by J. M. Guzmán, S. Singh, G. Rodríguez, and E. A. Pantelides. Oxford: Oxford University Press.

Frejka, Tomas and W. Ward Kingkade. 2001. "U.S. Fertility in International Comparison: An Exploration to Aid Projections." In *The Direction of Fertility in the United States,* edited by U.S. Census Bureau. Washington: Council of Professional Associations on Federal Statistics.

Frey, William H. 1995. "The New Geography of Population Shifts." In *State of the Union: American in the 1990s, Volume Two: Social Trends,* edited by R. Farley. New York: Russell Sage Foundation.

———. 2004. "The Fading of City-Suburb and Metro-Nonmetro Distinctions in the United States." In *New Forms of Urbanization: Beyond the Urban-Rural Dichotomy,* edited by A. G. Champion and G. Hugo. London: Ashgate.

Frey, William H. and Reynolds Farley. 1996. "Latino, Asian, and Black Segregation in U.S. Metropolitan Areas: Are Multiethnic Metros Different?" *Demography* 33 (1):35–50.

Frey, William H. and Kao-Lee Liaw. 1998. "The Impact of Recent Immigration on Population Redistribution Within the United States." In *The Immigration Debate: Studies of Economic, Demographic and Fiscal Effects of Immigration,* edited by J. P. Smith and B. Edmonston. Washington: National Academy Press.

Friedberg, Leora. 1998. "Did Unilateral Divorce Raise Divorce Rates? Evidence From Panel Data." *American Economic Review* 88:608–27.

Friedlander, Dov. 1983. "Demographic Responses and Socioeconomic Structure: Population Processes in England and Wales in the Nineteenth Century." *Demography* 20 (3):249–72.

Frier, Bruce W. 1999. "Roman Demography." In *Life, Death, and Entertainment in the Roman Empire,* edited by D. S. Potter and D. J. Mattingly. Ann Arbor: University of Michigan Press.

Fries, James F. 1980. "Aging, Natural Death, and the Compression of Morbidity." *New England Journal of Medicine* 303 (3):130–35.

Frisch, Rose. 1978. "Population, Food Intake, and Fertility." *Science* 199:22–30.

Frisch, Rose E. 2002. *Female Fertility and The Body Fat Connection.* Chicago: University of Chicago Press.

Fuguitt, Glenn and D. Brown. 1990. "Residential Preferences and Population Redistribution in 1972–88." *Demography* 27:289–600.

Fuguitt, Glenn and James Zuiches. 1975. "Residential Preferences and Population Distribution." *Demography* 27:589–600.

Furstenberg, Frank. 1998. "Review Symposium: Families According to Research: Relative Risks: What Is The Family Doing To Our Children?" *Contemporary Sociology* 27 (3):223–225.

Furstenberg, Frank and Andrew J. Cherlin. 1991. *Divided Families: What Happens to Children When Parents Part?* Cambridge: Harvard University Press.

Gadalla, Saad. 1978. *Is There Hope? Fertility and Family Planning in a Rural Community in Egypt.* Chapel Hill: Carolina Population Center, University of North Carolina.

Gage, T. B. 1994. "Population Variation in Cause of Death: Level, Gender, and Period Effects." *Demography* 31 (2):271–94.

Galle, Omer, Walter Gove, and J. McPherson. 1972. "Population Density and Pathology: What Are The Relations for Man." *Science* 176:23–30.

Galloway, P. 1984. *Long Term Fluctuations in Climate and Population in the Pre-Industrial Era.* Berkeley: University of California.

Gao, F., E. Bailes, D. L. Robertson, Y. Chen, C. M. Rodenburg, S. F. Michael, L. B. Cummins, L. O. Arthur, M. Peeters, G. M. Shaw, P. M. Sharp, and B. H. Hahn. 1999. "Origin of HIV-1 in the Chimpanzee Pan Troglodytes Troglodytes." *Nature* 397 (6718):436–41.

Garcia y Griego, Manuel, John R. Weeks, and Roberto Ham-Chande. 1990. "Mexico." In *Handbook on International Migration,* edited by W. J. Serow, C. B. Nam, D. F. Sly, and R. H. Weller. New York: Greenwood Press.

Garreau, Joel. 1991. *Edge City: Life on the New Frontier.* New York: Doubleday.

Gartner, R. 1990. "The Victims of Homicide: A Temporal and Cross-National Comparison." *American Sociological Review* 55:92–106.

Garza, Gustavo. 2004. "The Transformation of the Urban System in Mexico." In *New Forms of Urbanization:*

Beyond the Urban-Rural Dichotomy, edited by A. G. Champion and G. Hugo. London: Ashgate.

Gates, Gary J. and Freya L. Sonenstein. 2000. "Heterosexual Genital Sexual Activity Among Adolescent Males: 1988 and 1995." *Family Planning Perspectives* 32 (6):295–297.

Gelbspan, Ross. 1997. *The Heat Is On.* Reading, MA: Addison-Wesley.

Gendell, Murray. 1998. "Trends in Retirement Age in Four Countries, 1965–95." *Monthly Labor Review* (August):20–30.

Gendell, Murray and Jacob S. Siegel. 1996. "Trends in Retirement Age in the United States, 1955–1993, by Sex and Race." *Journal of Gerontology: Social Sciences* 51B:S132–S139.

Gibson, Campbell. 1976. "The U.S. Fertility Decline, 1961–1975: The Contribution of Changes in Marital Status and Marital Fertility." *Family Planning Perspectives* 8:249–52.

Gilbar, G. 1994. "Population Growth and Family Planning in Egypt, 1985–92." *Middle East Contemporary Survey* 16:335–48.

Gillis, J. R., L. A. Tilly, and D. Levine, Ed. 1992. *The European Experience of Declining Fertility: A Quiet Revolution 1850–1970.* Oxford: Basil Blackwell.

Glass, D. V. 1953. *Introduction to Malthus.* New York: Wiley.

Global Aging Report. 1997. "Guaranteeing the Rights of Older People: China Takes a Great Leap Forward." *Global Aging Report* 2 (5):3.

Gober, Patricia. 1993. "Americans on the Move." *Population Bulletin* 48 (3).

———. 1999. "Settlement Dynamics and Internal Migration of the U.S. Foreign-Born Population." In *Migration and Restructuring in the United States,* edited by K. Pandit and S. D. Withers. New York: Rowman & Littlefield.

———. 2000. "Immigration and North American Cities." *Urban Geography* 21 (1):83–90.

Godfrey, Brian J. and Yu Zhou. 1999. "Ranking World Cities: Multinational Corporations and the Global Urban Hierarchy." *Urban Geography* 20 (3):268–281.

Godwin, William. 1793 (1946). *Enquiry Concerning Political Justice and Its Influences on Morals and Happiness.* Toronto: University of Toronto Press.

Goldman, Noreen. 1984. "Changes in Widowhood and Divorce and Expected Durations of Marriage." *Demography* 21 (3):297–308.

Goldscheider, Calvin. 1971. *Population, Modernization, and Social Structure.* Boston: Little, Brown.

———. 1989. "The Demographic Embeddedness of the Arab-Jewish Conflict in Israeli Society." *Middle East Review* 21 (3):15–24.

———. 1996. *Israel's Changing Society: Population, Ethnicity, and Development.* Boulder, CO: Westview Press.

Goldscheider, Calvin and William Mosher. 1991. "Patterns of Contraceptive Use in the United States: The Importance of Religious Factors." *Studies in Family Planning* 22 (2):102–15.

Goldscheider, Frances. 1995. "Interpolating Demography With Families and Households." *Demography* 32 (3):471–480.

Goldscheider, Frances and Calvin Goldscheider. 1993. *Leaving Home Before Marriage: Ethnicity, Familism, and Generational Relationships.* Madison: University of Wisconsin Press.

———. 1994. "Leaving and Returning Home in 20th Century America." *Population Bulletin* 48 (4).

Goldscheider, Frances and Calvin Goldscheider. 1999. "Changes in Returning Home in the United States, 1925–1985." *Social Forces* 78 (2):695–720.

Goldscheider, Frances and Linda Waite. 1991. *New Families, No Families: The Transformation of the American Home.* Berkeley and Los Angeles: University of California Press.

Goldstein, Alice and Wong Feng, Ed. 1996. *China: The Many Facets of Demographic Change.* Boulder, CO: Westview Press.

Goldstein, S. and A. Goldstein. 1981. "The Impact of Migration on Fertility: An 'Own Children' Analysis For Thailand." *Population Studies* 35 (2):265–84.

Goldstone, Jack A. 1986. "State Breakdown in the English Revolution: A New Synthesis." *American Sociological Review* 92:257–322.

———. 1991. *Revolution and Rebellion in the Early Modern World.* Berkeley: University of California Press.

Goliber, Thomas. 1985. "Sub-Saharan Africa: Population Pressures on Development." *Population Bulletin* 40 (1).

Goode, William J. 1993. *World Changes in Divorce Patterns.* New Haven: Yale University Press.

Goodkind, D. M. 1995. "Vietnam's One-or-Two Child Policy in Action." *Population and Development Review* 21 (1):85–112.

Goodkind, Daniel. 1997. "The Vietnamese Double Marriage Squeeze." *International Migration Review* 31 (1):108–127.

Gordon, Charles and Charles F. Longino, Jr. 2000. "Age Structure and Social Structure." *Contemporary Sociology* 29 (5):699–703.

Gordon, M. 1975. *Agriculture and Population.* Washington: Government Printing Office.

Gottlieb, M. S., H. M. Schanker, P. T. Fan, A. Saxon, J. D. Weisman, and I. Pozalski. 1981. "Pneumocystis Pneumonia—Los Angeles." *Mortality and Morbidity Weekly Report* 30 (June 5):250–2.

Goudie, Andrew. 1990. *The Human Impact on the Natural Environment, 3rd Edition.* Cambridge: MIT Press.

Goudsmit, Jaap. 1997. *Viral Sex: The Nature of AIDS.* New York: Oxford University Press.

Gould, Peter. 1993. *The Slow Plague: A Geography of the AIDS Pandemic.* Oxford: Blackwell.

Gould, W. T. S. and A. M. Findlay, Ed. 1994. *Population Migration and the Changing World Order.* New York: Chichester.

Gove, Walter. 1973. "Sex, Marital Status, and Mortality." *American Journal of Sociology* 79 (1):45–67.

Grabill, W., Clyde Kiser, and Pascal Whelpton. 1958. *The Fertility of American Women.* New York: Wiley.

Graebner, W. 1980. *A History of Retirement.* New Haven: Yale University Press.

Graefe, Deborah R. and Daniel L. Lichter. 1999. "Life Course Transitions of American Children: Parental Cohabitation, Marriage, and Single Motherhood." *Demography* 36 (2):205–217.

Graunt, John. 1662 (1939). *Natural and Political Observations Made Upon the Bills of Mortality, Edited with an Introduction by Walter F. Willcox.* Baltimore: Johns Hopkins University Press.

Green, A. E., T. Hogarth, and R. E. Shackleton. 1999. "Longer Distance Commuting as a Substitute for Migration in Britain: A Review of Trends, Issues and Implications." *International Journal of Population Geography* 5 (1):49–67.

Greenhalgh, Susan. 1986. "Shifts in China's Population Policy 1984–86: Views From the Central, Provincial, and Local Levels." *Population and Development Review* 12 (3):491–516.

Greenhalgh, Susan, Z. Chuzhu, and L. Nan. 1994. "Restraining Population Growth in Three Chinese Villages, 1988–93." *Population and Development Review* 20 (2):365–96.

Gregorio, David I., Stephen J. Walsh, and Deborah Paturzo. 1997. "The Effects of Occupation-Based Social Position on Mortality in a Large American Cohort." *American Journal of Public Health* 87 (9):1472–5.

Grieco, Elizabeth M. and Rachel C. Cassidy. 2001. "Overview of Race and Hispanic Origin: 2000." *Census 2000 Brief* (C2KBR/01-1). Washington, DC.

Grodsky, Eric and Devah Pager. 2001. "The Structure of Disadvantage: Individual and Occupational Determinants of the Black-White Wage Gap." *American Sociological Review* 66 (4):542–567.

Grossbard-Schechtman, Shoshana. 1985. "Marriage Squeezes and the Marriage Market." In *Contemporary Marriage: Comparative Perspectives on a Changing Institution,* edited by K. Davis and S. Grossbard-Schechtman. New York: Russell Sage Foundation.

——. 1993. *On the Economics of Marriage: A Theory of Marriage, Labor, and Divorce.* Boulder, CO: Westview Press.

Guest, Avery M., Gunnar Almgren, and Jon M. Hussey. 1998. "The Ecology of Race and Socioeconomic Distress: Infant and Working-Age Mortality in Chicago." *Demography* 35 (1):23–34.

Guihati, K. 1977. "Compulsory Sterilization: The Change in India's Population Policy." *Science* 195:1300–1305.

Guillard, Achille. 1855. *Éléments de Statistique Humaine ou Démographie Comparée.* Paris: Guillaumin et Cie Libraires.

Guilmoto, Christophe Z. 1998. "Institutions and Migrations: Short-Term Versus Long-Term Moves in Rural West Africa." *Population Studies* 52:85–103.

Gupta, Geeta Rao. 1998. "Claiming the Future." In *The Progress of Nations: 1998,* edited by UNICEF. New York: United Nations Childrens Fund.

Gustafsson, Bjorn and Mats Johannson. 1999. "In Search of Smoking Guns: What Makes Income Inequality Vary Over Time in Different Countries." *American Sociological Review* 64:585–605.

Guttentag, M. and P. Secord. 1983. *Too Many Women?* Beverly Hills: Sage.

Guz, D. and J. Hobcraft. 1991. "Breastfeeding and Fertility: A Comparative Analysis." *Population Studies* 45:91–108.

Haas, W. H. and W. J. Serow. 1993. "Amenity Retirement Migration Process: A Model and Preliminary Evidence." *The Gerontologist* 33 (2):212–220.

Hacker, J. David. 2003. "Rethinking the 'Early' Decline of Marital Fertility in the United States." *Demography* 40 (4):605–620.

Hagan, John and Holly Foster. 2001. "Youth Violence and the End of Adolescence." *American Sociological Review* 66:874–899.

Hahn, B. A. 1993. "Marital Status and Women's Health: The Effects of Economic Marital Adjustments." *Journal of Marriage and the Family* 55:495–504.

Hall, D. R. and J. Z. Zhao. 1995. "Cohabitation and Divorce in Canada: Testing the Selectivity Hypothesis." *Journal of Marriage and the Family* 57:421–27.

Hall, Ray and Paul White. 1995. "Population Change on the Eve of the Twenty-First Century." In *Europe's Population: Towards the Next Century,* edited by R. Hall and P. White. London: University College London Press.

Hampshire, S. 1955. "Introduction." In *Sketch for an Historical Picture of the Progress of the Human Mind,* edited by M. J. A. N. d. C. Condorcet. London: Weidenfield and Nicholson.

Han, Sun Sheng and Shue Tuck Wong. 1994. "The Influence of Chinese Reform and Pre-Reform Policies on Urban Growth in the 1980s." *Urban Geography* 15 (6):537–64.

Handwerker, P. 1986. *Culture and Reproduction: An Anthropological Critique of Demographic Transition Theory.* Boulder, CO: Westview Press.

Hank, Karsten and Hans-Peter Kohler. 2000. "Gender Preferences for Children in Europe: Empirical Results from 17 FFS Countries." *Demographic Research* 2 (1).

Hansen, J. 1970. *The Population Explosion: How Sociologists View It.* New York: Pathfinder Press.

Hansen, Kristin. 1994. "Geographic Mobility: March 1992 to March 1993." *Current Population Reports* 20–481.

Haq, Muhammad. 1984. "Age at Menarche and the Related Issue: A Pilot Study on Urban School Girls." *Family Planning Perspectives* 13 (6):559–67.

Hardin, Garrett. 1968. "The Tragedy of the Commons." *Science* 162:1243–48.

Harevan, Tamara K. and Peter Uhlenberg. 1995. "Transition to Widowhood and Family Support Systems in the Twentieth Century Northeastern United States." In *Aging in the Past: Demography, Society, and Old Age,* edited by D. Kertzer and P. Laslett. Berkeley: University of California Press.

Harkavy, Oscar. 1995. *Curbing Population Growth: An Insider's Perspective on the Population Movement.* New York: Plenum Press.

Harlan, J. 1976. "The Plants and Animals That Nourish Man." *Scientific American* 235 (3):88–97.

Harris, Marvin and Eric B. Ross. 1987. *Death, Sex, and Fertility: Population Regulation in Preindustrial and Developing Societies.* New York: Columbia University Press.

Harrison, Paul. 1993. *The Third Revolution: Population, Environment and a Sustainable World.* London: Penguin Books.

Hartley, Leslie Poles. 1967. *The Go-Between.* New York: Stein and Day.

Hartmann, Betsy. 1995. *Reproductive Rights and Wrongs: The Global Politics of Population Control.* Boston: South End Press.

Harvey, W. 1986. "Homicide Among Young Black Adults: Life in the Subculture of Exasperation." In *Homicide Among Black Americans,* edited by D. Hawkins. New York: University Press of America.

Hatcher, R. A., J. Trussell, F. Stewart, G. K. Stewart, D. Kowal, F. Guest, W. Cates Jr., and M. S. Policar. 1994. *Contraceptive Technology, 16th edition.* New York: Irvington.

Hatcher, Robert. 1998. "Choosing a Contraceptive." *http://www.emory.edu/familyplanning/contraception.cfm,* accessed 2001.

Hatton, T. J. and J. G. Williamson. 1994. "What Drove the Mass Migrations From Europe in the Late Nineteenth Century?" *Population and Development Review* 20 (3):533–59.

Haub, Carl. 1995. "How many people have ever lived on earth?" *Population Today* 23 (2):4–5.

Haug, Werner. 2000. "National and Immigrant Minorities: Problems of Measurement and Definition." *Genus* LVI (1–2):133–147.

Hauser, R. and P. Featherman. 1974. "Socioeconomic Achievements of U.S. Men, 1962 to 1972." *Science* 185:325–31.

Havanon, N., J. Knodel, and W. Sittitrai. 1992. "The Impact of Family Size on Wealth Accumulation in Rural Thailand." *Population Studies* 46 (1):37–52.

Haveman, Robert, Karen Holden, Kathryn Wilson, and Barbara Wolfe. 2003. "Social Security, Age of Retirement, and Economic Well-Being: Intertemporal Patterns Among Retired-Worker Beneficiaries." *Demography* 40 (2):369–394.

Hawley, Amos. 1972. "Population Density and the City." *Demography* 91 521-30.

Hayward, Mark D., Eileen M. Crimmins, Toni P. Miles, and Yu Yang. 2000. "The Signficance of Socioeconomic Status in Explaining the Racial Gap in Chronic Health Conditions." *American Sociological Review* 65:910–930.

Hecht, Jacqueline. 1987. "Johann Peter Süssmilch: A German Prophet in Foreign Countries." *Population Studies* 41:31–58.

Heer, David and Amyra Grossbard-Schechtman. 1981. "The Impact of the Female Marriage Squeeze and the Contraceptive Revolution on Sex Roles and the Women's Liberation Movement in the United States, 1960 to 1975." *Journal of Marriage and the Family* 43 (1):49–65.

Heilig, G., T. Büttner, and W. Lutz. 1990. "Germany's Population: Turbulent Past, Uncertain Future." *Population Bulletin* 45 (4).

Heligman, L., N. Chen, and O. Babakol. 1993. "Shifts in the Structure of Population and Deaths in Less Developed Regions." In *The Epidemiological Transition: Policy and Planning Implications for Developing Countries,* edited by J. N. Gribble and S. H. Preston. Washington: National Academy Press.

Henry, Louis. 1961. "Some Data on Natural Fertility." *Eugenics Quarterly* 8:81–91.

Henshaw, Stanley K., Susheela Singh, and Taylor Haas. 1999. "The Incidence of Abortion Worldwide." *International Family Planning Perspectives* 25 (Supplement):30–38.

Herbst, Sharon Tyler. 1995. *The New Food Lover's Companion, 2nd Edition.* New York: Barron's Educational Services.

Herlihy, David and Christiane Klapisch-Zuber. 1985. *Tuscans and Their Families: A Study of the Florentine Catasto of 1427.* New Haven: Yale University Press.

Herman, Tom. 1999. "The Big Shift." *The Wall Street Journal.* 29 November. P. R24.

Hern, Warren. 1976. "Knowledge and Use of Herbal Contraception in a Peruvian Amazon Village." *Human Organization* 35:9–19.

——. 1977. "High Fertility in a Peruvian Amazon Indian Village." *Human Ecology* 5:355–67.

——. 1992. "Polygyny and Fertility Among the Shipibo of the Peruvian Amazon." *Population Studies* 46:53–64.

Hernandez, Donald. 1993. *America's Children: Resources From Family, Government and the Economy.* New York: Russell Sage Foundation.

Herskind, A. M., M. McGue, N. V. Holm, T. I. A. Sorensen, B. Harvald, and J. W. Vaupel. 1996. "The Heritability of Human Longevity: A Population-Based

Study of 2,872 Danish Twin Pairs Born 1870–1900." *Human Genetics* 97:319–323.

Heuser, Robert. 1976. *Fertility Tables For Birth Cohorts by Color: United States, 1917–73.* Rockville, MD: National Center for Health Statistics.

Heymann, D., J. Chin, and J. Mann. 1990. "A Global Overview of AIDS." In *Heterosexual Transmission of AIDS,* edited by N. Alexander, H. Gabelnick, and J. Spieler. New York: Wiley-Liss.

Higgins, Matthew and Jeffrey G. Williamson. 1997. "Age Structure Dynamics in Asia and Dependence on Foreign Capital." *Population and Development Review* 23 (2):261–294.

Hill, D. and M. Kent. 1988. *Election Demographics.* Washington: Population Reference Bureau.

Hill, Mark E., Samuel H. Preston, and Ira Rosenwaike. 2000. "Age Reporting Among White Americans Aged 85+: Results of a Record Linkage Study." *Demography* 37 (2):175–186.

Himes, Norman E. 1976. *Medical History of Contraception.* New York: Schocken Books.

Himmelfarb, Gertrude. 1984. *The Idea of Poverty: England in the Early Industrial Age.* New York: Alfred A. Knopf.

Hinde, Thomas. 1995. *The Domesday Book: England's Heritage, Then & Now.* London: Tiger Books International.

Hirschman, A. 1958. *The Strategy of Economic Development.* New Haven: Yale University Press.

Hirschman, Charles. 1999. "Theories of International Migration and Immigration: A Preliminary Reconnaissance of Ideal Types." In *The Handbook of International Migration: The American Experience,* edited by C. Hirschman, P. Kasinitz, and J. DeWind. New York: Russell Sage Foundation.

Hirschman, Charles, Richard Alba, and Reynolds Farley. 2000. "The Meaning and Measurement of Race in the U.S. Census: Glimpses into the Future." *Demography* 37 (3):381–393.

Hirschman, Charles and M. Butler. 1981. "Trends and Differentials in Breast Feeding: An Update." *Demography* 18 (1):39–54.

Hobbs, Frank B. and Bonnie L. Damon. 1996. "65+ in the United States." *Current Population Reports* 23–190.

Hobbs, Frank and Nicole Stoops. 2002. *Demographic Trends in the 20th Century.* Washington: U.S. Census Bureau.

Hobcraft, John. 1996. "Fertility in England and Wales: A Fifty Year Perspective." *Population Studies* 50 (3):485–524.

Hockings, Paul. 1999. *Kindreds of the Earth: Badaga Household Structure and Demography.* Walnut Creek, CA: Alta Mira Press.

Hodell, David, Jason Curtis, and Mark Brenner. 1995. "Possible Role of Climate in the Collapse of Classic Mayan Civilization." *Nature* 375:391–394.

Hoerning, Roxanne. 1996. "American Honda Jump-Starts Sales Geographically." *Business Geographics* (March):24–26.

Hogan, Timothy D. and Donald N. Steinnes. 1998. "A Logistic Model of the Seasonal Migration Decision for Elderly Households in Arizona and Minnesota." *The Gerontologist* 38 (2):152–158.

Hohm, Charles. 1975. "Social Security and Fertility: An International Perspective." *Demography* 12 (4): 629–644.

Holdren, John. 1990. "Energy in Transition." *Scientific American* 263 (3):156–63.

Hollerbach, Paula. 1980. "Recent Trends in Fertility, Abortion, and Contraception in Cuba." *International Family Planning Perspectives* 6 (3):97–106.

Hollingsworth, Thomas Henry. 1969. *Historical Demography.* Ithaca: Cornell University Press.

Hollman, Frederick W., Tammany J. Mulder, and Jeffrey E. Kallan. 2000. "Methodology and Assumptions for the Population Projections of the United States: 1999 to 2100." U.S. Bureau of the Census Population Division Working Paper No. 38, Washington.

Horsley, G. H. R. 1987. *New Documents Illustrating Early Christianity: A Review of the Greek Inscriptions and Papyri Published in 1979.* Sydney, Australia: Macquarie University.

Hout, Michael. 1984. "Occupational Mobility of Black Men: 1962 to 1973." *American Sociological Review* 49 (3):308–22.

Howell, Nancy. 1979. *Demography of the Dobe Kung.* New York: Academic Press.

Hsu, Mei-Ling. 1994. "The Expansion of the Chinese Urban System, 1953–1990." *Urban Geography* 15 (6):514–36.

Hu, Y. and N. Goldman. 1990. "Mortality Differentials by Marital Status: An International Comparison." *Demography* 27:233–50.

Hughes, H. L. 1993. "Metropolitan Structure and the Suburban Hierarchy." *American Sociological Review* 58:417–33.

Hugo, Graeme. 1999. *Gender and Migrations in Asian Countries.* Liège: International Union for the Scientific Study of Populations.

Hume, David. 1752 (1963). "Of the Populousness of Ancient Nations." In *Essays: Moral, Political and Literary,* edited by D. Hume. London: Oxford.

Hummer, Robert A., Richard G. Rogers, Charles B. Nam, and Felicia B. LeClere. 1999. "Race/Ethnicity, Nativity, and U.S. Adult Mortality." *Social Science Quarterly* 80 (1):136–153.

Hummert, M. L., T. A. Garstka, J. L. Shaner, and S. Strahm. 1995. "Judgments About Stereotypes of the Elderly." *Research on Aging* 17 (2):168–89.

Hunt, Michael E. 1984. *Retirement Communities: An American Original.* New York: Haworth Press.

Huntington, Samuel P. 1996. *The Clash of Civilizations and the Remaking of the World Order.* New York: Simon & Schuster.

Hutchinson, E. P. 1967. *The Population Debate: The Development of Conflicting Theories up to 1900.* Boston: Houghton Mifflin.

Huttman, Elizabeth, Wim Blauw, and Juliet Saltman, Ed. 1991. *Urban Housing Segregation of Minorities in Western Europe and the United States.* Durham, NC: Duke University Press.

Hutzler, Charles and Susan V. Lawrence. 2003. "China Acts to Lower Obstacles to Urban Migration." *The Wall Street Journal.* 22 January. P. A2.

Huzel, J. 1969. "Malthus, the Poor Law, and Population in Early Nineteenth-Century England." *Economic History Review* 22:430–52.

———. 1980. "The Demographic Impact of the Old Poor Law: More Reflexions on Malthus." *Economic History Review* 33:367–81.

———. 1984. "Parson Malthus and the Pelican Inn Protocol: A Reply to Professor Levine." *Historical Method* 17:25–27.

Iacocca, Lee (with W. Novak). 1984. *Iacocca: An Autobiography.* New York: Bantam Books.

Ibrahim, Saad Eddin and Barbara Lethem Ibrahim. 1998. "Egypt's Population Policy: The Long March of State and Civil Society." In *Do Population Policies Matter? Fertility and Politics in Egypt, India, Kenya, and Mexico,* edited by A. Jain. New York: Population Council.

India Registrar General and Census Commissioner. 2001. "Census of India." *http://www.censusindia.net/,* accessed 2001.

INEGI. 2001a. "Caracteristicas sociodemograficas por tamano de localidad." *http://www.archive.org,* accessed 2001.

———. 2001b. "Censo de Poblacion 2000: Resultados Definitivos; Porcentaje de Poblacion de 65 y mas anos, por Entidad Federativa." *http://www.inegi.gob.mx,* accessed 2004.

———. 2003. "Porcentagje de defunciones generales por sexo y causa, 1990–2001." INEGI: *www.inegi.gob.mx/est/contenidos/espanol/tematicos/mediano/med.asp?t=mpob45&c=3222,* accessed 2004.

Inglehart, Ronald and Wayne E. Baker. 2000. "Modernization, Cultural Change, and the Persistence of Traditional Values." *American Sociological Review* 65:19–51.

International Labour Office. 2003. "Stop Child Trafficking!" *World of Work* 47 (June):4–6.

International Labour Organization. 2003. "Facts on Women at Work." ILO: *http://www.ilo.org/dyn/gender/,* accessed 2004.

Ireland Central Statistics Office. 1978. *Statistical Abstract of Ireland, 1976.* Dublin: Stationary Office.

Isaacs, S. and R. Cook. 1984. *Laws and Policies Affecting Fertility: A Decade of Change.* Baltimore: Johns Hopkins University.

Islam, M. Mazharul and M. Mosleh Uddin. 2001. "Female Circumcision in Sudan: Future Prospects and Strategies for Eradication." *International Family Planning Perspectives* 27 (2):71–76.

Israel Central Bureau of Statistics. 2001. "Statistical Abstract of Israel 2000." *http://www.archive.org,* accessed 2002.

Issawi, C. 1987. *An Arab Philosophy of History: Selections from the Prolegomena of Ibn Khaldun of Tunis (1332–1406).* Princeton: Princeton University Press.

Iversen, Roberta Rehner, Frank F. Furstenberg, Jr., and Alisa A. Belzer. 1999. "How Much Do We Count? Interpretation and Error-Making in the Decennial Census." *Demography* 36 (1):121–134.

Jacobsen, Joyce P. and Laurence M. Levin. 1997. "Marriage and Migration: Comparing Gains and Losses from MIgration for Couples and Singles." *Social Science Quarterly* 78 (3):688–699.

Jaffe, F. 1971. "Toward the Reduction of Unwanted Pregnancy." *Science* 174:119–27.

Jain, Anrudh. 1998. "Population Policies That Matter." In *Do Population Policies Matter? Fertility and Politics in Egypt, India, Kenya, and Mexico,* edited by A. Jain. New York: Population Council.

Jain, Anrudh and Arjun Adlakha. 1982. "Preliminary Estimates of Fertility Decline in India During the 1970s." *Population and Development Review* 8 (3):589–606.

James, Patricia. 1979. *Population Malthus: His Life and Times.* London: Routledge & Kegan Paul.

Jasso, Guillermina, Douglas S. Massey, Mark R. Rosenzweig, and James P. Smith. 2000. "The New Immigrant Survey Pilot (NIS-P): Overview and New Findings About U.S. Legal Immigrants at Admission." *Demography* 37 (1):127–138.

Jejeebhoy, Shireen J. 1998. "Associations Between Wife-Beating and Fetal and Infant Death: Impressions from a Survey in Rural India." *Studies in Family Planning* 29 (3):300–308.

Jenkins, J. Craig and Stephen Scanlan. 2001. "Food Security in Less Developed Countries, 1970 to 1990." *American Sociological Review* 66 (October):718–744.

Jeune, Bernard and Axel Skytthe. 2001. "Centenarians in Denmark in the Past and the Present." *Population: An English Selection* 13 (1):75–94.

Johnson, Allen W. and Timothy Earle. 1987. *The Evolution of Human Societies: From Foraging Group to Agrarian Societies.* Stanford: Stanford University Press.

Johnson, Ian. 1997. "China's New Containment Policy: Fighting the Rise of Megacities." *The Wall Street Journal* 11 December. P. A20.

Johnson, Kay. 1996. "The Politics of the Revival of Infant Abandonment in China, with Special Reference to Hunan." *Population and Development Review* 22 77–98.

Johnson, Norman J., Paul D. Sorlie, and Eric Backlund. 1999. "The Impact of Specific Occupation on Mortality in the U.S. National Longitudinal Mortality Study." *Demography* 36 (3):355–367.

Jones, Colin. 2002. *The Great Nation: France From Louis XV to Napolean 1715–99.* New York: Columbia University Press.

Jones, R. 1988. "A Behavioral Model for Breastfeeding Effects on Return to Menses Postpartum in Javanese Women." *Social Biology* 35:307–23.

Jordan, Miriam. 1996. "Marketing Gurus Say: In India, Think Cheap, Lose the Cold Cereal." *The Wall Street Journal,* 11 October, P. B5.

——. 1999. "In India's Toothpaste War, an Oddly Familiar Smile." *The Wall Street Journal.* 15 April. P. B1.

Judis, John B. and Ruy Teixeira. 2002. *The Emerging Democratic Majority.* New York: Scribner.

Kahn, J. and W. Mason. 1987. "Political Alienation, Cohort Size, and the Easterlin Hypothesis." *American Sociological Review* 52:155–69.

Kalbach, Warren E. and Wayne W. McVey. 1979. *The Demographic Bases of Canadian Society, Second Edition.* Toronto: McGraw-Hill Ryerson.

Kalleberg, Arne L., Barbara F. Reskin, and Ken Hudson. 2000. "Bad Jobs in America: Standard and Nonstandard Employment Relations and Job Quality in the United States." *American Sociological Review* 65: 256–278.

Kalmijn, Matthijs. 1991. "Shifting Boundaries: Trends in Religious and Educational Homogamy." *American Sociological Review* 56:786–800.

Kanaiaupuni, Shawn Malia and Katherine M. Donato. 1999. "Migradollars and Mortality: The Effects of Migration on Infant Survival in Mexico." *Demography* 36 (3):339–353.

Kannisto, V. 2001. "Mode and Dispersion of the Length of Life." *Population: An English Selection* 13 (1):159–172.

Kaplan, C. and T. Van Valey. 1980. *Census 80: Continuing the Factfinder Tradition.* Washington: U.S. Bureau of the Census.

Kaplan, David. 1994. "Population and Politics in a Plural Society: The Changing Geography of Canada's Linguistic Groups." *Annals of the Association of American Geographers* 84 (1):46–67.

Kaplan, Robert D. 2000. *The Coming Anarchy: Shattering the Dreams of the Post Cold War.* New York: Vintage Books.

Kapolowitz, P. B. and S. E. Oberfield. 1999. "Reexamination of the Age Limit for Defining When Puberty is Precocious in Girls in the United States: Implications for Evaluation and Treatment." *Pediatrics* 104 (4):936–41.

Kasarda, John. 1995. "Industrial Restructuring and the Changing Location of Jobs." In *State of the Union: America in the 1990s, Volume One: Economic Trends,* edited by R. Farley. New York: Russell Sage Foundation.

Kazemi, F. 1980. *Poverty and Revolution in Iran.* New York: New York University Press.

Keely, Charles. 1971. "Effects of the Immigration Act of 1965 on Selected Population Characteristics of Immigrants to the United States." *Demography* 8:157–70.

Keister, Lisa A. 2003. "Sharing the Wealth: The Effect of Siblings on Adults' Wealth Ownership." *Demography* 40 (3):521–542.

Keith, V. and D. Smith. 1988. "The Current Differential in Black and White Life Expectancy." *Demography* 25:625–32.

Kemper, Robert Van. 1977. *Migration and Adaptation.* Beverly Hills: Sage.

——. 1991. "Migracion Sin Fronteras: El Caso del Pueblo de Tzintzuntzan, Michoacan, 1945–1990." In *Migración y Fronteras,* edited by S. M. d. Antropología. Mexico, DF: Mesa Redonda.

——. 1996. "Migration and Adaptation: Tzintzuntzenos in Mexico City and Beyond." In *Urban Life: Readings in Urban Anthropology, 3rd Edition,* edited by G. Gmelch and W. P. Zenner. Prospect Heights, IL: Waveland Press.

Kemper, Robert Van and George Foster. 1975. "Urbanization in Mexico: The View From Tzintzuntzan." *Latin American Urban Research* 5:53–75.

Kennedy, K., R. Rivera, and A. S. McNeilly. 1989. "Consensus Statement on the Use of Breastfeeding as a Family Planning Method." *Contraception* 39:477–96.

Kephart, W. 1982. *Extraordinary Groups: The Sociology of Unconventional Life-Styles, 2nd Edition.* New York: St. Martin's Press.

Kertzer, David I. 1993. *Sacrificed for Honor: Italian Infant Abandonment and the Politics of Reproductive Control.* Boston: Beacon Press.

Kertzer, David I. 1995. "Toward a Historical Demography of Aging." In *Aging in the Past: Demography, Society, and Old Age,* edited by D. I. Kertzer and P. Laslett. Berkeley: University of California Press.

Kertzer, David I. and Dennis P. Hogan. 1989. *Family, Political Economy, and Demographic Change: The Transformation of Life in Casalecchio, Italy, 1861–1921.* Madison: University of Wisconsin Press.

Keyfitz, Nathan. 1966. "How Many People Have Ever Lived on the Earth?" *Demography* 3 (2):581–82.

——. 1968. *Introduction to the Mathematics of Population.* Reading, MA: Addison-Wesley.

——. 1971. "On the Momentum of Population Growth." *Demography* 8 (1):71–80.

——. 1972. "Population Theory and Doctrine: A Historical Survey." In *Readings in Population,* edited by W. Petersen. New York: MacMillan.

——. 1973. "Population Theory." Chapter III in *The Determinants and Consequences of Population Trends, Volume I,* edited by United Nations. New York: United Nations.

——. 1982. "Can Knowledge Improve Forecasts?" *Population and Development Review* 8 (4):729–51.

——. 1985. *Applied Mathematical Demography, 2nd Edition.* New York: Springer-Verlag.

Khan, M. and C. Prasad. 1985. "A Comparison of 1970 and 1980 Survey Findings on Family Planning in India." *Studies in Family Planning* 16 (6):312–20.

Khawaja, Marwan. 2000. "The Recent Rise in Palestinian Fertility: Permanent or Transient?" *Population Studies* 54 (3):331–346.

Kilman, Scott. 2001. "Food Industry Shuns Bioengineered Sugar." *The Wall Street Journal,* 27 April, P. A7.

Kim, Y. J. and Robert Schoen. 1997. "Population Momentum Expresses Population Aging." *Demography* 34 (3):421–28.

Kinsella, Kevin. 1995. "Aging and the Family: Present and Future Demographic Issues." In *Handbook of Aging and the Family,* edited by R. Blieszner and V. Hilkevitch Bedford. Westport, CT: Greenwood Press.

Kinstler, S. and Louis G. Pol. 1990. "The Marketing Implications of Migration on Hospital and Physician Supply and Demand in the United States." *Journal of Hospital Marketing* 4 (1):119–41.

Kishor, S. 1993. ""May God Give Sons to All": Gender and Child Mortality in India." *American Sociological Review* 58:247–65.

Kisker, E. and N. Goldman. 1987. "Perils of Single Life and Benefits of Marriage." *Social Biology* 34 (3–4):135–52.

Kitagawa, E. and P. Hauser. 1973. *Differential Mortality in the United States: A Study in Socioeconomic Epidemiology.* Cambridge: Harvard University Press.

Klatsky, A. L. and G. D. Friedman. 1995. "Annotation: Alcohol and Longevity." *American Journal of Public Health* 85 (1):16–18.

Kline, Jennie, Zena Stein, and Mervyn Susser. 1989. *Conception to Birth: Epidemiology of Prenatal Development,* Edited by B. MacMahon. New York: Oxford University Press.

Knodel, John. 1970. "Two and a Half Centuries of Demographic History in a Bavarian Village." *Population Studies* 24:353–69.

Knodel, John and Etienne van de Walle. 1979. "Lessons From the Past: Policy Implications of Historical Fertility Studies." *Population and Development Review* 5:217–45.

Knorr-Held, L. and E. Rainer. 2001. "Projections of Lung Cancer Mortality in West Germany: A Case Study in Bayesian Prediction." *Biostatistics* 2:109–129.

Knowlton, Charles. 1832 (1981). "Fruits of Philosophy: The Private Companion of Young Married People." In *"A Dirty, Filthy Book",* reprinted in S. Chandrasekhar. Berkeley and Los Angeles: University of California Press.

Knutson, Mark. 2000. "Charles K. Ponzi Website." *http://www.mark-knutson.com,* accessed 2000.

Komlos, J. 1989. "The Age at Menarch in Vienna: The Relationship Between Nutrition and Fertility." *Historical Methods* 22:158–63.

Konner, M. and C. Worthman. 1980. "Nursing Frequency, Gonadal Function, and Birth Spacing Among Kung Hunter-Gatherers." *Science* 207:788–91.

Koonin, Lisa M., Lilo T. Strauss, Camaryn E. Chrisman, and Wilda Y. Parker. 2000. "Abortion Surveillance—United States, 1997." *Mortality and Morbidity Weekly Report* 49 (SS11):1–44.

Korenromp, Eline L., Brian G. Williams, Eleanor Gouwa, Christopher Dye, and Robert W. Snow. 2003. "Measurement of Trends in Childhood Malaria Mortality in Africa: An Assessment of Progress Toward Targets Based on Verbal Autopsy." *The Lancet Infectious Diseases* 3 (6):349–358.

Krach, Constance A. and Victoria A. Velkoff. 1999. "Centenarians in the United States: 1990." *Current Population Reports* (23–199RV).

Kralt, John. 1990. "Ethnic Origins in the Canadian Census, 1871–1986." In *Ethnic Demography: Canadian Immigrant, Racial and Cultural Variations,* edited by S. S. Halli, F. Trovato, and L. Driedger. Ottawa: Carleton University Press.

Kraly, Ellen Percy and Robert Warren. 1992. "Estimates of Long-Term Immigration to the United States: Moving US Statistics Toward United Nations Concepts." *Demography* 29 (4):613–26.

Kranhold, Kathryn. 1999. "Food Woos 'Echo Boomers' With Live TV." *The Wall Street Journal.* 13 August. P. B8.

Kreager, Philip. 1986. "Demographic Regimes as Cultural Systems." In *The State of Population Theory: Forward From Malthus,* edited by D. C. a. R. Schofield. Oxford: Basil Blackwell.

——. 1993. "Histories of Demography: A Review Article." *Population Studies* 47 (3):519–39.

Kreider, Rose M. and Jason M. Fields. 2002. "Number, Timing, and Duration of Marriages and Divorces: 1996." *Current Population Reports* (70–80).

Krissman, Fred. 2000. "Immigrant Labor Recruitment: U.S. Agribusiness and Undocumented Migration from Mexico." In *Immigration Research for a New Century: Multidisciplinary Perspectives,* edited by N. Foner, R. G. Rumbaut, and S. J. Gold. New York: Russell Sage Foundation.

Kritz, M. and J. M. Nogle. 1994. "Nativity Concentration and Internal Migration Among the Foreign-Born." *Demography* 31 (3):509–24.

Kuijsten, Anton. 1999. "Households, Families and Kin Networks." In *Population Issues: An Interdisciplinary Focus,* edited by L. J. G. van Wissen and P. A. Dykstra. New York: Kluwer Academic/Plenum.

Kunitz, Stephen J. 1994. *Disease and Social Diversity: The European Impact on the Health of Non-Europeans.* New York: Oxford University Press.

Kunstler, J. H. 1993. *The Geography of Nowhere: The Rise and Decline of America's Man-Made Landscape.* New York: Simon & Schuster.

Landers, John. 1993. *Death and the Metropolis: Studies in the Demographic History of London 1670–1830.* Cambridge: Cambridge University Press.

Langford, C. and P. Storey. 1993. "Sex Differentials in Mortality Early in the Twentieth Century: Sri Lanka and India Compared." *Population and Development Review* 19 (2):263–82.

Larsen, Ulla. 1995. "Trends in Infertility in Cameroon and Nigeria." *International Family Planning Perspectives* 21 (4):138–142.

Larsen, Ulla and Sharon Yan. 2000. "Does Female Circumcision Affect Infertility and Fertility? A Study of the Central African Republic, Cote d'Ivoire, and Tanzania." *Demography* 37 (3):313–321.

Laslett, Peter. 1971. *The World We Have Lost.* London: Routledge & Kegan Paul.

——. 1991. *A Fresh Map of Life: The Emergence of the Third Age.* Cambridge: Harvard University Press.

——. 1995. "Necessary Knowledge: Age and Aging in the Societies of the Past." In *Aging in the Past: Demography, Society, and Old Age,* edited by D. Kertzer and P. Laslett. Berkeley: University of California Press.

Lauderdale, Diane S. 2001. "Education and Survival: Birth Cohort, Period, and Age Effects." *Demography* 38 (4):551–561.

Laws, G. 1994. "Implications of Demographic Changes for Urban Policy and Planning." *Urban Geography* 15 (1):90–100.

Laxton, Paul. 1987. *London Bills of Mortality.* Cambridge, UK: Chadwyk-Healey.

Leasure, J. William. 1962. "Factors Involved in the Decline of Fertility in Spain: 1900–1950." Doctoral Thesis, Economics, Princeton University.

——. 1982. "L' Baisse de la Fecondité aux États-Unis de 1800 a 1860." *Population* 3 607–22.

——. 1989. "A Hypothesis About the Decline of Fertility: Evidence From the United States." *European Journal of Population* 5:105–17.

Lee, Everett. 1966. "A Theory of Migration." *Demography* 3:47–57.

Lee, Richard B. 1972. "Population Growth and the Beginnings of Sedentary Life Among the Kung Bushmen." In *Population Growth: Anthropological Implications,* edited by B. Spooner. Cambridge: MIT Press.

Lee, Ronald. 1976. "Demographic Forecasting and the Easterlin Hypothesis." *Population and Development Review* 2:459–68.

——. 1987. "Population Dynamics in Humans and Other Animals." *Demography* 24:443–65.

——. 2000. "Long-Term Population Projections and the U.S. Social Security System." *Population and Development Review* 26 (1):137–144.

Lee, Ronald, Andrew Mason, and Tim Miller. 2001. "Saving, Wealth, and Population." In *Population Matters: Demographic Change, Economic Growth, and Poverty in the Developing World,* edited by N. Birdsall, A. C. Kelley, and S. W. Sinding. New York: Oxford University Press.

Légaré, Jacques and Laurent Martel. 2003. "Living Arrangements of Older Persons in the Early Nineties: An International Comparison." *Genus* LIX (1): 85–103.

Leibenstein, Harvey. 1957. *Economic Backwardness and Economic Growth.* New York: John Wiley.

Lengyel-Cook, M. and R. Repetto. 1982. "The Relevance of the Developing Countries to Demographic Transition Theory: Further Lessons from the Hungarian Experience." *Population Studies* 36 (1):105–28.

Leonard, Bill. 2000. "Introduction to Redistricting." California State Assembly Republican Caucus: *http://www.archive.org,* accessed 2001.

Leroux, C. 1984. "Ellis Island." *San Diego Union-Tribune,* 26 May, P. D1.

Lesthaeghe, Ron J. 1977. *The Decline of Belgian Fertility, 1800–1970.* Princeton: Princeton University Press.

——. 1998. "On Theory Development: Applications to the Study of Family Formation." *Population and Development Review* 24 (1):1–14.

Lesthaeghe, Ron J. and J. Surkyn. 1988. "Cultural Dynamics and Economic Theories of Fertility Change." *Population and Development Review* 14:1–45.

Lesthaeghe, Ron and Paul Willems. 1999. "Is Low Fertility a Temporary Phenomenon in the European Union?" *Population and Development Review* 25 (2):211–228.

Leung, J. 1983. "Singapore's Lee Tells Educated Women: Get Married and Have Families." *The Wall Street Journal,* 26 August, P. A13.

Levitt, Peggy, Josh DeWind, and Steven Vertovec. 2003. "International Perspectives on Transnational Migration: An Introduction." *International Migration Review* 37 (3):565–575.

Levy, S., G. R. Frerichs, and H. L. Gordon, Ed. 1994. *The Dartnell Marketing Manager's Handbook.* Chicago: Dartnell.

Lewis, Bernard. 1995. *The Middle East: A Brief History of the Last 2,000 Years.* New York: Scribner.

Lexis, W. 1875. *Einleitung ein die Theorie der Bevölkerungs-Statistik.* Strasbourg: Trubner.

Li, Nan and Shripad Tuljapurkar. 1999. "Population Momentum for Gradual Demographic Transitions." *Population Studies* 53:255–262.

Liesman, Steve. 1999. "Driven to Distraction, Saudi Women May Soon Take to the Wheel." *The Wall Street Journal.* New York. 1 March. P. A1.

Lightbourne, R., S. Singh, and C. Green. 1982. "The World Fertility Survey." *Population Bulletin* 37 (1).

Lillard, Lee A. and Constantijn W. A. Panis. 1996. "Marital Status and Mortality: The Role of Health." *Demography* 33 (3):313–327.

Lilly, John. 2001. "Bookshelf: A Week in Hell; Review of *Can't Take My Eyes Off of You* by Jack Lechner." *The Wall Street Journal*. 3 January, P. A15.

Linder, Forrest. 1959. "World Demographic Data." In *The Study of Population: An Inventory and Appraisal,* edited by P. M. Hauser and O. D. Duncan. Chicago: University of Chicago Press.

Lindh, Thomas and Bo Malmberg. 1999. "Age Structure Effects and Growth in the OECD, 1950–1990." *Journal of Population Economics* 12:431–449.

Lingappa J. R., L. C. McDonald, P. Simone, U. D. Parashar. 2004. "Wresting SARS from Uncertainty." *Emerging Infectious Diseases* 10 (2):*http://www.cdc.gov/ncidod/EID/vol1ch02/03-1032.htm*

Livi-Bacci, Massimo. 1991. *Population and Nutrition: An Essay on European Demographic History*. Cambridge: Cambridge University Press.

Livi-Bacci, Massimo. 2000. *The Population of Europe.* Oxford: Blackwell.

Livi-Bacci, Massimo. 2001. *A Concise History of World Population, Third Edition*. Malden, MA: Blackwell Publishers.

Lloyd, Cynthia B. and Serguey Ivanov. 1988. "The Effects of Improved Child Survival on Family Planning Practice and Fertility." *Studies in Family Planning* 19 (3):141–161.

Loaeza Tovar, Enrique M. and Susan Martin. 1997. *Migration Between Mexico & the United States: A Report of the Binational Study of Migration*. Washington: U.S. Commission on Immigration Reform.

Loftin, Colin and Sally K. Ward. 1983. "A Spatial Autocorrelation Model of the Effects of Population Density on Fertility." *American Sociological Review* 48: 121–128.

Logan, John R. and Richard D. Alba. 1993. "Locational Returns to Human Capital: Minority Access to Suburban Community Resources." *Demography* 30 (2): 243–68.

Logan, John R., Richard D. Alba, Tom NcNulty, and Brian Fisher. 1996. "Making a Place in the Metropolis: Locational Attainment in Cities and Suburbs." *Demography* 33 (4):443–453.

London, Bruce. 1987. "Structural Determinants of Third World Urban Change: An Ecological and Political Economic Analysis." *American Sociological Review* 52:28–43.

Long, Larry. 1988. *Migration and Residential Mobility in the United States*. New York: Russell Sage Foundation.

——. 1991. "Residential Mobility Differences Among Developed Countries." *International Regional Science Review* 14 (2):133–47.

Long, Larry and Alfred Nucci. 1996. "Convergence or Divergence in Urban-Rural Fertility Differences in the United States." In *Paper Presented at the Annual Meeting of the Population Association of America*. New Orleans.

Longley, Paul and Michael Batty. 1996. "Analysis, Modelling, Forecasting, and GIS Technology." In *Spatial Analysis: Modelling in a GIS Environment,* edited by P. Longley and M. Batty. New York: John Wiley.

Lopez, A. D. 1983. "The Sex Mortality Differential in Developed Countries." In *Sex Differentials in Mortality: Trends, Determinants and Consequences,* edited by A. D. Lopez and L. T. Ruzick. Canberra: Australian National University.

Lorimer, Frank. 1959. "The Development of Demography." In *The Study of Population,* edited by P. M. Hauser and O. D. Duncan. Chicago: University of Chicago Press.

Lowenstein, Roger. 2001. "Runaway Gravy Train: Executive Pay is Out of Control, and Shareholders are Tied to the Tracks." *SmartMoney,* July, *http://www.smartmoney.com/mag/ind/cfm?story=magazine-july01*

Lublin, Joann S. 1998. "Even Top Women Earn Less, New Study Finds." *The Wall Street Journal*. 10 November. P. 18.

Lutz, Wolfgang, Brian C. O'Neill, and Sergei Scherbov. 2003. "Europe's Population at a Turning Point." *Science* 299:1991–1992.

Lutz, Wolfgang, Warren Sanderson, and Sergei Scherbov. 2001. "The End of Wordl Population Growth." *Nature* 412:543–545.

Lutz, Wolfgang and Mahendra Shah. 2002. "Population in Sustainable Development." Global Science Panel on Population & Environment: *http://www.iiasa.ac.at/Admin/INF/hague/GSP_final_statement.pdf,* accessed 2003.

Lynch, Scott M. and J. Scott Brown. 2001. "Reconsidering Mortality Compression and Deceleration: An Alternative Model of Mortality Rates." *Demography* 38 (1):79–95.

Mɕller, Valerie. 1997. "AIDS: The 'Grandmother's Disease' of South Africa." *Ageing and Society* 17 (4):461–63.

Ma, Laurence J. C. and Carolyn Cartier, Ed. 2003. *The Chinese Diaspora: Space, Place, Mobility, and Identity.* Lanham, MD: Rowman & Littlefield.

Mabry, J. Beth and Vern L. Bengston. 2001. "Review of 'Ageing and Popular Culture' by Andrew Blaikie." *Contemporary Sociology* 30 (1):22–24.

Mackie, Gerry. 1996. "Ending Footbinding and Infibulation: A Convention Account." *American Sociological Review* 61:999–1017.

Macunovich, Diane J. 2000. "Relative Cohort Size: Source of a Unifying Theory of Global Fertility Transition?" *Population and Development Review* 26 (2): 235–261.

Magretta, Joan. 2002. *What Management Is: How it Works and Why It's Everyone's Business*. New York: Free Press.

Makinwa-Adebusoye, P. 1994. "Report of the Seminar on Women and Demographic Change in Sub-Saharan Africa, Dakar, Senegal, 3-6 March 1993." *IUSSP Newsletter* January–April P. 43–48.

Malthus, Thomas Robert. 1798 (1965). *An Essay on Population.* New York: Augustus Kelley.

——. 1872 (1971). *An Essay on the Principle of Population, Seventh Edition.* London: Reeves & Turner.

Mandelbaum, D. 1974. *Human Fertility in India.* Berkeley: University of California Press.

Manson, Gary A. and Richard E. Groop. 2000. "U.S. Intercounty Migration in the 1990s: People and Income Move Down the Urban Hierarchy." *Professional Geographer* 52 (3):493–504.

Manton, Kenneth G. and Kenneth C. Land. 2000. "Active Life Expectancy Estimates for the U.S. Elderly Population: A Multidimensional Continuous-Mixture Model of Functional Change Applied to Completed Cohorts, 1982–1996." *Demography* 37 (3):253–266.

Manton, Kenneth G., Eric Stallard, and Larry Corder. 1997. "Changes in the Age Dependence of Mortality and Disability: Cohort and Other Determinants." *Demography* 34 (1):135–157.

Mare, Robert D. 1991. "Five Decades of Educational Assortative Mating." *American Sociological Review* 56:15–32.

Marini, M. 1984. "Women's Educational Attainment and Parenthood." *American Sociological Review* 49 (4): 491–511.

Marini, Margaret Mooney and Pi-Ling Fan. 1997. "The Gender Gap in Earnings at Career Entry." *American Sociological Review* 62:588–604.

Markham, John P. and John I. Gilderbloom. 1998. "Housing Quality Among the Elderly: A Decade of Changes." *International Journal of Aging and Human Development* 46 (1):71–90.

Martin, Joyce A., Brady E. Hamilton, Paul D. Sutton, Stephanie J. Ventura, Fay Menacker, and Martha L. Munson. 2003. "Births: Final Data for 2002." *National Vital Statistics Reports* 52 (10).

Martin, Joyce A., Brady E. Hamilton, Stephanie J. Ventura, Fay Menacker, and Melissa M. Park. 2002. "Births: Final Data for 2000." *National Vital Statistics Reports* 50 (5).

Martin, Philip L. 1993. "Comparative Migration Policies." *International Migration Review* 28 (1):164–70.

Martine, George. 1996. "Brazil's Fertility Decline, 1965–95: A Fresh Look at Key Factors." *Population and Development Review* 22 (1):47–75.

Marx, Karl. 1890 (1906). *Capital: A Critique of Political Economy, translated from the Third German Edition by Samuel Moore and Edward Aveling and Edited by Frederich Engels.* New York: Modern Library.

Masnick, George. 1981. "The Continuity of Birth-Expectation Data With Historical Trends in Cohort Parity Distributions: Implications for Fertility in the 1980s."

In *Predicting Fertility,* edited by G. Henderson and P. Placek. Lexington, MA: Lexington Books.

Mason, Andrew. 1988. "Saving, Economic Growth, and Demographic Change." *Population and Development Review* 14 (1):113–44.

Mason, Karen Oppenheim. 1997. "Explaining Fertility Transitions." *Demography* 34 (4):443–454.

Massey, D., R. Alarcon, J. Durand, and H. Gonzalez. 1987. *Return to Atzlan: The Social Processes of International Migration from Western Mexico.* Berkeley: University of California Press.

Massey, Douglas. 1996a. "The Age of Extremes: Concentrated Affluence and Poverty in the Twenty-First Century." *Demography* 33 (4):395–412.

——. 1996b. "The False Legacy of the 1965 Immigration Act." *World on the Move: Newsletter of the Section on International Migration of the American Sociological Association* 2 (2):2–3.

——. 2002. "A Brief History of Human Society: The Origin and Role of Emotion in Social Life." *American Sociological Review* 67:1–29.

Massey, Douglas, J. Arango, G. Hugo, A. Kouaouci, A. Pellegrino, and J. E. Taylor. 1993. "Theories of International Migration: A Review and Appraisal." *Population and Development Review* 19 (3):431–66.

——. 1994. "An Evaluation of International Migration Theory: The North American Case." *Population and Development Review* 20 (4):699–752.

Massey, Douglas and Nancy A. Denton. 1993. *American Apartheid: Segregation and the Making of the Underclass.* Cambridge: Harvard University Press.

Massey, Douglas, Jorge Durand, and Nolan J. Malone. 2002. *Beyond Smoke and Mirrors: Mexican Immigration in an Era of Economic Integration.* New York: Russell Sage Foundation.

Massey, Douglas and Kristin Espinosa. 1997. "What's Driving Mexico-U.S. Migration? A Theoretical, Empirical, and Policy Analysis." *American Journal of Sociology* 102 (4):939–99.

Mathews, P., M. McCarthy, M. Young, and N. McWhirter. 1995. *The Guinness Book of World Records 1995.* New York: Bantam Books.

Matthews, William. 1980. "Inheritance Tax." *Vegetable Box* 4 (1):36.

Maume, D. 1983. "Metropolitan Hierarchy and Income Distribution: A Comparison of Explanations." *Urban Affairs Quarterly* 18:413–429.

Maxwell, N. 1977. "Medical Secrets of the Amazon." *America* 29:2–8.

Mayer, Peter. 1999. "India's Falling Sex Ratios." *Population and Development Review* 25 (2):323–344.

Mayhew, B. and R. Levinger. 1976. "Size and the Density of Interaction in Human Aggregates." *American Journal of Sociology* 82 (1):86–109.

McCaa, Robert. 1994. "Child Marriage and Complex Families Among the Nahuas of Ancient Mexico." *Latin American Population History* (26):2–11.

McCall, Leslie. 2000. "Gender and the New Inequality: Explaining the College/Non-College Wage Gap." *American Sociological Review* 65:234–255.

McDaniel, Anthony. 1992. "Extreme Mortality in Nineteenth Century Africa: The Case of Liberian Immigrants." *Demography* 29 (4):581–94.

McDaniel, Antonio. 1995. *Swing Low, Sweet Chariot: The Mortality Cost of Colonizing Liberia in the Nineteenth Century.* Chicago: University of Chicago Press.

———. 1996. "Fertility and Racial Stratification." In *Fertility in the United States: New Patterns, New Theories,* edited by J. B. Casterline, R. D. Lee, and K. Foote. New York: Population Council.

McDonald, P. 1993. "Fertility Transition Hypothesis." In *The Revolution in Asian Fertility: Dimensions, Causes, and Implications,* edited by R. Leete and I. Alam. Oxford: Clarendon Press.

McDonald, Peter. 2000. "Gender Equity in Theories of Fertility Transition." *Population and Development Review* 26 (3):427–439.

McDonough, Peggy, Greg J. Duncan, David Williams, and James House. 1997. "Income Dynamics and Adult Mortality in the United States, 1972 through 1989." *American Journal of Public Health* 87 (9):1476–83.

McEvedy, Colin and Richard Jones. 1978. *Atlas of World Population History.* New York: Penguin Books.

McFalls, J. and M. McFall. 1984. *Disease and Fertility.* New York: Academic Press.

McFarlan, D., N. McWhirter, M. McCarthey, and M. Young. 1991. *The Guinness Book of World Records.* New York: Bantam Books.

McGinnis, J. M. and W. H. Foege. 1993. "Actual Causes of Death in the United States." *Journal of the American Medical Association* 270 (18):2207–212.

McKeown, Thomas. 1976. *The Modern Rise of Population.* London: Edward Arnold.

McKeown, T. 1988. *The Origins of Human Disease.* Oxford: Basil Blackwell.

McKeown, T. and R. Record. 1962. "Reasons for the Decline of Mortality in England and Wales during the 19th Century." *Population Studies* 16 (2):94–122.

McLanahan, Sara and L. Casper. 1995. "Growing Diversity and Inequality in the American Family." In *State of the Union: America in the 1990s, Volume Two: Social Trends,* edited by R. Farley. New York: Russell Sage Foundation.

McLanahan, Sara and Gary Sandefur. 1994. *Growing Up With a Single Parent: What Hurts, What Helps.* Cambridge: Harvard University Press.

McNeill, William H. 1976. *Plagues and People.* New York: Doubleday.

McNicoll, Geoffrey. 1999. "Population Weights in the International Order." *Population and Development Review* 25 (3):411–442.

McWhirter, N. 1983. *The Guinness Book of World Records.* New York: Bantam Books.

Meadows, D. 1974. *Dynamics of Growth in a Finite World.* Cambridge, MA: Wright-Allen Press.

Meadows, D. H., D. L. Meadows, J. Randers, and W. Bherens III. 1972. *The Limits to Growth.* New York: American Library.

Meek, R. 1971. *Marx and Engels on the Population Bomb.* Berkeley, CA: Ramparts Press.

Menchik, P. 1993. "Economic Status as a Determinant of Mortality Among Black and White Older Men: Does Poverty Kill?" *Population Studies* 47:427–36.

Mendoza, Cristobal. 1999. "African Employment in Iberian Agriculture and Construction: A Cross-Border Comparative Approach." King's College Department of Geography Occasional Paper No. 47, London.

Mermelstein, R., B. Miller, T. Prhaska, V. Benson, and J. F. Van Nostrand. 1993. "Measures of Health." In *Health Data on Older Americans: United States, 1992, Vital and Health Statistics, Series 3: Analytic and Epidemiological Studies, No. 27,* edited by J. F. Van Nostrand, S. Furner, and R. Suzman. Hyattsville, MD: Public Health Service.

Merrick, Thomas and S. Tordella. 1988. "Demographics: People and Markets." *Population Bulletin* 43 (1).

Mesquida, Christian G. and Neil I. Wiener. 1999. "Male Age Composition and Severity of Conflicts." *Politics and the Life Sciences* 18 (2):181–189.

Messersmith, Lisa J., Thomas T. Kane, Adetanwa I. Odebiyi, and Alfred A. Adewuyi. 2000. "Who's at Risk? Men's STD Experience and Condom Use in Southwest Nigeria." *Studies in Family Planning* 31 (3):203–216.

Metchnikoff, E. 1908. *Prolongation of Life.* New York: Putnam.

Meyers, R. 1978. "An Investigation of an Alleged Centenarian." *Demography* 15 (2):235–37.

Mhloyi, Marvelous. 1994. *Status of Women, Population and Development.* Liège: International Union for the Scientific Study of Population.

Middleton, Nick. 1999. *The Global Casino: An Introduction to Environmental Issues, 2nd Edition.* London: Arnold.

Mier y Terán, Martha. 1991. "El Gran Cambio Demográfico." *Demos* 4:4–5.

Migration News. 1998. "Canada: Immigration, Diversity Up." *Migration News* 5 (1).

———. 2000. "Malaysia, Indonesia." *Migration News* 7 (11).

———. 2001a. "Scandinavia." *Migration News* 8 (1).

———. 2001b. "South Asia." *Migration News* 8 (1).

Mill, John Stuart. 1848 (1929). *Principles of Political Economy.* London: Longmans & Green.

——. 1873 (1924). *Autobiography*. London: Oxford University Press.

Miller, G. Tyler. 2004. *Living in the Environment: Principles, Connections, and Solutions, 13th Edition*. Belmont, CA: Brooks/Cole Thomson Learning.

Millman, Joel. 2000. "U.S. Contestants in November Scrutinize Power of Mexican Expatriates in Their Nation's Politics." *The Wall Street Journal*. 30 June. P. A16.

——. 2003. "Gangs Plague Central America." *The Wall Street Journal*, 11 December, p. A14.

Millman, Joel and Carlta Vitzthum. 2003. "Changing Tide: Europe Becomes New Destination for Latino Workers." *The Wall Street Journal*. 12 September. P. A1.

Minino, A. M. and B. L. Smith. 2001. "Deaths: Preliminary Data for 2000." *National Vital Statistics Reports of the National Center for Health Statistics* 49 (12).

Moch, L. P. 1992. *Moving Europeans: Migration in Western Europe Since 1650*. Bloomington: Indiana University Press.

Mogelonsky, M. 1995. "Revenge of the Fats." *American Demographics* (July):10.

Montgomery, Mark and Barney Cohen, Ed. 1998. *From Death to Birth: Mortality Decline and Reproductive Change*. Washington: National Academy Press.

Montgomery, Mark R., Richard Stren, Barney Cohen, and Holly Reed, Ed. 2003. *Cities Transformed: Demographic Change and Its Implications in the Developing World*. Washington: National Research Council.

Morelos, José B. 1994. "La Mortalidad en México: Hechos y Consensos." In *La Población en el Desarrollo Contemporáneo de México*, edited by F. Alba and G. Cabrera. Mexico City: El Colegio de Mexico.

Morgan, S. Philip. 1996. "Characteristic Features of Modern American Fertility." In *Fertility in the United States: New Patterns, New Theories*, edited by J. B. Casterline, R. D. Lee, and K. A. Foote. New York: Population Council.

——. 2001. "Should Fertility Intentions Inform Fertility Forecasts?" In *The Direction of Fertility in the United States*, edited by U.S. Census Bureau. Washington: Council of Professional Associations on Federal Statistics.

Morrill, Richard, John Cromartie, and Gary Hart. 1999. "Metropolitan, Urban, and Rural Commuting Areas: Toward a Better Depiction of the United States Settlement System." *Urban Geography* 20 (8):727–748.

Morris, M., A. D. Bernhardt, and M. S. Handcock. 1994. "Economic Inequality: New Methods for New Trends." *American Sociological Review* 59:205–19.

Mosher, Steven W. 1983. *Broken Earth: The Rural Chinese*. New York: Free Press.

Mosher, W. and W. Pratt. 1990. "Fecundity and Infertility in the United States, 1965–88." *Advance Data From Vital and Health Statistics of the National Center for Health Statistics, No. 192*.

Mouw, Ted. 2000. "Job Relocation and the Racial Gap in Unemployment in Detroit and Chicago, 1980 to 1990." *American Sociological Review* 65:730–753.

Muhuri, P. K. and S. H. Preston. 1991. "Effects of Family Composition on Mortality Differentials by Sex Among Children in Matlab, Bangladesh." *Population and Development Review* 17 (3):415–34.

Mulder, Tammany J. 2001. "Accuracy of the U.S. Census Bureau National Population Projections and Their Respective Components of Change." In *The Direction of Fertility in the United States*, edited by U.S. Census Bureau. Washington: Council of Professional Associations on Federal Statistics.

Muller, Peter O. 1997. "The Suburban Transformation of the Globalizing American City." *Annals of the American Association of Political and Social Science* 551:44–58.

Mumford, Lewis. 1968. "The City: Focus and Function." In *International Encyclopedia of the Social Sciences*, edited by D. Sills. New York: Macmillan.

Muñoz-Peréz, Francisco and France Prioux. 2000. "Children Born Outside Marriage in France and Their Parents: Recognitions and Legitimations Since 1965." *Population: An English Selection* 12:139–196.

Murdock, Steve H. and David R. Ellis. 1991. *Applied Demography: An Introduction to Basic Concepts, Methods, and Data*. Boulder, CO: Westview Press.

Murphy, Sherry L. 2000. "Deaths: Final Data for 1998." *National Vital Statistics Reports* 48 (11).

Murray, Christopher J. L. and Alan D. Lopez. 1996. *The Global Burden of Disease: A Comprehensive Assessment of Mortality and Disability From Diseases, Injuries, and Risk Factors in 1990 and Projected to 2020*. Cambridge: Harvard University Press.

Mutchler, Jan E., Jeffrey A. Burr, Amy M. Pienta, and Michael P. Massagli. 1997. "Pathways to Labor Force Exit: Work Transitions and Work Instability." *Journal of Gerontology: Social Sciences* 52B:S4–S12.

Myrdal, Gunnar. 1957. *Economic Theory and Underdeveloped Areas*. London: G. Duckworth.

Nag, Moni. 1962. *Factors Affecting Human Fertility in Non-Industrial Societies: A Cross-Cultural Study*. New Haven: Yale University Press.

Nalin, David, Richard Cash, and R. Islam. 1968. "Oral Maintenance Therapy for Cholera in Adults." *Lancet* (2):370–372.

National Center for Health Statistics. 1994. *Health United States 1993*. Hyattsville, MD: Public Health Service.

——. 2002. *Health, United States, 2002*. Hyattsville, MD: National Center for Health Statistics.

——. 2003. *Health, United States, 2003*. Hyattsville, MD: National Center for Health Statistics.

National Committee on the Status of Women. 1975. *Status of Women in India*. New Delhi: Indian Council of Social Science Research.

National Research Council. 1993. *A Census That Mirrors America: Interim Report*. Washington: National Academy Press.

Netanyahu, Benjamin. 1983. "West Bank Time Bomb? A Demographic Fallacy." *The Wall Street Journal*. 18 October. P. 30.

Nevins, Joseph. 2002. *Operation Gatekeeper: The Rise of the "Illegal Alien" and the Making of the U.S.-Mexico Boundary*. New York: Routledge.

Newburger, Eric C. and Andrea E. Curry. 2000. "Educational Attainment in the United States (Update)." *Current Population Reports* (20–536).

Newport, Frank and Lydia Saad. 1992. "Economy Weighs Heavily on American Minds." *Gallup Poll Monthly* (319):30–31.

Newsweek. 1974. "How to Ease the Hunger Pangs." 11 November, p. 15.

Nicholls, W. H. 1970. "Development in Agrarian Economies: The Role of Agricultural Surplus, Population Pressures, and Systems of Land Tenure." In *Subsistence Agriculture and Economic Development*, edited by C. Wharton. Chicago: Aldine.

Nickerson, J. 1975. *Homage to Malthus*. Port Washington, NY: National University Publications.

Nielsen, J. 1978. *Sex in Society: Perspectives on Stratification*. Belmont, CA: Wadsworth.

Nizard, A. and F. Muñoz-Pérez. 1994. "Alcohol, Smoking and Mortality in France since 1950: An Evaluation of the Number of Deaths in 1986 due to Alcohol and Tobacco Consumption." *Population: An English Selection* 6:159–94.

Nonaka, K., T. Miura, and K. Peter. 1994. "Recent Fertility Decline in Dariusleut Hutterites: An Extension of Eaton and Mayer's Hutterite Fertility." *Human Biology* 66:411–420.

Nordberg, O. 1983. "Counting on Wall Street." *American Demographics* 5 (8):22–25.

Notestein, Frank W. 1945. "Population—The Long View." In *Food for the World*, edited by T. W. Schultz. Chicago: University of Chicago Press.

Nu'Man, Fareed. 1992. *The Muslim Population in the United States: A Brief Statement*. Washington: American Muslim Council.

O'Brien, Robert M. 2000. "Age Period Cohort Characteristic Models." *Social Science Research* 29:123–139.

O'Connell, Martin and C. Rogers. 1983. "Assessing Cohort Birth Expectations Data From the Current Population Survey, 1971–1981." *Demography* 29 (3): 369–84.

O'Donnell, James J. 2001. "Augustine: Christianity and Society." University of Pennsylvania: *http://ccat.sas.upenn.edu/jod/augustine.html*, accessed 2001.

Oeppen, J. 1993. "Back Projection and Inverse Projection: Members of a Wider Class of Constrained Projection Models." *Population Studies* 47 (2):245–68.

Office of Immigration Statistics. 2003. *2002 Yearbook of Immigration Statistics*. Washington: U.S. Department of Homeland Security.

Ogawa, Naohira and Robert D. Retherford. 1997. "Shifting Costs of Caring for the Elderly Back to Families in Japan: Will it Work?" *Population and Development Review* 23 (1):41–58.

Ohlin, G. 1976. "Economic Theory Confronts Population Growth." In *Economic Factors in Population Growth*, edited by A. Coale. New York: John Wiley.

Okolo, Abraham. 1999. "The Nigerian Census: Problems and Prospects." *The American Statistician* 53 (4): 321–325.

Olausson, Petra Otterblad, Bengt Haglund, Gunilla Ringback Weitoft, and Cnattingius Sven. 2001. "Teenage Childbearing and Long-Term Socioeconomic Consequences: A Case Study in Sweden." *Family Planning Perspectives* 33 (2):70–74.

Oliver, S. 1991. "Look Homeward." *Forbes* (1 April): 55–56.

Olshansky, S. and B. Ault. 1986. "The Fourth Stage of the Epidemiologic Transition." *The Milbank Quarterly* 64 (3):355–391.

Olshansky, S. Jay and Bruce A. Carnes. 1997. "Ever Since Gompertz." *Demography* 34 (1):1–15.

Olshansky, S. Jay, Leonard Hayflick, and Bruce A. Carnes. 2002. "No Truth to the Fountain of Youth." *Scientific American* 286 (6):92–96.

Omran, Abdel. 1971. "The Epidemiological Transition: A Theory of the Epidemiology of Population Change." *Milbank Memorial Fund Quarterly* 49:509–38.

———. 1977. "Epidemiologic Transition in the United States." *Population Bulletin* 32 (2).

Omran, Abdel R. 1973. "The Mortality Profile." In *Egypt: Population Problems and Prospects*, edited by A. R. Omran. Chapel Hill: Carolina Population Center.

Omran, Abdel R. and Farzaneh Roudi. 1993. "The Middle East Population Puzzle." *Population Bulletin* 48 (1).

Ong, Li Lian. 2003. *The Big Mac Index : Applications of Purchasing Power Parity*. New York: Palgrave Macmillan.

Oppenheimer, Valerie K. 1967. "The Interaction of Demand and Supply and Its Effect on the Female Labour Force in the U.S." *Population Studies* 21 (3):239–59.

———. 1994. "Women's Rising Employment and the Future of the Family in Industrializing Societies." *Population and Development Review* 20 (2):293–342.

O'Rand, Angela M. and John C. Henretta. 1999. *Age and Inequality: Diverse Pathways Through Later Life*. Boulder, CO: Westview Press.

ORC Macro. 2000. "New Directions: DHS Surveys Incorporate Geographic Data." *DHS+ Dimensions* 2 (1):1–2.

Oren, Michael B. 2002. *Six Days of War: June 1967 and the Making of the Modern Middle East.* New York: Oxford University Press.

Organization for Economic Co-operation and Development. 2003. "Ageing Societies and the Looming Pension Crisis." OECD: *http://www.oecd.org/document/59/0,2340,en_2649_37435_2512699_1_1_1_37435,00.html*, accessed 2004.

Orshansky, Mollie. 1969. "How Poverty is Measured." *Monthly Labor Review* 92 (2):37–41.

Ortiz de Montellano, Bernard R. 1975. "Empirical Aztec Medicine." *Science* 188:215–20.

Orzechowski, Shawna and Peter Sepiella. 2003. "Net Worth and Asset Ownership of Households: 1998 and 2000." *Current Population Reports* (70–88).

Osgood, D. Wayne, Janet K. Wilson, Patrick M. O'Malley, Jerald G. Bachman, and Lloyd D. Johnston. 1996. "Routine Activities and Individual Deviant Behavior." *American Sociological Review* 61:635–55.

Overbeek, Johannes. 1980. *Population and Canadian Society.* Toronto: Butterworths.

Palloni, Alberto. 2001. "Increment-Decrement Life Tables." In *Demography: Measuring and Modeling Population Processes,* edited by S. H. Preston, P. Heuveline, and M. Guillot. Oxford: Blackwell.

Palmore, James and Robert Gardner. 1983. *Measuring Mortality, Fertility, and Natural Increase.* Honolulu: East-West Population Institute.

Pampel, Fred C. 1993. "Relative Cohort Size and Fertility: The Socio-Political Context of the Easterlin Effect." *American Sociological Review* 58:496–514.

———. 1996. "Cohort Size and Age-Specific Suicide Rates: A Contingent Relationship." *Demography* 33:341–55.

———. 1998. "National Context, Social Change, and Sex Differences in Suicide Rates." *American Sociological Review* 63:744–758.

Pandit, Kavita and Suzanne Davies Withers. 1999. "Introduction: Geographic Perspectives on Migration and Restructuring." In *Migration and Restructuring in the United States: A Geographic Perspective,* edited by K. Pandit and S. D. Withers. New York: Rowman & Littlefield.

Pappas, G., S. Queen, W. Hadden, and G. Fisher. 1993. "The Increasing Disparity in Mortality Between Socioeconomic Groups in the United States, 1960 and 1986." *The New England Journal of Medicine* 329 (2):103–09.

Park, C. B. and N. Cho. 1995. "Consequences of Son Preference in a Low-Fertility Society: Imbalance of the Sex Ratio at Birth." *Population and Development Review* 21 (1):59–84.

Passel, Jeffrey S., Randy Capps, and Michael Fix. 2004. *Undocumented Immigrants: Facts and Figures.* Washington: Urban Institute Immigration Studies Program.

Pathak, K. B., Griffith Feeney, and Norman Y. Luther. 1998. *Accelerating India's Fertility Decline: The Role of Temporary Contraceptive Methods.* Honolulu: East-West Center Program on Population.

Paulson, Richard J., Robert Boostanfar, Peyman Saadat, and Eliran Mor. 2002. "Pregnancy in the Sixth Decade of Life: Obstetric Outcomes in Women of Advanced Reproductive Age." *Journal of the American Medical Association* 288 (18):2320–2323.

Pebley, Ann, H. Delgado, and E. Brinemann. 1979. "Fertility Desires and Child Mortality Experiences Among Guatemalan Women." *Studies in Family Planning* 11 (4):129–36.

Peng, Peiyun. 1996. "Population and Development in China." In *The Population Situation in China: The Insiders' View,* edited by China Population Association and the State Family Planning Commission of China. Liège, Belgium: International Union for the Scientific Study of Population.

Pennisi, Elizabeth. 2001. "Malaria's Beginnings: On the Heels of Hoes?" *Science* 293 (5529):416–417.

Peter, K. 1987. *The Dynamics of Hutterite Society.* Edmonton, Canada: University of Alberta Press.

Petersen, William. 1975. *Population.* New York: MacMillan.

———. 1979. *Malthus.* Cambridge: Harvard University Press.

Phillips, David. 1974. "The Influence of Suggestion on Suicide." *American Sociological Review* 39:340–54.

———. 1977. "Motor Vehicle Fatalities Increase Just After Publicized Suicide Stories." *Science* 196:1464–65.

———. 1978. "Airplane Accident Fatalities Increase Just After Newspaper Stories About Murder and Suicide." *Science* 201:748–50.

———. 1983. "The Impact of Mass Media Violence on U.S. Homicides." *American Sociological Review* 48 (4):560–68.

———. 1993. "Psychology and Survival." *Lancet* 342: 1142–45.

Phillips, David and D. Smith. 1990. "Postponement of Death Until Symbolically Meaningful Occasions." *Journal of the American Medical Association* 263 (14):1947–51.

Phillips, Kevin. 1969. *The Emerging Republican Majority.* New Rochelle, NY: Arlington House.

Piccinino, Linda J. and William D. Mosher. 1998. "Trends in Contraceptive Use in the United States: 1982–1995." *Family Planning Perspectives* 30 (1): 4–10.

Pimentel, David, X. Huang, A. Cardova, and M. Pimentel. 1997. "Impact of Population Growth on Food Supplies and Environment." *Population and Environment* 19 (1):9–14.

Pimentel, David, R. Terhune, R. Dyson-Hudson, S. Rochereau, R. Samis, E. Smith, D Denman, D. Reifschneider, and M. Shepard. 1976. "Land Degradation: Effects on Food and Energy Resources." *Science* 194:149–55.

Plane, David and Peter Rogerson. 1994. *The Geographical Analysis of Population: With Applications to Planning and Business*. New York: John Wiley.

Plato. 1960 (360 BC). *The Laws*. New York: Dutton.

Poednak, A. P. 1989. *Racial and Ethnic Differences in Disease*. New York: Oxford University Press.

Pol, Louis G. and Richard K. Thomas. 2001. *The Demography of Health and Health Care, Second Edition*. New York: Kluwer Academic/Plenum.

Poleman, Thomas. 1975. "World Food: A Perspective." *Science* 188:510–18.

Pollack, R. A. and Susan Cotts Watkins. 1993. "Cultural and Economic Approaches to Fertility: Proper Marriage or *Mésalliance*." *Population and Development Review* 19 (3):467–96.

Pope, Hugh. 2001. "Today's Saudi Arabia Thirsts for Dollars." *The Wall Street Journal*. 21 June. P. A12.

Popkin, Barry M. 1993. "Nutritional Patterns and Transitions." *Population and Development Review* 19:138–157.

Popkin, Barry M. 2002. "The Dynamics of the Dietary Transition in the Developing World." In *The Nutrition Transition: Diet and Disease in the Developing World*, edited by B. Caballero and B. M. Popkin. San Diego: Academic Press.

Population Action Council. 1985. "India Seeking Two-Child Family Norm by 2000." *Popline* 7 (5):1–2.

Population Action International. 1990. *Cities: Life in the World's 100 Largest Metropolitan Areas*. Washington: Population Action International.

Population Information Program. 1985. "Periodic Abstinence." *Population Reports of the Johns Hopkins University* 1 (3).

Population Reference Bureau. 2003. *2003 World Population Data Sheet*. Washington: Population Reference Bureau.

Portes, Alejandro. 1995. "Children of Immigrants: Segmented Assimilation and its Determinants." In *The Economic Sociology of Immigration: Essays on Networks, Ethnicity and Entrepreneurship*, edited by A. Portes. New York: Russell Sage Foundation.

Portes, Alejandro and Rubén G. Rumbaut. 1990. *Immigrant America: A Portrait*. Berkeley: University of California Press.

——. 2001. *Legacies: The Stories of the Immigrant Second Generation*. Berkeley and Los Angeles: University of California Press.

Postel, Sandra. 1996. "Forging a Sustainable Water Strategy." In *State of the World 1996*, edited by Lester Brown and associates. New York: Norton.

Poston, Dudley. 1986. "Patterns of Contraceptive Use in China." *Studies in Family Planning* 17:217–27.

Poston, Dudley. 1990. "Apportioning the Congress: A Primer." *Population Today* 18 (7/8):6–8.

Poston, Dudley and B. Gu. 1987. "Socioeconomic Development, Family Planning, and Fertility in China." *Demography* 24 (4):531–52.

Poston, Dudley, M. X. Mao, and M-Y. Yu. 1994. "The Global Distribution of the Overseas Chinese Around 1990." *Population and Development Review* 20 (3):631–46.

Potter, L. 1991. "Socioeconomic Determinants of White and Black Males' Life Expectancy Differentials, 1980." *Demography* 28:303–21.

Potter, R. 1992. *Urbanisation in the Third World*. Oxford: Oxford University Press.

Potts, Malcolm and Roger Short. 1999. *Ever Since Adam and Eve: The Evolution of Human Sexuality*. Cambridge: Cambridge University Press.

Prema, K. and M. Ravindranath. 1982. "The Effect of Breastfeeding Supplements on the Return of Fertility." *Studies in Family Planning* 13 (10):293–96.

Preston, Samuel 1970. *Older Male Mortality and Cigarette Smoking*. Berkeley: Institute of International Studies, University of California.

Preston, Samuel. 1978. *The Effects of Infant and Child Mortality on Fertility*. New York: Academic Press.

——. 1986. "Are the Economic Consequences of Population Growth a Sound Basis for Population Policy?" In *World Population and U.S. Policy: The Choices Ahead*, edited by J. Menken. New York: Norton.

Preston, Samuel and Michel Guillot. 1997. "Population Dynamics in an Age of Declining Fertility." *Genus* LIII (3–4):15–31.

Preston, Samuel and M. R. Haines. 1991. *Fatal Years: Child Mortality in Late Nineteenth-Century America*. Princeton: Princeton University Press.

Preston, Samuel, Patrick Heuveline, and Michel Guillot. 2001. *Demography: Measuring and Modeling Population Processes*. Oxford: Blackwell.

Price, Daniel O. 1947. "A Check on Underenumeration in the 1940 Census." *American Sociological Review* 12 (1):44–49.

Proctor, Bernadette D. and Joseph Dalaker. 2003. "Poverty in the United States: 2002." *Current Population Reports* (60–222).

Pumain, Denise. 2004. "An Evolutionary Approach to Settlement Systems." In *New Forms of Urbanization: Beyond the Urban-Rural Dichotomy*, edited by A. G. Champion and G. Hugo. London: Ashgate.

Pumphrey, G. 1940. *The Story of Liverpool's Public Service*. London: Hodden & Stoughton.

Qiao, Xiaochun. 2001. "From Decline of Fertility to Transition of Age Structrue: Ageing and its Policy Implications in China." *Genus* LVII (1):57–81.

Quinn, T. 1987. "The Global Epidemiology of the Acquired Immunodeficiency Syndrome." In *Report of the Surgeon General's Workshop on Children With HIV Infections and Their Families*, edited by B. Silverman and A. Waddell. Washington: U.S. Department of Health and Human Services.

Rajulton, F. 1991. "Migrability: A Diffusion Model of Migration." *Genus* XLVII (1–2):31–48.

Raley, R. Kelly and Larry Bumpass. 2003. "The Topography of the Divorce Plateau." Demographic Research 8-8: *http://www.demographic-research.org*, accessed 2004.

Randall, Sara. 1996. "Whose Reality? Local Perceptions of Fertility Versus Demographic Analysis." *Population Studies* 50 (2):221–234.

Rao, Vijayendra. 1993. "Dowry 'Inflation' in Rural India: A Statistical Investigation." *Population Studies* 47 (2):283–93.

Ravenstein, E. 1889. "The Laws of Migration." *Journal of the Royal Statistical Society* 52:241–301.

Reagan, Patricia B. and Randall J. Olsen. 2000. "You Can Go Home Again: Evidence From Longitudinal Data." *Demography* 37 (3):339–350.

Reed, Holly, John Haaga, and Charles Keely, Ed. 1998. *The Demography of Forced Migration: Summary of a Workshop*. Washington: National Academy Press.

Reed, Holly, Marjorie A. Koblinsky, and W. Henry Mosley, Ed. 2000. *The Consequences of Maternal Morbidity and Maternal Mortality: Report of a Workshop*. Washington: National Academy Press.

Rees, William E. 1996. "Revisiting Carrying Capacity: Area-Based Indicators of Sustainability." *Population and Environment* 17 (3):195–215.

Rees, William E. and Mathis Wackernagel. 1994. "Ecological Footprints and Appropriated Carrying Capacity: Measuring the Natural Capital Requirements of the Human Economy." In *Investing in Natural Capacity: The Ecological Economics Approach to Sustainability*, edited by A. Jansson, M. Hammer, C. Folke, and R. Costanza. Washington: Island Press.

Reher, D. and R. Schofield. 1993. *Old and New Methods in Historical Demography*. Oxford: Clarendon Press.

Reider, Ian and Susan Pick. 1992. "El Aborto: Quién Debe Tomar la Decisión." *Demos* 5:35–36.

Reissman, L. 1964. *The Urban Process: Cities in Industrial Societies*. New York: Free Press.

Rendell, Michael S. 2000. "Entry or Exit? A Transition–Probability Approach to Explaining the High Prevalence of Single Motherhood Among Black Women." *Demography* 36 (3):369–376.

Retherford, Robert D. 1975. *The Changing Sex Differentials in Mortality*. Westport, CT: Greenwood Press.

Revelle, Roger. 1984. "The Effects of Population Growth on Renewable Resources." In *Population Resources, Environment, and Development*, edited by United Nations. New York: United Nations.

Riche, Martha. 1997. "Should the Census Bureau Use 'Statistical Sampling' in Census 2000?" *Insight,* 18 August, p. 1.

Riddle, J. M. 1992. *Contraception and Abortion From the Ancient World to the Renaissance*. Cambridge: Harvard University Press.

Riley, Matilda White. 1976a. "Age Strata in Social Systems." In *Handbook of Aging and the Social Sciences*, edited by R. Binstock and E. Shanas. New York: Van Nostrand Reinhold.

———. 1976b. "Social Gerontology and the Age Stratification of Society." In *Aging in America*, edited by C. S. Kart and B. Manard. Port Washington, NY: Alfred.

Riley, Nancy and Robert W. Gardner. 1996. *China's Population: A Review of the Literature*. Liège: International Union for the Scientific Study of Population.

Rindfuss, Ronald R., S. Philip Morgan, and Kate Offutt. 1996. "Education and the Changing Age Pattern of American Fertility: 1963–1989." *Demography* 33 (3):277–290.

Ritchey, P. and C. Stokes. 1972. "Residence Background, Migration, and Fertility." *Demography* 9:217–30.

Rives, Bill. 1997. "Strategic Financial Planning for Hospitals: Demographic Considerations." In *Demographics: A Casebook for Business and Government*, edited by H. J. Kintner, T. W. Merrick, P. A. Morrison, and P. R. Voss. Santa Monica, CA: RAND.

Robey, B. 1983. "Achtung! Here Comes the Census." *American Demographics* 5 (10):2–4.

Robey, Bryant, John A. Ross, and Indu Bhushan. 1996. "Meeting Unmet Need: New Strategies." *Population Reports of the Population Information Program of the The Johns Hopkins University* Series J (No. 43).

Robinson, J. Gregory, Kirsten K. West, and Arjun Adlakha. 2002. "Coverage of the Population in Census 2000: Results from Demographic Analysis." *Population Research and Policy Review* 21:19–38.

Robinson, Robert V. and Elton F. Jackson. 2001. "Is Trust in Others Declining in America? An Age-Period-Cohort Analysis." *Social Science Research* 30:117–145.

Robinson, Warren C. 1997. "The Economic Theory of Fertility Over Three Decades." *Population Studies* 51:63–74.

Rodgers, Joseph Lee, Hans-Peter Kohler, Kirsten Ohm Kyvik, and Kaare Christensen. 2001. "Behavior Genetic Modeling of Human Fertility: Findings From a Contemporary Danish Twin Study." *Demography* 38 (1):29–42.

Rogers, Andrei. 1992. "Introduction." In *Elderly Migration and Population Redistribution: A Comparative Study,* edited by A. Rogers. London: Bellhaven Press.

Rogers, Everett M. 1995. *Diffusion of Innovations, Fourth Edition*. New York: Free Press.

Rogers, Richard G. 1992. "Living and Dying in the U.S.A.: Sociodemographic Determinants of Death Among Blacks and Whites." *Demography* 29:287–303.

Rogers, Richard G., Robert A. Hummer, and Charles B. Nam. 2000. *Living and Dying in the USA: Behavioral,*

Health, and Social Differentials of Adult Mortality. San Diego: Academic Press.

Rogers, Richard G. and Eve Powell-Griner. 1991. "Life Expectancies of Cigarette Smokers and Nonsmokers in the United States." *Social Science and Medicine* 32:1151–59.

Roof, Wade Clark. 1999. *Spiritual Marketplace: Baby Boomers and the Remaking of American Religion.* Princeton: Princeton University Press.

Rosenhause, S. 1976. "India Taking Drastic Birth Control Step." *Los Angeles Times,* 25 September, p. A2.

Rosental, Paul-Andre. 2003. "The Novelty of an Old Genre: Louis Henry and the Founding of Historical Demography." *Population-E* 58 (3):97–130.

Rosenwaike, Ira. 1979. "A New Evaluation of the United States Census Data on the Extreme Aged." *Demography* 16 (2):279–88.

——, Ed. 1991. *Mortality of Hispanic Populations.* New York: Greenwood Press.

Ross, Catherine E. and Chia-ling Wu. 1995. "The Links Between Education and Health." *American Sociological Review* 60:719–745.

Ross, John A. and E. Frankenberg. 1993. *Findings From Two Decades of Family Planning Research.* New York: Population Council.

Rossi, Peter. 1955. *Why Families Move.* New York: Free Press.

Ruggles, Steven. 1994a. "The Origins of African-American Family Structure." *American Sociological Review* 59:136–51.

——. 1994b. "The Transformation of American Family Structure." *American Historical Review* 99 (1):103–28.

——. 1997. "The Rise of Divorce and Separation in the United States, 1880–1990." *Demography* 34 (4): 455–466.

Rumbaut, Rubén G. 1991. "Passages to America: Perspectives on the New Immigration." In *The Recentering of America: American Society in Transition,* edited by A. Wolfe. Berkeley: University of California Press.

——. 1994. "Origins and Destinies: Immigration to the United States Since World War II." *Sociological Forum* 9(4):583–621.

——. 1995a. "The New Immigration." *Contemporary Sociology* 24 (4):307–11.

——. 1995b. "Vietnamese, Laotian, and Cambodian Americans." In *Asian Americans: Contemporary Trends and Issues,* edited by P. G. Min. Beverly Hills: Sage.

——. 1997. "Assimilation and Its Discontents: Between Rhetoric and Reality." *International Migration Review* 31 (4):923–60.

Russell, Bertrand. 1951. *New Hopes for a Changing World.* New York: Simon & Schuster.

Russell, Cheryl. 1999. "Been There, Done That!" *American Demographics:* January 1999. *http://www.americandemographics.com/,* accessed 2001.

Ryder, Norman. 1964. "Notes on the Concept of a Population." *American Journal of Sociology* 69 (5):447–63.

——. 1965. "The Cohort as a Concept in the Study of Social Change." *American Sociological Review* 30 (6):843–61.

——. 1987. "Reconsideration of a Model of Family Demography." In *Family Demography: Methods and Their Applications,* edited by J. Bongaarts, T. K. Burch, and K. W. Wachter. Oxford: Clarendon Press.

——. 1990. "What is Going to Happen to American Fertility?" *Population and Development Review* 16 (3):433–54.

Sadik, Nafis. 1994. "The Cairo Consensus: Changing the World by Choice." *Populi* 21 (9):4–5.

Sagan, Carl. 1989. "The Secret of the Persian Chessboard." *Parade Magazine.* 5 February, P. 14.

Sakamoto, Arthur, Huei-Hsia Wu, and Jessie M. Tzeng. 2000. "The Declining Significance of Race Among American Men During the Latter Half of the Twentieth Century." *Demography* 37 (1):41–51.

Salaff, J. and A. Wong. 1978. "Are Disincentives Coercive? The View From Singapore." *International Family Planning Perspectives* 4 (2):50–55.

Salant, Tanis, Christine Brenner, Nadia Rubaii-Barrett, and John R. Weeks. 2001. "Illegal Immigrants in U.S.–Mexico Border Counties: Costs of Law Enforcement, Criminal Justice, and Emergency Medical Services." San Diego: Border Counties Coalition.

Salter, C., H. B. Johnston, and N. Hengen. 1997. "Care for Postabortion Complications: Saving Women's Lives." *Population Reports* Series L (10).

Samara, R., B. Bucker, and A. Tsui. 1996. *Understanding How Family Planning Programs Work: Findings From Five Years of Evaluation Research.* Chapel Hill: University of North Carolina, Carolina Population Center.

Sampson, Robert J. and John H. Laub. 1990. "Crime and Deviance Over the Life Course: The Salience of Adult Social Bonds." *American Sociological Review* 55: 609–27.

Sampson, Robert J. and Jeffrey D. Morenoff. 2003. "Spatial (Dis)Advantage and Homicide in Chicago Neighborhoods." In *Spatially Integrated Social Science,* edited by M. F. Goodchild and D. G. Janelle. New York: Oxford University Press.

Samuel, T. John. 1966. "The Development of India's Policy of Population Control." *Milbank Memorial Fund Quarterly* 44:49–67.

Samuel, T. John. 1994. *Quebec Separatism is Dead: Demography is Destiny.* Toronto: John Samuel.

San Diego Union-Tribune. 2003. "Plague Found in 3 Squirrels." 21 August, p. B2.

Sanchez-Albornoz, N. 1974. *The Population of Latin America: A History.* Berkeley: University of California Press.

——. 1988. *Españoles Hacia America: La Emigración en Masa, 1880–1930.* Madrid: Alianza Editorial.

Sanderson, Stephen K. 1995. *Social Transformations: A General Theory of Historical Development.* Cambridge, MA: Blackwell.

Sanger, Margaret. 1938. *Margaret Sanger: An Autobiography.* New York: Norton.

Santelli, John S., Laura Duberstein Lindberg, Joyce Abma, Clea Sucoff McNeely, and Michael Resnick. 2000. "Adolescent Sexual Behavior: Estimates and Trends From Four Nationally Representative Surveys." *Family Planning Perspectives* 32 (4):156–165.

Sardon, Jean-Paul. 2002. "Recent Demographic Trends in the Developed Countries." *Population: English Edition* 57 (1):111–156.

Sassen, Saskia. 2001. *The Global City: New York, London, Tokyo, 2nd Edition.* Princeton: Princeton University Press.

Sauvy, Alfred. 1969. *General Theory of Population.* New York: Basic Books.

Saywell, Trish. 2003. "Cupid the Bureaucrat? Singapore Tries to Play Matchmaker." *The Wall Street Journal,* 30 January, p. A8.

Schachter, Jason. 2001a. "Geographical Mobility: March 1999 to March 2000." *Current Population Reports* (20–538).

———. 2001b. "Why People Move: Exploring the March 2000 Current Population Survey." *Current Population Reports* (23–204).

Schapera, Isaac. 1941. *Married Life in an African Tribe.* New York: Sheridan House.

Schiller, N. G., L. Basch, and C. S. Blanc. 1995. "From Immigrant to Transmigrant: Theorizing Transnational Migration." *Anthropological Quarterly* 68 (1):48–63.

Schinaia, N., A. Ghirardini, F. Chiarotti, P. Mannucci, and The Italian Group. 1991. "Progress to AIDS among Italian HIV-Seropositive Haemophiliacs." *AIDS* 5 (4):385–91.

Schneider, Paula. 2003. "Content and Data Quality in Census 2000." U.S. Census Bureau Census 2000 Testing, Experimentation, and Evaluation Program, Topic Report Series, No. 12, Washington: U.S. Census Bureau.

Schnore, Leo, C. Andre, and H. Sharp. 1976. "Black Suburbanization 1930–1970." In *The Changing Face of the Suburbs,* edited by B. Schwartz. Chicago: University of Chicago Press.

Schoen, Robert. 1983. "Measuring the Tightness of a Marriage Squeeze." *Demography* 20 (1):61–78.

Schoen, Robert and Young J. Kim. 1998. "Momentum Under a Gradual Approach to Zero Growth." *Population Studies* 52:295–299.

Schoen, Robert, Young J. Kim, Constance A. Nathanson, Jason Fields, and Nan Marie Astone. 1997. "Why Do Americans Want Children?" *Population and Development Review* 23 (2):333–358.

Schoen, Robert and R.M. Weinick. 1993. "The Slowing Metabolism of Marriage: Figures From 1988 U.S. Marital Status Life Tables." *Demography* 30 (4):737–746.

Schoenbaum, Micheal and Timothy Waidmann. 1997. "Race, Socioeconomic Status, and Health: Accounting for Race Differences in Health." *The Journal of Gerontology Series B* 52B (Special Issue) 61–73.

Schofield, R. and D. Coleman. 1986. "Introduction: The State of Population Theory." In *The State of Population Theory: Forward From Malthus,* edited by D. Coleman and R. Schofield. Oxford: Basil Blackwell.

Schofield, Roger and David Reher. 1991. "The Decline of Mortality in Europe." In *The Decline of Mortality in Europe,* edited by R. Schofield, D. Reher, and A. Bideau. Oxford: Clarendon Press.

Schreyer, Edward R. 1980. "Foreward." In *Through the '80s, Thinking Globally, Acting Locally,* edited by F. Feather. Washington: World Future Society.

Schultz, Ellen E. 2000. "'Phased Retirement' Option for Workers Is Mainly a Boon for Their Employers." *The Wall Street Journal,* 27 July, p. B1.

Schuman, H. and J. Scott. 1990. "Generational Memories." *ISR Newsletter* 16 (6):4–5.

Schumpeter, Joseph. 1942. *Capitalism, Socialism, and Democracy, Second Edition.* New York: Harper & Row.

Scrimshaw, Susan. 1978. "Infant Mortality and Behavior in the Regulation of Family Size." *Population and Development Review* 4 (3):383–403.

Seager, Joni, Ed. 1990. *The State of the Earth Atlas.* New York: Simon & Schuster.

———. 1993. *Earth Follies: Coming to Feminist Terms with the Global Environmental Crisis.* New York: Routledge.

Seitz, John L. 1995. *Global Issues: An Introduction.* Cambridge, MA: Blackwell.

Sell, R. and G. DeJong. 1983. "Deciding Whether to Move: Mobility, Wishful Thinking and Adjustment." *Sociology and Social Research* 67 (2):146–65.

Serbanescu, F., L. Morris, P. Stupp, and A. Stanescu. 1995. "The Impact of Recent Policy Changes on Fertility, Abortion, and Contraceptive Use in Romania." *Studies in Family Planning* 26 (2):76–87.

Shafik, Nemat Talaat and Ian Johnson. 2003. "Letter from the Vice President of Environmentally and Socially Sustainable Development and from the Vice President of Infrastructure." In *Environment Matters at the World Bank: Annual Review 2003,* edited by the World Bank. Washington: World Bank.

Shelton, Beth Anne. 2000. "Understanding the Distribution of Housework Between Husbands and Wives." In *The Ties that Bind: Perspectives on Marriage and Cohabitation,* edited by L. J. Waite. New York: Aldine de Gruyter.

Shiffman, Jeremy. 2000. "Can Poor Countries Surmount High Maternal Mortality?" *Studies in Family Planning* 31 (4):274–289.

Shiller, Robert. 2000. *Irrational Exuberance*. Princeton: Princeton University Press.

Shkolnikov, Vladimir, France Meslé, and Jacques Vallin. 1996. "Health Crisis in Russia: Part I, Recent Trends in Life Expectancy and Causes of Death From 1970 to 1993." *Population: An English Selection* 8:123–154.

Short, James F., Jr. 1997. *Poverty, Ethnicity, and Violent Crime*. Boulder, CO: Westview Press.

Short, Susan E., Ma Linmao, and Yu Wentao. 2000. "Birth Planning and Sterilization in China." *Population Studies* 54:279–291.

Shryock, Henry. 1964. *Population Mobility Within the United States*. Chicago: University of Chicago Press.

Shryock, Henry, Jacob Siegel, et al., condensed by E. Stockwell. 1976. *The Methods and Materials of Demography*. New York: Academic Press.

Shweder, Richard A. 1997. "It's Called Poor Health for a Reason." *The New York Times*, 9 March , P. A23.

Siegel, J. 1993. *A Generation of Change: A Profile of America's Older Population*. New York: Russell Sage Foundation.

Siegel, Jacob S. 2002. *Applied Demography: Applications to Business, Government, Law, and Public Policy*. San Diego: Academic Press.

Simcox, David. 1999. *Ending Illegal Immigration: Make it Unprofitable*. Washington: Negative Population Growth.

Simmel, Georg. 1905. "The Metropolis and Mental Life." In *Classic Essays on the Culture of Cities*, edited by R. Sennet. New York: Appleton-Century-Crofts.

Simmons, George C., C. Smucker, S. Bernstein, and E. Jensen. 1982. "Post-Neonatal Mortality in Rural India: Implications of an Economic Model." *Demography* 19 (3):371–89.

Simmons, I.G. 1993. *Environmental History: A Concise Introduction*. Oxford: Basil Blackwell.

Simmons, L. 1960. "Aging in Preindustrial Societies." In *Handbook of Gerontology*, edited by C. Tibbetts. Chicago: University of Chicago Press.

Simon, Julian. 1981. *The Ultimate Resource*. Princeton: Princeton University Press.

Simon, Julian L. 1992. *Population and Development in Poor Countries: Selected Essays*. Princeton: Princeton University Press.

Simons, Marlise. 2000. "Between Migrants and Spain: The Sea that Kills." *The New York Times,* 30 March, p. A2.

Singer, Max. 1999. "The Population Surprise." *Atlantic Monthly*. August. p. 22–25.

Singh, Susheela and Renee Samara. 1996. "Early Marriage Among Women in Developing Countries." *International Family Planning Perspectives* 22 (4):148–157.

Sivin, I. 1989. "IUDs are Contraceptives, not Abortifacients: A Comment on Research and Belief." *Studies in Family Planning* 20:355–59.

Skinner, G. William. 1997. "Family Systems and Demographic Processes." In *Anthropological Demography: Toward a New Synthesis,* edited by D. Kertzer and T. Fricke. Chicago: University of Chicago Press.

Smail, J. Kenneth. 1997. "Population Growth Seems to Affect Everything But Is Seldom Held Responsible for Anything." *Politics and the Life Sciences* 16 (2): 231–236.

Smart Growth Network. 2001. "About Smart Growth." *http://www.smartgrowth.org/about/default.asp*, accessed 2001.

Smelser, Neil J. 1997. "The Rational and the Ambivalent in the Social Sciences: 1997 Presidential Address." *American Sociological Review* 63:1–16.

Smith, Adam. 1776. *An Inquiry into the Nature and Causes of the Wealth of Nations: http://socserv.mcmaster.ca/econ*, accessed 2003.

Smith, C. A. and M. Pratt. 1993. "Cardiovascular Disease." In *Chronic Disease Epidemiology and Control,* edited by R. C. Brownson, P. L. Remington, and J. R. Davis. Washington: American Public Health Association.

Smith, D. S. 1978. "Mortality and Family in the Colonial Chesapeake." *Journal of Interdisciplinary History* 8 (3):403–27.

Smith, David P. 1992. *Formal Demography*. New York: Plenum.

Smith, G., M. Shipley, and G. Rose. 1990. "Magnitude and Causes of Socioeconomic Differentials in Mortality: Further Evidence From the Whitehall Study." *British Medical Journal* 301:429–32.

Smith, James P. 1997. "Wealth Inequality Among Older Americans." *The Journal of Gerontology Series B* 52B (Special Issue) 74–81.

Smith, K. R. and N. J. Waitzman. 1994. "Double Jeopardy: Interaction Effects of Marital and Poverty Status on the Risk of Mortality." *Demography* 31 (3): 487–507.

Smith, Michael P. and Luis E. Guarnizo. 1998. *Transnationalism From Below*. New Brunswick: Transaction.

Smith, Stanley, Jeff Tayman, and David Swanson. 2001. *Population Projections for States and Local Areas: Methodology and Analysis*. New York: Plenum.

Smits, Jeroen, Wout Ultee, and Jan Lammers. 1998. "Educational Homogamy in 65 Countries: An Explanation of Differences in Openness Using Country-Level Explanatory Variables." *American Sociological Review* 63:264–285.

Snipp, C. M. 1989. *American Indian: The First of This Land*. New York: Russell Sage Foundation.

Snow, John. 1936. *Snow on Cholera*. London: Oxford University Press.

Sobotka, Tomas. 2003. "Re-Emerging Diversity: Rapid Fertility Changes in Central and Eastern Europe After the Collapse of the Communist Regimes." *Population-E* 58 (4–5):451–486.

Society of Actuaries. 2000. *The RP-2000 Mortality Tables,* Schaumberg, IL: Society of Actuaries, *http://www.soa.org,* accessed 2001.

Sontag, Deborah. 2000. "Gaza Adding Children at an Unrivaled Rate." *The New York Times,* 24 February, p. A1.

Sörenson, A. and H. Trappe. 1995. "The Persistence of Gender Equality in Earnings in the German Democratic Republic." *American Sociological Review* 60: 398–406.

South, Scott J. and Glenna Spitze. 1994. "Housework in Marital and Nonmarital Households." *American Sociological Review* 59:327–47.

Spar, Edward J. 1998. "Demographics: End of Year (D.C. Year, That Is) Wrap-Up." *Business Geographics.* January, p. 38–39.

Spence, A. 1989. *Biology of Human Aging.* Englewood Cliffs, NJ: Prentice-Hall.

Spengler, Joseph. 1974. *Population Change, Modernization, and Welfare.* Englewood Cliffs, NJ: Prentice-Hall.

———. 1979. *France Faces Depopulation: Postlude Edition, 1936–1976.* Durham, NC: Duke University Press.

Spooner, Brian, Ed. 1972. *Population Growth: Anthropological Implications.* Cambridge: MIT Press.

Srinivasan, K., P. H. Reddy, and K. N. M. Raju. 1978. "From One Generation to the Next: Changes in Fertility, Family Size Preferences, and Family Planning in an Indian State Between 1951 and 1975." *Studies in Family Planning* 9 (10–11):258–71.

Stack, Steven. 1987. "Celebrities and Suicide: A Taxonomy and Analysis." *American Sociological Review* 52:401–12.

Stangeland, Charles E. 1904. *Pre-Malthusian Doctrines of Population.* New York: Columbia University Press.

Staples, R. 1986. "The Masculine Way of Violence." In *Homicide Among Black Americans,* edited by D. Hawkins. New York: University Press of America.

Stark, O. and R. Lucas. 1988. "Migration, Remittances, and the Family." *Economic Development and Cultural Change* 36:465–81.

Starr, Paul. 1987. "The Sociology of Official Statistics." In *The Politics of Numbers,* edited by W. Alonso and P. Starr. New York: Russell Sage Foundation.

Statistics Bureau of Japan. 1998. *Results of the First Basic Complete Tabulation of the 1995 Population Census.* Tokyo: *http://www.state.go.jp*

Statistics Canada. 1995. "Census 1996." *Focus for the Future* 10 (1):1–3.

———. 2001a. "Canadian Statistics—Population by Age Group, Canada, the Provinces and Territories."

http://www.statcan.ca/english/Pgdb/People/Population/demo31b.htm, accessed

———. 2001b. "Glossary." *http://www.archive.org,* accessed

———. 2003a. "2001 Census." *http://www.archive.org,* accessed 2003.

———. 2003b. "Persons in Low Income Before Tax." *http://www.archive.org,* accessed 2004.

———. 2003c. "Selected Leading Causes of Death by Sex." *http://www.archive.org,* accessed 2004.

Statistika Centralbyran (Sweden). 1983. *Pehr Wargentin: den Svenska Statistikens Fader.* Stockholm: Statistika Centralbyran.

Steffensmeier, Darrell and Miles Harer. 1987. "Is the Crime Rate Really Falling? An 'Aging' U.S. Population and Its Impact on the Nation's Crime Rate." *Journal of Research in Crime and Delinquency* 24:23–48.

Stephen, Elizabeth Hervey and Anjani Chandra. 2000. "Use of Infertility Services in the United States: 1995." *Family Planning Perspectives* 32 (3):132–137.

Stephenson, G. 1964. *A History of American Immigration.* New York: Russell and Russell.

Stewart, J., B. Popkin, D. Guilkey, J. Akin, L. Adair, and W. Flieger. 1991. "Influences on the Extent of Breast-Feeding: A Prospective Study in the Philippines." *Demography* 28 (2):181–200.

Stolnitz, George J. 1964. "The Demographic Transition: From High to Low Birth Rates and Death Rates." In *Population: The Vital Revolution,* edited by R. Freedman. Garden City: Anchor Books.

Stone, L. 1975. "On the Interaction of Mobility Dimensions in Theory on Migration Decisions." *Canadian Review of Sociology and Anthropology* 12:95–100.

Straus, M. 1983. "Societal Morphogenesis and Intrafamily Violence in Cross-Cultural Perspective." In *International Perspectives on Family Violence,* edited by R. Gelles and C. Cornell. Lexington, MA: Heath.

Stycos, J. Mayone. 1971. *Ideology, Faith and Family Planning in Latin America.* New York: McGraw-Hill.

Sung, K. 1995. "Measures and Dimensions of Filial Piety in Korea." *The Gerontologist* 35 (2):240–47.

Sussman, G. 1977. "Parisian Infants and Norman Wet Nurses in the Early Nineteenth Century: A Statistical Study." *Journal of Interdisciplinary History* 7 (4): 637–53.

Sutherland, I. 1963. "John Graunt: A Tercentenary Tribute." *Royal Statistical Society Journal, Series A* 126:536–37, reprinted in K. Kammeyer, ed. 1975. *Population: Selected Essays and Research.* Chicago: Rand McNally.

Suzman, Richard M., D. P. Willis, and Kenneth G. Manton. 1992. *The Oldest Old.* New York: Oxford University Press.

Szreter, Simon. 1996. *Fertility, Class, Gender in Britain 1860 to 1940.* Cambridge: Cambridge University Press.

Szreter, Simon and Eilidh Garrett. 2000. "Reproduction, Compositional Demography, and Economic Growth: Family Planning in England Long Before the Fertility Decline." *Population and Development Review* 26 (1):45–80.

Ta-k'un, W. 1960. "A Critique of Neo-Malthusian Theory." Excerpted in *Population and Development Review* (1979) 5 (4):699–707.

Tarmugi, Abdullah. 1999. "United Nations General Assembly Plenary, 2 July 1999." *http://www.archive.org*, accessed 2001.

Taueber, C. 1993. "Sixty-Five Plus in America." *Current Population Reports* (23–178RV).

Taylor, W., J. S. Marks, J. R. Livengood, and J. P. Koplan. 1993. "Current Issues and Challenges in Chronic Disease Control." In *Chronic Disease Epidemiology and Control,* edited by R. C. Brownson, P. L. Remington, and J. R. Davis. Washington: American Public Health Association.

Teitelbaum, M. and J. Winter. 1988. "Introduction." In *Population and Resources in the Western Intellectual Tradition,* edited by M. Teitelbaum and J. Winter. New York: Population Council.

Teitelbaum, Michael. 1975. "Relevance of Demographic Transition for Developing Countries." *Science* 188:420–25.

Thatcher, Roger. 2001. "The Demography of Centenarians in England and Wales." *Population: An English Selection* 13 (1):139–156.

The Alan Guttmacher Institute. 1998. *Into a New World: Young Women's Sexual and Reproductive Lives.* New York: Alan Guttmacher Institute.

The Economist. 1994. "Little Match Girls." 15 January, p. 38.

——. 1995a. "Eastern European Demography: of Death and Dying." 7 January, p. 41–42.

——. 1995b. "The Most Expensive Slum in the World." 6 May, p. 35–36.

——. 1996. "Cities: Fiction and Fact." 8 June, p. 44–45.

——. 1998a. "China: The X-Files." 14 February, p. 42–43.

——. 1998b. "How AIDS Began." (7 February):p. 81–2.

—— 2000a. "Over the Sea to Spain." 12 August, p. 40.

—— 2000b. "Who Wants to Live For Ever?" 23 December, p. 23.

——. 2001. "Slave-ships in the 21st Century?" 21 April : p. 41.

——. 2003a. "Afghanistan: The Pains of Motherhood." 22 November. P. 42.

—— 2003b. "The Great Leap Forward: Demand for Cars in China is Accelerating at a Remarkable Rate." 1 February, p. 53–54.

The Economist. 2003. "The Sucking Sound from the East." 26 July, p. 35–36.

The Environmental Fund. 1981. "The Perils of Vanishing Farmland." *The Other Side* No. 23.

The Presbyterian Panel. 1998. "Background Survey for the 1997–1999 Presbyterian Panel." Louisville, KY: Presbyterian Church.

Thomas, Hugh. 1997. *The Slave Trade: The Story of the Atlantic Slave Trade: 1440–1870.* New York: Simon & Schuster.

Thompson, Warren. 1929. "Population." *American Journal of Sociology* 34 (6):959–75.

Tietze, Christopher and S. Lewit. 1977. "Legal Abortion." *Scientific American* 236 (1):21–27.

Tobias, A. 1979. "The Only Article on Inflation You Need to Read: We're Getting Poorer, But We Can Do Something About It." *Esquire* 92 (5):. p. 49–55.

Tobler, Waldo. 1995. "Migration: Ravenstein, Thornthwaite, and Beyond." *Urban Geography* 16 (4): 327–343.

Tolnay, S. 1989. "A New Look at the Effect of Venereal Disease on Black Fertility: The Deep South in 1940." *Demography* 26:679–90.

Torrieri, Nancy and Holland Jennifer. 2000. "Census Scene, Part 3: Enabling an Annual Census." *Geospatial Solutions* (November):38–42.

Toulemon, Laurent. 1997. "Cohabitation Is Here To Stay." *Population: An English Selection* 9:11–46.

Trafzer, Clifford E. 1997. *Death Stalks the Yakama.* East Lansing: Michigan State University Press.

Tu, Ping. 2000. "Trends and Regional Differentials in Fertility Transition." In *The Changing Population of China,* edited by Z. Peng and G. Zhigang. Malden, MA: Blackwell.

Turke, P. 1989. "Evolution and the Demand for Children." *Population and Development Review* 15:61–90.

Turner, R. K., D. Pearce, and I. Bateman. 1993. *Environmental Economics: An Elementary Introduction.* Baltimore: Johns Hopkins University Press.

Tyner, James A. 1999. "The Geopolitics of Eugenics and the Exclusion of Philippine Immigrants From the United States." *The Geographical Review* 89 (1):54–73.

U.K. Public Records Office. 2001. "Domesday Book." *http://www.archive.org*, accessed 2001.

U.S. Border Patrol. 2003. "Southwest Border Apprehensions." *http://uscis.gov/graphics/shared/aboutus/statistics/msrsep03/SWBORD.HTM*, accessed 2004.

U.S. Census Bureau. 1975. *Historical Statistics of the United States.* Washington: Government Printing Office.

——. 1978a. *Country Demographic Profiles: India.* Washington: Government Printing Office.

——. 1978b. "History and Organization." Washington: U.S. Bureau of Census.

——. 1979a. "Country Demographic Profiles: Mexico." Washington, DC: U.S. Bureau of Census.

——. 1979b. "Fertility of American Women: June 1978." *Current Population Reports* Series P–20 (No. 341).

——. 1993. *1990 Census of Population and Housing, Technical Documentation.* Washington: Government Printing Office.

——. 1997. *Report to the Congress: The Plan for Census 2000, Revised and Reissued* (August). Washington: U.S. Bureau of Census.

——. 1999. "Census Bureau Looks Back Over a Century of Accomplishments." *http://www.census.gov/Press-Release/www/1999/cb99-250.html,* accessed 2003.

——. 2000a. "Census 2000 Geographic Terms and Concepts." *http://www.census.gov/geo/www/tiger/glossry2.pdf,* accessed 2003.

——. 2000b. "Computing Apportionment." *http://www.census.gov/population/www/censusdata/apportion-ment/computing.html,* accessed 2001.

——. 2000c. "Congressional Apportionment—Who's Counted." *http://www.census.gov/population/www/censusdata/apportionment/who.html,* accessed 2001.

——. 2000d. *Population Projections of the United States by Age, Sex, Race, Hispanic Origin, and Nativity: 1999 to 2100. http://www.census.gov/population/www/projections/natproj.html.*

——. 2001a. "International Data Base." *http://www.census.gov/ipc/www,* accessed 2001.

——. 2001b. *Statistical Abstract of the United States, 2000.* Washington: Government Printing Office.

——. 2001c. "Urban and Rural Classification Census 2000 Urban and Rural Criteria." *http://www.census.gov/geo/www/ua/ua_2k.html,* accessed 2001.

—— 2001d. "Urban Area Criteria for Census 2000—Proposed Criteria." *Federal Register,* (60):17017–17033, 28 March.

——. 2003a. "Census 2000." *http://factfinder.census.gov,* accessed 2003.

——. 2003b. "Current Population Survey, 2003 Annual Social and Economic Supplement." *http://ferret.bls.census.gov/macro/032003/hhinc/new02_001.htm,* accessed 2004.

——. 2003c. "Educational Attainment in the United States: March 2002, Detailed Tables (PPL-169)." *http://www.census.gov/population/www/socdemo/education/ppl-169.html,* accessed 2004.

U.S. Department of Housing and Urban Development. 1999. *Housing Our Elders: A Report Card on the Housing Conditions and Needs of Older Americans.* Washington: U.S. Department of Housing and Urban Development, Office of Policy Development and Research.

U.S. Department of Justice. 1971. *Crime in the United States, 1970.* Washington: Federal Bureau of Investigation.

——. 1981. *Crime in the United States, 1980.* Washington: Federal Bureau of Investigation.

——. 2000. *Crime in the United States, 1999.* Washington: Federal Bureau of Investigation.

U.S. Immigration and Naturalization Service. 1991. *An Immigrant Nation: United States Regulation of Immigration, 1798–1991.* Washington: Government Printing Office.

——. 2000. *1998 Statistical Yearbook of the Immigration and Naturalization Service.* Washington: Immigration and Naturalization Service, U.S. Department of Justice.

——. 2002. *2000 Statistical Yearbook of the Immigration and Naturalization Service.* Washington: Immigration and Naturalization Service, U.S. Department of Justice.

U.S. Office of Management and Budget. 2000. "Standards for Defining Metropolitan and Micropolitan Statistical Areas." *Federal Register* 65 (249):82227–82238.

U.S. Social Security Administration. 2001a. "Annual Statistical Supplement, 2001." *http://www.ssa.gov,* accessed 2001.

——. 2001b. "Considering Retiring? Consider the Options." *http://www.ssa.gov,* accessed 2001.

——. 2003. "Status of the Social Security and Medicare Programs: A Summary of the 2003 Annual Reports." *http://www.ssa.gov/OACT/TRSUM/trsummary.html,* accessed 2004.

Udry, J. Richard. 1994. "The Nature of Gender." *Demography* 31 (4):561–74.

——. 2000. "Biological Limits of Gender Construction." *American Sociological Review* 65:443–457.

UNAIDS. 2004. "AIDS Epidemic Update 2003." Joint United Nations Programme on HIV/AIDS: *http://www.unaids.org/en/Resources/Publications/Corporate+publications/aids+epidemic+update+-+december+2003.asp,* accessed 2004.

UNAIDS/WHO. 2000. *AIDS Epidemic Update: December 2000.* Geneva: World Health Organization.

Underwood, Carol. 2000. "Islamic Precepts and Family Planning: The Perceptions of Jordanian Religious Leaders and Their Constituents." *International Family Planning Perspectives* 26 (3):110–117.

United Nations. 1953. *Causes and Consequences of Population Trends.* New York: United Nations.

——. 1958. *Principles and Recommendations for National Population Censuses.* New York: United Nations.

——. 1973. *The Determinants and Consequences of Population Trends: New Summary of Findings on Interaction of Demographic, Economic and Social Factors, Volume I.* New York: United Nations.

——. 1979. *Trends and Characteristics of International Migration Since 1950.* New York: United Nations.

——. 1989. *Trends in Population Policy.* New York: United Nations.

——. 1991. *Demographic Yearbook 1989.* New York: United Nations.

——. 1996a. *Demographic Yearbook 1994.* New York: United Nations.

——. 1996b. *Urban Agglomerations 1996.* New York: United Nations.

——. 2000. *Replacement Migration: Is It a Solution to Declining and Ageing Populations?* New York: Population Division of the United Nations.

——. 2001. "Human Development Report 2001: Making New Technologies Work for Human Development." United Nations Development Program: *http://hdr.undp.org/reports/global/2001/en/,* accessed 2003.

United Nations Development Programme. 2003. "Human Development Report 2003: Millenium Development Goals: A Compact Among Nations to End Human Poverty." *http://www.undp.org/hdr2003/,* accessed 2003.

United Nations Development Programme. 1997. *Human Development Report 1997.* New York: Oxford University Press.

United Nations Division for the Advancement of Women. 2004. "The Convention on the Elimination of All Forms of Discrimination against Women (CEDAW)." *http://www.un.org/womenwatch/daw/cedaw/,* accessed 2004.

United Nations Population Division. 1993. *The Sex and Age Distribution of the World Populations: The 1992 Revision.* New York: United Nations.

——. 1994. *World Population Growth from Year 0 to Stabilization.* Mimeograph, 7 June. New York: United Nations.

——. 2000. *World Marriage Patterns 2000.* New York: United Nations.

——. 2001. "World Contraceptive Use, 2001." *http://www.un.org/esa/population/publications/contraceptive2001/WallChart_CP2001_web.xls,* accessed 2004.

——. 2002a. *International Migration 2002.* New York: United Nations.

——. 2002b. *International Migration Report 2002.* New York: United Nations.

——. 2002c. *World Urbanization Prospects: The 2001 Revision, Data Tables and Highlights.* New York: United Nations.

——. 2003a. *Impact of AIDS.* New York: United Nations.

——. 2003b. "Partnership and Reproductive Behaviour in Low-Fertility Countries." *http://www.un.org/esa/population/publications/reprobehavior/partrepro.pdf,* accessed 2004.

United Nations Population Fund. 1987. *1986 Report by the Executive Director.* New York: United Nations.

——. 1991. *Population and the Environment: The Challenges Ahead.* New York: United Nations.

United Nations Statistics Division. 2003. "Population and Housing Census Dates." *http://unstats.un.org/unsd/demographic/census/cendate,* accessed 2003.

United States Citizenship and Immigration Services. 2004. "Glossary & Acronyms." *http://uscis.gov/graphics/glossary.htm,* accessed 2004.

Universal Press Syndicate. 1982. ""Dear Abby"." 3 March.

Upadhyay, U. D. and B. Robey. 1999. "Why Family Planning Matters." *Population Reports* Series J (49).

Vallin, Jacques. 1999. *Mortalite, Sexe et Genre,* edited by A. Pinnelli. Liege: International Union for the Scientific Study of Population.

——. 2002. "The End of the Demographic Transition: Relief or Concern?" *Population and Development Review* 28 (1):105–120.

van de Kaa, Dirk J. 1987. "Europe's Second Demographic Transition." *Population Bulletin* 42 (1).

van de Walle, Etienne. 1992. "Fertility Transition, Conscious Choice and Numeracy." *Demography* 29: 487–502.

——. 2000. "'Marvellous Secrets': Birth Control in European Short Fiction, 1150–1650." *Population Studies* 54:321–330.

van de Walle, Etienne and John Knodel. 1967. "Demographic Transition and Fertility Decline: The European Case." In *International Population Conference.* Sydney, Australia.

——. 1980. "Europe's Fertility Transition: New Evidence and Lessons for Today's Developing World." *Population Bulletin* 34 (6).

Vanderpool, Chris. 1995. *Integrated Natural Resource Systems.* East Lansing: Michigan Agricultural Experiment Station, Michigan State University.

Vandeschrick, Christopher. 2001. "The Lexis Diagram, a Misnomer." *Demographic Research* 4 (3).

VanLandingham, Mark J., Somboon Suprasert, Werasit Sittitrai, Chayan Vaddhanaphuti, and Nancy Grandjean. 1993. "Sexual Activity Among Never-Married Men in Northern Thailand." *Demography* 30:297–314.

Velkoff, Victoria A. and Arjun Adlakha. 1998. "Women of the World: Women's Health in India." Washington: U.S.Census Bureau.

Ventura, Stephanie J., William D. Mosher, Sally C. Curtin, Joyce Abma, and Stanley K. Henshaw. 2000. "Trends in Pregnancies and Pregnancy Rates by Outcome: Estimates for the United States, 1976–96." *Vital and Health Statistics* 21 (56).

Veyne, P. 1987. "The Roman Empire." In *A History of Private Life, Volume 1: From Pagan Rome to Byzantium,* edited by P. Veyne. Cambridge: Harvard University Press.

Visaria, Pravin and Vijaylaxmi Chari. 1998. "India's Population Policy and Family Planning Program: Yesterday, Today, and Tomorrow." In *Do Population Policies Matter? Fertility and Politics in Egypt, India, Kenya, and Mexico,* edited by A. Jain. New York: Population Council.

Viscusi, W. K. 1979. *Welfare of the Elderly: An Economic Analysis and Policy Prescription.* New York: John Wiley.

Vishnevsky, Anatoly and Vladimir Shkolnikov. 1999. "Russian Mortality: Past Negative Trends and Recent Improvements." In *Population Under Duress: The Geodemography of Post-Soviet Russia,* edited by G. J. Demko, G. Ioffe, and Z. Zayonchkovskaya. Boulder, CO: Westview Press.

Vlassoff, C. 1990. "The Value of Sons in an Indian Village." *Population Studies* 44:5–20.

Wackernagel, Mathis and William E. Rees. 1996. *Our Ecological Footprint: Reducing Human Impact on the Earth.* Philadelphia: New Society.

Waite, Linda, Frances K. Goldscheider, and C. Witsberger. 1986. "Nonfamily Living and The Erosion of Traditional Family Orientations Among Young Adults." *American Sociological Review* 41 (2):235–51.

Waite, Linda J. 2000. "The Family as a Social Organization: Key Ideas for the Twenty-First Century." *Contemporary Sociology* 20 (3):463–469.

Waite, Linda J. and Maggie Gallagher. 2000. *The Case for Marriage.* New York: Doubleday.

Waldman, Peter. 1991. "As Acreage Erodes, Palestinians Feel Pressure to Deal." *The Wall Street Journal.* 10 May. P. A9.

Waldron, I. 1983. "The Role of Genetic and Biological Factors in Sex Differences in Mortality." In *Sex Differentials in Mortality: Trends, Determinants and Consequences,* edited by A. D. Lopez and L. T. Ruzicka. Canberra: Australian National University.

———. 1986. "What Do We Know About Causes of Sex Differences in Mortality? A Review of the Literature." *Population Bulletin of the United States* 18–1995: 59–76.

Walker, Chip. 1995. "Meet the New Vegetarian." *American Demographics* 17 (1):9.

Wall, Richard, Jean Robin, and Peter Laslett, Ed. 1983. *Family Forms in Historic Europe.* New York: Cambridge University Press.

Wallace, Paul. 2001. *Agequake: Riding the Demographic Rollercoaster Shaking Business, Finance and Our World.* London: Nicholas Brealey.

Wallace, Robert. 1761 (1969). *A Dissertation on the Numbers of Mankind, in Ancient and Modern Times, 1st and 2nd editions, revised and corrected.* New York: Kelley.

Wallerstein, Immanuel. 1974. *The Modern World System.* New York: Academic Press.

———. 1976. "Modernization: Requiescat in Pace." In *The Uses of Controversy in Sociology,* edited by L. A. Coser and O. N. Larsen. New York: Free Press.

Wallis, C. 1995. "How to Live to be 120." *Time,* 6 March, p. 85.

Walsh, J. 1974. "U.N. Conference: Topping Any Agenda Is the Question of Development." *Science* 185:1144.

———. 1975. "U.S. Agribusiness and Agricultural Trends." *Science* 188:531–34.

Ware, Helen. 1975. "The Limits of Acceptable Family Size in Western Nigeria." *Journal of Biosocial Science* 7:273–96.

Warren, C., F. Hiyari, P. Wingo, A. Abel-Aziz, and L. Morris. 1990. "Fertility and Family Planning in Jordan: Results From the 1985 Jordan Husband's Fertility Survey." *Studies in Family Planning* 21:33–39.

Warren, Carol A. B. 1998. "Aging and Identity in Premodern Times." *Research on Aging* 20 (1):11–35.

Warren, Robert and Ellen Percy Kraly. 1985. *The Elusive Exodus: Emigration from the United States.* Washington: Population Reference Bureau.

Wasserman, I. M. 1984. "Imitation and Suicide: A Reexamination of the Werther Effect." *American Sociological Review* 49:427–36.

Watkins, Susan Cotts. 1986. "Conclusion." In *The Decline of Fertility in Europe,* edited by A. Coale and S. C. Watkins. Princeton: Princeton University Press.

———. 1991. *From Provinces Into Nations: Demographic Integration in Western Europe, 1870–1960.* Princeton: Princeton University Press.

Wattenberg, Ben J. 1997. "The Population Explosion is Over." *The New York Times Magazine,* 28 November.

Weber, Adna. 1899. *The Growth of Cities in the Nineteenth Century.* New York: Columbia University Press.

Weeks, Gregory. 2003. *The Military and Politics in Postauthoritarian Chile.* Tuscaloosa: University of Alabama Press.

Weeks, John R. 1970. "Urban and Rural Natural Increase in Chile." *Milbank Memorial Fund Quarterly* XLVIII (1):71–89.

———. 1988. "The Demography of Islamic Nations." *Population Bulletin* 43 (4):1–53.

———. 1996. "Vignettes of PAA History—The Beginnings." *PAA Affairs* 29 (4):3–5.

———. 2004. "Using Remote Sensing and Geographic Information Systems to Identify the Underlying Properties of Urban Environments." In *New Forms of Urbanization: Conceptualizing and Measuring Human Settlement in the Twenty-first Century,* edited by A. G. Champion and G. Hugo. London: Ashgate.

Weeks, John R. and Karyl Fuller. 1996. "Population Distribution and Migration: Components of Housing Demand and Supply." In *Human Settlements Habitat: Proceedings and Recommendations of the International Symposium on Human Settlements,* edited by D. Wozniak, T. Shuman, A. Roet, and M. Garret. San Diego: International Institute for Human Resource Development, San Diego State University.

Weeks, John R., M. Saad Gadalla, Tarek Rashed, James Stanforth, and Allan G. Hill. 2000. "Spatial Variability in Fertility in Menoufia, Egypt, Assessed Through the Application of Remote Sensing and GIS Technologies." *Environment and Planning A* 32:695–714.

Weeks, John R., Arthur Getis, Allan G. Hill, M. Saad Gadalla, and Tarek Rashed. 2004. "The Fertility Transition in Egypt: Intra-Urban Patterns in Cairo." *Annals of the Association of American Geographers* 94 (1): 74–93.

Weeks, John R., Rubén G. Rumbaut, Claire Brindis, Carol Korenbrot, and D. Minkler. 1989. "High Fertility Among Indochinese Refugees." *Public Health Reports* 104:143–50.

Weg, Ruth. 1975. "Changing Physiology of Aging: Normal and Pathological." In *Aging: Scientific Perspectives and Social Issues,* edited by D. Woodruff and J. Birreu. New York: Van Nostrand.

Weiss, K. H. 1973. "Demographic Models for Anthropology." *American Antiquity* 38 (2: Part II).

Weiss, M. 1988. *The Clustering of America.* New York: Harper & Row.

Weiss, Michael J. 2001. *The Clustered World.* New York: Little Brown.

Weller, Robert, John Macisco, and G. Martine. 1971. "Relative Importance of the Components of Urban Growth in Latin America." *Demography* 8 (2): 225–32.

Wells, R. 1971. "Family Size and Fertility Control in Eighteenth-Century America: A Study of Quaker Families." *Population Studies* 25 (1):73–82.

———. 1982. *Revolutions in Americans' Lives.* Westport, CT: Greenwood Press.

———. 1985. *Uncle Sam's Family: Issues in and Perspective on American Demographic History.* Albany: State University of New York Press.

Wendelken, Sandra. 1996. "Religiously Mapping." *Business Geographics* (September):54.

Westerhof, Gerben J., Anne E. Barrett, and Nardi Steverink. 2003. "Forever Young?: A Comparison of Age Identities in the United States and Germany." *Research on Aging* 25 (4):366–383.

Westing, Arther. 1981. "A Note on How Many Humans Have Ever Lived." *BioScience* 31 (7):523–24.

Westley, Sidney B. 2002. "Assessing Women's Well-Being in Asia." *Asia-Pacific Population & Policy, East-West Center* 61.

Westoff, Charles. 1978. "Marriage and Fertility in the Developed Countries." *Scientific American* 239 (6):51–57.

———. 1990. "Reproductive Intentions and Fertility Rates." *International Family Planning Perspectives* 16:84–89.

———. 1994. "What's the World's Priority Task? Finally, Control Population." *The New York Times Magazine* 6 February.

———. 1997. "Population Growth: Large Problem, Low Visibility." *Politics & Life Sciences* 16 (2):226–227.

Westoff, Charles. 2003. *Trends in Marriage and Early Childbearing in Developing Countries, DHS Comparative Reports No. 5.* Calverton, MD: ORC Macro.

Westoff, Charles and Akinrinola Bankole. 1996. "The Potential Demographic Significance of Unmet Need." *International Family Planning Perspectives* 22 (1): 16–20.

Westoff, Charles and Ronald Rindfuss. 1974. "Sex Preselection in the United States: Some Implications." *Science* 184:633–36.

Westoff, Charles and Norman Ryder. 1977. "The Predictive Validity of Reproductive Intentions." *Demography* 14 (4):431–54.

Wharton, C. and R. Blackburn. 1988. "Lower Dose Pills." *Population Reports, Population Information Program, John Hopkins University* Series A, (Number 7).

White, Andrew A. and Keith F. Rust, Ed. 1997. *Preparing for the 2000 Census: Interim Report II.* Washington: National Academy Press.

White, Katherine J. Curtis. 2002. "Declining Fertility Among North American Hutterites: The Use of Birth Control Within a Darisusleut Colony." *Social Biology* 49 (1–2):58–73.

Whitmore, Thomas M. 1992. *Disease and Death in Early Colonial Mexico.* Boulder, CO: Westview Press.

Wiatrowski, William J. 2001. "Changing Retirement Age: Ups and Downs." *Monthly Labor Review* 124 (4): 3–12.

Wierzbicki, Susan. 2003. *Beyond the Immigrant Enclave: Network Change and Assimilation.* New York: LFB Scholarly Publishing.

Willcox, Walter F. 1936. *Natural and Political Observations Made Upon the Bills of Mortality by John Graunt.* Baltimore: Johns Hopkins Press.

Williams, J. D. 1994. "Using Demographic Analysis to Fix Political Geography." *Applied Demography* 8 (3):1–3.

Williams, L. and Basil Zimmer. 1990. "The Changing Influence of Religion on U.S. Fertility." *Demography* 27 (3):475–81.

Williams, Naomi and Chris Galley. 1995. "Urban-Rural Differentials in Infant Mortality in Victorian England." *Population Studies* 49:401–420.

Wilmoth, John R. and Shiro Horiuchi. 1999. "Rectangularization Revisited: Variability of Age at Death Within Human Populations." *Demography* 36 (4):475–495.

Wilson, Allan C. and Rebecca L. Cann. 1992. "The Recent African Genesis of Humans." *Scientific American* 266 (4):68-74.

Wilson, Thomas. 1991. "Urbanism, Migration, and Tolerance: A Reassessment." *American Sociological Review* 56:117–123.

Wilson, W. 1978. *The Declining Significance of Race.* Chicago: University of Chicago Press.

Winter, John G., Ed. 1936. *Papyri in the University of Michigan Collection: Miscellaneous Papyri.* Ann Arbor: University of Michigan Press.

Wirth, Louis. 1938. "Urbanism as a Way of Life." *American Journal of Sociology* 44:3–24.

Wise, David A. 1997. "Retirement Against the Demographic Trends: More Older People Living Longer, Working Less, and Saving Less." *Demography* 34 (1):83–95.

Wolffsohn, M. 1987. *Israel: Polity, Society, Economy, 1882–1986: An Introductory Handbook.* Atlantic Highlands, NJ: Humanities Press.

Wonacott, Peter, Jeanne Whalen, and Bhushan Bahree. 2003. "China's Growing Thirst for Oil Remakes the Global Market." *The Wall Street Journal.* 3 December. P. A1.

Wood, J. W., R. J. Ferrell, and S. N. DeWitte-Aviña. 2003. "Was the Black Death Yersinial Plague?" *The Lancet Infectious Diseases* 3 (6):327–328.

Woodrow-Lafield, Karen. 1996. "Emigration from the USA: Multiplicity Survey Evidence." *Population Research and Policy Review* 15 (2):171–99.

World Bank. 1994a. *Population and Development: Implications for the World Bank.* Washington: World Bank.

——. 1994b. *World Development Report 1994: Infrastructure for Development.* New York: Oxford University Press.

——. 1995. *Monitoring Environmental Progress: A Report on Work in Progress.* Washington: World Bank.

——. 1998. "West Africa: Community-Based Natural Resource Management." *Findings From Economic Management and Social Policy* 107 (March).

——. 1999. "Gender and Law: Eastern Africa Speaks." *Human Development* (126).

——. 2000. *World Development Report 2000/2001: Attacking Poverty.* New York: Oxford University Press.

——. 2003a. "Data & Statistics/Technical Notes." *http://www.worldbank.org/data/databytopic/top-notes.html,* accessed 2003.

——. 2003b. "Global Poverty Monitoring." *http://www.worldbank.org,* accessed 2004.

——. 2003c. "World Development Report 2003: Sustainable Development in a Dynamic World." *http://econ.worldbank.org/wdr/wdr2003,* accessed 2003.

World Commission on Environment and Development. 1987. *Our Common Future.* New York: Oxford University Press.

World Health Organization. 2001a. "Global Estimates of Maternal Mortality for 1995: Results of an In-Depth Review, Analysis and Estimation Strategy." *http://www.archive.org,* accessed 2001.

——. 2001b. "World Health Statistics Annual 1997-1999 (on-line edition)." *http://www.who.int/whosis,* accessed 2001.

——. 2002. "Suicide Prevention." *http://www.who.int/mental_health/prevention/suicide/suicideprevent/en,* accessed 2004.

——. 2003. "Severe Acute Respiratory Syndrome (SARS): Report by the Secretariate." *http://www.who.int/gb/EB_WHA/PDF/EB113/eeb11333.pdf,* accessed 2004.

——. 2004a. "Avian Influenza A (N5N1)—Update 17." *http://www.who.int/csr/don/2004_02_04/en/,* accessed 2004.

——. 2004b. "WHO UNICEF Coverage Estimates of Vaccines and Biologicals." *http://www.who.int/vaccines/globalsummary/timeseries/tswucoveragemcv.htm,* accessed 2004.

World Population Conference. 1927. *Proceedings of the World Population Conference, held at the Salle Centrale, Geneva, August 29th to September 3rd, 1927, Edited by Mrs. Margaret Sanger.* London: E. Arnold.

World Resources Institute. 1998. *World Resources 1998–99.* New York: Oxford University Press.

——. 2000. *World Resources 2000/2001.* New York: Oxford University Press.

Wright, R. 1989. "The Easterlin Hypothesis and European Fertility Rates." *Population and Development Review* 15:107–22.

Wrigley, E. A. 1974. *Population and History.* New York: McGraw-Hill.

Wrigley, E. A. 1987. *People, Cities and Wealth.* Oxford: Blackwell.

——. 1988. "The Limits to Growth." In *Population and Resources in Western Intellectual Tradition,* edited by M. T. a. J. Winter. New York: Population Council.

Wrigley, E. A. and R. S. Schofield. 1981. *The Population History of England, 1541–1871: A Reconstruction.* Cambridge: Harvard University Press.

Yablonsky, Lewis. 1997. *Criminology: Crime and Criminality, Fourth Edition.* New York: Prentice-Hall.

Yamaguchi, K. and L. R. Ferguson. 1995. "The Stopping and Spacing of Childbirths and Their Birth-History Predictors: Rational-Choice Theory and Event-History Analysis." *American Sociological Review* 60:272–98.

Yamani, Mai. 2000. *Changed Identities: The Challenge of the New Generation.* Washington: Brookings Institution.

Yap, Mui Teng. 1995. "Singapore's 'Three or More' Policy: The First Five Years." *Asia-Pacific Population Journal* 10 (4):39–52.

York, Richard, Eugene A. Rosa, and Thomas Dietz. 2003. "Footprints on the Earth: The Environmental Consequences of Modernity." *American Sociological Review* 68:279–300.

Young, T. K. 1994. *The Health of Native Americans: Toward a Biocultural Epidemiology.* New York: Oxford University Press.

Zarate, A. and A. de Zarate. 1975. "On the Reconciliation of Research Findings of Migrant-Nonmigrant Fertility Differentials in Urban Areas." *International Migration Review* 9 (2):115–56.

Zelinsky, Wilbur. 1971. "The Hypothesis of the Mobility Transition." *The Geographical Review* 61 (2):219–49.

Zeng, Yi and Linda George. 2000. "Family Dynamics of 63 Million (in 1990) to More Than 330 Million (in 2050) Elders in China." *Demographic Research* 2 (5).

Zhao, Z. 1994. "Demographic Conditions and Multigeneration Households in Chinese History: Results From Genealogical Research and Microsimulation." *Population Studies* 48 (3):413–26.

Zhou, Yixing and Laurence J. C. Ma. 2000. "Economic Restructuring and Suburbanization in China." *Urban Geography* 21 (3):205–236.

Zhu, Yu. 2004. "Changing Urbanization Processes and *In Situ* Rural-Urban Transformation: Reflections on China's Settlement Definitions." In *New Forms of Urbanization: Beyond the Urban-Rural Dichotomy*, edited by A. G. Champion and G. Hugo. London: Ashgate.

Zinsser, Hans. 1935. *Rats, Lice and History.* Boston: Little, Brown.

Zipf, George K. 1949. *Human Behavior and the Principle of Least Effort.* Reading, MA: Addison-Wesley.

Zlaff, Vivian. 1973. "Ethnic Segregation in Urban Israel." *Demography* 11 (2):161–84.

Zlotnick, H. 1994. "Expert Group Meeting on Population Distribution and Migration." *International Migration Review* 28 (1):171–204.

Zuberi, Tukufu. 2001. *Thicker Than Blood: How Racial Statistics Lie.* Minneapolis: University of Minnesota Press.

GEOGRAPHIC INDEX

status of women, 410, 413, 445
violence against women, 249
Indonesia, 47
food production, 512
migration, 276, 569
population size and growth, 64
Iran, 7–10
Ireland, 221
Israel, 8, 9, 10, 267, 268
ethnicity in, 442
fertility, 247, 442
migration, 298
population policy, 548
racial-ethnic mortality differential
in, 196
Italy, 112
family and households, 406
fertility, 257–258
immigration to, 569
labor force participation, 426
life expectancy, 184
migration, 583
older population, 372
population size and growth, 58, 59

Japan, 47
abortions in, 231
age structure, 344, 572
agricultural production, 508
carbon dioxide emissions, 521
fertility, 258
GNI PPP of, 497
infant mortality rates, 155
irrigation, 513
life expectancy, 187
migration, 304, 583
older population, 371, 372, 386,
394
per-person income, 500
population registers, 135
population size and growth, 50,
62–63
sex ratios, 386
Tokyo, 533
urbanization, 457
Jordan
fertility, 251, 252, 266–268
per-person income, 500
refugees, 310
urban-rural fertility differential,
462

Kazakhstan, 300
Kenya
fertility, 237, 251, 252
status of women, 410
Korea, 393
age structure, 572
fertility, 239
sex ratios, 327
Kuwait, 521

Latin America
See also specific countries
abortions in, 236
fertility, 245, 254
impact of declining mortality,
342–343
migration, 287, 300, 303, 304
mortality decline in, 185
older population, 382
per-person income, 500
population policy, 548
urban-rural fertility differential,
462
Latvia
age structure, 348–349
fertility, 209
Lebanon, 114
Libya, 295
Lithuania, 172
Liverpool, 191
London, 74, 132, 140, 191, 456

Malaysia, 393, 569
Mali, 462
Manchester, 191
Mexico
abortions in, 232
accidental deaths in, 169
age structure, 329, 330, 361–362,
571
carbon dioxide emissions, 521
census of, 123–124, 131
colonias, 486
crude death rate (CDR) in, 172
degenerative diseases in, 167,
168
demographic surveys in, 137
ecological footprint, 534–535
education, 422
ethnicity in, 441–442
family and households, 408
fertility, 207, 208, 209, 211, 234,
235, 251, 252, 266
food production, 511
gender differential in mortality in,
159
Guadalajara, 473
infant mortality rates, 155
labor force participation,
426–427
life expectancy, 183–184
maternal mortality, 161
metropolitan areas, 468
Mexico City, 468, 473, 475,
486
migration, 11, 295, 296–299, 300,
475, 571, 585
Monterrey, 473
older population, 373, 385–386,
396
per-person income, 501

population projection for,
337–338
population size and growth,
56–57
sex ratios, 385–386
suicide rates, 169
top causes of death in, 162
Tzintzuntzan, 474–476
urban definition, 455
urban transition, 474–476
Middle East, 7–10. See also specific
countries
migration, 295
population size and growth,
59–60

Netherlands, 113
immigration to, 11
population registers, 135
New Zealand
fertility, 254
migration, 285
population size and growth, 64
Nicaragua, 57
Niger, 209, 221
Nigeria
age structure, 329, 330, 331
census of, 113–114
economic development and population growth, 506
fertility, 237
infecundity rates in, 204
Lagos, 487
population size and growth, 47,
50, 60
poverty, 435
North Africa
See also specific countries
age structure, 571
education, 423
fertility, 254
life expectancy, 190
population size and growth,
59–60
North America
See also specific countries
age at marriage, 411, 414–415
age structure, 571
agricultural production, 508
divorce rates, 420
education, 423
gender gap in life expectancy in,
160
migration, 287
older population, 371
population momentum, 354
population policy, 548
population size and growth,
50–56
son preference, 239
Norway, 135

SUBJECT INDEX

For region and country specific information please see Geographic Index.